Companion to Biochemistry

Selected Topics for Further Study

List of Contributors

N. J. M. Birdsall	National Institute for Medical Research, Mill Hill, London.
A. T. Bull	Biological Laboratory, University of Kent, Canterbury.
B. F. C. Clark	M.R.C. Laboratory of Molecular Biology, University of Cambridge.
J. F. Collins	Department of Molecular Biology, University of Edinburgh.
A. d'Albis	M.R.C. Biophysics Research Unit, King's College, University of London.
A. P. Dawson	School of Biological Sciences, University of East Anglia.
W. B. Gratzer	M.R.C. Biophysics Research Unit, King's College, University of London.
S. E. Halford	Molecular Enzymology Laboratory, University of Bristol.
J. Hindley	Department of Biochemistry, Medical School, University of Bristol.
A. G. Lee	National Institute for Medical Research, Mill Hill, London.
P. M. Meadow	Department of Biochemistry, University College, University of London.
J. C. Metcalfe	National Institute for Medical Research, Mill Hill, London.*
A. J. Munro	Department of Pathology, University of Cambridge.
A. A. Newton	Department of Biochemistry, University of Cambridge.
B. A. Newton	M.R.C. Biochemical Parasitology Unit, Molteno Institute, University of Cambridge.
G. W. Offer	Department of Biophysics, King's College, University of London.
C. A. Pasternak	Department of Biochemistry, University of Oxford.
M. C. Perry	Department of Biochemistry, Chelsea College of Science and Technology, University of London.
K. Roberts	Department of Ultrastructural Studies, John Innes Institute, Norwich.
M. J. Selwyn	School of Biological Sciences, University of East Anglia.
T. F. Slater	Department of Biochemistry, Brunel University.
H. Smith	Department of Microbiology, University of Birmingham.
J. O. Thomas	Department of Biochemistry, University of Cambridge.
K. F. Tipton	Department of Biochemistry, University of Cambridge.
P. J. Winterburn	Department of Biochemistry, University College, Cardiff.

* *now at* Department of Pharmacology, University of Cambridge.

Companion to Biochemistry
Selected Topics for Further Study

Edited by

Alan T. Bull
Reader in Biology, University of Kent at Canterbury

John R. Lagnado
Lecturer in Biochemistry, Bedford College, University of London

Jean O. Thomas
University Lecturer in Biochemistry, Fellow of New Hall, Cambridge

Keith F. Tipton
University Lecturer in Biochemistry, Fellow of King's College, Cambridge

Longman

Longman
1724-1974

Longman Group Limited
London

Associated companies, branches and representatives
throughout the world

© Longman Group Limited 1974

ISBN 0 582 46004 2

Library of Congress Catalog Card Number
73-89498

Printed in Great Britain by
William Clowes & Sons Limited
London, Colchester and Beccles

Contents

Acknowledgements

We are grateful to the following for their advice and helpful comments: J. R. Coggins, H. B. F. Dixon, K. R. F. Elliott, M. D. Houslay, J. Kinderlerer, R. D. Kornberg, R. N. Perham, P. H. Rubery, C. J. R. Thorne. We also wish to thank Mr. Robert Welham of Longman Group Limited for his help and advice during the preparation of this book.

Preface

Advanced students of biochemistry frequently complain of a dearth of books suitable for their use. The basic texts generally do not take the subject far enough and may hardly touch on certain topics. On the other hand, reviews and monographs are written with a different purpose and for a different audience; therefore, they are not always satisfactory as a teaching medium. They are often too detailed, too expensive and, in some subjects, appear too infrequently. It is hoped that the *Companion to Biochemistry* may fill this gap between basic text and specialist review by providing a collection of articles on topics suitable for advanced courses in biochemistry. These articles, written with the student in mind, assume a familiarity with material in the basic texts and, building on this, provide accounts, at approximately the level of the final-year undergraduate, of twenty or so areas of biochemistry.

We realise that the choice of topics must, necessarily, be a personal one. We have tried to select topics that are either poorly treated in the textbooks, or poorly understood by the average final-year undergraduate. We have also included some topics that are not at present regarded as central to the teaching of biochemistry but which may become increasingly important. We are aware that some subjects have been omitted but many of these have been adequately covered in recent publications. We have allowed the type of presentation to vary in the hope that this would allow authors to develop their topics according to their own special interests.

We hope that the *Companion to Biochemistry* will be useful. If it is, we would like to prepare new editions from time to time and would welcome suggestions from both students and teachers for further topics to be covered and for ways of improving the book.

A. T. B.
J. R. L.
J. O. T.
K. F. T.

1
Protein Biosynthesis

B. F. C. Clark
MRC Laboratory

1 General Feature

1.1 Introduction

The explanation of protein biosynthesis in molecular terms is a central problem in molecular biology. Only since 1961 has significant progress been made by biochemists in elucidating the molecular mechanisms of protein biosynthesis and its relation to genetic information. Most of our new knowledge has come from experiments using bacterial cell extracts — *in vitro* systems. Similar eukaryotic cell extracts which can actively synthesise proteins are only now becoming a reality. Experiments using these new *in vitro* systems should confirm the general belief that the molecular mechanisms as detailed for bacteria are general throughout living systems. However, the control of protein synthesis in multicellular organisms is more complex than in prokaryotes. Furthermore, because of the compartments that exist in eukaryotic cells, different control mechanisms can occur in different parts of the same cell. For instance, mitochondrial proteins are thought to be initiated by the same mechanism as bacterial ones and differently from those of the eukaryotic cytoplasm.

The purpose of this chapter is to describe generally accepted facts about the molecular mechanisms involved in protein biosynthesis. This is not to say that we know everything about protein synthesis, as will be clear from the later sections. After a brief description of the assorted components used to make proteins and the information required for directing the machinery, each component and its interactions will be dealt with in more detail consonant with our current knowledge. I hope to be able to separate hard facts from new ideas which may be somewhat speculative.

1.2 Direction in which proteins are made

In the present state of our knowledge it appears that we can simplify the problem of protein biosynthesis to an attempt to understand the mechanism of polypeptide chain formation. This is because the spatial or tertiary structure of proteins seems to depend upon the primary structure, or the linear arrangement, of the constituent amino acids. Thus, when the amino acids have been put together in the correct order the polypeptide chain automatically twists and folds into the correct spatial structure, or conformation, for a particular biological function.

Proteins are made in the cell from the amino (N-terminal) end towards the carboxy (C-terminal) end. The peptide bond is formed between two amino acids by the elimination of water. This reaction does not proceed spontaneously under physiological conditions so the cell activates the two amino acids and also gives instructions for which amino acids are to be joined. In biological material a very intricate machinery has been evolved to control and direct peptide bond formation.

1.3 Rate of peptide bond synthesis

The best estimate for the rate of peptide bond formation in bacteria is about ten per second. For example, the rate of addition of amino acids for the synthesis of β-galactosidase *in vivo* is fifteen per second whereas a very active *in vitro* system for synthesising bacteriophage T4-coded lysozyme adds six amino acids per second. The rate for animal cells appears to be a little slower since globin is made *in vivo* at the rate of two amino acids added per second.

1.4 Adaptor molecules are attached to amino acids

In the cell the amino acids are activated enzymically by joining them to adaptor molecules called transfer ribonucleic acids (tRNAs). tRNA is sometimes called soluble ribonucleic acid (sRNA) but it is better to reserve this more general term for all RNA (irrespective of function) of low molecular weight and soluble in 1M salt solution. The adaptor molecules constitute a heterogeneous set of small RNA molecules each about 80 nucleotides long with a molecular weight of about 25 000. All tRNA molecules have the sequence CpCpA at one end (the 3'-OH end) and thus have a terminal *cis* vicinal diol group, the point where the amino acid is attached. More structural details will be given in section 4.

1.5 Activation

The overall activation reaction is directed by an activating enzyme as shown in Fig. 1.1. This activation, as described in section 3, is thought to occur in two separable steps. Overall, in the presence of ATP the amino acid is joined by the enzyme to one of the hydroxyl groups of the terminal adenosine of the tRNA in an ester linkage. The product, aminoacyl-tRNA, is the activated intermediate in protein biosynthesis. Biochemists do not usually try to distinguish between the two hydroxyl groups on the terminal adenosine because they are chemically equivalent in that the amino acid can migrate between them more rapidly than the rate of peptide bond formation.

Two aminoacyl-tRNAs could react with each other with the formation of a peptide bond if they were in the correct juxtaposition. However, in solution there would be strong competition by water for the hydrolysis of the aminoacyl-tRNA. It has been apparent since the later 1950s that protein biosynthesis occurs at special sites in the cell. In fact, it occurs on special cellular ribonucleoprotein particles called ribosomes whose molecular weight in bacteria is about $2 \cdot 7 \times 10^6$ daltons.

1.6 Ribosomal sites

There is good evidence that there are two ribosomal sites for the aminoacyl-tRNAs when a peptide bond is being formed, as in Fig. 1.2. So far we know very little about the detailed structure of these sites and the evidence for them is largely circumstantial. The sites position the peptidyl-tRNA (P-site for $tRNA_n$ in Fig. 1.2) and aminoacyl-tRNA (A-site for $tRNA_{n+1}$ in Fig. 1.2) during the propagation or *elongation* phase of protein biosynthesis. The concept of

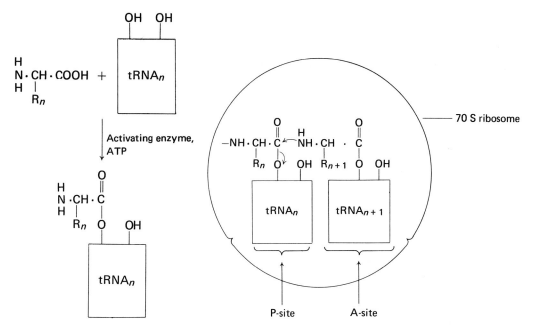

Fig. 1.1 Activation of amino acids.   Fig. 1.2 Peptide bond formation on the ribosome.

two sites may need some revision for special stages of protein synthesis such as starting, but this is not completely clear at present. The bacterial ribosomal unit is called a 70 S particle because of its sedimentation properties in the ultracentrifuge. It is composed of about 60 per cent RNA and 40 per cent protein. There are some 18 000 such ribosomes in the bacterial cell.

1.7 Peptide bond formation

Although a peptide bond might be expected to form spontaneously between two aminoacyl-tRNAs in the correct ribosomal sites, there is sound evidence available to indicate that the formation is brought about enzymically; the enzyme, *peptidyl transferase*, is an integral part of the ribosome. Supernatant protein factors have also been implicated in the elongation phase of biosynthesis. The EF-Tu factor is concerned with the binding of aminoacyl-tRNA to the ribosome whilst the EF-G factor is involved in the transfer of new peptidyl-tRNA from one site to the other. Superficial ribosomal proteins may be involved too, as described later. The special steps of initiation and termination of protein biosynthesis also require special factors called *initiation factors* and *release factors*, respectively.

1.8 Genetic programming

Now we come to the controlling element in protein synthesis. The programming of the order of amino acids during protein biosynthesis is carried out by high molecular weight RNA (usually $>10^5$ daltons) called messenger RNA (mRNA), which is directly related to the genetic material, deoxyribonucleic acid (DNA). In addition to ensuring hereditary continuity the genetic material contains information for the synthesis of proteins so there must be a relation between the *four* constituent nucleotides and the *twenty* constituent amino acids of proteins. This relationship is the genetic code. The experiments designed to solve the code have yielded much important information on the mechanism of protein biosynthesis (see section 2).

4

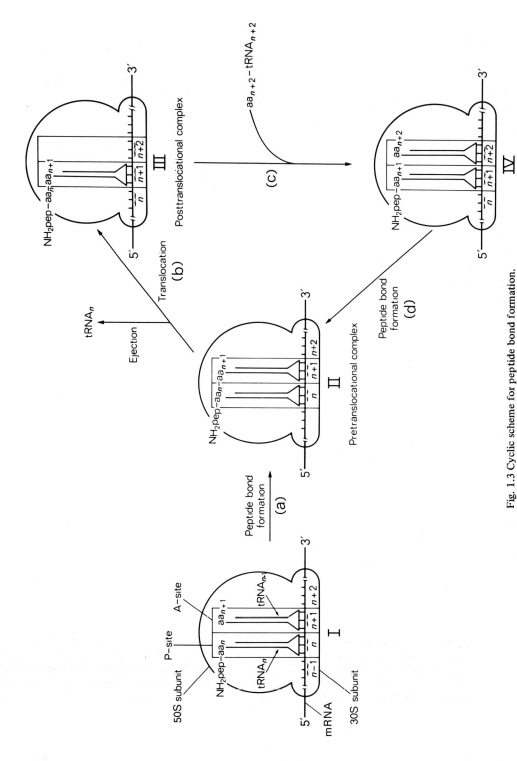

Fig. 1.3 Cyclic scheme for peptide bond formation.

1.9 Relative movement of mRNA and the ribosome

The generally accepted scheme for peptide chain elongation which involves relative movement of the mRNA and ribosome is shown as a cyclic scheme for peptide bond formation in Fig. 1.3. This scheme is simplified in that various superficial factors interacting with the tRNA or ribosome are omitted and that a static view of the ribosome involving special tRNA binding sites is assumed. At present there is not enough evidence to suggest that we drop the latter view and think in terms of a more dynamic situation involving activated binding states.

The cyclic scheme shown is really self-explanatory, starting with a situation in I with a growing peptide attached to $tRNA_n$ in the peptidyl-tRNA binding site (P-site) of a 70 S ribosome decoding codon n of the mRNA, and an aminoacyl-$tRNA_{n+1}$ decoding codon $n + 1$ in the aminoacyl-tRNA binding site (A-site). The mRNA is bound to the 30 S subunit and the tRNA stretches across both 30 S and 50 S subunits. The peptide bond is made by the enzyme peptidyl transferase on the 50 S subunit in step (a) leaving, in II, an uncharged $tRNA_n$ in the P-site and a new peptide extended by one amino acid, aa_{n+1}, attached to $tRNA_{n+1}$ in the A-site. Movement of the $tRNA_{n+1}$ and mRNA now occurs in step (b) to free the A-site (III) for a new incoming aminoacyl-$tRNA_{n+2}$ in step (c). Step (b), involving movement of $tRNA_{n+1}$ with concomitant ejection of $tRNA_n$, is usually called *translocation*. When the new aminoacyl-$tRNA_{n+2}$ is bound in the A-site as in IV, the ribosome is back to a state equivalent to I ready for a new round of peptide bond formation and translocation giving the cyclic feature to the scheme.

2 The Genetic Code

2.1 Concepts

The well-known double helical structure of DNA, first proposed by Watson and Crick, specifies base complementarity between the two chains, which run in opposite directions, i.e. they are antiparallel. The direction of the chain is referred to in terms of the direction of the sugar-phosphate links (see Fig. 1.4 where DNA is represented in two dimensions). These two concepts of base-pairing and polarity of the nucleic acid chain are fundamental to the understanding of protein biosynthesis.

Experiments using bacterial systems have established that there is collinearity between the genetic message on DNA and protein formation, i.e. the order of amino acids in the protein corresponds to the order of bases in the gene—sometimes called the *Sequence Hypothesis*. This collinearity is achieved in the cell by a two-step process shown in Fig. 1.4 which summarises the transmission of genetic information.

First one strand of the DNA is copied to give mRNA with the aid of an enzyme called RNA polymerase. This step is called transcription. Then the mRNA is translated into protein. This concept of a one-way flow in genetic information from DNA→RNA→protein has been called the *Central Dogma* especially by Crick. (For a detailed discussion of recent minor qualifications see Crick 1970.)

The mRNA is copied from a strand of DNA to give a complementary and antiparallel product just like the other strand of DNA (except that uracil replaces thymine). mRNA is synthesised from the 5'-phosphate end (or 5'-end) in units to the diol end (or 3'-end). As already stated, proteins are synthesised from the N-terminal end to the C-terminal end. Although the first evidence for this fact came from *in vivo* experiments with mammalian

6

systems (Dintzis 1961), it is well substantiated in bacterial cell-free systems. During protein biosynthesis the mRNA is translated from 5′-end to the 3′-end, i.e. in the same direction as its synthesis (see Fig. 1.4). Thus there is a directional relationship between the order of addition of amino acids during protein synthesis and the sequence of nucleotides in mRNA. This is known as the *polarity of reading of the genetic message.*

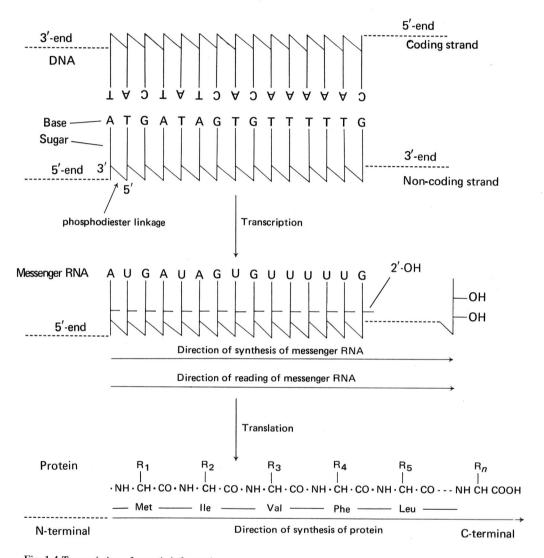

Fig. 1.4 Transmission of genetic information.

2.2 Solving the code

Clearly the most direct and convincing way of deciphering the genetic code would be to compare the nucleotide and amino acid sequences, respectively, of a piece of DNA and its specified protein. This problem was quite insoluble in the 1950s and the whole field of ingenious theoreticians who tried to make inspired guesses about the genetic code was hampered by a lack of experimental evidence. In 1960 the horizon began to brighten with the implication that the newly discovered messenger RNA, a complement of genetic material, actually carried direct information for the specification of a protein sequence. Thus, in biochemical terms the genetic code refers to the relationship of the nucleotide sequence of mRNA to the amino acid sequence of its relevant protein. The direct experimental elucidation of this relationship became a distinct possibility with the breakthrough experiment of Nirenberg and Matthaei (1961). They were able to programme the synthesis of a polypeptide containing only one amino acid by the addition of a synthetic polyribonucleotide containing only one type of nucleotide to a bacterial cell extract.

2.3 General properties

At about the same time the other most important contribution to our knowledge about the general nature of the genetic code came from *in vivo* genetic studies by Crick and his collaborators (Crick *et al.* 1961). The elegant genetic work was confirmed by *in vitro* biochemical studies several years later with the advent of synthetic mRNA of chemically defined nucleotide sequences. Additional evidence on the general nature of the genetic code came from studies of the amino acid sequence changes in variant human haemoglobins and in mutant tobacco mosaic virus (TMV) coat proteins where the changes were induced by chemical means. The general nature of the genetic code from all of these lines of evidence and from later biochemical experiments can be summarised as follows:

1. The genetic code is triplet in nature, i.e. a sequence of three nucleotides in mRNA specifies one amino acid in a protein. The group of three nucleotides used to specify an amino acid is called a codon. This property was suggested by theoretical considerations. By 1961 biochemists knew that mRNA contained only four nucleotides which have to be related to the twenty essential amino acids found in proteins. It was obvious that out of possible arrangements (4^1, 4^2 and 4^3) for the nucleotides, 4^3 (= 64) would be required to specify all twenty amino acids. Thus, in the early 1960s an arrangement of three nucleotides to specify an amino acid was assumed by the biochemists. This assumption dictated the next feature of the code which was found first by the geneticists.

2. The genetic code is degenerate. In other words one amino acid (of which there are twenty) can be specified by more than one codon (of which there are sixty-four possibilities).

3. The genetic code is non-overlapping and sequential. Thus the nucleotides are read off in groups of three sequentially from a fixed point. The codons are contiguous so that under normal conditions of translation there are no stretches of meaningless codons. Genetic evidence strongly favours the non-overlapping nature of the code since point mutation (a single base change) causes an alteration of only one amino acid in the protein.

2.4 The cell-free system

For several years before the cell-free experiments carried out by Nirenberg and Matthaei, research biochemists had been breaking cells from various source materials and attempting to reassemble a mixture of components active in synthesising proteins. Nirenberg and Matthaei were successful in constructing a relatively stable cell extract active for the synthesis of proteins and in directing the synthesis of a protein-like product by translation of a synthetic polyribonucleotide acting as mRNA; the composition of such a polypeptide-synthesising cell extract is shown in Fig. 1.5. They thus opened the way for an experimental investigation to elucidate the genetic code. During the rapid expansion of this field, to which Ochoa and his group also made a significant contribution, the mechanism of protein biosynthesis and the components involved were gradually elucidated. The separation of the crude cell extract (following the scheme shown in Fig. 1.6) into identifiable components showed that aminoacyl-tRNA is an intermediate in protein biosynthesis and confirmed the role of the ribosome as the site for peptide bond formation. For example, when polyuridylate was added to S-100 the presence of ribosomes was needed to make polyphenylalanine from added

$[^{14}C]$-Phe

> *E. coli* S-30 + mercaptoethanol, Mg^{2+}, K^+, *Tris* buffer pH 7.8, ATP, phosphoenol pyruvate, pyruvate kinase, poly U. (The S-30 cell extract is made as shown in Fig. 1.6.)

$[^{14}C]$-polyPhe

Fig. 1.5 Components of cell-free system for synthesis of polyPhe.

phenylalanine. An S-100 extract is defined according to the production scheme in Fig. 1.6 in terms of the centrifugal force used in the preparation. In addition, radioactively labelled polyPhe could be made from added $[^{14}C]$-phenylalanyl-tRNA. Thus, Phe-tRNA is an intermediate in this cell-free system. Polyribonucleotides containing more than one base component in random sequence were tested to see which other amino acids could be incorporated. Possible codon nucleotide sequences contained in these synthetic polymers were calculated from the input ratios of nucleoside diphosphate substrates during synthesis of the polymers by the enzyme polynucleotide phosphorylase. The correlation of the calculated codon frequency with the proportions of the different amino acids incorporated into the polypeptides synthesised permitted codons to be assigned to particular amino acids. In later experiments, the base compositions of the polymers were determined experimentally. This gave more reliable assignments since it was found that the experimentally determined base compositions did not always agree with the calculated figures. These experiments with heteropolymers permitted the assignment of codons only in terms of base composition (rather than base sequence) whereas homopolymers gave unambiguous results, e.g. UUU was clearly the codon for Phe. After a period of intensive effort the two main groups in the field, making use of a large variety of copolymers containing combinations of up to all four nucleotides, showed that most of the possible codons could be assigned to amino acids. Although the results using the copolymer-stimulated system were not always clear cut, and it was realised that minor effects should be treated with scepticism, the assignments were on the whole correct. One error, obvious to us now, was the assignment of the codon AAU to Lys (Speyer *et al.* 1962).

Some additional results were obtained using polyribonucleotides with ends of defined nucleotide sequence generated using the enzyme polynucleotide phosphorylase to add particular nucleotides on to a polynucleotide primer. Ochoa's group (Salas *et al.* 1965) established the direction of reading of the mRNA in protein synthesis using a block polymer of this type (Fig. 1.7). Although the effect obtained with poly ($A_{20}C$) was not absolutely convincing, the conclusion was confirmed by other methods in later years, most prettily by Streisinger and his colleagues, who determined amino acid sequences within a mutated piece of the T4 bacteriophage-induced lysozyme (Terzaghi *et al.* 1966). Difficulties arise from using block polymers such as poly ($A_{20}C$) due to the presence of nucleases (nucleic acid degradative enzymes) in the *E. coli* cell extracts. Ochoa's group overcame this problem by

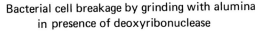

Bacterial cell breakage by grinding with alumina
in presence of deoxyribonuclease

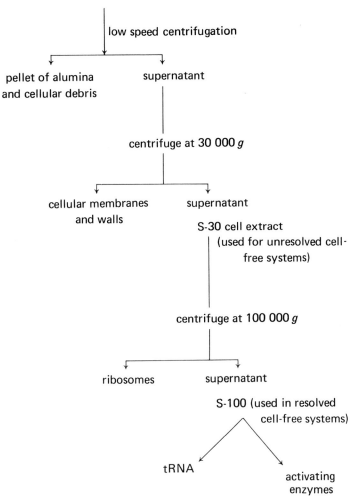

Fig. 1.6 Resolution of the bacterial cell-free system.

constructing their cell-free system from a combination of carefully washed *E. coli* ribosomes and a ribosomal supernatant from a class of nuclease-free bacteria — *Lactobacillus arabinosus*. Nowadays there are even available *E. coli* strains (e.g. MRE 600) which lack RNase I.

In 1963 it was clear that the use of random copolymers as messengers would not give the sequences of nucleotides within codons. Further progress in this field needed the isolation of a natural or synthetic mRNA of known nucleotide sequence which could be directly compared with the amino acid sequence of the polypeptide produced by translation of the mRNA in a cell-free system. The chemical synthesis of oligodeoxynucleotides of known sequence, together with the possibility of transcribing these into synthetic mRNA of known sequence, was the first step in this direction. This approach has been used uniquely by Khorana's group to confirm and establish experimentally (Khorana 1968) most of the features of protein biosynthesis related to the transfer of genetic information, but this group was not the first to elucidate the nucleotide sequences of codons.

10

Biochemical evidence

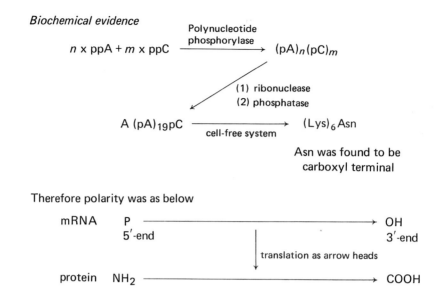

Fig. 1.7 Polarity of reading of the genetic message.

2.5 Triplet binding assay

Precise assignment of nucleotide sequences to codons came in another major breakthrough contributed to molecular biology by Nirenberg's group (Nirenberg and Leder 1964) and it came at a time when the technology involved for any predictable progress was so complicated as to be a great dampener on the enthusiasm of the biochemists in the field. By 1964 it was known that synthetic polyribonucleotides stimulated the binding of aminoacyl-tRNA to ribosomes and the complex of these three components could be identified by the method of ultra-centrifugation through sucrose density gradients. Unfortunately, this method of analysis is very tedious. Nirenberg and his colleagues devised a quick and ingenious way of isolating the aminoacyl-tRNA coupled with ribosomes and mRNA using a membrane filter made of nitrocellulose. Ribosomes were found to adsorb to these filters and so did some polyribonucleotides. However, aminoacyl-tRNA tagged by a radioactive label on the amino acid passed through the filter unless it was bound to the ribosome. In particular, Nirenberg and Leder (1964) were able to induce the binding of phenylalanyl-tRNA (Phe-tRNA) to ribosomes by poly U but not by other homopolymers, the bound complex being adsorbed on the nitrocellulose filter. Subsequently use of polyribonucleotide copolymers as synthetic messengers showed that the species of aminoacyl-tRNAs which became bound to ribosomes by the copolymers were derived from the same amino acids that were incorporated by the copolymers in the cell-free system. These experiments also demonstrated the involvement of the ribosome:messenger:aminoacyl-tRNA complex as a coded intermediate in protein biosynthesis.

Even nucleotide triplets, such as UpUpU, stimulated the formation of the complex, providing direct confirmation of the triplet nature of the genetic code. Nirenberg and his co-workers realised that trinucleotides of known sequences were easier to synthesise than mRNA. Their work, which started with the assignment of the heterobase triplet GpUpU (see Fig. 1.8 for the components in the triplet binding assay) to valine, rapidly progressed so that about fifty out of the sixty-four codon sequences were assigned, directly and convincingly, using this method.

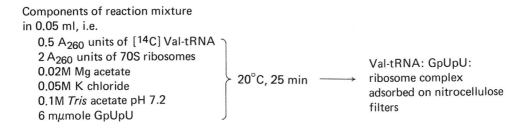

Fig. 1.8 The triplet-induced binding assay.

In some cases the binding assay gives unreliable results because the effect observed is small; the effect can sometimes be increased by using a purified tRNA species. This fact encouraged the development of methods for fractionation of mixed bacterial tRNAs so that as many tRNAs as possible, each specifically chargeable with one amino acid, could be tested in the binding assay. Indeed, fractionation has sometimes yielded several tRNA species each of which is chargeable with the same amino acid yet is usually bound by a different triplet (or set of triplets) (see section 2.8), giving direct proof of the degeneracy of the genetic code. In practice the tRNA species were seldom purified to homogeneity since it was found that a partial purification would suffice to show an effect. When the binding complexes with available triplets were still not stable enough to be detected by this method, many codons could be determined and others confirmed using the cell-free system programmed by synthetic mRNA of defined sequences as described above. Further strong support for the accuracy of the *in vitro* system for codon assignments comes from biochemical analysis of altered polypeptides produced from *in vivo* mutagenic events.

2.6 RNAs of repeating sequences

Some of the best experimental evidence for the assignment of codon sequences has come from Khorana's laboratory using synthetic mRNAs of known nucleotide sequences (Khorana 1968). The problem of relating the nucleotide sequence of mRNA directly with the amino acid sequence of its polypeptide product was solved by an admirable combination of organic chemical synthesis and enzymic synthesis. At first the reiterative copying mechanism of the enzyme RNA polymerase was employed to make long chain polyribonucleotides by copying sequentially from a series of complementary oligodeoxyribonucleotides of about twelve units long. Although this method worked when the oligodeoxyribonucleotide contained a repeating sequence of two different nucleotides, the yield was so low as to be almost negligible when a repeating sequence of three different nucleotides was used. Fortunately, the difficulty was overcome by using the scheme outlined in Fig. 1.9. The enzyme DNA polymerase extends a short double-helical oligodeoxyribonucleotide preparation of known repeating nucleotide sequence as exemplified in the diagram. This chain extension is probably accomplished by one chain slipping relative to the other (in this case, by two nucleotides) yet maintaining base pairing, thus exposing ends on opposite chains. The enzyme then incorporates nucleotides from deoxynucleoside triphosphates to fill in the gaps at the ends of the double helix. The long chain double-stranded synthetic DNA made in this way constitutes a synthetic gene of known

12

sequence which can be used many times since it is easily recovered from the cell-free system. A polypeptide, whose amino acid sequence is determined by analysis of its radioactive constituents, is synthesised in a two-stage reaction in the second stage of which the cell-free extract is added to the mRNA synthesised by RNA polymerase using the synthetic DNA as template. Figure 1.9 makes it clear that to make the specified polyribonucleotide, poly (U-G),

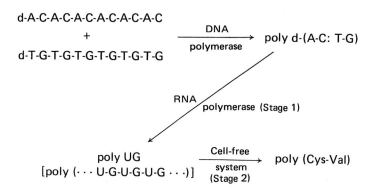

Fig. 1.9 Use of synthesised complementary polydeoxyribonucleotides for directing the two stage cell-free synthesis of defined polypeptides.

containing U and G in a repeating sequence, only the triphosphates containing U and G are added to the reaction to ensure the copying of only one strand of the DNA. Since the polypeptide in the example shown contained a repeating cysteine-valine sequence, and UGU was known to be a codon for cysteine from the binding assay described above, it was possible to assign the other possible codon, GUG, from the repeating UG sequence -U-G-U-G-U-G-, to valine. This assignment was in fact made before it was confirmed by the triplet binding assay.

2.7 Special features of the code

By the end of 1966 the elucidation of the genetic code was complete. The full assignment of codon sequences is shown in Fig. 1.10, a diagram based on an arrangement first proposed by Crick. As predicted by the geneticists there are very few nonsense codons. Genetic and biochemical studies have indicated that the codons UGA, UAG and UAA are signals for polypeptide chain termination. These codons do not normally signify amino acids but can be decoded by suppressor tRNAs; this will be described in section 9. It is likely that these codons signify chain termination in eukaryotic as well as bacterial cells.

Polypeptide chain initiation differs strikingly from termination in that the assigned codons for initiation also signify an amino acid when the codon occurs internally in the reading of messenger RNA. The results from *in vitro* experiments identify AUG and GUG as initiation codons whereas they also signify methionine and valine, respectively, for internal translation. Whether the triplet alone is a complete initiation signal will be discussed later (section 7). So far, although AUG has been identified as an initiation codon in many natural mRNA sequences GUG has been found only once (see section 7.15). The significance of this finding is unclear at present. Termination signals are not usually decoded by a special tRNA whereas initiation signals are decoded by a special type of initiator tRNA, formylmethionyl-tRNA in bacteria and mitochondria and a special non-formylated methionyl-tRNA in eukaryotic cytoplasm.

SECOND LETTER

		U	C	A	G	
FIRST LETTER	U	UUU ⎱ Phe UUC ⎰ UUA ⎱ Leu UUG ⎰	UCU ⎱ UCC ⎟ Ser UCA ⎟ UCG ⎰	UAU ⎱ Tyr UAC ⎰ UAA ⎱ Ter UAG ⎰	UGU ⎱ Cys UGC ⎰ UGA Ter UGG Trp	U C A G
	C	CUU ⎱ CUC ⎟ Leu CUA ⎟ CUG ⎰	CCU ⎱ CCC ⎟ Pro CCA ⎟ CCG ⎰	CAU ⎱ His CAC ⎰ CAA ⎱ Gln CAG ⎰	CGU ⎱ CGC ⎟ Arg CGA ⎟ CGG ⎰	U C A G
	A	AUU ⎱ AUC ⎟ Ile AUA ⎰ AUG Met	ACU ⎱ ACC ⎟ Thr ACA ⎟ ACG ⎰	AAU ⎱ Asn AAC ⎰ AAA ⎱ Lys AAG ⎰	AGU ⎱ Ser AGC ⎰ AGA ⎱ Arg AGG ⎰	U C A G
	G	GUU ⎱ GUC ⎟ Val GUA ⎟ GUG ⎰	GCU ⎱ GCC ⎟ Ala GCA ⎟ GCG ⎰	GAU ⎱ Asp GAC ⎰ GAA ⎱ Glu GAG ⎰	GGU ⎱ GGC ⎟ Gly GGA ⎟ GGG ⎰	U C A G

THIRD LETTER

Fig. 1.10 Codon assignments. (After Crick 1966.) Ter indicates a termination (or nonsense) codon.

Although the genetic code was elucidated by *in vitro* experiments, the correctness of the codon sequences has been confirmed by analysis of mutant proteins and recently by actual sequencing of the complete coat protein cistron of a naturally occurring messenger RNA, from bacteriophage MS2 (Min Jou *et al.* 1972). Where amino acid sequence analyses of mutant proteins have identified amino acid replacements, the codons for these amino acids are related by a single base change arising from a single point mutation in the gene.

Efforts to prove the universality of the bacterial genetic code stimulated the search for eukaryotic cell-free systems active for synthesising proteins. Results obtained from such systems support the contention that the code is indeed universal. Furthermore, now that messenger RNA sequences are being determined it is possible to decide which codon of a degenerate set is used for a particular amino acid and how often. Different frequencies of codon usage may turn out to be significant in the regulation of protein synthesis since certain aminoacyl-tRNAs occur in controllable small amounts.

Although few mRNA sequences are at present known some codons, such as AUA (isoleucine) and UUA (leucine) appear to be used infrequently. In fact AUA had not been found at all in natural mRNA until recent work attempting to determine the sequence of the whole MS2 bacteriophage RNA identified this codon in the A-protein gene (Contreras *et al.* 1973).

Various attempts have been made by theoreticians to find underlying rules in the codon assignments shown in Fig. 1.10. By observation it is seen that in general the left-hand side of the table contains the more non-polar (hydrophobic) amino acids. The origin of the genetic code is related to the origin of life. The difficulty of theorising on this problem is well illustrated by the papers of Crick (1968) and Orgel (1968).

2.8 Wobble hypothesis

The most helpful rationalisation of patterns occurring in the genetic code has been made by Crick in his 'wobble hypothesis' (Crick 1966). This is derived from knowledge of the role of tRNA in the transfer of information from mRNA to protein.

An elegant experiment by Chapeville and his colleagues (Chapeville *et al.* 1962) in the days when random copolymers were used as synthetic mRNA showed that once the amino acid was attached to a tRNA, it could be converted into another amino acid and still be coded for by the original amino acid codon. Thus, the tRNA was the recognition feature for transfer of coding information. It was predicted that the tRNA would recognise the mRNA by the same base-pairing mechanism that operates in double-stranded DNA, and that since the code is triplet in nature a sequence of three nucleotides in the tRNA might be its decoding feature. This prediction turned out to be correct and the position of the three nucleotides, the anticodon, in the sequences of tRNAs has been well established (see section 4).

Crick noticed that amino acid codons could be grouped in sets according to differences in only the base of the third position of the codon. For example, UU (U or C) coded Phe, UU (A or G) coded Leu and it was thought that in a yeast system GC (U, C or A) coded Ala. A quick incursion into the realms of model building enabled him to suggest that certain non Watson–Crick base pairs (i.e. other than G·C and A·U) were possible, with only slight distortion of a double-helix. He proposed that such a slight distortion could be possible during the interaction of mRNA with tRNA, so that the new rules of pairing of the third base [3′-end] of the codon with the first base [5′-end] of the anticodon (notice that codon and anticodon are complementary and antiparallel) were devised as shown in Fig. 1.11. Even the first two tRNA sequences published supported the hypothesis: they gave the anticodon assignments IGC (5′→3′) for alanine tRNA and IGA for serine tRNA. Although these tRNAs came from yeast the Ala-tRNA from yeast was coded on bacterial ribosomes in the triplet binding assay by GC (U, C or A) consistent with the anticodon IGC. Determinations of nucleotide sequences of anticodons of other bacterial tRNAs show a similar pattern of coding relationships consistent with the wobble hypothesis. An important aspect of the hypothesis is that it readily explains the existence of a high degree of *ambiguity* within the multiple species of tRNA, i.e. having more than one codon for the same tRNA. For example, two species of phenylalanyl-tRNA could

Third position of codon		First position of anticodon
A or G	————————	U
G	————————	C
U	————————	A
U or C	————————	G
U,C or A	————————	I

I is inosine, a minor base found in yeast tRNA. It
can be formed by deamination of the 6-NH_2 group of A.

Fig. 1.11 Predicted base pairing for the third position of a codon with tRNA. 'The Wobble Hypothesis.' (From Crick 1966.)

have anticodons GAA and AAA with codons UU (U or C) and UUU respectively. The details of the mechanism whereby UUU might, for example, be translated by one species of tRNA in preference to the other are unknown.

3 Amino acid activating enzymes

3.1 Introduction

In the late 1950s it was found that amino acids could be enzymically bound to a then uncharacterised class of RNA by a soluble protein fraction from rat liver. The protein fraction, which was soluble at pH 7, was isolated by precipitation at pH 5. This so-called pH 5 enzyme fraction was used in the early studies of amino acid activation in protein biosynthesis. The name 'activating enzyme' is used here because it relates its role to the protein biosynthetic reactions under discussion. Alternative names for this enzyme activity which catalyses the linkage of amino acid to tRNA are aminoacyl-tRNA synthetase, and the systematic, but cumbersome, name — amino acid : tRNA ligase (AMP). Up to the present bacterial, bovine pancreas and yeast enzymes have been obtained in a pure state.

Amino acid activating enzymes constitute up to 10 per cent of the cell's protein. This is equivalent to 1000–5000 molecules per cell. (These figures are, of course, approximate since the amount may vary under different growth conditions and the molecular weights of most of the enzymes are not known.)

At present, in bacterial cells at least, there appears to be only one activating enzyme for each of the twenty naturally-occurring amino acids; glutaminyl- and glutamyl-tRNA (and asparaginyl- and aspartyl-tRNA) are made by different enzymes. Modifications such as formylation of methionine for polypeptide chain initiation occurs after charging the tRNA with the amino acid; there is no enzyme capable of charging a tRNA with formylmethionine. Other modifications of amino acids in proteins, such as acetylation, are thought to occur after the protein has been made.

In general it is assumed that only twenty activating enzymes are found in all biological material. However, in a few cases multiple species of a particular amino acid activating enzyme have been detected. For instance, two species of phenylalanine activating enzyme have been isolated from *Neurospora crassa,* a fungus. The existence of more than one activating enzyme for the one amino acid in eukaryotic cells may be due to compartmentation where it is likely that one enzyme will be cytoplasmic and one mitochondrial.

Since there is only one activating enzyme for charging several different tRNAs *(isoaccepting species)* with the same amino acid it was hoped that a knowledge of the several tRNA primary structures would reveal how the activating enzyme recognised them. Unfortunately, however, these studies have not been very helpful (see section 3.2 and section 4).

3.2 Structure

How the activating enzyme distinguishes one tRNA from another is still a mystery in spite of much physico-chemical investigation. What is really needed, of course, is an X-ray crystallo-graphic study of a complex of an activating enzyme and its specific (or cognate) tRNA. Such a complex has not been crystallised although there are now available for X-ray analysis crystals of both tRNAs and activating enzymes. The tRNA crystals will be described in section 4. The lysine activating enzyme from yeast was the first to be crystallised but the crystals were not suitable for high resolution X-ray analysis. At present good crystals are available for yeast leucine activating enzyme of mol. wt. 120 000 daltons, *Bacillus stearothermophilus* tyrosine activating enzyme (mol. wt. 95 000) and a fragment of *E. coli* methionine activating enzyme of mol. wt. 66 000 containing the part of the enzyme active for loading the tRNA with methionine.

Although any classification of the activating enzymes is very tentative at present they can be considered as falling into three classes: I consists of monomeric enzymes containing one polypeptide chain (α) of mol. wt. about 110–120 000 daltons, e.g. yeast leucine, *E. coli* valine and isoleucine activating enzymes; II describes dimeric (α_2) enzymes of total molecular weight ranging from about 100 000 to 180 000, e.g. *B. stearothermophilus* tryptophan and tyrosine, *E. coli* proline and tryptophan activating enzymes (mol. wt. about 100 000) and *E. coli* methionine activating enzyme (mol. wt. about 180 000); III consists of enzymes containing four subunits, e.g. $\alpha_2\beta_2$, *E. coli* or yeast phenylalanine activating enzyme (mol. wt. 280 000).

Several activating enzyme species are now available in amounts up to 1 gram. Determination of the sequence of the *E. coli* methionine activating enzyme is now under way, and others will doubtless soon be attempted.

3.3 Function and reaction mechanism

Activating enzymes catalyse the esterification of an amino acid to tRNA. This is usually referred to as 'loading' or 'charging' the tRNA. Studies on the mechanism of the charging reaction have yielded the generally acceptable separation of the overall reaction into two distinct steps as in Fig. 1.12: step I is the activation reaction and step II is the transfer reaction. An aminoacyl-adenylate formed in step I remains enzyme-bound and transfers its amino acid to tRNA in step II. Either step can be used to assay the enzyme activity.

The overall charging reaction requires the presence of magnesium ions and satisfies its energy requirements by the splitting of ATP to AMP and inorganic pyrophosphate, PP_i. Different activating enzymes have different pH optima. Although some of the activating enzymes work best at a pH between 8 and 9 the charging of the tRNA is usually carried out at about pH 7·5 because of the instability of the aminoacyl-ester bond formed (Fig. 1.12) even under mildly alkaline conditions.

Both the steps shown in Fig. 1.12 are reversible. The equilibrium constant for the overall reaction varies between 0·3 and 0·7 depending on the enzyme. This shows that the aminoacyl-ester bond is of high potential energy, comparable with that of the pyrophosphate bond in ATP. In step I of the charging reaction the amino acid is joined to the 5'-phosphate of AMP in an acyl phosphate (a mixed acid anhydride) linkage. This very reactive acyl phosphate, as an enzyme complex (as shown in the diagram), transfers its acyl group in step II to another adenosine residue – the terminal adenosine of the tRNA – in an aminoacyl-ester linkage with a hydroxyl of the terminal 2', 3'-diol group. The point of attachment is generally taken to be the 3'-OH but since aminoacyl migration can occur 10^4 times per second between these two hydroxyl groups – nearly 1000 times faster than peptide bond formation – it is meaningless to distinguish between them for amino acid attachment to tRNA. Actually, very strong evidence has recently been obtained that charging attaches the amino acid to the 2'-OH group of the tRNA and that migration to the 3'-OH group is necessary before peptide bond formation can occur. The significance of this migration is unclear.

A word of caution must be inserted here about the generally accepted two-step reaction. It has recently been attacked as artifactual by Loftfield (1972). He has produced evidence that a

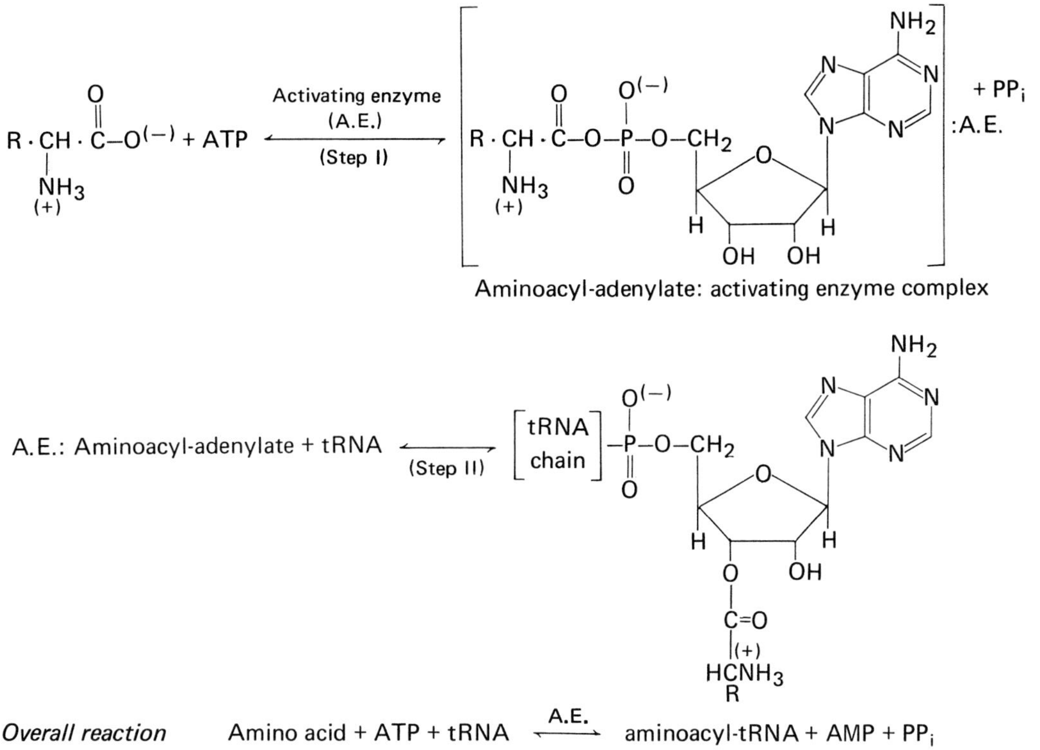

Fig. 1.12 Charging of tRNA with an amino acid in a two-step enzymic process.

concerted one step reaction is more likely to occur *in vivo*. At present therefore it is prudent to retain an open mind about the *in vivo* situation.

3.4 Assay

The activating enzyme is assayed for ability to charge tRNA with an amino acid in a buffered solution containing magnesium (approx. 10 mM); the amino acid, ATP and tRNA are in excess so that the initial rate of esterification of the amino acid to tRNA is dependent upon the amount of enzyme. The aminoacyl-tRNA, radioactively labelled in the amino acid, is precipitated with cold trichloroacetic acid, collected by filtration on cellulose nitrate or cellulose acetate filters and the dried radioactive precipitate is quantitated either by liquid scintillation counting or by gas flow counting. (Obviously the same scheme can be used to assay for the presence of tRNA if the activating enzyme is added in an excess to make the amount of tRNA rate-limiting.)

  Since both stages of the overall charging reaction are reversible it is possible to discharge an aminoacyl-tRNA (i.e. remove the amino acid) by the addition of AMP and pyrophosphate in the presence of the activating enzyme under charging conditions lacking ATP.

  The two steps of the charging reaction can be assayed separately. For instance, the formation of the aminoacyl-adenylate bound to the enzyme can be conveniently assayed by its back reaction. Radioactive pyrophosphate supplied in the absence of tRNA will equilibrate with unlabelled ATP so that radioactive ATP is obtained by the breakdown of the aminoacyl-adenylate: enzyme complex. The radioactive ATP can be conveniently isolated by its adsorption on charcoal. Since the rate of aminoacyl-adenylate formation is the rate-determining step in the exchange reaction, the amino acid dependent incorporation of radioactive pyrophosphate into ATP measures the rate of aminoacyl-adenylate formation. However, the rate determining step in the overall reaction is the second transfer step. Furthermore, depending on the enzyme the ratio of the two rates varies from 10 to a 100.

3.5 Accuracy

The two steps in the charging reaction clearly permit the activating enzyme to carry out two specific checks for ensuring accuracy in protein biosynthesis.

First, the correct amino acid is recognised by the enzyme in the formation of the aminoacyl-adenylate complex. Amino acid analogues have only a low chance of competing in this reaction. For example, the methionine activating enzyme forms an ethionyl-adenylate complex at one-hundredth the rate of that of the methionyl-adenylate complex.

Even if the activating enzyme were to make a mistake in choosing its amino acid, the second specificity check, for the correct tRNA, is very accurate. Although in general the activating enzymes do not recognise a wrong amino acid and join it to AMP there are exceptions, the first being discovered by Berg and his colleagues (Berg *et al.* 1961). For example, bacterial isoleucine activating enzyme will form valyl-adenylate albeit with a lower affinity of binding than for the isoleucyl derivative. However, the accuracy of protein synthesis is maintained in the transfer reaction, since the isoleucine enzyme will not form valyl-tRNA from the adenylate complex.

3.6 Recognition of tRNA

How the activating enzyme recognises its cognate tRNA is unknown and is the subject of much intensive research. Comparison of the primary structures of tRNAs charged by the same enzyme has not given an answer. Examples of this approach will be found in the section on tRNA structure and function (section 4). The most detailed information comes from a study of single base changes in mutant tRNAs but how such base-changes alter the tertiary structural recognition points cannot yet be elucidated. Information on the three-dimensional structure of the enzyme and tRNA is clearly needed before a proper interpretation is possible.

When the recognition site of a tRNA for its activating enzyme is discussed an uncertainty arises. There is no guarantee that every tRNA is recognised in a similar fashion by some property of a similarly located set of nucleotides. Perhaps the recognition of particular tRNAs by their specific activating enzymes has evolved differently, so that by now there are different classes of tRNA recognition features. If this is correct no generalisation about the recognition process will be possible when details of recognition of one tRNA by its activating enzyme are known.

4 The Adaptor

4.1 Introduction

In the late 1950s it was realised, especially by Crick, that there seemed to be no simple way in which nucleic acids could programme the synthesis of protein by direct structural interactions with amino acids. Thus, Crick proposed his *Adaptor Hypothesis* whereby an adaptor molecule (at that time vaguely specified) would intervene between the amino acid and the nucleic acid which carried information for directing the amino acid sequence. Shortly afterwards, in 1957,

Hoagland discovered in a rat extract a type of RNA which could bind amino acids specifically. This discovery of what was soon recognised as the missing adaptor molecule stimulated an ever expanding amount of research into the relationship between structure and function of tRNA.

4.2 Characterisation

In addition to decoding genetic information carried by messenger RNA, tRNAs function by carrying esterified and activated amino acids to the ribosomal site for peptide bond formation during protein biosynthesis.

Although tRNA and rRNA (ribosomal RNA) make up about 20 and 80 per cent of total RNA whilst the less stable *(in vivo)* mRNA accounts for about 2 per cent, as little as 1 per cent of the total DNA functions as template for the synthesis of both tRNA and rRNA, in roughly equal amounts. Evidence for the genome content of tRNA and rRNA comes from hybridisation studies. (Hybridisation involves the formation of stable double-strands, containing single strands of DNA gene with its specific complementary RNA.)

The total tRNA is estimated to make up about one per cent of the bacterial cell's dry weight. In a rapidly growing bacterial cell there are of the order of 4×10^5 tRNA molecules of perhaps fifty different types (the exact number of species is not known). There is no definite estimate of how much of the tRNA in the cell is charged with an amino acid, but it is likely that the catalytic activity of the activating enzymes is capable of keeping the cellular tRNAs fully charged for peptide bond formation if there are sufficient free amino acids available.

4.3 Isolation

Transfer RNA is usually extracted from bacterial cells with buffered aqueous phenol and can be obtained almost free of large RNA and DNA which are presumably trapped inside the cells. Addition of ethyl alcohol to the aqueous layer precipitates the tRNA; the phenol layer contains proteins and cell debris. Several further precipitations from salt solutions, then an extraction of the tRNA into methoxyethanol from phosphate buffer, to remove polysaccharides and traces of DNA, and a final ethanol precipitation from an aqueous organic mixture yields a tRNA preparation which can be used for functional tests or further purified by column chromatography to give the individual tRNA species.

4.4 Multiplicity

Since tRNA as isolated can be charged with many amino acids in a non-competitive way it is obviously a mixture of many tRNA species, each capable of charging with a particular amino acid. In addition, the genetic code studies made it clear that several tRNAs can be charged with the same amino acid. Thus the multiplicity of tRNA species is such that whilst the precise number in bacterial cells is unknown about fifty species are generally thought to be present.

4.5 Function in the cell-free system

The direct participation of tRNA as a carrier of the amino acid in protein synthesis was first shown in studies using a bacterial cell-free system; a synthetic mRNA, poly U, stimulated the transfer of phenylalanine from Phe-tRNA into polyPhe (see section 2).

In 1962 a group of biochemists (Chapeville, *et al.* 1962) proved that the interaction of the aminoacyl-tRNA with a specific coding sequence in messenger RNA is independent of the amino acid in the complex. Thus, once the amino acid is attached to the tRNA by its specific activating enzyme it plays no part in messenger recognition; the information is carried by the tRNA adaptor. In a classical experiment Chapeville and his co-workers (Fig. 1.13) altered the

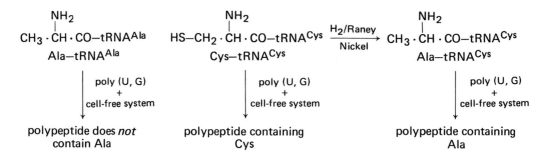

Fig. 1.13 An experiment to show that alteration of the amino acid part of aminoacyl-tRNA does not change the decoding properties of the tRNA.

cysteine residue of charged Cys-tRNA^{Cys} to alanine by treatment with Raney nickel, giving Ala-tRNA^{Cys}. (Uncharged tRNA which is capable of being charged with cysteine is referred to as tRNA^{Cys} whereas the full name for the charged tRNA is Cys-tRNA^{Cys}. If there is more than one isoaccepting species they are usually designated by subscripts.) From earlier studies using the cell-free system it was known that poly (U, G) random copolymer, stimulated the incorporation of cysteine but not of alanine. However, when the synthetic Ala-tRNA^{Cys} was added to the cell-free system programmed by poly (U, G) then Ala was incorporated into the polypeptide. Thus, the alteration of the amino acid did not alter the coding response of the tRNA to which the amino acid was attached. This was confirmed in a mammalian system in which cell-free extracts of rabbit reticulocytes supplemented with the altered Cys-tRNA^{Cys} (i.e. Ala-tRNA^{Cys}) synthesised rabbit haemoglobin containing alanine at positions where cysteine normally occurs, again showing the unique interaction of tRNA with a messenger RNA independent of the nature of the amino acid attached. This has also been confirmed using tRNAs deliberately charged with the wrong amino acid.

4.6 Fractionation of tRNA species

Pure tRNAs were required for two reasons: to provide pure aminoacyl-tRNAs for work on the genetic code (as already described in section 2) and so that determination of their nucleotide sequences (primary structures) would be possible.

Since all tRNA molecules have similar properties and are all about eighty nucleotides long, a combination of fractionation methods is usually required to achieve complete purification of a single species. Countercurrent distribution which was used successfully in the isolation of the tRNA species (yeast tRNA^{Ala}) whose structure was the first to be elucidated, was notable among the early methods used (see Doctor 1971). Indeed it is still used but usually in combination with some type of column chromatography.

Although I shall be discussing the purification of unlabelled tRNA, some of the chromatographic methods, suitably scaled down, can be used to purify of the order of half a milligram of [^{32}P]-labelled tRNA needed for the rapid sequencing methods recently developed in Sanger's laboratory (see Barrell 1971).

The most widely applicable column chromatographic methods are three systems involving: 1, an inert diatomaceous earth called chromosorb (Kelmers et al. 1971); 2, DEAE-Sephadex (Nishimura 1971); and 3, benzoylated DEAE-cellulose (Roy et al. 1971). The first of these systems uses chromosorb impregnated with a solution of a quaternary amine in isoamylacetate constituting a stationary phase. The tRNA species separate when applied to such a column in a buffered salt gradient. This system clearly resembles countercurrent distribution with the organic phase held stationary and is usually called 'reversed phase partition chromatography'. Although the system has a high resolving power and has been used to purify phenylalanine tRNA from *E. coli* there is a strong suspicion that it can inactivate some tRNA species. In this respect the straightforward ion-exchange chromatography employed in the second system has advantages; the tRNA is eluted with an increasing salt gradient. The third system is an invention of Tener (1967) (see Roy et al. 1971). The benzoylation of DEAE-cellulose (BD-cellulose) makes the adsorbing material less polar and more aromatic in character resulting in a stronger affinity for molecules with aromatic character. This is more marked when the tRNA is charged, i.e. in the form of aminoacyl-tRNA. Clearly, if any of the aminoacyl-tRNAs are modified to have aromatic character then they also will be more strongly adsorbed to the column. This has become the basis of a general method for isolation of a particular aminoacyl-tRNA from a mixture of tRNAs. The mixture is charged with one amino acid which is then modified by the addition of a phenoxyacetyl group. The modified aminoacyl-tRNA is then the last to be eluted from the BD-cellulose column.

4.7 Test of purity

The most reliable test of purity of a tRNA is proof of a unique nucleotide sequence. However, this is not usually feasible, although a two-dimensional paper electrophoretic fingerprint of an enzymically digested sample of [^{32}P]-labelled tRNA can be carried out fairly rapidly for this purpose. Unlabelled tRNA is generally estimated by means of a charging test. If the molar ratio of amino acid to tRNA approaches unity when the tRNA is enzymically charged under maximum conditions then the tRNA is taken to be pure. Of course, this assumes that the sample does not contain isoaccepting species. This possibility is checked by column chromatography using several different systems.

4.8 Sequence determination and minor bases

Methods of determining the nucleotide sequence of RNA are only briefly outlined here (for more details see Barrell 1971). First of all, the pure tRNA species is completely digested with T-1 RNase (which splits the RNA after Gp residues) and also separately with pancreatic RNase (which splits RNA after pyrimidine nucleotides). The nucleotide sequences of the short (usually < ten units) oligonucleotides produced are determined by the use of other RNA splitting enzymes with different specificities, usually combinations of venom phosphodiesterase and spleen phosphodiesterase and perhaps U-2 RNase. Analysis of the complete T-1 and pancreatic RNase products (usually called the primary digestion products) often shows that some of the oligonucleotides overlap. The task of overlapping short fragments is made easier by the fact that tRNA contains several inherent markers in the form of minor (rare) bases which are different from the usual four. (Incidentally, many of these minor bases are peculiar to tRNA. They are not found in messenger RNA and only a few methylated bases occur in ribosomal RNA in

22

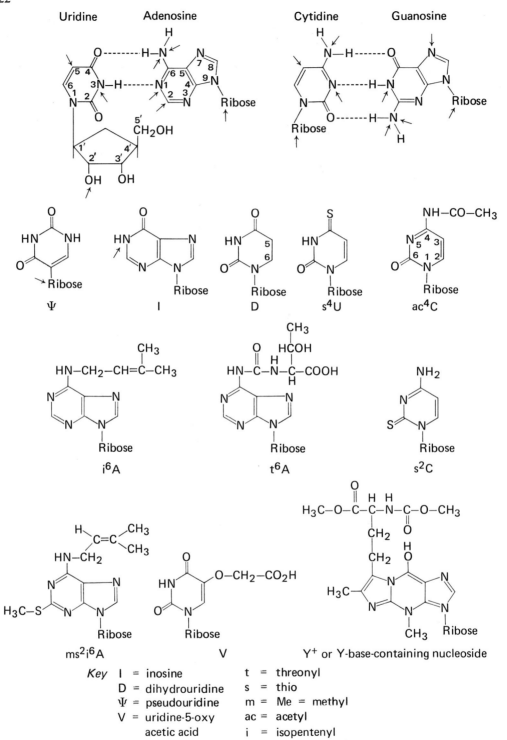

Fig. 1.14 Some minor nucleosides found in tRNA. The normal four bases are given in base pairs for reference. The arrows indicate positions in which methyl group substitutions have been found, in, for example, 2'-O-methyl guanosine (Gm), 2'-O-methyl cytidine (Cm), 5-methyl cytidine (m⁵C).

addition to the usual four.) A partial listing of tRNA minor base structures is given in Fig. 1.14. The full list is continually being expanded. Very little is known about the enzymic mechanisms producing these rare bases (see section 4.10). Partial digestion of tRNA with either T-1 RNase or pancreatic RNase under mild conditions gives large fragments of tRNA which can in turn be degraded further in a primary digestion (see above) to give the constituent oligonucleotides which can then be put in order. This method of overlapping small fragments to obtain long sequences, analogous to the methodology for determining the sequence of proteins, has thus led to complete primary structures of many tRNAs.

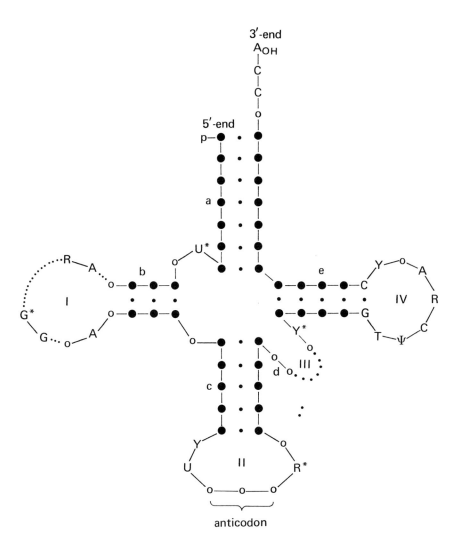

Fig. 1.15 General clover-leaf structure for tRNA. Full circles are H-bonded bases in base pairs. Open circles are bases not in clover-leaf base pairs. H-bonds in base pairs are represented by big dots. R = purine base. Y = pyrimidine base. * indicates that the nucleotide may be modified. Base-paired regions (stems) are numbered a to e and non base-paired bases are in loops I to IV. The dotted part of loops I and III indicate variation on number of nucleotides.

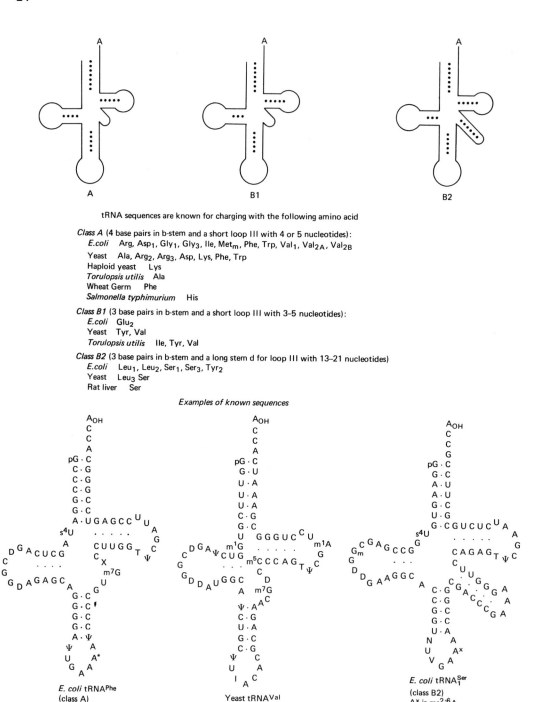

tRNA sequences are known for charging with the following amino acid

Class A (4 base pairs in b-stem and a short loop III with 4 or 5 nucleotides):
E.coli   Arg, Asp_1, Gly_1, Gly_3, Ile, Met_m, Phe, Trp, Val_1, Val_{2A}, Val_{2B}
Yeast   Ala, Arg_2, Arg_3, Asp, Lys, Phe, Trp
Haploid yeast   Lys
Torulopsis utilis   Ala
Wheat Germ   Phe
Salmonella typhimurium   His

Class B1 (3 base pairs in b-stem and a short loop III with 3–5 nucleotides):
E.coli Glu_2
Yeast   Tyr, Val
Torulopsis utilis   Ile, Tyr, Val

Class B2 (3 base pairs in b-stem and a long stem d for loop III with 13–21 nucleotides)
E.coli Leu_1, Leu_2, Ser_1, Ser_3, Tyr_2
Yeast Leu_3 Ser
Rat liver   Ser

Examples of known sequences

E. coli $tRNA^{Phe}$
(class A)
[A^* is ms^2i6A]
[X is unknown]

Yeast $tRNA^{Val}$
(class B1)

E. coli $tRNA^{Ser}_1$
(class B2)
A^x is ms^2i6A
V is U-5 oxyacetic acid
N is unknown

Fig. 1.16 Classes of tRNA secondary structure.

4.9 Structural properties

The sedimentation coefficient of tRNA (about eighty nucleotides long and mol. wt. about 25 000) determined by sucrose density gradient centrifugation is 4 S. At least thirty-five tRNA sequences are known. All tRNAs start with a free phosphate at the 5'-end of the polynucleotide and end with a common sequence CpCpA at the 3'-end. The 3'-adenosine has a free *cis* vicinal 2',3'-diol to which the amino acid becomes attached as already described. The remarkable feature of the known primary structures is that they can all be fitted to a base-paired secondary structure — the 'clover leaf' structure (Fig. 1.15) — first proposed by Holley for yeast alanine tRNA (Holley *et al.* 1965). Such a clover leaf structure will accommodate the constant features of most known tRNA sequences (Fig. 1.15). The exceptions so far are (i) yeast and *E. coli* initiator methionine tRNAs, (ii) tRNAs involved in cell wall metabolism rather than protein synthesis, (iii) a *Salmonella typhimurium* histidine tRNA and (iv) a bacteriophage-coded leucine tRNA.

Most regions of the tRNA structure are remarkably constant, e.g. stems a, c, and e have 7, 5 and 5 base pairs respectively and loops II and IV each contain 7 non-base-paired nucleotides. The variable regions are confined to stems b and d and loops I and III. The loops are often referred to by trivial names, e.g. loop I as the D-loop, since it usually contains some dihydro U bases; loop II as the anticodon loop; loop III as the variable 'finger' and loop IV as the GTΨC-containing loop, in which most of the nucleotides are constant. Stem a is often called the amino acid stem since the amino acid is attached to its 3'-terminal OH. The secondary structures of tRNA may be conveniently classified as A or B depending on whether the number of base pairs in stem b is 4 or 3 respectively; then a further subdivision can be made into B1 and B2 where these types differ by the number of nucleotides in stem d and loop III (see Fig. 1.16). Examples of this classification are listed in Fig. 1.16.

Although very little is known yet about the detailed three-dimensional structure of tRNA there is much evidence supporting the clover leaf arrangement in two dimensions. Physico-chemical studies measuring ultra-violet absorption changes with increasing temperature (melting curves), nuclear magnetic resonance studies identifying base-paired protons, Raman and infra-red spectral studies and tritium exchange studies have all given evidence for base-paired helical regions in tRNA. Light-scattering measurements and low angle X-ray scattering in solution have suggested that the tRNA molecule is long and thin. Many other experiments involving susceptibility of parts of the structure to enzymic cleavage, or to chemical reagents thought to be specific for single-stranded regions have confirmed the availability of the putative bases of loops I and II for modification. Chemical modification has been used to investigate the folding of the tRNA molecule but suffers from the drawback that reaction at one site may cause partial unfolding and give misleading results. However, one fact that seems certain from experiments of this type is that loop IV is somehow buried in the tertiary structure so that it is not available for reaction unless the molecule is unfolded by heating. The only unambiguous way of obtaining tertiary structural details is X-ray analysis. Since 1968 (Clark *et al.* 1968) many different tRNA species have been crystallised. However, only crystals of yeast tRNA[Phe] have so far yielded high resolution X-ray diffraction patterns (Kim *et al.* 1971). Because the X-ray work has progressed slowly there have been many ingenious attempts at building models for the tertiary structure of tRNA based on the available biochemical and physico-chemical evidence. The most detailed model is that due to Levitt (1969). The overall folding plan is shown in Fig. 1.17. However, it is very hard to visualise structures from such drawings. For example, the space-filling model of such a structure built for bacterial initiator tRNA is shown in Fig. 1.18.

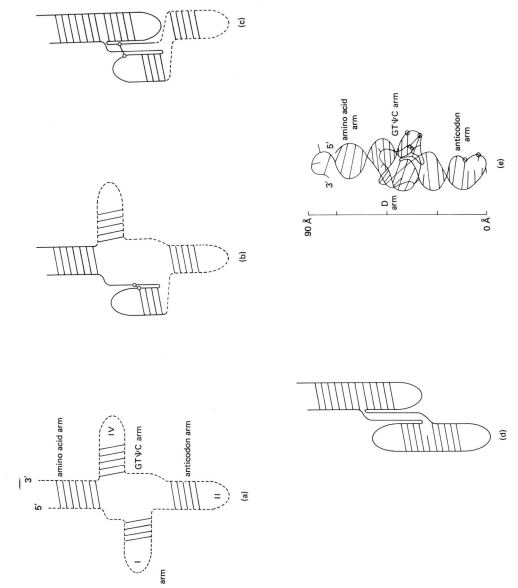

Fig. 1.17 A schematic diagram showing how indirect evidence can be used to define a tertiary structure of transfer RNA.

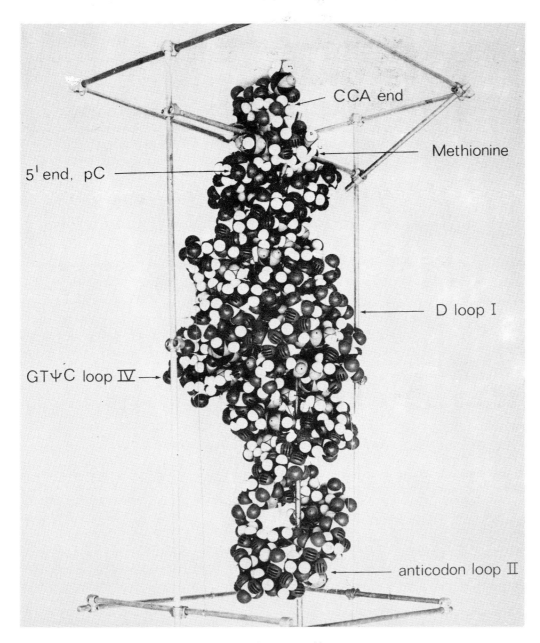

Fig. 1.18 Space filling model proposed for structure of tRNA$_f^{Met}$ by M. Levitt (1969). The dimensions are roughly 90 x 35 x 25Å.

Briefly, the model is constructed in the following way (see Fig. 1.17): (an arm is used to describe a stem plus a loop).

(a) Clover leaf secondary structure (supported by thirty-five independent tRNA sequences).
(b) Positioning of two arms necessitated by the photoreaction between thio U and a C residue nearby.

(c) Positioning of a third arm to bring together for hydrogen-bonding the only nucleotides whose sequence change is coordinated (other than those in normal base-paired regions) from an examination of the thirty-five structures.

(d) Positioning of the fourth arm to give a long thin molecule.

(e) Formation of base-pairs between loops at the centre of the molecule to give a precise tertiary structure. The circled bases are some of the constant ones in the normal tRNA species.

Very recently Kim and his colleagues (Kim *et al.* 1973) have obtained a tentative outline for the phosphate backbone of yeast tRNA^{Phe} from their X-ray crystallographic studies. A very different overall shape from Levitt's model emerges from their present data. Although the stems

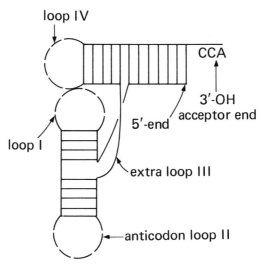

Fig. 1.19 Proposed arrangement of clover leaf in projection of tertiary structure given by X-ray analysis to 4 Å resolution. (Taken from Kim *et al.* 1973.)

are stacked as in Levitt's model (i.e. a on e, and b on c, as in Fig. 1.17(d)) the two resulting pieces of double helix are positioned at 90° to each other as in Fig. 1.19. The total structure should be known in the next year or so.

4.10 Relating structure to function of tRNA

4.10.1 Use of fragments and decoding.

Some ingenious attempts have been made to relate the structure of tRNA to its function. One obvious method has been to test tRNA fragments for particular functions, but this direct approach has not been particularly successful, probably because the native tertiary structure is also very important for function. The use of tRNA fragments to locate the activating-enzyme binding site on the tRNA has thus been unsuccessful. However, a fragment of the initiator tRNA containing the anticodon loop (Fig. 1.20, see also section 7.16) appears to behave independently in that it can be specifically bound to ribosomes by the initiator coding triplets just like the intact formylmethionyl-tRNA^{Met}. One might therefore expect the anticodon loop to protrude from the bulk of the tRNA to enable a base pairing interaction with messenger RNA. This anticodon loop region has an ordered structure which is a strict requirement for its

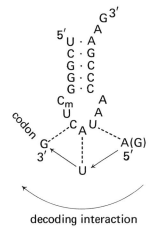

decoding interaction

Fig. 1.20 Triplet coding of an anticodon-containing fragment of bacterial initiator tRNA.

success in binding to ribosomes. If the side of the base-paired stem is removed, the residual linear fragment does not show any specific binding.

The anticodon sequence was unambiguously located by examination of the first few tRNA primary structures for triplets complementary to the appropriate codons. It was also located directly by genetic studies which can give suppressor tRNAs that function by a simple base change in their anticodon (section 9.4), and using the anticodon fragment (section 7.16).

4.10.2 Peptidyl transferase.

Other work using oligonucleotide fragments charged with amino acids has given information about the substrate requirements for the peptide bond forming enzyme, peptidyl transferase (see section 8.6, for details). Aminoacyl oligonucleotides containing the terminal sequence CpCpA or longer are active substrates for the enzyme indicating that the enzyme appears to recognise the -CpCpA end of the tRNA in the peptidyl-tRNA site. Shorter fragments such as CpA-amino acid are inactive.

4.10.3 Activating enzyme recognition.

1 Complex formation.
The association constant for the complex of tRNA and its activating enzyme is about 10^{-7}M — not a very strong binding but detectable by a variety of physical techniques such as gel filtration, sucrose gradient centrifugation, equilibrium dialysis, fluorescence quenching of tryptophan residues in the enzyme, and adsorption on membrane (nitrocellulose) filters. The last technique causes denaturation of the enzyme, which is less likely with the other methods.

The foolproof way of elucidating the specific recognition of a particular tRNA by its cognate activating enzyme would be to solve the three-dimensional structure of a crystal form of the tRNA-activating enzyme complex by X-ray analysis. However, no complex has yet been crystallised. In the meantime several indirect methods have been tried without real success as described below.

2 Comparison of primary structures of isoaccepting species.
Comparison of the primary structures of two or more tRNAs chargeable by the same enzyme might be expected to reveal features involved in enzyme recognition. The first such sequences

30

available for comparison were the two methionine tRNA sequences (Fig. 1.21). Both the bacterial initiator tRNA, (tRNA$_f^{Met}$), and the normal methionine tRNA, (tRNA$_m^{Met}$), are charged by the same activating enzyme and form enzyme complexes of about the same stability. However, detailed study of the sequences did not shed light on the features

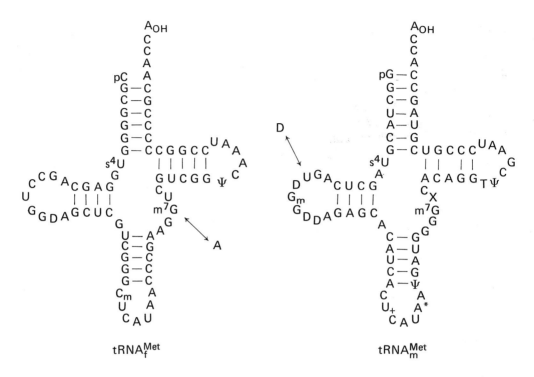

Fig. 1.21 Primary structures of bacterial methionine tRNAs arranged in 'clover-leaf' forms. ↔ signifies that some of the molecules have one base substituted for another, probably reflecting transcription of different genes. † and * are positions of base modifications. Base pairing is shown here by dashes.

responsible for recognition. Similar structural comparisons are now possible for bacterial valine and leucine tRNAs respectively. The different methionine and valine tRNAs have the same anticodons and this might be held to be important in recognition. Some of the leucine tRNAs however have different anticodons, to recognise leucine codons that differ in the first base. Either one cannot generalise recognition features from these examples or else there may turn out to be different classes of recognition.

3 Mischarging.

An extension of the comparative structural approach also gives little further information. This involves mischarging a set of tRNAs of known primary structure with a pure activating enzyme and then comparing the primary structures. The best example is the use of yeast phenylalanine activating enzyme to charge yeast, wheat germ, and *E. coli* phenylalanine tRNAs and *E. coli* valine, alanine and lysine tRNAs with phenylalanine. From these studies it is proposed that the a, b and c stems are involved in specificity. Time will tell. One disadvantage of this approach is that the conditions of charging are somewhat abnormal so that subtle changes in conformation of the tRNA could occur, leading to erroneous interpretations of features determining the charging.

Heterologous mischarging can also be obtained under abnormal charging conditions using dimethylsulphoxide. For instance, yeast valine activating enzyme will mischarge many tRNAs under these conditions. The effect of dimethylsulphoxide on tRNA structures is not clear however, so it is unwise to read too much into structural comparisons obtained in this way.

4 Inactivation by reagents.

Excision of fragments of the structure, or modifications with chemical reagents, may likewise perturb the three-dimensional structure of a tRNA. Using the excision technique some Russian workers have, however, suggested that the 5'-dinucleotide and the second and third bases of the anticodon are not necessary for enzyme recognition.

5 Mutagenesis.

More reliable information comes from single base changes in tRNAs that affect enzyme recognition. Such genetic studies are in progress. It is likely that one base change in a tyrosine tRNA changes it so that it now becomes chargeable by the glutamine activating enzyme. However, interpretation is still dogged by the possibility of a base affecting the specific folding of the tRNA rather than a recognition point.

In conclusion, apart from the obvious common recognition of all tRNAs because of the common CpCpA 3'-end, we know little about features responsible for specificity.

4.10.4 Conformational changes.

When the conformation of tRNA is somehow changed it may no longer be chargeable. Sometimes the chargeability can be recovered showing that the 'denaturation' is reversible, e.g. in the cases of *E. coli* tryptophan tRNA and yeast leucine tRNA. This is not yet interpretable in terms of detailed structural changes.

4.10.5 Protein synthesis factors.

A number of protein factors found either in the supernatant of a cell extract or loosely bound to the ribosome interact specifically with tRNA. In fact any aminoacyl-tRNA except initiator tRNA is carried to the ribosomal A-site in the form of a ternary complex with an elongation factor, EF-Tu, and GTP (see section 8.4). There is also a special set of bacterial enzymes concerned with recognising only the initiator tRNA, fMet-tRNA$_f$. For instance, initiation factor IF-2 is thought to carry the initiator tRNA in the form of a ternary complex with GTP to the initiation site on the ribosomal subunit (see section 7.12). A transformylase enzyme specifically recognises charged Met-tRNA$_f$ and formylates it.

Details of the specific structural features involved in the above interactions are still unknown.

4.10.6 CCA repair enzyme

This enzyme, of widespread occurrence, also known as CCA pyrophosphorylase, has no assigned function in the cell. However it is a useful enzyme for labelling and modifying the ends of tRNA. It has been purified from *E. coli,* yeast and rat liver; the molecular weight of the yeast enzyme is about 70 000 daltons.

In the presence of pyrophosphate the enzyme pyrophosphorolyses the 3'-end-pCpCpA of all tRNA molecules giving two moles of CTP (pppC), and one mole of ATP (pppA) per mole of tRNA. However the enzyme is usually used *in vitro* for completing a tRNA molecule when the terminal -pA, -pCpA or -pCpCpA are removed by other enzymic or chemical means. If the substrate ATP or CTP for this process contains a radioactive label in the pA or pC part then the end of the tRNA can be radioactively tagged for biochemical studies of protein synthesis. The precise *in vivo* function of the -CCA repair enzyme is unclear, but it seems likely that it adds CCA to a precursor species to make the final tRNA product. The prevalence of the enzyme in the cell will, however, presumably ensure that any damaged ends are repaired so that the tRNA can be charged with an amino acid.

Shortened tRNA as substrates for the -CCA repair enzyme can be produced by the reverse action of repair, as above, or using the enzyme venom phosphodiesterase which cleaves 5'-nucleotides stepwise from 3'-ends of polynucleotides. It is, however, difficult to control each enzyme step precisely and a chemical degradation method is normally used to provide the best substrate for the CCA repair enzyme. This makes use of the fact that there is only one *cis* vicinal diol in the tRNA molecule, in the sugar of the 3'-terminal adenosine. A *cis* diol system is cleaved with periodate; the ring opens, and two aldehyde groups are formed. The unstable oxidised nucleoside residue is readily eliminated in the presence of a nucleophilic base such as cyclohexylamine leaving a terminal -pCpCp on the tRNA. The end phosphate is removed with the enzyme alkaline phosphatase giving the -CCA repair enzyme substrate, tRNA-pCpC. The periodate-base-phosphatase treatment can be repeated to remove nucleotides stepwise. Useful information on the specificity of the -CCA repair enzyme can be obtained by testing the incorporation of nucleotide analogues into tRNA substrates produced in this way.

4.10.7 Acylaminoacyl-tRNA hydrolase.

This enzyme of common occurrence appears to be a scavenging enzyme, perhaps for the hydrolysis of peptidyl-tRNAs which have by chance fallen off the ribosome. All blocked (i.e. acylated) aminoacyl-tRNAs and short peptidyl-tRNAs, except formylmethionyl-tRNA, are hydrolysed at the amino acid-tRNA ester linkage.

4.10.8 Modifying enzymes

All tRNAs contain a variety of minor bases which are modifications of the normal four bases. These are introduced into the tRNA molecule by special enzymes after transcription. What determines the site and type of modification is at present completely unknown. Some modifications like those producing T and Ψ occur in all normal tRNAs. The role of the minor bases is unknown. One of their possible functions is to break up the secondary structure to allow proper folding. Because each modification is made by a separate enzyme, tRNA maturation is probably linked to many other metabolic pathways, possibly leading to involvement in hitherto unspecified control mechanisms.

Little progress has been made in purifying the enzymes involved in modification of tRNA. Although methylated bases (e.g. 7-methyl-G, m^7G, and 5-methyl-C, m^5C), quite often occur at constant points in the primary structure if they occur at all, they are by no means always present. Little is known about the detailed mechanism of methylases even though they have been recognised for a long time. In fact, although there is a large number of methylating activities it is not even clear whether one enzyme can carry out more than one specific methylation, or whether there is a separate enzyme for each modification.

4.11 Precursor tRNA

A transfer RNA is made in the cell by DNA-dependent RNA polymerase which copies a length of DNA (a gene) specifying a particular tRNA. An *E. coli* chromosome contains different genes for fifty or so different tRNA molecules but multiple copies of each gene may exist and the numbers of copies may vary for different tRNAs. About 0·5 per cent of the chromosome appears to code for tRNA genes.

Genes for tRNA are probably all longer in primary structure than mature tRNAs, analogous to primary gene products for mRNA and rRNA. The role of the extra pieces of RNA in precursor tRNA is not yet known. Precursor tRNAs are not usually observed in cell extracts since they have very short half-lives. Only one tRNA precursor has been studied in structural detail so far, that for a special bacterial tyrosine tRNA whose gene can be carried on a transducing bacteriophage.

The primary transcription product of the *E. coli* tyrosine tRNA gene (i.e. the precursor) has been sequenced by Altman and Smith (1971). It contains extra nucleotides at both ends which are cleaved to give the mature species. An enzyme activity, tentatively called the RNase P enzyme, which splits off the forty-one extra nucleotides from the 5′-end of the precursor during the maturation process, has been detected in proteins removed from ribosomes by washing with salt solution. Maturation of the 3′-end of the molecule which is only two bases longer in the precursor is still a mystery. The precursor as isolated might in fact be derived from a still earlier precursor longer at the 3′-end; the presence of a 5′-triphosphate, pppG, at the 5′-end however confirms that this end of the precursor is in fact the end of the primary transcription product.

Two possible forms of the precursor secondary structure are shown in Fig. 1.22. During transcription it is likely that form I is the preferred structure while later when the clover leaf of the tRNA is possible the second form II will be stabler.

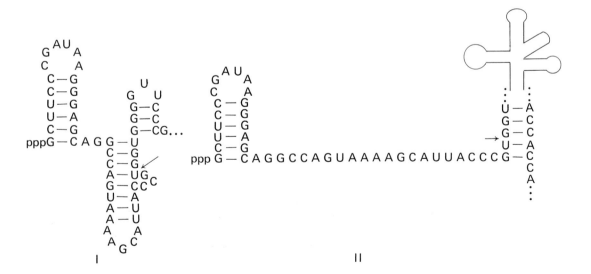

Fig. 1.22 Possible configuration for tyrosine precursor. The hydrogen-bonded loops shown are thought to be stable. The arrows indicate the beginning of the 5′-end of the tRNA moiety and a cleavage point. In the mature tRNA, none of the last four nucleotides shown at the 3′-end of structure II is involved in hydrogen bonding. (Taken from Altman and Smith 1971.)

4.12 Synthesis of tRNA genes

An elegant combination of synthetic organic chemistry and enzymology has been used by Khorana and his colleagues (Khorana *et al.* 1972) to synthesise a piece of double-stranded DNA corresponding to the sequence of yeast alanine tRNA without modifications of normal bases. Currently a gene for bacterial tyrosine tRNA is also being made. Whether these gene pieces will be useful in transcription studies is now unclear since the extra bits needed to complete genes for precursor tRNAs are not identified.

5 Ribosomes

5.1 Sites of protein synthesis

Once again most of the detailed information about ribosomes comes from studies using bacterial cell-free systems, although it has recently been possible to isolate very active ribosomes from extracts of mammalian cells, such as reticulocytes, liver and ascites tumour cells, and plant cells such as wheat germ.

Experiments in the late 1950s, using bacteria infected with a DNA-containing virus, T2 bacteriophage, showed that ribosomes were not active in carrying information to direct the synthesis of proteins but were themselves the sites of protein biosynthesis. After infection of the cell with T2 bacteriophage, no new ribosomal RNA was made so the rRNA could not function as a template which carried information for a bacteriophage protein's amino acid sequence. It was already known at that time that radioactive amino acids supplied to a bacterial cell could be incorporated into polypeptide chains on the surface of ribosomes. In the early 1960s it was found that after bacteriophage infection new messenger RNA was made in the bacterial cell and this mRNA carried information for the synthesis of bacteriophage-specific proteins.

Protein synthesis never appears to occur free in solution but only on ribosomal surfaces. The ribosomal structure is designed to orient two charged tRNA molecules specified by mRNA, in such a way as to permit the formation of a peptide bond between a polypeptide and an incoming amino acid.

Both prokaryotic and eukaryotic cells contain ribosomes composed of two subunits, one of which is about double the size of the other. In bacteria the ribosomes and their subunits have sedimentation constants of 70 S, and 50 S and 30 S respectively. In each case the ratio of the content of RNA to protein is 63:37. The ribosomes found in eukaryotic cytoplasm seem to have a higher protein content and larger RNA molecules, and have sedimentation coefficients of 80 S, 60 S and 40 S respectively for the single ribosome and its subunits. Ribosomes in eukaryotic cells appear to be rigidly ordered in a cellular membrane called the endoplasmic reticulum and this makes their isolation with functional integrity more difficult. Indeed ribosomes from mammalian cells when isolated often contain extraneous membrane material. From a functional aspect 80 S ribosomes appear to differ in some details from 70 S ribosomes. One clear example is their mode of involvement in polypeptide chain initiation (see section 7). An interesting feature of eukaryotic cells is the occurrence in organelles such as mitochondria and choroplasts of smaller ribosomes, very similar in properties to those of bacteria.

In rapidly growing (log phase) bacterial cells (generation time of about 20 min) there are of the order of $(15-18) \times 10^3$ ribosomes. Each ribosome has a molecular weight of about $2 \cdot 7 \times 10^6$ **and** has dimensions about $150 \times 150 \times 200$ Å. Since an average bacterial cell 2 μ long and 1 μ

across has a molecular weight of about 10^{12}, it can be estimated that in total the ribosomes make up nearly a quarter of the total cellular mass (i.e. the cell's dry weight). Thus a good proportion of the cell is devoted to the business of making proteins. Only one protein is made at a time on a ribosome. When the protein is finished the ribosome is released and can take part in the synthesis of another type of protein as programmed by a newly attached mRNA. Before the attachment of new mRNA the 70 S or 80 S ribosome (monosome) is dissociated into subunits (see section 7).

5.2 Polyribosomes

Depending on the ionic environment, ribosomes can exist in a variety of associated or dissociated states which may or may not be related to functional states within the cell. In particular, Mg^{2+} ions favour association whilst K^+ ions favour dissociation. In the native state one 50 S particle and one 30 S particle combine giving a single 70 S particle. Higher sedimenting states found in the cell are due to several monomeric 70 S particles attached to a strand of mRNA. Indeed if a cell is very carefully broken it is possible to isolate most of the

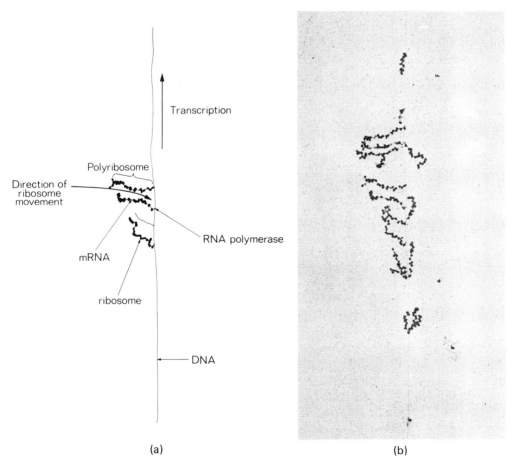

(a)   (b)

Fig. 1.23 (a) Diagram explaining electron photomicrograph shown in (b). (b) Electron photomicrograph by O. L. Miller Jr. and B. A. Hamkalo (e.g. see Miller 1973) to visualise genes of *Escherichia coli* in action; **magnification** is about 34 000 x.

ribosomes in polyribosomal formations with several ribosomes (all making the same proteins) strung along a single-stranded mRNA like beads on a string. It has been established, in both bacterial and mammalian cells, that the functional formations for protein synthesis are polyribosomes, e.g. for the synthesis of globin in reticulocytes they are quite often of the order of 200 S, containing five ribosomes on a single mRNA. These polyribosome formations have been confirmed by electron microscopy. Indeed the elegant studies of Miller and his colleagues (e.g. Miller 1973, see Fig. 1.23) show that often coupling of transcription and translation occurs in the cells; polyribosomes showing translational activity can be seen to be attached to new mRNA being transcribed off DNA. For the synthesis of proteins of mol. wt. 30 000–50 000 daltons (300–500 amino acids) about twelve to twenty ribosomes will be attached to the mRNA molecules. At maximal utilisation of mRNA length, taking the diameter of the ribosome to be 200 Å, there could be one ribosome per sixty nucleotides on the mRNA. In this context, when the mRNA attached to a ribosome is digested with a ribonuclease, the ribosome is found to protect a stretch of about thirty nucleotides from enzymic degradation. This must be a measure of the amount of mRNA which is not exposed.

The ability of a single mRNA to function simultaneously on several ribosomes helps to explain why a cell needs so little mRNA (only 1–2 per cent of total RNA). Polyribosomes are clearly an efficient and economical way of using mRNAs.

5.3 Subunits

Preparation of 70 S ribosomes has been described in section 2.4. When bacterial cells are broken by grinding with acid-washed alumina the polyribosomes are nearly all degraded to 70 S ribosomes. This is also partly due to the fact that the ribosome adsorbs a membrane ribonuclease (RNase I) and also has associated with it another ribonuclease (RNase II). These ribonucleases cleave the mRNA during the isolation of the ribosomes, which need to be washed very carefully to remove the ribonucleases before being used in a cell-free system. Another way to overcome this problem is to use ribosomes from ribonuclease-free bacteria or to use ribonuclease inhibitors such as bentonite (a diatomaceous earth).

Whether 70 S ribosomes exist free in the cell is debatable. There is reasonable evidence for the existence of about 10–20 per cent of the ribosomes in the 70 S form, whereas about 70 per cent are incorporated into polyribosomes. The remaining 10–20 per cent (but this appears to vary with the bacterial strain) exist in their subunit form.

It is known that 50 S and 30 S subunits recycle through a round of protein synthesis, i.e. after one protein chain has been synthesised using a 70 S particle the subunits can dissociate. This agrees with the idea that the 30 S subunit joins alone to mRNA before the 50 S unit when synthesis of polypeptide chains is started (see section 7 on initiation).

Native subunits appear to function differently from artificially formed subunits obtained by dissociating 70 S particles in the presence of low magnesium-ion concentration (10^{-4}M). The artificial or *derived* subunits are characteristically different from the native subunits in that under the usual Mg^{2+} concentrations ((10–15) $\times$ 10^{-3}M) used in cell free systems they are reassociated to give 70 S particles. Derived 30 S subunits apparently can be converted into native subunits by the addition of special proteins called initiation factors. It is unclear whether derived and native 50 S subunits are different.

5.4 Ribosomal RNA

The ribosomal RNA is an integral part of ribosomal structure unlike mRNA because when rRNA is removed the ribosomal structure and hence the functional properties are destroyed.

Ribosomes normally have ribonucleases adsorbed on their surfaces. The ribonucleases survive extraction with phenol in the separation of ribosomal protein from the RNA and will then degrade the RNA. Hence strict precautions must be taken to inactivate any possible ribonuclease present when ribosomal RNA (rRNA) is isolated. Ribonuclease II does not appear

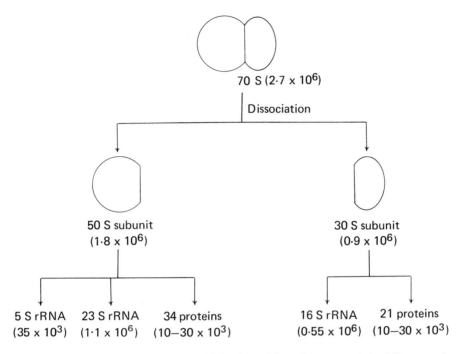

Fig. 1.24 Composition of a 70 S ribosome. (Molecular weights of components in daltons are shown in brackets.)

to be too much of a problem as it is only loosely bound and can be removed if the ribosomes are centrifuged through a sucrose gradient during preparation. When ribosomes are extracted with a mixture of phenol and water in the presence of a ribonuclease inhibitor such as bentonite the RNA is usually extracted into the aqueous phase. Bacterial ribosomes extracted carefully in this way release equimolar amounts of 23 S, 16 S and 5 S RNA components. If the subunits are separated before their RNA is extracted it is found that both the 23 S and 5 S come from the 50 S subunit, whereas only the 16 S RNA comes from the 30 S subunit (see Fig. 1.24).

The molecular weights of the bacterial rRNA species are about 10^6, 5.5×10^5 and 3.5×10^4 daltons for 23 S, 16 S and 5 S respectively. Hybridisation experiments show that about 0.4 per cent of the bacterial chromosome codes for rRNA. Thus, there are about six genes for each of 23 S, 16 S and 5 S rRNAs. *E. coli* rRNA genes are closely linked and some have been shown to be adjacent, with 5 S, 23 S and 16 S in tandem formation. Although it seemed likely that the precursor rRNA could, therefore, contain the three rRNAs with perhaps spacer sequences linking them, there is good electron microscopic evidence for at least the 16 S and 23 S rRNAs being transcribed in distinct pieces off the adjacent genes. This contrasts with the situation in mammalian cells; rRNA genes are transcribed as one unit in a special part of the nucleus, the nucleolus, to give a large 45 S RNA. The 45 S RNA is degraded in a series of steps yielding the final 18 S and 28 S rRNAs for the 40 S and 60 S subunits respectively. The 5 S rRNA appears to be transcribed separately. There is another small RNA, a 5.8 S rRNA, which is also part of

38

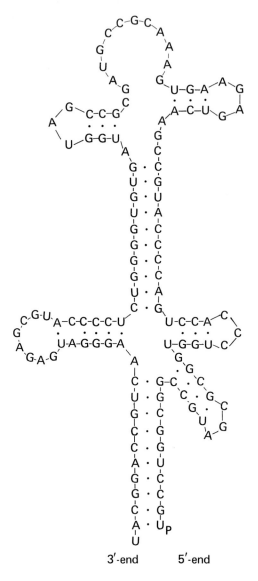

Fig. 1.25 *E. coli* 5 S RNA arranged in one possible secondary structural form. (Taken from Boedtker and Kelling 1967.) Dots indicate hydrogen bonds.

the mammalian 60 S subunit. Its function is as yet uncertain. Eukaryotic cells usually contain of the order of 50–250 genes for the 45S type RNA (e.g. yeast contains 140). However in special cases as in amphibian cells, gene amplification raises the number to the order of thousands (see also Chapter 7; section 2.3).

The rRNAs are largely single strands but appear to contain double-stranded regions as a result of base-pairing. The 23 S and 16 S RNA contain small amounts of methylated bases although the proportion of methylated bases is much smaller than for tRNA. The 5 S RNA does not appear to be covalently bound to the 23 S RNA in the 50 S particle, nor does it contain methylated bases. This raised the possibility that it has a different function from the longer RNAs, which have structural importance. Other functions have not been determined for the larger RNAs. Some proposals have been, for example, that 5 S RNA is involved in polypeptide chain termination or is a structural part of the A-site for binding aminoacyl-tRNA but the actual function remains undetermined.

5 S RNA is 120 nucleotides long and was the first [^{32}P]-labelled RNA to be sequenced by Sanger and his colleagues (Sanger et al. 1965). The primary structure allows base pairing to occur in a number of possible ways (see for example Fig. 1.25) giving a certain amount of secondary structure to the molecule in base-paired regions (Boedtker and Kelling 1967). Since an unequivocal arrangement has not been determined it is not possible to predict anything about tertiary structure. Available evidence so far suggests that the larger rRNAs also contain many regions of secondary structure. How the various ribosomal proteins interact with the RNA structures is not yet known; nor is the location or conformation of the tRNA or mRNA binding sites.

5.5 Ribosomal proteins

So far it has not been possible to treat a ribosomal subunit in such a way as to remove one protein at a time. Instead, proteins are stripped off the RNA in batches by washing with solutions (e.g. LiCl, CsCl) of increasing ionic strength; this also gives rise to undegraded RNA. The proteins can then be separated by electrophoresis on polyacrylamide gels or by chromatography on carboxymethyl-cellulose columns. The immense job of determining the structure of the ribosome is being attacked with tremendous energy by several laboratories.

Electron microscopic methods are not yet sufficiently well-developed to yield structural details of the ribosome. Likewise the ribosome is so large as to place potential X-ray analysis of the three-dimensional structure very much in the future. Anyway, attempts to crystallise ribosomes have not been successful: microcrystals of ribosomes found in cooled chick embryo cells are too small. Perhaps the most we can hope for will be a crude picture of how the various ribosomal components are related rather than ever to attempt to deduce information at an atomic level.

At present it is generally agreed that twenty-one different proteins, ranging in molecular weight from 10×10^3 to 30×10^3 daltons have been identified as components of the 30 S subunit. These proteins are numbered S1 to S21, S indicating small subunit. Since different proteins are recovered in different yields after a salt wash to remove superficial proteins, it is often difficult to tell whether a particular protein is a true structural component of the ribosome or not. The proteins from the 50 S (large) subunit are now identified and numbered L1 to L34; they are less well characterised than the 30 S subunit proteins. So far only two of the thirty-four proteins have been shown tentatively to be identical, confirming ribosomal asymmetry; proteins L7 and L12 apparently contain the same amino acid sequences but differ in that the N-terminal residue of L7 is acetylated. The role of the acetyl group is unclear but these proteins have been implicated in the structure of ribosomal A-site so much interest is being shown in them. These proteins are described further in section 8. Some of the other protein components have their N-terminal ends blocked – probably by formyl groups. The primary structures of many ribosomal proteins are now being investigated.

5.6 Structure and function

Attempts are now being made to assign functions to particular ribosomal components. One approach is to try to show where particular proteins associate with ribosomal RNAs; the ribosome is either digested partially to obtain pieces of RNA protected by a protein or proteins, or specific proteins are bound to naked rRNA to find which sequences are protected from nuclease digestion. Of course, identification of the protein's location on the rRNA calls for knowledge of the rRNA sequence but sequence determinations of bacterial rRNAs are well under way; the 16 S rRNA sequence should be completed in the next couple of years.

A chemical approach to the elucidation of three-dimensional structure is to cross-link proteins, or a protein and a nucleic acid, and then identify the components which are in a juxtaposition defined by the chemical reagent. This approach has been only recently applied because of the lack of suitable bifunctional reagents.

Valuable information on the packing of the proteins in the ribosome has come from a study of the reconstruction of subunits from their component parts. By testing the effects on assembly of omitting particular components it is possible to deduce roughly the order of protein-nucleic acid interactions for folding to proceed. Our knowledge of the total structure is, however, still sketchy.

The most likely explanation for the existence of two subunits in the functional ribosome for peptide bond formation is that two subunits are necessary for movement along the mRNA. Indeed it is possible to construct models whereby an oscillation between the subunits causes the mRNA to move over the ribosome. Perhaps the ribosome is not such a passive character as is suggested by referring to it as the site for protein biosynthesis.

5.7 Reconstitution of ribosomes

The major achievements in reconstitution of active subunits have been by Nomura and his colleagues (see Nomura 1973); they succeeded most readily in the case of the *E. coli* 30 S subunit. This was done by incubating the components in suitably buffered conditions in the presence of a sulphydryl-containing compound. High ionic strength gives optimal reconstitution at the elevated temperature of 40°C. At this temperature any undesired base pairs in the isolated RNA are broken and the proper positioning of the protein species can occur. Presumably an RNA-protein aggregate is formed which folds automatically into the appropriate 30 S subunit spatial structure. No external influences such as enzymes appear to be needed.

Reconstitution experiments may also identify proteins involved in ribosomal functions, such as peptide bond formation and mRNA and tRNA binding sites. Reconstitution is a specific process: 16 S rRNA cannot be replaced by 23 S rRNA. Assembly also requires all the inner-lying proteins to be present.

Reconstitution of the 50 S subunit has proved to be more difficult; perhaps a precursor rather than the mature rRNA is needed. For some unknown reason the reconstitution of the 50 S subunit from *E. coli* needs also the presence of 30 S subunit components, although the 50 S subunit of *Bacillus stearothermophilus* has been reconstituted from only 50 S components. The peptidyl transferase activity of 50 S ribosomal subunits will be described in section 8.

6 Messenger RNA

6.1 Discovery

In 1956 Volkin and Astrachan (Volkin and Astrachan 1956) identified a rapidly radioactively labelled RNA fraction in cells infected by DNA bacteriophages. An interpretation of their results became apparent to Brenner and his colleagues who showed that in bacteriophage-infected cells the stable ribosomes did not carry information for protein synthesis (Brenner *et al.* 1961). These experiments introduced the idea of the ribosome as a mere carrier for information-carrying template RNA which was a rapidly labelled species. Jacob and Monod (1961) introduced the name messenger RNA (mRNA) for this somewhat hypothetical template RNA. Such a concept was already needed to explain the situation in nucleated cells where information had to be transferred from DNA in the nucleus to the ribosomal sites for protein synthesis in the cytoplasm. Experimental proof of the existence of mRNA came from the use of a synthetic mRNA, poly U, by Nirenberg and Matthaei in 1961, when they were able to programme the synthesis of polyphenylalanine in a cell-free system (see section 2).

6.2 Function

Messenger RNA is the intermediary in the transfer of information from DNA to the protein-synthesising sites on the ribosome. The sequence of nucleotides in mRNA directly programmes the order of amino acids in a protein synthesised on a ribosome. The elucidation of the genetic code revealed that mRNA is translated into protein by a process in which a sequence of three nucleotides (a triplet) signifies one amino acid (see section 2).

Although mRNA comprises only about 3 per cent of total cellular RNA, it was thought until recently that in bacteria up to 99 per cent of the total DNA (genome) is transcribed into mRNA. This apparent contradiction is resolved because the function of mRNA involves greater molar quantities of rRNA and tRNA and although tRNA and rRNA are quite stable and have low turnover rates, in general bacterial mRNA is relatively unstable. In contrast, mRNA of eukaryotic cells is thought to be much more stable. For example, the half-life of the mRNA in rat liver is about 5 hr. at 30°C whereas that of bacterial mRNA is a few minutes. Recent research on chromosome structure has revealed that varying amounts, perhaps only 10 per cent, of the eukaryotic DNA codes for mRNA although it is still likely that 50 per cent or more of the bacterial genome is transcribed into mRNA.

There are about 1000 mRNA molecules in the bacterial cell but they vary greatly in chain length, reflecting the different proteins needed by the cell. The average *E. coli* polypeptide chain contains 300–500 amino acids so that the average size of the mRNA is 900–1500 nucleotides. However, some mRNAs carry information for more than one polypeptide chain. Because of this, they are called *polycistronic* mRNAs and clearly contain at least enough nucleotides to signify the appropriate proteins. They may also contain some sequences of nucleotides used as signals for starting and stopping chains or for defining ribosomal attachment points. Not much is known so far about these possible additional sequences. Usually polycistronic messengers are translated into polypeptides which have related functions. For example, ten enzymes in the pathway needed to synthesise the amino acid histidine are programmed by a polycistronic mRNA containing about 12 000 nucleotides, i.e. having a molecular weight of about 4×10^6 daltons. Studies of the sequence of certain regions of viral polycistronic mRNA are described in Chapter 7.

6.3 Synthesis

Although double-stranded DNA acts as a template for the enzyme RNA polymerase, which makes all types of RNA, only one strand of the DNA appears to be transcribed for a given gene. The DNA strand which is copied in a complementary and antiparallel fashion is known as the sense (or coding) strand. The role of the other DNA strand is not known. It is also unclear whether the RNA polymerase makes one strand of polycistronic mRNA by reading double-stranded DNA, or whether the enzyme separates the DNA strands before copying one of them. The direction of synthesis is from the 5'-phosphate end to the 3'-diol end. The mRNA is also read (or decoded) in groups of three nucleotides (triplets) from the 5'-end to 3'-end from the special starting point which has not been completely defined (see section 6.7 on phage RNA).

6.4 Degradation

At least two exonucleases have been implicated in the breakdown of RNA. Both of them, polynucleotide phosphorylase and RNase II, act by degrading the RNA stepwise from the diol (3'-) end. Since this is the opposite direction to that of synthesis, then if these were the true mRNA degradative enzymes fragments of unfinished mRNA would appear in the cell unless there were some means of protecting the RNA being synthesised from degradation. However, neither enzyme is thought to be actually involved in the degradation of mRNA at least in the first steps.

There is evidence that *in vivo* an mRNA is degraded by an unidentified enzyme from the 5'-end to the 3'-end, i.e. in the same direction in which it is translated. It is likely that nascent mRNA is protected from degradation by addition of ribosomes; if a ribosome fails to bind to the beginning of the mRNA degradation could begin. Degradation of a long polycistronic mRNA probably starts before translation is complete—an added hindrance to the biochemist aiming to isolate an intact mRNA. Coupling of translation and transcription can clearly occur as evidenced by the extraordinary electron photomicrograph in Fig. 1.23. Thus, the mRNA is translatable before it is completely synthesised.

6.5 Identification of mRNA

Rapid labelling or complementarity by hybridisation with DNA are no longer sufficient to identify a species as mRNA; it is also necessary to check its function in a cell-free system. Of the gene transcription products only mRNA is endowed with the property of carrying template information for programming the synthesis of a specific protein. Thus, only mRNA is capable of converting 70 S ribosomes to polyribosomes yielding a functional unit for actively synthesising protein.

The ultimate proof of an mRNA's character would be the determination of its nucleotide sequence, permitting a clear correlation with a known protein's amino acid sequence. It is probably safe to say that the extreme technical difficulties associated with this structural test will mean that only functional tests will be possible for some time.

Some sequences are known at both ends of whole bacteriophage RNA, as well as in between and at ends of cistrons; recently the whole of the cistron sequence coding for the coat-protein of MS2 has been determined by Fiers and his colleagues (Min Jou *et al.* 1972). These results are described in Chapter 7.

6.6 Natural mRNA

Synthetic polyribonucleotides can function adequately as mRNA in bacterial cell-free systems in spite of not having proper starting or attachment signals. This functional property has been instrumental in elucidating many biochemical features of the genetic code. Given the proper signals, however, the ribosomes will function preferentially with accurate specificity in translation. Special conditions usually involving a high Mg^{2+} concentration are necessary for the translation of synthetic mRNA.

Bacteriophage mRNA stands out as being the most readily available natural messenger for studies on protein biosynthesis. A bacterial cell-free system supplied with RNA extracted from bacteriophage (e.g. f2, MS2, R17 or Qβ) containing single-stranded RNA will synthesise protein which is characteristic of the bacteriophage. This natural mRNA can be extracted from the bacteriophage in such a way that its integrity with respect to starting signals, ribosomal attachment points and secondary structure, is maintained. Although single-stranded RNA from plant viruses such as TMV (tobacco mosaic virus), will make polypeptides in a bacterial cell-free system, the products have not been characterised properly as viral proteins; translation of these viral RNAs probably starts in the wrong places so that incorrect proteins are made. A fair test of a plant virus RNA's ability to act as messenger is currently hampered by the lack of very active plant cell extracts. In contrast it has recently been possible to assemble mammalian cell extracts which are very active in synthesising proteins. Single-stranded animal virus RNAs will work as mRNAs in mammalian cell extracts, e.g. encephalomyocarditis (emc) virus RNA can be translated actively in an ascites tumour cell extract. Although animal viral messengers are not as well characterised as bacteriophage messengers, several mammalian mRNAs have now been isolated and translated *in vitro* whereas attempts to isolate bacterial mRNAs in pure form have been unsuccessful.

Part of the problem is the sensitivity of large RNA, such as mRNA, to breakdown during isolation. Even though cells are extracted in the presence of phenol, and every care taken to separate the RNA under mild conditions, a small number of breaks in the RNA chain by physical or biochemical means can easily destroy the integrity of mRNA. Purification of a single bacterial mRNA is further complicated since there are so few copies of a specific mRNA (less than ten per cell) available at a time. Also polycistronic bacterial mRNA may be so long (of the order of 10^4 nucleotides) that degradation occurs before complete translation, so isolation of a pure species will be impossible. Nevertheless, mRNAs which at least have complete cistrons have been detected in bacterial extracts where transcription and translation are coupled. In some cases large DNA pieces containing complete genes, such as those isolated from T-even bacteriophage DNA or bacteriophage DNA carrying bacterial genes, can be transcribed and translated *in vitro*. Then the mRNA product, present in very low amount, can be detected by a biological assay. For instance bacteriophage T4-coded enzymes such as lysozyme and deoxycytidine monophosphate hydroxymethylase, and bacterial alkaline phosphatase and β-galactosidase, can all be synthesised *in vitro*.

To persuade the bacterial cell to make more of a particular mRNA special genetic tricks like bacteriophage infection are needed. This is not necessary for those animal cells which have specialised functions and which make a large amount of one protein. For instance reticulocytes which make mostly haemoglobin are a source of globin mRNA which has recently been isolated in a form capable of being actively translated in animal cell extracts. This type of mRNA has also been tested in a possibly more convincingly native system—oocytes. The oocyte is a complete cell which is large enough to be injected with a mRNA preparation to test its translation. The resulting translation occurs in an almost completely *in vivo* system, so in many ways a more reliable test of proper translation is obtained than in a cell-free system. The list of isolated and tested mammalian mRNAs is rapidly being extended. In addition to globin mRNA, eye lens α-crystallin mRNA, myosin mRNA, ovalbumin mRNA and myeloma protein mRNA have been identified. It is of interest that all eukaryotic mRNAs so far identified are monocistronic. If this proves to be a general property of eukaryotic mRNAs (non-viral in origin) it will contrast sharply with the bacterial situation.

6.7 Bacteriophage RNA as mRNA

Since large amounts of bacteriophage RNA can be isolated readily in an intact form from virus particles, it has become the most widely used mRNA for *in vitro* studies. The R17 bacteriophage, like the closely related f2, MS2 and M12 bacteriophages, contains a single-stranded RNA of about 3300 nucleotides. *In vivo*, and *in vitro*, R17 RNA codes for the synthesis of only three proteins. They have been identified as (1) the maturation or A protein, a minor component of the particle, (2) the coat-protein, a major structural component of the virus particle and, (3) the RNA synthetase or replicase, not found in the virus but made before bacteriophage replication in the bacterial cell. The order listed is also the structural gene order from the 5′-end of the bacteriophage RNA (see also Chapter 7, section 3.6). The proteins are not synthesised in equivalent amounts *in vivo* or *in vitro*. For each A protein molecule about three synthetase and twenty coat-protein molecules are made.

When R17 is used to stimulate the incorporation of radioactive amino acids into polypeptides in a bacterial cell-free system, the products can be readily fractionated and identified. The smallest is the coat-protein (129 amino acids) of known amino acid sequence. Thus translation of the coat-protein cistron of known sequence (Min Jou *et al.* 1972) in bacterial extracts has given us a system to confirm the basic protein biosynthetic mechanisms deduced using synthetic polynucleotides. Information about initiation and termination signals for proteins synthesis has come from use of this type of viral RNA as mRNA. The results will be described in sections 7 and 9 on punctuation.

6.8 Attachment of mRNA to the ribosome

Whereas ribosomes can attach at several points on long artificial mRNA such as synthetic polynucleotides, each natural monocistronic mRNA presumably has one specific attachment point which is defined by a nucleotide sequence or secondary structural restriction on the mRNA. The attachment point corresponds to, or is close to, the sequence of nucleotides designating polypeptide chain initiation. In the case of polycistronic mRNA there are several specific attachment points, namely at the beginning of each cistron. The idea that all ribosomes are first attached to the cistron near the 5′-phosphate end before rolling along to the other cistrons is probably wrong. The mechanism that translates different cistrons of the polycistronic mRNA at different rates is not known.

The functioning of several ribosomes simultaneously on the same mRNA gives rise to polyribosomes. A ribosome moves along the mRNA from the 5′-end to the 3′-end in some undefined way without losing contact with it; perhaps the ribosomal subunits are responsible for permitting such relative motion. Degradation by ribonuclease of mRNA with attached ribosomes indicates that the ribosome protects a stretch of about thirty nucleotides in the mRNA.

During initiation of protein biosynthesis the mRNA is attached to the 30 S subunit, probably under the influence of a protein factor and the initiator tRNA (see section 7). The exact location in the 30 S subunit is unknown. For translation to proceed the 50 S subunit must be joined to the 30 S subunit.

6.9 Control of translation

Bacteriophage RNAs contain a high degree of secondary structure (see also Chapter 7, sections 3.8 and 3.9). For example, 70 per cent of the coat-protein cistron sequence of MS2 RNA can be arranged in base pairs producing many loops. Tertiary structure arises from folding the secondary structure in three dimensions. This folded bacteriophage RNA clearly is translated only at exposed regions. However, once a ribosome is attached properly it probably opens up the rest of the structure. Alternatively it is possible that a special protein factor, a type of initiation factor (section 7), will expose the RNA at a specific initiation site to start ribosomal translation. To what extent tertiary structure or special proteins are involved in specifying RNA translation is currently being investigated. It is also possible that the high degree of organised structure in bacteriophage RNA is related to its packaging into virus particles.

In the case of mammalian cells special proteins may take part in binding to mRNA, perhaps even in precursor form, for transport from the nucleus to the cytoplasm. The role of these proteins is surmised to be protection against nuclease digestion. The question whether specific initiation factor proteins discriminate between various mammalian mRNAs is still open. However, very recently a purified factor has been shown to select a viral RNA (emc) for translation in mammalian cells. Should this finding turn out to be widespread, it means that classes of initiation factor will select classes of mRNA comparable with a mechanism postulated for control of cellular differentiation.

6.10 Precursor mRNA

Very little is known about the structure of possible precursor mRNAs in bacterial or mammalian cells. Some mRNAs have been tentatively identified as part of much larger RNA species found in mammalian cell nuclei. The larger diverse RNA species in the nucleus are called heterogeneous nuclear RNA which is probably associated with protein transport factors (see also Chapter 7, section 2.5). The nuclear precursor, together with its postulated transport protein, has been called an *informofer*. Before the RNA is translated as mRNA it must be stripped of its transport protein and cleaved to its functional size. Most mammalian mRNAs so far identified have an unexplained structural feature; they contain poly A sequences of up to about 80 units long near the 3'-ends. These poly A sequences appear to be added to a precursor by a special enzyme before translation. Although the role of the poly A has not been explained, its presence has proved useful in the isolation of mammalian mRNAs. RNAs containing poly A sequences can be bound tightly to cellulose columns into which poly dT sequences have been incorporated, and can thus be separated from other RNAs. A key problem remaining in structural studies on mammalian mRNA is how to obtain material radioactively labelled to a specific activity high enough for the Sanger fingerprinting techniques.

7 Polypeptide chain initiation

7.1 Discovery of an initiator tRNA

The mechanism by which protein synthesis begins both in bacterial and in eukaryotic cells is now well understood. Proteins start in a slightly different way in mammalian or plant cells, and bacterial cells—at least as far as cytoplasmic synthesis is concerned; there is, however, growing evidence that in cytoplasmic organelles, such as mitochondria, proteins may start by the same mechanism as in bacteria. Polypeptide chain initiation has attracted much interest because it is generally thought that this would be the most obvious point of control of protein biosynthesis.

The process of protein biosynthesis has been studied most informatively in the past using extracts of *E. coli* (see section 2). Synthetic polyribonucleotides containing a variety of different bases were shown to behave as artificial messenger RNA in bacterial cell-free systems with reasonable and comparable efficiencies; if a starting signal existed it would be expected that only the polyribonucleotide containing the starting signal would lead to the formation of polypeptides. In addition, cell-free systems programmed with a synthetic messenger (e.g. poly A) gave polypeptide products (e.g. oligolysines) containing no modified end units. Thus, for some time during the early period in the elucidation of the genetic code it was assumed that no special starting signal was needed in the mRNA and that no special protein factor was involved in starting the polypeptide chain. The assumption was made that any tRNA specified by the first codon in a messenger RNA would be able to initiate protein synthesis. This simple view of the mechanism of polypeptide chain initiation arose as a lucky artefact of the cell-free system which allows chains to start erroneously under special conditions. Even the finding that there appeared to be a restriction in the N-terminal positions of *E. coli* proteins to methionine, alanine and serine (about 45, 30 and 15 per cent respectively) did not immediately alter this view. Yet the results were especially interesting since methionine is a rare amino acid amounting to only about 3 per cent of the total amino acids in *E. coli* proteins.

The currently accepted mechanism of polypeptide chain initiation in bacteria arises from the discovery in 1964 by Marcker and Sanger (Marcker and Sanger 1964) of the formation of *N*-formylmethionyl-tRNA *in vivo* and *in vitro*. The terminology indicates that the formyl group (HCO-) is attached to the amino group of the methionine residue as illlustrated in the Fig. 1.26(b) where the methionine is shown attached to the terminal adenosine of the tRNA. In *N*-formylmethionyl-tRNA the amino group is thus blocked and hence no longer available for formation of a peptide bond; *N*-formylmethionine is thus restricted to the N-terminal position in the polypeptide chain. There was the immediate suspicion that this particular tRNA was

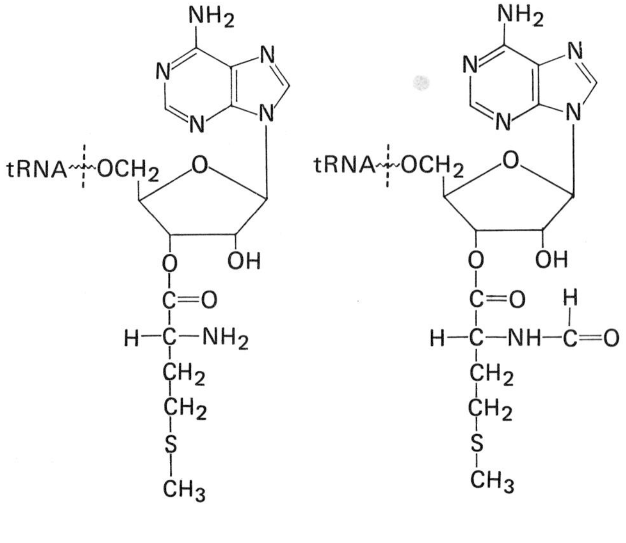

Fig. 1.26 Methionyl-adenosine (a) and formylmethionyl-adenosine (b) at the 3′-end of tRNA; the dotted lines indicate the points of cleavage of Met-tRNA and fMet-tRNA by pancreatic ribonuclease.

somehow involved in chain initiation. We were able to show that this hypothesis was correct, using a bacterial cell-free system (Clark and Marcker 1966).

The discovery of the initiator tRNA came from a study of the esterification of methionine to its specific tRNA in *E. coli*. Radioactive methionine (in this case labelled with ^{35}S) was incorporated into Met-tRNA with a crude *E. coli* extract containing the methionine activating enzyme. Two radioactive products resulted from digestion of the charged tRNA by pancreatic ribonuclease; one was the expected methionyl-adenosine (Fig. 1.26(a)) and the other was less positively charged than Met-A at pH 3·5. The unexpected product was identified chemically as formylmethionyl-adenosine (Fig. 1.26(b)). The fMet-A could also be detected in charged tRNA isolated from growing cells, showing that its formation in the cell-free system is not an artefact.

The maximum amount of formylation of methionyl-tRNA was determined to be about 60–70 per cent of the total methionine accepting activity. This finding suggested that there were two species of methionyl-tRNA in *E. coli* in a ratio of about 70:30, and that only the more abundant species could be formylated.

7.2 Two classes of Met-tRNA

Direct confirmation of this suggestion came when we separated two classes of methionine-accepting tRNAs from each other by counter-current distribution. The different methionine-accepting tRNAs (tRNAs^{Met}) are now usually separated by column chromatography on DEAE-Sephadex, or benzoylated DEAE-cellulose. One species, tRNA$_m^{Met}$, when charged to give methionyl-tRNA$_m^{Met}$ did not accept formyl groups, while the other species, tRNA$_f^{Met}$, could be completely converted into formylmethionyl-tRNA$_f$. Thus *E. coli* tRNA contains two different methionine-accepting tRNAs of which only one can then be converted into the formyl derivative. In abbreviated form the charged species are termed Met-tRNA$_m$ and Met-tRNA$_f$. When Met-tRNA$_f$ is formylated it becomes fMet-tRNA$_f$. Two classes of tRNA^{Met} have also been identified in the cytoplasm of eukaryotic cells. However the initiator species, again referred to as tRNA$_f^{Met}$, is functional in eukaryotic systems as Met-tRNA$_f$ i.e. unformylated.

7.3 The mechanism of formylation

There are two possible ways in which fMet-tRNA$_f$ could be formed: (1) by formylation of methionine to give formylmethionine after which the latter becomes attached to tRNA$_f^{Met}$, (2) by formylation after the formation of Met-tRNA$_f$. Marcker's work (Marcker 1965) distinguished between these two possibilities and showed conclusively that the formylation of the amino group of methionine only happens after methionine has become attached to tRNA$_f^{Met}$. There is no activating enzyme for charging with formylmethionine. The formylation reaction is catalysed by an enzyme which requires N^{10}-formyl tetrahydrofolic acid as the

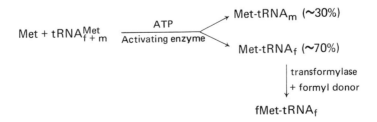

Fig. 1.27 Formation of fMet-tRNA$_f$

source of formyl groups; this formyl donor is the normal cofactor in the cell for reactions involving transfer of formyl groups. Such reactions are catalysed by a class of enzymes known as transformylases. The transformylase catalysing the formation of fMet-tRNA$_f$ has been purified and shown to be specific for Met-tRNA$_f$. Despite a very thorough search no other formylated aminoacyl-tRNA has been found. The transformylase does not even recognise methionine on Met-tRNA$_m$ and it therefore follows that the enzyme must recognise the structure of tRNA$_f$. This is in contrast with the properties of the methionyl-tRNA synthetase which attaches methionine to both forms of tRNA^{Met} (Fig. 1.27) and therefore probably recognises some common feature of both tRNAs. The reactions leading to the formation of fMet-tRNA$_f$ are summarised in Fig. 1.27.

7.4 Evidence for formylmethionyl-tRNA as a chain initiator

The role of fMet-tRNA$_f$ as an initiator in protein biosynthesis was first shown using bacterial cell-free systems programmed with synthetic polynucleotides of random nucleotide sequence. Only the random undefined polynucleotides containing the bases U, A and G, poly (U, A, G) or U and G, poly (U, G) stimulated the incorporation of methionine into polypeptides and these products were found to contain N-terminal *N*-formylmethionine. When fractionated methionine-accepting species of tRNA were used it was found that poly (U, A, G) incorporated formylmethionine from fMet-tRNA$_f$ into N-terminal positions, while methionine from Met-tRNA$_m$ was incorporated into internal positions. When poly (U, G) was used as a messenger only fMet-tRNA$_f$ and not Met-tRNA$_m$ allowed the incorporation of methionine into protein. Furthermore, it was established that this methionine became incorporated into N-terminal positions as formylmethionine. However, the experiments most clearly revealing the function of tRNA$_f$ were done under conditions where formylation of Met-tRNA$_f$ was prevented. Under such conditions it was demonstrated that Met-tRNA$_f$, even when not formylated, in the presence of poly (U, A, G) or poly (U, G) still incorporated methionine into N-terminal positions (Fig. 1.28). This unexpected result clearly indicated that tRNA$_f^{Met}$ was especially adapted to function as a chain initiator. Figure 1.28 also shows the results of programming the incorporation of methionine into polypeptide products by synthetic mRNAs of known repeating sequences such as ApUpGpApUpG, poly (A-U-G) and UpGpUpGpUpG, poly (U-G). These observations therefore showed that tRNA$_f^{Met}$ could function as a polypeptide chain initiator in a bacterial system programmed with a synthetic messenger RNA and indicated the codon assignments (also indicated by the triplet binding assay) shown in Fig. 1.28 (Clark and

Synthetic messenger	Source of Met		Position of Met in polypeptide		Codons used
	Met-tRNA$_m$	Met-tRNA$_f$ fMet-tRNA$_f$	internal	N-terminal	
random poly (U, G)	—	+	—	+	GUG
random poly (U, A, G)	+	+	+	+	AUG, GUG
poly (U-G)	—	+	—	+	GUG
poly (A-U-G)	+	+	+	+	AUG

Fig. 1.28 Position of methionine in polypeptides synthesised in a cell-free system programmed with various synthetic polynucleotides.

Marcker 1966). Whether either of the codons AUG or GUG assigned to the initiator tRNA is sufficient as a signal for initiation will be discussed later.

That fMet-tRNA could function as an initiator *in vitro* was confirmed in a most striking way by work using bacteriophage RNA (e.g. R17 or f2 RNA) as a natural mRNA in a cell-free system. In such systems the most abundant product directed by bacteriophage RNA is the coat-protein. When such an RNA was used in the cell-free system it was found that formylmethionine became incorporated into the N-terminal position. This was a most surprising result since it is known that the N-terminal amino acid of the coat-protein from intact virus particles is alanine. However, alanine was found to be the amino acid next to formylmethionine in the coat-protein synthesised in the cell-free system. It is now known that an enzyme removes the formylmethionine from the newly synthesised coat-protein *in vivo*.

7.5 Deformylation

The enzyme that removes the formyl group from growing polypeptides in prokaryotes is called the deformylase. This enzyme, although detected in *E. coli, Bacillus stearothermophilus* and *Bacillus subtilis*, has not been very well purified because of its instability in cell extracts. So far it is known that the enzyme can remove the formyl group from substrates such as formylmethionylphenylalanyl-tRNA and formylmethionyl-puromycin (an analogue to be described later) but not from formylmethionine itself.

The high proportion of methionine end groups ($\sim$ 45 per cent) in *E. coli* proteins can thus be explained by an enzymic removal of the formyl group from the terminal formylmethionine. At present no satisfactory explanation can be given for the observed high frequency of N-terminal alanine and serine in *E. coli* proteins.

7.6 Initiator tRNA binding site on the ribosome: specification of sites by puromycin

One explanation of the ability peculiar to Met-tRNA$_f$ to incorporate methionine only into an N-terminal position, whether formylated or not, is that the tRNA part of Met-tRNA$_f$ has a particular conformation that allows it to fit directly into the P-site on the ribosome, in contrast with all other aminoacyl-tRNAs which bind to the A-site (see Fig. 1.29, where a model is presented for the binding of charged tRNAs just before the first peptide bond formation).

The special binding properties of fMet-tRNA$_f$, by which the initiator binds to the ribosome more firmly than other aminoacyl-tRNAs under the direction of triplets or of synthetic polynucleotides, favour such a mechanism involving a site for the initiator distinct from other tRNAs. Strong evidence comes from experiments using the antibiotic puromycin (see also section 8).

Puromycin, because of its structural similarity to the terminal end of aminoacyl-tRNA (Fig. 1.30), inhibits protein biosynthesis by substituting for an aminoacyl-tRNA in the A-site. The polypeptide is thereby displaced from polypeptidyl-tRNA as polypeptidyl-puromycin. For example, addition of puromycin to polyphenylalanyl-tRNA or polylysyl-tRNA bound to ribosomes by poly U or poly A, respectively, yields the appropriate polypeptidyl-puromycin. A normal peptide bond is formed, but the polypeptidyl-puromycin is stable and ejected because puromycin does not contain the unstable activated carboxylic acid ester group (i.e. the normal linkage of an amino acid carboxyl group to the tRNA) nor presumably does it contain enough of the tRNA-like structure to bind to the ribosomal site and be translocated.

In a study of a model system for the formation of the first peptide bond it was found that when Met-tRNA$_f$, whether formylated or not, was bound to ribosomes by the triplet ApUpG,

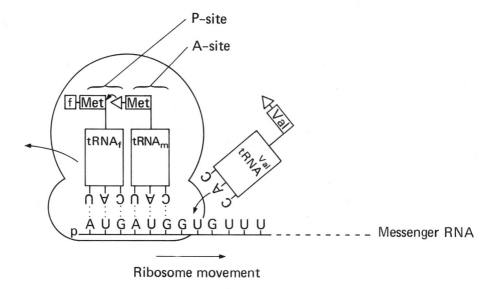

Fig. 1.29 Initiation on the 70 S ribosome.

addition of puromycin gave rise to the peptide derivatives formylmethionyl-puromycin or methionyl-puromycin respectively. Yet Met-tRNA$_m$, when bound to ribosomes with the same triplet codon ApUpG was found to be completely unreactive with puromycin in agreement with the behaviour of all other aminoacyl-tRNAs. These experimental results confirmed that there were at least two sites on the ribosome, one where binding of an aminoacyl-tRNA leads to formation of a peptide bond with puromycin and another where no peptide bond is formed with puromycin. The former reactive site is generally accepted as being the P-site and the latter the

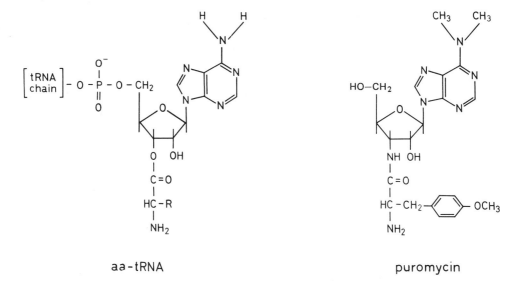

Fig. 1.30 Structural similarity of the 3'-end of aminoacyl-tRNA (especially of tyrosyl-tRNA) to the antibiotic, puromycin.

A-site. These experiments further demonstrated that the tRNA part of Met-tRNA$_f$, whether the methionine is formylated or not, is able to enter the P-site directly and it is this that confers upon it its special function as a chain initiator. In the light of this hypothesis the dual functions of the codons AUG and GUG also can be understood. The distinction of their dual role is made at the level of the tRNA. In chain initiation Met-tRNA$_f$ recognises AUG or GUG in the beginning of a messenger RNA and becomes bound in the P-site. For polypeptide elongation the incoming aminoacyl-tRNA will always be brought to the A-site and in this site AUG codes for methionine and GUG for valine. Although the above model appears to be basically sound, the actual specification of the binding sites became more complicated with the discovery of protein factors called 'initiation factors'. The relationship of the peptidyl-tRNA binding site to the initiator site is also not completely understood, but at the initiation step on 70 S ribosomes they appear to be identical (see sections 7.8 and 7.11).

7.7 Involvement of initiation factors and GTP

In 1966 it was realised that although natural mRNA such as bacteriophage f2 RNA, could be translated in a crude extract (S-30) of *E. coli*, resolution and purification of the crude components led to a loss of natural mRNA activity in a recombined system. However, the recombined system including salt-washed ribosomes, S-100 supernatant and aminoacyl-tRNA still permitted the translation of synthetic mRNAs (such as poly U and poly A) with reasonable efficiency. A search for the lost factors needed to translate natural mRNA resulted in the discovery of the initiation factors by three independent groups of workers. Although such factors could be obtained from the ribosomal supernatant of *E. coli* cells, they are usually prepared by washing crude ribosomes with a relatively high molarity salt solution such as 0·5M or 1·0M NH$_4$Cl. The three types of bacterial initiation factor so far described are now systematically called IF-1, IF-2 and IF-3. When they were first discovered they were called F1, F2 and F3, or A, C and B respectively, by different groups of workers. I have emphasised 'types' of initiation factor since different species of both IF-2 and IF-3 have been found in *E. coli*. The situation in mammalian cells is much less well characterised and is an area of intense current research. So far it looks as if initiation factors will be found in eukaryotic cells with similar properties to those in bacterial cells although they need not necessarily be found on the ribosome. The three bacterial factors have been separated from each other and purified to homogeneity but the early work used crude preparations including all three. For optimal function of the initiation factors GTP is required.

The early characterisation experiments investigated the translation of a variety of mRNAs in a resolved cell-free system to which initiation factors were added. Significantly translation of bacteriophage RNA was observed to occur efficiently only when initiation factors were added. Also, synthetic mRNAs which contained the initiator codon AUG near the 5'-phosphate end were translated at low magnesium ion concentration (5 mM) only in the presence of initiation factors and GTP. Synthetic mRNA not containing an initiator codon was not translated at low magnesium concentration even in the presence of initiation factors. Furthermore, the artificial mRNAs were all translated at higher magnesium ion concentration (>10 mM) and under these conditions the translation of mRNA containing an initiator codon could start incorrectly without a requirement for the initiator tRNA, fMet-tRNA$_f$. These results suggested that natural polypeptide chain initiation occurred at low magnesium ion concentration with a strict requirement for initiation factors, GTP, and the involvement of the initiator tRNA. Further developments came from an investigation of the part played by initiation factors in the formation of the coded initiator tRNA-ribosomal complex.

7.8 The 70 S ribosomal initiation complex

Again at low magnesium concentration (5 mM) it was possible to bind only the initiator RNA to the 70 S ribosomes in the presence of initiation factors, GTP and the initiator codon —whether present as a triplet (ApUpG or GpUpG) or in a synthetic polyribonucleotide, or in a natural mRNA such as f2 RNA.

In the initiation complex the initiator was found to be bound in the P-site as shown by its release with puromycin. Furthermore the conformation of the fMet-tRNA$_f$ is specifically recognised by the initiation factors at a magnesium ion concentration of 5mM since uncharged tRNA$_f^{Met}$ and unformylated Met-tRNA$_f$ do not form complexes under these conditions.

7.9 Use of a GTP analogue

Another interesting feature of the initiation complex was discovered by replacing GTP by its analogue, GMP-PCP. This analogue has the P-O-P of the β-γ pyrophosphate link replaced by P-CH$_2$-P so that the usual γ-phosphate removal associated with peptide bond elongation is prevented. Although formation of the initiation complex takes place when GTP is replaced by GMP-PCP, the bound initiator tRNA is not released by puromycin. The interpretation of this effect is not clear cut. Clearly there can be no hydrolysis of GTP during the formation of the initiation complex as in peptide bond elongation (section 8) and the actual role of GTP is still unclear. If GTP is involved in some allosteric way by priming the interaction of the initiation factor with the tRNA, perhaps the GMP-PCP is capable of giving a modified complex whose stereochemistry is incorrect for reaction with puromycin on the 50 S subunit.

Alternatively, Bretscher (1968a) proposed a mechanism in terms of hybrid tRNA binding sites to account for the above result. Since the usual two-site hypothesis involves sites extending from the 30 S to 50 S ribosomal subunits, the sites can be considered to be composed of 30 S and 50 S partial sites. He proposed that during the formation of the initiation complex the initiator tRNA ends up in the total P-site only via an intermediate complex where a dislocated hybrid of a 50 S A-site and 30 S P-site is involved. Thus, it is possible that GMP-PCP might allow the hybrid site complex to form but it would prevent a subsequent relocation to a total P-site which could require the hydrolysis of GTP, in a manner similar to polypeptide chain translocation (see section 8).

Another possible explanation comes from the recent finding that GTP hydrolysis has been identified with release of initiation factor IF-2 from the ribosome; without this release the formation of the first peptide bond will not be possible. Thus, it is possible that the GTP analogue locks the ribosome and prevents peptide bond formation.

7.10 The role of the formyl group

It is known that the formyl group on fMet-tRNA$_f$ is not necessary for placing methionine in a terminal position in polypeptides synthesised in a prokaryotic cell-free system at high magnesium concentration, or under normal conditions in eukaryotic cells, so why a formyl group should occur at all is not clear. The available evidence suggests that it is involved somehow in the formation of the first peptide bond and the formation of the ribosomal initiation complex. When the rate of protein synthesis was measured in a cell-free system where formylation did, or did not, occur, it was clear that in the presence of formylated Met-tRNA$_f$ protein synthesis was considerably faster than when the Met-tRNA$_f$ was not formylated. This rate effect was especially marked at the beginning of the incubation suggesting that the step affected was the formation of the first peptide bond. In the puromycin experiments (described in section 7.6) where only one peptide bond is formed, it was also noted that the formation of formylmethionyl-puromycin was much faster than the formation of methionyl-puromycin. Furthermore it seems (see section 8) that Met-tRNA$_f$ when formylated is a better substrate for the enzyme responsible for peptide bond formation than is the unformylated species.

The involvement of initiation factors and the requirements for the formyl group in polypeptide chain initiation under conditions of low magnesium ion concentration suggested another indirect role for the formyl group. There is good evidence that formylation of Met-tRNA$_f$ alters the structure of the tRNA$_f^{Met}$ in such a way that it can form a specific complex at low magnesium ion concentration with initiation factors and GTP in the P-site of the 70 S ribosome. The present ideas about the functions of particular initiation factors amplify this latter role of the formyl group. It is likely that the initiator tRNA is brought to the ribosome (in fact the subunit 30 S as we shall see in sections 7.11 and 7.12) as a complex with IF-2 and GTP. This initiation factor, IF-2, does not recognise the non-formylated Met-tRNA$_f$ at low magnesium ion concentration so it is possible that again the formylation specifies the recognition by the initiation factor. However, it is not yet certain that this recognition occurs away from the ribosome as we recently suggested (Rudland *et al.* 1971).

Since bacteria are thought to make proteins faster than eukaryotic cells we can speculate that formylation may have evolved to speed up protein synthesis by permitting the first peptide bond to be made faster.

7.11 Role of ribosomal subunits in initiation

Early studies of cell-free systems using electron microscopy showed that mRNA molecules are bound to 30 S subunits, which had been formed by subjecting 70 S ribosomes to low magnesium ion concentrations(10^{-4} M) followed by sucrose gradient fractionation. Since it was also known that polypeptidyl-tRNA is attached to 50 S subunits (Gilbert 1963) a division of function between the two subunits was a possibility.

Nomura first proposed, as a result of some modest binding effects, that the initiator tRNA forms a complex with a 30 S ribosomal subunit and mRNA before the larger 50 S subunit joins them to expose the aminoacyl-tRNA binding site. Recently there has been much support for the involvement of the 30 S subunit in this way. Studies of polyribosome metabolism have shown that there is an abundance of polyribosomes and subunits but few monosomes (perhaps only 10 per cent) in lysates of *E. coli*. The rates at which newly formed, radioactively labelled 30 S and 50 S subunits and mRNA entered, and were released from, polyribosomes led to the conclusion that the free subunits regularly joined (coupled) from a pool to form a monosome on mRNA. It was assumed that the couple then moved along the mRNA and dissociated after reaching the end. All of the subunits exchanged with polyribosomes. Further, *in vivo* and *in vitro* studies using heavy isotopic labels have proved that ribosomal subunits exchange with one another over several rounds of protein synthesis. These studies, together with results of experiments elucidating the functions of the initiation factors (section 7.12) have led to the generally accepted idea that polypeptide chain initiation starts by interaction of mRNA and initiator tRNA on a free 30 S subunit. It is also thought, though not yet proved, that chain termination involves the removal of the intact 70 S ribosome which is later dissociated prior to another chain initiation; there is controversial evidence for a protein dissociation factor being concerned before initiation but whether the dissociation factor is an initiation or an anti-reassociation factor is not clear. Finally there is much evidence from studies of the formation of the coded initiator tRNA: 30 S complex and its dependence on initiation factors that the 30 S subunit is indeed involved in natural chain initiation.

7.12 Function of each initiation factor

Protein chain initiation depends on the formation of a ribosomal initiation complex prior to peptide bond formation. The formation of the initiation complex requires at least three macromolecular recognition processes. They are (a) the selection of the initiator tRNA (in bacteria, fMet-tRNA$_f$) from all other aminoacyl-tRNAs, (b) selection of the correct initiation codon AUG (or GUG) on the mRNA corresponding to the amino terminal methionine residue of the programmed protein, (c) interaction of mRNA and initiator tRNA with a 30 S ribosomal subunit not yet engaged in protein synthesis. The specificity of these recognition processes appear to be controlled by the initiation factors. Process (a) appears to be directed by IF-2 and process (b) by IF-3; process (c) occurs as a consequence of processes (a) and (b) but appears to be stabilised by IF-1. However, the precise function of each of the initiation factors is still in dispute; all that is certain is that initiation factors are required to initiate protein synthesis directed by natural mRNA in *E. coli* extracts. Some of the properties of bacterial initiation factors taken from the confusing multitude of published papers are summarised below.

All of the initiation factors appear to be in limiting amounts with less than one molecule per ribosome in the cell and their amounts apparently depend on the cell growth conditions. Thus, they are possible regulatory factors in protein synthesis. Another general feature of the initiation factors is that they appear to be localised on the 30 S subunit.

7.12.1 Initiation factor IF-1

IF-1 has been purified to homogeneity from *E. coli* cells. It is a small protein with a molecular weight of about 9000 daltons and is readily separated from the other types of initiation factors which can be adsorbed to a DEAE-cellulose column under conditions where IF-1 is not retained. So far IF-1 has not been separated into components and its function is not certain, although it is generally agreed that IF-1 is somehow concerned with the stabilisation of the coded ribosomal initiation complex formed with IF-2, GTP and the initiator tRNA. However an alternative view, which has recently been gaining ground, is that IF-1 is the active dissociation factor for dissociating the pool of 70 S monosomes to yield 30 S subunits for the initiation process. A third recent proposal that it is analogous to elongation factor (EF-Ts) (see section 8), and that its role is to recycle IF-2 by releasing it from an IF-2:GDP complex, seems to have lost support. Although IF-1 is known to form a complex with IF-3 the function of the complex is not yet clear.

7.12.2 Initiation factor IF-2

IF-2 is much larger than IF-1. So far two distinct species of IF-2 (IF-2a and IF-2b) have been identified but appear to have identical functions. IF-2a has been purified and has a molecular weight of about 80 000, but IF-2b with a molecular weight of approximately 100 000 has not been well purified.

Like IF-1, IF-2 also binds to IF-3, but the function of this complex is also unclear. There is similar doubt about the reports that IF-2 directs the binding of mRNA to 30 S subunits. However, IF-2 is undoubtedly concerned with binding the initiator tRNA to bacterial ribosomes, in the presence of GTP and mRNA. There is reasonable support for the idea that IF-2 plays the specific role of selecting the initiator tRNA for transport to the 30 S subunit. This would be analogous to the function of the elongation factor EF-Tu (see section 8), which takes all other aminoacyl-tRNAs, except the initiator, as a ternary complex with GTP to the 70 S ribosomal A-site. In comparison (as in Fig. 1.31) IF-2, which can form a very weak

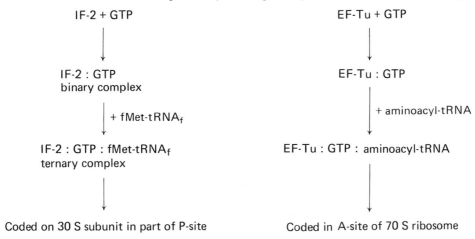

Fig. 1.31 Comparison of the roles of IF-2 and EF-Tu.

complex with GTP, forms a ternary complex with GTP and fMet-tRNA$_f$ but no other aminoacyl-tRNAs for coding in the initiation site of the 30 S subunit. Possibly IF-1 stabilises this complex on the 30 S subunit.

The GTP analogue, GMP-PCP, can substitute for GTP in the formation of the 30 S initiation complex. The large 50 S subunit joins to the 30 S initiation complex apparently without involvement of GTP hydrolysis to complete the 70 S initiation complex with the initiator tRNA in the 'P' site — not quite the P-site apparently (see following comments). Recent research suggests that the IF-2:GDP complex after GTP hydrolysis has to be removed before the initiation complex with the P-site filled is properly set for the first peptide bond formation. Thus GMP-PCP may prevent peptide bond formation by locking the initiator tRNA in the inactive 'P'-site condition (see Fig. 1.32 which summarises the roles of the initiation factors).

7.12.3 Initiation factor IF-3

Although IF-1 and IF-2 are sufficient to permit formation of the initiation complex and translation using synthetic polynucleotides that contain an initiation codon, a third class of initiation factor (IF-3) is necessary for initiation complex formation and translation with a natural mRNA, at least if the mRNA is in the native state; IF-3 is not required if the natural mRNA (e.g. phage RNA) is first denatured with formaldehyde. So far several different species of IF-3 have been identified after much controversy. The most well documented species of IF-3 is one purified from *E. coli*, which has a molecular weight of 22 500 daltons.

The precise function of IF-3 is still unknown although it is well established that it is necessary for initiation to occur on natural mRNA. The proposal that IF-3 directs the initial binding of mRNA to the 30 S subunit prior to the fMet-tRNA$_f$ binding is without experimental proof. Indeed the detailed mechanism of binding of mRNA to the ribosomes during the

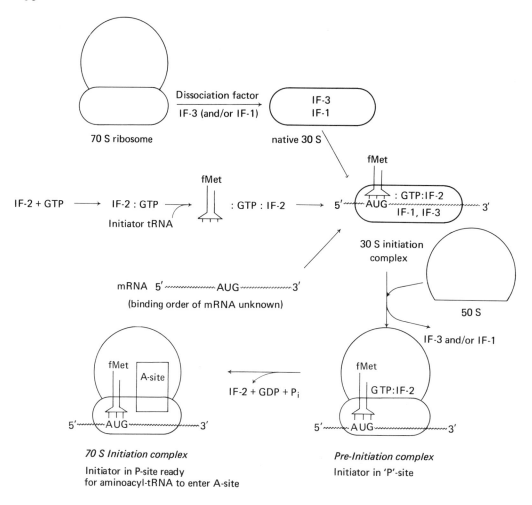

Fig. 1.32 Scheme for steps so far detected in polypeptide chain initiation.

initiation process, and whether it occurs before, during or after binding of the initiator tRNA is not known. There may be a curious difference in the prokaryotic and eukaryotic systems on this point. In bacteria the binding of the initiator tRNA seems to be mRNA-coded so that binding of the mRNA appears to take place before or at the same time as the tRNA. In contrast, binding of a mammalian (rabbit reticulocyte) initiator tRNA is apparently non-coded so here the mRNA can occur after the initiator tRNA binding. This discrepancy is under active investigation.

IF-3 appears to increase the stability of the coded ribosomal initiation complex. Thus a currently favoured role for IF-3 is that it may recognise a special initiation signal or feature of mRNA structure to allow coding of the initiator tRNA; for example, IF-3 might open up a folded region of mRNA to expose the initiating triplet. Many studies have tried to prove this point — so far without success. The idea is supported by the finding of multiple species of IF-3 in bacterial extracts. Differential recognition of different initiation sites by several classes of IF-3 raises the interesting idea that control of protein synthesis through messenger selection could occur in this way (see section 6.9). The present evidence, although slightly favouring messenger selection, is frankly inconclusive. So far we know that one IF-3 activity preferentially selects the initiation site for the coat-protein cistron of R17 phage RNA, and another the initiation site of the late mRNA of phage T4. The hypothesis that selection by an appropriate IF-3 species is a reasonable explanation of the mechanism of switching mRNA read-out that occurs during bacteriophage infection has not yet been proved. The situation in eukaryotic cells will be described in the section 7.13.

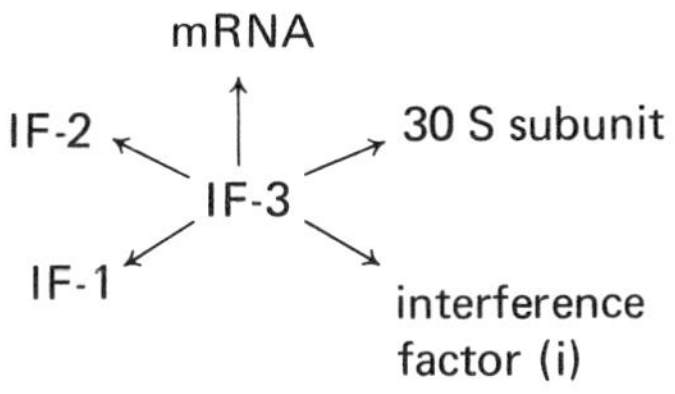

Fig. 1.33 Postulated interations of IF-3.

Bacterial IF-3 may also be involved in two other aspects of initiation. It may have a dissociating effect on bacterial 70 S ribosomes, and it interacts with another recently discovered set of protein factors called interference factors. The multiple interactions of IF-3 are summarised in Fig. 1.33 without a real knowledge of which are natural phenomena.

7.12.4 Dissociation factor

Several groups of workers have shown that addition of IF-3 in pure form to 70 S ribosomes dissociates them into their subunits. This property has been taken to be an active function in protein synthesis necessarily preceding initiation. However, growing evidence now indicates that IF-1 is really the dissociation factor, and that IF-3 just artifactually loosens the binding of the subunits, perhaps by removing magnesium thereby causing dissociation of 70 S couples. The situation is confused by the fact that IF-3 and IF-1 form a complex with each other. Certainly dissociation occurs before initiation, and although there is a good chance that either IF-1 or IF-3 is the active factor, recent work suggests that a new factor with anti-reassociation activity is the real one.

7.12.5 Interference factors

The gene order of phage RNA such as R17 RNA and selection of the second gene, the coat-protein cistron, by an IF-3 have already been described (see section 6.7). The discrimination by the IF-3 is not however absolute; some A protein (perhaps 10 per cent of the amount of coat-protein) is also made, but only very small amounts of the synthetase. Because of the lack of complete discrimination it has been suggested that the multiplicity of species of IF-3 is an artefact of the cell-free system, in which the real initiation factor is split into subunits, and that isolation of the real macromolecular initiation factor would result in absolute discrimination in selection of message. It remains to be seen whether this is true. What we do know from recent work is that it is possible to change the specificity of mRNA selection of an IF-3. Revel and his colleagues (Groner *et al.* 1972) looked for a factor which would inhibit the translation of MS2 phage RNA (almost identical with R17 phage) directed with a mixture of IF-3 species. They were able to isolate from bacterial extracts a protein factor of molecular weight about 74 000 with the function of inhibiting coat-protein synthesis and allowing synthesis of the phage synthetase. This factor is called interference factor, i. Apparently i interacts with an IF-3 and prevents it from taking part in initiation complex formation at the coat-protein cistron initiation site. A whole class of interference factors which specifically inhibit the messenger selection of IF-3 has recently been found. Their existence is a clear example of translational control.

A remarkable property of factor i has recently been identified. It has been shown to be identical with, and can replace subunit I of, $Q\beta$ phage synthetase which is composed of 4 main subunits, only one of which (subunit II) is coded by the phage RNA. As will be described in section 8, subunits III and IV have been identified as bacterial elongation factors EF-Tu and EF-Ts respectively. The significance of this finding which links RNA synthesis with protein synthesis is not clear.

The current suggested roles of the initiation factors are summarised in the overall scheme for the initiation step shown in Fig. 1.32. Experiments which have measured the length of mRNA protected by the ribosome from ribonuclease digestion in these complexes have indicated that there is no movement of the ribosome along the mRNA during the change of the initiator binding state from the 'P'-site to the P-site.

Before concluding this section one other factor should be mentioned. Does the discharged initiator tRNA after first peptide bond formation fall off the ribosome or is it ejected by an ejection factor?

7.12.6 Ejection of $tRNA_f^{Met}$

An activity distinct from the initiation factors has recently been identified with a tRNA-ejection activity. It was isolated in an impure form from a crude initiation factor preparation. The sparse evidence available suggests that this postulated ejection factor specifically recognises uncharged $tRNA_f^{Met}$ on the ribosome and removes it with a GTP requirement. It is not yet known if the GTP is hydrolysed in the process.

7.13 Eukaryotic initiation factors

Our present knowledge about the existence and function of eukaryotic initiation factors is much less detailed than for bacterial factors. Certainly it is likely that similar classes of factors exist, although the identification of eukaryotic IF-3 type has been controversial. The confusion has come from the difficult problem of isolating pure eukaryotic initiation factors in an active state. Four different initiation factors in impure form have been isolated from reticulocyte extracts. They have been defined as M1, resembling IF-2; M2a; aGTPase activity; M2b similar to IF-1, and M3 similar to IF-3.

Very recently an IF-2 type factor has been isolated from reticulocytes and HeLa cells which has been shown to form a complex with initiator tRNA (Met-tRNA$_f$ in eukaryotic cells) and probably GTP, but the binding to the small (40 S) ribosomal subunit does not appear to be mRNA coded. This apparent difference from the bacterial situation is not yet understood.

The real controversy, however, has centred upon the function or even the existence of IF-3 type factors in eukaryotic cells. Although such an mRNA selection factor has for some time been thought to exist in controlling myosin synthesis in muscle cells, studies of translation of various pure mRNAs in oocytes have favoured lack of mRNA discrimination. It is now thought that the oocyte has the possible disadvantage of containing many classes of IF-3 since it is not a differentiated cell. Appropriate pure mRNAs injected into oocytes will indeed programme the synthesis of globin and eye lens α-crystallin with high efficiency. Very recently the hypothesis which demands the existence of discriminatory IF-3 factors has gained support from the purification of an IF-3 factor from ascites tumour cell extracts. It selects the initiation site of a viral RNA, encephalomyocarditis virus RNA, and discriminates against other eukaryotic mRNAs such as globin mRNA.

7.14 Universality of methionine initiator tRNAs

The presence of fMet-tRNA$_f$ has been detected throughout the prokaryotic cell world and in eukaryotic organelles. Thus, it is possible to generalise that where protein synthesis is fast and occurring on small, 70 S-like monosomes fMet-tRNA is used as an initiator, together with its appropriate initiation factors. For example fMet-tRNA$_f$ has been found in extracts of *B. subtilis* and *B. stearothermophilus* and in chloroplasts from *Euglena gracilis* and mitochondria from yeast and rat liver. Furthermore *E. coli* Met-tRNA$_f$ can be formylated by extracts of *Lactobacillus leichmanii*, *Pseudomonas*, *Streptomyces antibioticus* and *Clostridium tetramorphus*. The only exception to the generalisation that fMet-tRNA$_f$ is the prokaryotic initiator has been found in *Streptococcus faecalis R* which cannot synthesise the formyl donor (formyl tetrahydrofolate) and can be grown on a folate free medium. The cell extracts of *S. faecalis R* are incapable of formylating *E. coli* Met-tRNA$_f$ so it is likely that in this bacterial strain Met-tRNA$_f$ is the initiator. The significance of this finding and its relation to the situation in eukaryotic cells is unclear.

The initiating species in eukaryotic cell cytoplasms has been shown to be Met-tRNA$_f$. The lack of formylation appears to be due to the absence of transformylase activity in eukaryotic cell cytoplasms, since eukaryotic initiator tRNAs such as yeast, guinea pig, liver, reticulocyte and ascites tumour Met-tRNA$_f$ can be formylated with the bacterial transformylase enzyme. However, further discrimination by eukaryotic initiation factors would probably exclude the formylated species from being an efficient initiator tRNA on 80 S ribosomes, an idea which has been confirmed using reticulocyte cell extracts.

In conclusion, although some differences in the detailed mechanism of protein initiation does exist for prokaryotic and eukaryotic cells in each case the initiator tRNA is a methionine tRNA.

7.15 The signal for initiation

Although the coding properties of the initiator tRNA and the nucleotide sequences around the initiating codon of several different cistrons are now known, no generalisation can be made about the exact nature of the starting signal for protein synthesis. *In vitro* studies show that the bacterial initiator tRNA is coded by both AUG and GUG but until recently only AUG had been found in the limited number of known mRNA cistron starting sequences. However GUG has now been found once—at the start of the A-protein cistron in MS2 RNA. Thus the importance of GUG as a natural starting codon is not clear.

There is some evidence that AUG alone is not enough for the starting signal and the signal could well include, or have close to it, a sequence or structure responsible for binding the 30 S subunit to the mRNA. The sequences so far determined for the initiation sites on mRNA have been for bacteriophage RNA where it has been possible to correlate determined nucleotide sequences with known amino acid sequences of the bacteriophage proteins. Small pieces of RNA suitable for sequence determination have been isolated from experiments in which 70 S initiation complexes were digested with pancreatic ribonuclease; the initiator tRNA located the ribosome at the initiation site on the mRNA and the ribosome protected the mRNA against

	fMet	Arg	Ala	Phe	Ser

A-protein: AUUCCUAGGAGGUUUGACCU · AUG · CGA · GCU · UUU · AGU ·G

	fMet	Ala	Ser	Asn	Phe

Coat-protein: AGAG(C)CCUCAACCGGGGUUUGAAGC · AUG · GCU · UCU · AAC ·UUU

	fMet	Ser	Lys	Thr	Thr	Lys

Synthetase: AAACAUGAGGAUUACCC · AUG · UCG · AAG · ACA · ACA · AAG

Fig. 1.34 R17 initiation sites with assigned amino acids.

digestion around the appropriate initiation signal. This work is discussed in more detail in chapter 7. Some examples of the determined sequences (Steitz 1969) are shown in Fig. 1.34. The coat-protein and A-protein cistron starting sequences can be arranged in secondary structure to give hairpin loops but the synthetase starting sequence cannot. It is possible, as already stated, that an IF-3 has something to do with recognising and unfolding such sites containing secondary structure.

7.16 Dual coding and primary structure of initiator tRNA

It is unlikely that two different initiator tRNAs are coded by AUG and GUG. At present, we believe the dual coding to be a function of the decoding characteristics of the initiator tRNA in the P-site on the ribosome. Strong evidence for this comes from an interpretation of the

	(f)Met-$tRNA_f$	Met-$tRNA_m$
Methionyl-tRNA synthetase	+	+
Transformylase	+	−
Codons	AUG GUG	AUG
Source of methionine	amino-terminal positions	internal positions
Initiation factor, IF-2a	+	−
Transfer factor, EF-Tu	−	+
Puromycin reactivity	+	−
Acylaminoacyl-tRNA hydrolase	−	+
Binding of tRNA to 30 S ribosomes	strong	weak

The plus sign indicates that reaction occurs with that particular tRNA.

Fig. 1.35 Summary of biological properties of methionine tRNAs.

function of tRNA$_f^{Met}$ in terms of its primary structure. Although it is possible to separate two initiator tRNA species with identical functions their primary structures differ only in one position outside the anticodon loop and, not surprisingly, they are both coded by AUG and GUG. Furthermore, a radioactive [^{32}P]-nonadecanucleotide containing the anticodon CAU has been isolated from partial enzymic digests of tRNA$_f^{Met}$ (see Fig. 1.20). This anticodon loop fragment is coded by both AUG and GUG at the same site as the initiator tRNA, fMet-tRNA, on 30 S subunits. Thus, the dual coding observed for polypeptide chain initiation is a special feature of the structure of the initiator tRNA of which the anticodon loop is a part; protein factors are not required to explain this phenomenon.

7.17 Functions of methionine tRNAs in terms of primary structures

It was hoped that a comparison of the primary structures of the two types of methionine tRNA, tRNA$_f^{Met}$ and tRNA$_m^{Met}$ would reveal functional parts of the molecule. Figure 1.35 summarises the different reactions that these two classes of methionine tRNA undergo. However, even knowing the structures (shown in Fig. 1.21) an interpretation of function in terms of primary structure is not possible (see also Section 4.10.3, para *1*). The recent elucidation of the primary structure of a eukaryotic initiator RNA, yeast tRNA$_f^{Met}$, also does not allow any convincing explanation of its function. Current work is aimed at interpreting the function of the initiator tRNA in terms of its spatial structure.

7.18 Model systems for polypeptide chain initiation

Many of the properties of natural polypeptide chain initiation have been mimicked by *N*-acetyphenylalanyl-tRNA. When coded by poly U, and in the presence of initiation factors and GTP at low magnesium concentrations, this analogue will bind to a puromycin-sensitive site (like initiator tRNA). However, the analogy should not be taken too far, since the analogue is not as efficient an initiator as fMet-tRNA$_f$ nor does it form such a stable ribosomal initiation complex. Indeed the analogue cannot form a stable complex directed by its triplet, UpUpU, in contrast to the stable initiator-tRNA complex directed by ApUpG. Other blocked aminoacyl-tRNAs are worse initiators than *N*-acetylphenylalanyl-tRNA. The explanation of this presumably is that the spatial structure of *N*-acetylphenylalanyl-tRNA fortuitously resembles that of fMet-tRNA$_f$: this would possibly be the reason for the efficient artefactual translation of poly U in cell-free systems.

8 Polypeptide Chain Elongation

8.1 General features

After polypeptide chain initiation a series of peptide chain elongation steps occurs before the process of chain termination releases the completed protein. Elongation is defined as the addition of amino acids one at a time to a growing polypeptide in a sequence programmed by mRNA. The overall process of chain elongation has been already described briefly in Section 1.9. This simple version of elongation is complicated by the involvement of cell supernatant factors now called elongation factors. So far functions have been described for elongation factor EF-T, which consists of two functionally separate subunits EF-Tu and EF-Ts, and elongation factor EF-G. Elongation factors used to be called transfer factors.

The mechanism of polypeptide chain elongation is generally represented as occurring in three steps. For convenience, and in the absence of strong contradictory evidence, elongation will be described in terms of two ribosomal sites. Step 1, which occurs after the 70 S initiation complex is formed or during polypeptide chain growth is described by (c) in Fig. 1.3 and involves the codon-directed binding of aminoacyl-tRNA to a vacant ribosomal A-site adjacent to an occupied P-site. Step 2 is the peptide bond-forming step (d) in Fig. 1.3. In this step there is peptidyl transfer from the peptidyl-tRNA of fMet-tRNA$_f$ in the P-site to aminoacyl-tRNA in the A-site. Step 3 is concerned with a translocation of the peptidyl-tRNA back into the P-site to permit a new incoming aminoacyl-tRNA to be bound in the A-site, with concomitant ejection of the discharged tRNA from the P-site. This translocation, step (b) in Fig. 1.3, involves relative movement of the ribosome and mRNA. EF-T is concerned with Step 1 and EF-G is somehow concerned with Step 3. Both of these protein factors need GTP hydrolysis in order to function.

8.2 Protein factors for elongation

Most of our knowledge about the polymerisation of amino acids on the ribosome has come from a study of model systems which in large part have been derived from the development of the bacterial cell-free system (see Section 2). Lipmann's group was the first to show that protein factors as well as ribosomes were required for the polymerisation of amino acids by components of the cell-free system (Nishizuka and Lipmann 1966). They used poly U to programme a resolved bacterial cell-free system for the synthesis of polyphenylalanine from Phe-tRNA and were able to isolate from the ribosomal supernatant fluid two complementary protein fractions which participated in the condensation of the activated amino acids. These protein fractions were required for polyphenylalanine synthesis in the presence of both ribosomal subunits and GTP. Similar fractionation of mammalian cell-free systems by other workers have implicated less well characterised protein fractions, called transfer factors, in poly-U-directed polyPhe synthesis by reticulocyte and rat liver components. More progress, as usual, has been made with the bacterial system and these will now be described in detail.

In the early studies one of the complementary protein fractions was shown to be coincident with a ribosome-dependent GTPase, an enzyme which hydrolysed GTP to inorganic phosphate (P_i) and GDP. Thus the activity now known as EF-G was readily assayed as in Fig. 1.36. Since it did not incorporate GMP from GTP into polymer it was clearly different from an RNA polymerase type of reaction. Although the GTPase activity was shown to be linked to protein synthesis, when the stoicheiometry of the reaction was determined there was an abundant excess (> 50-fold) of GTP hydrolysis over the formation of peptide bonds. This embarrassment was reduced (although it is still with us) when further work resulted in a better purification of the protein fractions and of the ribosomes. Although both these polymerising protein fractions were shown to be found in the ribosomal supernatant the one associated with GTPase activity

$$\gamma\text{-}[^{32}P]\,GTP \xrightarrow[\substack{\text{washed ribosomes} \\ \text{+ EF-G fraction}}]{\text{pH 7·4 buffer, 10 mM-MgCl}_2} GDP + {}^{32}P_i$$

determined as
phosphomolybdate
complex and extracted
into organic phase

Fig. 1.36 Assay for EF-G factor.

usually adhered strongly to ribosomes and contaminated them. It needed a very careful and detailed fractionation of supernatant proteins, and rigorous washing of the ribosomes to progress in the characterisation of these polymerisation factors. By a procedure involving fractional precipitation, gel adsorption and several column chromatography steps using DEAE-Sephadex and hydroxylapatite, the two complementary protein fractions EF-T and EF-G (originally called T and G) could, however, be separated, and they have since also been crystallised.

$$[^{14}C] \text{ Phe-tRNA} \xrightarrow[\substack{\text{GTP, saturating EF-G factor} \\ \text{+ enzyme fraction containing EF-T}}]{\substack{\text{pH 7·4, 10 mM-MgCl}_2 \text{ salt buffer,} \\ \text{mercaptoethanol,} \\ \text{poly U, washed ribosomes,}}} [^{14}C] \text{ polyPhe}$$

Fig. 1.37 Assay for EF-T factor.

In the early fractionations EF-G was assayed as described above (Fig. 1.36). Saturating amounts of EF-G were needed to check for the polymerisation stimulation or transfer function of the EF-T activity, as shown in Fig. 1.37. The ribosomes were checked for G contamination by the assay for GTPase activity. This did not require added poly U or Phe-tRNA. The use of GMP-PCP (the analogue of GTP in which the oxygen bridge between the β- and γ-phosphates is replaced by a methylene bridge as described earlier) confirmed that the hydrolysis of GTP was really required in the polymerisation of phenylalanine from Phe-tRNA; when GTP was replaced by GMP-PCP in this reaction there was complete inhibition of polymerisation.

One point about EF-T and EF-G very significant for future structural studies is the relatively large amount of these proteins in the bacterial cell, which means that it should be possible to prepare them on a reasonable scale. About 1 per cent of the cell supernatant protein is EF-T and about 2 per cent is EF-G: these amounts can even increase a few fold when protein synthesis is occurring rapidly. It seems that there is about one molecule of EF-T and one molecule of EF-G per ribosome in the cell. Both EF-T and EF-G have molecular weights of about 80 000 but EF-T splits into subunits to function in the elongation step as will be described in Section 8.4.

8.3 Hydrolysis of GTP

It is very difficult to determine accurately the number of GTP molecules split during one round of peptide bond formation and translocation. Those workers who see the whole process as a concerted mechanism believe that only one GTP is split. However, when each step in elongation is studied *in vitro* (as we shall see in the next few sections) two molecules of GTP are split for one complete round of amino acid addition to the growing polypeptide.

8.4 Step 1 of elongation requires EF-Tu

Elongation factor, EF-T, has been shown to be composed of two components called EF-Tu and EF-Ts in equimolar amounts. These components have different functions in polypeptide chain elongation. In fact only EF-Tu appears to be involved in step 1 of elongation in which aminoacyl-tRNA is brought to the 70 S ribosomal A-site; EF-Ts is needed only to recycle EF-Tu in an active form. The two components have recently been purified on a large scale separately, in contrast to early procedures where they were isolated together as EF-T. The key to obtaining active EF-Tu is to stabilise it, as first performed by Weissbach and colleagues, by purifying it in the presence of its cofactor product GDP (Miller and Weissbach 1970) with which it forms a tightly bound complex (see below). The molecular weights of EF-Tu and EF-Ts are approximately 45 000 and 35 000 respectively. The functions of these components have been elucidated in *in vitro* experiments and are summarised in Fig 1.38.

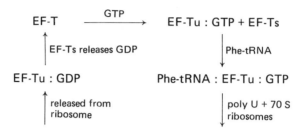

Decoding of mRNA by ternary complex bound in A-site of 70 S ribosome.

Fig. 1.38 Role of elongation factors EF-Tu and EF-Ts in step 1 of elongation.

When GTP is mixed with EF-T a binary complex of EF-Tu:GTP is formed, displacing EF-Ts. The dissociation constant of this complex is about 3×10^{-7}M in the presence of 10 mM magnesium ions. The complex binds to nitrocellulose filters so the presence of EF-Tu in cell extracts can easily be detected if the complex is made with radioactive GTP. This is a more convenient assay for following EF-Tu during its preparation than the previous one described for EF-T in Section 8.2. If any aminoacyl-tRNA (except the initiator) is added to the binary complex, the ternary complex aminoacyl-tRNA:GTP:EF-Tu is formed. This is shown in Fig. 1.38 for Phe-tRNA. The ternary complex does not bind to nitrocellulose filters so addition of the aminoacyl-tRNA to the cell extract before filtration prevents an equivalent molar amount of EF-Tu from being adsorbed by the filter. It is this ternary complex and not the aminoacyl-tRNA alone which binds to the 70 S ribosomal A-site during step 1 of the polypeptide elongation process (see step (c) of Fig. 1.3). After this binding step, which illustrates the transfer function of the factor, GTP is hydrolysed presumably prior to peptide bond formation, and EF-Tu is released from the ribosome probably in a tight complex with the GDP product. The timing of the GTP hydrolysis and release of GDP in relation to peptide bond formation and translocation is unclear. What is known is that EF-Tu forms a binary complex with GDP which is 100 times more tightly associated than is the GTP complex; the dissociation constant of EF-Tu:GDP in the presence of 10 mM magnesium ions is 3×10^{-9}M. The stable EF-Tu:GDP complex does not react with aminoacyl-tRNA and regeneration of the EF-Tu:GTP complex for participation in another round of chain elongation would need very high concentrations of GTP. However EF-Ts catalyses the substitution of GTP for GDP in an EF-Tu:GDP binary complex by first forming EF-T from EF-Tu and EF-Ts, and expelling GDP. Then the EF-T complex (dissociation constant about 5×10^{-5}M) is readily dissociated by GTP to start another round of aminoacyl-tRNA binding. An understanding of the detailed molecular workings of EF-Tu should be possible in the foreseeable future because large crystals of the EF-Tu:GDP complex suitable for X-ray analysis have recently been produced.

There is a distinct similarity between the formation of the EF-Tu factor:GTP:aminoacyl-tRNA complex and that of the initiation factor:GTP:initiator-tRNA complex (see Section 7). Both complexes require the presence of GTP in their assembly, and in each case the GTP in the complex can be replaced by the analogue GMP-PCP in support of the idea that no hydrolysis of GTP occurs in the formation of this complex.

Many of the details of the elongation factor-stimulated transfer of aminoacyl-tRNA to the ribosome were first worked out by Lengyel and his colleagues (Skoultchi *et al.* 1970). They used *in vitro* systems containing stabler elongation factors from the heat stable bacterium, *Bacillus stearothermophilus.* The elongation factors from *B. stearothermophilus* corresponding to *E. coli* EF-Tu and EF-Ts are termed S_3 and S_1. The other one, S_2 is equivalent to EF-G.

8.5 Step 2 requires peptidyl transferase

Step 2 of polypeptide chain elongation involves the formation of a peptide bond. Since both the peptidyl-tRNA and aminoacyl-tRNA are highly reactive species it might be thought that they would react spontaneously to form the peptide bond. In this respect the ribosome might play a passive part in providing the appropriate sites to permit the reactive species to approach each other. However, it is known that the formation of the peptide bond on the ribosome is catalysed by an enzyme, peptidyl transferase, which is a ribosomal component. It is not clear what happens to the excess of energy released during peptide bond formation and during GTP hydrolysis. The energy is not needed for bond formation but may be needed to actuate possible conformational changes of interacting macromolecules during chain elongation and for the relative movement of the ribosome and mRNA.

Model systems have been devised to obtain information about the enzymically catalysed peptide bond formation and especially about the substrate specificities at the ribosome catalytic site. The use of the aminoacyl-tRNA analogue puromycin has already been described in Section 7.6, with reference to a definition of the P- and A-sites on the ribosome. Peptide bond formation can be studied by adding puromycin to a system containing ribosome-bound polypeptide complex. Puromycyl-polypeptides are formed with release of the polypeptide from the ribosome in a reaction which has enzymic characteristics; GTP is not required for the puromycin release reaction. That this model system really does appear to portray correctly a normal peptide bond formation was confirmed by the demonstration that the formation of a single peptide bond from peptidyl-tRNA and aminoacyl-tRNA does not require GTP.

There is much evidence that the catalytic centre of peptidyl transferase is on the 50 S unit. For instance polypeptidyl-tRNA can be isolated attached to 50 S subunits after dissociation of the 70 S ribosomes. PolyPhe-tRNA thus attached to 50 S ribosomal subunits can be shown to react with puromycin in a manner analogous to peptide bond formation but independent of mRNA or the 30 S subunit.

8.6 Substrate requirements at the P-site

Details about the substrate specificity at the catalytic site in the neighbourhood of the P-site have been determined in a curious but useful system developed by Monro (1967). The aminoacyl-tRNA is replaced as before by puromycin (Pm) but the peptidyl-tRNA is replaced by a terminal oligonucleotide fragment of initiator tRNA with its attached formylmethionine (i.e. CpApApCpCpA-fMet). In Section 7.6 it was shown that the initiator tRNA, fMet-tRNA, appears from its reaction with puromycin to be located in the P-site. The reaction of this fragment with puromycin is catalysed by washed 50 S ribosomal subunits and is dependent upon the presence of monovalent cations, divalent cations and ethanol, whereas 30 S subunits, soluble protein factors, GTP and mRNA are not required. Thus, this fragment reaction is similar to the enzymically catalysed peptide bond formation and is summarised in Fig. 1.39. The curious requirement for ethanol or methanol probably reflects some stabilisation of the fragment in the P-site. When other aminoacyl-tRNA fragments, blocked and unblocked, and of various lengths, were tested in the ethanol-dependent reaction on 50 S subunits, it was clear that the enzyme requires an acylaminoacyl-tRNA or acylaminoacyl-tRNA fragment as a substrate in the P-site and recognises at least the terminal CpCpA.

$$\text{CpApApCpCpA-fMet + Pm} \xrightarrow{\text{50 S, ethanol}} \text{fMet.Pm}$$

Fig. 1.39 Fragment reaction locating peptidyl transferase.

The intact initiator tRNA, fMet-tRNA$_f$, has also been bound to 50 S ribosomal subunits in the presence of ethanol, and a reaction with aminoacyl-tRNA similar to the fragment reaction has been characterised. So there is good evidence that this ethanol-dependent fragment reaction with Pm is similar to peptide bond formation.

8.7 Substrate requirements at the A-site

The substrate requirements of the peptidyl transferase at the A-site have been studied using chemically synthesised analogues of the aminoacyl-adenosine end of aminoacyl-tRNA (Rychlik et al. 1969). When a polypeptidyl-tRNA such as polyLys-tRNA was bound to the P-site, a reaction occurred with phenylalanyl-adenosine (Phe-A). If the adenosine was replaced by I and C there was a decrease in reactivity, and complete loss of activity if A was replaced by G or U. The 2'-hydroxyl group of the sugar was required possibly for binding the terminal A to the ribosome, since Phe-deoxy A did not work. The aminoacyl-nucleoside cannot be replaced by other aminoacyl esters or the amide. Thus, there is a strictly specific requirement for an aminoacyl ester of adenosine with a neighbouring (cis vicinal) hydroxyl group. Further characterisation of the peptidyl transferase will come from its clear identification as one or more of the ribosome's constituent proteins.

8.8 Step 3 of polypeptide chain elongation involves EF-G

When it became clear that actual peptide bond formation did not require the hydrolysis of GTP another role for the EF-G factor was sought. The stage of polypeptide chain elongation in which more energy was required was clearly translocation, and EF-G has indeed been found to be involved in the poorly understood translocation step (step (b) of Fig. 1.3). EF-G and GTP are together required for the movement of peptidyl-tRNA from the 70 S ribosomal A-site to the P-site. Concomitantly the discharged tRNA is ejected from the P-site, and there is evidence that EF-G is also concerned at this point. This can be distinguished from the special case of ejection of initiator tRNA after the first peptide bond formation (see section 7.12). The hydrolysis of GTP is thought to provide the driving energy for the translocation process.

The involvement of EF-G in translocation has been neatly shown by experiments with a model cell-free system as summarised in Fig. 1.40; this indicates the binding of aminoacyl-tRNA and initiator tRNA to ribosomes by a short synthetic mRNA and specified protein factors. The results of using as mRNA a hexanucleotide containing an initiator codon are shown in experiment (a) of Fig. 1.40. EF-G factor was found not to be required for dipeptide synthesis, and coded binding of aminoacyl-tRNA in the A-site required EF-T factor and GTP. (This experiment did not use purified EF-Tu.) When these findings are compared with those of experiment (b) of Fig. 1.40 which illustrates the use of nonanucleotide as mRNA, the involvement of the EF-G factor became clear. The translation of the third codon stringently requires the presence of EF-G factor and involves hydrolysis of GTP. If the EF-G factor was

Factors added		Product bound to ribosomes
(a) Using as mRNA ApUpGpUpUpU		
1. Initiation	$\xrightarrow[\text{+ GTP + Phe-tRNA}]{\text{ribosomes + fMet-tRNA}}$	fMet-tRNA
2. Initiation + EF-T	$\xrightarrow{\text{as in (a)1}}$	fMet.Phe-tRNA
(b) Using as mRNA ApUpGpUpUpUpUpUpU		
1. Initiation	$\xrightarrow{\text{as in (a)1}}$	fMet-tRNA
2. Initiation + EF-T	$\xrightarrow{\text{as in (a)1}}$	fMet.Phe-tRNA
3. Initiation + EF-T + EF-G	$\xrightarrow{\text{as in (a)1}}$	fMet.Phe.Phe-tRNA

Fig. 1.40 Experiment showing function of factor EF-G.

omitted, even in the presence of EF-T factor and GTP, only the dipeptide was synthesised. This result suggests that the hydrolysis of GTP is concerned in the translocation of the mRNA during the synthesis of the tripeptide.

These and other results tempted workers to call the EF-G a translocase but there is no evidence that EF-G, a GTPase activity, is responsible for moving the new peptidyl-tRNA from the A-site to the P-site. This moving activity will probably be called the translocase but it has not yet been properly identified. Another possible translocase, protein L12, or its acetylated derivative, protein L7, from the 50 S subunit will be discussed later.

8.9 Inhibition of EF-G by fusidic acid

The sterol drug fusidic acid (FA) inhibits the uncoupled GTPase activity manifested by EF-G only in the presence of ribosomes and at the translocation step — more evidence that the ribosome-linked GTPase activity functions in translocation. Further details about the effect of various drugs such as fusidic acid on protein synthesis components will be found in reviews listed in the Bibliography. Fusidic acid appears to inhibit translocation according to the scheme outlined in Fig. 1.41. Bacterial EF-G forms an unstable intermediate with 70 S ribosomes, or the 50 S subunit, and GTP. The unstable intermediate can be stabilised in a non-functional state by using the non-hydrolysable GTP analogue, GMP-PCP. Normally the GTP is hydrolysed permitting the isolation of a stable complex of EF-G, GDP and 50 S subunit. This complex necessarily dissociates during translocation to allow another round of translocation. Although fusidic acid actually stimulates the formation of the ternary ribosomal complex using purified components this affords a ready explanation of its inhibition of translocation. It stimulates the ribosomal binding by forming a non-dissociable complex with the translocation product complex EF-G:50 S:GDP. Thus fusidic acid prevents the dissociation of this ternary complex and locks the ribosome into a non-functional state after one round of translocation as shown in Fig. 1.41. Lack of dissociation of the normal ternary complex at 0°C may also be the reason why EF-G will not function in translocation at 0°C.

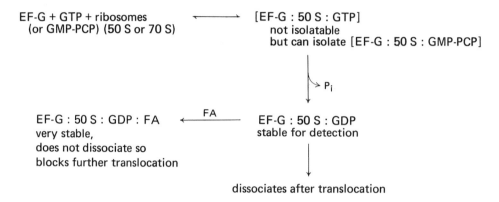

Fig. 1.41 How fusidic acid inhibits the translocation step.

Although no mutants of initiation factors, release factors or EF-T have so far been found, two types of mutants of EF-G are known. One occurs in a temperature-sensitive *E. coli* mutant (at elevated temperatures it ceases to function). The other type occurs in a fusidic acid-resistant mutant which presumably has an altered EF-G no longer able to form a complex with the fusidic acid.

8.10 Ribosomal sites for binding EF-T and EF-G: use of thiostrepton

Strong evidence has been gathered recently for a common site or overlapping site of action on the ribosome for both EF-Tu and EF-G functions. The evidence comes from at least three types of experiment.

The first set of experiments involved the use of fusidic acid. As described above FA binds to ribosomes with EF-G and GDP. Normally FA does not inhibit the coded binding of the aminoacyl-tRNA:GTP:EF-Tu ternary complex to the 70 S ribosome. However if FA is added with EF-G and GTP and allowed to bind to ribosomes then subsequent binding of the EF-Tu ternary complex is prevented. This competition for binding suggests a common site of action or at least close, or overlapping, sites which are not independent of each other.

The second set of experiments also used an antibiotic, in this case, thiostrepton. Thiostrepton inhibits the ribosomal binding activities of both EF-Tu and EF-G. It appears to work by binding irreversibly to the 50 S ribosomal site for EF-Tu and EF-G. Thiostrepton also prevents GTP hydrolysis associated with both EF-Tu and EF-G binding. Although these experiments seem to be clear evidence of a common site for EF-Tu and EF-G activity associated with the ribosomal A-site, the possibility remains that thiostrepton, the EF-Tu ternary complex and EF-G:GDP bind to different, but reciprocally interacting, sites.

The third set of experiments is very recent and complete interpretation will be a future problem. It is possible to remove from salt-washed ribosomes a protein required for both EF-Tu and EF-G ribosomal associated functions. The removal conditions involve a further high salt (1M ammonium chloride) wash in the presence of 40 per cent ethanol. These treated ribosomes are inactive in protein synthesis and do not bind EF-Tu and EF-G complexes. At least five proteins have been identified in this ethanolic wash supernatant and these can be fractionated. When two very similar acidic proteins (formerly called A_1 and A_2) known as L7 and L12 are added back to the ethanol-treated ribosomes, the ability to bind elongation factor complexes is restored. These proteins have been identified as structural protein constituents of the 50 S subunits, and the evidence seems to indicate that they form part of a common or overlapping ribosomal site for the EF-Tu and EF-G functions.

Proteins L7 and L12 have been intensively studied recently in the belief that they play an important role in translocation. They have a molecular weight of about 13 800 daltons and differ only in that the N-terminal residue of L7 is acetylated; the amino acid sequence for the *E. coli* protein is now known. The finding that L7 and L12 are the same is the only example known so far that the ribosome contains more than one copy of a particular protein. The acetylated protein, L7, is inactive in the test of restoring activity to the ethanol treated ribosomes, whereas the non-acetylated species L12 is active. Although ethanol-treated ribosomes minus L7 and L12 proteins are inactive in elongation factor binding, they still bind thiostrepton. Thus it is likely on this evidence that the thiostrepton binding to the ribosome is related to but not identical with the EF-Tu and EF-G binding sites.

8.11 Eukaryotic elongation factors

Elongation factor activities similar to the bacterial ones probably also exist in eukaryotic cells. However, in spite of intense current research active eukaryotic elongation factors have not yet been isolated in pure form for detailed studies. Present indications are that factors corresponding to EF-Tu and EF-G called EF-1 and EF-2 occur in a brain extract. No evidence of an EF-Ts-like factor has been detected. Although EF-2 seems to have a molecular weight similar to EF-G, EF-1 is much larger (mol. wt., 186 000 daltons) than EF-Tu. So far it has not been possible to form a ternary complex of EF-1, GTP and aminoacyl-tRNA. Both EF-1 and EF-2 give complexes with GTP. This distinguishes EF-2 from EF-G because EF-G forms a complex with GTP only in the presence of ribosomes.

Another interesting difference between EF-2 and EF-G is the inhibition of the eukaryotic factor by diphtheria toxin. This toxin, in the presence of NAD as cofactor, inhibits translocation in eukaryotic cells as well as the GTPase activity of EF-2, thus implicating EF-2 in the translocation mechanism.

8.12 Stringent factor and magic spots: a control mechanism?

When normal bacterial strains (genetically defined as RC^+) are deprived of a particular essential amino acid during their growth, not only does protein synthesis stop but RNA synthesis is also substantially reduced. Such *'stringent'* control of RNA synthesis suggests that the regulation of RNA synthesis is linked to protein synthesis in bacteria. It is also possible to identify mutant *E. coli* strains (RC^-) where this coupling is altered so that RNA synthesis is under *'relaxed'* control and is not reduced when a required amino acid is removed from the growth medium.

Some indication of the way in which this control is exercised was obtained by measuring the amount of charged tRNA in stringent cells under various growth conditions. During exponential growth of an *E. coli* culture of RC^+ cells, the degree of aminoacylation of each tRNA is of the order of 80 per cent; in contrast, during amino acid starvation stringent strains contain only 10–40 per cent of the aminoacylated-tRNA corresponding to the missing amino acid. Further, it appeared that, at least in *E. coli*, the synthesis of stable RNAs (rRNA and tRNA) was preferentially reduced relative to that of mRNA during the stringent response to amino acid starvation. Thus it seemed likely that the stringent response exemplifies regulation of a special class of RNA. Recent work has given further indication of how this could occur. Not only does uncharged tRNA accumulate during the stringent response, but two unusual guanosine nucleotide derivatives also appear to be made in large amounts (Cashel and Gallant 1969). It now seems that the formation of these derivatives, called for convenience MS1 and MS2, is part of the stringent response. MS stands for the trivial name of 'magic spot' from their original appearance on chromatograms. MS1 and MS2 are ppGpp (guanosine tetraphosphate) and pppGpp (guanosine pentaphosphate) with phosphates attached to the 5′ and 3′-hydroxyl groups. MS1 and MS2 are thus analogues of GDP and GTP, respectively.

How these compounds are degraded is unknown, but their synthetic pathway has recently been elucidated and some indication of their role is being suggested by experiments with *in vitro* systems. Using purified components of cell-free systems made from stringent cells it has been possible to synthesise MS1 and MS2 *in vitro*. Their synthesis requires washed ribosomes, a factor from the high salt wash of ribosomes (as for the initiation factor preparation), GDP or GTP, and ATP; GDP produces MS1 and GTP, MS2. The synthesising factor has been called the stringent factor and much effort is being expended to identify its role in protein synthesis. It appears not to be any of the initiation or elongation factors.

If carefully purified ribosomes are used, then synthesis of MS1 or MS2 by stringent factor requires uncharged tRNA in addition to the components listed above, and, further, the uncharged tRNA must be bound to the ribosome with its coding mRNA. Thus there now appear to be two pathways for binding tRNA to the ribosomal A-site depending on the cellular growth conditions. Under normal conditions the aminoacyl-tRNA is synthesised and forms a complex with EF-Tu and GTP and decodes the mRNA during peptide bond formation. However, if the tRNA is not charged it binds directly to the ribosome decoding the mRNA and setting off a repetitive reaction synthesising MS1 or MS2. The relation between these pathways is not yet understood.

Some light has recently been thrown on the link between the stringent response and control of RNA synthesis. *In vitro* experiments have shown that MS1 inhibits synthesis of RNA by bacterial RNA polymerase. The control of initiation of the stable classes of RNA has been tentatively assigned to a transitory subunit psi, ψ, of RNA polymerase. Somewhat fortuitously, it was discovered that ψ is composed of subunits III and IV of the isolatable RNA-synthesising enzyme, Qβ phage synthetase (or replicase). Since MS1 inhibits and binds to both ψ and EF-T, subunits III and IV have been compared with EF-Tu and EF-Ts and have been shown to be identical with them (Blumenthal *et al.* 1972). Thus elongation factors from protein synthesis have been identified as also probably being concerned with initiation of DNA transcription by RNA polymerase. At any rate the elongation factors play a role in synthesising phage RNA. It is worth remembering that subunit I of Qβ synthetase is also the interference factor i which is implicated in control of translation (see section 7).

At present it is not possible to predict how this postulated coupling of protein synthesis and RNA synthesis will turn out. It seems that as more and more details of cellular metabolism are elucidated more proteins with multifunctional roles will be found.

8.13 Conformation changes in ribosomal complexes during elongation?

Ultracentrifugation studies have recently been made of ribosomal-tRNA complexes at different stages of polypeptide chain elongation. Two main types of complexes were used, one called the pretranslocational complex (equivalent to state II of Fig. 1.3) and the other called the posttranslocational complex (i.e. state III of Fig. 1.3). At high centrifugation rates different complexes were found to have reproducibly different sedimentation values which were interpreted as arising from different conformational states of the ribosome. However at lower speeds the different complexes sedimented together. Consequently, the difference in sedimentation velocities shown in high-speed centrifugation is no proof of a difference in conformational states for the complexes; the different tRNA positions in the different complexes could have different stabilising effects on the coupling of the subunits.

9 Polypeptide Chain Termination

9.1 Definition

When homopolyribonucleotides such as poly U or poly A are used to programme the synthesis of a polypeptide in a cell-free system, nearly all of the polypeptide product can be isolated attached to tRNA. The small amount of chemical hydrolysis of the ester linkage in peptidyl-tRNA at physiological conditions is nowhere near adequate to explain the rate at which the bacterial cell completes proteins so there must be a mechanism for hydrolysing polypeptidyl-tRNA. It is now known that mRNA contains signals for polypeptide chain termination; the codons UAA, UAG and UGA have been shown to be chain-terminating codons by genetic and biochemical methods. Whether natural chain termination *in vivo* involves a structural signal involving more than the terminator codons is not known; *in vitro* a terminator codon suffices. Polypeptide chain termination is defined as the mRNA-programmed release of free polypeptide from polypeptidyl-tRNA growing on the ribosomal surface. The first significant information about polypeptide chain termination came from genetic studies mainly in the laboratories of Brenner and Garen (see Garen 1968).

9.2 Genetic studies elucidating terminator codons

The replacement of an mRNA codon for a particular amino acid by another codon specific for another amino acid is called a missense mutation. This type of mutation is observed only if the new amino acid does not inactivate the protein's function. Alternatively, the change to a codon which does not correspond to any amino acid is called a nonsense mutation. The fact that there is extensive degeneracy in the genetic code suggests that there are few nonsense codons.

9.2.1 Nonsense mutations

The discovery in DNA-containing bacteriophage of nonsense mutations causing the premature termination of polypeptide chains helped in the identification of terminator codons. At first, the nonsense mutants were thought to act by interrupting the reading of the genetic message. A class of nonsense mutants called amber mutants in the gene for the head protein of bacteriophage T4 was used (as noted in section 2) to prove the collinearity of gene and protein. These amber mutants of phage-T4 produced N-terminal fragments of coat-protein in bacterial cells by a premature termination at the growing end of the polypeptide. Since the fragments were not attached to tRNA when they were isolated it appeared that the mechanism for their production was similar to that of normal polypeptide chain termination. It could be shown that the short polypeptides did not arise from mRNA fragments but were instead a result of a mutation in the mRNA.

The amber mutants were found to be suppressible mutants; bacteriophages containing them can grow on permissive (su⁺) bacterial hosts but not on non-permissive (su⁻) hosts. Although amber mutants could be found in a variety of bacteriophage genes the same permissive host could suppress all of these mutants. Amber mutants are thus defined by the suppressing host; see for example Fig. 1.45. Further knowledge has permitted their definition in terms of a common mechanism for their suppression. Other nonsense mutants, called ochre mutants, have different properties from amber mutants. The transmission coefficient, a measure of completed polypeptide chains in suppressed translation of the nonsense mutant compared with the total number of starts, in the case of amber mutants is quite high (30–65 per cent, depending on the suppressing host) whereas it is usually very low (about 5 per cent) for ochre mutants (see later sections 9.2 and 9.6 for possible explanations). Because of this low transmission coefficient the completed polypeptides produced by suppressors of ochre mutants are in such small amounts that they are extremely difficult to detect biochemically. Ochre suppressors suppress amber mutants as well but the converse is not true.

9.2.2 Amber and ochre codons

Two types of experiments showed that amber and ochre mutations were related to specific codons. One involved investigation of the effect of the mutagen hydroxylamine on the production and reversion of amber and ochre mutants of the rII gene of the bacteriophage T4. The other involved a study of the amino acids inserted into the bacteriophage head protein and specified by codons related to the amber codon.

The latter approach involved connecting the amber codon to amino acids with known codons by specified mutations. The wild-type bacteriophage was shown to contain glutamine at the place of termination in the amber mutant. The amber codon, which was related to glutamine (codon CAG) by biochemical studies and tryptophan (UGG) by mutagenically induced transitions, was concluded to be UAG; a transition describes an alteration of a pyrimidine to the other pyrimidine or a purine to the other purine. This was confirmed from mutations to other codons by transversions; a transversion describes the mutation of a pyrimidine to a purine or vice versa. A similar approach using even more extensive connections to amino acid replacements was used in analyses of the bacterial enzyme alkaline phosphatase and its many

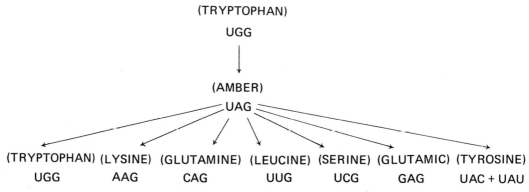

Fig. 1.42 Amber codon's relatedness to amino acid codons.

isolatable mutants as shown in Fig. 1.42. Genetic evidence fixed the ochre codon as UAA since it can be obtained by a transition from the amber codon UAG.

9.3 Chain terminating signals

Studies of the head protein fragments caused by amber mutants of T4 bacteriophage showed that the mutants lead to efficient termination of the polypeptide chain. This was suggestive evidence that amber and ochre codons code for proper chain termination. Furthermore, since the ochre termination was much more difficult to suppress than the amber (see section 9.2),

```
                Coat-protein                                              Synthetase
                                                                          fMet  Ser
       Ala Asn Ser Gly  Ile  Tyr
 f₂ (G)CA AAC UCC GG C AUC UAC UAA UAG A C G CCG GCC AUU CAA ACA UG
 R17 (G)CA AAC UCC GG U AUC UAC UAA UAG A U G CCG GCC AUU CAA ACA UGA GGA UUA CCC AUG UCG
```

Fig. 1.43 Intercistronic nucleotide sequence for bacteriophages R17 and f2 RNAs. Differences between the two phages are blocked. The termination codons are underlined. (Adapted from Nichols and Robertson 1971.)

UAA was considered to be the normal chain terminating codon whereas UAG could be a rare chain terminator. Recent genetic studies have implicated another codon, UGA, as a terminator codon, also by characterisation of a third type of suppressible nonsense mutation. These suppressors are now known as UGA suppressors and were previously called opal suppressors. No bacterial aminoacyl-tRNAs which can be coded by UAA, UAG or by UGA have so far been found although some eukaryotic aminoacyl-tRNAs, apparently able to decode chain terminator codons, have been described; these unusual tRNAs may be suppressor tRNAs in eukaryotic cells.

Very little is known about *in vivo* chain termination signals. The only evidence available so far again comes from sequence studies on phage RNA; the nucleotide sequences at the ends of the coat-protein cistrons in two phage, R17 and f2, are known. Both sequences (Fig. 1.43) show that codons for the last amino acid of the protein is followed by not just one terminator codon but two in tandem (Nichols and Robertson 1971). Recent studies make it unlikely that this double codon sequence UAAUAG is a universal termination signal. One explanation of this double terminator sequence is that it might protect against spontaneous mutations capable of converting a single termination codon into a sense codon. The double terminator codon makes it very unlikely indeed that a chance mutation could suppress termination.

9.4 Mechanism of bacterial suppression

Studies of bacterial suppression have been useful in attempts to elucidate the mechanism of chain termination. When an amber mutant is suppressed a mechanism is provided by the permissive host (su$^+$) for competing with polypeptide chain termination. The propagated chain results in a complete protein and has an amino acid inserted at the site of termination in the non-permissive host (su$^-$). The mechanism of bacterial suppression has been elucidated most clearly using the su$_I^+$ and su$_{III}^+$ suppressors. Consideration of the translation of mRNA where chain termination could be suppressed led to three possibilities: altered ribosomes, altered activating enzymes, or altered tRNAs could be involved in such a way as to prevent chain termination.

When tRNA from su$^+$ cells was loaded with amino acids using su$^+$ activating enzymes, the resulting charged su$^+$ aminoacyl-tRNAs could be completely discharged by activating enzymes

74

from su⁻ cells. If a chain terminating tRNA had been charged by su⁺ activating enzymes then the su⁻ activating enzymes would not have been able to discharge it. Thus, the second possibility (altered activating enzymes) could not be responsible for bacterial suppression.

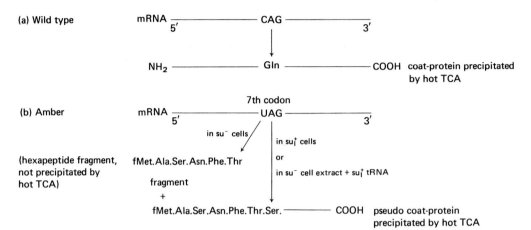

Fig. 1.44 An amber mutant of an RNA bacteriophage as mRNA to indicate mechanism of suppression.

The involvement of tRNA in bacterial suppression was demonstrated best by *in vitro* studies of protein synthesis directed by bacteriophage RNA which contained an amber codon. The cell-free system is described in Fig. 1.44. It is possible to isolate coat-protein amber mutants of the RNA-containing bacteriophage f2 or R17. In either case the amber codon causes chain termination in su⁻ cells after the sixth amino acid from the N-terminal end of the coat-protein. The use of RNA extracted from these amber mutant-containing bacteriophage (referred to below as amber RNA) as natural mRNA *in vitro* allows the construction of the most reasonable system for a biochemical study of chain termination; these aspects will be described later. Although the N-terminal fragment made *in vitro* by the amber codon-containing RNA starts with formylmethionine this amino acid is later removed *in vivo* since the completed coat-protein starts with alanine (see Section 7).

The system outlined in Fig. 1.44 can be assayed quite simply. Since the N-terminal hexapeptide is soluble in hot trichloroacetic acid (TCA) (Fig. 1.44(b)) whereas the complete coat-protein is not (Fig. 1.44(a)), the amount of radioactive formylmethionine in a hot TCA precipitate from such a cell-free system programmed by amber RNA is a measure of the suppression in that system. When a cell extract from su⁻ cells is programmed with amber RNA very little coat-protein is made (Fig. 1.44(b)) compared with that made from wild-type RNA (Fig. 1.44(a)). Thus, various components of an su⁺ cell extract can be added to the su⁻ system to check which causes suppression. In this way su⁺ᵢ tRNA was shown to be the active component from a fractionated extract and inserted serine at the seventh position of the polypeptide, as expected from the classification shown in Fig. 1.45. The su⁺ᵢᵢᵢ tRNA has also been shown to insert tyrosine, as expected. Clearly the su⁺ tRNA which competes with chain termination must be decoding the UAG codon in the mRNA. Attempts to purify serine su⁺ tRNA and tyrosine su⁺ᵢᵢᵢ tRNA have been only partly successful. In both cases the purified fractions were bound slightly to ribosomes in the triplet binding assay by the amber codon UAG. Unfortunately, the binding was so weak as to require confirmation by the activity of the purified fraction in the more sensitive and reliable *in vitro* assay already described.

Although it was established that tRNA was the suppressing agent, further details of the mechanism of suppression called for a determination of the primary structure of su⁺ tRNA. The mutation su⁻ to su⁺ clearly requires a change in a tRNA. This could be brought about either by a change in an existing su⁻ amino acid tRNA gene so that the newly expressed tRNA could read a UAG codon, and still be charged by the activating enzyme or, alternatively, by modification

Bacterial suppressor	Amino acid inserted
su⁺ₗ	Ser
su⁺ₗₗ	Gln
su⁺ₗₗₗ	Tyr

Fig. 1.45 Classification of amber suppressors determined using *in vivo* systems.

of a chain terminator tRNA for UAG so that it will accept an amino acid. If the suppressor gene is a structural gene for a tRNA and the mutation is in the anticodon, the su⁺ tRNA for each of the suppressor strains will not recognise its former su⁻ codon. Thus, it would be expected to be a dispensable or redundant tRNA in the su⁻ cell. This has been shown to be the mechanism of bacterial suppression for the su⁺ₗₗₗ class. Nowadays the idea of redundant tRNAs is no longer viewed with suspicion, so that the mechanism proposed on the basis of a knowledge of the

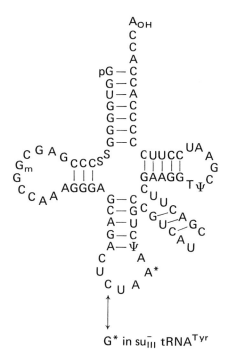

Fig. 1.46 Nucleotide sequence of su⁺ₗₗₗ tRNA Tyr. The arrow shows the base substitution for converting su⁻ and su⁺ tRNA. s⁴U = S.

primary structures of su⁻ and su⁺ tyrosine tRNAs is accepted as a general feature of bacterial cells. Smith and his colleagues were able to isolate pure su^+_{III} tyrosine tRNA on a radioactive scale after a clever piece of genetic surgery which allowed them to make bacterial cells which generated more copies than usual of this tRNA. They determined the nucleotide sequence of the suppressor tyrosine tRNA (Goodman *et al.* 1968) and found the anticodon CUA (see Fig. 1.46). The primary structure of su⁻ tyrosine tRNA, which was a revertant (a spontaneous change in the backward direction) of the su⁺ tyrosine tRNA, was next determined. This su⁻ tRNA was found to be a minor component of the normal tyrosine tRNA of the su⁻ cells, and differed from the su⁺ tRNA only in the anticodon as shown (Fig. 1.46). Thus suppressor cells contain a single base mutation from a G-like residue (G*) in the anticodon of a wild-type tRNA^{Tyr} to a C, causing a loss of ability to decode a tyrosine codon (UAU or UAC) but producing decoding of the chain terminating UAG.

9.5 Mechanism of polypeptide chain termination

Early experiments which investigated chain termination using a cell-free system programmed by synthetic copolymers of unknown sequence as mRNA suffered from severe technical disadvantages. The product polypeptidyl-tRNA could not easily be separated from any free polypeptide produced by a chain terminating mechanism, and the situation was further complicated by the lability of the polypeptidyl-tRNA to hydrolysis (yielding free polypeptide). Fairly reliable assays were, however, devised for separating polypeptides containing mainly phenylalanine or lysine from their respective polypeptidyl-tRNAs. Thus synthetic copolymers containing mainly U or A residues could be tested for their chain terminating ability in a cell-free system. As with the homopolymers, most of the polypeptide product is attached to tRNA when a cell-free system is programmed with copolymers such as poly (U, C), poly (U, G) and poly (A, C). However, poly (U, A) was found to give about 50 per cent of the product as free polypeptide. This was the first biochemical support for the suggestion from genetic studies that UAA is a terminating codon. An interesting control experiment indicated that absolute nonsense did not result in chain termination. Absolute nonsense is obtained by using a base which does not base pair in the normal way. Thus a copolymer containing mainly U and small amounts of xanthosine, X, which does not base pair, gave polyPhe-tRNA with no released polypeptide; the small proportion of X allowed fairly long stretches of polyPhe to be made before blocking further translation.

The use of synthetic polynucleotides of defined repeating sequences by Khorana and his colleagues provided additional evidence that the chain terminating codons identified by genetic studies were correct. Polymers of the triplets UAG, UAA or UGA are not translated into polypeptides in the cell-free system.

The most satisfactory mRNA available for studying chain termination has been the coat-protein amber mutant of bacteriophage RNA which has already been described in section 9.4 (see Fig. 1.44). Clearly, when amber RNA directs the synthesis of polypeptide *in vitro* UAG, the amber codon, is translated as a chain terminating signal. Polypeptide chain termination in the system directed by amber RNA has been demonstrated to be an active process by the work of Bretscher (1968b). He first purified the six tRNA species corresponding to the amino acids preceding the amber codon position in the amber RNA. When these charged tRNA species were added to a resolved cell-free system programmed by amber RNA the free hexapeptide was obtained. Since the chain terminating mechanism was not affected by the absence of the extra tRNA species it was unlikely that a chain terminating tRNA is required, as expected from the work on suppression. Furthermore, when the amber RNA was replaced by wild-type RNA in the presence of the six purified tRNA species, hexapeptidyl-tRNA was the product. Thus, the hexapeptidyl-tRNA which was chemically stable under the conditions of the experiment needs an active process for the release of the hexapeptide and the glutamine codon CAG in the wild-type RNA is not misread as the chain terminator UAG.

9.6 Release factors

Ganoza (1966) was the first to observe that the supernatant of an *E. coli* cell extract contained protein factors which were connected with polypeptide chain termination. She studied the formation of free polypeptides when a cell-free system was programmed with the copolymer, poly (U, A), of random nucleotide sequence. When the cell-free system was resolved into its components, and various supernatant protein fractions were added, the release of polypeptide from peptidyl-tRNA was found to depend on the presence of a particular supernatant factor. In an extension of her earlier work she also used synthetic polynucleotides, such as $AUGU_{14}A_n$, containing a defined sequence and starting with a chain initiator codon. Since some of this class of molecules contained UAA the synthetic mRNA caused chain termination. This termination could be shown to be dependent upon the presence of her crude supernatant release factor.

Capecchi (1967) was able to obtain more reliable information about the release factor by using a modified cell-free system programmed by the amber R17 RNA described in Fig. 1.44. All amino acids were removed during the preparation of this system so that peptide synthesis could be controlled by the successive addition of amino acids contained in the N-terminal fragment, fMet.Ala.Ser.Asn.Phe.Thr., made by the amber RNA. Capecchi purified a protein release factor (RF factor) of molecular weight about 40 000 which was essential for release of the hexapeptide from the tRNA in the mRNA-directed peptidyl-tRNA: ribosome complex. The hexapeptide was not released from the tRNA in the presence of RF factor if ribosomes were absent; this property distinguishes the RF factor from another protein fraction with hydrolase activity which has been isolated from supernatant proteins and which hydrolyses ribosome-free peptidyl-tRNA in solution. Although this hydrolysis does not appear to be involved in chain

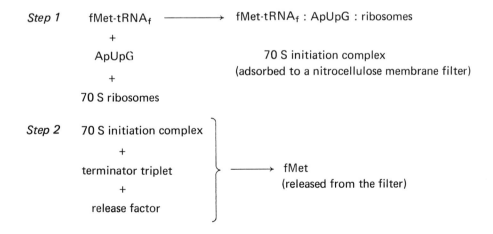

Step 1 fMet-tRNA$_f$ ⟶ fMet-tRNA$_f$: ApUpG : ribosomes

  +

ApUpG 70 S initiation complex

  + (adsorbed to a nitrocellulose membrane filter)

70 S ribosomes

Step 2 70 S initiation complex

  +

terminator triplet ⟶ fMet

  + (released from the filter)

release factor

Fig. 1.47 Rapid assay for polypeptide chain termination.

termination, it could be a scavenging enzyme which hydrolyses any growing peptidyl-tRNA released in error from the ribosome.

Further properties of the RF factor came from work by Caskey and his colleagues (Caskey *et al.* 1968). They set up a simple assay for chain termination as shown in Fig. 1.47. Using only triplets in the well-known triplet-directed tRNA binding assay they were able to translate initiator and terminator codons sequentially. In their assay they first bound initiator tRNA, fMet-tRNA, to 70 S ribosomes with the triplet ApUpG to form a stable 70 S initiation complex (step 1 of Fig. 1.47); this was done at high magnesium ion concentration (10 mM) to avoid the requirement for initiation factors. The methionine was radioactively labelled so that the location of the formylmethionine could easily be followed. The release of free formylmethionine from the ribosomal initiation complex (step 2 of Fig. 1.47) was shown to require the presence of a terminator triplet and the release factor but nothing else; no chain terminator-tRNA is involved. Assuming that the initiator codon is in the P-site under these conditions, it appears that the terminator codon is decoded in the A-site on the ribosome. It was shown that all three proposed terminator triplets (UpApA, UpApG and UpGpA) worked in this assay for termination. The RF factor can in fact be resolved into two release factors, RF-1 and RF-2, which have been found to work in conjunction with different terminator codon sets. RF-1 responds to both UAA and UAG whilst RF-2 responds to UAA and UGA. The fact that both release factors work with UAA is probably why UAA appeared to cause more efficient termination in the earlier genetic studies. The way in which the protein release factor can decode an mRNA codon and hydrolyse peptidyl-tRNA is a stimulating unsolved problem since the RF factor certainly does not appear to contain any RNA. The latest estimates show that there is only about one release factor molecule per thirty ribosomes in the bacterial cell; this is therefore another possible point in translational control. There are about 500 molecules of RF-1 of mol. wt. 44 000 daltons and about 700 molecules of RF-2 of mol. wt. 47 000 in the *E. coli* cell.

70 S ribosomes + RF + radioactive terminator triplet $\longrightarrow$ 70 S ribosomes : RF : terminator triplet (complex adsorbed on nitrocellulose filters)

Fig. 1.48 Assay for terminator codon.

A partial reaction, as shown in Fig. 1.48, has been useful for studying the codon recognition properties of release factors. The presence of a particular RF directs the binding of radioactive triplet to 70 S ribosomes forming a complex which can conveniently be measured since it is adsorbed to nitrocellulose filters. This partial assay also led to the discovery of a third release factor, RF-3, formerly called factor S. RF-3 stimulates the binding of RF-1 or RF-2 to ribosomes and in doing so RF-3 also binds to the ribosomal complex. RF-3 does not recognise the codon nor does it form a complex with RF-1 or RF-2 alone. It seems that RF-3 may function by permitting formation of the ribosomal complex with a lower concentration of terminator triplet than would otherwise be possible.

A somewhat unexpected property of the bacterial termination scheme is the lack of requirement for GTP. An unexplained functional relation between GTP and RF-3 is suggested by *in vitro* studies. RF-3 stimulates the formation of RF:terminator triplet:70 S ribosome complexes which seem to be actively dissociated by GTP and GDP. Although RF-3 also becomes incorporated into the ribosomal complex it dissociates from it upon the addition of GTP or GDP. In contrast to these currently rather confusing results with GTP in the bacterial system, it has been established that GTP is actively involved in the termination reaction in eukaryotic cells such as reticulocytes.

9.7 Mammalian release factor

Release factor activity was first detected in rabbit reticulocyte extracts using a modification of the formylmethionine release assay described for the bacterial system (Fig. 1.47). Although the assay used bacterial initiator tRNA, in the eukaryotic systems, in contrast with the bacterial ones, the intermediate complex (fMet-tRNA: 80 S reticulocyte ribosomes) was formed without the addition of the ApUpG triplets. The probable normal non-coded binding of eukaryotic initiator tRNAs to ribosomal subunits has already been mentioned in section 7. The release of formylmethionine from the fMet-tRNA:80 S ribosome complex was shown by Caskey and his colleagues (Goldstein *et al.* 1970) to require a protein release factor, a polynucleotide containing U and A residues, and GTP. This requirement for GTP could not be replaced by GDP or the GTP analogue, GMP-PCP suggesting hydrolysis of GTP during eukaryotic termination. Thus in striking contrast to the situation so far elucidated for the bacterial release factor function, the mammalian release factor function requires GTP and appears to involve its hydrolysis. It has recently been shown that the tetranucleotides UpApApA, UpApGpA, UpGpApA and UpApGpG (with a terminator codon in the first three base positions) are active templates for causing mammalian RF activity whereas terminator triplets are not; this reveals another slight difference from the bacterial situation. In fact, oligonucleotides longer than three residues appear to be necessary for forming a proper coding complex with RF and 80 S ribosomes. The one mammalian RF contains information for coding all three terminator codons unlike the bacterial situation. It is large (mol. wt. about 150 000 daltons) and contains subunits, although it appears to function only as the large complex, which is considered to be equivalent to a complex of the three bacterial RF activities. Similar mammalian RF activities have also been detected in extracts from guinea pig liver and Chinese hamster liver.

9.8 Role of GTP in protein biosynthesis

GTP hydrolysis in mammalian RF activity seems to be related to terminator codon recognition because although GTP hydrolysis is stimulated by UpApApA it does not occur with ApApApA. In further support of this idea, GTP hydrolysis during release factor activity can occur on ribosomes not carrying nascent peptidyl-tRNA and is not inhibited by antibiotics which inhibit the hydrolysis of peptidyl-tRNA. Thus it is unlikely that GTP hydrolysis is directly linked to the peptidyl-tRNA hydrolysis step of polypeptide chain termination.

The binding of reticulocyte RF to ribosomes is stimulated by both GTP and its analogue GMP-PCP but not by GDP indicating a guanine nucleotide requirement analogous to (a) the bacterial IF-2-dependent binding of initiator tRNA during polypeptide chain and (b) the EF-Tu binding of aminoacyl-tRNA during polypeptide chain elongation. In all three cases (initiation, elongation and termination) the protein factor-stimulated *binding* can use GTP or GMP-PCP but completion of all three protein synthetic events as detailed in *in vitro* experiments requires phosphate hydrolysis from the specific nucleotide derivative, GTP. Since the information on the bacterial polypeptide chain termination event is strangely out of step with the analogies above, it is tentatively assumed that a GTP requirement in this event has been artefactually masked so far by the nature of the *in vitro* experiments.

When the total count is made of the number of GTP splits for peptide bond formation the *in vitro* experiments give the answer two for the elongation step and one each tentatively for initiation and termination as described in earlier sections. GTP is not required for the actual peptide bond-forming step by the peptidyl transferase. Yet it does appear to be hydrolysed for the translocation of peptidyl-tRNA during protein synthesis (section 8.8), a step involving EF-G. We also have the alternative viewpoint that *in vivo* a concerted mechanism during the elongation step may require only one GTP split.

The relevance of the newly discovered requirement for GTP in the synthesis of magic spot compounds (section 8.12) to a role in protein biosynthesis is not yet understood.

9.9 Peptidyl-tRNA hydrolysis; its relation to peptidyl transferase and site of release activity

Elucidation of the detailed mechanism of peptidyl-tRNA hydrolysis during polypeptide chain termination will be difficult because of the involvement of the ribosome whose structure will be of unresolved complexity for some time. Ribosomal binding of an RF is not sufficient to trigger the hydrolysis of peptidyl-tRNA but there is reasonable evidence that the RF-1 or RF-2 participates in the hydrolysis activity. It is not yet clear whether the hydrolysis is due to a cooperative interaction of RF-1 or RF-2 and a ribosomal constituent. It has, however, been shown that the ribosome must be in an active form (i.e. able to form peptide bonds) for the release activity to be possible.

Many antibiotics such as tetracycline, streptomycin, sparsomycin, chloramphenicol, amicetin and lincocin are inhibitors of codon-directed peptidyl-tRNA hydrolysis in *E. coli*. The development of assays for evaluating terminator codon recognition and peptidyl-tRNA hydrolysis independently have enabled the site of action of such antibiotics to be located. Tetracycline and streptomycin inhibit codon recognition whereas amicetin, lincocin, chloramphenol and sparsomycin inhibit release of fMet in the quick assay (see Fig. 1.47) for peptidyl-tRNA hydrolysis without significantly affecting codon recognition. This latter class of antibiotics has been shown also to inhibit the peptidyl transferase activity of the 50 S ribosomal subunit. A different set of antibiotics is known to inhibit both the peptidyl transferase and RF-mediated peptidyl-tRNA hydrolysis on reticulocyte ribosomes. These experiments suggest that the hydrolytic event in peptidyl-tRNA hydrolysis might perhaps be related to peptide bond formation and might even be another facet of the peptidyl transferase activity.

The evidence on the ribosomal site of action of the release activity also agrees with this idea. It is thought that for release to occur the peptidyl-tRNA must be sensitive to puromycin and thus be in the P-site. To obtain this situation a translocation event must occur after the last peptide bond formation prior to termination. The activity of the release factor at the A-site suggests that water could be used at this site as a substrate by the peptidyl transferase to hydrolyse the peptidyl-tRNA bound in the P-site.

Work by Caskey and his colleagues (Caskey *et al.* 1971) using the partial reactions shown in Fig. 1.49, has supported the postulate that peptidyl transferase plays a role in peptide chain termination. A complex of the initiator tRNA with bacterial or reticulocyte ribosomes is first made as in reaction A, then the fMet is transferred to a series of substrates as in reactions B, C and D, which have been shown to be characteristic of the peptidyl transferase activity. Reaction B gives evidence of a model reaction for peptide bond formation whereas reaction C shows that the same enzyme can form an ethyl ester with the appropriate substrates, uncharged tRNA and ethanol. However, if ethanol is replaced by acetone (reaction D) fMet is produced by hydrolysis, a reaction also characteristic of the release activity.

Since peptidyl transferase can clearly hydrolyse peptidyl-tRNA, it is highly likely that this activity is somehow involved in normal polypeptide chain termination. Supporting this possibility are some studies with antibiotics in which peptide bond formation has been inhibited (lincocin for *E. coli* ribosomes and anisomycin for reticulocyte ribosomes) and in

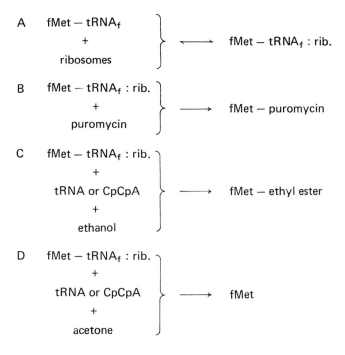

Fig. 1.49 Model reactions as assays for peptidyl transferase or termination. Intermediate complexes used in each case may be reticulocyte derived (i.e. fMet-tRNA$_f$:80 S rib. as shown) or bacterial (fMet-tRNA$_f$:ApUpG:70 S rib.)

which there has been simultaneous stimulation of peptidyl-tRNA hydrolysis with acetone. Thus the currently held view is that the RF (RF-1 or RF-2) activity may modify the peptidyl transferase, or restrict the choice of nucleophilic substrate, so that a hydrolytic event is promoted.

How the last discharged tRNA is ejected from the ribosome is not known; perhaps the elongation factor EF-G or a special factor is required. Quite probably the final step in termination is the release of the 70 S or 80 S ribosome from mRNA. At present it is unlikely that dissociation into subunits occurs during termination. What releases the ribosome from mRNA and whether termination is a point in control of protein biosynthesis are also unclear at this time.

Bibliography

General Reading — books and review articles

BOSCH, L. ed. (1972) *The Mechanism of Protein Synthesis and its Regulation.* North Holland Research Monographs Frontiers of Biology. Vol. 27. Excellent chapters for recent work covering sections 3 — 9, the best starting point for the advanced student and research worker.

DAVIDSON, J. N. (1972) *The Biochemistry of the Nucleic Acids,* 7th Edn. Chapman and Hall. Good background reading especially for nucleic acid structure and function.

82

LEWIN, B. M. (1970) *The Molecular Basis of Gene Expression,* Wiley – Interscience.
   Reasonable general background and a good source of references but a bit dated.
WATSON, J. D. (1970) *Molecular Biology of the Gene,* 2nd edn., W. A. Benjamin, Inc.
   Excellent basic text for protein synthesis and related molecular genetics.
The Molecular Basis of Life, readings from the *Scientific American,* W. A. Freeman and Co.
   Very readable introduction to the field.

Relevant articles in series such as:

   Annual Reviews of Biochemistry, Vols. 1 – 42.
   Progress in Nucleic Acid Research and Molecular Biology, eds. J. N. Davidson and W. E.
   Cohn, Academic Press, Vols. 1 – 12.
   Progress in Biophysics and Molecular Biology, Vols. 1 – 26.

For experimental details about procedures two series can be recommended:

   Procedures in Nucleic Acid Research, eds. G. L. Cantoni and D. R. Davies, Harper and Row,
   Vol. 1, 1966; Vol. 2, 1971.
   Methods in Enzymology, Academic Press. Special volumes on nucleic acids and protein
   synthesis, eds. K. Moldave and L. Grossman, Vol. XIIA, 1967, Vol. XIIB, 1968, Vol. XX,
   1971 and Vol. XXI, 1971.

Review Articles

Section 2
CRICK, F. H. C. (1966) The genetic code III. *Scient. Am.,* **215,** Oct. pp. 55 – 62.
Cold Spring Harb. Symp. Quant. Biol. (1966), 'The genetic code', Vol. XXXI – contains many
   excellent papers by leading workers in the field.
STRETTON, A. O. W. (1965) 'The genetic code.' *Brit. Med. Bull.* **21,** 229 – 35. A good source
   of references for the early genetic work.

Section 3
LOFTFIELD, R. (1972), 'The mechanism of aminoacylation of transfer RNA.' *Prog. Nucl.
   Acid Res. Molec. Biol.* **12,** 87 – 128.
NEIDHARDT, F. C. (1966) 'Roles of amino acid activating enzymes in cellular physiology'.
   Bact. Rev., **30,** 701 – 19.
PETERSON, P. J. (1966) 'Amino acid selection in protein biosynthesis'. *Biol. Rev.,* **42,**
   552 – 613.

Section 4
ARNOTT, S. (1971) 'The structure of transfer RNA.' *Prog. Biophys. Molec. Biol.,* **22,**
   179 – 213.
CRAMER, F. (1971) 'Three-dimensional structure of tRNA.' *Prog. Nucl. Acid Res. Molec.
   Biol.,* **11,** 391 – 421.
SMITH, J. D. (1972) 'Genetics of transfer RNA.' *Ann. Rev. Gen.,* **6,** 235 – 56.
ZACHAU, H. G. (1969), 'Transfer ribonucleic acids.' *Angew. Chem. Internat. Edit.,* **8,**
   711 – 26. A good source of references for earlier work.

Section 5

KURLAND, C. G. (1972) 'Structure and function of the bacterial Ribosome.' *Ann. Rev. Biochem., 41*, 377 – 408.

NOMURA, M. (1973) 'Assembly of bacterial ribosomes.' *Science, 179*, 864 – 73.

PESTKA, S. (1971) 'Inhibitors of ribosome functions.' *Ann. Rev. Microbiol., 25*, 487 – 562.

Section 6

MILLER, O. L. (1973) 'The visualization of genes in action.' *Scient. Am. 228*, Mar. pp. 34 – 42.

SINGER, M. F. and LEDER, P. (1966) 'Messenger RNA: an evaluation.' *Ann. Rev. Biochem., 35*, 195 – 230.

Section 7

CLARK, B. F. C. and MARCKER, K. A. (1968) 'How proteins start.' *Scient. Am., 218*, Jan. pp. 36 – 42.

Cold Spring Harb. Symp. Quant. Biol. (1969). Relevant papers in: 'The mechanism of protein synthesis,' Vol. XXXIV.

LUCAS-LENARD, J. and LIPMANN, F. (1971) 'Protein biosynthesis'. *Ann. Rev. Biochem., 40*, 409 – 48. Also contains background information for sections 8 and 9.

Section 8

LENGYEL, P. and SÖLL, D. (1969) 'Mechanism of protein synthesis.' *Bact. Rev., 33*, 264 – 301.

LIPMANN, F. (1969) 'Polypeptide chain elongation in protein biosynthesis.' *Science, 164*, 1024 – 31.

Section 9

GAREN, A. (1968) 'Sense and nonsense in the genetic code.' *Science 160*, 149 – 59.

References for the text

ALTMAN, S. and SMITH, J. D. (1971) 'Tyrosine tRNA precursor molecule polynucleotide sequence'. *Nature New Biol. 233*, 35 – 39.

BARRELL, B. G. (1971) 'Fractionation and Sequence Analysis of Radioactive Nucleotides'. *Procedures Nucl. Acid Res. 2*, 751 – 79.

BERG, P., BERGMANN, F. H., OFENGAND, E. J. and DIECKMANN, M. (1961) 'The enzymic synthesis of amino acyl derivatives of ribonucleic acid.' *J. Biol. Chem., 236*, 1726 – 34

BLUMENTHAL, T., LANDERS, T. A. and WEBER, K. (1972) 'Bacteriophage Qβ replicase contains the protein biosynthesis elongation factors EF Tu and EF Ts.' *Proc. Nat. Acad. Sci. U.S.A., 69*, 1313 – 17.

BOEDTKER, H. and KELLING, D. G. (1967) 'The ordered structure of 5 S RNA'. *Biochem. Biophys. Res. Commun, 29*, 758 – 66.

BRENNER, S., JACOB, F. and MESELSON, M. (1961) 'An unstable intermediate carrying information from genes to ribosomes for protein synthesis'. *Nature, Lond., 190*, 576 – 81.

BRETSCHER, M. S. (1968a) 'Translocation in protein synthesis: a hybrid structure model'. *Nature, Lond., 218*, 675 – 77.

84

BRETSCHER, M. S., (1968b) 'Polypeptide chain termination: an active process'. *J. Mol. Biol.* **34**, 131 – 36.

BROWNLEE, G. G., SANGER, F. and BARRELL, B. G. (1967) 'Nucleotide Sequence of 5 S-ribosomal RNA from *Escherichia coli*'. *Nature, Lond.*, **215**, 735 – 36.

CAPECCHI, M. R. (1967) 'Polypeptide chain termination *in vitro:* isolation of a release factor'. *Proc. Nat. Acad. Sci. U.S.A.*, **58**, 1144 – 51.

CASHEL, M. and GALLANT, J. (1969) 'Two compounds implicated in the function of the RC gene of *Escherichia coli*'. *Nature, Lond*, **221**, 838 – 41.

CASKEY, C. T., TOMPKINS, R., SCOLNICK, E., CARYK, T. and NIRENBERG, M. (1968) 'Sequential translation of trinucleotide codons for the initiation and termination of protein synthesis.' *Science,* **162**, 135 – 38.

CASKEY, C. T., BEAUDET, A., SCOLNICK, E. and ROSMAN, M. (1971) 'Peptidyl Transferase hydrolysis of fMet-tRNA.' *Proc. Nat. Acad. Sci. U.S.A.*, **68**, 3163 – 67.

CHAPEVILLE, F., LIPMANN, F., VON EHRENSTEIN, G., WEISBLUM, B., RAY, W. J. and BENZER, S. (1962) 'On the role of soluble ribonucleic acid in coding for amino acids.' *Proc. Nat. Acad. Sci. U.S.A.*, **48**, 1086 – 92.

CLARK B. F. C. and MARCKER, K. A. (1966) 'The role of *N*-formylmethionyl-sRNA in protein biosynthesis.' *J. Mol. Biol.*, **17**, 394 – 406.

CLARK, B. F. C., DOCTOR, B. P., HOLMES, K. C., KLUG, A., MARCKER, K. A., MORRIS, S. J. and PARADIES, H. H. (1968) 'Crystallization of transfer RNA'. *Nature, Lond.*, **219**, 1222 – 24.

CONTRERAS, R., YAEBAERT, M., MIN JOU and FIERS, W. (1973) 'Bacteriophage MS2 RNA: nucleotide sequence of the end of the A-protein gene and the intercistronic region'. *Nature New Biol.*, **241**, 99 – 101.

CRICK, F. H. C. (1966) 'Codon-anticodon pairing: The wobble hypothesis'. *J. Mol. Biol.*, **19**, 548 – 55.

CRICK, F. H. C. (1968) 'The origin of the genetic code'. *J. Mol. Biol.*, **38**, 367 – 79.

CRICK, F. H. C. (1970) 'Central dogma of molecular biology'. *Nature, Lond.*, **227**, 561 – 63.

CRICK, F. H. C., BARNETT, L., BRENNER, S. and WATTS-TOBIN, R. J. (1961) 'General nature of the genetic code for proteins'. *Nature, Lond.*, **192**, 1227 – 32.

DINTZIS, H. M. (1961) 'Assembly of the peptide chains of hemoglobin.' *Proc. Nat. Acad. Sci. U.S.A.* **47**, 247 – 61.

DOCTOR, B. P. (1971) 'Countercurrent distribution of transfer ribonucleic acid'. *Procedures Nucl. Acid Res.*, **2**, 588 – 607.

GANOZA, M. C. (1966) 'Polypeptide chain termination in cell-free extracts of *E. coli*'. *Cold Spring Harb. Symp. Quant. Biol.*, **31**, 273 – 78.

GAREN, A. (1968) 'Sense and nonsense in the genetic code'. *Science,* **160**, 149 – 59.

GILBERT, W. (1963) 'Polypeptide synthesis in *Escherichia coli*. II, The polypeptide chain and s-RNA'. *J. Mol. Biol.*, **6**, 389 – 403.

GOLDSTEIN, J. L., BEAUDET, A. L. and CASKEY, C. T. (1970) 'Peptide chain termination with mammalian release factor.' *Proc. Nat. Acad. Sci. U.S.A.*, **67**, 99 – 106.

GOODMAN, H. M., ABELSON, J., LANDY, A., BRENNER, S. and SMITH, J. D. (1968) 'Amber suppression: a nucleotide change in the anticodon of a tyrosine transfer RNA'. *Nature, Lond.*, **217**, 1019 – 24.

GRONER, Y., POLLACK, Y., BERISSI, H. and REVEL, M. (1972) 'Cistron specific translational control protein in *Escherichia coli*'. *Nature New Biol.*, **239**, 16 – 19.

HOLLEY, R. W., APGAR, J., EVERETT, G. A., MADISON, J. T., MARQUISEE, M., PENSWICK, J. R. and ZAMIR, A. (1965) 'Structure of a ribonucleic acid'. *Science,* **147**, 1462 – 65.

JACOB, F. and MONOD, J. (1961) 'Genetic regulatory mechanisms in the synthesis of proteins'. *J. Mol. Biol.*, **3**, 318 − 56.

KELMERS, A. D., WEEREN, H. O., WEISS, J. F., PEARSON, R. L., STULBERG, M. P. and NOVELLI, G. D. (1971) 'Reversed-phase chromatography systems for transfer ribonucleic acids − preparatory scale methods'. *Methods in Enzymology*, **XX**, 9 − 34.

KHORANA, H. G. (1968) 'Synthesis in the study of nucleic acids'. *Biochem. J.* **109**, 709 − 25.

KHORANA, H. G., AGARWAL, K. L., BUCHI, H., CARUTHERS, M. H., GUPTA, N. K., KLEPPE, K., KUMAR, A., OHTSUKA, E., RAJBHANDARY, U. L., VOND DE SANDE, J. H., SGARAMELLA, V., TERAO, T., WEBER, H. and YAMADA, T. (1972) 'Total synthesis of the structural gene for an alanine transfer ribonucleic acid from yeast'. *J. Mol. Biol.*, **72**, 209 − 17.

KIM, S. H., QUIGLEY, C. J., SUDDATH, F. L. and RICH, A. (1971) 'High resolution X-ray diffraction patterns of crystalline transfer RNA that show helical regions'. *Proc. Nat. Acad. Sci. U.S.A.*, **68**, 841 − 45.

KIM, S. H., QUIGLEY, G. J., SUDDATH, F. L., McPHERSON, A., SNEDEN, D., KIM, J. J., WEINZIERL, J. and RICH, A. (1973) 'Three-dimensional structure of yeast phenylalanine transfer RNA: folding of the polynucleotide chain'. *Science*, **179**, 285 − 88.

LEVITT, M. (1969) 'Detailed molecular model for transfer ribonucleic acid'. *Nature, Lond.*, **224**, 759 − 63.

LOFTFIELD, R. B. (1972) 'The mechanism of aminoacylation of transfer RNA'. in *Prog. Nucl. Acid Res. Molec. Biol.* **12**, 87 − 128.

MARCKER, K. A., and SANGER, F. (1964) '*N*-formylmethionyl − sRNA.' *J. Mol. Biol.*, **8**, 835 − 40.

MARCKER, K. A. (1965) 'The formation of *N*-formylmethionyl − sRNA.' *J. Mol. Biol.*, **14**, 63 − 70.

MILLER, D. L. and WEISSBACH, H. (1970), 'Studies on the purification and properties of factor Tu from *E. coli.*' *Arch. Biochem. Biophys.* **141**, 26 − 37.

MILLER, Jr., O. L. (1973) 'The visualization of genes in action.' *Scient. Am.*, **228**, March pp. 34,− 42.

MIN JOU, W., HAEGEMAN, G., YSEBAERT, M. and FIERS, W. (1972) 'Nucleotide sequence of the gene coding for the bacteriophage MS2 coat protein.' *Nature, Lond.*, **237**, 82 − 88.

MONRO, R. E. (1967) 'Catalysis of peptide bond formation by 50 S ribosomal subunits from *Escherichia coli.*' *J. Mol. Biol.*, **26**, 147 − 51.

NICHOLS, J. and ROBERTSON, H. (1971) 'Sequences of RNA fragments from the bacteriophage f2 coat protein cistron which differ from their R17 counterparts'. *Biochim. Biophys. Acta*, **228**, 676 − 81.

NIRENBERG, M. W. and MATTHAEI, J. H. (1961) 'The dependence of cell-free protein synthesis in *E. coli* upon naturally occurring or synthetic polyribonucleotides.' *Proc. Nat. Acad. Sci. U.S.A.*, **47**, 1588 − 602.

NIRENBERG, M. W. and LEDER, P. (1964) 'RNA codewords and protein synthesis.' *Science*, **145**, 1399 − 407.

NISHIMURA, S. (1971) 'Fractionation of transfer RNA by DEAE − Sephadex A-50 column chromatography.' *Procedures Nucl. Acid Res.*, **2**, 542 − 64.

NISHIMURA, S. (1972) 'Minor components in transfer RNA: their characterization, location and function in *Prog. Nucl. Acid Res. Molec. Biol.* **12**, 49 − 85.

NISHIZUKA, Y. and LIPMANN, F. (1966) 'Comparison of guanosine triphosphate split and polypeptide synthesis with a purified *E. coli* system.' *Proc. Nat. Acad. Sci. U.S.A.*, **55**, 212 − 19.

NOMURA, M. (1973). 'Assembly of bacterial ribosomes'. *Science*, **179**, 864 − 73.

ORGEL, L. E. (1968) 'Evolution of the genetic apparatus.' *J. Mol. Biol.*, **38**, 381 – 93.

ROY, K. L., BLOOM, A. and SÖLL, D. (1971) 'tRNA separations using benzoylated DEAE – cellulose.' *Procedures Nucl. Acid. Res.*, **2**, 524 – 41.

RUDLAND, P. S., WHYBROW, W. A. and CLARK, B. F. C. (1971) 'Recognition of bacterial initiator tRNA by an initiation factor.' *Nature New Biol.*, **231**, 76 – 78.

RYCHLIK, I. ČERNA, J., CHLADEK, S., ŽEMLIČKA, J. and HALADOVÁ, Z. (1969) 'Substrate specificity of ribosomal peptidyl transferase: 2' (3')-*O*-aminoacyl nucleosides as acceptors of the peptide chain on the amino acid site.' *J. Mol. Biol.*, **43**, 13 – 24.

SALAS, M., SMITH, M. A., STANLEY, W. M., WAHBA, A. J. and OCHOA, S. (1965) 'Direction of reading of the genetic message.' *J. Biol. Chem.*, **240**, 3988 – 95.

SKOULTCHI, A., ONO, Y., WATERSON, J. and LENGYEL, P. (1970) 'Peptide chain elongation: indications for the binding of an amino acid polymerization factor, guanosine 5'-triphosphate – aminoacyl transfer ribonucleic acid complex to the messenger – ribosome complex'. *Biochemistry*, **9**, 508 – 14.

SPEYER, J. F., LENGYEL, P., BASILIO, C. and OCHOA, S. (1962) 'Synthetic polynucleotides and the amino acid code, IV.' *Proc. Nat Acad. Sci. U.S.A.*, **48**, 441 – 48.

STEITZ, J. A. (1969) 'Polypeptide chain initiation: nucleotide sequences of the three ribosomal binding sites in bacteriophage R17 RNA.' *Nature, Lond.*, **224**, 957 – 64.

TERZAGHI, E., OKADA, Y., STEISINGER, G., EMRICH, J., INOUYE, M. and TSUGITA, A., (1966) 'Change in a sequence of amino acids in phage T4 lysozyme by acridine-induced mutations.' *Proc. Nat. Acad. Sci. U.S.A.*, **56**, 500 – 7.

VOLKIN, E. and ASTRACHAN, L. (1956) 'Phosphorus incorporation in *Escherichia coli* ribonucleic acid after infection with bacteriophage T2.' *Virology*, **2**, 149 – 61.

2
Chemical Modification of Proteins

Jean O. Thomas
Department of Biochemistry, University of Cambridge

1 Introduction

A preoccupation with the actual structure of proteins is justified if one considers the variety of functions that proteins perform in the cell. For example, keratin of hair, and collagen of connective tissue are structural proteins, enzymes perform a catalytic role, antibodies a protective role, haemoglobin transports oxygen, ferritin stores iron, and some proteins and polypeptides are hormones. The biological function of a protein is a direct consequence of its three-dimensional structure and this in turn is determined by its amino acid sequence, i.e. its primary structure. To obtain even an inkling of how proteins carry out their individual functions one is thus led to look at all levels of their structure. One might ask how the chain is folded up (e.g. which groups are 'buried' and which 'exposed'), which groups in the protein are involved in maintaining the folded structure and in the association of subunits in multimeric proteins, and which groups are involved in the recognition and binding of small molecules (e.g. antigens by antibodies, substrates by enzymes, hormones by receptors).

X-ray crystallography is potentially able to reveal the positions of almost all the atoms in a protein molecule and would thus seem to be the ideal way to establish the structure. So far, however, only thirty or so protein structures have been solved in this way, mainly because of the difficulty in obtaining most proteins in crystal forms suitable for crystallographic work; the additional problem of obtaining isomorphous derivatives containing heavy atoms is discussed in section 9.

The picture obtained by X-ray crystallography is, of course, a static one. While there is evidence that this is representative of the structure of the protein in solution, there is also evidence for subtle changes in the structure of proteins in solution relevant to their functions. These changes in structure are often manifested in changes in chemical reactivity of the proteins in solution. A real grasp of the way in which a protein might function can, in fact, come only from the combined results of a number of approaches.

This chapter is intended to summarise the numerous ways in which chemical modification contributes to our knowledge of the structure and action of proteins. Examples have been selected to illustrate the most important points; no attempt has been made to catalogue all the possible chemical reactions of all the modifiable residues in proteins. The chemical principles underlying the modification reactions are on the whole straightforward, usually amounting to no more than nucleophilic substitution and addition reactions. From time to time throughout the chapter details of chemistry will be indicated but otherwise the objective will be to obtain a broad coverage.

Reference to individual papers has been kept to a minimum. The review by Freedman (1971) is a useful general survey of the field and those by Singer (1967), Cohen (1968), Spande *et al.* (1970), Stark (1970) and Shaw (1970) deal with particular aspects of the chemical modification of proteins and will be mentioned again later. A broad coverage of the literature dealing with primary structure and chemical modification of proteins, annually since 1968, is to be found in chapters of that title in a series of Specialist Reports (ed. Young). Papers in a symposium volume, *Chemical Reactivity and Biological Role of Functional Groups in Enzymes* (ed. Smellie 1970), illustrate some particular problems of protein structure and function currently being studied with the aid of chemical modification. A book by Means and Feeney (1971) is a handy source of references on all aspects of chemical modification of proteins.

2 The scope of the problem

Proteins are complex organic molecules to which an unique linear amino acid sequence imparts an unique pattern of folding of the polypeptide chain in three dimensions. Elucidation of the structure of proteins is concerned with analysis of the linear arrangement of amino acids (the primary structure) and of the spatial arrangement of the polypeptide chain (the three-dimensional structure). Chemical modification is a major tool in both approaches. Different aims, however, call for different strategies. In studies of primary structure, whether chemical modification is being used to facilitate peptide fractionation or more directly in sequence analysis, any influence of the folding of the polypeptide chain on the reactivities of particular amino acid side-chains is deliberately removed. In contrast, the use of chemical modification in studies of tertiary and quaternary structure hinges on the fact that each protein reacts in an unique way with a given reagent because of the particular reactivity conferred on the amino acid side-chains by that three-dimensional structure. Careful study of this reactivity with different reagents can enable some conclusions to be drawn about the arrangement in space of the polypeptide chain. Studies of this sort are useful and of interest when taken in conjunction with X-ray crystallographic analysis of protein structure. In the absence of crystallographic work chemical evidence is invaluable.

The chemical reactivity of proteins arises from the reactivity of the side-chains of the sulphur-containing amino acids (cysteine and methionine), the basic amino acids (lysine, histidine and arginine), the acidic amino acids (aspartic and glutamic acids) and the activated aromatic amino acids (tyrosine and tryptophan). These side-chains (with the exception of the side-chain of tryptophan) can function as sulphur, nitrogen or oxygen nucleophiles in addition and displacement reactions involving various reagents; sulphur is, in general, more reactive as a nucleophile than nitrogen, and nitrogen is more reactive than oxygen; the hydroxy side-chains of serine and threonine are, on the whole, so unreactive as not to be of use in chemical modification, but they sometimes lead to unwelcome side-reactions during modification of the more nucleophilic groups. Nucleophilic reactions of amino acid side-chains are usually strongly pH dependent. The reactive species is unprotonated (uncharged in the case of basic amino acids and methionine; anionic in the case of cysteine, tyrosine and the acidic amino acids) and pH thus determines the amount of reactive form present. Restriction of nucleophilic attack to one particular type of amino acid side-chain through control of pH is discussed later (section 6.4). Nucleophilic amino acid side-chains are often investigated using acylating or alkylating reagents such as acyl- or alkyl-halides which suffer nucleophilic displacement of the halogen (iodide and bromide are better leaving groups than chloride) as in Fig. 2.1(a); alkylation may also take place through nucleophilic addition across a double bond in the reagent (e.g. as in the reaction of thiol groups with N-ethyl maleimide (Fig. 2.1(b)).

(a) Protein—XH + YR → Protein—XR + YH
Protein—XH + YCOR → Protein—XCOR + YH
(X = S, NH or O; Y = halogen)

(b) Protein—SH +

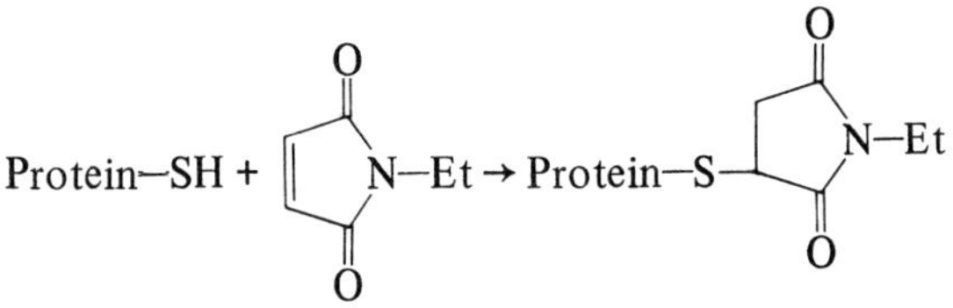

Fig. 2.1 Nucleophilic displacement (a), and nucleophilic addition (b) by protein side-chains.

Some side-chains are susceptible to oxidation; those of cysteine and methionine are most commonly oxidised to cystine and methionine sulphoxide, respectively, and those of histidine, tyrosine and tryptophan are oxidised in the ring systems. Photo-oxidation is a popular tool for modification of such residues. The aromatic rings of tyrosine, histidine and tryptophan are subject also to substitution under certain conditions, e.g. iodine will react with histidine and tyrosine giving, initially, 3-iodo derivatives; tryptophan however is likely to be oxidised instead and destroyed.

Whatever its ultimate aim chemical modification of proteins calls for a simple way of estimating accurately the stoicheiometry of reaction. This is usually achieved by taking advantage of a change in the absorption spectrum of the protein which may occur on modification, or by using a reagent which is labelled either radioactively, or with a chromophoric substituent and measuring the radioactivity or amount of chromophore incorporated. If the modified amino acid is stable to 6N-HCl, or indeed if it is completely destroyed, amino acid analysis can be invaluable in estimating the number of such residues in the protein molecule. Peptide mapping is also of use in establishing the number of residues modified.

The subsequent sections of this chapter will take up in turn the involvement of chemical modification in studies of the primary structures (section 3) and tertiary and quaternary structures (sections 4–7) of proteins. The discussion of primary structure indicates how protein chains are often treated chemically at the outset to facilitate handling and how chemistry plays a part at every subsequent stage through chain cleavage and purification of peptides to the steps in which the sequences of the peptides are actually determined. The discussion of three-dimensional structure is concerned with specific chemical modification of the active centres of enzymes and other proteins, and reaction with 'group-specific' reagents for the purpose of identifying 'buried' and 'exposed' side-chains and thus mapping the topography of the molecule. The marriage of physical and chemical methods in the study of protein structure is seen in the covalent attachment of spectroscopic probes at specific positions in the molecule, and in the preparation of protein derivatives containing heavy atoms for X-ray crystallographic studies (section 9).

3 Studies of primary structure

The individual steps in the determination of the amino acid sequence of a protein are now well established although a particular protein may require a different combination of these. Table 2.1

Table 2.1   Chemical modification in primary structure studies

1. Amino acid analysis
2. Determination of total half-cystine, cysteine and total tryptophan
3. Cleavage of intra- or inter-molecular disulphide bridges
4. Determination of N- and C-terminal amino acids
5. Chain cleavage: chemical cleavage
                     restriction or enhancement of enzymic cleavage
6. Peptide fractionation and purification: diagonal methods
7. Determination of peptide sequences: Edman degradation
                                         mass spectrometry

outlines the strategy in a typical case; chemical modification has proved indispensable at each stage. Methods and techniques used in protein sequence determination have been summarised in a book by Schroeder (1968).

3.1 Preliminaries

Disulphide bonds in proteins are likely to undergo disulphide exchange with one another or with cysteinyl residues, and are therefore routinely cleaved and blocked as a first step in sequence analysis; separate experiments are carried out at a later stage to determine the correct pairing of cysteinyl residues in bridges. The bonds may be cleaved either reductively (Fig. 2.2(a)) with a thiol (such as 2-mercaptoethanol) and then stabilised by carboxymethylation, or oxidatively (Fig. 2.2(b)) with performic acid. Both S-carboxymethylcysteine and cysteic acid

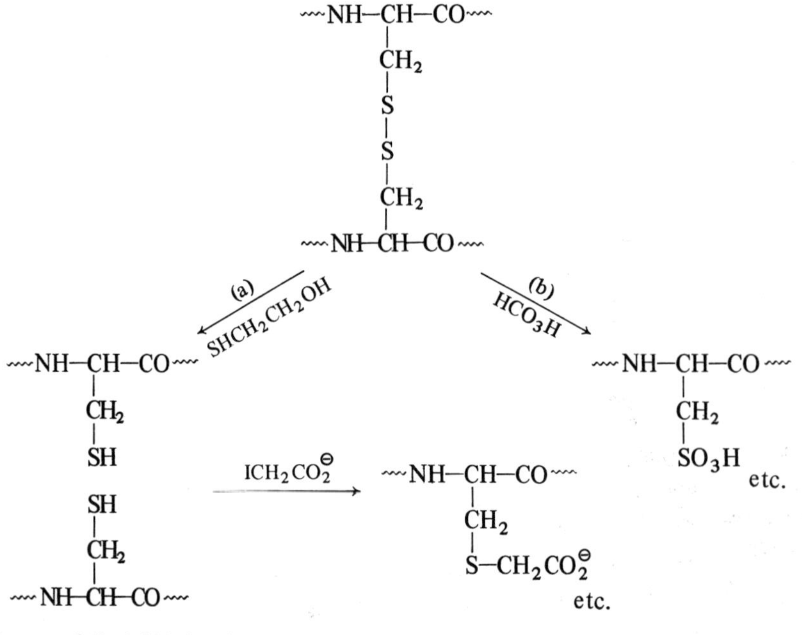

Fig. 2.2 Cleavage of disulphide bonds reductively (a), oxidatively (b).

are stable to further handling. These procedures do not damage the peptide linkages of the protein, although performic acid does destroy tryptophan and oxidises methionine to its sulphone.

If the protein contains cysteine (whether or not it contains disulphide bonds) it is subjected to one of the procedures illustrated in Fig. 2.2 to avoid forming the complex mixture of products that would otherwise arise through partial oxidation of the thiol groups during handling.

The total amount of half-cystine (i.e. cysteine, whether or not it is involved in a disulphide bond) in a protein may be estimated accurately by automatic amino acid analysis, either as S-carboxymethylcysteine or as cysteic acid. To obtain the *cysteine* content the native protein is unfolded in a denaturant (e.g. concentrated urea or guanidine solution) and treated

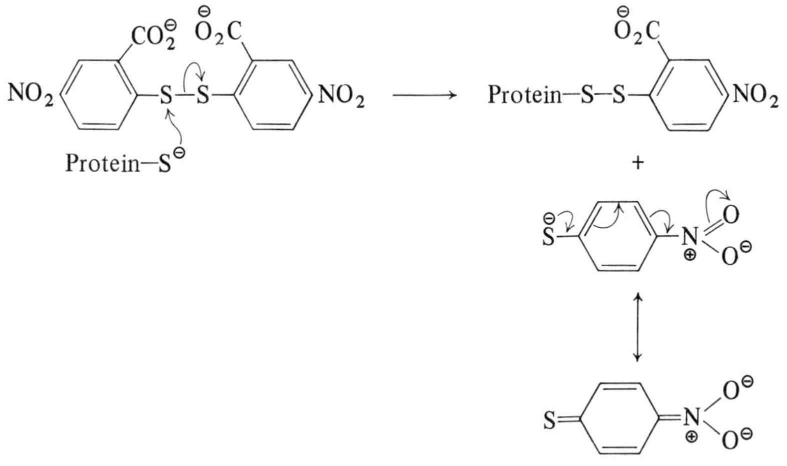

Fig. 2.3 Reaction of protein thiol groups with 5,5'-dithiobis-(2-nitrobenzoic acid), DTNB.

with 5,5'-dithiobis-(2-nitrobenzoic acid) (DTNB). In the reaction at pH 8, the anionic forms of protein thiol groups attack the disulphide bond of the reagent expelling a coloured, resonance-stabilised, nitrothiophenolate anion (Fig. 2.3) which may be estimated quantitatively from its absorption at 412 nm.

Since the indole side chain is easily destroyed, the tryptophan content of a protein is also estimated at the outset of sequence analysis. Tryptophan is destroyed in conventional acid hydrolysates but can be estimated by alkylation with 2-hydroxy-5-nitrobenzyl bromide. The reagent is rapidly hydrolysed in aqueous solution ($t_{1/2} < 1$ min) and only nucleophiles as reactive as the side-chain of cysteine and the indole nucleus react before the reagent is destroyed. Alkylation is specific for tryptophan under acid conditions because thiol groups are unionised (Barman and Koshland 1967), and may be determined from the change in absorbance at 410 nm. Careful control of the reaction conditions is necessary because of the potential mixture of substitution products. The reagent is also useful as a chromophoric probe for the environments of tryptophan residues in proteins (cf. section 10.1).

3.2 Determination of N- and C-terminal amino acid residues

3.2.1 N-Terminal residues

The two reagents most widely used for labelling the free α-amino group in peptides and proteins are fluoro-2,4-dinitrobenzene (Dnp-F) and dansyl chloride (Dns-Cl). They act similarly, both undergoing nucleophilic attack by amino groups at pH 8, and the linkages generated are stable to acid under conditions where the peptide bonds are all cleaved (Fig. 2.4). The labelled amino acid can be detected and identified by virtue of either its colour (in the case of the Dnp derivative) or its fluorescence (for the Dns derivative); the latter method is about 1000 times more sensitive.

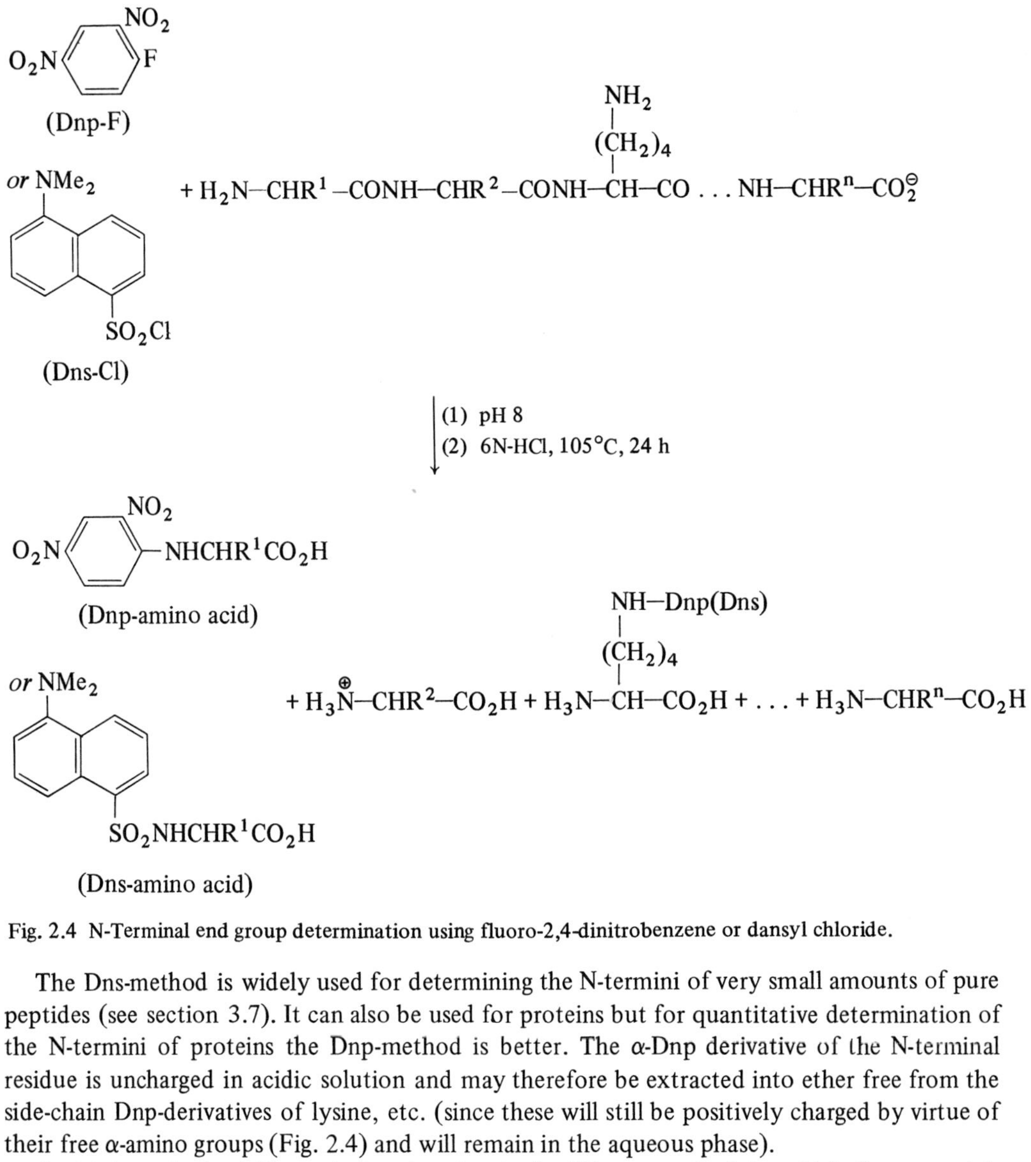

Fig. 2.4 N-Terminal end group determination using fluoro-2,4-dinitrobenzene or dansyl chloride.

The Dns-method is widely used for determining the N-termini of very small amounts of pure peptides (see section 3.7). It can also be used for proteins but for quantitative determination of the N-termini of proteins the Dnp-method is better. The α-Dnp derivative of the N-terminal residue is uncharged in acidic solution and may therefore be extracted into ether free from the side-chain Dnp-derivatives of lysine, etc. (since these will still be positively charged by virtue of their free α-amino groups (Fig. 2.4) and will remain in the aqueous phase).

A third method of N-terminal estimation is the Edman method in which the reagent is phenylisothiocyanate (Ph·NCS) and the N-terminal amino acid is identified as its phenylthiohydantoin. In contrast with the two previous methods the N-terminal amino acid is cleaved selectively without affecting the other peptide bonds, and the Edman method is, in fact, most often used for sequential degradation of peptides and proteins from the N-terminus (see section 3.7).

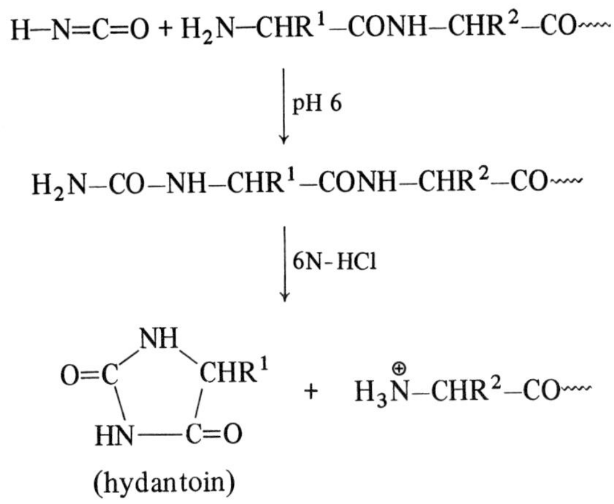

$$H-N=C=O + H_2N-CHR^1-CONH-CHR^2-CO\cdots$$

$$\downarrow pH\ 6$$

$$H_2N-CO-NH-CHR^1-CONH-CHR^2-CO\cdots$$

$$\downarrow 6N\text{-}HCl$$

$$
\begin{matrix}
& NH & \\
O=C & \diagdown & CHR^1 \\
\diagdown & & \diagup \\
HN & \!\!\!\!\!\!\!\!\! - C=O &
\end{matrix}
\qquad + \quad H_3\overset{\oplus}{N}-CHR^2-CO\cdots
$$

(hydantoin)

Fig. 2.5 Cyanate method for N-terminal estimation.

The related 'cyanate' (cyanic acid) method (Fig. 2.5) is used only infrequently for identifying the N-terminus in proteins; the N-terminal amino acid is cleaved and identified as its hydantoin.

3.2.2 C-Terminal residues

In contrast with the amino terminus, the carboxy terminus of a protein is relatively inert and not amenable to chemical modification. Good chemical methods for determining the C-terminal amino acids in peptides and proteins are consequently sadly lacking and enzymic methods (using carboxypeptidases) are more satisfactory. Of the chemical methods available, probably the most widely used is hydrazinolysis (Fig. 2.6). In the basic reaction mixture all the peptide

$$H_2N-CHR^1-CONH-CHR^2-CO\ldots NH-CHR^n-CO_2H$$

$$\downarrow N_2H_4$$

$$H_2N-CHR^1-CONHNH_2 + H_2N-CHR^2-CONHNH_2 + \ldots + H_2N-CHR^n-CO_2H$$

Fig. 2.6 Hydrazinolysis method for C-terminal estimation.

bonds are converted into the corresponding hydrazides whereas the resonance-stabilised C-terminal carboxylate ion is inert and remains unchanged; the unique C-terminal amino acid can then be separated from the mixture of hydrazides for identification. (Even here the method depends on the fact that carboxylate groups are less reactive than amide bonds, themselves generally unreactive.) Disadvantages of the method are low yields, nasty side-reactions, and the problems of identifying C-terminal asparagine (Asn) and glutamine (Gln) because of the difficulty in separating, for example, Asp-β-hydrazide (arising from C-terminal Asn) from Asp-α-hydrazide (from internal Asp).

A second non-enzymic method for C-terminal residues is gaining currency. It involves incorporation of tritium into the C-terminal carboxy group by exchange with the active hydrogen of the oxazolone that is formed by cyclisation in the presence of acetic anhydride (Fig. 2.7). The tritiated amino acid can be indentified in a total hydrolysate from its elution

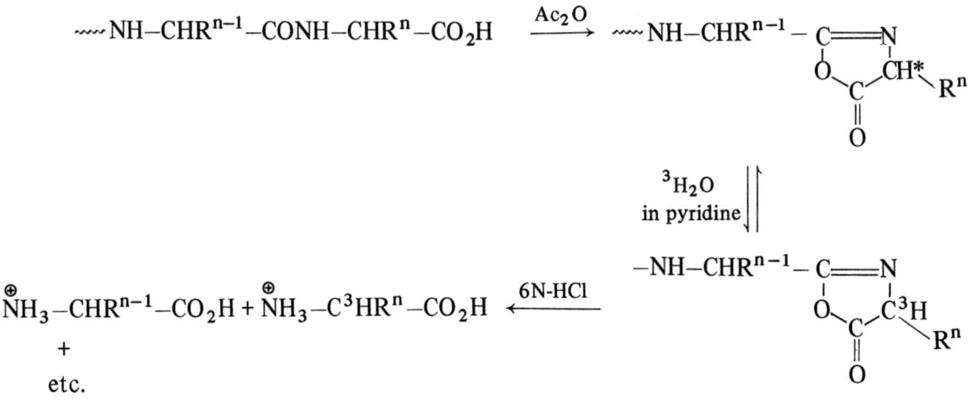

$$\sim\sim NH-CHR^{n-1}-CONH-CHR^{n}-CO_2H \xrightarrow{Ac_2O} \sim\sim NH-CHR^{n-1}-C{=\!=\!=}N$$

(structure with O, C, CH^*, R^n, C, O)

$$\downarrow {}^{3}H_2O \text{ in pyridine}$$

$$-NH-CHR^{n-1}-C{=\!=\!=}N$$

(structure with O, C, $C^{3}H$, R^n, C, O)

$$\overset{\oplus}{N}H_3-CHR^{n-1}-CO_2H + \overset{\oplus}{N}H_3-C^3HR^n-CO_2H \xleftarrow{\text{6N-HCl}}$$

$$+$$

etc.

Fig. 2.7 C-Terminal estimation by tritium exchange; the active hydrogen is indicated by an asterisk.

position in ion-exchange chromatography. This method, however, also gives ambiguous results in certain cases.

In a third method described recently (Hamada and Yonemitsu 1973) (cf. Chibnall and Rees 1951) the terminal group is esterified and the protein then reduced in aqueous solution with borohydride. After total acid hydrolysis the C-terminal amino acid will appear as the corresponding amino alcohol and can be identified by ion-exchange chromatography. C-Terminal Asn is easily identified by this method as β-amino-γ-butyrolactone; Asp would appear as aspartidiol.

3.3 Chain cleavage: chemical methods

The specificity of enzymic cleavage of polypeptide chains has been exploited with great success in the analysis of protein sequences, to give unique fragments of the original chain in high yield. Among the most widely used enzymes are trypsin (which cleaves on the C-terminal side of lysine and arginine), chymotrypsin (which cleaves on the C-terminal side of tyrosine, phenylalanine and tryptophan; and more slowly after leucine and isoleucine), pepsin, thermolysin and papain (which have lower specificities). Analysis of two different enzymic digests gives two sets of fragments, each overlapping the cleavage points of the other, so that the fragments may be ordered and the original sequence reconstructed.

A number of chemical methods have been reported for the specific cleavage of certain peptide bonds in model peptides (see the review by Spande et al. 1970) but in the majority of cases the yields are so low that the methods are of little use for cleavage of polypeptide chains in sequence analysis. Chemical cleavage is most useful for bonds formed by the rarer amino acids in proteins since only a small number of fragments is then generated and formidable problems of fractionation are avoided. One method, cleavage at methionyl bonds with cyanogen bromide, has proved to be specific and to give high yields of cleavage products and is now very frequently used for initial dissection of the protein chain. A method for cleaving tryptophyl bonds is also described below.

3.3.1 Cleavage at methionyl bonds with cyanogen bromide

The chemistry of the reaction is shown in Fig. 2.8(a), with (i) as the final step; there is a concerted nucleophilic displacement of bromide by sulphur and of sulphur by the carbonyl oxygen. Cleavage occurs specifically at methionyl peptide bonds and, since methionine is one of

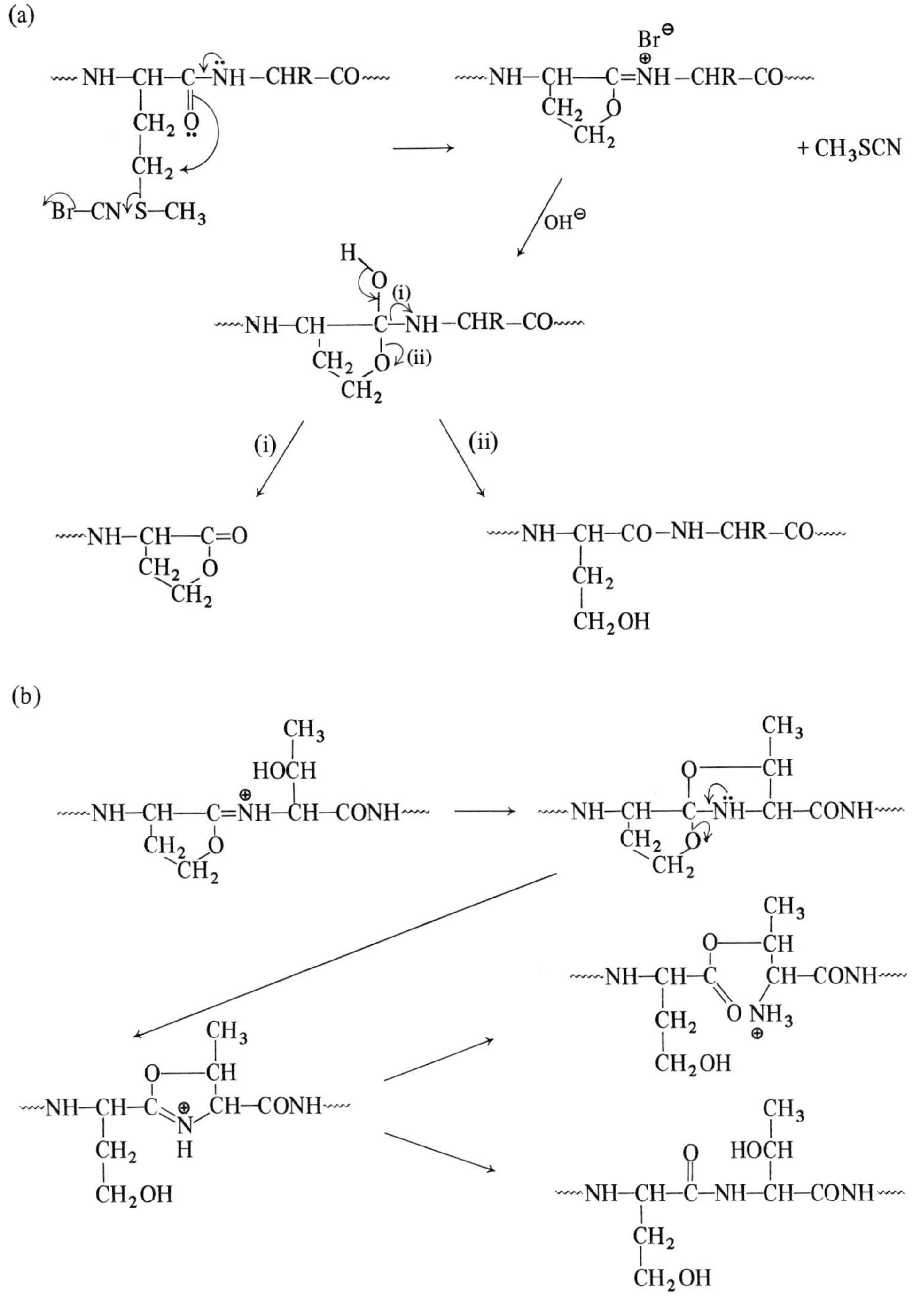

(a)

(b)

Fig. 2.8   (a) Cleavage (i), and non-cleavage (ii), of methionyl bonds with cyanogen bromide. (b) Reaction with cyanogen bromide of methionyl bonds in the sequence -Met-Thr-.

the 'rarer' amino acids in most protein chains, cyanogen bromide cleavage leads to a small number of comparatively large fragments (e.g. glyceraldehyde 3-phosphate dehydrogenase from pig muscle has nine methionine residues per subunit of molecular weight 36 000, and thus gives ten 'cyanogen bromide fragments' which can be separated and purified). The C-terminal residue

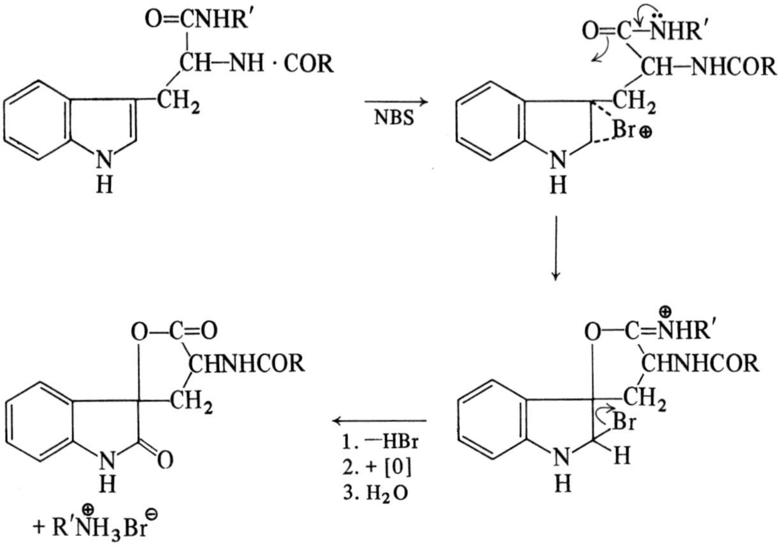

Fig. 2.9   Cleavage of (a) tryptophyl, (b) tyrosyl peptide bonds with N-bromosuccinimide.

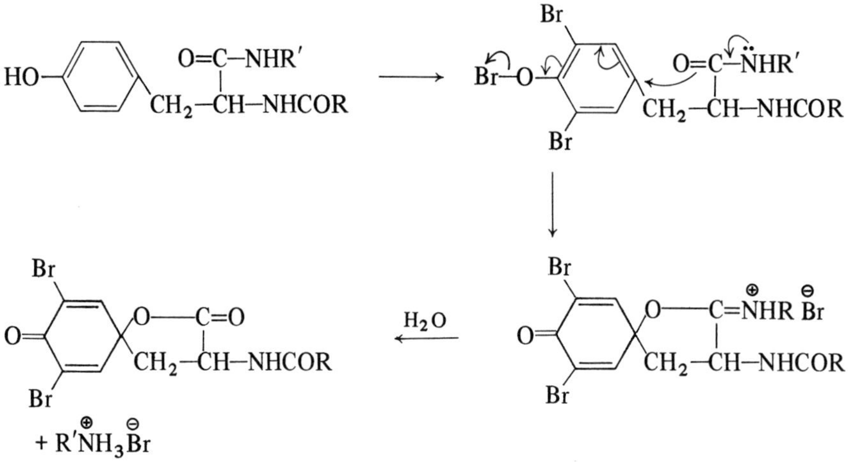

of all fragments except the original C-terminal one is homoserine lactone. In certain cases, the peculiar characteristics of some proteins have meant that cyanogen bromide has been the most suitable agent for chain cleavage, e.g. when the basic amino acids, potential points of attack by trypsin, are few and concentrated in one region of the chain (as in pepsin), or when the molecule is particularly large and calls for an initial dissection into a number of manageable pieces (as in the immunoglobulins).

Although the cyanogen bromide method is well proven, there have been a few recent reports of its failure. Where methionine occurs in the sequence -Met-Ser- or -Met-Thr- the yield of cleavage is often low although the methionine is completely converted into homoserine. This may be due to formation of an iminolactone involving the hydroxy side-chain (Fig. 2.8(b)). Even in the absence of serine or threonine adjacent to methionine cleavage may be incomplete; possibly the cyclic imidate does not cleave as in (i) (Fig. 2.8(a)) but the homoserine lactone ring instead opens as in (ii).

3.3.2 Cleavage at tryptophyl and tyrosyl bonds with N-bromosuccinimide (NBS)

Cleavage of protein chains with N-bromosuccinimide is sometimes useful although far less widely used than cleavage with cyanogen bromide. NBS at pH 3·4 will cleave tryptophyl peptide bonds (Fig. 2.9(a)) in proteins in reasonable yield, but its reactivity extends also to tyrosyl peptide bonds (Fig. 2.9(b)) (although this can be blocked by O-dinitrophenylation or other suitable substitution) and, to a lesser extent and at higher temperature, to histidyl bonds. Conversely, if tryptophan residues are blocked by reaction with 2-hydroxy-5-nitrobenzyl bromide, cleavage can be restricted to tyrosyl peptide bonds; on occasions specificity can be effected merely through control of pH. It is always pleasing to come across proteins that are particularly suited to certain reagents, e.g. the lysine-rich histone F_1 from rabbit thymus has one tyrosine and no tryptophan, histidine or sulphur-containing amino acid residues and is tailor-made for cleavage with N-bromosuccinimide; in this case cleavage at the tyrosyl bond is virtually quantitative (Rall and Cole 1970).

The mechanism of cleavage at tryptophyl and tyrosyl bonds with NBS is similar to that of cleavage with CNBr at methionyl bonds, namely nucleophilic attack by the carbonyl oxygen of the amide bond to be cleaved, with displacement of a good leaving group. In the case of methionine this was the -SCH₃ group, and in the case of tryptophan and tyrosine is a bromine atom, since the first stage of reaction is bromination as shown in Fig. 2.9. Nucleophilic displacement at tryptophan generates a spirodioxindole-γ-lactone (I) and at tyrosine a spirodienone lactone (II) with chain cleavage in both cases. Further details of this, and of the cleavage of histidyl bonds, may be found in the review by Spande *et al.* (1970).

3.4 Restriction of enzymic cleavage

It will have become evident from the previous section that one of the main advantages of enzymic over chemical cleavage is its specificity, e.g. trypsin cleaves lysyl and arginyl peptide bonds only. If the specificity can be further increased by restricting attack to only one of these two then the method becomes even more attractive.

There are various ways of 'blocking' lysyl residues by chemical modification to restrict tryptic attack to arginyl bonds; for reasons to be described the converse is unfortunately not true. The reactivity of lysine residues makes restriction of tryptic attack relatively easy, in a way that is not possible for other enzymes such as chymotrypsin, which cleaves after aromatic and hydrophobic (i.e. 'unmodifiable') amino acids. The specificity of trypsin for amino acid residues with basic side-chains is explained in the X-ray crystallographic model of trypsin by the presence of a negatively charged aspartic acid side-chain at the bottom of the 'hole' in the molecule that accommodates the substrate side-chain. Accordingly, modification of lysyl side-chains with reagents that either remove the positive charge, or reverse it, or even preserve it but extend its distance from the polypeptide backbone, will prevent tryptic attack. A typical sequence of events using lysine-blocking in sequence analysis of proteins is outlined in Table 2.2. The requirements of an ideal blocking group can be summarised as follows: it must be easily introduced, stable to the conditions of tryptic digestion and peptide fractionation, and easily removed so that each purified peptide can be redigested with trypsin at the lysyl bonds to

Table 2.2   Strategy of lysine-blocking in sequence analysis

1. Introduction of lysine-blocking groups
2. Digestion with trypsin and purification of peptides with C-terminal arginine
3. Removal of lysine-blocking groups from purified C-terminal arginine peptides
4. Digestion with trypsin (cleavage at lysyl bonds) and purification of the peptides
5. Sequence analysis of the peptides; partial ordering of peptides within each blocked fragment putting arginine at the C-terminus in each case.

give fragments representing the sequence between a pair of arginine residues. The advantages of lysine-blocking are that the mixture of peptides from a tryptic digest is less complex than it would otherwise be, and that some ordering of the peptides with C-terminal lysine is achieved. A few examples will show that some reagents satisfy the criteria for good blocking groups better than others.

3.4.1 Modifications that remove the charge

N-Acetyl groups are easily introduced by acylation of amino groups with acetic anhydride but are impossible to remove selectively; they are thus never used for lysine blocking in sequence analysis. On the other hand, trifluoroacetylation, e.g. by reaction of the protein with ethyl trifluorothioacetate at pH 9.5 (Fig. 2.10), may be reversed under mildly alkaline conditions, usu-

$$\text{Protein}-NH_2 \;+\; CF_3COSCH_2CH_3 \;\longrightarrow\; \text{Protein}-NH-COCF_3$$

1M-piperidine, 0°C

Fig. 2.10   Trifluoroacetylation of protein amino groups with ethyl thiotrifluoroacetate.

ally by treatment with 1M-piperidine at 0°C. The lability of the *N*-trifluoroacetyl group arises, of course, from the presence of three strongly electron-withdrawing fluorine atoms which make the carbonyl group extremely susceptible to nucleophilic attack. In fact, care must be taken to carry out all manipulations with trifluoroacetyl peptides at moderate pH (around pH 8). A disadvantage of trifluoroacetylation is the insolubility of some peptides after removal of the charge on the amino group, which makes fractionation difficult. The properties of the *N*-trifluoroacetyl group form the basis of a 'diagonal method' for selective purification of lysine-containing peptides and will be discussed later (section 3.6).

3.4.2 Modifications that preserve the charge

Imidoesters serve to prevent tryptic attack even though the modified side-chain still carries a positive charge (Fig. 2.11), but the difficulty of removal (which requires high pH values) makes them unsuitable for lysine-blocking in sequence analysis. Imidoesters are used in excess for modification of the protein at pH 8 to offset the competing hydrolysis of the reagent.

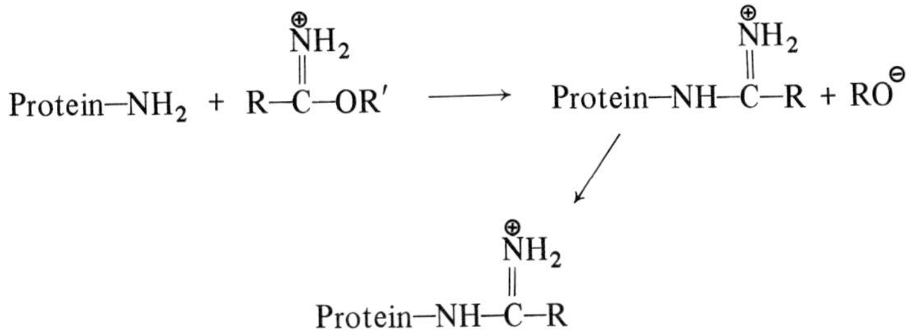

$$\text{Protein}-\text{NH}_2 + \text{R}-\overset{\overset{\oplus}{\text{NH}_2}}{\underset{\|}{\text{C}}}-\text{OR}' \longrightarrow \text{Protein}-\text{NH}-\overset{\overset{\oplus}{\text{NH}_2}}{\underset{\|}{\text{C}}}-\text{R} + \text{RO}^{\ominus}$$

$$\text{Protein}-\text{NH}-\overset{\overset{\oplus}{\text{NH}_2}}{\underset{\|}{\text{C}}}-\text{R}$$

Fig. 2.11 Amidination of protein amino groups with imidoesters.

ε-Amino groups in proteins can be converted to guanidino groups by reaction with O-methylisourea (Fig. 2.12); lysine is thus converted into homoarginine. It was recently shown that homoarginyl bonds can, in fact, act as poor substrates for trypsin. An interesting case in this connection is that of trypsin inhibitors, small proteins of 50–60 amino-acid residues, which

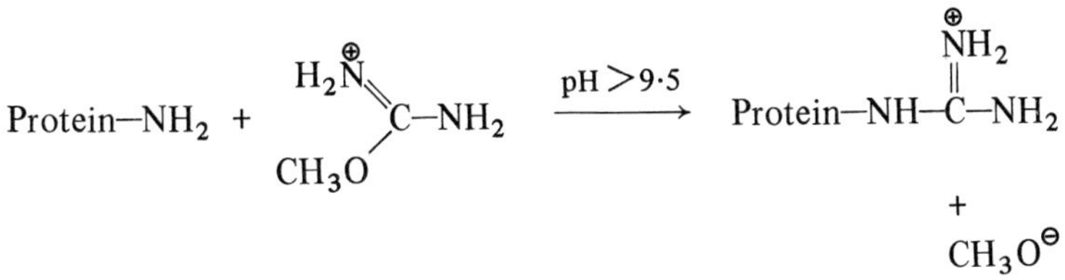

$$\text{Protein}-\text{NH}_2 + \begin{array}{c}\overset{\oplus}{\text{H}_2\text{N}}\\ \\ \text{CH}_3\text{O}\end{array}\!\!\!\!\!\text{C}-\text{NH}_2 \xrightarrow{\text{pH} > 9.5} \text{Protein}-\text{NH}-\overset{\overset{\oplus}{\text{NH}_2}}{\underset{\|}{\text{C}}}-\text{NH}_2$$

$$+ \ \text{CH}_3\text{O}^{\ominus}$$

Fig. 2.12 Guanidination of protein amino groups with O-methylisourea.

fall into two classes depending on whether they have an arginyl or a lysyl bond at the 'active site'. These bonds are cleaved in the enzyme–inhibitor complex and it has been assumed that this cleavage is a prerequisite for inhibition. Thus chemical modification of an inhibitor with a lysine blocking reagent that leaves the inhibitory activity unaffected has been taken to imply that the inhibitor has an arginyl active site. However homoarginyl bonds can in fact be cleaved by trypsin, so guanidination should not be used to distinguish between lysyl and arginyl active sites.

3.4.3 Modifications that reverse the charge

Acylation of amino groups in proteins with cyclic anhydrides of dicarboxylic acids replaces the positively charged amino group with a negatively charged derivative. The first reagent to be used for this purpose was succinic anhydride (Fig. 2.13(a)). Succinylation will prevent tryptic cleavage at lysyl bonds but is of limited usefulness in sequence work because it is irreversible.

The anhydrides of maleic acid and its derivatives are, however, free from this disadvantage and have consequently gained wide currency as reversible lysine-blocking reagents (Butler et al. 1969; Dixon and Perham 1968) (Fig. 2.13(b)). The ease of removal of the blocking groups in mildly acidic conditions (pH 3·5) is attributed to general acid catalysis of amide bond hydrolysis by the intramolecular carboxyl group, which the rigidity of the double bond in the reagent (contrast the succinyl group) directs into an orientation favourable for catalysis; the mechanism proposed by Kirby and Lancaster (1972) is shown in Fig. 2.13(b). Maleyl blocking groups can

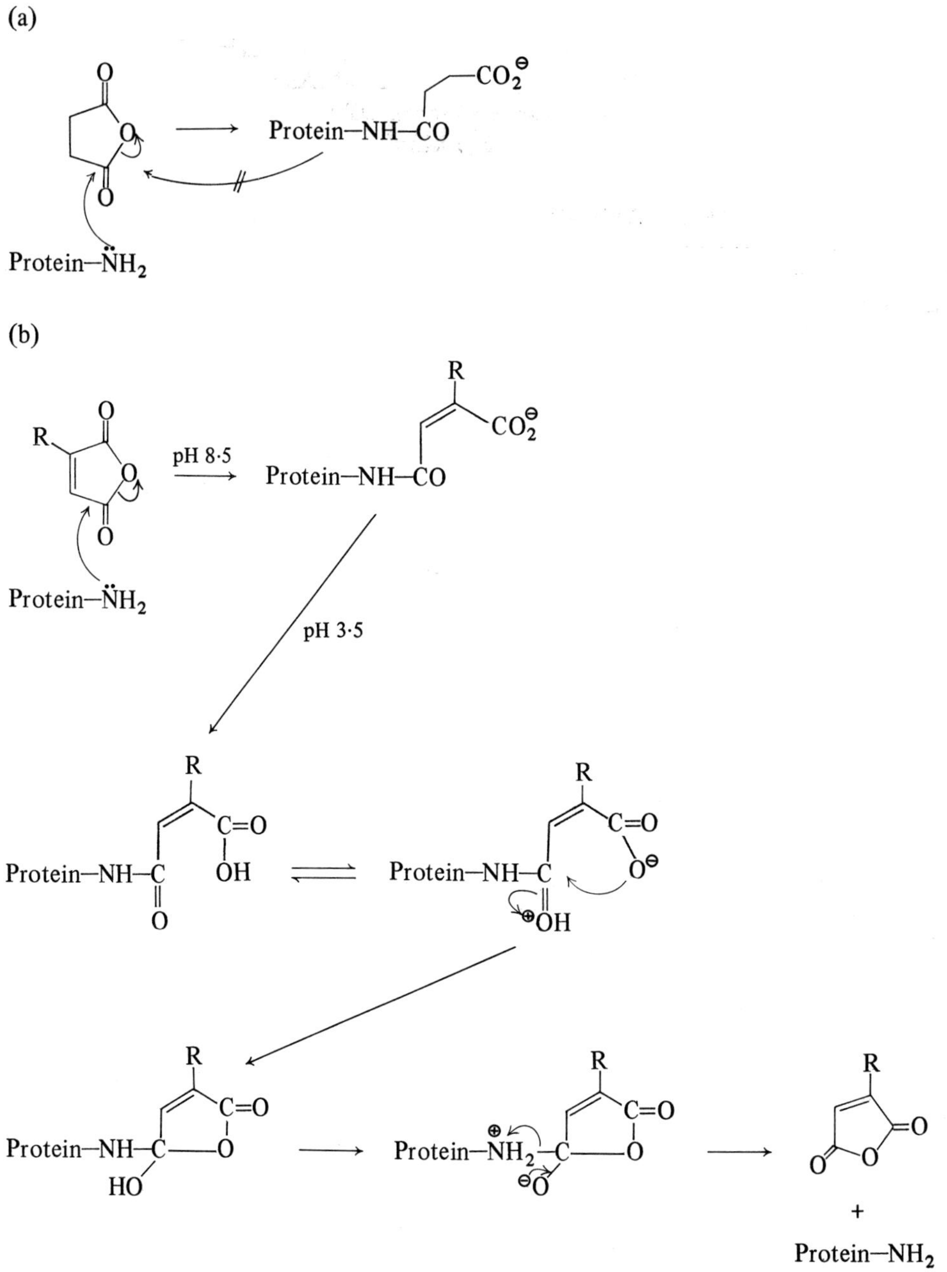

100

(a)

(b)

Fig. 2.13 Acylation of protein amino groups by (a) succinylation, (b) maleylation (R = H) or citraconylation (R = CH$_3$).

thus be removed by incubation of the protein at pH 3·5 overnight at 60°C and methyl maleyl (citraconyl) groups at 20°C. The additional methyl group of citraconic anhydride presumably directs the carboxyl group even more firmly into the position where it can be of assistance so that the blocking group is more easily removed; two methyl groups (in dimethyl maleic anhydride) enhance the facilitated hydrolysis to such an extent that the modification is reversed in 5 min at 20°C at pH 3·5, and appreciably even at neutral pH, but the consequent difficulties of manipulating the blocked peptides make this reagent much less useful than citraconic or maleic anhydride. Various other anhydrides have been introduced (e.g. tetra-fluorosuccinic) but have so far been used only rarely. The effect of the electron-withdrawing fluorine atoms on the stability of the *N*-acyl bond means that this blocking group (unlike the succinyl) can be removed at pH 9·5 and 0°C (cf. trifluoroacetyl and acetyl).

Another use of maleylation and citraconylation in studies of the primary structure of proteins is solubilisation of large peptides, e.g. those obtained by cleavage of protein chains with cyanogen bromide (see section 3.3), to render them amenable to column fractionation methods. The negative charges introduced by maleylation or citraconylation presumably cause unfolding and repulsion of chains and hence disaggregation. Succinylation has the same effect but its irreversibility precludes its use on cyanogen bromide fragments destined for sequence analysis.

The tertiary and quaternary structures of proteins are likewise disrupted by reaction with the dicarboxylic acid anhydrides and this, together with the reversibility of citraconylation and maleylation, is being exploited in investigations of the role of lysine residues in the maintenance of quaternary structure (cf. section 7).

3.5 Enhancement of enzymic cleavage

It is possible to restrict tryptic attack on a protein as described in section 3.4. In some cases, however, it is advantageous to *increase* the number of points of proteolysis. In particular the side-chains of cysteine (or reduced cystine) may be alkylated with ethyleneimine so that they masquerade (Fig. 2.14) as lysine residues (with which they are isosteric) and the *S*-(β-amino-

$$\text{Protein—SH} + \triangleright\text{NH} \longrightarrow \text{Protein—S—CH}_2\text{CH}_2\text{NH}_2$$

Fig. 2.14   *S*-Aminoethylation of proteins with ethyleneimine.

ethyl) cysteinyl bonds can then be cleaved by trypsin. Comparable tricks are not known for other enzymes or other reagents. Aminoethylation is irreversible but has proved invaluable on occasion in dealing with what is known as the 'core' problem, e.g. where a large and insoluble tryptic peptide occurs because two consecutive basic residues in the amino acid sequence are far apart. It has been widely used for this purpose on the 'cores' obtained from the β-chain in studies of haemoglobin variants, and renders all the peptides suitable for 'fingerprinting' analysis by paper electrophoresis and chromatography.

An impressive use of chemical modification to direct enzymic cleavage has been reported (Slobin and Singer 1968) for the immunoglobulins. These are large four-chain molecules having intra- and inter-molecular disulphide bonds which would give a complex mixture of peptides on cleavage with trypsin. The molecule can, however, be dissected into large pieces (useful, *inter alia*, for studies of sequence homology, etc.) by first acylating lysine residues, then reducing the disulphide bridges and aminoethylating, and finally blocking the arginine residues (see below). Tryptic digestion is then restricted to the initial half-cystines (now *S*-aminoethylated).

3.6 Chemical modification in peptide fractionation and purification

The solubilisation of mixtures of large peptides (e.g. CNBr peptides) by citraconylation or maleylation has already been mentioned (section 3.4).

3.6.1 Diagonal techniques: principles

Diagonal techniques provide a way of purifying peptides containing particular types of amino acids, usually by paper electrophoresis. The idea of diagonal electrophoresis originated with work on the photo-oxidation of histidine peptides and the consequent changes in electrophoretic mobility of the peptides, but did not really achieve prominence until the introduction in 1963 of the method for disulphide-bridged peptides described below. Diagonal techniques have been reviewed by Hartley (1970) and by Perham in vol. 1 (1969) of the series edited by Young.

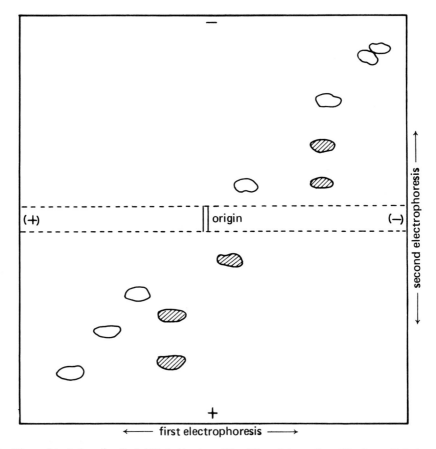

Fig. 2.15   Diagonal technique for disulphide-bridged peptides. The mixture of peptides is applied at the 'origin'. The dotted lines indicate the strip of paper on which the first electrophoresis is carried out, sewn on to a fresh sheet of paper for the second electrophoretic step. The hatched areas indicate the off-diagonal peptides.

'Diagonal separations' are based on the following principles: if a mixture of peptides is subjected to electrophoresis on paper first in one dimension and then at right angles under identical conditions, the peptides will lie on the diagonal of the paper since they move the same distance in both electrophoretic steps. If, however, some of the peptides are modified between the two electrophoretic steps so that they suffer a charge change, they will have an altered mobility in the second electrophoretic step and will thus lie off the diagonal. If the modification is specific for certain types of amino acid side-chains then peptides containing these residues may be purified selectively from complex mixtures. The method works most conveniently if the modification can be carried out on the paper, e.g. by exposing the paper to volatile reagents in a closed vessel.

3.6.2 Cystine-containing peptides

Performic acid vapour will oxidise disulphide bonds and generate two residues of cysteic acid, which carries a negative charge at pH 6·5:

$$R-S-S-R' + HCO_3H \rightarrow RSO_3H + R'SO_3H$$

The cysteic acid peptides derived from the initial disulphide-bridged peptide will thus lie off the electrophoretic diagonal (e.g. at pH 6·5) (Fig. 2.15). Two off-diagonal peptides are usually obtained from an inter-chain disulphide bridge and also from an intra-chain bridge since enzymic digestion often occurs between the two half-cystines; it may thus not be possible to distinguish inter- from intra-chain bridges solely from the number of off-diagonal peptides.

The performic acid method, the first practical application of the diagonal technique, has been particularly useful in making the correct assignment of disulphide bridges in proteins whose primary structure is known, and for identifying the disulphide-bridged peptides in complex molecules such as the immunoglobulins and the family of serine proteases (see Hartley 1970). In the serine proteases the method has provided an easy way of studying sequence homologies in the disulphide-bridge regions between members of the family (e.g. trypsin, chymotrypsin and elastase) and between the enzymes from different species. The disulphide-bridged peptides are usually generated using pepsin to cleave the protein chain, since the acidic conditions used for the digestion minimise the undesirable disulphide exchange reactions which tend to occur at higher pH values such as that used for tryptic digests (pH 8).

3.6.3 Cysteine-containing peptides

Air-oxidation of thiol groups, particularly on paper, precludes direct application of the performic acid technique to cysteine peptides, but if the thiol groups are first modified by disulphide exchange with cystamine they can then be treated in the same way as cystine

$$R-SH + \begin{matrix} SCH_2CH_2NH_2 \\ | \\ SCH_2CH_2NH_2 \end{matrix} \longrightarrow \begin{matrix} R-S-S-CH_2CH_2NH_2 \\ + \\ HSCH_2CH_2NH_2 \end{matrix} \xrightarrow[\text{etc.}]{HCO_3H,}$$

Fig. 2.16 Exchange of protein thiol groups with cystamine.

peptides (Fig. 2.16). This method has been used in analysis of the amino acid sequences around the thiol groups in the muscle protein myosin.

3.6.4 An alternative method for thiol groups and disulphide bonds

When S-carboxymethylated thiol groups (or reduced disulphide bonds) are converted to the sulphone derivatives by oxidation with performic acid, the pK_a of the carboxy group is lowered. and this change can be detected on electrophoresis at pH 3·5. This is the basis of an alternative diagonal method for thiol groups and disulphide bonds. In contrast with the method described above (3.6.2), it will not, of course, give the actual pairing of cysteine residues in disulphide bonds. Half-cystine peptides from intra- and inter-chain disulphide bridges in immunoglobulins could, however, be distinguished by first using mild reduction to break the *inter*-chain linkages followed by carboxymethylation with radioactive iodoacetic acid, and then further reduction and non-radioactive carboxymethylation to alkylate the *intra*-chain half-cystine residues.

3.6.5 Lysine-containing peptides

Lysine-containing peptides can be isolated and purified by a diagonal method in which the peptides from a digest of *N*-trifluoroacetylated protein are subjected to two electrophoretic steps, in two dimensions at right angles, one before and one after exposure of the paper to ammonia vapour. The trifluoroacetyl groups are removed from lysine-containing peptides, generating a charge change such that the altered peptides lie off the diagonal in the direction of the cathode. Lysine occurs too commonly in most proteins for this method to be practicable but it was ideally suited to isolation of the C-terminal peptide of pepsin for sequence analysis since the few basic residues present in this enzyme are clustered near the C-terminus. The method is also useful for the lysine peptides in cyanogen bromide-generated fragments. Maleylation and citraconylation (section 3.4.3) are alternatives to trifluoroacetylation.

3.6.6 Methionine-containing peptides

In a diagonal procedure involving methionyl residues the charge change is between the positively charged sulphonium salt (generated by alkylation of methionine with iodoacetamide at acid pH values) and the neutral product obtained when this is heated at pH 7 and 100°C. These conditions cleave the chain so that the initial methionine residue appears as the C-terminal homoserine residue (actually released as the lactone) of the N-terminal half of the

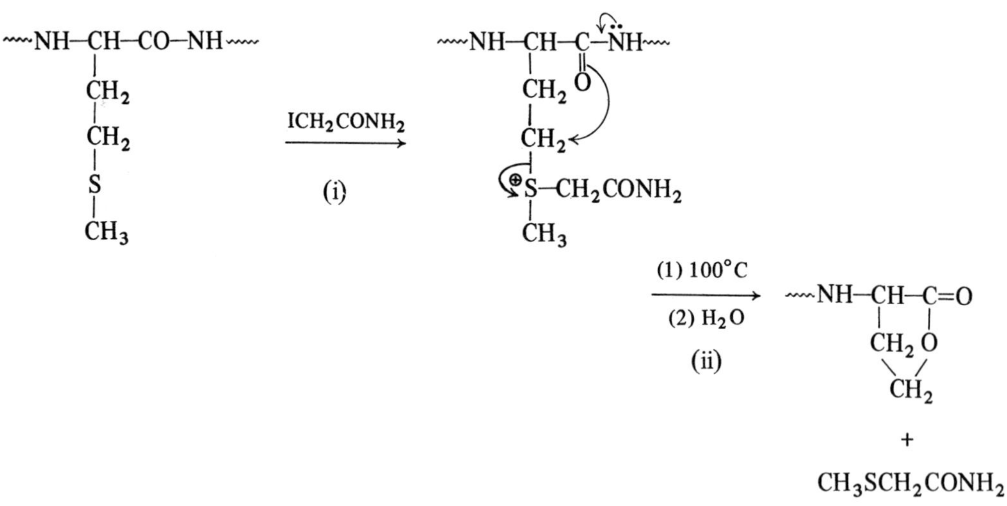

Fig. 2.17   Methionine diagonal technique; the charge change is generated in (ii).

chain (Fig. 2.17). The methionine diagonal method, so far not widely used, was of considerable value in determining the sequences around the four methionine residues in pepsin — a stubborn protein in which clustering of basic residues at the C-terminus leaves a peptide of about 300 residues resistant to tryptic attack.

3.6.7 Histidine-containing peptides

In a diagonal method for histidine peptides (Cruickshank *et al.* 1971) the charge change is generated by removal of a modifying group (dinitrophenyl) (cf. 3.6.5). Dinitrophenylation of histidine residues is carried out on the intact protein, after blocking lysine side-chains by citraconylation. The modified protein is then digested with pepsin in formic acid giving a mixture of peptides – from which the lysine blocking groups have been lost under the acid conditions. The two electrophoretic steps are then carried out, one before, one after exposure of the paper to 2-mercaptoethanol to remove *N*-Dnp groups by thiolysis, and the histidine peptides identified as those that move off the diagonal towards the cathode. The procedure is summarised in Fig. 2.18.

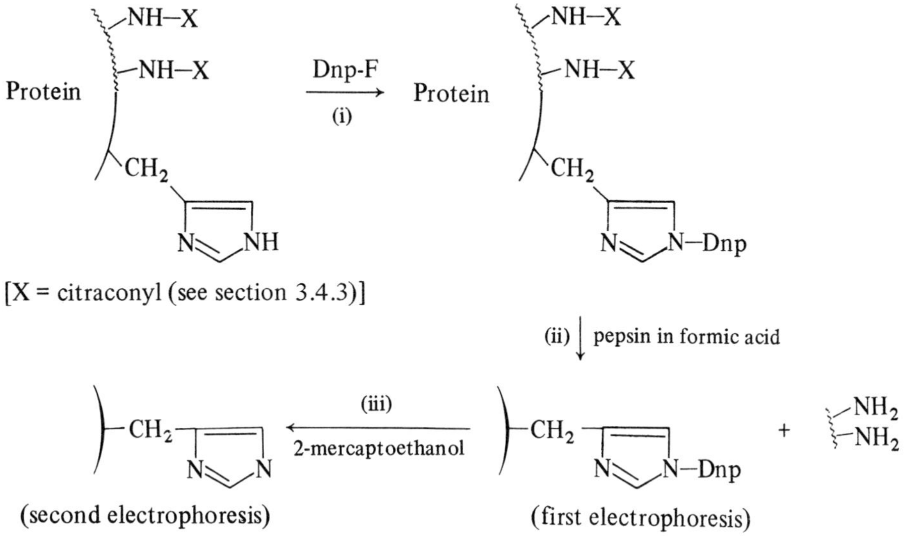

Protein $\xrightarrow[\text{(i)}]{\text{Dnp-F}}$ Protein

[X = citraconyl (see section 3.4.3)]

(ii) ↓ pepsin in formic acid

(iii) ← 2-mercaptoethanol

(second electrophoresis)   (first electrophoresis)

Fig. 2.18   Histidine diagonal technique; the charge change is generated in (iii).

The elegance of diagonal methods has no doubt been responsible for attempts to exploit the principle even further. For instance a 'column diagonal' procedure will permit identification and hence selective purification of phosphorylated peptides (e.g. those obtained from the active sites of enzymes such as phosphoglucomutase). The procedure is to pass a mixture of peptides through an anion-exchange column before, and after, treatment of the mixture with alkaline phosphatase; those, and only those, peptides from which a phosphate group can be removed are eluted in different (earlier) positions from the column after treatment with alkaline phosphatase. This was used recently to isolate the peptide containing the lysine residue that binds the cofactor pyridoxal phosphate as an imine (i.e. a Schiff base) in glutamate decarboxylase. The imine was stabilised by reduction and the peptides from a proteolytic digest were treated as above (Strausbauch and Fischer 1970); the peptide eluted early was the one that contained covalently bound pyridoxal phosphate.

3.7 Chemical modification in the actual sequence-determining steps

The method most commonly used for sequence determination of peptides is, of course, the Edman degradation (reviewed by Stark 1970), and the purpose of the chemical modification is to introduce an instability at the bond linking the first amino acid residue to the rest of the polypeptide chain so that this residue can then be selectively removed. The procedure is shown in Fig. 2.19. The unprotonated α-amino group first adds across the N=C bond of the reagent. The crucial instability arises from intra-molecular nucleophilic attack by the sulphur atom at the carbonyl group of the first peptide bond in acidic conditions (anhydrous, to avoid damage to the remainder of the polypeptide chain), forming a favourable five-membered ring as shown and eliminating the rest of the chain intact. The initial thiazolinone is extracted with an organic solvent and then rearranged in aqueous acid to the stable phenylthiohydantoin derivative of the N-terminal residue (i.e. the Pth-amino acid), which can be identified. Repeated many times on a progressively shorter peptide, this gives a method of sequential degradation from the N-terminus.

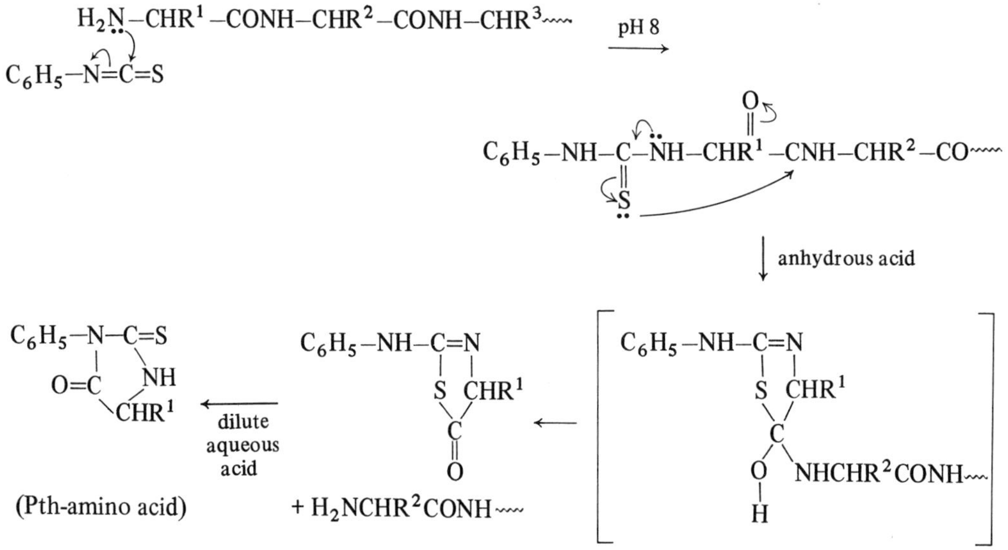

Fig. 2.19   Edman degradation of peptides and proteins.

It is inappropriate here to do more than recall briefly the variations on the way in which the Edman method is used in practice: the 'direct' method requiring rather a large amount of peptide (*ca.* 1 μmole) involves identification of the Pth-amino acid; the 'subtractive' method, which requires a little less, instead compares the amino acid composition of the peptide before and after the degradative cycle; the popular 'dansyl–Edman' method, which requires only about 20 nmoles of pure peptide for sequence analysis, uses the Edman cycle merely as a way of removing residues and identifies the newly exposed N-terminus each time by the ultra-sensitive dansyl method (see section 3.2.1), using a small sample removed after each degradative cycle.

At the moment the sequencer (sequenator), the machine introduced by Edman and Begg in 1967 for automatic determination of amino acid sequences of proteins (see Stark 1970), is still very much in its infancy and is expensive. It is capable of determining about 40–60 residues from the N-terminus of a protein. The chemistry of the process is exactly as described above for the manual method (Fig. 2.19). All operations are carried out in a small spinning cup so that the reactants are spread in a thin film on the walls. The thiazolinones generated are extracted with organic solvent and siphoned off into a fraction collector for identification — either as such, or after conversion to the phenylthiohydantoins, or to the original amino acid. Since it is essential that the sample stays in the cup during all the siphoning operations it is easy to see why the automatic method works best for whole proteins or very large peptides (e.g. cyanogen bromide fragments), and worst for small peptides which often extract with the amino acid derivative. This is especially true for small peptides containing lysine residues since the side-chain amino group is also modified by phenylisothiocyanate and the peptide is thus made more hydrophobic.

Two approaches are being taken to make small peptides suitable for analysis in the sequenator by the Edman method. One is to attach the peptide to a resin so that it remains insoluble throughout. The other, applicable only to lysine-containing peptides, involves treating the peptide first with a phenylisothiocyanate containing sulphonic acid groups in the aromatic ring so that the ϵ-amino groups are modified and the solubility of the peptide in organic solvents diminished, and then using the normal reagent for sequential degradation. This should also prove useful in reducing losses of small lysine-containing peptides analysed manually by the Edman method.

Several variations on the Edman method have been reported for the purpose of producing thiohydantoins with properties appropriate to particular methods of identification, e.g. methylthiohydantoins are volatile and can be easily identified by g.l.c.; p-bromophenyl-thiohydantoins give characteristic mass spectrometric fragmentation patterns; fluorescent thiohydantoins should make the sensitivity of detection of thiohydantoins equal to that of the dansyl–Edman method.

The Edman method is easily the most widely used method of establishing amino acid sequences; there is no comparable method for sequential degradation from the C-terminus. One method investigated for this purpose (see Stark 1970) involves generation of a thiohydantoin at the C-terminus by treatment first with acetic anhydride and then with ammonium thiocyanate. In practice, however, the method is limited and allows removal of only three or four residues. All amino acid sequences are in fact determined by sequential degradation from the N-terminus.

3.8 Chemical modification in mass spectrometric determination of amino-acid sequences

Mass spectrometry of peptides has been reviewed a number of times (see, for example, Shemyakin et al. 1970, and reviews cited therein).

Volatility is a prerequisite for mass spectrometry, and peptides are rendered highly involatile by intermolecular hydrogen bonding between amide groups. Volatilisation can be achieved if hydrogen bonding is eliminated by methylating all the amide bonds (permethylation) and by substituting the termini to remove their ionic character, e.g. by acylation of the N-terminus and esterification of the C-terminus (Fig. 2.20).

$$H_3\overset{\oplus}{N}-CHR^1-CONH-CHR^2 \cdots NH-CHR^{n-1}-CONH-CHR^n-CO_2^{\ominus}$$

(1) Ac_2O
(2) NaH/dimethylformamide
(3) MeI

$$CH_3CO-\underset{\underset{CH_3}{|}}{N}-CHR^1-\underset{\underset{CH_3}{|}}{CON}-CHR^2 \cdots \underset{\underset{CH_3}{|}}{N}-CHR^{n-1}-\underset{\underset{CH_3}{|}}{CON}-CHR^n-CO_2CH_3$$

Fig. 2.20   Modification of peptides for mass spectrometry.

The amide bond is not normally regarded as a reactive species and it owes not only its planar structure but also its inertness (i.e. lack of reactivity in carbonyl addition reactions) to resonance stabilisation:

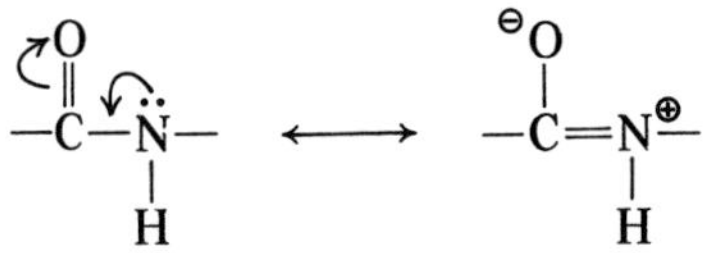

As a consequence of this resonance, however, the amide proton acquires a slight acidity and can be removed with a strong base, e.g. dimethylformamide, or the carbanions obtained from dimethylsulphoxide or dimethylacetamide by treatment with sodium hydride. Permethylation is then achieved with an excess of methyl iodide. Once this is done the sample (at least for peptides up to about 10-12 residues) is volatile and in the mass spectrometer will give a

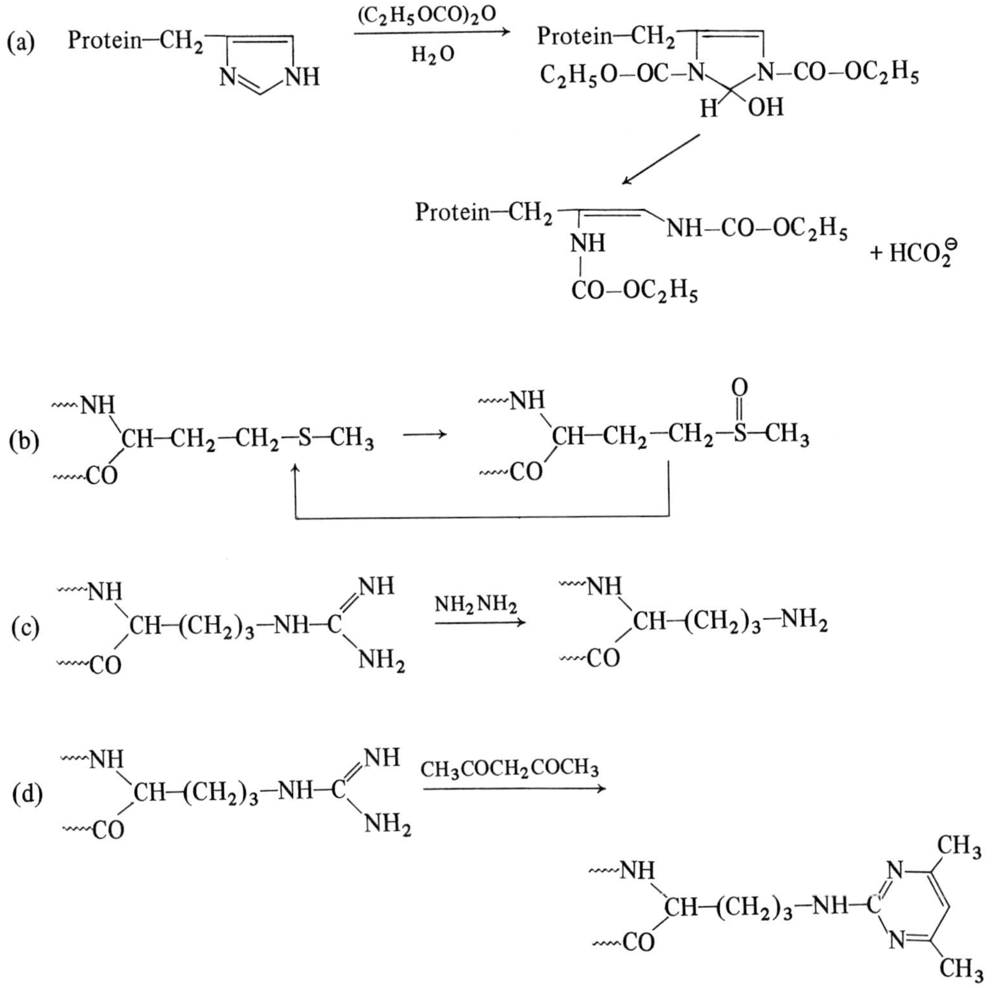

Fig. 2.21 Modification of (a) histidine, (b) methionine, (c) and (d) arginine side-chains for mass spectrometry.

fragmentation pattern that can be interpreted in terms of sequential breakdown of the parent ion, thus revealing the sequence of amino acid residues in the peptide.

Two problems must be overcome if the mass spectrometer is to supersede the conventional 'wet' methods of sequencing (e.g. the dansyl–Edman method). One is the amount of sample required (at present this is comparatively large) and the other is the difficulty posed by certain amino acid side-chains. Chemical modification is slowly providing a solution to the latter. The nucleophilic side-chains of cysteine, histidine, lysine, methionine and arginine all give involatile quaternary ammonium, or sulphonium, salts under the conditions for permethylation (although it is possible that this may be avoided in some cases under mild conditions). Various dodges have been employed to circumvent the problem. For instance lysine at the C-terminus of a tryptic peptide can be removed by digestion with carboxypeptidase B; cysteine-containing peptides will give satisfactory spectra either after desulphurisation with Raney nickel or as the S-carboxymethyl derivatives; histidine can be acylated with an excess of diethyl pyrocarbonate, $(EtO—CO)_2O$, so that the imidazole ring is broken and quaternisation avoided (Fig. 2.21(a)); methionine can be reversibly converted into its sulphoxide for protection during the permethylation step (Fig. 2.21(b)); arginine can be converted either into ornithine by treatment with hydrazine (Fig. 2.21(c)), or into ornithine derivatives by treatment with diketones, and these then give good spectra, e.g. acetylacetone converts arginyl residues into N-2-(4,6-dimethyl)-pyrimidyl ornithine residues (Fig. 2.21(d)).

Peptide mixtures are now being analysed by low resolution mass spectrometry (Morris et al. 1971) so that the problem of obtaining pure peptides for sequence analysis, and hence of having to deal with very small quantities, may be partly over.

4 Studies of tertiary and quaternary structures: the scope of the problem

The remainder of this chapter will be concerned with chemical modification of native proteins. The features to be studied can be broadly classed as follows:

> Active centres of enzymes and other proteins
> The topography of protein molecules
> Subunit interactions in multimeric proteins
> The relationship between primary structure and biological activity of peptides and proteins

The role that chemical modification plays in physical studies of protein structure is exemplified by the preparation of heavy-atom derivatives for X-ray analysis and the specific attachment of fluorescent probes and spin-labels.

5 The active site

5.1 Introduction

Specific reaction at the active centre of an enzyme can arise in a number of ways, of which the most generally applicable is the use of a reagent that binds at the substrate-binding site (sections 5.2, 5.3). Some enzymes, however, have particularly reactive and accessible groups at the active centre (section 5.4) while others, through some quirk of their structure, bind specifically reagents that are ordinarily non-specific (section 5.5), and both situations may lead to unique reaction at the active centre.

In addition to what one might call conventional, straightforward, affinity labelling there are other approaches to modification of the active centres of enzymes, less routinely used and less well tried. They illustrate nicely the ingenuity applied to the problem and will be considered in sections 5.7–5.10.

A group labelled by a substrate analogue is clearly in the vicinity of the active centre, but may or may not be directly involved in catalysis. There are instances where side-chains near catalytic groups can be modified without any direct effect on activity, e.g. when chymotrypsin is inactivated by photo-oxidation, a histidine and a methionine residue are destroyed but kinetic analysis indicates that only histidine is catalytically essential and modification of methionine, also in the active centre region, gives an enzyme which is still partly active (Koshland et al. 1962). On the other hand, two groups are thought to be involved in the photo-oxidative inactivation of phosphoglucomutase: a fast-reacting histidine and a fast-reacting methionine. Even if inhibition is accompanied by modification of only one group, steric effects (see below), and charge and conformational effects must be discounted before the groups modified can be unambiguously assigned a role in catalysis. Optical rotatory dispersion (o.r.d.) and circular dichroism (c.d.) spectra are useful indicators of gross conformational changes (see Chapter 4).

5.2 Affinity labelling

The best, and most obvious way of specifically labelling the active centre of an enzyme is to turn enzymic specificity to advantage by designing an inhibitor with a close structural resemblance to a normal substrate; specificity of binding is thus assured and, once bound (and thus present in a high local concentration), a suitably designed reagent may then react at, or near, the active site. An affinity label is really a competitive inhibitor into which is incorporated some 'warhead' that leads to irreversible (covalent) reaction with the enzyme. Such inhibitors are often activated alkylating agents, such as α-keto alkyl bromides or iodides, or acylating

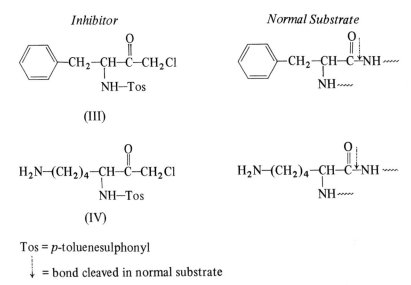

Tos = p-toluenesulphonyl

↓ = bond cleaved in normal substrate

Fig. 2.22   Active-site-directed inhibitors of chymotrypsin (III) and trypsin (IV).

agents such as acyl bromides or chlorides. Alkylation and acylation of the enzyme both occur by nucleophilic attack on the reagent by an amino acid side-chain (Fig. 2.1(a)). Once the covalent linkage has been generated at the active site, and provided that it is stable, the protein may be denatured and digested enzymically, and the 'active-site peptide' containing the modified residue may be identified using some convenient feature of the inhibitor such as radioactivity, or spectroscopic or fluorescent properties. Isolation of 'active-site peptides' from an enzyme from several sources has made it possible to compare amino acid sequences in these important regions and to look for amino acid sequence homology and evidence of evolutionary relationships.

A standard text by Baker (1967) and a review by Singer (1967) discuss fully the principles of design of active-site-directed irreversible inhibitors. The subject has also been well reviewed by Shaw (1970). Once the organic chemical problems of synthesis of the inhibitor have been overcome, reaction with the enzyme is usually straightforward – provided, of course, that there is a nucleophilic side-chain suitably positioned for attack on the bound reagent.

Some examples will illustrate the sort of substrate analogues used for affinity labelling. Bromo- and chloro-methyl ketones are useful aklylating groups to incorporate into substrates and pseudosubstrates. For instance tosylphenylalanyl chloromethyl ketone (III) and tosyl-lysine chloromethyl ketone (IV) (Fig. 2.22) are irreversible inhibitors of chymotrypsin and trypsin, respectively, because the amino acid side-chains serve to direct them into the normal substrate-binding site. In general, halomethyl ketone derivatives of amino acids are easily made and might be useful, for example, in studies of the active centres of the enzymes of amino acid metabolism.

Alkylating warheads are often conveniently introduced into substrates, substrate analogues or competitive inhibitors by bromoacetylation of an amino, hydroxy or sulphydryl group in the molecule (Fig. 2.23). Some examples are shown in Table 2.3.

$$R-YH_2 \xrightarrow[\substack{\text{or} \\ BrCH_2COBr}]{(BrCH_2CO)_2O} R-Y-CO-CH_2Br \xrightarrow{Protein-XH} R-Y-CO-CH_2-X-Protein$$
$$+ HBr$$

X,Y = NH, O or S
R−YH$_2$ = substrate, substrate analogue or competitive inhibitor

Fig. 2.23 Bromoacetyl derivatives of substrates, substrate analogues or competitive inhibitors as affinity labels.

A nice illustration of the mapping of active centre regions is the affinity labelling of an antibody directed against the dinitrophenyl (Dnp) group. Two affinity labels based on the Dnp hapten were used: bromoacetyl Dnp-ethylene diamine (V) and bromoacetyl N^ϵ-Dnp-lysine (VI). The former labelled only light chains of the antibody and the latter only heavy chains (Haimovitch, Givol and Eisen 1970). This shows that both types of chain in the antibody contribute to the antibody combining-site; it also illustrates that two active-site-directed

Table 2.3 Bromoacetyl derivatives of substrates as affinity labels

Enzyme	Affinity label	Residue labelled
Carboxypeptidase A$_\gamma$	N-Bromoacetyl-N-Me-L-Phe	Glu-270
β-Galactosidase	N-Bromoacetyl-β-D-galactopyranosylamine	Met
Carnitine acetyl transferase	O-Bromoacetylcarnitine	His

$$Dnp-NH-CH_2-CH_2-NH-CO-CH_2Br$$

$$Dnp-NH-(CH_2)_4-\underset{\underset{CO_2H}{|}}{CH}-NH-CO-CH_2Br$$

(V)   (VI)

reagents may attack in different positions (see below). The earliest affinity labelling experiments for antibody combining sites were carried out with aromatic diazonium salts related to the natural hapten and led to modification of specific tyrosine residues in the antibody (Fig. 2.24) (see Singer 1967).

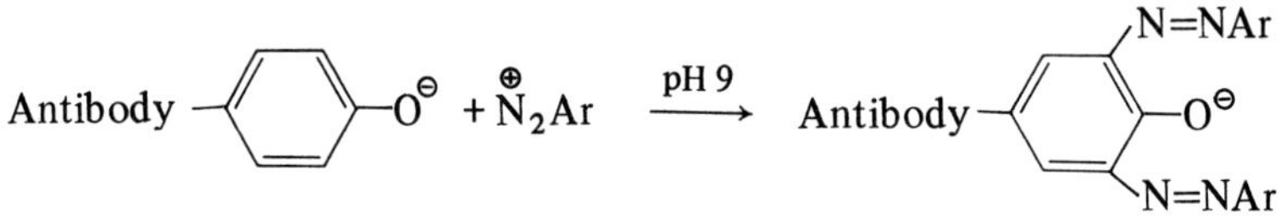

Ar = aryl radical

Fig. 2.24   Affinity labelling of antibodies with diazonium salts of haptens.

Diazoacetyl compounds are also useful as affinity labels. They are subject to nucleophilic attack by carboxy groups at pH values of about 5, giving esters (Fig. 2.25); at this pH only thiol groups, of all the side-chains in the protein, are sufficiently nucleophilic to generate side reactions. Various diazoacetyl compounds (e.g. N-diazoacetylnorleucine methyl ester, methyl diazoacetate) will label an aspartic acid residue in the active centre of pepsin, in a reaction which is facilitated by metal ions (e.g. Cu^{2+}, Ag^+). The diazo compounds are not acting as substrate analogues and it has been suggested that specific reaction occurs because a second carboxy group nearby binds and orients the positively charged complex of a metal ion with a carbene (from the diazo compound) such that the first carboxy group is attacked.

$$R-CO-CH_2N_2 \xrightarrow{\text{Protein}-CO_2H} R-CO-CH_2O-CO-Protein$$

Fig. 2.25   Reaction of diazoacetyl compounds with protein carboxy groups.

The exact position of the alkylating or acylating group in an active-site-directed inhibitor will clearly determine the site of reaction with the protein. This is illustrated in the irreversible inhibition of luciferase from fireflies. The inhibitor 2-cyano-6-chlorobenzothiazole (VII), an analogue of luceriferin (VIII), arylates the enzyme, which displaces the 6-chlorine atom. Although the site of reaction has not yet been determined, it is known that there is no reaction

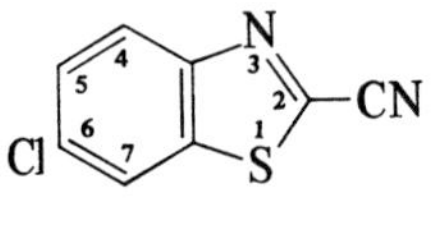

(VII) 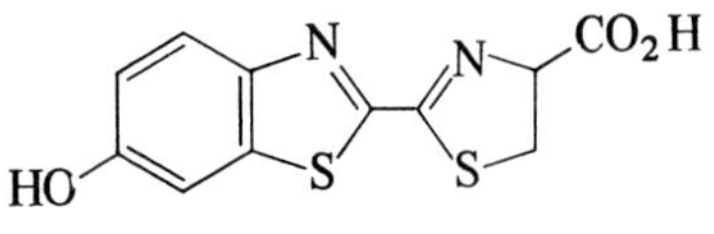   (VIII)

at the reactive thiol groups in the enzyme. These are in fact known to be near the binding site for the carboxy group of luciferin, i.e. far from the 6-position (Lee and McElroy 1971). Inhibition of chymotrypsin with different affinity labels is an interesting example of labelling of different groups in the vicinity of the active centre. For instance, tosylphenylalanine chloromethyl ketone (III) reacts with histidine-57, diphenylcarbamyl chloride (IX) with serine-195, and 3-phenoxy-1,2-epoxypropane (X) with methionine-192 (see Hartley 1964). The X-ray model for chymotrypsin confirms that these residues are at, or near, the active site.

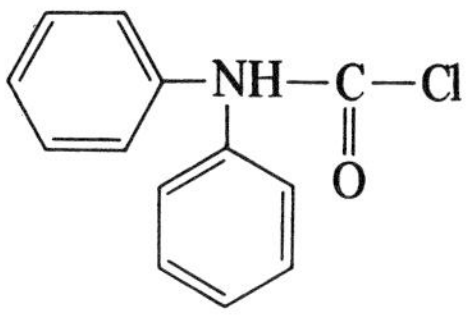

(IX)

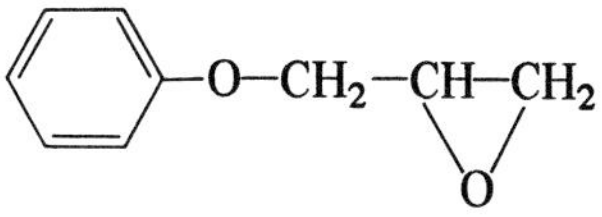

(X)

When bulky groups are used as affinity labels there lurks a possibility that inhibition may be due to steric blockage of the active site rather than to modification of a catalytically important group, and this should be borne in mind when assigning roles in catalysis to groups modified by affinity labelling. Steric effects are less likely, but not of course excluded, when the modifying group is small. For example, methyl p-nitrobenzenesulphonate inhibits chymotrypsin irreversibly, but steric effects are unlikely since the bulky group is eliminated (Fig. 2.26) and the

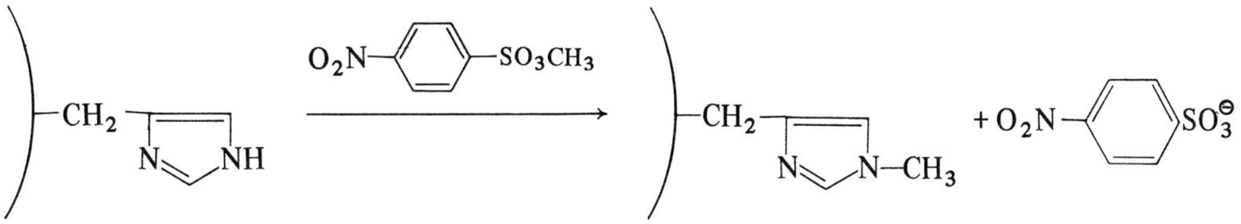

Fig. 2.26   Methylation of the catalytic histidine residue in chymotrypsin with methyl p-nitrobenzenesulphonate.

enzyme is methylated at N-3 of the histidine at the active centre (Nakagawa and Bender 1970), suggesting this residue to be essential for catalysis. Inhibition of chymotrypsin by reaction of methionine-192 with (X) is clearly partly steric since this methionine can be carboxymethylated without affecting V_{max} for the enzyme, although K_m is increased.

It might be expected that if the structure of the transition state in an enzymic reaction were known, and an analogue carrying a reactive substituent could be synthesised, this would be more likely to label a group actually involved in the catalytic mechanism. This was the rationale behind the use of glycidol (2,3-epoxypropanol) phosphate (XI) as an inhibitor of the enzyme

(XI)

triosephosphate isomerase. The inhibitor is based on the putative transition state in the conversion of dihydroxyacetone phosphate to glyceraldehyde 3-phosphate. A single carboxygroup in the enzyme was labelled as an ester. It may unfortunately be less easy to make a reasonable guess as to the structure of the transition states in other enzymic reactions, but in principle the approach is a powerful one.

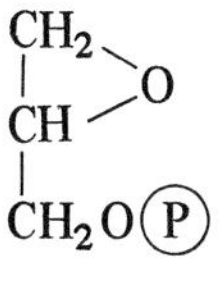

5.3 Labelling of the active site with substrate

The best site-specific reagent for an enzyme is, of course, its substrate, but the bonds in the enzyme–substrate complex are transitory and of no use in chemical labelling studies. In some instances, however, it is possible to stabilise these bonds and 'freeze' the complex. The best examples are enzymes such as aldolases which have a lysine residue at the active site, and which function by forming imine linkages with carbonyl groups in substrates. The imine can be stabilised by reduction with borohydride and the substrate is then covalently bound to the active site as a label.

Certain other enzymes can be labelled at the active site by substrate in a rather different way. For example, phosphoglucomutase is the intermediate carrier of the phosphate group when glucose-1-phosphate is isomerised to glucose-6-phosphate; the phosphorylated enzyme is easily formed in the presence of either of the sugar phosphates and is stable in the absence of a phosphate acceptor. ^{32}P-labelled substrate is thus a convenient affinity label.

5.4 Reaction at the active site because of hyper-reactive groups

Some well studied inhibitors that react specifically with the active sites of enzymes bear no immediately obvious relation to the normal substrates of the enzymes, e.g. di-isopropylfluoro-phosphate (DFP) (XII) as a specific inhibitor of the serine proteases and iodoacetate (XIII) as

$$(CH_3)_2CHO \diagdown \underset{(CH_3)_2CHO \diagup}{P} \diagup\hspace{-0.3em}\diagdown \underset{F}{\overset{O}{}} \qquad\qquad ICH_2CO_2^\ominus$$

$$\text{(XII)} \qquad\qquad\qquad \text{(XIII)}$$

an inhibitor of glyceraldehyde 3-phosphate dehydrogenase (GPDH). In these cases specific binding is probably relatively unimportant (although the bulky isopropyl groups could occupy the 'hole' in chymotrypsin into which the aromatic side-chains of normal substrates bind; and the negative charge of iodoacetate could direct it into the site in GPDH that binds the phosphate group of glyceraldehyde 3-phosphate). Instead the specificity of reaction is probably due to the considerably enhanced reactivity of the active centre hydroxy and thiol groups respectively. In support of this, one may recall that DFP is also a potent inhibitor of trypsin and subtilisin, which have considerably different binding sites from chymotrypsin, consistent with their different specificities; and that iodoacetamide is a potent inhibitor of glyceraldehyde 3-phosphate dehydrogenase, acting at the same thiol group as iodoacetate.

Other examples of 'hyper-reactive' side-chains in proteins (see Shaw 1970) are the ϵ-amino group of lysine-41 in bovine ribonuclease, which of all the lysine residues (many of which must be on the surface) reacts by far the most rapidly with fluoro-2,4-dinitrobenzene; and cysteine-25 in papain. Hyper-reactive groups have abnormally low pK_a values, e.g. lysine-41 in ribonuclease has an apparent pK of 8·8, in contrast with the value of about 10 expected for ϵ-amino groups; cysteine-25 in papain has a pK of 8·5 in contrast with the normal value (9–10) for thiol groups. Speculations that such hyper-reactivity effects are due to specific side-chain interactions within the three-dimensional structure of the protein have been considerably strengthened by X-ray crystallographic models for the proteins. For instance, the Asp/His/Ser 'charge-relay system' in chymotrypsin (Blow *et al.* 1969) accounts for the enhanced nucleo-philicity of the serine hydroxy-group; and the hyper-reactivity of lysine-41 in ribonuclease may be due to a neighbouring arginine side-chain (cf. Shaw 1970).

5.5 Reaction at the active centre through specific adsorption

Group-specific reagents will sometimes react selectively with side-chains at the active centre not because they resemble substrates (section 5.2, 5.3), nor because of hyper-reactive side-chains (section 5.4), but rather as a consequence of specific adsorption of the reagent. This results from complementary structural features (e.g. charged regions) in the reagent and the enzyme. For instance, the rapid alkylation of a histidine residue in ribonuclease by iodoacetic acid, but not iodoacetamide, can be accounted for by adsorption of the reagent to a positively charged group in the enzyme, possibly the normal phosphate-binding site; consistent with this is reversible inhibition of the enzyme by sulphate or phosphate anions. In fact, careful analysis showed that carboxymethylation of ribonuclease at pH 5·5 gave two monosubstituted products. Peptide mapping showed that one was carboxymethylated at histidine-12 and the other at histidine-119. The modifications were mutually exclusive and led to the suggestion that the two histidine residues were close together in the active centre, a conclusion that has since been supported by the X-ray crystallographic model for ribonuclease; chemical and physical studies of pancreatic ribonuclease have been reviewed by Richards and Wyckoff (1971).

5.6 Labelling of active sites with group-specific reagents

Groups at the active centre may be labelled with quite general group-specific reagents without relying on hyper-reactivity (section 5.4) or specific adsorption (section 5.5). This is done using a competitive inhibitor to protect the active centre reversibly while all other reactive groups are treated with reagent, and then labelling with, for example, a radioactive form of the reagent after removal of the inhibitor. A peptide containing the modified residue can then be identified in enzymic digests in the usual way and analysed.

This approach was successfully applied to trypsin; benzamidine (a competitive inhibitor) was used for reversible protection of aspartic acid-177 at the ionic binding site while carboxyl groups were coupled to glycineamide using a water-soluble carbodiimide. When benzamidine was removed only aspartic acid-177 was labelled by radioactive glycineamide (Eyl and Inagami 1971). Reversible protection is also useful for labelling particular proteins *in situ* (e.g. in membranes) (see section 5.10).

5.7 Photoaffinity labelling

This approach has been developed by Westheimer and his colleagues (e.g. Hexter and Westheimer 1971) for enzymes, such as chymotrypsin, that work through a covalent acyl-enzyme intermediate. Substrates of the enzyme carrying a diazo function in the acyl group (e.g. diazoacetyl esters) give rise to a diazoacetyl group covalently linked to the catalytic centre

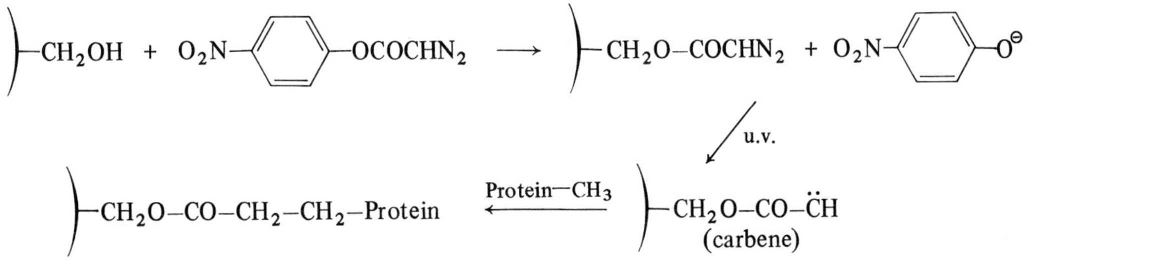

Fig. 2.27   Photoaffinity labelling of an enzyme that forms an acyl-enzyme intermediate.

(i.e. an *O*-diazoacetyl enzyme). Irradiation leads to photolysis of the diazo group, generating at the active site a reactive carbene (Fig. 2.27) which will immediately attack neighbouring side-chains. Since attack is non-specific, occurring at almost all amino-acid side-chains (i.e. alkyl as well as the commonly reactive nucleophiles such as amino and thiol groups), carbenes are useful tools for exploring groups at the active site that would otherwise be inaccessible to chemical methods of investigation. This more than offsets the disadvantages resulting from a mixture of products. When ^{14}C-diazoacetyl chymotrypsin was photolysed, *O*-[^{14}C]-carboxymethylserine, *N*-[^{14}C]-carboxymethylhistidine and *O*-[^{14}C]-carboxymethyltyrosine were found.

Photochemical generation of reactive intermediates provides an elegant approach to the study of antibody combining sites. Haptens are used which are aryl azides or diazoketones, both of which are quite stable. Photolysis of the antibody–hapten complex, however, gives rise to highly reactive nitrenes (Fleet, Porter and Knowles 1969) or carbenes (Converse and Richards 1969) respectively, which will react at the antibody combining site. Both nitrenes and carbenes can insert into C–H bonds.

5.8 Active-site-directed photo-oxidation

There are numerous examples of photo-oxidation of proteins by irradiation in solution in the presence of a suitable photo-sensitising dye such as methylene blue or rose bengal. The dyes absorb light of a suitable wavelength and in the energised state are able to transfer the energy direct to 'exposed' photo-oxidisable residues (methionine, histidine, tryptophan, tyrosine, cysteine and cystine) in the protein. If the rate of inactivation of the enzyme is equal to the rate of loss of an amino acid residue, this residue is implicated in the catalytic mechanism. (Photo-oxidation of chymotrypsin has been reviewed by Koshland *et al.* 1962.) However, the results obtained by this method are often not clear-cut and should be interpreted with caution.

Scoffone and his co-workers (for example, Scoffone *et al.* in Smellie 1970) have shown that it is possible to confine photo-oxidation to the active centre. The trick is simply to use a group at the active centre as the photosensitiser. For instance, the haem group in cytochrome *c* will serve this purpose and only histidine-18 and methionine-80 are modified on irradiation of ferricytochrome *c* at pH 8·2. A dinitrophenyl group attached to lysine-41 in ribonuclease (cf. section 5.4) will serve the same purpose, and leads to specific photo-oxidation of methionine-30, histidine-12 and tyrosine-97, consistent with the proximity of these groups to lysine-41 in the X-ray model. So far specific photo-oxidation has been confined to studies of proteins whose three-dimensional structure is known, but the agreement of the results with the X-ray models inspires confidence in the approach.

5.9 Dynamic labelling of active centres

A somewhat esoteric but elegant approach to the study of active centres involves measurement of the reactivity of side-chains towards group-specific reagents, not in the native enzyme itself but in the enzyme–substrate complex (Christen 1970). For example, tetranitromethane reacts with an essential tyrosine residue of aspartate aminotransferase only in the presence of the substrate pair, glutamate and α-ketoglutarate. This is still a novel approach but one that holds great promise for analysis of dynamic processes during catalysis. It nicely complements the approach of using substrate deliberately to protect catalytic groups at the active centre while other manipulations are being performed (cf. section 5.6).

5.10 Active centre labelling of proteins *in situ*

A particular protein present in a mixture can be labelled specifically by taking advantage of the specificity of interaction between an enzyme and its substrate, or a hormone or neuro-transmitter and its receptor, and using alkylating or acylating analogues of the substrate, hormone or neurotransmitter. Alkylating analogues of acetylcholine, e.g. (XIV), have recently been used to label the cholinergic receptor in intact membranes; (XIV) is a maleimide derivative

$$\text{N}-\overset{O}{\underset{O}{\parallel\!\!\!\parallel}}-\text{CH}_2-\overset{\oplus}{\text{N}}(\text{CH}_3)_3 \quad \text{I}^{\ominus}$$

(XIV)

which alkylates thiol groups. Despite specific interactions, the results obtained by labelling *in situ* using specifically designed inhibitors are not always clear-cut, and non-specific labelling of many proteins can occur (cf. Chavin 1971). Reversible protection with a competitive inhibitor will often overcome this problem and will also enable group-specific reagents to be used for *in situ* labelling (cf. section 5.6). Exploration of receptor sites *in situ* is, however, a very specialised approach that requires a good deal of care and skill.

6 The topography of a protein molecule

6.1 Introduction

X-ray crystallography shows that, in general, the hydrophobic side-chains (e.g. of valine, leucine and phenylalanine) are directed towards the interior of the protein and free of water whereas the hydrophilic side-chains (e.g. of aspartic acid, lysine and serine) are on the outside and exposed to water (although the relative accessibility of a particular type of amino acid side-chain varies from one protein to another (Lee and Richards 1971)). Numerous studies involving chemical modification of proteins have been concerned with identification of residues on the 'outside' of a protein molecule. Such residues may be essential for binding of substrates, or for catalysis, or neither. The reagents normally used in such studies are termed 'group-specific' and have recently been reviewed by Stark (1970). No attempt will be made here to duplicate that treatment; instead we shall concentrate on closer examination of the approach in principle, its advantages and its limitations.

First, is it valid to think in terms of a definite surface for a protein? The convolutions in the 'surface' are such that in addition to being completely buried, or completely exposed to the solvent, groups 'at the surface' may be masked slightly because of a shallow folding and the presence of neighbouring amino acid side-chains. (For a detailed discussion see the review by Kronman and Robbins 1970.) Differential reactivity of amino acid side-chains of a particular kind can be due to some of the side-chains being more 'buried' than others. It also depends, however, on the size of the reagent and on its polarity; small hydrophobic reagents are more likely to be able to penetrate into the protein molecule and to react with groups that are quite distant from the surface. Thus the most reactive groups are not necessarily the most 'exposed'; examples of this type of behaviour are given below. Despite such pitfalls, it is fair to say that

118

group-specific labelling has yielded much useful information about the three-dimensional structure of proteins. It has often complemented X-ray crystallographic analysis of the protein structure, and it has been particularly informative in cases where X-ray analysis is not at hand. It is, of course, much more useful if the primary structure of the protein is known so that 'reactive' residues can be identified within a known sequence as a first step to outlining the shape of the molecule (albeit very roughly) in the absence of precise X-ray information. The residues modified are identified by locating labelled peptides in enzymic digests using some property such as colour or radioactivity, and determining their sequences.

6.2 Stoicheiometry

The stoicheiometry of reaction of a group-specific reagent with a protein in solution may be established in a number of ways, depending on the nature of the reagent. Amino acid analysis is useful if the modified amino acid is either completely unharmed or completely destroyed by hydrolysis in 6N-HCl (e.g. *S*-carboxymethylcysteine from reaction of thiol groups with iodoacetic acid is stable and easily quantitated); alternatively chromophoric changes occurring on reaction may be used for spectroscopic quantitation, e.g. when ϵ-amino groups are converted to trinitrophenyl derivatives by reaction with 2,4,6-trinitrobenzenesulphonic acid, or when tyrosine is nitrated by tetranitromethane. If the reagent is radioactively labelled, e.g. 2-[^{14}C]-iodoacetate commonly used for thiol groups, then incorporation of reactivity is measured.

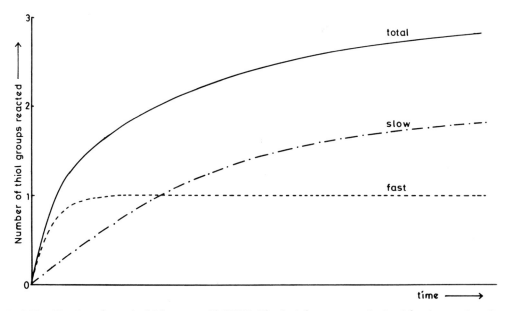

Fig. 2.28   Titration of protein thiol groups with DTNB. The dotted curves are calculated for the reaction of individual groups, where a single thiol group (. . . . . .) reacts ten times as fast as two others (.—.—.—.). The continuous curve is the observed time course of reaction of the three groups in the protein.

In such measurements of stoicheiometry, e.g. the reaction of thiol groups with 5,5'-dithiobis-(2-nitrobenzoic acid), differential reactivity of groups may be recognised. Figure 2.28 illustrates what might be expected for a protein with one thiol group that reacts rapidly and two others that react ten times more slowly, the rest of the thiol groups being unreactive. It is clear from the figure, however, that it may be difficult to recognise immediately that one group is considerably more reactive than the other two.

6.3 The nature of the reagent

The reactivity of a particular side-chain in a protein towards different reagents may vary in a way that is not due to any intrinsic properties of the group as displayed in model compounds. This is because reagents of different size, shape and charge respond to different features of the protein environment around the reactive group (cf. section 6.1). Thus, for instance, in rabbit muscle aldolase, where only three of the eight cysteine side-chains per polypeptide subunit can be modified, two cysteine residues react relatively rapidly with iodoacetamide or N-ethyl-maleimide and one reacts slowly. With 5,5'-dithiobis-(2-nitrobenzoic acid), however, one cysteine residue reacts rapidly and two react slowly (see the paper by Perham and Anderson in Smellie 1970). The same three thiol groups are involved in each case. It is possible, however, that as well as reacting at different rates different reagents may react at different sites (cf. section 5.2). A further point to note is that reaction at one group may affect the reactivity of others, perhaps making accessible groups that are in fact not so in the native structure.

The location of tyrosine side-chains has often been the subject of study. The aromatic ring is hydrophobic and would thus be expected to be buried while the phenolic hydroxy-group may either be directed 'inwards' or exposed to the surrounding medium depending on its state of ionisation. This, and the fact that some group specific reagents used for tyrosine residues substitute into the ring (e.g. iodine, tetranitromethane) while others modify the hydroxy-group (e.g. N-acetylimidazole and cyanuric fluoride) will explain why different estimates of the number of exposed tyrosine residues in a particular protein are frequently found in the literature.

The point made earlier (section 6.1) about the dangers of equating accessibility of side-chains with location on the 'surface' of a protein is best illustrated with reference to tyrosine residues. Tetranitromethane has been widely used in recent years as a probe for exposed tyrosine residues in a number of proteins. When the X-ray crystallographic model of cytochrome c become available, however, the easily nitratable residues in this case were shown to be buried in a hydrophobic environment; the same appears to be true for one or two other proteins. Similar effects may apply in the use of iodine. One should thus be very careful in interpreting accessibility to these (and possibly other) reagents in terms of the topography of the protein molecule.

6.4 Some considerations in the use of group-specific reagents

If the side-chain can ionise then the most nucleophilic species is the form that predominates at high pH, i.e. the unprotonated amino group, or the phenolate form of the tyrosine side-chain. One might thus expect that suitable control of the pH would confine reaction to side-chains of one particular type. Unfortunately this is only partly true, since the pK values of amino acid side-chains lie within a fairly narrow range (Table 2.4) and when one class is, for example, half-ionised, several others will also exist to greater or lesser extents in the unprotonated form. For instance, at pH 8·5 sulphydryl groups will be approximately 50% ionised, histidine residues completely unprotonated and a small proportion of the amino groups will also be unprotonated; they are therefore all potential candidates for nucleophilic attack on the modifying reagents. On the other hand, nucleophiles differ in their intrinsic reactivity (see section 2) and a reactive species present in low concentration at a particular pH will compete effectively with a less reactive species present in higher concentration. The ionised sulphydryl side-chain is such a good nucleophile that exposed thiol groups should be blocked in some way if reaction is to be confined to other less reactive nucleophiles.

Table 2.4   pK values of ionisable groups in proteins

Amino acid	Ionising group	pK observed in proteins
N-Terminal	α-NH$_2$	7·5–8·5
C-Terminal	α-CO$_2$H	3·1–3·8
Histidine	Imidazole	6·0–7·4
Lysine	ϵ-NH$_2$	9·4–10·6
Arginine	Guanidino	>12·0
Aspartic acid	β-CO$_2$H	3·0–4·7
Glutamic acid	γ-CO$_2$H	3·0–4·7
Tyrosine	Phenolic-OH	9·6–10·0
Cysteine	—SH	8·5–9·5

Despite what has just been said, suitable choice of pH can sometimes be sufficient to restrict reaction to a particular type of side-chain, e.g. in acid solution all potential nucleophiles are fully protonated and thus inactive – all except methionine whose sulphur atom becomes protonated only at low pH. Under these conditions (e.g. pH 2–3), therefore, reagents such as iodoacetic acid which alkylate cysteine side-chains at neutral pH will react solely with methionine. A further point to remember, however, is that change in pH may itself alter the conformation of a protein and lead to differences in group reactivity.

Hyper-reactivity of side-chains and specific adsorption of reagents leading to specific site modification with group specific reagents have already been discussed (sections 5.4 and 5.5).

6.5 Some group-specific modifications

Other reviews should be consulted for a comprehensive list of group-specific reagents and details of their use (see Means and Feeney 1971; Stark 1970, and reviews cited by them). It will be sufficient here to mention a few of those commonly used to measure 'exposed' residues in proteins; some will already have been encountered in this and previous sections.

Amino groups are generally acylated at pH about 8 by nucleophilic attack on reagents such as the anhydrides used in lysine blocking (see section 3.4.3). Alternatively they can function as nucleophiles in displacement reactions, as they do when amino groups are modified with fluoro-2,4-dinitrobenzene or dansyl chloride in N-terminal determinations (see section 3.2.1), or with 2,4,6-trinitrobenzenesulphonic acid.

The high pH required to generate the nucleophilic unprotonated species of the guanidino side-chain of arginine (pK_a 12-13) leads to unfolding of proteins. This is the case in reaction with 1,3- or 1,2-diketones (Fig. 2.29). Modification of arginine is, however, possible under mild conditions using the trimeric self-condensation product of 2,3-butanedione (diacetyl) (XVa), but since the reaction is irreversible it has no attraction as a device for arginine blocking in sequence studies (see section 3.4; cf. Table 2.2). Another reagent has, however, recently become a candidate for this; phenylglyoxal (XVb) reacts with arginine residues (and α-amino groups, but not ϵ-amino groups, which have a high pK_a) under mild conditions close to neutral pH. The product is unstable and breaks down slowly under mild conditions to regenerate arginine

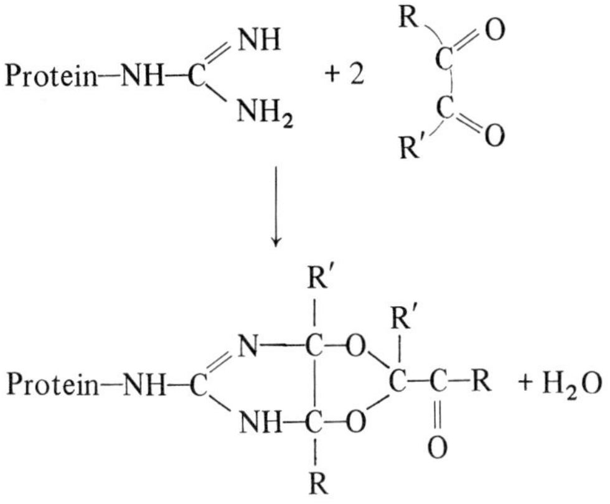

Fig. 2.29 Reaction of arginine side-chains with 1,2-diketones.

residues. It is thus potentially useful, not only in sequencing work but also in studies of the role of arginine residues in biological activity.

It is usually difficult to effect specific modification of histidine residues other than by affinity-labelling. Photo-oxidation is often exploited but methionine, tyrosine, tryptophan, cystine and cysteine can also react. Diethyl pyrocarbonate $((C_2H_5CO)_2O)$ at slightly acid pH has found some favour (cf. Fig. 2.21(a)) and should not react under these conditions with lysine residues; diazo-1H-tetrazole (XVI), another reagent used to modify histidine residues, reacts also with tyrosine to give a coloured product. Disubstitution of both histidine and tyrosine occurs.

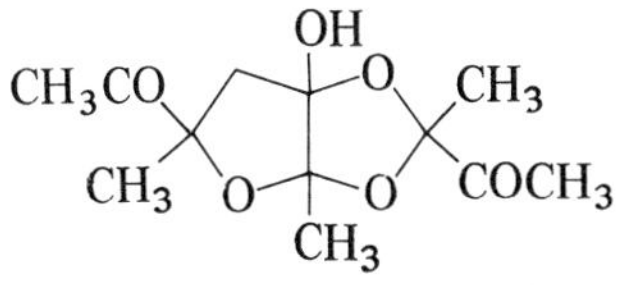

(XVa) 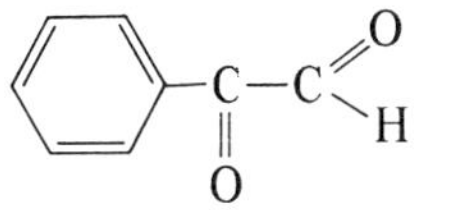   (XVb) 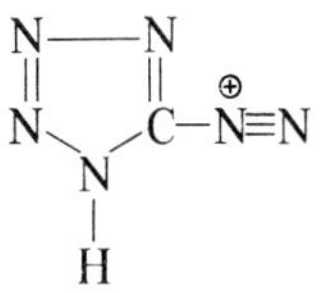   (XVI)

The high reactivity of exposed thiol groups has been mentioned many times already. Alkylation with haloacetates by nucleophilic displacement, or with N-ethylmaleimide by nucleophilic addition, or with DTNB by nucleophilic displacement have already been discussed.

Aspartic and glutamic acids can be coupled to nucleophiles (e.g. amines) using water soluble carbodiimides; tyrosine can be nitrated or iodinated (cf. section 6.3); and tryptophan can be alkylated with 2-hydroxy-5-nitrobenzyl bromide (section 3.1).

6.6 The relative reactivity of 'exposed' functional groups

6.6.1 Competitive labelling of amino groups

It is often difficult to assign degrees of reactivity to particular groups within a class that react readily. If one exposed group is considerably more reactive than other exposed ones, and the reaction of the protein can be monitored spectroscopically, then it may be possible to distinguish the reaction of one group kinetically and subsequently to identify it by peptide mapping. However, where differences in reactivity are less extreme it is tricky to decide which residue amongst a particular group is most reactive. The method of competitive labelling of amino groups overrides this difficulty. The procedure is as follows:

A large excess of protein is treated briefly with radioactive (^{14}C- or ^{3}H-) acetic anhydride in the presence of a large amount of competing nucleophile (e.g. phenylalanine); it is then allowed to react to completion with unlabelled acetic anhydride. The acetyl phenylalanine formed is extracted and its specific activity is determined. The specific activities of lysine-containing peptides in an enzymic digest are then measured and, if the amino acid sequence of the protein is known, relative reactivities can be assigned to individual amino groups. By measuring the rates of reaction as a function of pH, pK values for the amino groups may also be deduced. The ϵ-amino groups of all the lysine residues in elastase that become acetylated are found to have pK values within the range expected for primary amines. The N-terminal valine residue however has a pK of 9·7, consistent with crystallographic evidence which shows that this residue forms an internal ion-pair with aspartic acid-194 and is buried (see Hartley 1970). It is possible that measurements of the reactivities of amino groups by competitive labelling will reveal also small local changes in the conformation of proteins with changes in pH.

6.6.2 Use of double-labelling

A related approach has been used to determine the relative reactivity of tyrosine residues in proteins.

The extent of labelling of a particular tyrosine residue with radioactive iodine does not give an accurate estimate of its reactivity since a hydrophobic environment favours the introduction of a second iodine atom faster than the first. Thus the most extensively substituted tyrosines are those that are most buried, illustrating the point made earlier (section 6.3).

The double-label technique of Pressman and his co-workers does not entirely obviate this difficulty, but it does enable different tyrosine residues to be compared semi-quantitatively. A sample of the protein iodinated to a low level with ^{125}I-iodine and another sample labelled to a higher level with ^{131}I-iodine are mixed, digested enzymically, and the radioactive peptides are analysed both for the ratio ^{125}I:^{131}I and for the amount of di-iodotyrosine. A highly reactive residue will have either a high ratio of ^{125}I:^{131}I in monoiodotyrosine, or a large amount of ^{125}I-label — whether in mono- or di-iodotyrosine. This, of course, says nothing about the degree of *exposure*. However, the amount of di-iodotyrosine obtained from a particular residue is an indication of the hydrophobicity of its environment.

Like the method of competitive labelling of amino groups, double-labelling with iodine enables the most reactive group to be identified, and different tyrosines to be compared; it could presumably also act as an index of local conformational changes under different conditions. When applied by Pressman and his colleagues (Seon *et al*. 1971) to the tyrosine residues of seven Bence–Jones proteins (immunoglobulin light chains) it demonstrated similar reactivities and microenvironments for corresponding tyrosine residues in each, implying that the proteins have similar three-dimensional structures.

6.7 Mapping the topography of a protein with cross-linking agents

If two side-chains in a protein (preferably one of the defined amino acid sequence) can be linked intramolecularly with a small bifunctional reagent, the span length of the reagent defines the maximum distance apart for the two residues in the three-dimensional structure. In the absence of X-ray studies such chemical information may be all the structural evidence that there is. One should nonetheless be cautious about categorical assignment of intramolecular distances on this sort of evidence, because of the danger of spatial distortion of the protein through the constraints imposed by the cross-linking agent. The protein groups usually cross-linked are the nucleophilic cysteine, lysine or histidine side-chains. Bifunctional reagents in chemical modification of proteins have been reviewed by Wold (1967).

A bifunctional reagent first reacts monofunctionally either relatively with a nucleophile which is relatively reactive (cf. section 5.4) or near a site which for some reason absorbs the reagent (cf. section 5.5) and then 'searches' for another nucleophile within striking distance. If the reagent does not react rapidly with a second group in the protein and is instead hydrolysed by water then reaction with the protein is monofunctional. This must, of course, be taken into account when the stoicheiometry of reaction of the (e.g. radioactive) reagent with the protein is being calculated, and in order to obtain an accurate estimation of the number of actual *cross-links*, both the number of moles of reagent incorporated and the number of potentially reactive groups remaining after reaction with the cross-linking reagent must be measured (e.g. the number of remaining amino groups could be estimated spectrophotometrically using trinitrobenzenesulphonic acid).

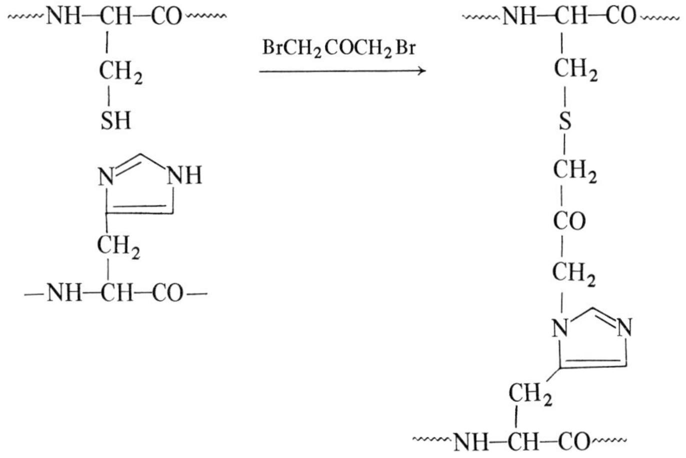

Fig. 2.30   Cross-linking of cysteine-25 and histidine-158 in papain with dibromoacetone.

The sort of information to be gained from the cross-linking approach is nicely illustrated by experiments on the plant proteases papain, ficin and stem bromelain. Papain contains a single cysteine residue (which is catalytically active) and two histidine residues, one of which is implicated by kinetic studies as being involved in catalysis. Dibromoacetone will cross-link cysteine-25 to histidine-158 (Fig. 2.30) suggesting that these residues are within 5 Å of each other in the three-dimensional structure. This is confirmed by the X-ray model for papain. Dibromoacetone will also inactivate ficin and stem bromelain which have a catalytically essential thiol group. Treatment with radioactive reagent and isolation of the cross-linked peptide shows that in each case this cysteine residue is cross-linked to a histidine residue (Husain and Lowe 1970). There are two consequences of the experiment. First, it may be concluded that the three-dimensional structures of the three enzymes are probably very similar in the region of the active centre and, secondly, the amino acid sequences of the cross-linked peptides from the three different sources may be compared. This shows that the sequences

(a)	Papain	-Asn-Gln-Gly-Ser-Cys-Gly-Ser-Cys-Trp-
	Ficin	-Gln-Gln-Gly-Gln-Cys-Gly-Ser-Cys-Trp-
	Stem bromelain	-Asn-Gln-Asp-Pro-Cys-Gly-Ala-Cys-Trp-

(b)	Papain	-His-Ala-Val-Ala-Ala-
	Ficin	-His-Ala-Val-Ala-Leu-
	Stem bromelain	-His-Ala-Val-Thr-Ala-

Fig. 2.31   Amino acid sequences near the catalytic cysteine (a), and histidine (b), residues in three plant proteases.

around the histidine and cysteine residues are homologous (Fig. 2.31) and that the three proteases thus constitute a 'family' of enzymes, analogous with the serine proteases.

An elegant experiment carried out on staphylococcal nuclease by Anfinsen and his co-workers (Cuatrecasas *et al.* 1969) is worth considering in some detail, since it illustrates several points of interest (Fig. 2.32). Nitration with tetranitromethane in the presence of a competitive inhibitor (deoxythymidine $3',5'$-diphosphate) resulted in specific modification of tyrosine-115 and, in the absence of the inhibitor, of tyrosine-85. The nitro groups of the 3-nitrotyrosine residues formed were reduced with dithionite to give aromatic 3-amino groups with pK values characteristically much lower (below 4·7) than those of ordinary protein amino groups. The reagent 1,5-difluoro-2,4-dinitrobenzene (XVII) reacted selectively and mono-functionally at pH 5 with aminotyrosine-115 (or −85), and bifunctional reaction was then achieved by raising the pH to 9·4 to allow the other end of the reagent to attack a group in the protein. Cross-links were formed between tyrosine-115 and lysine-136 (Fig. 2.32), and tyrosine-85 and lysine-116. Another cross-linking reagent of different span, p,p'-difluoro-m,m'-dinitrophenylsulphone (XVIII), linked tyrosine-115 with lysine-53 and tyrosine-85 with tyrosine-115. These results show that tyrosines-85 and 115 are nearby in space, in accord with the X-ray crystallographic model and with earlier physicochemical results suggesting some co-operative action between these residues in catalysis.

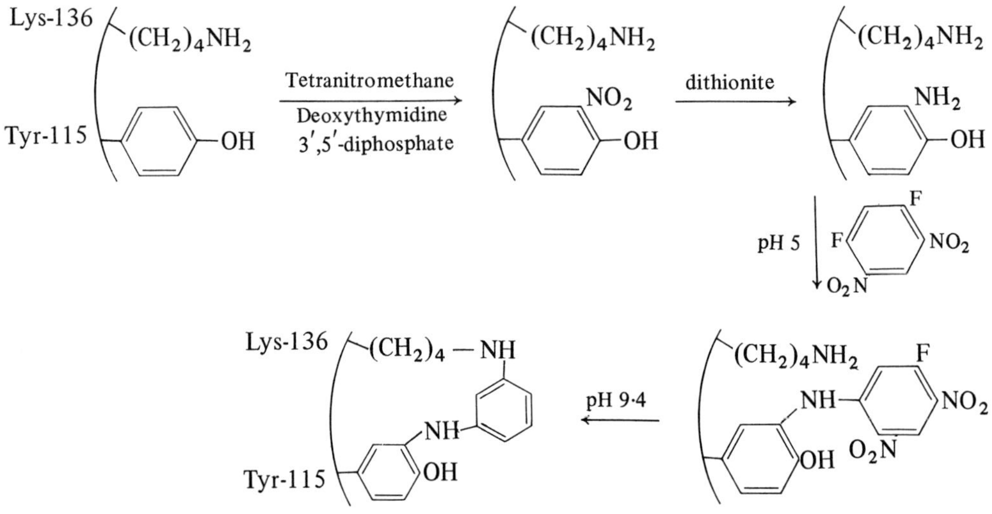

Fig. 2.32   Cross-linking of 3-aminotyrosine-115 and lysine-136 in staphylococcal nuclease with 1,5-difluoro-2,4-dinitrobenzene.

F NO₂ ... structures

$$\text{(XVII)} \qquad \text{(XVIII)}$$

(XVII) — 1,5-difluoro-2,4-dinitrobenzene (F, NO₂, F, NO₂ substituents on benzene ring)

(XVIII) — O₂N, F / S(=O)(=O) / O, NO₂, F bis-sulphone structure

The cross-linked proteins mentioned so far are monomeric, for which 'intra-molecular' is synonymous with intra-chain. In oligomeric proteins, however, intra-molecular cross-links can be formed between subunits, so that bifunctional reagents are also potentially capable of revealing something about the functional and spatial relationship between groups on different subunits. For instance, a bifunctional alkylating agent has been used to cross-link cysteine-93 of the β-chain to histidine-46 of the α-chain in haemoglobin. The haemoglobin modified in this way does not bind oxygen co-operatively showing that functional subunit interactions have been perturbed.

Glyceraldehyde 3-phosphate dehydrogenase is a tetrameric enzyme with one essential thiol group (the side-chain of cysteine-149) per subunit. This can be acetylated with acetyl phosphate below pH 7 to give a stable acyl enzyme; if the pH is raised to 8·5 the acetyl group is transferred to lysine-183. While it is unclear whether this lysine has any direct role in normal enzymic catalysis, it must lie fairly close to the active-site thiol group in the three-dimensional structure. It is of some interest to know whether the active site is formed essentially within one subunit (i.e. whether the cysteine and lysine residues cross-linked are in the same chain) or between subunits, from groups contributed by at least two polypeptide chains. The proximity of the cysteine and lysine residues in space has recently been directly confirmed by treatment of the enzyme with 1,5-difluoro-2,4-dinitrobenzene (XVII) (Shaltiel and Tauber-Finkelstein 1971) leading to isolation of a peptide containing lysine-183 cross-linked by the reagent through its ε-amino group to the thiol group of a peptide containing cysteine-149. Unfortunately the question of the number of subunits contributing to the active site is still not resolved because the molecular weight of the cross-linked enzyme under dissociating conditions was not measured.

It is likely that intra-subunit cross-linking will also be useful in studies of the way in which subunits interact to form functional enzymes. For instance, in tryptophan synthetase a cross-link introduced between cysteine-80 and cysteine-117 of the isolated α-subunit by bis(maleimidomethyl) ether (XIX) prevents association of the α-subunit with the β-subunit to

$$\text{(XIX)}$$

(XIX) — N—CH₂—O—CH₂—N bis(maleimidomethyl) ether

give the native functional complex, $\alpha_2\beta_2$. If the changes caused by cross-linking could be established this should shed some light on the nature of the subunit interactions.

In structural studies, it is desirable for ease of interpretation at the level of primary structure that at best one, at worst a few, cross-links be generated. For some other purposes, however,

$$\overset{\overset{\displaystyle \oplus}{\displaystyle NH_2}}{\underset{\displaystyle RO-C}{\|}}-(CH_2)_n-\overset{\overset{\displaystyle \oplus}{\displaystyle NH_2}}{\underset{\displaystyle C-OR}{\|}}$$

(XX)

this consideration is irrelevant. Bifunctional imidoesters (XX) which react with lysine residues to form amidines (cf. Fig. 2.11), are now finding wide use as cross-linking agents in studies of the subunit composition of oligomeric enzymes. Imidoesters are quite general group-specific reagents for lysine side-chains (which are exposed at the surface of most proteins) and can be used to generate inter-subunit (i.e. intramolecular) cross-links, provided that the protein is present in very low concentration (0·5–1·0 mg/ml) to minimise intermolecular cross-linking. If a protein with n subunits is dissociated after such treatment it is not converted completely to monomer (mol. wt. M) but other species are also present with molecular weights $2M$, $3M$, . . ., nM (if all the subunits are identical). On an analytical scale the species present after cross-linking are easily resolved by polyacrylamide gel electrophoresis in the presence of the detergent sodium dodecyl sulphate (Davis and Stark 1970) and the number of bands on the gel thus gives the subunit composition directly. The bifunctional imidoester usually used is dimethyl suberimidate (XX, $n = 6$, R = CH_3) but others are available with different span lengths.

It is possible that cross-linking will also shed some light on functional changes in the structure of oligomeric enzymes. Isopropylmalate synthase normally gives the bands expected of a tetrameric structure when cross-linked with dimethyl suberimidate; in the presence of a feedback inhibitor, however, the results of cross-linking indicate that tetramers are converted to dimers (and possibly monomers), suggesting that the feedback inhibitor might perhaps work *in vivo* by causing dissociation (Kohlaw and Boatman 1971).

7 Subunit interactions in multimeric proteins

The types of side-chains critical in association of one subunit with another in multimeric proteins may be investigated chemically using group-specific reagents; the sort of approach adopted is as follows: First it must be shown (most easily by sedimentation velocity ultracentrifugation) that the protein can be dissociated into subunits (e.g. by concentrated urea or guanidine hydrochloride solutions) and then reassociated with full recovery of activity. The dissociated protein is then modified with a particular group-specific reagent, the modification quantitated, and the reassociation of subunits investigated. Loss of ability to associate suggests that the groups modified are involved in the association process. The modification is then reversed and the unblocked protein subunits should reassociate as well as those of the native protein.

In practice, several reagents must be investigated before a structural role can be assigned to a particular type of side-chain since the bulk of the modifying group, or its charge, may be the reason why the subunits will not reassociate (rather than loss of the unmodified side-chain). Further, unless the reagent is absolutely specific for one type of side-chain there is no guarantee that groups modified in side-reactions (and subsequently unblocked) are not the ones responsible for the observed effects.

It is desirable in studies of this sort that the modification should be reversible; even so irreversible side-reactions sometimes occur. For instance, although aldolase, a tetramer, can be reversibly dissociated (e.g. in urea) activity is not recovered if the tetramer is dissociated by citraconylation of the lysine residues, then unblocked and exposed to conditions favourable for reassociation. The number of free thiol groups is found to be less after citraconylation and it is likely that irreversible alkylation which interferes with reassociation and refolding has occurred by addition of thiol groups across the double bond of the reagent (cf. Fig. 2.1(b)). Reaction of thiol groups is not, of course, a problem when citraconylation is used to restrict tryptic attack in sequence analysis (section 3.4.3) since thiol groups will already have been blocked (section 3.1).

The protein coat of tobacco mosaic virus consists of a large number of identical subunits each of molecular weight 17 000. These subunits will assemble spontaneously into virus-like rods in the absence of nucleic acid. When the two lysine residues in the protein subunit are trifluoroacetylated, the subunits will not assemble into rods but the system reverts to normal when the trifluoroacetyl groups are removed, suggesting a possible role for one or both of the lysine residues in self-assembly. When the lysine residues are both amidinated by reaction with imidoesters (e.g. methyl acetimidate, XXI), however, rod formation proceeds normally,

$$CH_3-\overset{\overset{\displaystyle \overset{\oplus}{N}H_2}{\|}}{C}-OCH_3$$

(XXI)

suggesting that the requirement is, in fact, solely for a positive charge. When the reactivity of the lysine residues is studied in the intact virus, only one of the two (lysine-68) will react with imidoesters suggesting that the other (lysine-53) is buried (see Perham and Richards 1968). This is surprising for a polar group and one might speculate that lysine-53 is involved in an internal ion-pair critical for structural assembly. Work by Fraenkel-Conrat and Colloms (1967) has implicated lysine-53 in the assembly process in the presence of RNA.

8 Elucidation of structure–function relationships

Much of what might be said under this heading has already found its way into the discussions of modifications of the active site (section 5). Affinity-labels of various sorts react at or near the active site and the possibility of steric interference, perturbation of the protein structure, and alteration of the local charge must be eliminated before inactivation can be attributed to loss of catalytic groups. Care in interpretation is also required when more than one residue is being attacked (cf. section 5.1). The role of particular groups in self-assembly has already been mentioned (section 7). What remains is to ask whether other groups in proteins – groups not involved in catalysis – can be modified without effect on biological function. Groups involved in binding (e.g. of enzyme to substrate) are of particular importance. There are various ways of tackling this problem for peptides and proteins, and chemical modification is one of them. For small peptides, such as the hormones oxytocin, vasopressin and bradykinin, which are easily synthesised chemically, slight variations in the structure can be introduced during the synthetic process. The binding properties and the biological potency of the variants can then be compared with those of the 'native' structure and roles assigned to the altered amino acid residues. For proteins this is less feasible; only one enzyme, ribonuclease, has been completely synthesised chemically and the task is no small undertaking. If certain residues in the amino acid sequence could be replaced by others, the effect on folding and on biological activity could be tested and this approach, a formidable one, is being taken in a few laboratories. The strategy is to link together a series of peptide fragments (made either by the classical 'manual' method, or by the solid-phase method (reviewed by Merrifield 1969)), since particular amino acids can be replaced relatively easily in small fragments. A semi-synthetic approach also being exploited combines synthetic peptides with others obtained from enzymic digests of the protein being synthesised. Well characterised proteins of known three-dimensional structure are being chosen, and the results are awaited with interest.

One comparatively simple system that has been well studied along these lines is the S-peptide (residues 1–20) and S-protein (residues 21–124) system generated by cleavage of ribonuclease A with subtilisin at the bond between residues 20 and 21. The S-peptide and S-protein are separately inactive but reassociate, non-covalently, to regenerate full ribonuclease activity. The effect of replacing various residues in the S-peptide on its ability to combine with S-protein and regenerate ribonuclease activity has been investigated by Hofmann, and by Scoffone, and their co-workers (see the review by Richards and Wyckoff 1971). The amino acid sequence of the S-peptide is given in Fig. 2.33. Residues 15–20 can be omitted without effect either on binding

<pre>
1   5   10   15   20
Lys-Glu-Thr-Ala-Ala-Ala-Lys-Phe-Glu-Arg-Gln-His-Met-Asp-Ser-Ser-Thr-Ser-Ala-Ala
</pre>

Fig. 2.33   Amino acid sequence of the S-peptide from bovine pancreatic ribonuclease.

or on the activity. If the sulphur of methionine-13 is modified either by oxidation or by alkylation, binding of S-peptide to S-protein is severely impaired, but without much effect on the activity of the resulting complex, suggesting that methionine is important for binding. Phenylalanine-8 is also important in binding and can be replaced without ill effect by tyrosine, but not by glycine, alanine, isoleucine or lysine. Histidine-12 is crucial for activity; photo-oxidation and iodination decrease both binding and activity, while carboxymethylation at N-3 or conversion of the imadazole to a pyrazolyl ring both affect activity without affecting binding.

If chemical modification or photo-oxidation inactivates a protein without affecting the binding of substrate, direct interference with the catalytic groups or their relative orientation is indicated. An interesting example that points up the precision of enzyme action is the conversion of the bacterial protease subtilisin (which shares the charge-relay system of the mammalian proteases, trypsin, chymotrypsin, etc.) into thiosubtilisin by the following sequence of events (Fig. 2.34). Phenylmethanesulphonyl fluoride reacts specifically with the active-site serine of subtilisin (and other serine proteases) as in (i), and then nucleophilic

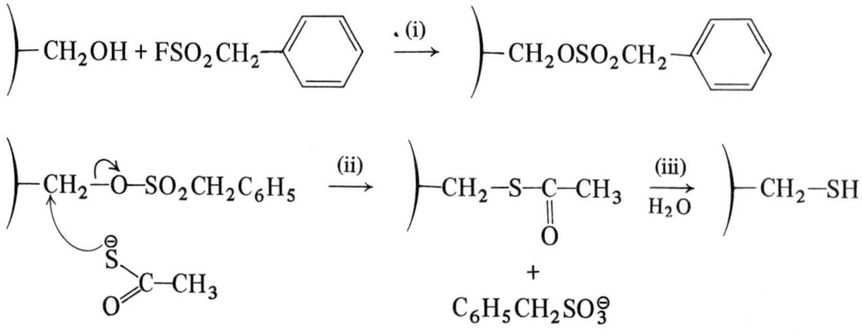

Fig. 2.34   Conversion of subtilisin to thiosubtilisin.

displacement of sulphonate by thioacetate as in (ii) gives a thiol ester which is hydrolysed, (iii) (Neet and Koshland 1966). The thiol enzyme is catalytically inactive towards normal substrates and pseudosubstrates (although it does retain some, probably non-specific, activity towards *p*-nitrophenyl acetate), but binding appears to be unimpaired. Loss of activity is somewhat surprising since an -SH group at the active site might have been expected to function more efficiently in catalysis than an -OH group because of its greater nucleophilicity, and one is led to conclude that the difference of 0·45 Å in the Van der Waals radii of the sulphur and oxygen atoms at the active site is critical. Obviously a precise orientation is required of the atoms that are to react. This type of approach to active-site modification is free from the criticism of steric hindrance that applies to some studies involving active-site-directed inhibitors (cf. section 5.2).

Chemical modification can reveal the importance of surface charge in binding enzymes to macromolecular substrates. The hydrolytic activity of the enzyme elastase against its natural substrate elastin was destroyed by reaction of its three ε-amino groups with maleic anhydride (converting positively charged groups to negatively charged ones, cf. 3.4.3) although the activity of the enzyme in the hydrolysis of small ester substrates was unchanged. The ε-amino groups may thus be involved in binding elastase to elastin. The implication that the surface of elastin has a net negative charge was supported by the demonstration that when the negative charge on a bacterial protease which did not digest elastin was diminished by modification of its carboxy groups with a water-soluble carbodiimide, the protease would now bind to elastin and showed some elastolytic activity (see Gertler 1971).

Specific modification of proteins is generally possible only by affinity labelling or if only one residue of a particular type, or reactivity, is present. The α-amino group is unique in a protein and can thus be modified selectively to investigate its role. It is possible (see, for example, Van Heyningen and Dixon 1967) to remove an exposed α-amino group without effect on the remainder of the protein structure by transamination of the α-amino group with glyoxylate; this gives an α-keto derivative of the protein. The α-ketoacyl residue can sometimes be removed, leaving the protein shorter by one residue by treatment with *o*-phenylene diamine in acid conditions, possibly through imine formation at one amino group of the reagent followed by intra-molecular attack by the other amino group on the first peptide bond.

If unambiguous conclusions are to be drawn from chemical modification experiments about the relation between structure and activity in proteins, then there must be evidence that the intended modification, and not concurrent reactions, is responsible for the observed change in activity; one of the best ways of showing this is to reverse the change (provided, of course, that the native protein is unaffected by the conditions for reversal) and demonstrate full regain of the original characteristics. Possible hazards are typified in a study of myoglobins (Habeeb and Atassi 1970) in which, of a series of cyclic anhydrides tested (cf. section 3.4.3), only citraconic anhydride gave full regain of immunochemical properties after blocking and deblocking, indicating some irreversible side-reactions with the other reagents. Since myoglobin has no thiol groups, a possible explanation is that the *O*-esters of hydroxy side-chains (whose formation undoubtedly accompanies *N*-acylation) are not as easily destroyed under the acid conditions used for deblocking as might be expected.

Chemical modification is not the only way in which proteins with specifically altered side-chains may be obtained; genetic variants afford the same opportunity. Haemoglobins constitute the best example in mammalian systems. A large number of so called 'abnormal' haemoglobins have been characterised which are allelic variants of the normal form, and which contain single amino acid changes corresponding to single base-changes in the genetic code. Since the three-dimensional structure of haemoglobin is known, the position in space of the substitution can be located. If the amino acid replaced is involved in subunit association, in binding the haem group, or in the structural changes occurring on oxygenation and deoxygenation, or if it is an 'internal' residue whose replacement disturbs folding of the polypeptide chain, the substitution is found to be critical and leads to loss or impairment of biological function, and to clinical symptoms. Conversely, variant haemoglobins with substitutions in surface residues (as seen in the X-ray model) can generally function normally. The 'molecular pathology' of human haemoglobin has been reviewed by Perutz and Lehmann (1968). This sort of approach could be a fruitful one for other proteins with genetic variants, serving essentially the same purpose as chemical modification with group-specific reagents in giving a broad indication of the distribution of side-chains in space (preferably with a protein of known primary structure).

Bacteria are relatively easily selected for mutations in well characterised proteins. The genetics of the tryptophan synthetase system of *E. coli* has been beautifully worked out in Yanofsky's laboratory (see Helinski and Yanofsky 1966) and permits some of the residues essential for a functional complex to be distinguished. The α and β subunits separately catalyse the formation of indole from indole-3-glycerol phosphate and the conversion of indole to tryptophan. When the subunits are combined the rates of these partial reactions are 30–100 times greater. Mutants defective in tryptophan synthetase exist in which one or other of the α or β subunits is defective so that the overall reaction is not catalysed, and the isolated altered subunit will also not catalyse the partial reaction but will still stimulate the other subunit, suggesting that the association between subunits has not been affected by the mutation. The effect of mutation at certain positions in the α subunit on its thermostability has also been investigated. Substitution of a particular glycine residue with arginine or valine gave a heat-labile protein whereas replacement by glutamic acid increased thermostability relative to that of the normal form.

9 Chemical modification to produce heavy atom derivatives for X-ray crystallography

In general the solution of protein structures by X-ray crystallography requires at least two isomorphous derivatives of the protein containing a heavy (electron-dense) atom so that the phases of the X-ray reflections can be calculated and the intensities of the spots on the X-ray film thereby translated into atomic coordinates. Each heavy atom should bind (covalently or otherwise) at a single site in the unit cell, with 100 per cent occupancy, and without affecting the crystal structure in any way. Each of these three requirements may be difficult to fulfil; the last is the most crucial. The preparation of isomorphous derivatives has been reviewed by Blake (1968).

The method usually tried first is to soak crystals of the protein in a solution of a heavy metal salt (e.g. uranyl acetate, sodium chloroplatinate) in the hope that there will turn out to be specific binding sites in the protein for the heavy metal ions. Uranyl ions appear to bind to carboxy groups; platinum, palladium, gold and mercury to nitrogen-containing side-chains, particularly histidine; and chloroplatinate to the sulphur of methionine and possibly cystine.

A less empirical approach is the use of enzyme-inhibitor complexes where the inhibitor contains a heavy atom. A drawback here, however, is that conformational changes often occur when enzymes bind inhibitors and these changes are particularly marked in the active site. The enzyme–inhibitor complex will thus not be truly isomorphous with the native enzyme. It may. however, be isomorphous with an enzyme-inhibitor complex containing the inhibitor (or a derivative of it) lacking the heavy atom, and can, of course, be used to solve the structure of the inhibited enzyme. For example, the pipsyl (*p*-iodobenzenesulphonyl) and tosyl (*p*-toluenesulphonyl) derivatives of α-chymotrypsin at serine-195 form an isomorphous pair, in which the heavy atom iodine replaces the methyl group, and were used to solve the three-dimensional structure of inhibited chymotrypsin.

A second systematic approach to the preparation of isomorphous derivatives is to aim for binding to particular functional groups in proteins that are known to react with heavy atoms. The most common are thiol groups, which react readily with mercurials (e.g. *p*-chloromercuri-benzoate, methylmercuric nitrate, etc.). The structure of haemoglobin, for instance, was solved entirely with derivatives obtained by this method.

Binding sites for heavy metals can be introduced into proteins by chemical modification. For example, picolinamidine derivatives of amino groups will chelate metals (Fig. 2.35). Thiol

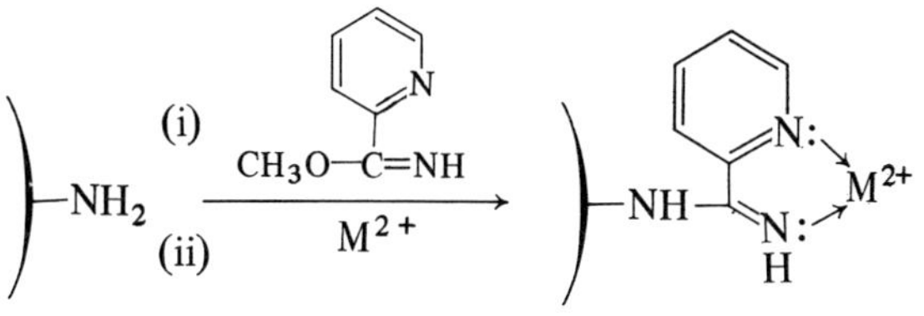

Fig. 2.35   Metal chelation by the picolinamidine group.

binding sites for mercurials may be introduced into proteins that lack cysteine residues by modification of amino groups. The difficulty is, of course, that most proteins contain a large number of lysine residues; these will be in exposed positions in the protein leading to several sites of modification and minimising the chances of obtaining a single-site heavy atom derivative on treatment with mercurials. This approach is thus best suited to, and indeed is invaluable for, proteins containing only one (or only one reactive) lysine residue, as illustrated below.

One way of modifying the ε-amino groups of lysine residues is by reaction of the protein in solution with a thiolactone (e.g. *N*-acetyl homocysteine thiolactone), as shown in Fig. 2.36(a). The lactone acylates the nucleophilic amino group, and the ring-opening generates a thiol group.

A second method is to generate a thiol ester by acylating amino groups with a cyclic anhydride carrying a thiol ester substituent as shown in Fig. 2.36(b); hydrolysis then gives a thiol group. In a third method (Fig. 2.36(c)), amino groups are acylated with the *N*-hydroxy-succinimide ester (an active ester) of a thiazolidine, and the thiazolidine ring is subsequently opened by treatment with mercuric ions. When this was tried on insulin (which has no thiol groups and a single lysine residue) three monosubstitution products were obtained, corresponding to reaction at the two α-amino groups and the single ε-amino group, but unfortunately the crystals of the only one that would crystallise then proceeded to disintegrate when exposed to mercuric ions. In fact, none of these three ingenious ways of introducing thiol groups into proteins has yet led to the successful production of a protein derivative suitable for X-ray work.

Tobacco mosaic virus protein has a single cysteine residue which appears to be buried, and two lysine residues only one of which will react with a variety of imidoesters in the intact virus (cf. section 7). This motivated the synthesis of an imidoester containing a thiol group (Perham and Thomas 1971) (Fig. 2.36(d)). The modification and the introduction of a mercury atom have been successfully carried out, and the position of the heavy atom in the three-dimensional structure has been located by low resolution X-ray work on oriented gels. The results from this and other such derivatives are leading towards a more precise knowledge of the molecular structure of a virus which has already been under study for over thirty years.

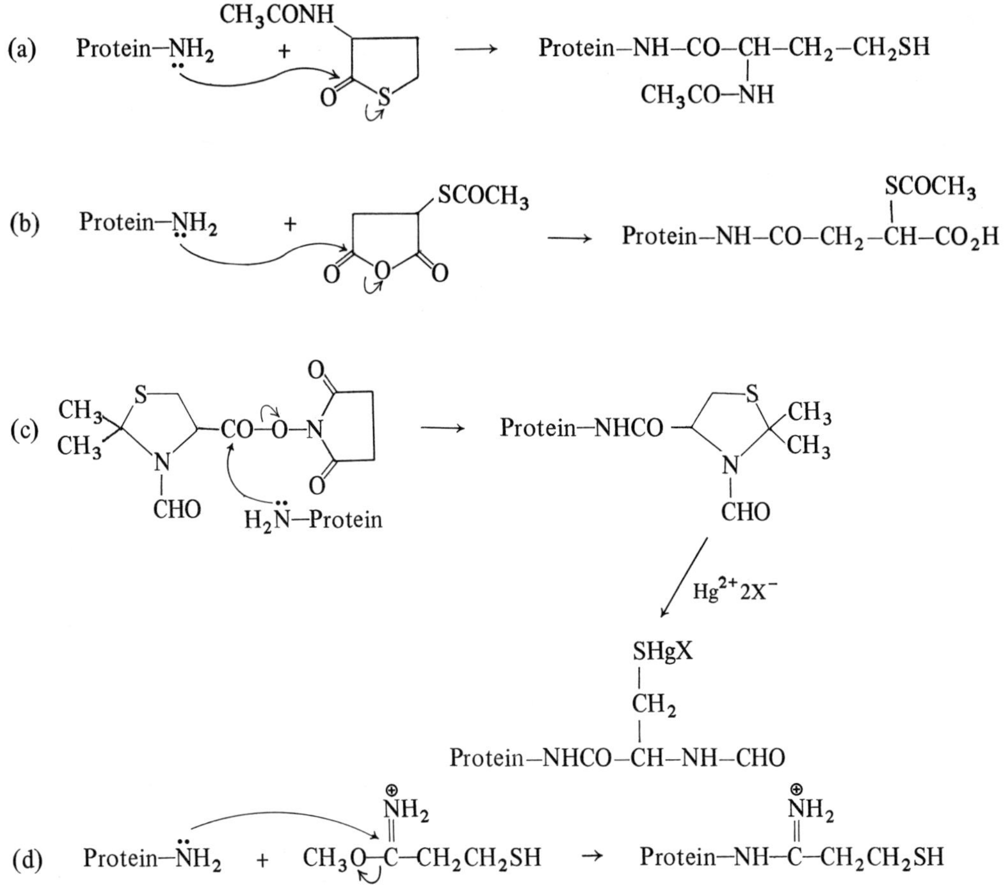

Fig. 2.36   Introduction of thiol groups into proteins by modification of amino groups.

As far as X-ray work goes the fewer the lysine residues in the protein the better the chance of success in generating unique sites for mercurials by introduction of thiol groups. However if the principle of deliberately introducing electron-dense atoms for X-ray work were to be extended to electron microscopy (e.g. of macromolecular assemblies, cell surfaces), a large number of lysine residues for conversion into thiol-containing groups would be advantageous.

10 Attachment of probes of conformation

Optical and n.m.r. methods for investigation of protein structure are dealt with elsewhere in this book (Chapters 3 and 4). Information about the local environment of particular groups in proteins can often be obtained by attaching to the protein small molecules with characteristic spectra which are sensitive to changes in environment. Such probes (often called reporter groups) may be chromophores or fluorophores, or free radicals (spin-labels). Some aspects of the use of attached probes, in particular chromophoric reporter groups, fluorescence probes and spin-labels will be considered in the remainder of this section.

Ideally only one molecule of the probe should be introduced into each protein molecule for easily interpretable results. Such specificity of attachment may be achieved in ways already described, i.e. using affinity labels with the required spectroscopic properties to study the active centre (cf. section 5.2); or turning to advantage the peculiarly high reactivity of certain groups at the surface for attachment of probes there (cf. section 5.4). A moment's reflection will suggest a possible limitation of the method: the probes themselves are often bulky ring-systems capable, one would suspect, of distorting the protein structure and thereby altering the very situation under study. Every effort must therefore be made to check that this has not occurred.

10.1 Chromophoric reporter groups

Difference spectroscopy has been used to study chromophores in hydrophobic environments as well as to monitor changes in protein conformation. Solvent perturbation studies (in organic–aqueous mixtures) carried out on ribonuclease dinitrophenylated specifically on lysine-41 (cf. section 5.4) showed that the Dnp-group is at least partly directed 'inwards' towards a hydrophobic environment. Mercurials are known to react with thiol groups, and a

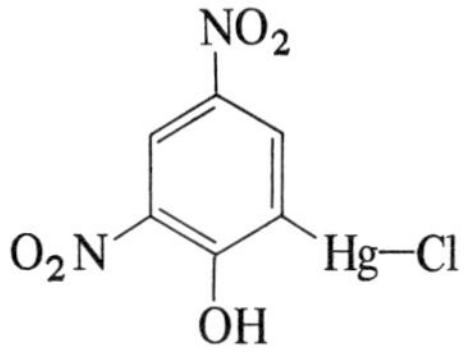

(XXII)

class of mercurials carrying chromophoric groups (e.g. chloromercurinitrophenols such as XXII) should thus provide a handle for studying the microenvironment of accessible thiol groups in proteins (see the paper by Gutfreund and McMurray, in Smellie 1970). Likewise 2-hydroxy-5-nitrobenzylbromide is an environmentally sensitive probe for tryptophan residues (section 3.1) Activation of zymogens is likely to involve conformational changes which may be monitored with a probe, e.g. the arsanilazo chromophore attached to tyrosine-248 in procarboxypeptidase showed large spectral changes on activation to carboxypeptidase.

10.2 Fluorescence probes

The use of covalently bound fluorescence probes, unlike the use of intrinsic fluorescence (i.e. of tryptophan residues or of a coenzyme such as NADH), introduces the danger of distortion by the probe itself as already mentioned.

The dansyl group has been used as a probe for amino groups; it is particularly useful in cases where for one reason or another it binds with a 1:1 stoicheiometry to the protein, e.g. dansyl chloride reacts covalently with a single lysine residue in heavy meromyosin with complete loss of ATPase activity. A newer fluorescence probe, which has various advantages over the dansyl group is NBD-Cl (7-chloro-4-nitrobenz-2-oxa-1,3-diazole) (XXIII) (see Birkett *et al.* in Smellie

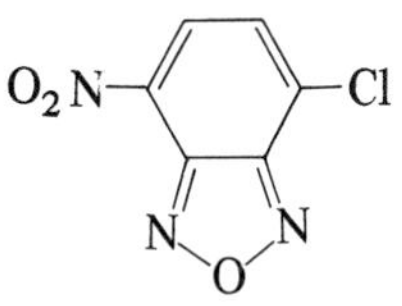

(XXIII)

1970). This can react with both thiol groups and ϵ-amino groups. Specific modification of one thiol group per subunit of rabbit muscle phosphorylase b can be achieved, and the properties of the bound probe can then be examined as a function of binding of the physiological substrates and effectors of the enzyme. (In this instance, the fact that the features of activation by AMP were unchanged was taken as evidence that the native structure was not grossly distorted.) The enzyme will also bind manganese ions, and further information can be obtained by studying the effect of substrates, activators etc., on the quenching of NBD fluorescence by Mn^{2+} ions. Conformational probes thus present a way of obtaining a dynamic picture of a protein in solution (after all, its physiological state) and reveals small changes in structure that the static picture built up by X-ray crystallography does not disclose.

A single dansyl group attached to trypsinogen has provided a means of monitoring conformational changes that occur on activation to give trypsin. Unique reaction was achieved by dansylation at low pH of an aminotyrosine residue generated by reduction of a nitrotyrosine residue, itself produced by selective nitration of a single tyrosine residue (cf. the cross-linking of aminotyrosine residues in staphylococcal nuclease, section 6.7).

10.3 Spin-labels

In general these are compounds of the type (XXIV) or (XXV) where choice of R varies with the protein under study. The nitroxide group is a free radical and the spin-label in free aqueous solution will show a characteristic sharp three-line e.s.r. (electron spin resonance) spectrum arising from interaction of the unpaired electron with the nitrogen nucleus. When this interaction is restricted, as in solutions of high viscosity or when the spin-label is attached to a macromolecule, the spectrum is broadened. Changes in the e.s.r. spectrum may thus be used to monitor conformational changes or asssociation–dissociation phenomena in proteins.

(XXIV)   (XXV)

As with other conformational probes the problem of interpreting the spectroscopic results is preceded by the problem of actually introducing the probe into a specific position, and again spin-labels based on substrate analogues and on group-specific reagents which react with abnormally reactive groups in particular proteins have been designed. Several of these are described in the reviews by Hamilton and McConnell (1968) and McConnell and McFarland (1970).

Spin-labels based on di-isopropylfluorophosphate (DFP) indicate differences in the active sites of chymotrypsin and acetylcholinesterase, the spin-label attached to chymotrypsin being much more strongly immobilised. Similarly a nitroxide spin-label attached at various distances from a dinitrophenyl group has been useful in defining the depth of the binding site of an anti-Dnp antibody; when the distance is large enough the spin-label is no longer immobilised in the cleft and can rotate freely. A maleimide-based spin-label introduced at a single thiol group in actin becomes severely immobilised when actin polymerises and will thus serve as a probe for actin–actin interactions.

In all the examples cited the conclusions drawn from the spin-labelling experiments are clear cut: the label is either immobilised or rotating freely in solution. It should be emphasised that this is where the usefulness of the spin-labelling method lies; in general it is not well suited to answering questions about subtle changes in protein conformation. In some instances, however, this has been done e.g. in an investigation of subunit interactions in haemoglobins. Positions 93 of the β chains are very close to the important contact regions between the α and β subunits, and iodoacetamide-based probes of type XXV (R = $-NHCOCH_2I$) attached specifically by alkylation to the cysteine residues in these positions are found to be sensitive to conformational changes occurring on oxygenation, when the subunits move closer together (see McConnell and McFarland 1970). Perturbation of the structure by the probes themselves must again be guarded against, and this is particularly important in experiments such as those described for haemoglobin where the probe is attached in a very critical region of the molecule; in such cases interpretation of e.s.r. spectra is stretched to its limits. In fact X-ray crystallographic evidence now suggests that introduction of a spin-label at cysteines β-93 does indeed perturb the structure. However the modified haemoglobin still binds oxygen co-operatively showing that the spin-label is a 'tolerable structural perturbation' (McConnell and McFarland 1970).

There are other ways of using spin-labels to investigate protein structure but these do not rely on chemical modification of the protein for attachment of probes and will not be discussed here. A typical approach might be to design a spin-label based on a coenzyme and then use interaction of the spin-label with neighbouring nuclei or a paramagnetic ion bound to the protein as a measure of distances within the molecule.

11 Conclusion

The aim has been to indicate some of the ways in which chemical modification of proteins has contributed to an understanding of their structure and function. Its usefulness in the sequential degradation of proteins and peptides, and in the determination of end-groups, and to restrict or promote enzymic attack is undisputed. Active-site-directed irreversible inhibitors provide an important way of labelling the active centres of enzymes in order that they may be mapped and compared for species differences. X-ray crystallography may often be dependent on the introduction of a heavy atom through chemical modification. The introduction of probes of conformation, such as fluorescence probes and spin-labels, may prove useful although the positions where they would be most informative in proteins are often those where they will cause most damage. In the investigation by chemical modification of structural features critical for biological activity the ideal system will involve a completely reversible reagent free of any side-effects, and this is difficult to achieve. One alternative to specific modification in studies of biological activity is the chemical synthesis of variants in which the amino acid sequence at certain positions is altered, but this is at present far less feasible for proteins than for biologically active peptides. Another approach is to use naturally occurring variants, ideal systems in which to study the effects on structure and activity of replacing certain amino acids. In particular, a set of well characterised defective proteins from bacterial mutants is a system worth exploiting as a way of unravelling structure–function relationships. Isolation of natural mutants and chemical modification are, in fact, complementary approaches to the study of structure and function in proteins. Certain amino acid replacements (e.g. cysteine to alanine) might be hard to come by as natural mutations while others (e.g. isoleucine to valine) are impossible to achieve by chemical modification (although accessible through chemical synthesis) and might be most easily obtained by natural mutation.

12 References

BAKER, B. R. (1967). *Design of Active-Site-Directed Irreversible Enzyme Inhibitors.* Wiley, New York.

BARMAN, T. E. and KOSHLAND, D. E. (1967). 'A colorimetric procedure for the quantitative determination of tryptophan residues in proteins.' *J. Biol. Chem.,* **242**, 5771–76.

BLAKE, C. C. F. (1968). 'The preparation of isomorphous derivatives'. *Adv. Protein Chem.,* **23**, 59–120.

BLOW, D. M., BIRKTOFT, J. J. and HARTLEY, B. S. (1969). 'Role of a buried acid group in the mechanism of action of chymotrypsin.' *Nature Lond.,* **221**, 337–340.

BUTLER, P. J. G., HARRIS, J. I., HARTLEY, B. S. and LEBERMAN, R. (1969). 'The use of maleic anhydride for the reversible blocking of amino groups in polypeptide chains.' *Biochem. J.,* **112**, 679–89.

CHAVIN, S. I. (1971). 'Isolation and study of functional membrane proteins.' *F.E.B.S. Letters,* **14**, 269–82.

CHIBNALL, A. C. and REES, M. W. (1951). 'The amide and free carboxyl groups of insulin.' *Biochem. J.,* **48**, xlvi.

CHRISTEN, Ph. (1970). 'Chemical approaches to intermediates of enzymic catalysis.' *Experientia,* **26**, 337–47.

COHEN, L. A. (1968). 'Group-specific reagents in protein chemistry.' *Ann. Rev. Biochem.* **37**, 695–726.

CONVERSE, C. A. and RICHARDS, F. F. (1969). 'Two stage photosensitive label for antibody combining sites.' *Biochemistry,* **8**, 4431–36.

CRUICKSHANK, W. H., RADHAKRISHNAN, T. M. and KAPLAN, H. (1971). 'A diagonal paper electrophoretic method for the selective isolation of histidyl peptides.' *Canad. J. Biochem.,* **49**, 1225–32.

CUATRECASAS, P., FUCHS, S. and ANFINSEN, C. B. (1969). 'Cross-linking of aminotyrosyl residues in the active site of staphylococcal nuclease.' *J. Biol. Chem.,* **244**, 406–12.

DAVIES, G. E. and STARK, G. R. (1970). 'Use of dimethylsuberimidate, a cross-linking reagent, in studying the subunit structure of oligomeric proteins.' *Proc. Natn. Acad. Sci. U.S.A.,* **66**, 651–56.

DIXON, H. B. F. and PERHAM, R. N. (1968). 'Reversible blocking of amino groups with citraconic anhydride.' *Biochem. J.,* **109**, 312–14.

EYL. A. W. and INAGAMI, T. (1971). 'Identification of essential carboxyl groups in the specific binding site of bovine trypsin by chemical modification.' *J. Biol. Chem.,* **246**, 738–46.

FLEET, G. W. J., PORTER, R. R. and KNOWLES, J. R. (1969). 'Affinity labelling of antibodies with aryl nitrene as reactive group.' *Nature Lond.,* **224**, 511–12.

FRAENKEL-CONRAT, H. and COLLOMS, M. (1967). 'Reactivity of tobacco mosaic virus and its protein toward acetic anhydride.' *Biochemistry,* **6**, 2740–45.

FREEDMAN, R. B. (1971). 'Applications of the chemical reactions of proteins in studies of their structure and function.' *Q. Rev. Biol.* **25**, 431–54.

GERTLER, A. (1971). 'Selective, reversible loss of elastolytic activity of elastase and subtilisin resulting from electrostatic changes due to maleylation.' *European J. Biochem.,* **23**, 36–40.

HABEEB, A. F. S. A. and ATASSI, M. Z. (1970). 'Enzymic and immunochemical properties of lysozyme. Evaluation of several amino group reversible blocking reagents.' *Biochemistry,* **9**, 4939–44.

HAIMOVITCH, J., GIVOL, D. and EISEN, H. N. (1970). 'Affinity labelling of the heavy and light chains of a myeloma protein with anti-2,4-dinitrophenyl activity.' *Proc. Natn. Acad. Sci. U.S.A.,* **67**, 1656–61.

HAMADA, T. and YONEMITSU, O. (1973). 'Selective reduction of peptide-ester groups in aqueous solution. IV. Application to carboxyl-terminal determination of proteins.' *Biochem. Biophys. Res. Commun.,* **50,** 1081-86.

HAMILTON, C. L. and McCONNELL, H. M. (1968). 'Spin labels'. In: *Structural Chemistry and Molecular Biology,* pp. 115-49. (Eds. A. Rich and N. Davidson.) W. H. Freeman, San Francisco.

HARTLEY, B. S. (1964). 'The structure and activity of chymotrypsin.' In *Structure and Activity of Enzymes,* pp. 47-60 (Eds. T. W. Goodwin, J. J. Harris and B. S. Hartley) Academic Press, London and New York.

HARTLEY, B. S. (1970). 'Strategy and tactics in protein chemistry.' *Biochem. J.,* **119,** 805-22.

HELINSKI, D. R. and YANOFSKY, C. (1966). 'Genetic control of protein structure.' In: *The Proteins* (2nd edn.) (Ed. H. Neurath), vol. 4, pp. 1-93. Academic Press, London and New York.

HEXTER, C. S. and WESTHEIMER, F. H. (1971). 'Intermolecular reaction during photolysis of diazoacetyl α-chymotrypsin.' *J. Biol. Chem.,* **246,** 3928-33.

HUSAIN, S. S. and LOWE, G. (1970). 'The amino-acid sequence around the active-site cysteine and histidine residues, and the buried cysteine residue in ficin'; 'The amino-acid sequence around the active-site cysteine and histidine residues of stem bromelain.' *Biochem. J.,* **117,** 333-40; 341-46.

KIRBY, A. J. and LANCASTER, P. W. (1972). 'Structure and efficiency in intramolecular and enzymic catalysis. Catalysis of amide hydrolysis by the carboxy group of substituted maleamic acids.' *J. Chem. Soc. Perkin II,* 1206-14.

KOHLAW, G. and BOATMAN, G. (1971). 'Cross-linking of salmonella isopropylmalate synthase with dimethyl suberimidate: evidence for antagonistic effects of leucine and acetyl-CoA on the quaternary structure.' *Biochem. Biophys. Res. Commun.,* **43,** 741-46.

KOSHLAND, D. E., STRUMEYER, D. H. and RAY, W. J. (1962). 'Amino acids involved in the action of chymotrypsin.' *Brookhaven Symp. Biol.,* **15,** 101-33.

KRONMAN, M. J. and ROBBINS, F. M. 'Buried and exposed groups in proteins.' In *Fine Structure of Proteins and Nucleic Acids.* Dekker, New York.

LEE, B. and RICHARDS, F. M. (1971). 'The interpretation of protein structures: estimation of static accessibility.' *J. Molec. Biol.,* **55,** 379-400.

LEE, R. T. and McELROY, W. O. (1971). 'Isolation and partial characterisation of a peptide derived from the luciferin binding site of firefly luciferase.' *Archs. Biochem. Biophys.,* **146,** 551-56.

McCONNELL, H. M. and McFARLAND, B. G. (1970). 'Physics and chemistry of spin labels.' *Q. Rev. Biophys.,* **3,** 91-136.

MEANS, G. E. and FEENEY, R. E. (1971). *Chemical Modification of Proteins.* Holden-Day, San Francisco, California.

MERRIFIELD, R. B. (1969). 'Solid-phase peptide synthesis.' *Adv. Enzymol.,* **32,** 221-96.

MORRIS, H. R., WILLIAMS, D. H. and AMBLER, R. P. (1971). 'Determination of the sequences of protein-derived peptides and peptide mixtures by mass spectrometry.' *Biochem. J.,* **125,** 189-201.

NAKAGAWA, Y. and BENDER, M. L. (1970). 'Methylation of histidine-57 in α-chymotrypsin by methyl p-nitrobenzenesulphonate. A new approach to enzyme modification.' *Biochemistry,* **9,** 259-67.

NEET, K. E. and KOSHLAND, D. E. (1966). 'The conversion of serine at the active site of subtilisin to cysteine: a "chemical mutation".' *Proc. Natn. Acad. Sci. U.S.A.,* **56,** 1606-11.

PERHAM, R. N. and RICHARDS, F. M. (1968). 'Reactivity and structural role of protein amino groups in tobacco mosaic virus.' *J. Molec. Biol.,* **33,** 795-807.

PERHAM, R. N. and THOMAS, J. O. (1971). 'Reaction of tobacco mosaic virus with a thiol-containing imidoester and a possible application to X-ray diffraction analysis. *J. Molec. Biol.*, **62**, 415–18.

PERUTZ, M. F. and LEHMANN, H. (1968). 'Molecular pathology of human haemoglobin'. *Nature Lond.*, **219**, 902–9.

RALL, S. C. and COLE, R. D. (1970). 'Quantitative cleavage of a protein with *N*-bromosuccinimide'. *J. Am. Chem. Soc.*, **92**, 1800–1.

RICHARDS, F. M. and WYCKOFF, H. W. (1971). 'Bovine pancreatic ribonuclease.' In *The Enzymes* (3rd edn.) (Ed. P. D. Boyer), vol. 4, pp. 647–806. Academic Press, New York and London.

SCHROEDER, W. A. (1968). *The Primary Structure of Proteins.* Harper and Row, New York and London.

SEON, B.-K., ROHOLT, O. A. and PRESSMAN, D. (1971). 'Common micro-environments of tyrosyl residues in several Bence–Jones proteins.' *Biochim. Biophys. Acta,* **229**, 396–406.

SHALTIEL, S. and TAUBER-FINKELSTEIN, M. (1971). 'Introduction of an intramolecular crosslink at the active site of glyceraldehyde 3-phosphate dehydrogenase.' *Biochem. Biophys. Res. Commun.,* **44**, 484–90.

SHAW, E. (1970). 'Selective chemical modification of proteins.' *Physiol. Rev.,* **50**, 244–96.

SHEMYAKIN, M. M., OVCHINNIKOV, YU. A., VINOGRADOVA, E. I., KIRYUSHKIN, A. A., FEIGINA, M. YU., ALDANOVA, N. A., ALAKHOV, YU. B., LIPKIN, V. M., MIROSHNIKOV, A. I., ROSINOV, B. V. and KAZARYAN, S. A. (1970). 'The rational use of mass spectrometry for amino acid sequence determination in peptides and extension of the possibilities of the method.' *F.E.B.S. Letters,* **7**, 8–12.

SINGER, S. J. (1967). 'Covalent labelling of active sites.' *Adv. Protein Chem.,* **22**, 1–54.

SLOBIN, L. I. and SINGER, S. J. (1968). 'The specific cleavage of immunoglobulin polypeptide chains at cysteinyl residues.' *J. Biol. Chem.,* **243**, 1777–86.

SMELLIE, R. M. S. (ed.) (1970). *Chemical Reactivity and Biological Role of Functional Groups in Enzymes.* Biochemical Society Symposia, Number 31. Academic Press, London and New York.

SPANDE, T. F., WITKOP, B., DEGANI, Y. and PATCHORNIK, A. (1970). 'Selective cleavage and modification of peptides and proteins.' *Adv. Protein Chem.,* **24**, 97–260.

STARK, G. R. (1970). 'Recent developments in chemical modification and sequential degradation of proteins.' *Adv. Protein Chem.,* **24**, 261–308.

STRAUSBAUCH, P. H. and FISCHER, E. H. (1970). 'Structure of the binding site of pyridoxal 5′-phosphate to *E. coli* glutamate decarboxylase.' *Biochemistry,* **9**, 233–38.

VAN HEYNINGEN, S. and DIXON, H. B. F. (1967). 'Scission of the N-terminal residue from a protein after transamination.' *Biochem. J.,* **104**, 63P.

WOLD, F. (1967). 'Bifunctional reagents.' *Meth. Enzymol.,* **11**, 617–40.

YOUNG, G. T. (ed). (1969–72). *Amino-acids, Peptides and Proteins,* vols. 1–4. Specialist Periodical Reports of the Chemical Society. The Chemical Society, London.

3
Nuclear Magnetic Resonance Spectroscopy of Proteins

J. C. Metcalfe, N. J. M. Birdsall and A. G. Lee
National Institute for Medical Research, Mill Hill, London

1 Introduction

Although the biochemical literature of the past few years has contained a rapidly increasing number of reports of nuclear magnetic resonance (NMR) experiments on macromolecular systems, it has remained rather difficult for the non-specialist to grasp the kind of information which this technique can provide and to relate it to the information obtained by more familiar spectroscopic methods. In this outline we will illustrate the present range of structural problems which can be studied by NMR, by selecting a few of the key experiments which have been performed with proteins.

It is helpful to start with a general description of the kind of information which can be obtained from NMR spectra. The observation of nuclear resonance depends on the absorption of radiofrequency radiation by nuclei with magnetic moments orientated in a strong magnetic field. The absorption of energy results in the reorientation of magnetic moments with respect to the applied field. An NMR spectrometer therefore has the same formal elements as a UV spectrometer: a source of radiation (in the radiofrequency (rf) range) a system to detect the absorption of energy by the sample (an rf receiver), and a display system. The only additional feature is that the difference in energy levels between ground and excited states of the nuclei is

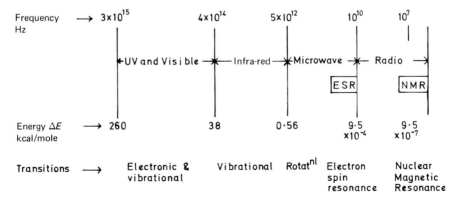

Fig. 3.1 The energies and frequencies of transitions in the electromagnetic spectrum (adapted from Kowalsky and Cohn 1964).

created by applying a high and extremely uniform magnetic field (10–60 kG), using an electromagnet or superconducting solenoid. The magnitude and the frequencies of transitions over the entire spectroscopic range are compared in Fig. 3.1. The most characteristic feature of NMR is the very small value of the energy absorbed in the transitions of nuclei to the excited state, of the order of 10^{-3} cal. Thus the frequency of radiation that the nucleus absorbs is 10^6 times smaller than the frequency of UV or visible spectroscopy. Consequently the appearance of the spectra depend on very slight variations in the electronic configuration of the molecule, so that they reflect with remarkable sensitivity the stereochemical structure of the molecule. We can regard the nucleus as a naturally occurring and non-perturbing probe of chemical structure and environment. For this reason NMR has long been established as the most powerful method for structural analysis of small molecules in organic chemistry. Applications to the study of macromolecular structures have been far fewer because the problems of data collection and interpretation become increasingly difficult as molecular structures become more complex, and radical new solutions are required, different in kind to those applied to small molecules.

The interpretation of NMR spectra in terms of structure and molecular motion is generally based on four parameters which characterise the absorption of the radiofrequency radiation.

1. The intensity of a resonance absorption line. This is directly proportional to the number of nuclei in a given chemical environment.
2. The precise value of the frequency (v) required for the absorption of energy by a nucleus is termed *the chemical shift*.
3. The splitting of resonances into mulitplets of fine structure, characterised by *coupling constants*.
4. The processes by which excited nuclei can return to the lower energy levels, or exchange energy with neighbouring nuclei of the same kind, are termed relaxation mechanisms and are characterised by *relaxation times*.

This outline applies to all forms of NMR spectroscopy, irrespective of the chemical nature of the observed nucleus, and the above parameters are discussed in the following section. The experiments described in this chapter will deal more specifically with proton (^{1}H) and ^{13}C magnetic resonance, since these are the most commonly occurring nuclei in proteins and allow virtually any chemical group within the structure to be examined. Other non-perturbing magnetic nuclei that could be used include ^{15}N, ^{17}O, ^{31}P, ^{2}H and ^{33}S. Artificially introduced nuclei may also be of use (e.g. ^{19}F), but these are essentially probe experiments, distinct from the main objective of exploiting the very high information content of NMR of intrinsic nuclei. For example, staphylococcal nuclease contains more than 855 protons each of which will give rise to a separate resonance or group of resonances, and for each of which *at least* three parameters may be measured (intensity; chemical shift; relaxation times). Similar amounts of data are contained in the ^{13}C spectrum. The immediate problem is that many of the hundreds of ^{1}H resonances will have similar chemical shifts and will overlap in the spectrum: for example, resonances from amino-acid residues of the same kind will overlap if they have a similar environment in the tertiary structure of the protein. The spectra have to be simplified to distinct resonances and assigned to specific residues within the protein sequence. The resonances may also be broadened by internal steric restrictions on the motion of the amino-acid residues imposed by the structure of the protein, and this further complicates the resolution of single resonances. In spite of these technical problems, which have, in principle, been largely overcome, the NMR spectrum permits the simultaneous observation of all chemical groups within the protein. In this sense the amount of data is comparable to X-ray diffraction, and the technique may be regarded as complementary to X-ray diffraction in that it is the only method at present capable of providing very detailed structural information on proteins in solution. It is also complementary in providing detailed motional information not accessible from X-ray diffraction, which gives an essentially static picture of a molecule. It does not, of course, provide coordinates defining the three-dimensional atomic structure of the protein, and only relatively coarse distance information has been obtained in the special cases where the technique can be applied as a 'spectroscopic ruler' to estimate distances in proteins.

2 Basic Concepts

The physical basis of the NMR technique has been described in several texts listed in the bibliography: here we simply try to make the vocabulary of NMR sufficiently familiar to allow adequate description of the experiments discussed in the following sections.

2.1 The resonance condition

An atomic nucleus which has a nuclear magnetic moment, μ, will also possess spin angular momentum. The magnetic moment will therefore precess about the direction of an applied magnetic field, H_0, in the same way that the spin axis of a gyroscope will precess about the earth's gravitational field (Fig. 3.2). For a given type of nucleus, e.g. a proton, the frequency of

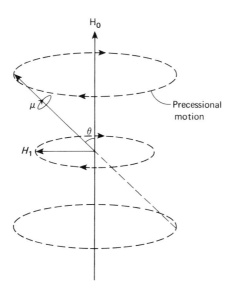

Fig. 3.2 The precession of the magnetic moment μ of a nucleus about the direction of the applied field, H_0. For nuclei with $I = \frac{1}{2}$ the angle θ is 54·9° and is independent of H_0. The energy of interaction of the magnetic moment in the direction of the fields is $-\mu H_0$ and in the opposite direction is $_+ \mu H_0$.

this precession depends only on the magnitude of the applied magnetic field. Because of the quantised nature of the spin angular moment of any magnetic nucleus (defined by the spin number $I = \frac{1}{2}, 1, 1\frac{1}{2}, \ldots$) the nuclear magnetic dipole can be considered to precess *at certain fixed orientations* about the applied magnetic field, and for protons and ^{13}C nuclei ($I = \frac{1}{2}$) there are only two such possible orientations, defined by the angle θ (Fig. 3.2). Each orientation corresponds to an energy of interaction between the magnetic moment of the dipole, μ, and the applied magnetic field given by $-\mu H_0$ (aligned with the field) and $_+\mu H_0$ (against the field). The energy levels of the two orientations for the proton and ^{13}C nuclei dipoles are therefore separated by $2\mu H_0$, which is the energy required to cause a transition between the two energy levels. The resonance frequency ν is then given by

$$h\nu = 2\mu H_0$$

where h is Planck's constant. The value of ν depends only on the strength of the magnetic moment of the nucleus, which is fixed, and the magnetic field which it experiences.

We can illustrate how we cause a transition of a nucleus between the two energy levels in the NMR experiment by considering the effect of a second field, H_1, very much weaker than H_0 rotating in a plane perpendicular to H_0 (Fig. 3.2). The H_1 field will only exert a significant torque on the nuclear dipole and reorientate it when the rate and sense of rotation of H_1 is exactly equal to the rate of precession of the nuclear dipole. At this resonance condition energy equivalent to $2\mu H_0$ is absorbed from the rotating field H_1.

The H_1 field induces transitions between the two orientations of the nuclear dipole with equal probability in *either direction*. Since the energy level separation $2\mu H_0$ is very small at equilibrium, there is only a slight excess of nuclei in the lower energy level (a few parts per million for protons at present H_0 field strengths and at room temperature). At low intensities, the H_1 field induces more upward transitions than downwards transitions, because of the excess of nuclei in the lower energy level, and there is a small net absorption of energy from H_1 which is observed as the resonance signal. The effect of H_1 is to tend to equalise the two populations of nuclei, and this is counteracted by the process of relaxation which returns excited nuclei to the lower energy level and restores the difference in populations of nuclei. However, as the intensity of H_1 is increased, the two energy levels eventually become equally populated. No net absorption of energy can then occur and the system is said to be *saturated*. This sets a limit to the intensity of H_1 which can be used to observe resonance, and therefore on the intensity of the observed signal which can be obtained from a given sample concentration.

To summarise: resonance occurs when a radiofrequency field H_1 is generated so that its magnetic vector rotates perpendicular to H_0 at the precise precession frequency of the nucleus. It is the absorption of energy from H_1 that is observed as the resonance signal. The low sensitivity of NMR is inherent in the small energy differences in the transitions.

2.2 Chemical shifts

All proton nuclei in a molecule do not undergo resonance at precisely the same frequency because the position of the resonance depends on the effective magnetic field H_{eff} experienced by the nucleus. This is generally less than the applied field, because the nuclei are shielded to differing extents by the magnetic fields of the surrounding electrons. That is, $H_{eff} = H_0 (1 - \sigma)$, where σ is the shielding constant. Thus the precise resonance frequency is a measure of the shielding of a particular nucleus, which reflects its precise electronic environment and therefore its chemical bonding. Nuclei in identical chemical environments experience the same effective magnetic fields and have the same resonance frequencies.

Resonances are measured relative to an internal standard dissolved in the same solution as the sample, or relative to an external standard, usually in a coaxial capillary. To enable separations between resonances to be compared at different field strengths, they are expressed as the dimensionless number, δ, in parts per million of the H_1 radiofrequency:

$$\delta = \frac{\Delta\nu}{\nu_{H_1}} \times 10^6$$

where $\Delta\nu$ is the chemical shift difference in Hz (c/s) between the standard and the observed resonance and ν_{H_1} is the H_1 radiofrequency.

2.3 Coupling constants

In molecules with magnetic nuclei in different chemical environments, the absorption lines are often split into multiplets. This splitting is caused by a coupling between the magnetic nuclei which takes place via the bonding electrons, and occurs because one nucleus senses the magnetic orientation of other nuclei. The coupling may be between like or unlike nuclei (homo- or heteronuclear coupling). For example, although the ^{12}C nucleus has no magnetic moment, the stable ^{13}C isotope has a spin of $I = \frac{1}{2}$ and will split the ^1H resonance in a ^{13}CH group into two peaks because of the two possible spin states for the ^{13}C nucleus. Since the interaction is mutual, the H resonance will also be split into two peaks with the same separation. The peak separation measured in Hz is termed the coupling constant J. The natural abundance of ^{13}C is only about 1 per cent so that generally the splitting of proton resonances is only just detectable in the proton spectra of organic molecules as a pair of satellite resonances corresponding to 1 per cent of the intensity of the parent proton resonance. If the ^{13}C content is enriched the satellite intensities increase proportionately at the expense of the parent resonance. Proton spectra show extensive ^1H–^1H homonuclear coupling because the abundance of the ^1H nucleus is over 99 per cent. Provided that the chemical shift separation of two sets of coupled proton nuclei is large compared with their coupling constants, the number of multiplets and their intensity ratios are simply defined by the number of coupled nuclei. For example, if n proton nuclei are coupled to a group of m proton nuclei, then their resonances are split into multiplets of $(m + 1)$ and $(n + 1)$ lines respectively: this is termed *first-order* coupling. Where the shift difference is not large compared with the coupling constants, complex *second-order* multiplets are obtained.

These complex splitting patterns can make the observation and assignment of spectra rather difficult. However, provided the resonances of two magnetic nuclei (A and X) are not too close, the coupling between them can be removed by applying a radiofrequency field to excite the X nucleus between its two spin states, while observing the A nucleus. Because of the rapid exchange induced between the two orientations of X, the A nucleus will only sense the average orientation, and the resonance due to the A nucleus will collapse to a single peak. This double resonance experiment is termed *decoupling* and can be used to map out systematically which chemical groups are interacting magnetically with each other in a molecule of unknown structure.

2.4 Relaxation processes

We consider now the magnetisation of a set of chemically identical nuclei in a population of identical molecules. At equilibrium these nuclei will have a net magnetisation M_0 oriented in the direction of H_0, due to the excess of nuclei in the lower energy level. If a radiofrequency field H_1 is applied which is sufficiently intense to equalise the populations of nuclei in the two energy levels, the net magnetisation in the direction of the H_0 field (by convention the z axis) is reduced to zero ($M_z = 0$). If H_1 is now switched off the nuclei will decay back to their original equilibrium distribution and M_z will increase from zero to M_0 by a first-order rate process, characterised by a half-time T_1. Thus if M_z at some time is less then M_0, it will approach M_0 exponentially with time such that

$$\frac{dM_z}{dt} = \frac{-(M_z - M_0)}{T_1}$$

T_1 is called the *spin lattice* relaxation time since it involves the dissipation of nuclear spin energy to other nuclei and electrons in the sample (called collectively the lattice). T_1 is also called the *longitudinal* relaxation time because it refers to magnetisation in the direction of H_0.

In the plane perpendicular to H_0 (the xy plane), the individual nuclear moments precess at random about H_0 at equilibrium in the absence of H_1, and there is not net magnetisation in the xy plane. When H_1 is applied, it causes the nuclei to precess in phase with each other in the xy plane, and can be visualised as a 'bunching' of the precessing nuclear moments. This creates a net magnetisation rotating in the xy plane (M_x and M_y) which decays to zero when H_1 is removed, and a second relaxation time T_2 is defined by the first-order rate process of this decay:

$$\frac{dM_x}{dt} = \frac{-M_x}{T_2} \qquad \frac{dM_y}{dt} = \frac{-M_y}{T_2}$$

T_2 is called the spin–spin relaxation time since it is concerned with equilibrium within the spin system and does not affect the magnitude of M_z. T_2 is also termed the *transverse* relaxation time. For liquids there is an explicit relationship between T_2 and the linewidth at half height of the resonances ($\Delta\nu_{\frac{1}{2}}$) given by

$$\Delta\nu_{\frac{1}{2}} = \frac{1}{\pi T_2}$$

Both relaxation processes are caused by time-dependent magnetic fields at the nuclei, and these fields arise from the random thermal motions of neighbouring nuclei which are always present, but the dependence of T_1 and T_2 processes on molecular motion are distinct. Spin–lattice relaxation is caused by fluctuating magnetic fields which occur at a frequency corresponding to the precession frequency, whereas spin–spin relaxation is strongly influenced by motions of low frequency which have little effect on spin–lattice relaxation. Slow motion of the nuclei therefore leads to short T_2 values and thus to broad lines. This is a considerable limitation in the study of proteins by NMR, since proteins tumble in solution at very slow rates (10^8–10^6 s^{-1}).

2.5 Fourier transform NMR and the measurement of relaxation times

Conventional NMR spectra are obtained either by sweeping through the range of resonance frequencies (for protons about 10 ppm of the H_1 frequency and about 200 ppm for ^{13}C), or, more conveniently, by making a linear sweep of the magnetic field, H_0 at a constant H_1 frequency. A Fourier transform technique of generating the spectrum has been introduced which has effectively increased the sensitivity by 1–2 orders of magnitude for complex spectra. In this technique a very short intense rf pulse (10–100 μs) is applied to the sample, which has

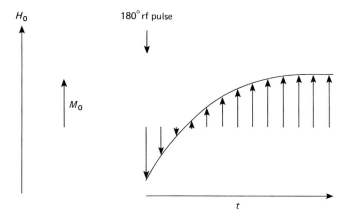

Fig. 3.3 Measurement of the spin–lattice relaxation time (T_1) by a pulse technique. The equilibrium nuclear magnetisation (M_0) is reorientated through 180° by a brief pulse and the decay back to the original orientation is measured as a function of time (t).

the effect of simultaneously exciting nuclei *over the whole resonance frequency range*. The resulting interferogram can then be Fourier transformed to give the characteristic absorption signals of a conventional NMR spectrum. The main advantage of this technique is that the short rf pulse which generates the whole spectrum can be repeated immediately after the relaxation processes have returned the spin system to its equilibrium state. This is often 10-100 times faster than the time taken to scan the spectrum by sweeping the field in the conventional way. The pulse can be repeated many times until the signal-to-noise ratio of the spectrum is adequate. This technique has made nuclei of low intrinsic sensitivity (e.g. ^{13}C) readily accessible to NMR spectroscopy. Another advantage is that the pulse technique is ideally suited to the measurement of both T_1 and T_2 relaxation times. These can be measured for all the resolved resonances in the spectrum.

The measurement of T_1 by a pulse technique is illustrated in Fig. 3.3. The nuclei are excited with a pulse of the precise duration and intensity necessary to reorientate the equilibrium nuclear magnetisation M_0 of the nuclei through 180°. This is an unstable orientation with respect to the applied field and the magnetisation will decay back to equilibrium with the

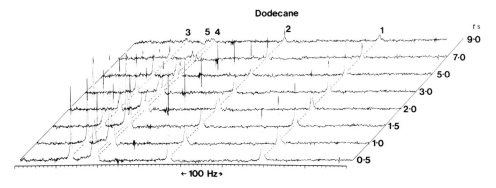

Fig. 3.4 ^{13}C NMR spectra of *n*-dodecane. The resonances numbered 1 to 5 correspond to carbons 1 + 12, 2 + 11, 3 + 10, 4 + 9 and 5 + 6 + 7 + 8, respectively. The spectra at increasing t values after the 180° pulse show decreasing intensities from which the T_1 values are calculated. The spectra are observed by using a 90° pulse which turns the magnetisation M_z into the xy plane. (Taken from Lee *et al.* 1973.)

characteristic half-time T_1. This decay process can be monitored by observing the residual spectra at increasing times, t, after the initial 180° pulse. The monitoring is performed by applying a second pulse of exactly 90°, which turns the residual magnetisation through 90° into the xy plane which contains the receiver coils. The spectra obtained from this type of experiment are illustrated in Fig. 3.4 for the ^{13}C relaxation of dodecane. Four of the non-equivalent pairs of carbon atoms in the chain are resolved, and the spectra at increasing t values show the time course of the decay after the initial 180° pulse ($t = 0$). By plotting the amplitude of the resonances as a function of t, the T_1 relaxation times can be calculated.

Using Fourier transform spectroscopy, at least 1·0 ml 10^{-4}–10^{-3}M equivalent protons, or 10^{-2}–10^{-3}M equivalent ^{13}C nuclei (enriched to 90 per cent from the natural abundance level of about 1 per cent) are required to allow good relaxation measurements in a time of a few hours, although of course higher concentrations greatly shorten the time required. These are relatively large amounts of material and sensitivity remains a considerable limitation on biological NMR, in spite of the introduction of the Fourier transform technique.

2.6 Chemical exchange

When a proton can exist in two chemically distinct environments, for example a carboxyl proton of an amino acid which is exchanging with H_2O protons, the spectrum observed depends on the rate of exchange of the proton between the two environments. If the rate of exchange is fast compared with the separation (in Hz) between the resonances, then a single symmetrical resonance is observed at a chemical shift which is the average of the shifts in the two environments, weighted for their relative lifetimes. If the exchange is slow compared with the chemical shift separation, separate resonances for the proton in each environment are observed with shifts corresponding to the two states, and intensities determined by their relative lifetimes. At intermediate rates of exchange, the resonance lineshapes may be complicated partially averaged spectra, which are substantially broader than the separate resonances. In this exchange broadened region, the exchange rate can often be calculated, whereas in the fast and slow exchange conditions, only lower and upper limits to the exchange rate can be set.

Similar conditions of fast, intermediate, and slow exchange also apply to the averaging of relaxation times of protons exchanging between more than one chemically distinct environment. For the fast exchange condition to apply, the proton must exchange between two environments at a rate fast compared with the relaxation rate in either environment.

It should be noted that in general the limits for the three exchange conditions for chemical shift and relaxation measurements will not coincide, since different NMR parameters are involved.

3 Amino Acids and Peptides

3.1 Amino-acid spectra

The effects of pH on the ^1H spectra of amino acids have been tabulated in detail. The spectra of alanine and serine shown in Fig. 3.5. provide good examples of spin–spin interactions. The chemical shifts of the ^1H spectra depend on the state of ionisation of the amino acid, and the effect of changes in ionisation is transmitted through the entire carbon chain in the aliphatic amino acids and the aliphatic portions of aromatic amino acids. The magnitude of the transmitted effect decreases with increasing distance from the titrating group. The effect of titrating groups in the side chain are also transmitted, as shown for the titration curves of histidine protons in Fig. 3.6. The histidine C_2H and C_4H titration curves are particularly important in the subsequent discussion of protein spectra.

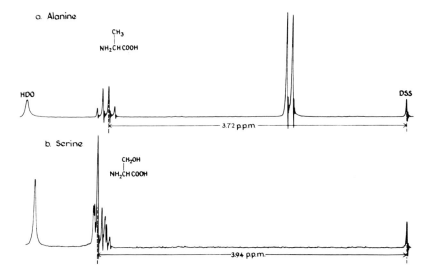

Fig. 3.5 ^1H spectra at 100 MHz of aqueous solutions of (a) alanine. The CH resonance is split into a quartet by the magnetic field of the CH_3 protons and the CH_3 is split into a doublet by the CH proton. This is first-order coupling, because the chemical shift between these two groups is much larger than the splitting caused by spin–spin interactions.

(b) Serine. The presence of the oxygen atom decreases the shielding of the neighbouring CH_2 and its resonance overlaps with the resonance of the CH proton. The chemical shift and the spin interaction are comparable and a complex pattern of second-order spin–spin interaction is observed. (Taken from Mandel 1965.)

The zwitterionic form of amino acids can be directly demonstrated by comparison of the area of the $\overset{+}{N}H_3$ with the side chain proton resonances, but protons attached to nitrogens which undergo rapid exchange with the solvent protons at neutral pH merge with the solvent signal. If the rate of the proton exchange processes fall in a suitable range, they can be followed by observing the shape and linewidth of the exchanging proton.

3.2 Conformation of amino acids

The magnitude of the coupling constant between two protons on adjacent aliphatic carbons is a function of the dihedral angle ϕ formed between the $H–C_\alpha–C_\beta$ and $C_\alpha–C_\beta–H$ bond pairs.

At present calculations define the dihedral angle ϕ with an accuracy of 5-10°. In some amino acids the magnitude of the coupling constants between protons on adjacent carbons was found to depend on the state of ionisation. This reflects differences in the preferred

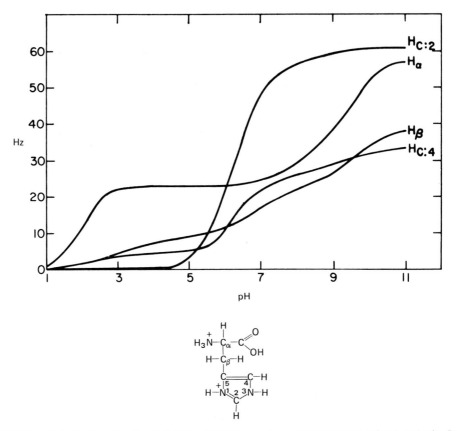

Fig. 3.6 The effect of pH on the chemical shifts of histidine protons at 60 MHz. The inflections in the $C_\alpha H$ curve correspond to the titration of the C_α amino and carboxyl groups while the $C_2 H$ curve reflects the protonation of N_1. The $C_\beta H_2$ and $C_4 H$ curves reflect the titration of all the ionisable groups to different extents. (Taken from McDonald and Phillips 1963.)

conformation about the C_α–C_β bond in the different ionised states. Thus in amino acids of the type

$$R-C_\beta H_2-C_\alpha H-COOH$$
$$\overset{|}{\underset{}{NH_2}}$$

it is generally assumed that of all the possible conformations obtained by rotation about the C_α–C_β bond, only the three staggered conformations occur to a significant extent. These three minimum energy conformations, called rotamers, are represented by:

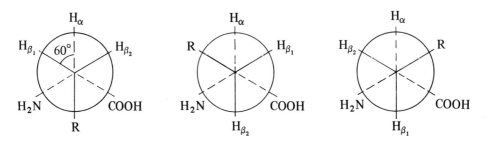

In many amino acids the two β protons are non-equivalent, which implies that rotation about the C_α–C_β bond is hindered so that the populations of the three rotamers about the C_α–C_β bond are not equal. The observed coupling constants between the two C_β protons and the C_α proton will then be the weighted average of the coupling constants in each of the rotamers. It is generally assumed that all *gauche* conformations of the C_β and C_α protons ($\phi = 60°$) in the three rotamers have the same coupling constant (J_{gauche}), and the *trans* conformations ($\phi = 180°$) have a distinct coupling constant (J_{trans}).

In principle, the relative lifetimes of each rotamer can be calculated provided that there are reliable values for J_{gauche} and J_{trans} and that the two non-equivalent gauche C_β protons are unequivocally assigned. The variation in estimates of J_{trans} and J_{gauche} obtained from simpler model compounds does not affect the calculated values of the lifetimes by more than 25 per cent, but the assignments of the C_β protons are only known unequivocally for a few amino acids. For example, the assignments for phenylalanine have been proved by stereospecific substitution of one of the C_β protons with deuterium. The changes in coupling constants, e.g. with pH and temperature, must reflect changes in rotamer populations. Several sets of relative rotamer populations in amino acids have been published (e.g. Pachler 1964).

3.3 Peptides

The changes in the chemical shifts which occur on forming a dipeptide between an amino acid and glycine have been summarised by Nakamura and Jardetzky (1967). The chemical shift of the glycine αCH_2 protons can be expressed as a sum of individual shifts, each representing a single chemical change. The shifts fall into two classes of differing magnitudes: first-order shifts of the αCH_2 glycine protons (0·4–0·7 ppm downfield) occur on titration of the N-terminal and the C-terminal groups and on the formation of an N-terminal peptide bond. All other shifts, e.g. due to charge on the neighbouring group and the formation of a C-terminal peptide bond, are relatively small second-order shifts ($< 0·2$ ppm). The shifts are additive to a very good approximation, and by adding together the appropriate shift terms, it is possible to predict the shift of the glycine αCH_2 protons in a peptide in any state of ionisation to within about 0·08 ppm.

The N-terminal and C-terminal titration shifts are transmitted to the side chain protons in the amino acid and to the αCH of the neighbouring residue but not farther. The length of the peptide chain does not affect the magnitude of the peptide bond formation shifts or the N-terminal and C-terminal shifts. *Thus with the exception of the titration of the terminal αNH_2 or COOH, there appear to be no primary structure effects on the chemical shifts of the side chain protons.* This implies that proton NMR is not generally of use for sequence analysis except for very short peptides.

3.4 α Helix formation in polypeptides

We can conclude from the above summary that in the absence of any secondary or tertiary structure, the NMR spectrum of a polypeptide is the superposition of the spectra of its constituent amino acids, after correcting for the effects of the formation of the peptide linkage. This makes it possible to detect secondary or tertiary structure by deviations from the predicted spectrum. The random coil to α helix transition in homopolymers, which can be manipulated by varying the solvent composition, provides a good model system to study the effect of secondary structure in the absence of tertiary structure. It is, of course, necessary to correct for any effects on shifts directly due to changes in the solvent composition.

For good examples of studies of the helix to random coil transition in poly-γ-benzyl-L-glutamate and poly-L-glutamic acid, see Markley *et al.* (1967) and Bradbury *et al.* (1968).

The αCH peptide resonance shows a substantial upfield shift on helix formation. In general this change in chemical shift in the transition region correlates closely with changes in the ORD spectra, but unlike the ORD spectra the chemical shift is rather insensitive to the nature of the side chain and probably gives a more reliable measure of the helix content. The NH peptide resonance generally shifts downfield on helix formation and this has been attributed to the dominant effect of the hydrogen bond in the helix. Both the NH and αCH resonances are broadened on helix formation, but the extent of line broadening falls off in the side chain with increasing distance from the peptide backbone. The broadening of the NH and αCH peptide resonances is much less than would be expected for the tumbling rate of a rigid helical rod of high molecular weight ($>$ 10 000). This may indicate that the individual residues in the polypeptide chain exchange between helical and random coil states even when the overall helix content is high. The side chain resonances generally show no change in shift on helix formation; any such changes in proteins must be determined by tertiary rather than secondary structure.

3.5 Peptide conformation analysis: valinomycin

Under favourable circumstances, NMR is capable of defining peptide conformation in solution, without reference to X-ray diffraction data. Small cyclic peptides in which the cyclic structure imposes a substantial conformational constraint have been extensively studied, and we take the work of Ivanov *et al.* (1969) on valinomycin as an example of this type of analysis. Valinomycin is a 36-membered cyclodepsipeptide:

$$
\left[
\begin{array}{cccc}
\overset{13}{CH(CH_3)_2} & \overset{14}{CH_3} & \overset{15}{CH(CH_3)_2} & \overset{16}{CH(CH_3)_2} \\
| & | & | & | \\
-NH-CH-CO-O-CH-CO-NH-CH-CO-O-CH-CO- \\
123456789101112
\end{array}
\right]_3
$$

which facilitates transport of potassium ions in bilayers and membranes by forming a 1 : 1 stoichiometric complex. The efficiency and high ion selectivity of the complexing reaction depends on the conformational state of valinomycin, which was determined by combining data from NMR, IR, ORD and dipole moment measurements. For example, the ORD curves of valinomycin vary with solvent polarity indicating an equilibrium between preferred conformers, which shifts with polarity. A critical difference in hydrogen bonding in two conformers was detected by comparing the NMR spectrum in the non-polar solvent CCl_4 with the spectra in perdeuterated dimethyl sulphoxide. The spectra show two pairs of doublets corresponding to two different types of NH signals ($N_{(1)}H$ and $N_{(7)}H$). In CCl_4 the signals are quite close together with chemical shifts which suggest that in this solvent the NH group form six intramolecular hydrogen bonds, whereas in $(CD_3)_2SO$ the $N_{(7)}H$ signal is shifted to low field while the $N_{(1)}H$ doublet remains practically unshifted, indicating that the three $N_{(7)}H$ protons are now hydrogen bonded to solvent molecules. Taken together with IR data the NMR spectra clearly indicate that in media of low polarity valinomycin exists in an equilibrium mixture of two forms, in which one form with all the amide groups forming six intramolecular H bonds is clearly predominant, and the other form with only three hydrogen bonds increases in importance in increasingly polar solvents. The data also show that the amide and ester groups must be in a *trans* conformation, and the only way in which this can be achieved with all six amide groups participating in mutual H-bonding is through carbonyl–amide hydrogen bonding in the 'direction of acylation' (Fig. 3.7). This constrains the conformation to a rigid framework of six condensed 10-membered rings formed by the hydrogen bonds with the whole resembling a bracelet about 8 Å in diameter and 4 Å high. However, the data still fit both the conformations A_1 and its inside-out counterpart A_2 which differ in the chirality of the ring and the orientation of the side chains (Fig. 3.8). The choice between these two conformations can be made unambiguously from the coupling constants in the amide linkages, which would lie in a

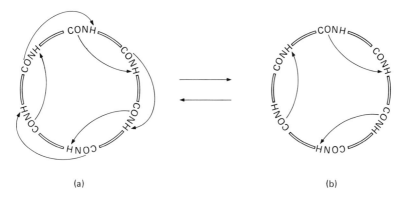

(a)   (b)

Fig. 3.7 The hydrogen bonding in the 'direction of acylation' in the two preferred conformers of valinomycin. (a) Six H-bond form which predominates in non-polar solvents. (b) Three H-bond form in polar solvents. (Taken from Ivanov *et al.* 1969.)

much lower range for A_2 than for A_1. The observed coupling constants not only exclude A_2 but show that one group of NH–C_αH fragments in the A_1 conformation are *gauche* oriented while the others are *cis* oriented. This implies that three of the ester groups are within the ring and three are outside. The coupling constants also showed that the L and D valyl side chains have *trans* oriented $C_{(8)}$H–$C_{(15)}$H and $C_{(2)}$H–$C_{(13)}$H protons, whereas the hydroxyisovaleryl side chains have *gauche* oriented $C_{(11)}$H–$C_{(16)}$H protons. We therefore arrive at the precisely defined A_1 bracelet conformation in Fig. 3.9(a) for valinomycin in CCl$_4$.

Analogous studies lead to the conformation of the valinomycin–K$^+$ complex in solution shown in Fig. 3.9(b). The hydrogen bonded framework of the A_1 conformation is retained in the complex, but all the ester carbonyls are now involved in ion dipole interactions with K$^+$. The conformation changes in that the three ester carbonyls which were directed outwards are now oriented towards the middle of the ring to form a hexagonal ring of oxygens around

A_1 A_2

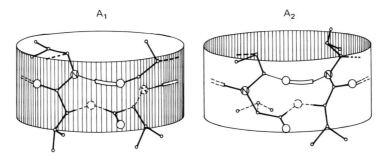

Fig. 3.8 The two possible 'bracelet' conformations of valinomycin consistent with six H-bonds and the amide and ester groups in a *trans* conformation. (Taken from Ivanov *et al.* 1969.)

152

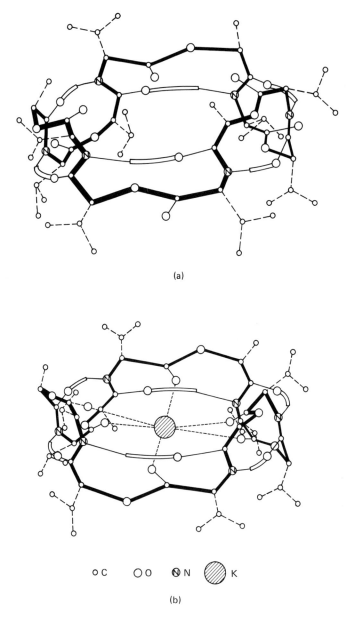

(a)

o C ⊙ O Ⓝ N ▨ K

(b)

Fig. 3.9 The conformations proposed for valinomycin (a) in CCl_4, (b) the K^+ complex in CCl_4. (Taken from Ivanov *et al.* 1969.)

the K^+ ion. The NMR spectra indicate a *gauche* orientation for all six NH—CH fragments as a consequence of reorientating the ester groups, consistent with the rigid symmetrical conformation of the complex shown in Fig. 3.9(b). A striking feature of the complex is that the K^+ ion and the H bonds are effectively shielded from solvent action by the hydrophobic branched side chains on the periphery of the molecule.

It is generally much more difficult to obtain conformational analysis of similar precision for linear peptides where the range of permissible conformations is potentially much greater.

4 Proteins

4.1 Ribonuclease

The first NMR spectrum of a protein, ribonuclease, was obtained by Saunders *et al.* (1957). The four broad peaks which were observed were accounted for approximately in terms of the spectra of the constituent amino acids. The spectrum of the *fully denatured* protein can be accounted for accurately by this procedure. The observation of broad, largely unresolved envelopes of resonances is expected since the linewidths are proportional to the inverse of the rate of tumbling and hence to the molecular weight, although local motional freedom of residues will sharpen the resonances. The observation of discrete resonances from single residues is the key to obtaining information from the NMR spectra and although the only general method by which peaks in the main body of the spectrum can be assigned is by selective isotopic substitution discussed later, the first detailed studies of proteins were concentrated on those resonances which fall outside the main broad envelopes. The most frequent examples are the C-2 protons of histidine residues, often observed as single resonances to the low field side of the main aromatic envelope. These histidine resonances were first examined in ribonuclease A which has also been extensively studied in other applications of NMR, and which we take for our main example. It has the advantages of low molecular weight and high solubility in the monomeric form. Its sequence is known, which is obviously essential for assignment, and its crystallographic structure has been determined.

4.2 Histidine assignment in ribunoclease

All four of the histidine C-2–H resonances of ribonuclease are resolved to low field of the main aromatic envelope (Fig. 3.10) These resonances shift on titration (Figs. 3.10 and 3.11) and the

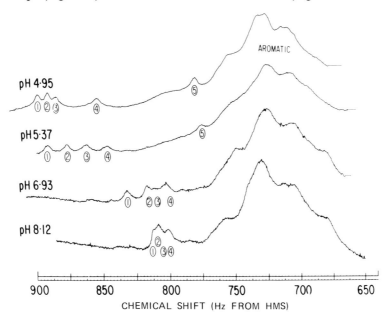

Fig. 3.10 ^{1}H NMR spectra at 100 MHz of the aromatic region of 0·012 M ribonuclease A. in deutero-acetate buffer at various pH values. Peaks 1–4 are C-2 imidazole peaks of the four histidine residues. Peak 5 is a C-4 imidazole resonance. The envelope labelled 'aromatic' includes three other C-4 imidazole peaks as well as peaks from six tyrosine and three phenylalanine residues. (Taken from Meadows *et al.* 1967.)

corresponding pK values are peak 1, 6·7 peak 2, 6·2; peak 3, 5·8; peak 4, 6·4. In addition, a peak attributable to a C-4 proton of histidine was observed with a pK of 6·7 which was therefore attributed to the same residue as peak 1 (Meadows *et al.* 1968).

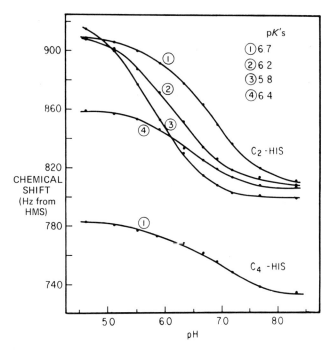

Fig. 3.11 Titration curves of histidine C-2 proton peaks and one histidine C-4 proton peak of ribonuclease A. The pK values are for 32°C in deutero-acetate buffer. (Taken from Meadows *et al.* 1968.)

The first approach to assigning these resonances to the histidines in ribonuclease (12, 48, 105 and 119 in the amino-acid sequence) was by chemical modification. Ribonuclease reacts with iodoacetamide to form two derivatives alkylated *either* at histidine 12 or at histidine 119. The titration curves of these derivatives showed that alkylation of a *single* histidine residue produces large changes in the pK value of *two* histidines, suggesting that either a conformational change has occurred, or that the two histidines are close together in the three-dimensional structure. Since the same two histidine peaks are affected by alkylation at either histidine 12 or 119 the latter explanation is more likely and is consistent with the crystal structure of the enzyme. Thus the chemical modification experiments allow the histidine peaks to be distinguished as two pairs; peaks 2 and 3 corresponding to histidines 12 and 119 at the binding site and peaks 1 to 4 to histidines 48 and 105.

To obtain a complete assignment a more subtle modification is required to affect the resonance of either histidine 12 or 199 but not both. This was achieved by selective deuteration of the C-2 proton of histidine 12. The subtilisin-cleaved enzyme, ribonuclease S, retains full enzymatic activity and is identical to ribonuclease A except that the peptide bond between residues 20 and 21 is cleaved. The histidine titration curves of ribonuclease S compared with native ribonuclease showed that peak 1 was quite unaffected and peaks 2 and 3 showed a slight increase in pK (by about 0·5 pH unit) but no change in shift of the protonated or unprotonated form. Peak 4, on the other hand, is shifted 0·02–0·03 ppm downfield in ribonuclease S compared to ribonuclease A. Histidine 48, which is 'buried' in the crystallographic structure of ribonuclease A, is less buried in ribonuclease S since histidine 48 is in the vicinity of the peptide bond between residues 20 and 21. It therefore seemed very probable that peak 4 could be assigned to histidine 48 on the basis of its abnormal chemical shift at low pH in ribonuclease A and the partial normalisation of this shift in ribonuclease S. Ribonuclease S was separated into the two components S peptide (residues 1-20) and S protein (residues 21-124), and the imidazole C-2 proton of histidine 12 was exchanged for deuterium by incubating the S peptide in D_2O until the C-2 proton was undetectable by NMR. The S peptide was recombined with S protein to reconstitute the enzymatically active ribonuclease S' which then showed only three histidine C-2 peaks. Comparison of the titration curves of ribonuclease S and ribonuclease S' indicated that either peak 1 or peak 2 was absent in ribonuclease S'. The chemical modification experiments had already shown that peak 1 could not be histidine 12, leaving the exchanged proton as that of peak 2. The full assignment is therefore:

Peak	pK	Histidine residue
1	6·7	105
2	6·2	12
3	5·8	119
4	6·4	48

This was the first time that the pK values of individual residues in a protein had been determined, and it also made possible the detailed interpretation of the NMR studies of inhibitor binding to ribonuclease described in the following section. It is also worth noting that although the two enzyme forms, ribonuclease A and S, have virtually identical structures, the titration curves of three of the four histidine C-2 protons differ in the two forms, and they are clearly a sensitive indicator of the conformational state of the enzyme.

4.3 Inhibitor binding to ribonuclease

When a complex is formed between a small molecule and a protein, changes may occur in the chemical shifts and relaxation times of the spectra of both the small molecule and the amino-acid residues at the protein binding site. We take the interaction of nucleotide inhibitors bound to ribonuclease as our main example.

Meadows and Jardetzky (1968) examined the interaction of ribonuclease with the specific inhibitor cytidine-3'-monophosphate (3'-CMP). At low concentrations of inhibitor histidine 119 was broadened and shifted downfield with histidine 12, while histidines 105 and 48 were unchanged. As the inhibitor to enzyme ratio approached equimolarity all four histidines were shifted although histidines 48 and 105 were least affected. The magnitude of the downfield shift of histidines 12 and 119 depended on the pH, and the histidine titration curves were measured in the presence of the inhibitor. The pK of histidine 12 increased from 6·2 in the free enzyme to 8·0 in the complex and for histidine 119 from 5·8 to 7·4. The affinity of 3'-CMP is maximal at pH 5·6 and at higher pH values the decrease in affinity follows closely the titration curve of the two active site histidine residues. On the low pH side however the decrease in affinity is not due to titration of the histidine but to the titration of the phosphate group of the inhibitor itself (pK 5·9). The enzyme-inhibitor interaction can be characterised by the two ionisation states of both enzyme and inhibitor, which fit the change in chemical shift of the active site histidines with increasing concentrations of 3'-CMP. At low inhibitor concentrations the data indicate that only the doubly charged form of 3'-CMP binds when both histidines are protonated, but at higher concentrations there may also be weaker binding by the singly charged form of the inhibitor. Binding of 3'-CMP in the dianionic form has recently been confirmed by ^{31}P NMR (Rüterjans et al. 1971).

The histidine 48 resonance in the presence of inhibitor is detectable only at low pH, since at higher pH it is severely broadened. It has been suggested that there is a conformational equilibrium affecting histidine 48 and that the position of this equilibrium is altered by inhibitor binding. The broadening observed in the presence of inhibitor probably arises from rapid interchange between conformational states of the protein which have different chemical

COMPLEX OF 3'-CMP WITH RNASE

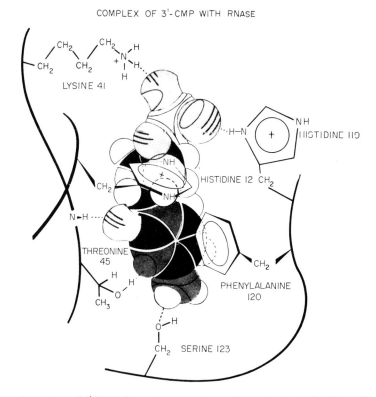

Fig. 3.12 Proposed structure of 3'-CMP ribonuclease complex at the active site cleft. (Taken from Meadows et al. 1969.)

shifts for the histidine 48 resonance: if the conformational change was slow, two separate histidine 48 resonances would be observed. Since histidine 48 is not at the active site, it is probably perturbed indirectly through neighbouring residues which are affected by inhibitor binding. The NMR data give an order of magnitude estimate for the conformational transition rate of about 60 s^{-1}.

In further detailed proton NMR studies of the binding of 2'-CMP, 3'-CMP and 5'-CMP to ribonuclease, the following conclusions are drawn (Meadows et al. 1969);

a. The pyrimidine rings of all three mononucleotides bind in the same way, close to an aromatic group on the enzyme, probably phenylalanine 120.

b. Both histidine 12 and histidine 119 are protonated in the mononucleotide–enzyme complex.

The phosphate group of both 2'-CMP and 3'-CMP binds specifically to histidine 119 but not to histidine 12. The phosphate group of 5'-CMP does not appear to bind preferentially to either of the active site histidines.

c. The ribose ring of the inhibitor in the 2'-CMP ribonuclease complex is in a different orientation from those of the other two complexes.

By combining this information with the X-ray diffraction structure of the active sites, structures were proposed for the three enzyme–inhibitor complexes, of which the 3'-CMP complex is shown in Fig. 3.12. The evidence for these structures is presented in full by Meadows et al. (1969) and the problem of the mechanism of action of ribonuclease is discussed by Roberts et al. (1969).

A recent paper by Benz et al. (1972) in which the T_1 relaxation times of all four histidines were measured, has demonstrated the importance of Fourier transform spectroscopy for proteins. On addition of 3'-CMP the T_1 values of the C-2 protons of histidines 119 and 48 decrease while those of histidines 12 and 105 remain unchanged, consistent with interaction of the phosphate group with histidine 119 but not histidine 12, and some change in the environment of histidine 48 possibly associated with the conformational change. Calculations show that the dominant relaxation mechanism for the histidine protons is probably an inter-residue proton dipolar interaction. This paper provides a good discussion of the problems of distinguishing motional changes caused by inhibitor binding and changes in proton–proton internuclear distances in the protein. The detailed analysis of the motion of side chains in a protein free of this ambiguity is in principle possible using ^{13}C NMR. For a ^{13}C–H group the relaxation of the ^{13}C nucleus is dominated by the ^{13}C-proton dipolar interaction, and any changes in ^{13}C T_1 values will generally depend directly on motional changes. The full potential of such measurements will only be realised when single assigned ^{13}C resonances can be studied, and such experiments are now feasible.

Earlier studies on other complexes, in which only the effects of complex formation on the spectra of the small molecules were examined should be mentioned here. These studies demonstrated that both chemical shifts and differential relaxation changes in the proton spectra of small molecules occurred on binding to proteins. Good prototype experiments were the analysis of the chemical shifts of the α- and β-anomers of N-acetyl-D-glucosamine on binding to lysozyme by Dalquist and Raftery (1968a, b), and the differential linewidth changes in the spectra of sulphonamides bound to bovine serum albumin, described by Jardetzky and Wade-Jardetzky (1965). The latter paper contains a good description of the criteria necessary to establish whether a complex fulfils the fast, intermediate, or slow exchange conditions for NMR relaxation measurements. Particularly detailed information can also be obtained from changes in the relaxation times of small molecules which bind to proteins containing a paramagnetic ion at the binding site. Paramagnetic ions provide powerful relaxation mechanisms depending on the proximity of the small molecule to the paramagnetic ion. Under favourable circumstances the distance of the paramagnetic ion from the ligand groups can be calculated, and under some exchange conditions the dissociation rate constant is obtained. A good recent example is Dwek et al. (1972) and there is an excellent review by Mildvan and Cohn (1970).

4.4 Unfolding of ribonuclease

Ribonuclease can refold to its native and enzymatically active tertiary structure, after being converted to a random coil by treatment with urea and mercaptoethanol. This indicates that all the information necessary for the formation of the native three-dimensional conformation is contained in the amino-acid sequence. Roberts and Benz (1972) have recently used NMR to study the unfolding of ribonuclease by a number of denaturants; by taking advantage of the increase in sensitivity from the Fourier transform technique, they were able to study the unfolding processes under conditions where they are wholly reversible. The results provide clear evidence for intermediates in the unfolding process between the well defined native and maximally unfolded states. The spectrum of the native enzyme shows four resolved histidine peaks (Fig. 3.10) but in the fully denatured state these are all superimposed. The residues are conveniently distributed through the primary sequence to act as markers for unfolding and we have already seen that their chemical shifts and titration curves are sensitive comformational indications of the enzyme. As the degree of unfolding was gradually increased, the intermediates were detected from the behaviour of the histidine resonances and from more qualitative observations on the main aromatic and aliphatic envelopes of resonances. The unfolding of ribonuclease by urea, guanidine hydrochloride, or heating at low pH appears to occur in the following sequence:

1. Dissociation of the N-terminal part of the chain from tertiary bonding to the rest of the protein to at least residue 12, but probably not as far as residue 14. This removes Phe 8 from the hydrophobic region near the active site cleft.
2. This facilitates the dissociation of the C-terminal part of the chain including Phe 120 and His 119, continuing backwards towards Cys 110.
3. More or less simultaneously the region Asp 14-Tyr 25 unfolds (affecting His 48) and then region 25-40 and 90-95.
4. Finally the β-structure involving residues 41-48, 61-85 and 96-110, which runs from one side of the molecule to the other, is disrupted together with its associated hydrophobic clusters.

It is important to realise that although a detailed and coherent picture is beginning to emerge the information is at present limited and speculative since it is derived mainly from 4 out of the 124 residues in the protein. The pathway of folding could be completely defined with information from all of the residues in the protein. This can be obtained in principle by isotopic substitution.

4.5 Selectively deuterated staphylococcal nuclease

Staphylococcal nuclease, in common with ribonuclease, has four histidines, but the spectrum in the histidine C-2-H region contains five peaks at most pH values. All these peaks change position with pH as would be expected for C-2 protons, but two of them have appreciably less intensity than the others. The most likely explanation is that these two peaks correspond to a single histidine which can exist in two states representing two conformations of the enzyme, and that the exchange rate of the histidine between these two states is slow compared with their chemical shift separation (< 24 s^{-1}). The addition of Ca^{2+} and/or the inhibitor thymidine-3′,5′-phosphate changes the pK of this histidine in only one of the two forms and presumably only binds to the enzyme in one of the two conformations (Markley et al. 1970). Although the inhibitor causes systematic changes in the main aromatic envelope of the nuclease [7 Tyr and 3 Phe residues], the envelope is relatively featureless and the changes are hard to interpret. The fully protonated protein spectrum is therefore not amenable to the kind of analysis applied to ribonuclease, and the problem was solved by reducing the number of proton resonance lines through selective deuteration of the protein (Markley et al. 1968).

Deuterated amino acids were obtained by hydrolysis of protein isolated from algae grown in 99 per cent D_2O, and the isotopic constitution of the amino acids was determined from their proton NMR spectra. Proton resonances were undetectable except for the 2,6-positions in the tyrosine ring, the 2-position in histidine, the β-position of aspartic acid and the γ-position of glutamic acid. The nuclease was isolated from the bacteria which had been grown in a medium containing the deuterated amino acids. In the first deuterated staphylococcal nuclease analogue, tryptophan and methionine were selected as the only fully protonated amino acids to be

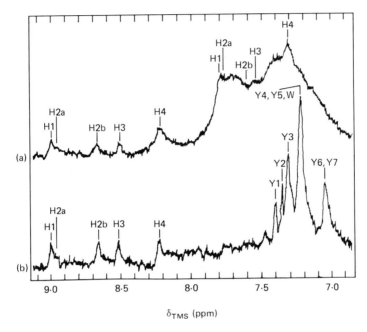

Fig. 3.13 ^{1}H spectra at 100 MHz of the aromatic region of protonated staphylococcal nuclease (a) and the selectively deuterated analogue of the enzyme (b) at 32°C; pH 8·0. Staphylococcal nuclease from the Foggi strain contains four histidine (S), seven tyrosine (Y), one tryptophane (W), and three phenylalanine residues. (Taken from Putter *et al.* 1970.)

present. The aromatic regions of the proton NMR spectra of the protonated and deuterated nucleases are compared in Fig. 3.13. The four C-2 histidine protons (S1–S4) in the deuterated enzyme correspond to the positions of the same protons in the protonated enzyme, strongly suggesting that the two forms of the enzyme have the same conformation. The tyrosine residues are protonated in the equivalent 2,6-positions and their resonances are therefore simplified to singlets in the deuterated enzyme, allowing the chemical shifts of the individual tyrosines to be determined (Y1–Y7). The peaks corresponding to the single tryptophan residue were also assigned. Similar spectral simplification of the aliphatic region was achieved. The four methionines are identified, of which only one was affected by the addition of inhibitor + Ca^{2+}. Extensive changes were also observed in the tyrosine peaks of the aromatic region.

Several other analogues were subsequently prepared containing different protonated amino acids (Putter *et al.* 1970) and a model of the binding site of the enzyme derived from the NMR experiments has been described. It is clear even from the limited number of analogues prepared that very detailed information can be obtained where the selective deuteration technique can be applied. The simplified proton spectra which are obtained also provide a method of following unfolding in much more detail than has been possible for ribonuclease, and the data again suggest that denaturation is not an all-or-none process but proceeds by stages in which partially denatured forms have appreciable stability (Putter et al. 1970; Jardetzky et al. 1971).

4.6 Other proteins

Two other interesting features of protein NMR spectra can only be mentioned briefly here. In some proteins, such as lysozyme, peaks shifted to unusually high field are observed, which shift downfield to the normal chemical shift range in the fully denatured state. These resonances have been attributed to methyl or methylene groups close to the faces of aromatic rings (ring current shifts). The residues giving these resonances can be tentatively assigned by reference to the crystal structure, and from perturbations induced by the binding of small molecules (McDonald and Phillips 1967).

Proteins containing haem prosthetic groups show resonances far to the high and low field of the normal range (e.g. Wüthrich et al. 1968). These resonances can be assigned to (a) protons of the porphyrin ring itself and (b) protons of the protein close to the porphyrin ring. The large shifts are due to the ring current of the porphyrin ring or to interaction with the paramagnetic iron atom of the haem. These resonances have been extensively examined in elegant studies of the perennial problem of haem-haem interactions in haemoglobin. One feature of special interest has been the use of mutant haemoglobins both as a method of assignment and to provide structures in which the quaternary structure is significantly altered (see for example, Lindstrom et al. 1972).

4.7 Conclusions

As an example of the isotopic substitution technique, the introduction of selective deuteration represents the most important biochemical advance achieved in the application of NMR spectroscopy to proteins. The introduction of Fourier transform spectroscopy has greatly simplified the technical problems in obtaining high resolution protein spectra rapidly or in dilute solution. It also allows the measurement of relaxation times for all the resolved resonances in the spectrum, which potentially contain important motional and distance information. The theoretical problems of interpretation are still substantial and ^{13}C data is likely to be particularly important.

Selective deuteration does not solve the problem of assignment, for which the only general solution is the insertion of labelled amino acid residues at known positions in the sequence of a protein (eventually by total synthesis). A number of approaches to assignment have been illustrated which frequently allow unambiguous assignments, but clearly require sophisticated biochemical manipulation of the protein.

Problems which now appear to be accessible to NMR are the identification of kinetic measurements with specific structural changes; detailed studies of interaction between small molecules and protein binding sites; and the pathways of folding and unfolding of proteins using ^{2}H and ^{13}C substituted residues. Recent work has also shown that ^{13}C NMR can be applied to examine lipid-protein interactions in intact biological membranes, by inserting ^{13}C-labelled phospholipids into the membranes. This significantly extends the range of macromolecular assemblies which can be studied by NMR and this work is reviewed by Lee et al. (1973).

5 Bibliography

5.1 General texts and reviews

EMSLEY, J. W., FEENEY, J. and SUTCLIFFE, L. H. (1966). *High Resolution Nuclear Magnetic Spectroscopy*, vols. I and II. Oxford, Pergamon Press. Recommended as a standard reference text for a more detailed discussion of NMR theory.

JACKMAN, L. M. (1959). *Application of Nuclear Magnetic Resonance Spectroscopy in Organic Chemistry*. Oxford, Pergamon Press. The early chapters of this book provide an excellent descriptive introduction to the elements of NMR theory.

KOWALSKY, A. and COHN, M. (1964). 'Application of nuclear magnetic resonance in biochemistry', *A. Rev. Biochem.*, **33**, 481-518.

LEE, A. G., BIRDSALL, N. J. M. and METCALFE, J. C. (1973). 'Nuclear magnetic resonance studies of biological membranes', *Chem. Br.*, **9**, 116-23. An account of recent work on lipids in bilayers and membranes, particularly ^{13}C relaxation measurements.

METCALFE, J. C. (1970). 'Nuclear magnetic resonance spectroscopy'. *The Physical Principles and Techniques of Protein Chemistry*, Part B. Chapter 14, pp. 275-363. New York and London, Academic Press.

MILDVAN, A. S. and COHN, M. (1970). 'Aspects of enzyme mechanisms studied by nuclear spin relaxation induced by paramagnetic probes', *Adv. Enzymol.*, **33**, 1-70. This is an important application of NMR to proteins not covered in this chapter.

ROBERTS, G. C. K. and JARDETZKY, O. (1970). 'NMR spectroscopy of amino acids, peptides, and proteins', *Adv. Protein Chem.* **24**, 447-545. A comprehensive and clear account of the state of the art up to June 1969.

5.2 Papers

BENZ, F. W., ROBERTS, G. C. K., FEENEY, J. and ISON, R. R. (1972). 'Proton spin-lattice relaxation studies of the histidine residues of pancreatic ribonuclease', *Biochim. Biophys. Acta*, **278**, 233-38.

BRADBURY, E. H., CRANE-ROBINSON, C., GOLDMAN, H. and RATTLE, H. W. E. (1968). 'Proton magnetic resonance and optical spectroscopic studies of water-soluble polypeptides: Poly-L-lysine HBr, Poly(L-glutamic acid), and Copoly(L-glutamic acid[42], L-lysine HBr[28], L-alanine[30])'. *Biopolymers*, **6**, 851-62.

DALQUIST, F. W. and RAFTERY, M. A. (1968a). 'A NMR study of association equilibria and enzyme-bound environments of N-acetyl-D-glucosamine anomers and lysozyme', *Biochemistry*, **7**, 3269-77.

DALQUIST, F. W. and RAFTERY, M. A. (1968b). 'An NMR study of enzyme inhibitor association. The use of pH and temperature effects to probe the binding environments', *Biochemistry*, **7**, 3277-80.

DWEK, R. A., RADDA, G. K., RICHARDS, R. E. and SALMON, A. G. (1972). 'Probes for the conformational transitions of phosphorylase a. Effect of ligands studied by proton-relaxation enhancement and chemical reactivities', *European J. Biochem.*, **29**, 509-14.

IVANOV, V. T., LAINE, I. A., ABDULAEV, N. D., SENYAVINA, L. B., POPOV, E. M., OVCHINNIKOV, Y. A. and SHEMYAKIN, M. M. (1969). 'The physicochemical basis of the functioning of biological membranes: the conformation of valinomycin and its K^+ complex in solution', *Biochem. Biophys. Res. Commun.*, **34**, 803-11.

JARDETZKY, O. and WADE-JADETZKY, N. G. (1965). 'On the mechanism of the binding of sulphonamides to albumin', *Mol. Pharmacol.*, **1**, 214-30.

JARDETZKY, O., THIELMANN, H., ARATA, Y., MARKLEY, J. L. and WILLIAMS, M. N. (1971). 'Tentative sequential method for the unfolding and refolding of staphylococcal nuclease at high pH', *Cold Spring Harb. Symp. Quant. Biol.* **36**, 257-61.

LINDSTROM, T. R., HO, C. and PISCIOTTA, A. V. (1972). 'NMR studies of haemoglobin M Milwaukee', *Nature New Biol.*, **237**, 263-64.

McDONALD, C. C. and PHILLIPS, W. D. (1963). 'A nuclear magnetic resonance study of structures of cobalt(II) − histidine complexes', *J. Am. Chem. Soc.*, **85**, 3736-42.

McDONALD, C. C. and PHILLIPS, W. D. (1967). 'Manifestations of the tertiary structures of proteins in high frequency NMR', *J. Am. Chem. Soc.*, **89**, 6333-41.

MANDEL, M. (1965). Proton magnetic resonance spectra of some proteins I. Ribonuclease, oxidised ribonuclease, lysozyme and cytochrome C', *J. Biol. Chem.*, **240**, 1586-92.

MARKLEY, J. L., MEADOWS, D. H. and JARDETZKY, O. (1967). 'Nuclear magnetic resonance studies of helix coil transitions in polyamino acids', *J. Molec. Biol.*, **27**, 25-40.

MARKLEY, J. L., PUTTER, I. and JARDETZKY, O. (1968). 'High resolution NMR spectra of selectively deuterated staphylococcal nuclease', *Science,* **161**, 1249-51.

MARKLEY, J. L., WILLIAMS, M. N. and JARDETZKY, O. (1970). 'NMR studies of the structure and binding sites of enzymes. XII. A conformational equilibrium in staphylococcal nuclease involving a histidine residue', *Proc. Natn. Acad. Sci. U.S.A.*, **65**, 645-51.

MEADOWS, D. H. and JARDETZKY, O. (1968) 'NMR studies of the structure and binding sites of enzymes'. IV. Cytidine 3'-monophosphate binding to ribonuclease', *Proc. Natn. Acad. Sci. U.S.A.*, **61**, 406-13.

MEADOWS, D. H., JARDETZKY, O., EPAND, R. M., RUTERJANS, H. H. and SCHERAGA, H. A. (1968). 'Assignment of the histidine peaks in the NMR spectrum of ribonuclease', *Proc. Natn. Acad. Sci. U.S.A.*, **60**, 766-72.

MEADOWS, D. H., MARKLEY, H. L., COHEN, J. S., JARDETZKY, O. (1967). 'NMR studies of the structure and binding sites of enzymes, I. Histidine residues', *Proc. Natn. Acad. Sci. U.S.A.*, **58**, 1307-13.

MEADOWS, D. H., ROBERTS, G. C. K. and JARDETZKY, O. (1969). 'NMR studies of the structure and binding sites of enzymes. VIII. Inhibitor binding to ribonuclease', *J. Molec. Biol.*, **45**, 491-511.

NAKAMURA, A. and JARDETZKY, O. (1967). 'Systematic analysis of chemical shifts in the NMR spectra of peptide chains. I. Glycine-containing dipeptides', *Proc. Natn. Acad. Sci. U.S.A.*, **58**, 2212-19.

PACHLER, K. G. R. (1964). 'NMR study of some α-amino acids. II. Rotational isomerism', *Spectrochim. Acta*, **20**, 581-7.

PUTTER, I., MARKLEY, J. L. and JARDETZKY, O. (1970). 'NMR studies of the structure and binding sites of enzymes. XI. Characterisation of selectively deuterated analogs of staphylococcal nuclease', *Proc. natn. Acad. Sci. U.S.A.*, **65**, 395-401.

ROBERTS, G. C. K., DENNIS, E. A., MEADOWS, D. H., COHEN, J. S. and JARDETZKY, O. (1969). 'The mechanism of action of ribonuclease', *Proc. Natn. Acad. Sci. U.S.A.*, **62**, 1151-58.

ROBERTS, G. C. K. and BENZ, F. W. (1973). 'Proton Fourier transform NMR studies of the unfolding of ribonuclease', *Ann. N.Y. Acad. Sci.* in press.

RUTERJANS, H. H., HAAR, H., MAURER, W. and THOMPSON, T. C. (1971) Abstr. 1st Eur. Biophys. Congr., Baden, Austria. 'NMR studies of the interaction of nucleotides with Ribonuclease A and T_1 using ^1H and ^{31}P magnetic resonance.'

SAUNDERS, M., WISHNIA, A. and KIRKWOOD, J. G. (1957). 'The NMR spectrum of ribonuclease', *J. Am. Chem. Soc.*, **79**, 3289-90.

WÜTHRICH, K., SHULMAN, R. G. and YAMANE, T. (1968). 'Proton magnetic resonance studies of human cyanomethemoglobin', *Proc. Natn. Acad. Sci. U.S.A.*, **61**, 1199-206.

4
Electronic Spectra and Optical Activity of Proteins

Anne d'Albis and W. B. Gratzer
King's College, London

1 Electronic absorption spectra

1.1 Basic principles: interaction of light with matter

It is not our object here to present spectroscopy-in-a-nutshell, but rather to recall some basic principles in a form which is useful for an understanding in simple physical terms of the phenomena which are of interest in the study of proteins. We shall assume familiarity with the basic concepts of molecular structure, and of electronic spectra. A number of good basic texts are available, such as Whiffen (1972), and at a somewhat more advanced level, Barrow (1962), Jaffe and Orchin (1962) and Baumann (1962), as well as Wiberg (1964). For useful descriptions of the theory of molecular spectra, at an essentially pictorial level, see also Robinson (1961) and Kasha (1961).

1.1.1 Nature of electronic transitions

The electric field of a light wave, passing through an assemblage of molecules, will induce a dipole moment in the electron cloud, of magnitude defined by a quantity, α, the polarisability of the molecule. The oscillating dipole set up by the oscillating electric field of the light generates a new secondary wave that accompanies the original light wave through the medium. The net result can be shown to amount to a modified wave, which is retarded relative to the incident wave (or to a wave *in vacuo*). The retardation depends on the length of the light path through the medium. The ratio of the velocity of the wave in the medium to that in a vacuum is the *refractive index* of the medium. It bears a simple relation to the polarisability. The polarisability, and therefore the refractive index, varies with the energy of the light, that is to say with its wavelength: the shorter the wavelength of the radiation, the higher its energy (which is proportional to the wave number, or reciprocal of the wavelength), and the larger the refractive index.

At some wavelength, however, the energy will correspond to that of an electronic transition: light is absorbed, and instead of merely oscillating at the frequency of the light wave (say 10^{15} Hz) electronic charge is redistributed in a long-term manner (say for a duration of 10^{-9}s). Where such a resonance condition is fulfilled it will not be surprising to find that due to the extraction of energy, the relation between refractive index and wavelength undergoes a convulsion. This is known as anomalous refractive index dispersion, and takes the form shown in Fig. 4.1. We shall return to this in considering optical rotation.

164

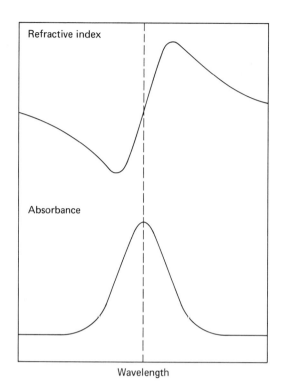

Fig. 4.1 Dispersion of refractive index. The upper curve shows how the refractive index changes in the region of the corresponding absorption band (lower curve).

1.1.2 Absorption bands: general principles

In order to understand the nature of the absorption spectrum it is necessary to consider the types of molecular orbital (for description see references cited above) that are involved in a given molecule or chromophore. (By *chromophore* one means the chemical unit, which may be a part of a large molecule, with its individual absorption characteristics, which it retains regardless of environment. Thus the phenyl, phenol and indole rings of phenylalanine, tyrosine and tryptophan are referred to as the aromatic chromophores of proteins, and have to a first approximation the same absorption spectra as the free amino acids.) In regard to the types of absorbing groups that are present in proteins, we need consider only σ-orbitals, π-orbitals, and non-bonding (lone pair) n-orbitals. For absorption bands in the accessible part of the ultraviolet (that is to say the wavelength range down to about 185 nm, in which one can work with modern commercial spectrophotometers, and aqueous solutions) one may even omit consideration of the ground-state σ-orbitals, which correspond to very strong chemical bonds, from which electrons are only promoted with great difficulty (that is to say at high energy) into other orbitals. The difference between bonding and anti-bonding molecular orbitals is that in the latter there is a node in the electron density distribution, at which the wave-function by convention changes sign. Thus bonding and anti-bonding σ-orbitals between two atoms (nuclei shown by a dot) may be written $\cdot + \cdot$ for σ and $+ \cdot \cdot -$ for σ^*. The latter is a repulsive state of high energy. The ground state, in which the molecule normally exists, contains no anti-bonding orbitals. In a conjugated π-electron system, conventionally described by writing alternating single and double bonds, as in the polyenes, the lobes of the atomic p-orbitals, standing out above and below the line of the nuclei, fuse into elongated clouds of charge. There is thus a cylinder of π electron density above the line of the nuclei, and one below. In benzene these cylinders become doughnuts above and below the plane of the ring. In the excited states (π^*) each cylinder of charge will be interrupted by one (for the lowest excited state) or more repulsive interactions, which lead in each case to a discontinuity in the electron density, otherwise described as a nodal plane, slicing through both cylinders. For instance in benzene each doughnut will in the first excited state be broken in half by a nodal plane perpendicular to the ring, cutting the doughnut at opposite ends of a diameter. The more nodes, the higher the energy, the greater the distance in energy terms of the excited state above the ground state, and the greater therefore the energy corresponding to the transition between them. Ultimately the energy will be so high that the transition will occur at wavelengths too short to observe by ordinary means. Figure 4.2 shows schematically the relative energies of the orbitals present in a simple molecule. The σ-electrons, as noted, occupy a deep potential well, the π-electrons come next, and the n-electrons, which are not engaged in a chemical bond, lie somewhat higher. This is the complement of ground-state orbitals. It will be apparent that the lowest-energy transition is that between the n-electrons, and the lowest excited state of the π-orbitals (π^*). Such a transition is referred to as an $n \rightarrow \pi^*$ transition. Next in order of increasing energy is the first $\pi \rightarrow \pi^*$ transition, and we then have the $n \rightarrow \sigma^*$ transition, but this is already of too high energy to concern us here.

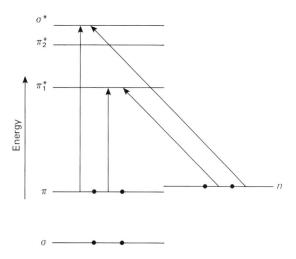

Fig. 4.2 Electronic energy levels, and transitions (arrows) for a typical simple molecule, containing σ-, π- and n-(lone pair) orbitals. The electron pairs in the filled ground-state orbitals are shown.

The diagram in Fig. 4.2 defines the wavelengths of the bands, but not their intensity. The latter, i.e. the transition probability, depends on a number of factors into which we cannot enter in detail here. One of them is the degree of spatial overlap between the orbitals involved. This is very low as between the n-electrons and the π-system (see above references for descriptions of the orbitals), and the intensity of such transitions is therefore always small. The next factor concerns the symmetry properties of the orbitals, which determines whether there is a directional shift of electronic charge. If there is not, the transition is forbidden. Forbiddeness, however, is a relative concept in large molecules, which afford complex environments, because the symmetry characteristics are perturbed by vicinal effects, and in practice it simply means that symmetry-forbidden transitions occur at much lower intensity than those which are allowed. The reader with a desire for a firmer grasp of these concepts is urged to consult some of the references that we have given.

We come now to a very important notion, that of the polarisation of electronic transitions. Since an electronic transition consists of the redistribution of electronic charge within the molecule, there must be a resultant vector, which represents the net magnitude and direction of this change. For example, imagine a transition from the ground state π-orbitals of benzene to the first excited state. The ground state orbitals are doughnuts of electron charge above and below the ring. In the first excited state, the two doughnuts are interrupted by a nodal plane perpendicular to the ring, giving a redistribution of charge in the plane of the ring. It will be clear that the vector which expresses the charge displacement must also then lie in the plane of

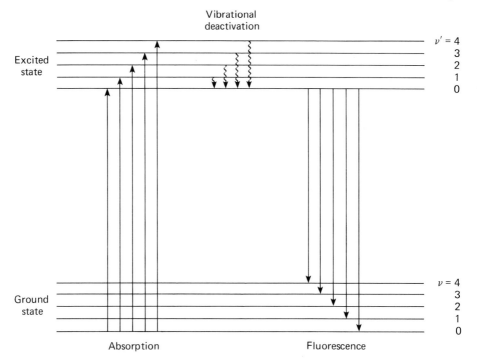

Fig. 4.3 Vibronic components of electronic transition. The arrows show a set of transitions (within the absorption band envelope) from the zero-vibrational level in the ground state to various vibrational levels in the first excited state, and a set of transitions (fluorescence) from the zero vibrational level in the first excited state to various vibrational levels in the ground state.

the ring. This vector is the *electronic transition moment*. If the electric field of the light is to induce a transition, it must have a component in the direction of the transition moment. If a sample can be procured (a crystal for example) in which all the molecules, and therefore the transition moment for a given transition in each, are oriented in the same direction, then plane-polarised light with the polarisation direction parallel to that of the transition moments will be maximally absorbed, whereas if polarised perpendicularly to this direction, it will not be absorbed at all. Use is made of this principle in flow dichroism of macromolecules for example, in which orientation is achieved by rapid streaming. Very large molecules, such as α-helical polypeptides, or DNA for instance, can also be oriented in solid films by stroking as the solution dries; the molecules tend to line up with their long (helix) axes along the direction of stroking, and in

DNA the planes of the bases therefore lie predominantly perpendicular to this axis. Since the bases are aromatic species with $\pi \rightarrow \pi^*$ transition moments in the plane of the rings, light polarised perpendicularly to the helix axis is much more strongly absorbed than light polarised along the axis.

We consider next the fine-structure of electronic absorption bands of complex molecules, and the effect of the environment, especially the solvent. Unlike atomic spectra, or spectra of simple molecules in the gas phase, the spectra in which we are interested show no sharp lines, but only rather broad absorption bands, that commonly contain irregularities in the form of poorly developed peaks or shoulders, and sometimes (see for example the absorption band of phenylalanine below), a series of peaks. Structure in such spectra may reflect the presence of two or more overlapping electronic absorption bands (a case in point is the long wavelength feature in the absorption band of tryptophan, also shown below), or, as in the case of phenylalanine, vibrational structure. Within each electronic state, there are a series of vibrational levels, and in principle transitions can occur between any vibrational level in the ground state, and any vibrational level in the excited state. In practice the number of vibrational bands within the envelope of the electronic absorption band is limited, first of all by the fact that at ordinary temperatures all the molecules will normally find themselves in the lowest vibrational state ($\nu = 0$). Secondly, there are allowed and forbidden vibrational transitions, depending on the symmetry properties of the vibrational wave-functions, and in particular the probability of individual vibrational transitions is determined by the Franck-Condon principle, which places a limitation on the internuclear spacings in the vibrational states between which electronic excitation can occur, and which we will not go into here, but which is expounded in the general references given above (see, e.g. Barrow 1962; also Reid 1957). A set of vibrational transitions within a transition between two electronic states is shown in Fig. 4.3.

1.1.3 Solvent effects on absorption spectra

Molecules in the gas phase show very highly developed vibrational structure, and often within each vibronic band they have a series of closely spaced rotational levels. In solution, the latter are not seen, and the vibrational structure too is broadened and smeared out. This results from interaction of the molecules with the solvent, and perhaps with each other, the formation of such complexes causing the breakdown of vibrational quantisation, and the local electric and magnetic fields of the surrounding molecules producing a multitude of substates (Stark and Zeeman interactions). The net result is that each vibronic band is broadened to form a kind of continuum of states, and the original fine-structure becomes much more sketchy. The solvent interactions become much stronger when one goes from non-polar to polar solvents, and the structure is blurred out still further (Fig. 4.4).

Another consequence of solvent interactions is that the energies of the ground and excited states are perturbed. In general they will not be perturbed to the same extent, and the energy gap between them will narrow or widen, the result being respectively a red- or blue-shift in the absorption band. Now in general excitation involves increased separation of charge, and the excited-state dipole moment will tend to be bigger than that of the ground state. The transition dipole will induce transient dipoles in the neighbouring solvent molecules, and will interact with them, causing the transition to be energetically facilitated, so that a red-shift will result (Fig. 4.5). In this case, the higher the solvent polarisability, the bigger the shift. When the solvent molecules are polar, however (i.e. possess a permanent dipole), the situation is more complex, for if on excitation, the dipole moment of the chromophore changes, the favoured configuration of solvent molecules around the solute ceases to be stable. The solvent molecules however need a time interval – long compared with that of the electronic transition – to orient

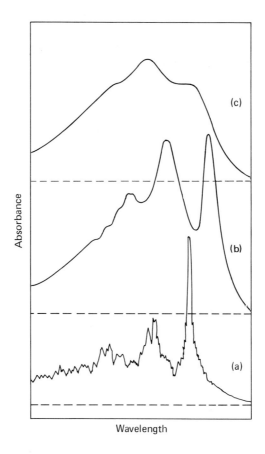

Fig. 4.4 Typical environmental effects on the absorption spectrum of an aromatic molecule: (a) vapour phase spectrum, showing rotational, as well as vibrational structure; (b) solution spectrum in a non-polar solvent, showing well-developed vibrational structure; (c) spectrum in a polar solvent, showing vestigial vibrational structure only.

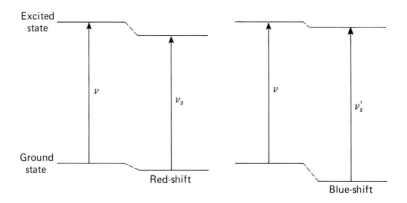

Fig. 4.5 Effect of solvent interactions on the energies of the ground and excited electronic states of a solute molecule, and the corresponding energies of the transition.

in such a way as to take account of the altered dipole moment of the solute, and the result is a destabilisation of the excited state as instantaneously attained. Depending on the energies involved this can lead to a blue-shift in a polar solvent. This as we shall see appears to be the case for the aromatic chromophores embedded in proteins.

A special case is presented by $n \to \pi^*$ transitions. The n-electrons, not being involved in a covalent bond, and being especially exposed, tend to interact most strongly of all with the solvent. With increasing solvent polarity therefore the ground (n) state is progressively more stabilised than the excited (π^*) state, and large blue-shifts result.

1.1.4 Experimental aspects of absorption spectra

An absorption band is defined by its position and intensity. The position is described by wavelength (λ) in nm, or by wave-number ($\bar{\nu} = 1/\lambda$) in cm^{-1}, the latter having the advantage that it is a measure of energy. The intensity is expressed in terms of the *molar absorptivity* (or extinction coefficient), ϵ, at any wavelength, generally that of the absorption maximum, or a specific (i.e. weight, rather than molar) absorptivity; for example $E_{1 cm}^{1 \%}$ is commonly used for proteins, and refers to a 1 per cent solution in 1 cm path. The intensity is more fundamentally described by the area of the band, $\int \epsilon \, d\bar{\nu}$ and the quantity $4.33 \times 10^{-9} \int \epsilon \, d\bar{\nu}$ is called the *oscillator strength* and denoted by f.

The *transmittance* of a solution is the ratio of light intensity after and before passage through the solution, i.e. $T = I/I_0$. The *absorbance* (optical density) is $A = -\log T$. This quantity is proportional to the concentration of the absorbing species (Beer's law). Since each successive thickness element through which the light passes takes out an equal proportion of the intensity incident on it, the intensity diminishes in a compound-interest, that is logarithmic, manner during its passage through the solution. The molar absorptivity is then the absorbance of a molar solution in a 1 cm path, and hence $A = \epsilon Cl$ where C is the molar concentration and l the path length in cm. This is a statement of the Lambert–Beer law, which also implies that the absorbance is proportional to path length. A very intense allowed transition might have $\epsilon \sim 10^5$ and $f \sim 1$, and forbidden transitions would range from about $\epsilon = 10^3$ to zero, or $f = 0.01$ to zero. Absorption spectra may be expressed by plots of A, or better, of ϵ, against λ or $\bar{\nu}$, or sometimes to provide a conspectus showing both strong and weak bands $\log \epsilon$ is plotted.

1.2 Absorption spectra of proteins in the near ultraviolet and visible

1.2.1 Aromatic amino acids

The most familiar part of the absorption spectrum of proteins occurs in the range 250–310 nm, and arises from the aromatic side chains, of tryptophan, tyrosine and phenylalanine. The absorption spectra of these amino acids are shown in Fig. 4.6. The indole chromophore of tryptophan gives rise to the strongest absorption band. It will be noted that at alkaline pH, the phenolic chromophore of tyrosine undergoes ionisation: $\phi OH \rightleftharpoons \phi O^- + H^+$, and this results in a large red-shift, and intensity enhancement of the spectrum. The extinction coefficient of a protein at, say, 280 nm, depends on the tyrosine and tryptophan content, and is used for concentration determination. Additional contributions, come from disulphide bonds, with a low-intensity peak around 250 nm, and histidine, with a band at about 220 nm, but this is not readily detected. For the most part it is only tyrosine and tryptophan that need to be considered.

170

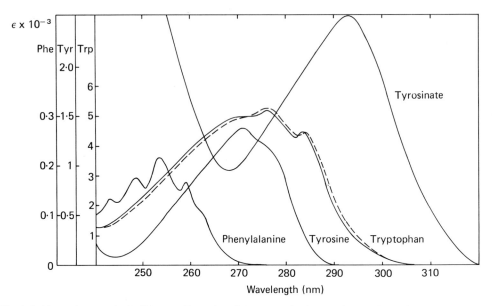

Fig. 4.6 Absorption spectra in the near-ultraviolet of the aromatic chromophores of proteins. Note that the intensity scales are different for the three spectra, that for phenylalanine being very low. The broken line indicates a red-shifted spectrum, brought about by some perturbation of the chromophore.

1.2.2 Difference spectra

As we have noted, a chromophore entering a different solvent environment will in general show perturbations in its spectrum, in particular a small wavelength shift. Such will be the case for an aromatic residue transferred from aqueous solution into the rather hydrocarbon-like environment inside a globular protein. This is exactly what happens when a protein folds up into its native conformation. Contrariwise, denaturation leads to exposure of these residues to the solvent. Transfer from an aqueous environment to one of a highly polarisable hydrocarbon nature, leads to a red-shift. In a broad absorption band, a small shift, of the order of 1 nm is not easily quantitated. One way is to measure the absorbance of the protein in one state (say the native form) against that in the denatured form at the same concentration. The result is a *difference spectrum*, the amplitude of which is an expression of the shift. For an idealised absorption band, the shape of the difference spectrum is shown in Fig. 4.7(a), and brief reflection will show that it is essentially (if the shift is small) the derivative of the absorption band. The bands of tryptophan and of tyrosine are not of course of this idealised shape, and give rise therefore to more complex, but characteristic difference spectra, as shown in Fig. 4.7(b). Such difference spectra may be generated experimentally by any effect that perturbs the energy levels of the chromophore (Fig. 4.5), e.g. by ionisation of a vicinal group (such as an α-carboxyl in an amino acid), by hydrolysis of a peptide to its amino acids or by a change in solvent polarity (Fig. 4.7(c)). The same general shape is produced by changing the environment of an aromatic side chain in a protein, most obviously by denaturation, and many studies on the thermodynamics and kinetics of protein folding and unfolding have been done in terms of the appearance or disappearance of such difference spectra (Fig. 4.7(c)). There are other ways in which the effect has been used. For example if, as is common, aromatic residues are present in the active site of an enzyme, a difference spectrum will often be generated by the attachment of a substrate or inhibitor. Again if aromatic residues are present in the contact regions of

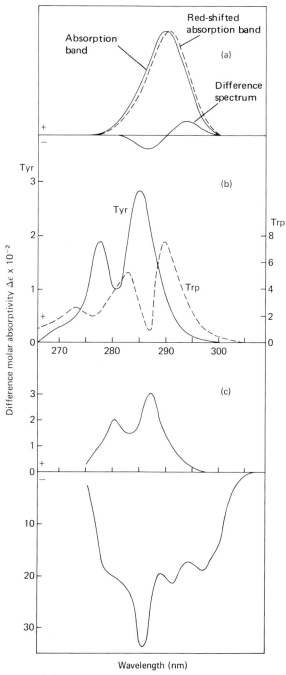

Fig. 4.7 Difference spectra: (a) difference spectrum for an idealised absorption band, undergoing a red-shift; (b) characteristic difference spectra of tyrosine and tryptophan (not on the same intensity scale), generated by addition of ethanol to an aqueous solution of the amino acid. The difference spectrum corresponds to a shift of the kind illustrated for tryptophan in Fig. 4.6. (c) Typical difference spectra encountered in proteins. The upper curve represents a red-shift induced by solvent perturbation of exposed residues in a protein containing tyrosine, but little tryptophan (bovine serum albumin); the lower curve shows the difference spectrum corresponding to the blue-shift that results from denaturation of a protein containing both tryptophan and tyrosine (trypsin).

protein molecules in association-dissociation equilibrium, any displacement of this equilibrium will result in the appearance of a difference spectrum. The difference method can therefore be used to follow (in equilibrium or kinetic experiments) a variety of equilibria involving ligand-binding, changes in tertiary, and in quaternary structure, and so forth.

Another kind of application is the determination of the number of tyrosine and tryptophan residues on the surface, and in the interior of a globular protein. Since a difference spectrum may be generated by a solvent change, the addition of a non-denaturing solvent (D_2O, sucrose solution or glycerol for example) to the protein solution will be expected to perturb those chromophores accessible to the medium, but not those in the interior, from which solvent is excluded. The size of the difference spectrum is calibrated with free tyrosine or tryptophan derivatives (or with the unfolded protein) and an estimate can be made of the fraction at the surface in the native molecule. Attempts to estimate the degree of exposure of partly buried chromophores, or the sizes of crevices housing some residues, have been made by comparing the effects of perturbants of different sizes.

Difference spectra of this kind are not to be confused with the spectroscopic change that results from the ionisation of tyrosine side chains (Fig. 4.6). Normal tyrosines have a pK_a of about 10, though in proteins this may be shifted by environmental factors, and tyrosines in the centre of a globular molecule may not titrate at all unless the protein is denatured. By following the absorbance at 295 nm, say, as a function of pH, it is possible to deduce the number of titratable tyrosines in a protein, and their pK's. This would be impossible by potentiometric titration, since the tyrosines will invariably be masked by the much more numerous lysines which titrate in the same range. For a full account of the aromatic spectra and difference spectra of proteins, see Wetlaufer (1962) and Donovan (1969).

1.2.3 Extrinsic chromophores

Of course absorption measurements are useful in a variety of other situations. The use of chromogenic substrates will be familiar to enzymologists: substrates exist, or can be designed which change their absorption spectrum when they are converted to the catalytic product. This provides good assay procedures. Many ligands also change their absorption spectra, usually to a rather modest extent, when they bind to a protein. If the perturbation to the spectrum can be measured and the absorbances corresponding to bound and unbound forms determined, the binding thermodynamics can often be evaluated. The design of such systems are also an essential part of fast kinetic experiments, using stopped-flow, or temperature-jump methods, for example, where the spectroscopic change serves as a marker for following the kinetics of

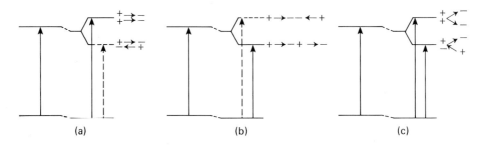

Fig. 4.8 Exciton resonance interaction of two identical chromophores, showing the two levels into which the excited state is split. The levels for the unperturbed chromophore are shown to the left in each case. Full arrows indicate allowed, and broken arrows forbidden transitions. (a) chromophores side-by-side, (b) head-to-tail, (c) at an angle to each other.

enzymic and other reactions. Sometimes a chromophore can be covalently introduced at a suitable point in a protein, and will be perturbed by events, such as those occurring in enzymic catalysis. Such chromophores have been flamboyantly described as reporter groups. In conjugated proteins (haemproteins, flavoproteins, metalloproteins, etc.) the absorption spectrum in the visible region has almost innumerable applications. It may be used to follow oxygen or carbon monoxide binding to respiratory proteins. It is often sensitive to changes in quaternary structure, and to the binding of other ligands. In favourable cases it may be used directly to give information about the state of the prosthetic group (distinguishing for instance between low-spin and high-spin ferric haem states, or between possible ligand-field symmetry states of the metal ion in metalloproteins).

1.3 Excitons and hypochromism

1.3.1 Resonance interactions

A phenomenon first recognised in the spectra of crystals is the distribution of excitation between a number of identical chromophores. This is a resonance phenomenon which one may view in at least two quite different ways: one may either think of the quantum of excitation as being smeared out over a number of close-lying chromophores, or one may envisage it as jumping rapidly back and forth from one to another. These views are strictly equivalent, and may be treated by time-independent or time-dependent quantum theory to give the same expressions (see for example Kasha 1963, McRae and Kasha 1964, and Förster 1960, for a particularly clear account). Consider now the case of a dimer. The transition moments are close together and for resonance, with a single quantum of excitation distributed between the two chromophores, the transition moments may be in phase or out of phase. These two configurations will give rise to two energy levels in the excited state, where previously there was one (Fig. 4.8). How this manifests itself in the absorption spectrum depends now on the geometry of the dimer. If the transition moments are parallel, the in-phase and out-of-phase arrangements are $\overset{+}{\underset{+}{\rightarrow}}\overset{-}{\underset{-}{\rightarrow}}$ and $\overset{+}{\underset{-}{\rightarrow}}\overset{-}{\underset{+}{\leftarrow}}$. Clearly the first of these represents a repulsive situation (increased energy), the second an attractive one (decreased energy). The second however has a resultant transition moment of zero, and is therefore forbidden. The allowed transition is from the ground state to an excited state of elevated energy, which results in a blue-shift in the absorption band. For transition moments arranged head-to-tail, we have $+ \rightarrow - + \rightarrow -$ and $+ \rightarrow - - \leftarrow +$. Now the in-phase arrangement is attractive, leading to lowered energy, and is allowed, and by analogy with the above argument, the result is a red-shift. In the general case the dimers will be at some angle to each other, and both the in-phase and out-of-phase combinations of the two transition moments will have a finite resultant, so that both states will give rise to an absorption component. In this general case, therefore, the result will be *splitting* of the absorption band.

The separation of the resonance components will be a measure of the interaction energy, which will be large if the transition moments are large, close together, and oriented at a favourable angle to each other. Weak interaction can lead to splitting or shifts which are small compared with the width of the absorption band, and may manifest itself in splitting in the strongest vibrational levels. If the interaction is very weak, there is transfer of excitation with no visible effect on the absorption spectrum; this is manifested in fluorescence, as we shall see, and is often referred to as Förster transfer.

To return now to the case of strong interaction: for three chromophores there will be three levels instead of two, and so on, and for an ordered polymer there will be in effect a continuum of levels making up a broad band in the excited state. For a card-stack of chromophores (extension of the parallel dimer considered above), the transition moments will be in phase only at the top of the exciton band, and it is to this position that transitions will occur.

1.3.2 Hypochromism and hyperchromism

Hypochromism is a phenomenon which is most familiar to workers in the nucleic acid field. In brief the term signifies that the absorption intensity is less than the sum of its parts. The bases in the free state, or in denatured DNA, have a much higher absorbance than in ordered, double-helical DNA. The effect is also important in proteins. Hypochromism arises in systems of parallel chromophores, and the opposite (hyperchromism) in chromophores aligned end-to-end (Fig. 4.9). In a simple intuitive way one might suppose that in an arrangement of the first

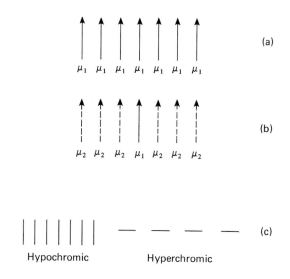

Fig. 4.9 Origins of hypo- and hyperchromism. Intensity changes result from interactions between different transitions (b), not between identical transitions (exciton interactions) (a). The card-stack arrangement of chromophores gives rise to hypochromism, the head-to-tail arrangement to hyperchromism (c).

kind, the excitation of one of the chromophores will lead to a displacement of charge in a direction which will inhibit (in consequence of electrostatic repulsion) the excitation of its neighbours in the stack, and vice versa for the second arrangement. This is not strictly correct, for one has to consider the interaction of any transition dipole with those of other electronic transitions in adjacent chromophores. Light oscillating with the frequency of a given absorption band will be absorbed, but will also set up oscillation of the electron clouds as described in section 1.1. The mode of the oscillations will be defined by the principal polarisabilities, one of which is associated with each electronic transition direction. These polarisation dipoles may oscillate in-phase or one cycle out-of-phase in relation to the exciting frequency. The polarisation dipoles in adjoining chromophores interact with the transition dipole of the given chromophore and may reinforce it or weaken it, to give hypo- or hyperchromism. The interaction is reciprocal, so that the intensity gained by some transitions through interaction with others will result in loss of intensity in the latter (though this may not be observable if they occur at inaccessibly low wavelengths). A given transition cannot steal intensity from itself. Thus the scheme shown in Fig. 4.9 must be interpreted in terms of the interaction of a transition dipole μ_1 in the given chromophore with those corresponding to other transitions, μ_2 say, in adjoining chromophores. This amounts to a dispersion force interaction.

The conservation of intensity within the exciton system may seem intuitively reasonable, since the expression that defines the intensity involves the wave-functions describing both the ground and the excited states, and the symmetry of the system (Fig. 4.8) is such that whatever applies to the transition from the ground state to one exciton state will apply in an equal and opposite sense to the transition to the other exciton state. A proper definition of the system requires that the effective ground state and first excited state be expressed in terms of the pure (unperturbed) state with a small contribution mixed in from the other states with which they interact in the array. When the intensity is evaluated from the wave-functions of the states so defined, terms involving like states (exciton interaction) vanish, while those involving interaction of unlike states remain finite. A clear treatment at this level may be found in Murrell (1964). Whether the intensity of a given transition increases or decreases depends on the relative alignment of the transition dipoles, in a predictable manner as shown in Fig. 4.9.

1.3.3 Far-ultraviolet absorption spectra of proteins and synthetic analogues

We come now to the manifestations of these phenomena in proteins, and to the relation of the absorption spectrum of a polypeptide chain to its secondary structure. The amide (e.g. peptide bond) chromophore has an absorption band at about 190 nm, a region that has been accessible with commercial spectrophotometers only in about the last ten years. Measurements present some problems, because almost everything (including water and most buffer ions, and even chloride and hydroxyl ions) begins to absorb at short wavelengths. Nevertheless, some useful information can be extracted from this region, for the peptide absorption is sensitive to conformation. When a polypeptide chain enters the α-helical conformation, some striking changes are observed in the absorption band. These are of two kinds: band splitting and

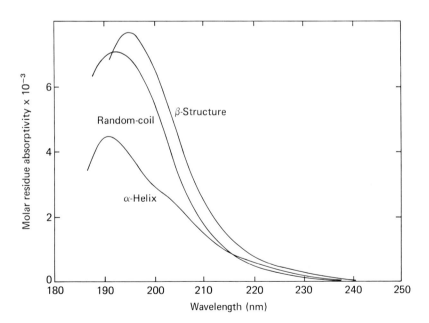

Fig. 4.10 Far-ultraviolet absorption spectra of polypeptide in solution in the α-helical, random-coil, and antiparallel-β conformations.

hypochromism. The chromophores in the α-helix form an exciton system, and the rules of Fig. 4.8 apply. Seen end-on the α-helix approximates to a square sided box, the peptide chromophores lying along the walls. The electronic transition moment of the strong $\pi \to \pi^*$ transition are tilted upwards at about 45° to the axis. One may therefore consider separately the resonance interactions amongst their axial, and amongst their radial resolved parts. It is easily enough shown that the former make up a head-to-tail system (Fig. 4.8), which results in a red-shift, whereas in the latter the phase relations that would lead to a shift cancel out, so that there is no change in the band position (for a pictorial version of the reasoning see for instance Kasha 1963, Gratzer 1967). The upshot is that the band is split into a component at 190 nm, polarised perpendicularly to the helix axis, and another at 205 nm, polarised along the axis. The appearance in solution is shown in Fig. 4.10.

It can be seen also that the absorption band of the α-helix is strongly hypochromic with respect to that of the random-coil. This forms the basis of a method of estimating the α-helix content of proteins and polypeptides, though in real proteins corrections must be applied for the absorption of side chains, especially those of aromatic amino acids, which have intense bands at short wavelengths. Denaturation, or other processes resulting in conformational changes can be followed in terms of the absorbance at short wavelengths, and so also can the hydrolysis of peptide bonds. The pleated-sheet, or β-conformation, as shown in Fig. 4.10, is *hyper*chromic in the region of peptide absorption. Proteins or polypeptides with unusual ordered conformations, such as collagen or polyproline, also show conformation-dependent effects in the peptide absorption band, and this is often useful for the observation of conformational transitions. (For a more detailed account of the nature and application of protein spectra in the low ultraviolet, see Gratzer 1967.)

2 Fluorescence

2.1 Excited states

2.1.1 What happens to excitation energy?

A molecule will remain in its excited state for perhaps $10^{-8}-10^{-9}$ s.* During this time a number of interesting things can happen. The molecule for example can decompose, or undergo a chemical reaction, since the excited state is apt to be more reactive. It can form a complex with a ground-state molecule of the same species (excimer) or of another species, such as the solvent (exciplex). It may ionise, since the pK of any ionising group will be changed on excitation. Most commonly the excitation energy is wholly or partly dissipated by thermal, collisional processes, and the molecule undergoes a radiationless return to its ground state; or the excitation is emitted as light in the transition to the ground state. This is *fluorescence*. Not all molecules are fluorescent, but aromatic species most commonly are.

2.1.2 Nature of fluorescence spectra

Fluorescence occurs always from the lowest excited state to the ground state. Excitation energy in higher electronic excited states is always degraded to the lowest electronic excited state

---

* We do not consider triplet states in which the spin of the excited electron is reversed. Transitions between singlet and triplet states are strongly forbidden, and the molecule is therefore trapped, once it is in its triplet state, which can have lifetimes of seconds. Phosphorescence is associated with triplet states, and is generally of interest only at low temperatures.

(internal conversion) before emission occurs. Moreover, the lifetime of the excited state is sufficient to allow the molecule to equilibrate in thermal (vibrational) terms with its surroundings, so that, as in absorption from the ground state, the molecule is always in its lowest vibrational state before emission. Thus any excess of vibrational energy that accompanies absorption to higher vibrational levels (Fig. 4.3) is lost by the time fluorescence occurs. This decreases the energy gap, as shown in Fig. 4.3, and the fluorescence band, is therefore always on the long-wavelength side of the absorption band. Furthermore, a rough mirror-image relation generally exists between the shapes of the absorption and fluorescence bands, because of the inverse pattern of vibronic components (Fig. 4.3), absorption proceeding from the lowest vibrational level of the ground state and fluorescence from the corresponding level in the excited state. The intensity of fluorescence is measured by the quantum yield (ratio of quanta put out in fluorescence to quanta absorbed in the excitation process), but for technical reasons this is tedious to determine, and fluorescence spectra generally appear on an arbitrary intensity scale.

Since the efficiency of populating the excited state depends on the amount of light absorbed, the highest fluorescent intensity will be elicited by irradiation at the absorption maximum. The trace of emitted intensity at given wavelength as a function of the wavelength of excitation is called the *excitation spectrum*, and follows the contour of the absorption spectrum. The wavelength distribution of the fluorescence, excited at fixed wavelength, is called the *emission spectrum*. As we have noted the latter lies to the red of the former.

2.1.3 Excitation transfer

Under the right circumstances another thing which can happen to excitation energy is transfer to another molecule, either of the same or of a different species. The acceptor molecule may emit (and if the two molecules are of different species, this is called sensitised fluorescence), or its energy may be degraded thermally. In this case the acceptor *quenches* the fluorescence of the donor. The mechanism of such an energy transfer, at least in the situations that interest us, involves a resonance between the transition dipoles, and depends on the degree of overlap of the emission band of the sensitiser and the absorption band of the emitter, and on the inverse of the sixth-power of the distance between the molecules, as well as on their relative orientation (parallel dipoles interacting most strongly). In practice this means that transfer can be appreciable over distances of the order of 100 Å. Now a typical globular protein might have a diameter of 50–100 Å, and the possibility of energy transfer between the chromophores within such a domain clearly exists. We shall later consider the manifestation of this effect, often referred to as Förster transfer, in proteins, and its uses.

2.1.4 Fluorescence polarisation

Suppose that an array of chromophores, in a crystal for example, is irradiated with light polarised in the direction of alignment of the transition moments. If the molecules do not rotate within the lifetime of the excited state, the fluorescent light will be polarised in the same direction. If the same molecules were randomly oriented, as in a solution, those molecules with their transition moment parallel to the plane of polarisation at any given instant will absorb maximally, and those perpendicular will not absorb at all. Again, if there were no reorientation within the excited state lifetime, the fluorescence, if it comes from those molecules that were excited directly by the incident light, would be polarised like the absorption. Contrariwise, if there were complete kinetic randomisation within the 10^{-9} or so seconds during which the excited state persists, the selective orientation of the excited molecules would be dissipated, and the emitted light would be unpolarised. (Also, if there are many energy transfer events within the same period, the polarisation will be progressively dissipated, and the emission will be wholly or partly depolarised.) Now a large molecule, the rotational diffusion of which is rather slow, especially so if the medium is viscous, may reorient quite slowly, or at any rate on a time scale comparable with that of the excited state duration. In such a case, the fluorescence depolarisation can be related to the rotational diffusion coefficient, and thus to the gross geometry of the molecule. More importantly, any change in the nature of the kinetic unit, such as might be occasioned by binding a ligand, by subunit association or dissociation, or by a conformational change, will be reflected in the fluorescence depolarisation. This is therefore a valuable, and rather easily applied method of following such effects. However the intrinsic chromophores of the molecule may not have an excited state lifetime in the right range, and in this case suitable chromophores (generally dye molecules) must first be covalently attached.

The rotational relaxation of a dye-labelled macromolecule can also be observed by a different technique, which, given the apparatus, is rapid and precise. This involves the observation of the relaxation process, in terms of polarisation of the fluorescence, through the lifetime of the excited state. Excitation is induced by a rapid light pulse, and the decay kinetics of the fluorescent intensity are observed, with a polarising prism in the emitted beam, which is oriented in one direction in one experiment and in the perpendicular direction for another. A comparison of the two profiles gives the change in anisotropy of fluorescence, as the molecules relax. Measurements are made on a time scale of nanoseconds.

2.2 Fluorescence of proteins and its applications

2.2.1 Fluorescence characteristics of the aromatic amino acids

Phenylalanine, tyrosine and tryptophan are all fluorescent. The emission spectra of the free amino acids are shown in Fig. 4.11. The intensity of the fluorescence depends, as we have noted, on the efficiency of the quenching processes that compete with it. Many non-fluorescent

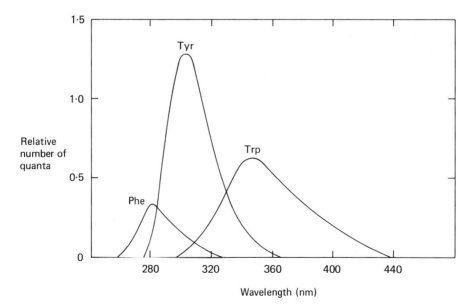

Fig. 4.11 Fluorescence spectra of the aromatic amino acids in aqueous solution (adapted from F. J. W. Teale and G. Weber (1957) *Biochem. J.*, **65**, 480, Fig. 7).

groups have the ability to quench, and the quantum yield therefore depends on the immediate surroundings of the chromophore. Thus, for example, large changes in intensity of the fluorescence of the free amino acids accompany changes in the ionisation state of the α-amino or α-carboxyl group. As might be imagined the quantum yields observed in different proteins vary over a wide range. Any gross changes in conformation, notably if the protein is denatured, and likewise accompanied by considerable changes in the fluorescent intensity.

2.2.2 Energy transfer and fluorescence of proteins

Since the linear dimension of a typical globular protein, or a subunit, is of the order of 50 Å, the separation of the aromatic residues is such that there is every likelihood that excitation energy transfer (see above) will occur. The spectral overlap requirement is fulfilled: the emission band of phenylalanine overlaps well with the absorption of tyrosine and of tryptophan, and the emission of tyrosine with the absorption of tryptophan. Thus, barring very unfavourable angles between transition moments, one may expect energy transfer in the direction phe → tyr → trp. This expectation is fully borne out. The contribution of phenylalanine, in terms of the fraction of light that phenylalanine residues, with their very low molar absorptivity, actually absorb is in practice negligible, and except in freakish proteins containing phenylalanine, but no tyrosine or tryptophan, its fluorescence is not observed. In general, the emission spectrum of proteins is to all intents that of tryptophan, though in some cases a contribution, in the form of a shoulder on the short-wavelength side, from tyrosine residues is seen. When a protein is unfolded, and becomes a rather loose coil, the average distance between the aromatic residues increases substantially, the probability of energy transfer diminishes, and the tyrosine contribution frequently becomes more prominent. Fluorescent intensity is therefore of value as a conformational marker.

The fluorescent intensity of a given species is modified by changes of solvent (for example higher viscosity makes for proportionately less radiationless decay), and is reduced by collision with quenchers added to the solution. The iodide ion, for example, is an effective quencher. This offers a means once more of examining the spectroscopic behaviour of external, exposed, aromatic residues. Moreover the association of a ligand with an aromatic residue at its binding site can change the fluorescence, and in favourable instances this can be used to measure binding processes. It may be remarked that when the solution is made alkaline, and there is appreciable ionisation of tyrosines, the tyrosinate chromophore can act as an energy acceptor from tryptophan, and since its own fluorescence is very weak, it will tend to quench the net emission from the protein. For more details concerning the intrinsic fluorescence of proteins and its uses, see Konev (1967) and Longworth (1971).

2.2.3 Extrinsic chromophores

Fluorescence in protein chemistry has perhaps been of the greatest value in situations involving cofactors, or other prosthetic groups, or artificially introduced tailor-made chromophores. The haem proteins represent a case in which the protein fluorescence is almost completely quenched. Energy transfer within the domain of a globular protein was first demonstrated in a classical experiment, involving photolysis of carboxyhaemoglobin. Carbon monoxide binds very strongly to the haem groups of haemoglobin. When exposed to a high intensity of light, the carboxyhaem group is excited, and the excited state decomposes, with the result that the carbon monoxide is lost. Irradiation at a wavelength which excites the aromatic residues of the protein has the same effect however and causes the carbon monoxide to fall off. The action spectrum of the photolysis (i.e. the wavelength-dependence of the quantum efficiency of the reaction) in fact follows the absorption spectrum of the carboxyhaemoglobin precisely. Since the haem group is such an effective quenching agent, observation of the protein fluorescence offers an extremely sensitive method of following the binding of haem to apoproteins. Even the binding of haemoglobin to its specific binding-proteins the serum haptoglobins can be followed in terms of the quenching of haptoglobin fluorescence in the haemoglobin–haptoglobin complex.

Many examples could be given of fluorescent cofactors. The fluorescence of NADH is useful for example in the assay of NAD^+-dependent dehydrogenases. The fluorescence of the pyridoxal chromophore, which is present in many proteins (e.g. glycogen phosphorylase), can be used as a sensitive marker for conformational events, such as the transition between allosteric states. An elegant way to study the interaction of the archetypal hapten, dinitro-phenyllysine, with its antibody also involves energy transfer. The dinitrophenyl group does not fluoresce, and is a strong quencher. When the hapten binds to the antibody, the tryptophan fluorescence of the latter is quenched to the extent of about 70 per cent. The fluorescent intensity is thus a reflection of the position of the binding equilibrium, and can therefore be used as a means of analysing the thermodynamics and kinetics of the interaction. Substrates or inhibitors modified by a fluorescing or quenching group have similarly been used to study interactions with enzymes.

In principle, energy transfer can also be applied to the measurements of actual distances between groups, or a change in such distances. The law that the transfer efficiency depends on the reciprocal of the sixth-power of the separation (see above) is very precisely obeyed. If now two groups, a sensitiser and an emitter, can be calibrated in a model system, to give the proportionality constant in the sixth-power equation, the extent of energy transfer can be directly interpreted in terms of distance (usually with the assumption of free rotation of the groups involved to allow the use of an average value for the relative orientation). Since many reasonably specific, fluorescent reagents are now available for the modification of reactive groups in proteins, and substrates or ligands can be similarly labelled to order, it is often possible to set up such a system, which will detect critical distances, or changes in geometry, during biologically important reactions. For example an estimate has been attempted of the angle between the prongs in immunoglobulin (a Y-shaped molecule, with the two antigen binding sites at the ends) by making a hybrid antibody, with one site specific for a hapten containing a sensitiser chromophore, the other for an emitter. The extent of energy transfer gives an estimate of the distance between them, and therefore of the angle of the Y in solution, since the lengths of the limbs are known from electron microscopy.

The fluorescent group need not for all purposes be covalently attached. Some substances are available (the best known being an anilinonaphthalenesulphonic acid) that bind to hydrophobic regions of proteins (e.g. the active sites of enzymes, which are in general hydrophobic cavities giving access to the interior), and in so doing acquire a greatly enhanced fluorescence. The displacement of such molecules by other ligands (such as substrates) is then accompanied by quenching of the fluorescence, and this again provides a good means of observing binding processes in many situations. A perusal of the current literature of protein chemistry will reveal many other applications of measurements of extrinsic fluorescence. Energy transfer is thought to be important in natural processes, especially photosynthesis. At low temperature, in frozen solution, proteins also display phosphorescence. Transfer distances involving triplet states are very short, and the suggestion has been made, though not yet explored, that this offers a means of measuring very small distances.

2.2.4 Fluorescence polarisation in proteins

Measurements of fluorescence polarisation of bound dyes of suitable excited-state lifetime have been widely used in protein chemistry. (The lifetimes of the aromatic amino acids in the excited state are too short to be comparable with rotational relaxation events of large molecules.) They can give a measure of changes in local rigidity, but more particularly they allow, as we have already mentioned, changes in the nature of the kinetic unit to be observed, especially in binding or dissociation equilibria. Again the attachment of haptens or antigens to labelled antibody can for example be followed. In principle rotational relaxation times, which can be determined, will give a measure of the dimensions of the equivalent hydrodynamic ellipsoid of the protein. A related method, as we have mentioned, with a number of advantages for extracting rotational diffusion coefficients, and detecting changes in the size or shape of the molecule, involves measurement (in the nanosecond time range) of the decay of emitted intensity, as the chromophores return to their ground state, of the two polarised components of the fluorescence. For a general account of work on fluorescent protein conjugates, see Dandliker and Portmann (1971).

3 Optical activity

3.1 Nature and origins of optical activity

3.1.1 Optical rotation

We shall consider here the nature and origins of optical rotation and circular dichroism. A simple qualitative explanation of the phenomena runs as follows: plane-polarised light, which is generated by passage of unpolarised light through a double-image prism, or a sheet of polaroid, consists of a wave, the electric vector of which oscillates in a plane (the plane of polarisation). It is convenient for the present purpose to visualise this as the resultant of two vectors, rotating at equal rates in opposite senses (Fig. 4.12). When they reinforce, at twelve o'clock, the resultant in the vertical plane will be maximal and positive. When they are at nine and three o'clock, the resultant is zero, and when they reinforce again at six o'clock, the resultant will be maximal and negative. Thus the resultant wave coming out of the plane of the paper performs a sinusoidal oscillation in the vertical plane. The two components are called the right- and left-circularly polarised rays, and clearly follow clockwise and counterclockwise screw motions. These components are not an abstraction, for circularly-polarised light can be generated in practice, as we shall see. Now in an optically active medium (a quartz crystal say) there is an intrinsic disymmetry: the lattice does not look the same to a left-handed, as to a right-handed ray travelling through it. Since with a finite refractive index, the velocity of the light will be retarded in the medium, it will not be surprising to find that the left- and right-handed components are not retarded to precisely the same extent. In other words, the refractive index of the material is different for the two rays, and the substance is birefringent. Thus after some number of rotations of the two circular components of Fig. 4.12(a), one will arrive at the situation shown in Fig. 4.12(b), where the screw described by the slower component lags behind that described by the faster. The resultant will now make an angle with the vertical plane, the angle of rotation, α. Thus the plane of polarisation is rotated in an optically active medium, and this is the phenomenon known as *optical rotation*, which is a measure of the difference between the refractive indices for left- and right-circularly polarised light n_l and n_r. (It

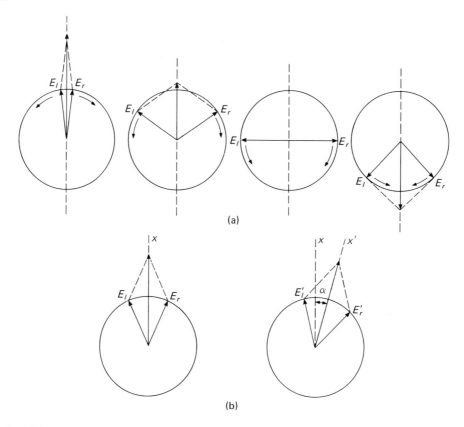

Fig. 4.12 (a) Plane-polarised light, broken down into its right- and left-circularly polarised components. The positions of the vectors of the latter and the resultant in the plane of polarisation are shown after successive time intervals. (b) Components and resultant before and after passage through an optically active medium, showing rotation of the plane of polarisation.

is easy in fact to show that $\alpha = (n_l - n_r)\,\pi/\lambda$ where λ is the wavelength. For a dextrorotatory substance α is positive, and for a laevorotatory material it is negative.)

The optical rotation can be measured at different wavelengths to give the *optical rotatory dispersion*. Just as the refractive index of a medium increases towards shorter wavelengths (see above), the refractive index difference, and hence the optical rotation, similarly increases. We have noted that in the region at which the resonance condition is fulfilled, that is to say at the wavelength of an absorption band, there is a convulsive change in the refractive index (Fig. 4.1), and again, as one would expect, the same is true of the refractive index difference, and thus the optical rotation. This manifestation (Fig. 4.13) is called the *Cotton effect*. A Cotton effect in which the rotation is positive on the long-wavelength side is a positive Cotton effect, and vice versa for negative rotation. A theoretical treatment gives rise to the *Drude equation*, which is however an approximation valid only for wavelengths well away from the Cotton effect. The Drude equation relates the specific optical rotation, $[\alpha]$, which we shall define later in terms of the concentration of optically active material, to the wavelength of measurement λ, and that of the centre of the Cotton effect λ_c:

$$[\alpha] = k_c/(\lambda^2 - \lambda_c^2)$$

where k_c is a constant. It will be seen that as λ approaches λ_c, $[\alpha]$ approaches infinity, so that

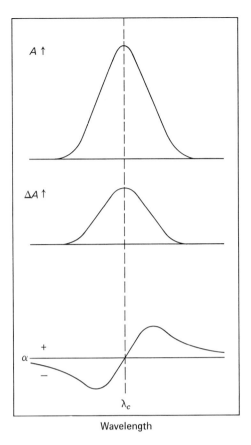

Wavelength

Fig. 4.13 Appearance of Cotton effect in circular dichroism and optical rotatory dispersion. The corresponding absorption band is also shown. The Cotton effect depicted here is a *positive* Cotton effect. In a negative Cotton effect the circular dichroism would be negative, and the optical rotatory dispersion would show a negative lobe at long wavelength and a positive lobe on the short wavelength side.

the condition $\lambda \gg \lambda_c$ must be met if the equation is to be applied. In fact a term such as the above may be expected to arise from each absorption band of the optically active chromophore, and the actual $[\alpha]$ will be the sum of many such terms. In practice, in the nature of the equation, if the Cotton effects are widely spaced, the contribution of the nearest (longest-wavelength) Cotton effect will dominate at a wavelength clear of all the Cotton effects (i.e. on the long-wavelength side of all absorption bands), and the Drude equation is most commonly obeyed well enough. The Drude constant, λ_c, which can be extracted from a plot of $[\alpha] \lambda^2$ against $[\alpha]$ for example, must then however be regarded as an empirical constant.

3.1.2 Circular dichroism

Since in an optically active medium the refractive indices associated with any transition are different for the left- and right-circularly polarised rays, it will not be surprising to find that this difference in the interaction with the light extends also to absorption. Indeed inasmuch as optical rotation measures the refractive index difference, and the refractive index of a material is related to its absorption in the manner indicated in Fig. 4.1, a precisely analogous relation should prevail between refractive index difference, i.e. optical rotation, and absorbance difference. Left- and right-circularly polarised light should then be absorbed with different

184

intensities $(A_l \neq A_r)$ and this is found to be the case (Fig. 4.13). This phenomenon is called circular dichroism, and can be measured directly. The absorbance difference $(A_l - A_r)$ in a band may be positive (positive Cotton effect) or negative (negative Cotton effect). The effect is small — a 1 per cent difference would be regarded as a large circular dichroism. It may be noted that if one of the circular components (see Fig. 4.12(a)) is preferentially absorbed, the two rotating vectors become of unequal length. The effect, it may readily be seen, is that the resultant describes an ellipse. The ellipticity of the light after passage through the medium is thus an alternative measure of the circular dichroism, and is in fact most commonly used to describe it. Circular dichroism (CD) and optical rotatory dispersion (ORD) are complementary phenomena, and may be analytically related (just like absorption and refractive index) by a set of relations, well known to physicists, the Kronig–Kramers transforms. Since the optical rotation at any wavelength is related to the circular dichroism over the entire wavelength range, and vice versa, the transformation of one into the other requires the numerical evaluation of an integral for each wavelength, and is scarcely feasible without a computer.

3.1.3 Physical basis of optical activity

Whereas this rather crudely intuitive view of optical activity will probably suffice as a basis for studying the effects observed in biological molecules, we will attempt a slightly more analytical approach, as a starting point for anyone wishing to pursue the subject in greater depth. We consider first a molecule in the form of a left- or right-handed screw. A real case of this kind is represented by hexahelicene, which consists of six fused benzene rings in an open circle. Because of steric hindrance at the open end this molecule is not planar, but twisted, and it may have a right- or left-handed screw-sense. Synthetic hexahelicene can be resolved into these two forms. The π-electron disposition is thus helical and electrons may be regarded as travelling in a helical path. Suppose now that the molecule is allowed to interact with plane-polarised light travelling in a direction perpendicular to the axis of the screw (Fig. 4.14). When the electric

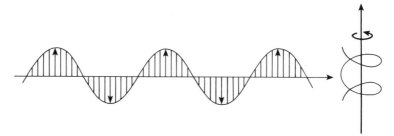

Fig. 4.14 Interaction of a plane polarised ray, propagated in the direction indicated, with a molecule in which the electrons are constrained to move in a helical path. Displacement of electronic charge has the components indicated.

vector of the incident polarised light is parallel with the screw axis of the molecule, it will cause a polarisation of electronic charge in this direction. That is to say an oscillating dipole will be induced up and down the helix axis. Furthermore, since the electron displacement is spiral, there will be a component of charge rotation around the axis, the result of which will be an induced magnetic dipole in the axial direction. It follows from Maxwell's equations that an oscillating magnetic dipole will emit radiation with a magnetic vector in the same plane as the dipole. (This magnetic dipole radiation is weak and its intensity is generally treated as negligible.) The new scattered wave combines with the incident radiation, and since the electric vector of the primary wave is perpendicular to its magnetic vector, and therefore to the electric vector of the secondary radiation, the resultant of the two is tilted out of the plane of the incident electric vector. This is in fact only half the story, because we must also consider the effect of the magnetic vector of the incident light on the helix. The magnetic field induces a current flux in the helix, in other words another contribution of induced electric dipole along the axis. The new oscillating electric dipole generates a secondary wave, which interacts, as before, with the electric vector of the primary radiation, causing rotation of the plane of polarisation. A full discussion, including a consideration of the effect of random orientation of the disymmetric molecules, is given by Kauzmann (1957). It is worth emphasising here, in regard to the orientation effect, that a left-handed screw remains a left-handed screw regardless of the direction from which one is viewing it.

Optical activity has been rigorously treated in quantum mechanical terms. The upshot is that the rotational strength of a transition depends on the vector product of the electric and magnetic transition moments: $R = \text{Im } (\mu m \cos \theta)$ where μ and m are these transition moments, and θ the angle between them. For R to be finite θ must differ from $90°$, which is its value in all symmetrical molecules. The magnetic dipole moment comes about from a rotational motion of electronic charge about an axis. The prefix Im indicates that one takes the imaginary part of this complex number, which arises merely from the fact that the electric and magnetic vectors are out of phase with each other. In hexahelicene, as we have shown, there is a rotational component to the charge displacement, with a component of electrostatic displacement along the same direction. This type of disymmetric chromophore is rather rare; more often the disymmetry is not a function of the chromophore *per se*, but of the disposition of groups around it. The most familiar example occurs in compounds containing an asymmetric carbon, with four different substituents, tetrahedrally disposed. Like a left- or right-handed helix this has an inherent disymmetry (in the sense that a left-hand helix is distinguishable from a right-hand helix, whichever way it is oriented). An important consideration is that different transitions in a molecule are associated with very different rotational strengths. Large magnetic moments may occur in very weak transitions, and $n \rightarrow \pi^*$ transitions (see above) are often associated with very large values. A component of the electric dipole in the same direction can be brought about by introduction into the $n \rightarrow \pi^*$ state of some of the character of other states with different symmetries. The result, at any rate, is that $n \rightarrow \pi^*$ transitions, which are often barely visible in the absorption spectrum, being frequently concealed in the long-wavelength tail of stronger bands, give rise in many cases to very prominent Cotton effects.

There are two general types of mechanism that are considered for generating a finite vector product of the electric and magnetic dipoles to fulfill the condition for optical activity. In the one (one-electron theory) the postulate is that the asymmetric field perturbs the symmetry of the electron displacement, so as to introduce a resultant electric moment into the magnetic transition. The other mechanism has been treated classically and quantum mechanically and is known as the coupled-oscillator model. Here different polarisabilities within the molecule interact with one another, their separation being smaller than, or comparable with, the wavelength of the light. Suppose two groups in the molecules are arranged at some angle to each other, and that a circularly-polarised ray, with the electric vector performing a right-handed corkscrew motion is propagated along the line joining the two groups. If the direction of the vector when it reaches the nearest group is represented by a full arrow, and that when it reaches the second group by a broken arrow (propagation into the plane of the paper), the arrangement will look like this: ↗ (the groups being separated by a small part of a wavelength). For left-handed circularly-polarised light the picture would be different ↖. These will also be the directions of the induced electric dipoles. The field resulting from the dipole indicated by the broken arrow will then induce a dipole in the other group, shown by a short

186

arrow, as follows in the two cases: ⌡ and ⌐ . The effect of the perturbation is clearly inverted when the screw-sense of the incident light is changed. For an assemblage of molecules randomly oriented, the two pictures would be mirror images, and one could therefore turn one into the other by reorientation. The two effects would therefore cancel and the result would be no optical activity. If on the other hand the molecule has no plane or centre of symmetry, it is not equivalent to its mirror image, and the two arrangements shown above are also not equivalent. These are the conditions for optical activity.

3.1.4 Exciton systems

A special case of optical activity occurs in disymmetric exciton systems. Suppose two identical chromophores, skewed relative to each other, but with their planes parallel and close together, undergo resonance interaction. Looking down on the chromophores, and denoting the top

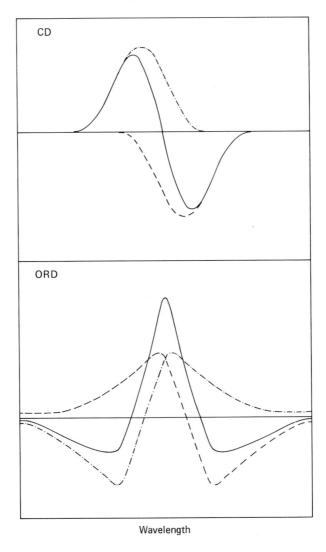

CD

ORD

Wavelength

Fig. 4.15 Optical activity of an exciton system, showing the combination (bold line) of equal and opposite positive and negative Cotton effects.

transition moment by a full, and the bottom transition moment by a broken line, we have the two possible phase relations: ✕ and ✕. In either case there is both rotational and translational displacement of charge. In the terms of the argument outlined above, these will combine in opposite senses to yield two identical but opposite rotational components R and $-R$. By reasoning represented in Fig. 4.8, the transitions corresponding to the two phase relationships will have different energies, and the net result in ORD and CD, will be as indicated in Fig. 4.15. This situation is expected for a helix in which exciton interactions between chromophores are prevalent. Non-exciton interactions between different transitions can also generate Cotton effects, but these are not characterised by the same symmetry property.

3.1.5 Practical considerations

The measurement of ORD outside the region of absorption is in instrumental terms easy. The observation of Cotton effects is more difficult, both in ORD and CD, for the light which is to be analysed is attenuated by the absorption of the sample. It is the ratio of optical activity to absorption that determines the difficulty or ease of observation. The measurement of circular dichroism requires the generation of circularly-polarised light, and this is done by passing plane-polarised light through a quarter-wave plate or other device, which retards one of the components shown in Fig. 4.12 by a quarter of a cycle. One can visualise that the resultant vector will describe either a left- or right-handed corkscrew motion, according to which one is retarded. In a good modern instrument absorbance differences of the order of 10^{-6} (generally on a peak sample absorbance of about 1) can be discriminated (and in ORD a rotation of some $0.001°$).

Since, as has been noted, CD and ORD are equivalent phenomena, the question arises which, given the choice, it is preferable to measure. The answer in general is CD, for it is on the whole more informative. The CD at any wavelength arises only from absorption bands that contribute finite absorbances at that wavelength. ORD on the other hand, since it is not confined to the region of absorption, contains a component of optical rotation from all Cotton effects in the molecule. One of the consequences is that a CD spectrum may be resolved just like an absorption spectrum into its components. This is clearly much more difficult (in fact generally impossible) for a set of such curiously shaped functions as ORD Cotton effects. This should be clear from Fig. 4.16. Also when a particular CD band arises from one kind of group in a complex molecule, such as a protein, its behaviour, which reflects the response of that particular group to its environment, can be observed more or less independently of contributions from other groups, as long as the overlap is not too extensive. This is not possible in ORD. On the other hand, ORD is often useful if the absorbance is very large, or if other absorbing but optically inactive species are present, for by operating at wavelengths outside the absorption region, observations can still be made. For instance protein denaturation reactions are often conveniently observed in this way.

Units Optical rotation is measured in degrees, α. The *specific rotation* is defined as $[\alpha] = 100\,\alpha/cd$ where c is the concentration in g/100 ml and d the optical path length, expressed for historical reasons, and deplorably, in dm. This archaism still persists. The molar rotation, written $[\phi]$ or $[m]$ is defined as $[m] = M[\alpha]/100$, where M is the molecular weight (or in polymers, such as proteins, generally the monomer-unit weight, so that $[m]$ is molar rotation with respect to residues, so as to be independent of the length of the polymer). There is a dependence of optical rotation on the refractive index of the medium, and sometimes a correction is applied (the Lorentz factor) to refer the optical rotation to vacuum conditions. Thus corrected, the molar rotation is generally written $[\phi']$ or $[m']$ and is $\{3/(n^2 + 2)\}\,[m]$.

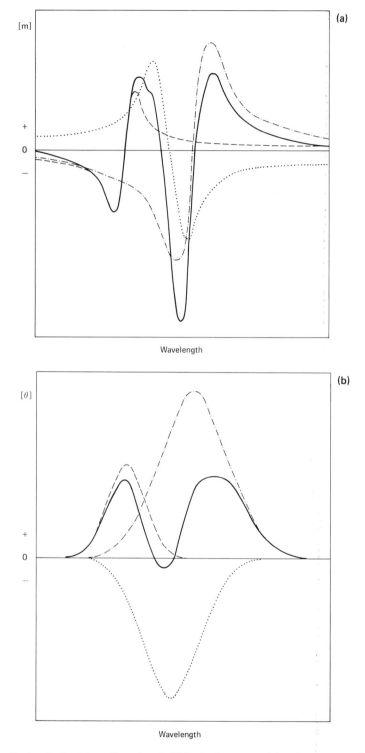

Fig. 4.16 (a) Resultant optical rotatory dispersion (bold line) of a system giving rise to two positive and one negative Cotton effects. (b) Resultant and component parts of circular dichroism of the same molecule.

Strictly speaking, since n, the refractive index, varies with wavelength, a different correction term should be applied at each wavelength. It is doubtful whether the correction has a high degree of quantitative validity. Circular dichroism is expressed as $\Delta A = A_l - A_r$ (the absorbances for left- and right-circularly polarised light). By analogy with absorbance (see above), the molar circular dichroism is $\Delta \epsilon = \Delta A/Cl$, where C is the molar concentration and l the path length in cm. The molar ellipticity (again commonly referred to mole residues of polymer) which is most often used to express the circular dichroism, is $[\theta] = 3300\ \Delta \epsilon$. Or if the circular dichroism, as measured, is expressed in ellipticity θ rather than absorbance difference, $[\theta] = 100\ \theta/Cl$. By analogy with the oscillator strength in absorbance (see above) the *rotational strength* of a band may be expressed in terms of the area under the circular dichroism curve, adjusted by a constant. Thus $R = 23 \times 10^{-40} \int \Delta \epsilon\ d\bar{\nu}$ (c.g.s. units).

3.2 Optical activity of proteins

3.2.1 Origins of optical activity in native proteins

There are three ways in which optical activity can in principle come about in a native protein. In the first place, the polypeptide chain is made up of L-residues, and optical activity must therefore be present in virtue of this alone. Secondly, an ordered spiral arrangement of residues, such as obtains in a right-handed α-helix, creates a new basis for optical activity. (This indeed was a situation envisaged by Pasteur, who adverted to the case of a right- or left-handed spiral staircase, made up of inherently symmetrical steps.) Thirdly an asymmetric distribution of charges or dipoles about a chromophore, resulting from the rigid tertiary structure, can also provoke optical activity. Let us first note that the intrinsic optical activity of chromophoric side chains, under the influence of the asymmetric α-carbon alone, is for practical purposes negligible. That is to say, in a random-coil polypeptide (a fully denatured protein for example), Cotton effects from the tyrosine and tryptophan side chains are just, but only just, observable, and can be neglected by comparison with the optical activity of the peptide chromophores.

3.2.2 Optical activity and chain conformation

The optical activity of the peptide chromophore is of especial interest, for it is strongly dependent on the conformation of the chain of which it is a part. It was known long before the observation of the Cotton effects became technically feasible that denaturation of proteins was accompanied by a large change of optical activity, and that whereas the ORD of the random-coil obeyed the one-term Drude law, that of the α-helix did not. The problem was theoretically treated by Moffitt, who deduced the exciton structure of the peptide absorption band (Fig. 4.8), and suggested that one Cotton effect should arise from each of the two exciton components. With two Cotton effects so close together, a single-term Drude law would not be obeyed. Instead one would write a Drude term for each Cotton effect. The resulting equation could be simplified for measurements at long wavelengths λ such that λ-λ_1 or λ-$\lambda_2 \gg \lambda_2$-λ_1, where λ_1 and λ_2 are the wavelengths of the two bands. One may then take a mean Drude constant, λ_0, half way between λ_1 and λ_2, and algebraic simplification now yields the celebrated Moffit equation:

$$[m'] = a_0 \left(\frac{\lambda^2}{\lambda^2 - \lambda_0^2}\right) + b_0 \left(\frac{\lambda^2}{\lambda^2 - \lambda_0^2}\right)^2$$

where a_0 and b_0 are constants and λ_0 is generally taken as 212 nm. In the random-coil, one term suffices to describe the ORD, and the second term disappears ($b_0 = 0$), whereas in the helix the second term is large (empirically b_0 about $-630°$), and is a consequence solely of the α-helical conformation. It was found that highly α-helical molecules obeyed the Moffitt equation, which was thought to provide a means of measuring the α-helix content of proteins. Operationally, and within limits, the Moffitt equation still stands, as do other two-term Drude relations (e.g. the Shechter–Blout equation). The Moffitt equation must however now be regarded as empirical, both because of a theoretical uncertainty that later came to light, and because the hypothesis concerning the underlying Cotton effects proved to be inadequate.

What was omitted from earlier considerations, was the existence of an $n \to \pi^*$ transition, corresponding to promotion of lone-pair electron of the peptide carbonyl oxygen atom into the π-system. This is a very weak band, which is barely detectable in the tail of the strong $\pi \to \pi^*$ absorption, at about 225 nm. It is nevertheless, as $n \to \pi^*$ transitions frequently are, associated with a large rotational strength, enhanced by perturbation by the electrostatic field of the helix. There are thus three observable Cotton effects in the low ultraviolet: one, which is negative, comes from the $n \to \pi^*$ transition, a second, also negative, from the first (parallel polarised) limb of the $\pi \to \pi^*$ transition, and a third, which is positive, from the second (perpendicularly polarised) $\pi \to \pi^*$ limb. The resulting ORD and CD curves are shown in Fig. 4.17.

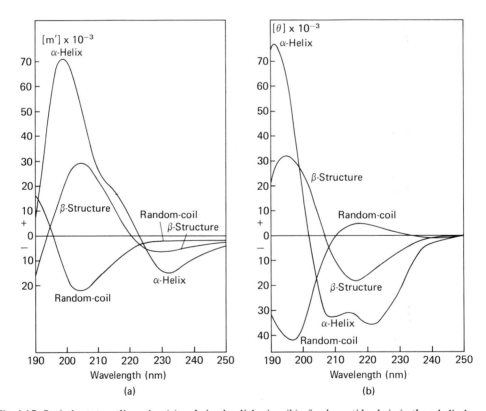

Fig. 4.17 Optical rotatory dispersion (a) and circular dichroism (b) of polypeptide chain in the α-helical, random-coil and β-conformations. Curves are based on the synthetic polypeptides, poly-L-glutamic acid and poly-L-lysine.

The ORD and CD of the random-coil presents a quite different appearance. The notion of a random-coil of course does not imply complete rotational freedom of the backbone, and certain ranges of values of the torsional angles are preferred, depending to some extent on the

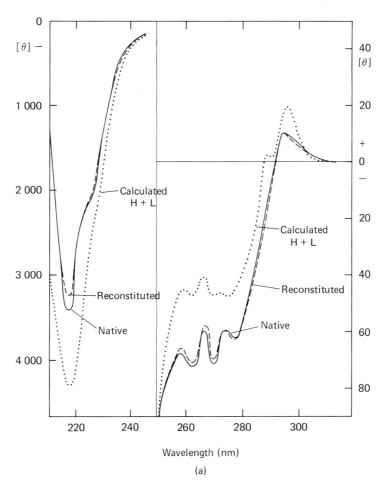

Fig. 4.18 Some circular dichroism spectra of proteins, illustrating applications of optical activity measurements.

(a) Effect of separation and recombination of heavy (H) and light (L) chains of immunogobulin, in the peptide and aromatic regions (taken from K. J. Dorrington and B. R. Smith (1972) *Biochim. Biophys. Acta,* **263,** 70.)

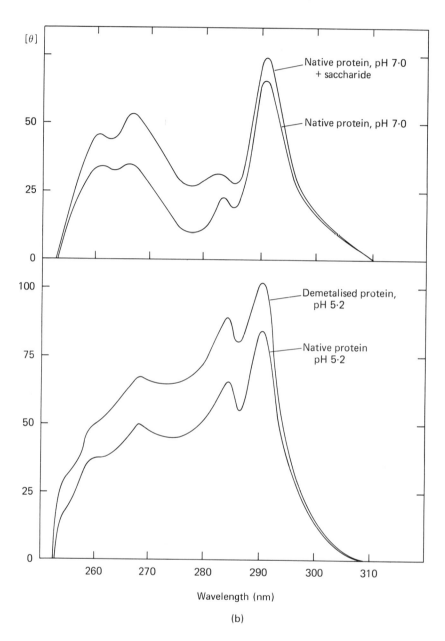

(b) Effect of binding of saccharide to, and removal of divalent cations from, concanavalin A (taken from data given by W. D. McCubbin, K. Oikawa and C. M. Kay (1971) *Biochem. Biophys. Res. Commun.*, **43**, 666 and M. N. Pflumm, J. L. Wang and G. M. Edelman (1971) *J. Biol. Chem.*, **246**, 4369).

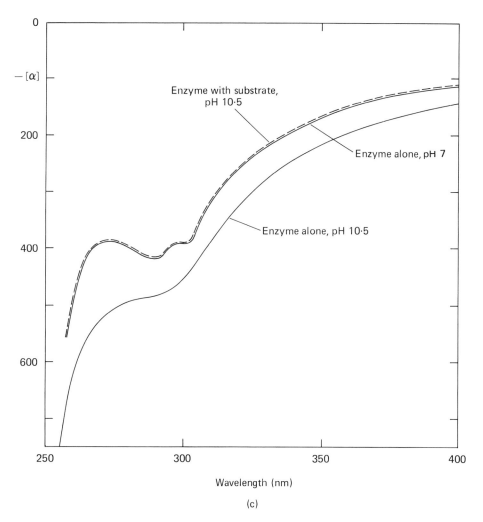

(c) Effect of alkaline denaturation on aromatic side-chain Cotton effects of trypsin, showing stabilising effect of substrate analogue (adapted from data given by J-J. Béchet and A. d'Albis (1969) *Biochim. Biophys. Acta.* **178**, 561).

bulk of the side chains. These affect the optical activity, which is not therefore necessarily quite the same in all polypeptide random-coils. A considerable degree of conformational tolerance is also acceptable in α-helices, and probably even more in β-structures. The curves shown in Fig. 4.17 are representative, though a certain variation exists (see e.g. Gratzer and Cowburn 1969, for summarised data). The ORD and CD curves of proteins by and large fit reasonably well to the expected shapes and magnitudes for a mixture of α-helix, β-structure and random-coil, but there are perturbations from other structural interactions in the rigid polypeptide chain, from likely differences between long and short helices, and from contributions of cystine and, as we shall see, aromatic, side chains. Unusual helices, such as the polyproline structures, collagen, etc., give their own characteristic systems of Cotton effects.

As a method of obtaining structural information about proteins then − a kind of poor man's crystallography − ORD and CD afford distinctly low-grade data. At the same time they offer a means of recognising α-helices (as in many fibrous proteins for example) and β-structures in proteins in dilute solution, and most of all they provide an incomparable method of observing conformational phenomena, to which in general the optical activity responds strongly. Applications along these lines are too numerous to mention.

3.2.3 Optical activity of side-chain chromophores

Although in the absence of structural perturbations the optical activity of side chains can be taken as negligible, this is by no means always, or even generally, so in native proteins. Perturbations can come about in a number of ways, and the implication is always that there is no free rotation about the αC–βC bond, otherwise asymmetric influences on the chromophore will tend to be averaged out. The perturbation may come about by exciton interaction between like chromophores (an example of this occurs in α-helical polytyrosine), from interaction between unlike chromophores, or from close-lying charges or dipoles. Complex systems of Cotton effects, often showing extensive vibrational fine structure, arise from tryptophan, tyrosine and phenylalanine, and disappear if the protein is denatured. Some examples are shown in Fig. 4.18. Some cases are even known in which the induced optical activity in aromatic residues is so large that it actually dominates the CD and ORD of the protein. At low temperatures, generally in glassy solvents in liquid nitrogen, the rotation of freely exposed chromophores on the surface of the protein is inhibited. This freezing-in of side-chain motion can cause large effects in the CD. Any events involving aromatic residues with large rotational strengths are likely to modify the measured CD. These could include conformational adjustments around the residues in question, ionisation changes in neighbouring groups or binding of ligands. Such changes in CD in the aromatic region have often been observed, and have led to inferences about sites of interactions, and have, moreover, been used to measure binding equilibria. Aromatic Cotton effects have been known to respond by changes in magnitude, and also inversions of sign in some systems. Such effects provide handles for the observation of highly localised events in globular proteins.

The disulphide chromophore generates optical activity near 250 nm. The sign of the Cotton effect depends, as has been shown with model compounds, on the screw-sense of the disulphide bond, since two stable configurations exist, which will transform into each other by rotation about the single bond. In native proteins rotational freedom of cystines is absent, and the net contribution, which in cystine-rich proteins may be by no means negligible, depends in sign and magnitude on the distribution of the bonds between the two permitted configurations, as well as on local perturbations by other groups.

At short wavelengths, especially in proteins of low helicity, in which the CD is not swamped by the helix contributions, many small Cotton effects are often seen. In some cases they doubtless arise from backbone conformational interactions, in others from side-chain chromophores. A good example is γ-globulin, in which their modification on separating the light and heavy chains, and the extent of their reappearance when these are recombined, has given important information about the relationship between the chains (see Fig. 4.18).

3.2.4 Optical activity of extrinsic chromophores

Prosthetic groups, often devoid of inherent optical activity, such as haem, commonly acquire optical activity as a result of the asymmetry of their environment when they are bound to the protein. The derivatives of haemoglobin all have systems of Cotton effects throughout the haem absorption bands, and these can be used to identify conformational states. Characteristic changes are caused by the interaction of α and β chains, which can be used to assess whether normal subunit interactions, such as are required for co-operative functions, are present. In some haemoglobin variants these interactions are faulty. Cofactors of all sorts acquire optical activity in proteins, and this again can be used to monitor changes in the immediate environment of the cofactor, that may occur independently of changes elsewhere.

The introduction of artificial chromophores has also frequently been of value. Modification of an enzyme at the active site, say by introduction of a dinitrophenyl group, which possesses an absorption band, and if rigidly held, a corresponding Cotton effect on the long-wavelength side of the intrinsic protein absorption, can for example be used to provide a means of studying interactions only in the vicinity of this residue. Nitration of tyrosines and the coupling of arsanilazo groups are among the many methods that have been used to introduce chromophores capable of individual observation into interesting regions of globular proteins. Large changes, including inversions, in extrinsic Cotton effects arising from such groups, in consequence of enzymically relevant events, can often be observed. Non-covalently bound dyes can also at times be used to follow changes in the conformation or ligand state of proteins.

Finally, the induced optical activity of ligands can be used to study binding reactions. Again chromophoric haptens provide a good example. Pharmacology is rich in examples of chromophoric biologically active molecules, which acquire optical activity when bound to their receptor proteins, and the attachment of which is measured by this means. Some good examples of phenomena involving extrinsic Cotton effects in proteins and their ligands are given in the following sources: Rockey *et al.* (1972); Vallee *et al.* (1971); Chignell and Chignell (1972); Blauer *et al.* (1972).

Bibliography

BARROW, G. M. (1962). *Introduction to Molecular Spectroscopy.* New York. McGraw-Hill.

BAUMANN, R. P. (1962). *Absorption Spectroscopy.* New York. John Wiley.

BÉCHET, J-J. and d'ALBIS, A. (1969). 'Binding of competitive inhibitors to the different pH-dependent forms of trypsin.' *Biochim. Biophys. Acta,* 178, 561–76.

BLAUER, G., HARMETZ, D. and SNIR, J. (1972). 'Optical properties of bilirubin-serum albumin complexes in aqueous solution. I. Dependence on pH.' *Biochim. Biophys. Acta,* 278, 68–88.

CHIGNELL, C. F. and CHIGNELL, D. A. (1972). 'The application of circular dichroism and optical rotatory dispersion to problems in pharmacology.' In *Methods in Pharmacology,* vol. 2, pp. 111–56. Ed. Chignell, C. F.

DANDLIKER, W. D. and PORTMANN, A. J. (1971). 'Fluorescent protein conjugates.' In *Excited States of Proteins and Nucleic Acids,* pp. 199–275. Eds. Steiner, R. F. and Weinryb, I. New York. Plenum Press.

196

DONOVAN, J. W. (1969). 'Ultraviolet absorption.' In *Physical Principles and Techniques of Protein Chemistry,* part A, pp. 101–70. Ed. Leach, S. J. New York. Academic Press.

DORRINGTON, K. J. and SMITH, B. R. (1972). 'Conformational changes accompanying the dissociation and association of immunoglobulin-G subunits.' *Biochim. Biophys. Acta,* **263**, 70–81.

FÖRSTER, T. (1960). 'Excitation transfer.' In *Comparative Effects of Radiation,* pp. 300–41. Eds. Burton, M., Kirby-Smith, J. S. and Magee, J. L. New York. John Wiley.

GRATZER, W. B. (1967). 'Ultraviolet absorption spectra of polypeptides.' In *Poly-α-Amino Acids,* pp. 177–238. New York. M. Dekker.

GRATZER, W. B. and COWBURN, D. A. (1969). 'Optical activity of biopolymers.' *Nature, Lond.* **222**, 426–31.

JAFFE, H. H. and ORCHIN, M. (1972). *Theory and Applications of Ultraviolet Spectroscopy.* New York. John Wiley.

KASHA, M. (1961). 'The nature and significance of n-π^* transitions.' In *Light and Life,* pp. 31–77. Eds. McElroy, W. D. and Glass, B. Baltimore. Johns Hopkins Press.

KASHA, M. (1963). 'Energy transfer mechanisms and the molecular exciton model for molecular aggregates.' *Radiat. Res.* **20**, 55–71.

KAUZMANN, W. (1957). *Quantum Chemistry.* New York. Academic Press.

KONEV, S. V. (1967). *Fluorescence and Phosphorescence of Proteins and Nucleic Acids.* New York. Plenum Press.

LONGWORTH, J. W. (1971). 'Luminescence of polypeptides and proteins. In *Excited States of Proteins and Nucleic Acids,* pp. 319–484 Eds. Steiner, R. F. and Weinryb, I. New York. Plenum Press.

McCUBBIN, W. D., OIKAWA, K. and KAY, C. M. (1971). 'Circular dichroism studies on concanavalin A.' *Biochem. Biophys. Res. Commun.,* **43**, 666–74.

McRAE, E. G. and KASHA, M. (1964). *Physical Processes in Radiation Biology.* New York. Academic Press.

MURRELL, J. M. (1964). *Theory of the Electronic Spectra of Organic Molecules.* London. Methuen.

PFLUMM, M. N., WANG, J. L. and EDELMAN, G. M. (1971). 'Conformational changes in concanavalin A.' *J. Biol. Chem.,* **246**, 4369–70.

REID, C. (1957). *Excited States in Chemistry and Biology.* London. Butterworth.

ROBINSON, G. W. (1961). 'Electronic excited states of simple molecules.' In *Light and Life,* pp. 11–30. Eds. McElroy, W. D. and Glass, B. Baltimore. Johns Hopkins Press.

ROCKEY, J. H., MONTGOMERY, P. C., UNDERDOWN, B. J. and DORRINGTON, K. J. (1972). 'Circular dichroism studies on the interactions of haptens with MOPC-315 and MOPC-460 mouse myeloma proteins and specific antibodies.' *Biochemistry,* **11**, 3172–81.

TEALE, F. J. W. and WEBER, G. (1957). 'Ultraviolet fluorescence of the aromatic amino acids.' *Biochem. J.,* **65**, 476–82.

VALLEE, B. L., RIORDAN, J. F., JOHANSEN, J. T. and LIVINGSTON, D. M. (1971). 'Spectrochemical probes for protein conformation and function.' *Cold Spring Harb. Symp. Quant. Biol.,* **36**, 517–31.

WETLAUFER, D. B. (1962). 'Ultraviolet spectra of proteins and amino acids.' *Adv. Protein Chem.,* **17**, 303–90.

WHIFFEN, D. H. (1972). *Spectroscopy.* 2nd ed. London. Longman.

WIBERG, K. B. (1964). *Physical Organic Chemistry.* New York. John Wiley.

5
Rapid Reaction Techniques

S. E. Halford
Molecular Enzymology Laboratory, University of Bristol.

1 Introduction

The contents of a letter cannot be revealed by looking at the envelope. Perhaps the envelope might indicate the arrival of the electricity bill but it still cannot disclose its cost. Likewise steady-state kinetics produce only limited information about enzyme mechanisms. This is due to the fact that kinetic experiments under steady-state conditions feature very low concentrations of enzyme; the only parameter of the progress of the reaction that can be measured is the overall rate of substrate utilisation or product formation since the steady state is defined by the invariance of the concentrations of intermediates. Of course the constants obtained by steady-state measurements are useful descriptions of an enzyme. But if the reactions of an enzyme with its substrate are studied at much higher concentrations of enzyme, individual intermediates in the pathway can be detected and the rates of their formation and decay measured. However the velocity of enzyme-catalysed reactions is linearly dependent upon the enzyme concentration so special techniques must be employed for the study of the rapid reactions at high enzyme concentrations.

One simple consideration will suffice to illustrate the value of rapid reaction techniques. The turnover number of enzyme reactions vary considerably from one to another though the majority fall within one order of magnitude of 100 s^{-1}. At this rate, the slowest step in the mechanism will have a half time of 7 ms or less. It is therefore necessary to observe the formation and decomposition of the intermediates on the time scale of milliseconds or less. The classical 'mix and shake' method or initiating a reaction can only be used profitably on reactions with half times of 10 s or more. Separate from the advantages of time resolution, the power of rapid reaction techniques also lies in their directness and simplicity: experimental observation replaces algebraic manipulations.

The techniques that have been most widely applied to the study of fast reactions in biological systems are the rapid mixing and the perturbation methods. The principles behind these methods are self-evident: rapid mixing devices are designed so as to minimise the time taken to mix two solutions while the perturbation methods depend on the response of a chemical equilibrium to an external perturbation. Only these two techniques and the type of information that they yield will be described in this chapter. Alternative approaches to the study of fast reactions include the spectroscopic methods which rely on the broadening of an absorption line of a chemical species when that species has a lifetime similar to the absorption frequency. The frequencies associated with nuclear magnetic resonance spectra are the most suitable for kinetic measurements but this technique is discussed elsewhere in this book (see Chapter 3) rather than here (see also Sykes and Scott 1972).

In this chapter, the theory and practice of rapid mixing and perturbation methods are discussed predominantly in terms of the elucidation of enzyme mechanisms. However this area of research is only one of the many divisions of biochemistry that have benefited from the application of rapid reaction techniques. For in the description of biological systems, kinetics may be considered an end in itself, in contrast to chemistry where kinetics is only a means to an end. The aim of kinetics in chemistry is exclusively the elucidation of reaction mechanisms. To achieve a full understanding, the dynamic properties of biological macromolecules must be related to all the static structural information that can be obtained but the biological significance of any system is its dynamic behaviour.

2 Elementary kinetics

The application of any kinetic technique must be preceded by an understanding of the basic theories of chemical kinetics. This section covers only the kinetics needed to comprehend this chapter. Fuller accounts of the theory can be found in most textbooks on physical chemistry or in specialised monographs (Frost and Pearson 1961), though most of the fundamental kinetics ever likely to be required by a biochemist are given in a concise account by Gutfreund (1972).

Kinetics are the analysis of a rate of change of concentration in terms of the concentrations of the reactants, rate constants being the proportionality constants in the derived expressions. For the reaction

$$A \xrightarrow{k} B$$

the rate equation is given by

$$-\frac{d}{dt}[A] = k[A] \tag{1}$$

where k is the rate constant.† The order of a reaction is determined by the power of the concentration term in the rate equation so the above reaction is first order. The units of a first order rate constant are reciprocal time, usually s^{-1}. Integration of equation (1) yields

$$\ln[A_0] - \ln[A] = kt$$

or

$$[A] = [A_0] e^{-kt}$$

Thus the progress curve of a first order reaction has the form of an exponential. The rate constant of a first order reaction can be evaluated from the gradient of a plot of $\ln[A]$ against time. Such a semi-log plot also has diagnostic value for if it is not linear, then the reaction is not first order.

---

† In multi-step reactions, forward-rate constants are written as $k_{+1}, k_{+2}, \ldots, k_{+n}$; the rate constants for the reverse of each of these steps as $k_{-1}, k_{-2}, \ldots, k_{-n}$. Concentrations of reactants are denoted by square brackets, $[A]$, with a suffix $[A_0]$ for that at zero time and a bar $[\overline{A}]$ for that at equilibrium. Dissociation constants for any step are given as K_n.

For the reversible reaction

$$A \underset{k_{-1}}{\overset{k_{+1}}{\rightleftharpoons}} B$$

the rate equation will be

$$\frac{d}{dt}[B] = k_{+1}[A] - k_{-1}[B] = k_{+1}([A_0] - [B]) - k_{-1}[B]$$

since $[A_0] = [A] + [B]$.
But at equilibrium,

$$k_{+1}[\overline{A}] = k_{-1}[\overline{B}]$$

$$\therefore \quad k_{+1}([A_0] - [\overline{B}]) = k_{-1}[\overline{B}]$$

Upon elimination of $[A_0]$,

$$\frac{d}{dt}[B] = k_{-1}[\overline{B}] + k_{+1}[\overline{B}] - k_{+1}[B] - k_{-1}[B]$$

$$= (k_{-1} + k_{+1})([\overline{B}] - [B])$$

This equation has the form of a first order expression but the apparent rate constant evaluated from the approach to equilibrium will in fact be the sum of the two first order rate constants. The ratio of the rate constants is given by the equilibrium constant though of course the equilibrium constant by itself provides no information about the absolute magnitudes of the rate constants.

The rate equation for the bimolecular reaction

$$A + A \xrightarrow{k} A_2$$

is given by

$$-\frac{d}{dt}[A] = k[A]^2 \tag{2}$$

$$\therefore \quad \frac{1}{[A]} - \frac{1}{[A_0]} = kt$$

The second power concentration term in equation (2) indicates that this reaction is second order. The molecularity of a reaction is defined by the number of molecules in the transition state complex. The rate constant can be evaluated from the gradient of a plot of $1/[A]$ against time and has the units of $M^{-1}\,s^{-1}$.

The half-time of a reaction ($t_{1/2}$) is defined by the time taken to complete one half of the reaction from any starting point. For a first order reaction,

$$\ln[A_0] - \ln\left[\frac{A_0}{2}\right] = kt_{1/2}$$

$$\therefore \quad t_{1/2} = \frac{\ln 2}{k} = \frac{0\cdot69}{k}$$

For a second order reaction,

$$\frac{2}{[A_0]} - \frac{1}{[A_0]} = kt_{1/2}$$

$$\therefore \quad t_{1/2} = \frac{1}{k[A_0]}$$

Thus the half-time of a first order reaction remains constant throughout the course of the reaction while in a second order reaction it will increase as the reaction proceeds towards completion.

For the more general type of bimolecular reaction,

$$A + B \xrightarrow{k} C$$

$$\frac{d}{dt}[C] = k[A] \cdot [B] \tag{3}$$

where k is again the second order rate constant. Under the condition that one of these reactants is in large excess over the other so that its concentration effectively remains constant throughout the reaction, equation (3) can be converted into a pseudo first order expression.
When

$$[A_0] \ll [B_0], \quad \frac{d}{dt}[C] = k_{app}[A]$$

where

$$k_{app} = k[B_0] \tag{4}$$

In this case, k_{app} will be a first order rate constant though the true second order rate constant can be evaluated from the linear dependence of k_{app} upon $[B_0]$.

However the fact that an observed rate constant is dependent upon the reactant concentration is not proof of the bimolecularity of the reaction. The majority of reactions encountered in enzyme kinetics involve many separate steps. In a multi-step reaction, the overall rate of product formation is determined by the rate of the slowest step in the pathway. Complications can ensue if there are two or more comparatively slow steps in the pathway; this situation will be further discussed in the next section where it will be shown that it results in the rate constant for product formation being a function of the several rate constants. If in a multi-step pathway, the rate-limiting step is unimolecular but is coupled to a bimolecular process, the rate constant measured from product formation will still vary with the concentrations of reactants. In the mechanism

$$A + B \underset{k_{-1}}{\overset{k_{+1}}{\rightleftharpoons}} C \xrightarrow{k_{+2}} D$$

$$\frac{d}{dt}[D] = k_{+2}[C].$$

Now assume that the initial pre-equilibrium is established much faster than the time taken for the second step of the reaction. Let also $[A_0] \ll [B_0]$. Hence the boundary conditions are

$$K_1 = \frac{[A] \cdot [B]}{[C]}$$

and by the conservation of mass

$$[A_0] = [A] + [C] + [D], \quad \text{and} \quad [B_0] \simeq [B]$$

By eliminating $[A]$ between these equations and then substituting $[D]$ in place of $[C]$ in the above rate equation, an expression can be obtained in which the only variable is $[D]$,

$$\frac{d}{dt}[D] = \frac{k_{+2}}{1 + K_1/[B_0]}([A_0] - [D])$$

This equation is now a first order rate equation for $[D]$, from which

$$k_{obs} = \frac{k_{+2}}{1 + K_1/[B_0]} \tag{5}$$

where k_{obs} is the rate constant measured from the production of D at a particular level of $[B_0]$. Equation (5) shows that k_{obs} will increase as $[B_0]$ is increased until $[B_0] \gg K_1$ whereupon k_{obs} will equal k_{+2} and be independent of $[B_0]$. The lack of variation of k_{obs} when $[B_0]$ is large enough to saturate $[A]$ permits one to use the concentration dependencies of the observed rate constants to distinguish between this mechanism and the alternative possibility of a bimolecular reaction under pseudo first order conditions. In the latter, the rate constant would be observed to increase indefinitely with increasing $[B_0]$ and not show a saturation effect (equation (4)).

Now consider the general solution to the rate equations of a reaction mechanism containing two consecutive steps,

$$A \xrightarrow{k_{+1}} B \xrightarrow{k_{+2}} C$$

This scheme is completely described by two rate equations for the mechanism contains only two independent concentration variables; given values for $[A]$ and $[B]$ at any one time, then only one value is possible for $[C]$ at that time. For linear mechanisms, the number of independent concentration variables is equal to the number of distinct steps in the pathway and one rate equation can be written down for each concentration variable. But certain cyclic mechanisms contain fewer independent concentration variables than steps in the pathway and a corresponding degeneracy is found among the rate equations. To return to our example.

$$\frac{d}{dt}[A] = -k_{+1}[A]$$

$$\therefore \quad [A] = [A_0]\, e^{-k_{+1}t}$$

$$\frac{d}{dt}[B] = k_{+1}[A] - k_{+2}[B]$$

$$= k_{+1}[A_0]e^{-k_{+1}t} - k_{+2}[B]$$

Integration by parts yields

$$[B] = k_{+1}[A_0]\, \frac{e^{-k_{+1}t} - e^{-k_{+2}t}}{k_{+2} - k_{+1}}$$

202

Now by the conservation of mass and if both $[B_0]$ and $[C_0] = 0$

$$[A_0] = [A] + [B] + [C]$$

$$\therefore \quad [C] = [A_0] \left\{ 1 + \frac{1}{k_{+1} - k_{+2}} \left(k_{+2}\, e^{-k_{+1}t} - k_{+1}\, e^{-k_{+2}t} \right) \right\}$$

For specific values of k_{+1} and k_{+2}, the concentration changes of each reactant with time are plotted out in Fig. 5.1. The rate equations for $[B]$ and $[C]$ are both complex expressions containing sums of exponentials in k_{+1} and k_{+2}. Thus both rate constants could be evaluated

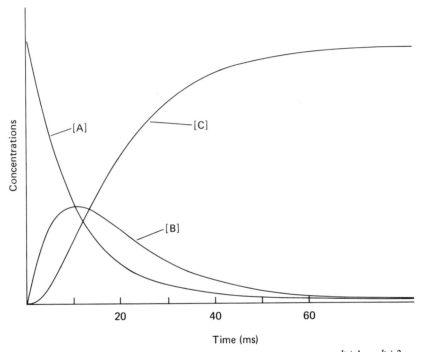

Fig. 5.1 Computer simulation of the concentration changes during the reaction $A \xrightarrow{k+1} B \xrightarrow{k+2} C$, with both k_{+1} and k_{+2} set at 100 s^{-1}. (Courtesy of Professor H. Gutfreund.) For the special case when $k_{+1} = k_{+2}$, $[C] = [A_0] \{1 - e^{-k_{+2}t} - k_{+2}t\, e^{-k_{+2}t}\}$

from records of the time-dependent concentration changes of either $[B]$ or $[C]$; the concentration of $[A]$ decays exponentially, determined solely by k_{+1}. One procedure for simplifying the rate equations is the steady-state assumption. While the concentration of $[B]$ is invariant with time,

$$\frac{d}{dt}[B] = k_{+1}[A] - k_{+2}[B] = 0$$

Under these conditions which arise whenever $[B]$ either fails to build up to any significant extent or as it passes through its maximum, the rate of disappearance of A is equal to the rate of formation of C. An alternative situation that yields a simple rate equation for $[C]$ is when either $k_{+1} \gg k_{+2}$ or $k_{+1} \ll k_{+2}$; in these cases, the appearance of C would follow an exponential progress curve determined solely by the smaller of the two rate constants. In more complicated mechanisms, however, explicit solutions for the integrated rate equations are sometimes unobtainable because there can be more unknowns than independent equations or, alternatively, the equations just cannot be integrated.

3 Rapid mixing techniques

3.1 Instrumentation

The first technique developed specifically for the study of fast reactions was the continuous flow method, originated by Roughton and Hartridge in 1923 in order to determine whether or not the combination of oxygen with haemoglobin was the rate-limiting step in the respiratory process (see Roughton 1963). Their apparatus is illustrated in Fig. 5.2. The two solutions are

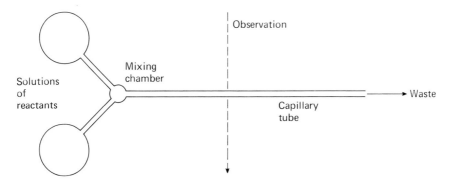

Fig. 5.2 Schematic diagram of the continuous flow apparatus of Roughton and Hartridge. The solutions are forced from their reservoirs into the observation tube by a constant head of gravitational pressure. The text covers details of the methods of observation that can be used to monitor the extent of reaction. When the apparatus is adapted for quenched flow experiments, the mixed solution is passed into the quenching solution instead of flowing down the waste pipe.

driven under pressure from their reservoirs through mixing jets and into a capillary tube across which is placed the observation system for monitoring the extent of the reaction. The age of the reaction mixture can be calculated from the flow rate and the distance between the mixing chamber and the observation point. At a flow rate of 10 ml/s and at a distance of 3 mm from the mixer, mixing was found to be completed within 1 ms. So long as the flow of solutions continues at a constant rate, the composition of the reaction mixture at any point downstream of the mixing chamber remains invariant with time. By adjusting either the flow rate or the distance between the mixing chamber and the observation point, it is possible to measure the extent of reaction at different time points after mixing.

A significant advance in rapid mixing techniques was the introduction of the stopped flow method developed principally by Chance and Gibson (see Chance 1963a). In this device (Fig. 5.3) the two reactant solutions are kept in small syringes until flushed into the mixing chamber by driving the syringe plungers forward; flow starts but is then suddenly stopped. This is achieved by passing the mixed solutions into a third syringe arranged so that its plunger is forced against a stopping barrier to prevent further flow and at the same time flicks a microswitch to trigger the recording device. During flow, the reaction mixture in the observation chamber is about 3 ms old but that portion of reaction mixture in the observation

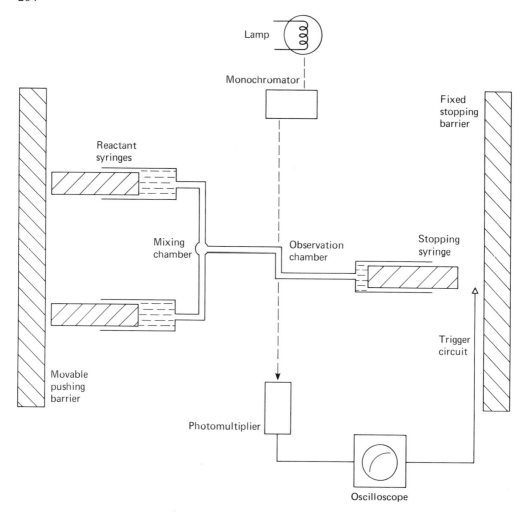

Fig. 5.3 Schematic diagram of a stopped flow apparatus set up to record extinction changes on an oscilloscope. The observation chamber is constructed as illustrated in order to optimise the optical light path.

chamber at the instant of flow arrest will remain there and age, thus permitting a continuous record of the complete time course of the reaction to be made. In brief, the operation of the stopped flow apparatus is the same as any conventional recording spectrophotometer except for the method of mixing the reactants.

The progress of a reaction is preferably monitored by some physical property that differs between the reactants and the products and can be directly related to the time-dependent concentration changes. An additional constraint in the stopped flow technique is that the response of the detection system must be faster than the reaction itself. When there exists an extinction difference between reactants and products in either ultra-violet or visible spectra, absorption optics can be used; monochromatic light is shone through the observation chamber and on to a photomultiplier whose output is recorded on a storage oscilloscope. The oscilloscope trace can be photographed for subsequent analysis. Reactions that do not involve any absorption changes but do proceed with either the liberation or uptake of protons from the solution can still be studied by spectrophotometric techniques provided that a pH indicator such as phenol red is added to the mixture. Fluoresence differences between reactants, intermediates and products have also provided suitable signals in the observation of many reactions by the stopped flow method. Fluoresence detection has the advantage of greater sensitivity over absorption optics. In the continuous flow technique, the response time of the detection system need not be faster than the reaction under study. For instance, thermocouples with a response time of 50 ms can be placed in the observation tube and provided that the flow rate is maintained for longer than 50 ms, the heat liberated in the reaction can be measured even at times less than 50 ms after mixing. In the last resort, reactions that do not possess any convenient method of observation can only be monitored by quenching the reaction mixture at various times after mixing and then chemically analysing its composition. This approach requires a modification of the continuous flow method known as the quenched flow technique; instead of observing the extent of reaction during flow down the capillary, the mixed solutions are passed through the capillary and into a quenching solution which prevents any further reaction. For most enzyme solutions, 6M perchloric acid is a suitable quench.

Though a wider range of reactions can be studied by the continuous flow method on account of its freedom from constraints upon the choice of detection systems, the stopped flow technique is in fact the more convenient to operate. Furthermore the determination of a complete reaction profile by continuous flow requires between 50 ml and 5 l of the reactant solutions while the comparable stopped flow experiment would use less than 2 ml of liquid. This saving in solution volume arises not only from the necessity of maintaining the flow rate for a fixed time in the continuous flow mode but also from the fact that the complete reaction profile is obtained in just one record from a single stopped flow experiment whereas each continuous flow experiment yields only the extent of reaction at one time point. Since many biochemical investigations are limited by the availability of the reactants, the stopped flow technique is now employed in biochemistry to the almost complete exclusion of continuous flow except in a few special cases where there is no alternative to quenched flow.

3.2 Transient kinetics of enzyme reactions

Let us start by considering a simple enzyme mechanism

$$E + S \underset{}{\overset{k_{+1}}{\rightleftharpoons}} ES \underset{}{\overset{k_{+2}}{\rightleftharpoons}} EP \underset{}{\overset{k_{+3}}{\rightleftharpoons}} E + P \tag{I}$$

The characterisation of such a scheme falls into two parts. First it is necessary to elucidate the physical mechanism. The mechanism is defined by the number and the type of intermediates and the rate constants of the individual steps. Secondly, the key events in the turnover, such as substrate recognition and catalysis, must be analysed in chemical terms. Both types of investigations benefit by the observation of the transient approach to the steady state. In the steady state, the enzyme will accumulate into the form of those intermediates that precede the rate-limiting step, but in the transient phase the concentrations of intermediates can build up to substantial levels even though these intermediates might not be present at significant concentrations in the steady state. For mechanism (I), this point has been illustrated by computer calculations on the time-dependent concentration changes of each intermediate as well as the product of the reaction (Fig. 5.4). Also shown in Fig. 5.4 are the differences in the transient phases caused by a change in the rate-limiting step; this behaviour is discussed in further detail below.

The study of transients in enzyme reactions can itself by subdivided into two fields, the transients of enzyme intermediates and the transients of reactants and products. In the latter, a

206

mathematical analysis of pre-steady state rate of product formation can reveal the rate constants for many of the individual steps in the mechanism. For this type of experiment, total product formation is observed and no differentiation is made between enzyme bound and free product. In contrast, the observation of transient intermediates requires that the intermediates be spectrally distinct. Consequently such experiments provide structural and chemical information about the separate events in the turnover as well as kinetic information. The

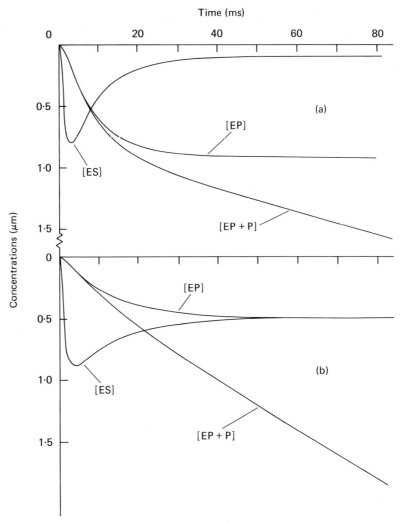

Fig. 5.4 Computer simulations of the concentration changes of enzyme-bound substrate [ES], enzyme-bound product [EP] and total product [EP + P] in the reaction (scheme I)

$$E + S \xrightleftharpoons{k_{+1}} ES \xrightleftharpoons{k_{+2}} EP \xrightleftharpoons{k_{+3}} E + P$$

after mixing together enzyme and substrate. (Courtesy of Professor H. Gutfreund.)

(a) Concentrations and rate constants assumed for the simulation were as follows: $[E_0] = 1.0 \ \mu M$, $[S_0] = 1.0 \ mM$, $k_{+1} = 10^7 \ M^{-1} \ s^{-1}$, $k_{+2} = 100 \ s^{-1}$, $k_{+3} = 10 \ s^{-1}$, and the rate constants for the reverse of each step as zero.

(b) As (a) in all respects except that $k_{+2} = k_{+3} = 40 \ s^{-1}$.

The rise in the concentration of [ES] and the consequent lag in [EP] production would be completed within the stopped flow mixing time (3 ms).

complementary nature of the transient kinetics of intermediates and those of product formation is demonstrated by Fig. 5.1, in which it was shown that the two rate constants in a simple consecutive reaction could be evaluated from records of the concentration changes of either the intermediate, [B], or the product, [C]. But without recourse to simplifying assumptions, one usually lacks explicit rate equations to describe the overall progress of a multi-step reaction or, in our case, the pre-steady state phase of an enzyme reaction. This might be considered a severe handicap to transient kinetics. However in practice, this handicap is almost never critical. But one skill required by the enzymologist is the ability to design experiments so as to ensure that the physical assumptions necessary for a mathematical solution are indeed valid under his particular conditions. The assumptions routinely involved are that the concentration of one reactant is in large excess over another or that one rate constant is much larger than another, and thus can easily be verified in the course of the experiment.

As an example of the derivation of a rate equation for transient product formation, consider scheme II

$$\text{E} + \text{AB} \underset{}{\overset{k_{+1}}{\rightleftharpoons}} \text{E} . \text{AB} \xrightarrow{k_{+2}} \underset{+ \text{B}}{\text{E} - \text{A}} \xrightarrow{k_{+3}} \text{E} + \text{A} \qquad \text{(II)}$$

In fact this reaction mechanism applies to many hydrolytic enzymes including the 'serine active site' proteases such as trypsin and elastase; the substrate, an ester AB, is hydrolysed to give an acyl-enzyme, $\text{E} - \text{A}$, with concomitant release of the alcohol, B, and the turnover is then completed by the deacylation of the acyl-enzyme. The reaction can be conveniently monitored by the release of the alcohol, B, provided that it is chromophoric. In this derivation, it will be assumed that [AB] is sufficiently large so that the bimolecular combination of E with AB is much faster than the subsequent steps and that the concentration of free enzyme is negligible. Hence,

$$[\text{E}_0] = [\text{E} . \text{AB}] + [\text{E} - \text{A}]$$

The rate of formation of $[\text{E} - \text{A}]$ is given by

$$\frac{\text{d}}{\text{d}t} [\text{E} - \text{A}] = k_{+2}[\text{E} . \text{AB}] - k_{+3}[\text{E} - \text{A}]$$

$$= k_{+2}[\text{E}_0] - (k_{+2} + k_{+3})[\text{E} - \text{A}]$$

By integration from the limit that $[\text{E} - \text{A}] = 0$ when $t = 0$,

$$[\text{E} - \text{A}] = \frac{k_{+2}[\text{E}_0]}{k_{+2} + k_{+3}} \{1 - e^{-(k_{+2} + k_{+3})t}\}$$

With the conservation of mass relationship given above,

$$[\text{E}_0] = \frac{k_{+2}[\text{E}_0]}{k_{+2} + k_{+3}} \{1 - e^{-(k_{+2} + k_{+3})t}\} + [\text{E} . \text{AB}]$$

The rate equation of the liberation of B in mechanism II is

$$\frac{\text{d}}{\text{d}t} [\text{B}] = k_{+2}[\text{E} . \text{AB}]$$

$$= \frac{k_{+2}[\text{E}_0]}{k_{+2} + k_{+3}} \{k_{+3} + k_{+2} e^{-(k_{+2} + k_{+3})t}\}$$

Integration, followed by evaluation of the constant from the boundary condition that [B] = 0 when t = 0, yields

$$[B] = [E_0]\left[\frac{k_{+2}k_{+3}t}{k_{+2} + k_{+3}} + \left(\frac{k_{+2}}{k_{+2} + k_{+3}}\right)^2 \{1 - e^{-(k_{+2} + k_{+3})t}\}\right] \tag{6}$$

The three components in equation (6) can be related to corresponding parts in the progress curves that illustrate total product formation (EP + P) in Fig. 5.4.

Upon completion of the transient phase, the rate of product formation becomes constant. In this steady state, the invariant rate of product formation (v) is given by the first term in equation (6)

$$v = \frac{[B]}{t} = [E_0]\frac{k_{+2} \cdot k_{+3}}{k_{+2} + k_{+3}} \tag{7}$$

After extrapolation of the steady-state back to zero time, it is possible to determine the amount of product liberated in the transient phase. Let this be π. From the second term in equation (6)

$$\pi = [E_0]\left(\frac{k_{+2}}{k_{+2} + k_{+3}}\right)^2 \tag{8}$$

Note that from equation (8), π = [E_0] when $k_{+2} \gg k_{+3}$ and that π = 0 when $k_{+2} \ll k_{+3}$. Thus the very existence of a transient phase locates the rate-limiting step in the reaction mechanism. Furthermore, it is not a trivial conclusion from equation (8) that the amplitude of transient product formation is equal to the enzyme concentration when $k_{+2} \gg k_{+3}$. In an oligomeric enzyme, [E_0] would represent the active site concentration and differences between the apparent number of active sites and the number of protein subunits can provide information on the nature of the subunit interactions. Transient product formation is only observed when the rate-limiting step occurs after the process that creates the spectrophotometric change with which the reaction is monitored; in mechanism II it was assumed that only the chemical transformation, E . AB → E − A + B, created the spectrophotometric change. This point is illustrated in Fig. 5.4, where a greater amplitude of transient product formation (EP + P) is observed in curve (A) when $k_{+2} > k_{+3}$ than in curve (B) where $k_{+2} = k_{+3}$.

The rate of product formation decays from the transient rate to the slower rate of the steady state in an exponential fashion. Therefore it is possible to calculate a first order rate constant of transient product formation (k_{obs}). The third term in equation (6) shows that

$$k_{obs} = k_{+2} + k_{+3} \tag{9}$$

By combination of equations (7) and (9), the individual rate constants, k_{+2} and k_{+3}, can be determined from the parameters that are measured from the rates of product formation, provided that [E_0] is known and the latter can be evaluated from the amplitude of transient product formation. The mathematical analysis of an enzyme mechanism such as that given above is indeed necessary because the measured parameters, k_{obs} and v, are both functions of sums and products of rate constants. To illustrate the importance of determining the rate constant for a single step in an enzyme mechanism, consider scheme II under the condition that $k_{+2} \gg k_{+3}$. Now the steady-state velocity will be determined solely by k_{+3} and provides no information as to the value of k_{+2}. But the rate of the step in which the enzyme hydrolyses the substrates is characterised by k_{+2} so if one seeks to analyse the mode of catalytic action of this enzyme, one has no option but to evaluate k_{+2}.

The theory of transient kinetics is just an extension of classical kinetics but the millisecond time scale of most transients in enzyme reactions requires the use of rapid reaction techniques such as stopped flow. The application of the transient kinetics of product formation in the elucidation of enzyme mechanisms can be demonstrated by some typical experiments on two different enzymes: horse liver alcohol dehydrogenase and *E. coli* alkaline phosphatase. In the reaction of alcohol dehydrogenase with ethanol and NAD, a transient phase of NADH production can be observed by the stopped flow technique prior to the steady state. This reaction can be followed by absorption spectrophotometry at 325 nm, a wavelength at which enzyme bound and free NADH have the same molar extinction coefficient. It had previously been shown for this enzyme that the steady-state rate is controlled by the dissociation of the product, NADH, from the enzyme. The theories of transient kinetics developed above are immediately applicable to two-substrate enzymes provided that the second non-chromophoric substrate is present at saturating level. From the amplitude of transient product formation, it

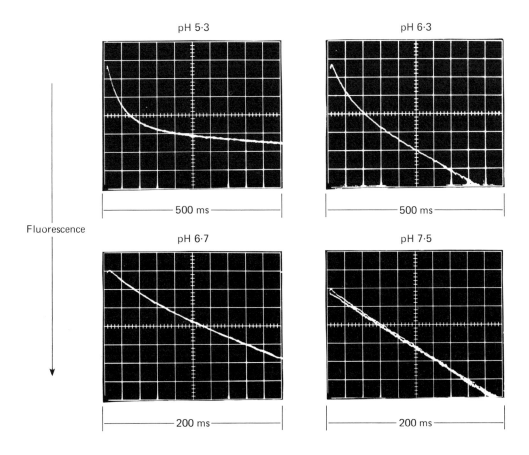

Fig. 5.5 Stopped flow fluorimeter records of the hydrolysis of 10 μM 4-methylumbelliferyl phosphate by 0·17 μM alkaline phosphatase in 0·1 M Tris-acetate buffers at various pH values. Both the final pH of the reaction mixtures and the time scale of the oscilloscope traces (horizontal axis) are noted by each photograph. For the fluorescence calibration, each division on the vertical axis corresponds to 0·05 μM 4-methylumbelliferone.

was concluded that the two active sites on this dimeric enzyme were functionally identical and acted independently of one another. At pH 7, the rate constant of transient NADH production was found to be 125 s^{-1} with ethanol as the substrate but when deuterated (d 6) ethanol was used instead, the transient rate constant fell to 25 s^{-1}, though identical steady-state rates had been measured for the two substrates. This isotope effect shows that the rate of the hydride transfer between ethanol and NAD on alcohol dehydrogenase can be measured in the transient phase. A full account of the transients of alcohol dehydrogenase is given by Shore and Gutfreund (1970).

In contrast to the comparatively simple mechanism of alcohol dehydrogenase, the analysis of the transients in the reactions of alkaline phosphatase have exposed a number of unexpected complexities. When read in sequence, three papers demonstrate how the mechanism of this enzyme has been slowly unravelled over the years through the continued application of rapid reaction techniques (Fernley and Walker 1966; Trentham and Gutfreund 1968; Halford 1971). Alkaline phosphatase hydrolyses a wide range of phosphate esters to yield the alcohol and inorganic phosphate. Figure 5.5 displays a montage of stopped flow records showing the increase in fluorescence due to the hydrolysis of 4-methylumbelliferyl phosphate by alkaline phosphatase at various pH values between 5·3 and 7·5. One product of this reaction, 4-methylumbelliferone, is intensely fluorescent at a wavelength well separated from protein fluorescence though the substrate itself is not fluorescent. The mechanism of alkaline phosphatase is similar to scheme II with B representing the product, 4-methylumbelliferone, and E − A a phosphoryl-enzyme.

Several aspects of the mechanism of alkaline phosphatase are revealed directly upon inspection of Fig. 5.5. Firstly, the difference between the rates of transient and steady-state product formation decreases with increasing pH. Quantitation of this effect with equations (7) and (9) shows that the rate constant k_{+2} is more or less independent of pH while k_{+3} increases linearly with increasing pH. At pH values below 6·5, $k_{+2} > k_{+3}$ while the converse, $k_{+2} < k_{+3}$, holds above pH 6·5. Consequently the steady-state rate is determined primarily by k_{+3} at low pH values and by k_{+2} at higher pH values (equation (7)). Secondly, the amplitude of transient product formation decreases with increasing pH. But given the different pH profiles of k_{+2} and k_{+3}, scheme II in fact demands the reduction in the transient amplitude (equation (8)). The amplitude of this transient is of further interest because at low pH values when $k_{+2} \gg k_{+3}$, only one mole of product is released per mole of enzyme though the enzyme is a dimer of two identical subunits. Among the possible reasons for the limitation on the amplitude could be partial denaturation of the protein or reversibility of the step, E . AB $\rightleftharpoons$ E − A + B. But as neither of these suggestions are applicable to alkaline phosphatase, it has been postulated that only one out of the two active sites operates at any one time though the nature of the subunit interactions that cause this behaviour has yet to be elucidated. Thirdly, alkaline phosphatase hydrolyses the wide range of phosphate esters at a uniform rate. The value of k_{+3} should not alter with different esters because this constant reflects the decomposition of a common intermediate, the phosphoryl-enzyme. But it was surprising to find that k_{+2} was also independent of the ester for the rate of the chemical cleavage was expected to be accelerated by a leaving group with a lower pK_a. Further experiments on the rate of formation of the phosphoryl-enzyme have shown that this process is in fact rate-limited by a conformation change of the protein and that the rate constant for the chemical step is much larger than the rate constant from transient product formation (see section 3.3).

The study of the transient kinetics of enzyme intermediates, as opposed to transient product formation, depends upon the detection of the intermediates. This can be achieved by alterations in the spectra of either the enzyme or the substrate. Several enzymes possess strong absorption bands in the visible spectrum and during their reactions with substrate, these absorption peaks can undergo either variations in intensity or shifts in the wavelength of maximal absorption. Perhaps the spectrophotometric characterisation of the transient intermediates within the turnover of the haemproteins, catalase and peroxidase, provide the most elegant example of this experimental approach (reviewed by Chance 1963b). Incidentally, the original reports of these experiments by Chance in 1940 constituted the first direct evidence for the existence of an enzyme–substrate complex. The extinction coefficient of catalase at 405 nm is reduced by some 40 per cent upon reaction with methyl hydrogen peroxide and is then returned to normal when that complex is reduced by the donor alcohol. Not only were the spectra of each intermediate determined but also their lifetimes. Furthermore, additional complexes between these enzymes and their substrates could be identified by their unique spectra but these were not found on the catalytic pathway and thus could be dismissed as abortive complexes. As an alternative to the absorption spectrophotometry of enzymes, temporal variations in protein fluorescence have also been used to monitor the formation and decay of transient intermediates. While protein fluorescence may be applicable to a wider range of enzymes, it is sometimes difficult to interpret the mechanistic significance of the observed signals. For instance, the quenching of protein fluorescence by ligand binding may not be linearly dependent upon the fractional saturation (Holbrook 1972).

In many enzymes, the active site is located within a hydrophobic pocket and thus the absorption spectrum of a substrate bound at the active site is often different from that free in aqueous solution. One development of this approach to the observation of transient intermediates has relied on the alterations to the spectrum of the dye, proflavin, as it binds at the active site of trypsin. Proflavin is a competitive inhibitor of trypsin. Upon addition of a

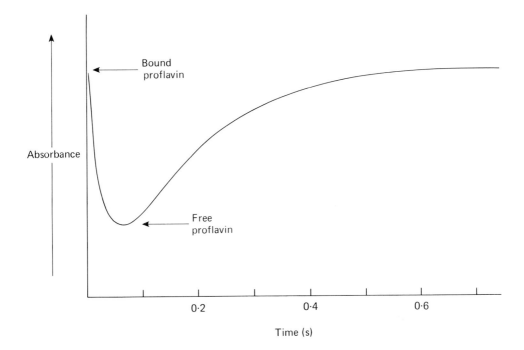

Fig. 5.6 Representation of a stopped flow spectrophotometer recording at 470 nm of the displacement of proflavin (0·2 mM) from trypsin (0·2 mM) by benzoyl-arginine ethyl ester (0·2 mM) in 0·1 M acetate buffer, pH 5·3.

non-chromophoric substrate to the complex of proflavin and trypsin, the displacement of the dye is recorded spectrophotometrically and when the products of the reaction dissociate, the dye will re-attach to the enzyme with the reversion of the spectral signal (Fig. 5.6). Such experimental records reveal the occupancy of the enzyme active site by those intermediates that bind to the enzyme more strongly than the dye itself. In this case (scheme II), the transient observed by the displacement of proflavin probably represents the formation and decomposition of the acyl-enzyme (Barman and Gutfreund 1966).

3.3 Partial reactions

By now the reader should be aware of the tendency of enzyme mechanisms to become rather complicated as new experimental techniques provide evidence for the existence of more and more intermediates. Hence it is often advantageous to first isolate a small section of the reaction pathway from the other events in the enzyme turnover and then examine that partial reaction by itself. The partial reactions that can be studied by rapid mixing techniques fall into three classes. In the first place, the substrate can be replaced by a substrate analogue that binds to the enzyme but does not undergo the chemical reaction of the substrate. Secondly, the reactions of a multi-substrate enzyme can be studied with an incomplete set of substrates. Thirdly, an enzyme can be permitted just a single turnover by the addition of a sub-stoichiometric level of substrate; provided that the enzyme concentration is in excess of both the substrate and the K_m for that substrate, all the substrate in the mixture should be utilised in one first order reaction. As in the discussion on transient kinetics, examples of partial reactions can be drawn from experiments on alkaline phosphatase and alcohol dehydrogenase.

One substrate analogue for alkaline phosphatase is 2-hydroxy-5-nitrobenzyl phosphonate,

$$CH_2PO_3H_2$$
—OH
$$O_2N—$$

On account of the methylene group adjacent to the phosphorus, this compound cannot be hydrolysed by alkaline phosphatase through it binds to the enzyme. Upon binding, the absorption spectrum of the chromophoric nitrophenyl group is perturbed. It could therefore be used to characterise details in the mechanism of substrate binding (Halford et al. 1969). From the concentration dependency of the rate constants measured from stopped flow records of the binding of this phosphonate to alkaline phosphatase (viz. equation (5)), it was concluded that binding took place in a two-step process

$$E + I \underset{\phantom{k_{+1}}}{\overset{k_{+1}}{\rightleftharpoons}} E.I \underset{\phantom{k_{+2}}}{\overset{k_{+2}}{\rightleftharpoons}} E*I$$

Moreover, the value thus obtained for the rate constant, k_{+2}, was in close agreement with the predicted value for the rate constant from a conformation change whose existence in the reaction with substrate had previously been postulated. This conformation change could be most readily observed with the substrate analogue for in the reaction with the true substrate, the formation of E*S is rate-limiting and its decomposition through the subsequent chemical step is very rapid so the concentration of this extra intermediate in the enzyme turnover is correspondingly small.

One partial reaction of alcohol dehydrogenase studied by Shore and Gutfreund (1970) was the displacement of NADH from the enzyme–NADH complex. This process could be followed spectrophometrically at 340 nm because at that wavelength there exists an extinction difference between enzyme bound and free NADH. The displacement of NADH was achieved by mixing in the stopped flow apparatus the enzyme–NADH complex with excess NAD.

$$E.NADH \rightleftharpoons E + NADH$$

$$E + NAD \rightleftharpoons E.NAD$$

At the high concentration of NAD, the rate of formation of E . NAD is limited by the dissociation of NADH from E . NADH. Hence the rate constant for NADH dissociation could be evaluated as $3 \cdot 0 \pm 0 \cdot 2$ s^{-1}. This value is in good numerical agreement with the turnover number of $3 \cdot 3$ s^{-1} measured from the steady-state rate of NADH production under the same conditions. These findings support the proposal that the dissociation of NADH is the rate-limiting step in the oxidation of ethanol by alcohol dehydrogenase.

Measurements of the rate constants for the dissociation of one ligand by displacement methods have now been extended to many systems. However the correlations between the rate constants obtained by this method and the overall mechanism of the enzyme under investigation are not always as direct as the above example. For instance, Harrigan and Trentham (1971) found that the thiol reagent, DTNB, reacted with the active site cysteine of glyceraldehyde 3-phosphate dehydrogenase only after the dissociation of enzyme bound NAD. But the rate constant of NAD dissociation obtained by its displacement with DTNB was only one-fifth of the turnover number from the steady state of NAD production. Therefore within the catalytic cycle, NAD must dissociate from a form of the enzyme that has been modified by the second ligand; in this case after acylation by 1,3 diphosphoglycerate. The paper by Harrigan and Trentham (1971) should be referred to for the kinetic theory of displacement reactions.

The other half of the kinetics of complex formation between enzyme and ligand is the measurement of the association rate constant.

$$E + L \xrightleftharpoons{k_{+1}} E.L$$

This problem is distinct from the analysis of the pathway of ligand binding which in many cases involves a subsequent rearrangement of the enzyme–ligand complex. However, experiments on these lines have often proved to be beyond the capability of stopped flow methods. The reason is that the majority of second order rate constants for the association of an enzyme with its ligand fall within the narrow range of 10^7–10^8 M^{-1} s^{-1} (see section 4.3). While the rate of a bimolecular reaction is determined by the concentration of reactants, the level at which the concentrations must be set in order to permit the resolution of the reaction on the millisecond time scale of stopped flow instrumentation is often lower than can be detected by spectrophotometric techniques. Furthermore, in reversible reactions, the first order rate constant of the reverse step also contributes to the rate of equilibration. In a few cases the second order rate constants have been evaluated from reactions at very low concentrations of enzyme and ligand, by monitoring the progress of the reaction with sensitive fluorescence detection. But in general, the study of extremely fast reactions requires the application of perturbation methods instead of mixing techniques.

4 Perturbation methods

4.1 Large perturbations and small relaxations

There is a limit to the speed with which two solutions can be mixed together. Despite considerable effort spent on the design of rapid mixing devices, any reduction below 1 ms for the time taken to mix two solutions has always introduced severe technical problems such as inhomogeneous mixing or cavitations during flow. Therefore new techniques are needed to extend the range of experimental observations to time intervals of less than 1 ms. As an alternative to mixing the reagents, the reaction can be initiated by the physical perturbation of a previously mixed set of reagents. The initiation of a new reaction from reagents that were unreactive prior to the perturbation requires a large change in the physical environment. Two common methods in which large perturbations are harnessed for the study of rapid reactions in solution are flash photolysis and pulse radiolysis. Descriptions of these and other perturbation methods, including references to the original work by Porter and others, can be found in Hague (1971).

Flash photolysis can be used for the study of reactions that are initiated by light. The principle of this method is that the reactants are subjected to a light flash of very high intensity in a region of the spectrum where at least one of the components of the reaction mixture is photosensitive. The fate of the excited component can be followed spectrophotometrically; the observation light beam is placed at right angles to the photolysis flash and is of much lower intensity than the flash. The time resolution of flash photolysis equipment is determined by the duration of the flash but at the same time the flash must be sufficiently intense to excite the chemical species. In conventional flash photolysis apparatus, this conflict between short duration and high intensity is resolved by obtaining the flash from the discharge of a bank of condensers through a tube filled with a rare gas such as krypton. Apparatus of this type has been used to follow photochemical events and reactions with half-times in the microsecond region. Since the advent of laser optics, flash techniques have been extended into the nanosecond and even picosecond ranges in order to observe the excited singlet states that are often the precursors of the triplet states which had been observed in the microsecond range. Pulse radiolysis is somewhat similar to flash photolysis except that in this case the perturbation is created by a pulse of ionising radiation instead of the light flash. The pulse, which may be of X-rays or an electron beam, usually lasts for a few microseconds. This technique has found its widest application in the study of the reactions of the hydrated electron, $[e(H_2O)_n]^-$.

Large perturbation methods have the advantage over all other rapid reaction techniques with respect to time resolution. In fact, the recent developments in flash photolysis can now be used to measure the rates of photochemically induced electronic excitations that are faster than any chemical reaction. However the applications of these methods to biochemistry are severely limited by the availability of suitable reactions that can be initiated by a particular perturbation. The main biological systems studied by flash photolysis include the primary events in photosynthesis and the photochemistry of nucleic acids. One other system in which flash photolysis has proved useful is the photo-dissociation of carbon monoxide from its complex with haemoglobin. This reaction has also been examined in a combined stopped flow-flash photolysis apparatus. To date, pulse radiolysis has been occasionally employed for investigations on the reactions of individual components from the electron transport chain though its application in this field may well become more widespread in the future.

However the perturbation methods for the study of fast reactions are not limited to large perturbations. Instead of initiating a reaction by a large perturbation, it is possible to apply a small perturbation to a mixture of reactants and products at equilibrium and then to monitor the response of the chemical system as it adjusts to the change in the equilibrium position caused by this perturbation. The self-adjustment of a molecular system to a new state of minimum free energy following a disturbance is known as a relaxation. A further characteristic of a relaxation is the time delay between the forcing function and the resultant self-adjustment, from which it is possible to determine a relaxation time. Relaxation processes have been known to physics for over a hundred years but not until the work of Eigen in 1954 was this approach used on chemical equilibria. Since then relaxation kinetics have been developed, principally by Eigen, into a whole new branch of chemical kinetics. In a chemical relaxation, the forcing function that shifts the equilibrium position is normally a variation in an intensive thermodynamic parameter such as temperature, pressure or electric field strength and the delay in the response is caused by the finite time taken for the reaction between the chemical species in that equilibrium. Consequently the rate constants in the equilibrium reaction can be calculated from the relaxation time of the chemical response (section 4.2). Of the many published reviews on relaxation methods, the best and most rigorous is by Eigen and De Maeyer (1963) while less detailed descriptions can be found in Eigen (1968) and Hague (1971).

There exist a great variety of relaxation techniques for chemical equilibria are sensitive to a number of different thermodynamic parameters that can be changed external to the system. Furthermore the forcing function can be applied either in a step-wise manner such as a jump from one temperature to another or as a periodic oscillation. In the latter case, the forced concentration changes will attempt to follow the periodic forcing function but when the frequency of the forcing function is similar to the rate at which the chemical equilibrium can respond to the external change, there will be established an interaction between the system and the forcing function through which energy will be absorbed. The absorption of energy by the system may be visualised as a phase lag between the periodicities of the forcing function and the chemical change. An example of an oscillatory method for relaxation kinetics is the ultrasonic technique. The passage of sound waves through a liquid will set up alternating regions of high and low pressure which are associated with slight increases and decreases of temperature. If the position of the chemical equilibrium is dependent upon either temperature or pressure, the disturbance caused by the sound waves may be used as the forcing function. Several experimental methods have been developed to measure the attenuation of sound waves in liquids (Eigen and De Maeyer 1963). Measurements of sound absorption by ultrasonic techniques can be used to evaluate relaxation times between 1 ms and 1 ns. On account of their excellent time resolution, these techniques are valuable for studies of very fast reactions such as proton transfer between acids and bases and the formation of coordination complexes on metal ions. While knowledge of these fundamental reactions is very necessary for biochemistry, the ultrasonic methods have no application to the reactions of biological macromolecules because their operation requires that the reactants constitute a significant molar fraction of the solution (i.e. $>10^{-2}$ M).

The relaxation methods that rely on step-wise perturbations are more pertinent to biochemical investigations. The most widely used of these techniques is the temperature-jump method. If a reaction is associated with a finite standard enthalpy change ΔH^0, the van't Hoff isochore

$$\frac{\mathrm{d}\ln K}{\mathrm{d}T} = \frac{\Delta H^0}{RT^2}$$

shows that its equilibrium constant will be dependent upon temperature. Consequently a sudden rise in temperature will alter the ratio of reactants to products and provided that this affects the extinction of the solution, the concentration changes may be followed by spectrophotometric means. The rise in temperature is normally obtained by one form of joule heating, the input of electrical energy. The energy stored in a high voltage capacitor can be suddenly dissipated into the solution by connecting the capacitor across the cell in which the equilibrium mixture is placed. The electrical discharge may be as high as 30 kV. The apparatus for this type of experiment is shown in Fig. 5.7. The cell containing the sample is constructed like a dumbell in order to produce a uniform field between the two electrodes so that the portion of solution in the observation light path is evenly heated. By this method, it is possible

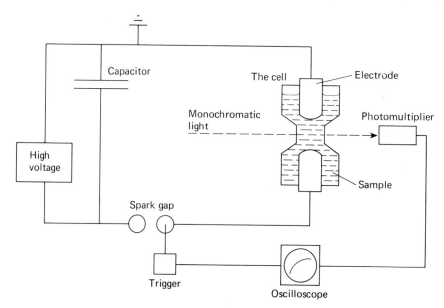

Fig. 5.7 Schematic diagram of a temperature-jump apparatus that utilises joule heating. Details of the design of the sample cell are shown. The chemical relaxation is monitored spectrophotometrically on an oscilloscope and the triggered spark gap ensures that the observation commences at the same time as the electrical energy is discharged through the cell.

to heat a solution of total volume 5 ml through 5°C with a heating time of 10 μs. The solution will remain at the elevated temperature for about a second before cooling down to the original temperature. Heating by electrical discharge can only be used on solutions of high conductivity. Non-conducting solutions may be heated rapidly by a pulse of light from a laser beam.

A step-wise perturbation to displace a chemical equilibrium can also be realised by a pressure-jump. The effect of pressure on an equilibrium constant is determined by the volume change of the reaction, ΔV^0, and is given quantitatively by the isotherm

$$\frac{d \ln K}{dP} = - \frac{\Delta V^0}{RT}$$

Experimentally, the pressure-jump can be obtained by first applying a pressure of about 100 atm to the sample placed in a specially designed vessel. The pressure is then reduced to the atmospheric level within 100 μs by puncturing a metal disc set in the wall of the sample container. At the present time, the pressure-jump method is not used as often as the temperature-jump on account of the inferior time resolution of the former technique and also because most reactions in solution are more sensitive to temperature changes than they are to pressure. However the pressure-jump method is particularly suitable for the study of reactions in aqueous solutions in which charged species are formed because the electrostriction of the solvent results in large volume changes. It should also be noted that if an aqueous solution is buffered by a compound that has a large heat of ionisation, a temperature-jump will shift the pH of that solution. Since proton transfer reactions are very rapid (section 4.3), the pH change will be coincident with the temperature-jump. Hence the equilibrium under study may be perturbed by both pH- and temperature-jumps operating simultaneously. Likewise pressure-jumps may be coupled to pH-jumps.

4.2 Relaxation kinetics

The relationship between a step-wise perturbation to a chemical equilibrium and the delayed chemical response can be further discussed by reference to Fig. 5.8. A temperature-jump that heats a solution of Tris buffer lowers the pH of that solution and thus protonates the phenol red indicator dissolved in the buffer; the protonation of phenol red is recorded by the absorbancy decrease at 570 nm (Fig. 5.8(a)). As proton transfer reactions are very fast, the delay between the forcing function and the chemical response is infinitesimal, and the equilibrium between the acidic and basic forms of phenol red adjusts to the changing temperature as rapidly as the temperature is itself changed. Consequently the rate of the spectral change in Fig. 5.8(a) only provides a measurement of the heating time of this temperature-jump instrument. On the other hand, when the same instrument is used to perturb an equilibrium mixture of an enzyme with an analogue of its substrate (Fig. 5.8(b)), the rate of the absorbancy increase at 430 nm due to the binding of the chromophoric substrate analogue is very much slower than that found in Fig. 5.8(a) for the heating time. In this system, the rate of the chemical relaxation is limited by the rate constants in the binding equilibrium and thus the relaxation lags behind the rise in temperature rather than keeping pace with it.

After a step-wise perturbation that causes a small disturbance to a chemical equilibrium, the relaxation to the new equilibrium position follows an exponential progress curve regardless of

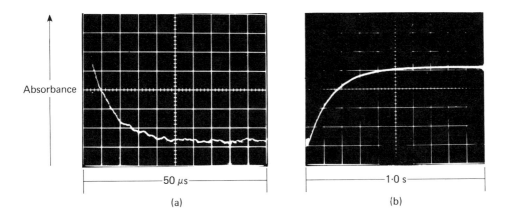

Absorbance

50 μs

(a)

1·0 s

(b)

Fig. 5.8 Spectrophotometric records of chemical relaxations following temperature-jumps. The constitution of the two chemical equilibria are
(a) 25 μM phenol red in 0·1 M Tris, 33 mM Na_2SO_4, pH 8·0.
(b) 60 μM (tetrameric) alkaline phosphatase and 0·20 mM.
2-hydroxy-5-nitrobenzyl phosphonate in 10 mM Tes, 0·1 M KCl, pH 8·3.
Note the difference in the time scales on the photographs of the two relaxations.

the kinetic order of the reaction. Let the difference in the equilibrium concentrations of one reactant before and after the perturbation be ΔC_0 and that between the final concentration and the level at time t during the relaxation be ΔC. The exponential is then described by the equation

$$[\Delta C] = [\Delta C_0]\, e^{-t/\tau}$$

where τ is the relaxation time. The reciprocal relaxation time is determined from the experimental data (viz. Fig. 5.8(b)) as the gradient of a semi-log plot of $\ln(\Delta C)$ versus time, an identical procedure to the evaluation of a first order rate constant. The above expression is in fact an integrated form of the fundamental equation of all relaxation processes,

$$\frac{\mathrm{d}}{\mathrm{d}t}[\Delta C] = -\frac{[\Delta C]}{\tau} \tag{10}$$

The differential equation was originally obtained from basic thermodynamics. It is the most convenient starting point for the derivation of the relationship between rate constants and relaxation times. An important restriction upon the validity of equation (10) is that the perturbation must be small, i.e. $[\Delta C_0] \ll [\overline{C}]$. It is this restriction that permits the linearisation of complex rate equations into simple exponentials.

The linearisation procedure may be demonstrated by the derivation of the relaxation equation for the reaction scheme

$$E + S \underset{k_{-1}}{\overset{k_{+1}}{\rightleftharpoons}} ES$$

Let the concentrations of the chemical species be (E), (S) and (ES) at time t during the relaxation and $(\overline{E})$, $(\overline{S})$ and $(\overline{ES})$ at the new equilibrium position after the perturbation, the deviations between the actual concentrations and their final values being δE, δS and δES.

$$(E) = (\overline{E} + \delta E) \qquad \text{where } \delta E \ll E$$
$$(S) = (\overline{S} + \delta S)$$
$$(ES) = (\overline{ES} + \delta ES)$$

The orthodox rate equation for this mechanism is

$$\frac{\mathrm{d}}{\mathrm{d}t}(ES) = k_{+1}(E)(S) - k_{-1}(ES)$$

$$\therefore \quad \frac{\mathrm{d}}{\mathrm{d}t}(\overline{ES} + \delta ES) = k_{+1}(\overline{E} + \delta E)(\overline{S} + \delta S) - k_{-1}(\overline{ES} + \delta ES)$$

But at the new equilibrium,

$$\frac{\mathrm{d}}{\mathrm{d}t}(\overline{ES}) = k_{+1}(\overline{E})(\overline{S}) - k_{-1}(\overline{ES}) = 0$$

$$\therefore \quad \frac{\mathrm{d}}{\mathrm{d}t}(\delta ES) = k_{+1}(\overline{E}\,\delta S + \overline{S}\,\delta E + \delta E\,\delta S) - k_{-1}(\delta ES)$$

Since the perturbation was small, product terms such as $\delta E\delta S$ can be ignored. Elsewhere, δE and δS can be substituted by $-\delta ES$, on account of the conservation relationship

$$\delta E = \delta S = -\delta ES$$

The rate equation has now been linearised to

$$\frac{d}{dt}(\delta ES) = -k_{+1}(\bar{E} + \bar{S})\,\delta ES - k_{-1}\,\delta ES$$

This expression has the same exponential form as equation (10), and by comparison with that equation

$$\frac{1}{\tau} = k_{-1} + k_{+1}(\bar{E} + \bar{S}) \tag{11}$$

The two rate constants in the equilibrium reaction can be determined from the dependence of $1/\tau$ upon $(\bar{E} + \bar{S})$; a graph of $1/\tau$ versus $(\bar{E} + \bar{S})$ should be a straight line with a gradient of k_{+1} and an intercept of k_{-1}. If the data does not yield a straight line, then the reaction under study must proceed by a different mechanism. In equation (11), the values of $(\bar{E})$ and $(\bar{S})$ refer to the equilibrium concentrations of the free reactants and differ from the total amount of these reagents added to the mixture by the concentration of $\overline{ES}$. The actual values of $(\bar{E})$ and $(\bar{S})$ may be calculated from the total concentrations if the dissociation constant K_1 is known. The dissociation constant also provides an independent check upon the ratio of the values obtained for k_{-1} and k_{+1}.

The relaxation equation for any single-step mechanism can be derived by the above method. But the full beauty of relaxation methods comes in their application to the study of multi-step equilibria. In this situation, the method itself splits the complicated mechanism into its elementary steps by separating these steps along the time axis. The relaxation of each part of the mechanism can then be described by single exponential. The term 'relaxation spectrum' has been introduced to describe the several relaxations arising from a multi-step equilibrium. The individual reciprocal relaxation times that comprise the relaxation spectrum are characteristic of the system in the same way that an NMR spectrum is characteristic of a particular molecule, though it is the concentration dependence of each reciprocal relaxation time that distinguishes one mechanism from another. However, the response of a multi-step equilibrium to a sudden perturbation in an external parameter does not proceed as a series of independent relaxations for the rapid re-equilibration of one step will affect the rate at which a second step will re-equilibrate. Mathematically, the relaxation times are the eigen values of a system of coupled differential equations. There will be one linear differential equation to describe the time course of re-equilibration of each independent concentration variable so there will be as many relaxation times as there are independent concentration variables. The problem is just the solution of the simultaneous differential equations in order to equate relaxation times with functions of rate constants.

The rate equations for the mechanism

$$E + S \underset{k_{-1}}{\overset{k_{+1}}{\rightleftarrows}} ES \underset{k_{-2}}{\overset{k_{+2}}{\rightleftarrows}} E^*S$$

are

$$\frac{d}{dt}(E) = k_{-1}(ES) - k_{+1}(E)(S)$$

$$\frac{d}{dt}(E^*S) = k_{+2}(ES) - k_{-2}(E^*S)$$

Linearisation and the use of the conservation relationships

$$\delta E + \delta ES + \delta E*S = 0$$

$$\delta E = \delta S$$

yields

$$\frac{d}{dt}(\delta E) = -[k_{-1} + k_{+1}(\bar{E} + \bar{S})]\,\delta E - k_{-1}\,\delta E*S$$

$$\frac{d}{dt}(\delta E*S) = -k_{+2}\,\delta E - (k_{+2} + k_{-2})\,\delta E*S$$

These equations must now be cast in the form

$$\frac{d}{dt}(\delta E) = -\frac{\delta E}{\tau}, \qquad \frac{d}{dt}(\delta E*S) = -\frac{\delta E*S}{\tau}$$

Insertion of these expressions into the linear rate equations produces

$$\frac{\delta E}{\tau} = [k_{-1} + k_{+1}(\bar{E} + \bar{S})]\,\delta E + k_{-1}\,\delta E*S$$

$$\frac{\delta E*S}{\tau} = k_{+2}\,\delta E + (k_{+2} + k_{-2})\,\delta E*S$$

$$\therefore \quad \begin{vmatrix} [k_{-1} + k_{+1}(\bar{E} + \bar{S})] - \dfrac{1}{\tau} & k_{-1} \\[2mm] k_{+2} & (k_{+2} + k_{-2}) - \dfrac{1}{\tau} \end{vmatrix} = 0$$

This may be written as the quadratic

$$\left(\frac{1}{\tau}\right)^2 - [k_{-1} + k_{+1}(\bar{E} + \bar{S}) + k_{+2} + k_{-2}]\,\frac{1}{\tau}$$

$$+ [k_{-1}k_{-2} + k_{+1}(k_{+2} + k_{-2})(\bar{E} + \bar{S})] = 0$$

For the solutions $1/\tau_1$ and $1/\tau_2$, the Gaussian would be

$$\left(\frac{1}{\tau} - \frac{1}{\tau_1}\right)\left(\frac{1}{\tau} - \frac{1}{\tau_2}\right) = 0$$

$$\therefore \quad \left(\frac{1}{\tau}\right)^2 - \left(\frac{1}{\tau_1} + \frac{1}{\tau_2}\right)\frac{1}{\tau} + \frac{1}{\tau_1}\frac{1}{\tau_2} = 0$$

By comparing coefficients with the previous quadratic

$$\frac{1}{\tau_1} + \frac{1}{\tau_2} = k_{-1} + k_{+1}(\bar{E} + \bar{S}) + k_{+2} + k_{-2}$$

and

$$\frac{1}{\tau_1}\frac{1}{\tau_2} = k_{-1}k_{-2} + k_{+1}(k_{+2} + k_{-2})(\bar{E} + \bar{S})$$

Hence all four rate constants in this mechanism can be determined from plots of the sum and the product of the reciprocal relaxation times against $(\bar{E} + \bar{S})$. But if the relaxation times are well separated, it is possible to 'uncouple' the equations in order to relate one relaxation time with a single step. If the mechanism consists of a fast binding step followed by a slower re-arrangement of the enzyme–substrate complex (i.e. $\tau_1 \ll \tau_2$), the relaxation of the binding step will be completed before the conformation change has effectively started to re-equilibrate. In this case the binding step can be treated as an isolated process, a single-step mechanism, but a concentration term must be included with the equation from the re-arrangement step in order to account for its coupling to the previous step.

$$\therefore \quad \frac{1}{\tau_1} = k_{-1} + k_{+1}(\bar{E} + \bar{S})$$

and

$$\frac{1}{\tau_2} = \frac{\tau_1}{\tau_1 \tau_2}$$

$$= \frac{k_{-1}k_{-2} + k_{+1}(k_{+2} + k_{-2})(\bar{E} + \bar{S})}{k_{-1} + k_{+1}(\bar{E} + \bar{S})}$$

$$\therefore \quad \frac{1}{\tau_2} = k_{-2} + \frac{k_{+2}}{1 + K_1/(\bar{E} + \bar{S})} \tag{12}$$

where $K_1 = k_{-1}/k_{+1}$. From equation (12), it can be seen that

$$\frac{1}{\tau_2} \to k_{-2} \qquad \text{when } (\bar{E} + \bar{S}) \ll K_1$$

$$\frac{1}{\tau_2} \to k_{-2} + k_{+2} \qquad (\bar{E} + \bar{S}) \gg K_1$$

A further increase in $(\bar{E} + \bar{S})$ above K_1 will not alter $1/\tau_2$. On the other hand, $1/\tau_1$ will increase linearly with $(\bar{E} + \bar{S})$. It is, however, possible to find only one relaxation in a given system even though the concentration dependence of its reciprocal relaxation time is characteristic of a two-step mechanism. For instance in the above two-step mechanism with $\tau_1 \ll \tau_2$, only the relaxation characterised by τ_2 would be observed spectrophotometrically when the unimolecular step involved a spectral change and the bimolecular step did not. But if only the faster initial binding produced the spectral change while ES and E*S had the same spectra, then the coupling between the two relaxations would ensure that both processes could be detected. Therefore it is important to measure the reciprocal relaxation times over a wide range of concentrations in order to determine whether or not they tend towards a constant value at the higher levels. Unfortunately, the biochemical literature contains several reports of the relaxation spectra of various enzyme reactions in which the reciprocal relaxation times were not measured over a sufficiently wide range of concentrations, with the result that the assignment of the mechanism to a particular reaction is either ambiguous or even on some occasions wrong.

The analysis of the relaxation spectra of very much more complicated mechanisms than the one considered here can be tackled by the same mathematical procedures, though the expressions tend to become rather cumbersome. Halford (1972) provides a list of relaxation equations from several mechanisms that might be encountered in studies of ligand binding. However it is comparatively simple to distinguish the previous mechanism of substrate binding followed by a conformation change from one in which the conformation change precedes the binding step

$$E \underset{k_{-1}}{\overset{k_{+1}}{\rightleftharpoons}} \underset{+S}{E^*} \underset{k_{-2}}{\overset{k_{+2}}{\rightleftharpoons}} E^*S$$

It can be shown for this scheme (see Gutfreund 1972, for the mathematics) that when the isomerisation is much slower than the binding process ($\tau_1 \gg \tau_2$) so that the relaxation equations may be uncoupled,

$$\frac{1}{\tau_2} = k_{-2} + k_{+2}(\bar{E}^* + \bar{S})$$

and

$$\frac{1}{\tau_1} = k_{+1} + \frac{k_{-1}}{1 + (\bar{S})/\{(\bar{E}^*) + K_2\}} \tag{13}$$

where $K_2 = k_{-2}/k_{+2}$. From equation (13),

$$\frac{1}{\tau_1} \to k_{+1} + k_{-1} \qquad \text{when } (\bar{S}) \ll (\bar{E}^*) + K_2$$

$$\frac{1}{\tau_1} \to k_{+1} \qquad \qquad (\bar{S}) \gg (\bar{E}^*) + K_2$$

A comparison of equations (12) and (13) shows that the reciprocal relaxation times from the isomerisation steps may be either increased or decreased by increasing the substrate concentration, depending upon whether that isomerisation occurs after or before the binding step. It is therefore possible to determine which pathway the reaction follows.

4.3 Relaxation spectra of enzyme reactions

For the elucidation of enzyme mechanisms, relaxation methods are complementary to the observation of transients by the stopped flow method. The two techniques should not be considered as rivals for the solution of the same problem even though certain experiments may be carried out with equal facility by either technique. Relaxation procedures can only be applied to reactions poised at an equilibrium position in which the concentrations of reactants are approximately equal to those of the products. Furthermore, in multi-step equilibria, the relaxation methods provide information about only those intermediates present in significant concentrations. On the other hand, intermediates of enzyme reactions whose concentrations are insignificant at either equilibrium or in the steady state may still be detected in the transient phase of that reaction after mixing the reactants together in a stopped-flow device (Fig. 5.4), and moreover the detection of the intermediates is totally independent of the final equilibrium position though it depends to some extent upon the location of the rate-limiting step within the mechanism. Consequently the stopped flow approach is usually preferred for studies on the overall reaction of an enzyme with its substrates. However in the field of partial reactions (see section 3.3), relaxation methods often have the advantage on account of both their superior time resolution and their ability to separate the individual steps of a complicated equilibrium into a series of discrete exponentials instead of yielding one rate equation containing sums and products of the exponentials.

One class of reactions that have been extensively studied by the temperature-jump technique is the combination of an enzyme with a specific ligand

$$E + L \underset{k_{-1}}{\overset{k_{+1}}{\rightleftharpoons}} EL$$

In nearly all reactions of this kind, the second order rate constants (k_{+1}) for the biomolecular processes have been found to fall within a surprisingly narrow range between 10^7 and 10^8 M^{-1} s^{-1}. However the first order rate constants (k_{-1}) in the direction of dissociation vary over several orders of magnitude, reflecting the different stabilities of enzyme–ligand complexes. The second order rate constants are only slightly slower than the values expected for a diffusion controlled reaction between a large enzyme molecule and the smaller substrate. While it is possible to discern a 'characteristic rate' for reactions of this sort, it should be remembered that there are a few exceptions to this rule. Other categories of reactions are associated with characteristic rates. For instance, the substitution of water from the inner coordination sphere of most metal ions by another ligand proceeds at a rate that is more or less independent of the nature of the substituting ligand but is characteristic of the particular metal ion (Diebler et al. 1969). In the few examples studied to date, the second order rate constants for the specific aggregation of two protein subunits have all been located between 10^5 and 10^6 M^{-1} s^{-1} (Gutfreund 1972), though many more reactions of this class must be analysed before they can be considered to have a characteristic rate.

In a diffusion controlled reaction, every collision between pairs of reacting molecules results in a reaction. The rate of a diffusion controlled reaction is therefore determined by the collision frequency rather than the activation energy of the transition state complex. The diffusion controlled rate represents the upper limit for a bimolecular process. Reactions involving protons are often diffusion controlled. In aqueous solution, the rate constants for the combination of two highly mobile species such as a proton with a base or a defect proton (OH^-) with an acid normally fall within the range of 10^{10}–10^{11} M^{-1} s^{-1} (Eigen and De Maeyer 1963). But in ice, proton transfer along a pre-formed hydrogen bond is no longer diffusion controlled and may be 100 times faster than the comparable reaction in aqueous solution. In the case of direct proton transfer between an acid and a base,

$$AH + B^- \rightleftharpoons A^- + BH$$

the reaction is usually diffusion controlled in the direction of proton transfer towards the better proton acceptor (the group with the higher pK_a) and the rate constant for the reverse reaction of proton transfer away from the better proton acceptor may then be calculated from the difference in their pK_a values. The knowledge of the fundamental reactions of protons that has been obtained by relaxation methods is most illuminating when it comes to the consideration of acid-base catalysis in enzyme mechanisms. When a diffusion controlled reaction involves an enzyme as one of the reacting partners, the rate constant is smaller than that expected from the maximal collision frequency, even after taking into account the relative immobility of the large protein molecule, because only a small fraction of the collisions will occur with a suitable orientation of the reactants.

In general, the mechanism of ligand binding to an enzyme contains several unimolecular steps in addition to the bimolecular combination. The bimolecular event is rapid and relatively unspecific. The unimolecular steps may be either conformation changes of the protein or distortion of the substrate or both. Their function is to confer additional specificity upon the enzyme reaction, to organise the enzyme active site so as to achieve efficient catalysis and in some cases to regulate the enzyme activity. Relaxation methods are ideally suited to the analysis of multi-step equilibria. One of the many systems in which the temperature-jump technique has been used to define the separate steps in ligand binding is the association of the substrate analogue, 2-hydroxy-5-nitrobenzyl phosphonate with alkaline phosphatase (see section 3.3; also Halford 1972). In this experiment, the different relaxations were separated by monitoring the chemical response to the perturbation of a given equilibrium mixture at either a fast or a slow time base setting on the oscilloscope. The reciprocal relaxation times were then studied as a function of the concentrations so that each relaxation could be related to either a conformation change or a binding step and the mechanism was thus elucidated. In the case of the co-operative binding of NAD to yeast glyceraldehyde-3-phosphate dehydrogenase, an enzyme composed of four identical subunits, Kirschner *et al.* (1966) observed three relaxations; two of these could be related to the binding of NAD to different conformations of the enzyme though the binding at all four sites of either tetrameric conformation could be described by a single pair of rate constants, while the third relaxation referred to the conformation change involving all four protein subunits in an all-or-none fashion. In other words, the relaxation spectrum fits neatly to the precepts of the 'Monod model' for co-operative ligand binding. The paper by Kirshner *et al.* (1966) provides the most attractive demonstration of the power of relaxation methods.

5 Finale

This chapter has been limited so far to the theory of rapid reaction techniques and their application to the study of the reactions of enzymes. Even within this limitation, no indication has been given to show that these techniques are currently employed on a wide range of enzymes. The examples that were provided to illustrate either the operation of experimental techniques or various points of kinetic theory were all selected from less than half a dozen enzymes. However, the review by Gutfreund (1971) should redress the balance for described in it are the results from rapid reaction investigations on many different enzymes. If one word could summarise the contribution of rapid reaction techniques to the elucidation of enzyme mechanisms, the word would be directness. In most physical techniques, there exists a gap between the experimental observation and the resultant conclusion that must be bridged by theory and numerical calculations. But with rapid reaction methods, the experimental observation of a concentration change against time, whether it be an intermediate, a transient or a relaxation, is often one and the same thing as the final conclusion. If one seeks to detect an intermediate, it is only logical to use the appropriate technique.

However it would also be a grave injustice to rapid reaction techniques if it were thought that their only application was on enzyme mechanisms. A list of systems of biological interest that have been studied by these methods would include among others, helix-coil transitions in polypeptides; the folding of globular proteins; the reactions of antibodies, carriers and transport proteins including haemoglobin; regulatory proteins such as the *lac* repressor; metabolic regulation; photosynthesis, electron transport and oxidative phosphorylation; base pairing and the formation of double helices from complementary single strands of DNA. In addition, some attention is now being turned away from the properties of isolated components and towards the dynamic behaviour of intact cell organelles or even whole cells of yeast or bacteria. Many of the rapid reaction techniques described here can be adapted for such purposes. Just one reference to the use of rapid reaction techniques to a problem outside the field of enzyme mechanisms shall be considered in detail.

Several antibiotics of medium molecular weight have been shown to act as specific carriers of alkali metal ions through membranes. Monactin, a cyclic 'poly-ether', is one compound of this type and has the ability to transport potassium ions across membranes while excluding sodium ions. However, monactin can form a complex with either K^+ or Na^+. The rates of complex formation of either alkali metal ion with monactin were found to be almost diffusion controlled; the second order rate constants for the combinations were both measured by relaxation techniques to be about 3.10^8 M^{-1} s^{-1} (Diebler *et al.* 1969). The rate constants for the dissociation of the metal ions from their complex with monactin were evaluated at 10^3 s^{-1} for K^+ and 3.10^5 s^{-1} for Na^+. Thus the half-times of the dissociation processes are approximately 700 and 2 μs for K^+ and Na^+ respectively. Now a value of 100 μs is a reasonable estimate for the time taken by the monactin complex to diffuse through a lipid bilayer of 100 Å thickness. Consequently the lifetime of the K^+–monactin complex is long enough for the complex to diffuse through the membrane but the Na^+–monactin complex will break down faster than it can be transported. While this may well be a naïve approach to the problem of selective permeability of membranes, it at least has the advantage of explaining complicated biological phenomena in terms of the simplest physical chemistry. And that is the whole point of the exercise.

6 References

The references cited below fall into three categories. For the record, each specific example that demonstrates the application of a rapid reaction technique in a particular problem is provided with the reference to the original literature. Review articles could be consulted for topics that are either beyond the scope of this chapter or are not covered in sufficient detail. However, the first suggestion for further reading must be one of the two books: the monograph by Hague (1971) is strongest on the methodology of the perturbation techniques and their results on chemical reactions but lacks an account of the transients of enzyme reactions; Gutfreund (1972) covers the biophysical chemistry of enzymes and is particularly good on transient and relaxation kinetics.

BARMAN, T. E. and GUTFREUND, H. (1966). 'Optical and chemical identification of kinetic steps in trypsin- and chymotrypsin-catalysed reactions.' *Biochem. J.*, **101**, 411–16.

CHANCE, B. (1963a). 'Rapid reactions.' in Friess *et al.*, 728–49.

CHANCE, B. (1963b). 'Enzyme kinetics in the transient state.' In Friess *et al.*, 1314–60.

DIEBLER, H., EIGEN, M., ILGENFRITZ, G., MAAS, G. and WINKLER, R. (1969). 'Kinetics and mechanism of reactions of main group metal ions with biological carriers.' *Pure Appl. Chem.*, **20**, 93–115.

226

EIGEN, M. (1968). 'New outlooks on physical enzymology.' *Quart. Rev. Biophys.*, **1**, 3–33.

EIGEN, M. and De MAEYER, L. (1963). 'Relaxation methods.' In Friess *et al.*, 895–1054.

FERNLEY, H. N. and WALKER, P. G. (1966). 'Phosphorylation of *Escherichia coli* alkaline phosphatase by substrate.' *Nature, Lond.* **212**, 1435–37.

FRIESS, S. L., LEWIS, E. S. and WEISSBERGER, A., Eds. (1963). *Techniques of Organic Chemistry*, Vol. 8. *Investigations of Rates and Mechanisms of Reactions*. New York. Interscience.

FROST, A. A. and PEARSON, R. G. (1961). *Kinetics and Mechanism*. New York. Interscience.

GUTFREUND, H. (1971). 'Transients and relaxation kinetics of enzyme reactions.' *Ann. Rev. Biochem.* **40**, 315–44.

GUTFREUND, H. (1972). *Enzymes: Physical Principles*. London. John Wiley.

HAGUE, D. N. (1971). *Fast Reactions*. London. John Wiley.

HALFORD, S. E. (1971). '*Escherichia coli* alkaline phosphatase: an analysis of transient kinetics.' *Biochem. J.*, **125**, 319–27.

HALFORD, S. E. (1972). '*Eschericia coli* alkaline phosphatase: relaxation spectra of ligand binding.' *Biochem. J.*, **126**, 727–38.

HALFORD, S. E., BENNETT, N. G., TRENTHAM, D. R. and GUTFREUND, H. (1969). 'A substrate-induced conformation change in the reaction of alkaline phosphatase from *Escherichia coli*.' *Biochem. J.*, **114**, 243–51.

HARRIGAN, P. J. and TRENTHAM, D. R. (1971). 'Reactions of D-glyceraldehyde-3-phosphate dehydrogenase with chromophoric thiol reagents.' *Biochem. J.*, **124**, 573–80.

HOLBROOK, J. J. (1972). 'Protein fluorescence of lactic dehydrogenase.' *Biochem. J.*, **128**, 921–31.

KIRSCHNER, K., EIGEN, M., BITTMAN, R. and VOIGHT, B. (1966). 'The binding of nicotinamide-adenine dinucleotide to yeast D-glyceraldehyde-3-phosphate dehydrogenase: temperature-jump relaxation studies on the mechanism of an allosteric enzyme.' *Proc. Natn. Acad. Sci. U.S.A.*, **56**, 1661–67.

ROUGHTON, F. J. W. (1963). 'Rapid reactions.' In Friess *et al.*, 704–27.

SHORE, J. D. and GUTFREUND, H. (1970). 'Transients in the reactions of liver alcohol dehydrogenase.' *Biochemistry*, **9**, 4655–59.

SYKES, B. D. and SCOTT, M. D. (1972). 'NMR studies of the dynamic aspects of molecular structure and interaction in biological systems.' *Ann. Rev. Biophys. Bioeng.*, **1**, 27–50.

TRENTHAM, D. R. and GUTFREUND, H. (1968). 'The kinetics of the reaction of nitrophenyl phosphates with alkaline phosphatase from *Escherichia coli*.' *Biochem. J.*, **106**, 455–60.

6
Enzyme Kinetics

K. F. Tipton

Department of Biochemistry, University of Cambridge

1 Introduction

The study of enzyme kinetics is frequently regarded as being an esoteric pursuit which has little relevance to the general study of biochemistry. In many cases kinetic studies have been divorced from their biochemical context and have been regarded as being an end in themselves. There is no doubt that a kinetic study of an enzyme-catalysed reaction *per se* can make a valid contribution to biochemistry but the real importance of kinetic studies lies in a much wider context. In studies of the operation and control of metabolic pathways it is necessary to know how efficiently each individual enzyme in the system will be operating with the steady-state levels of metabolites that are available to it in the cell and how the enzymes will respond to any changes in these metabolite levels. Such studies, of course, will necessitate a detailed knowledge of the kinetics of each enzyme in the system. If, on the other hand, studies are aimed at working out the detailed chemical mechanism by which the enzyme works, a thorough knowledge of the kinetics of the system again represents an essential starting point. For example it is important to know the answer to such questions as whether, in a two-substrate reaction, both substrates are bound to the enzyme at the same time and whether it is possible to study the binding of one substrate to the enzyme in the absence of the other. Again the answers to these questions can frequently be provided most easily by kinetic studies. Thus for metabolic studies and for any detailed study of the mechanism of enzyme action an understanding of the kinetics of the reaction catalysed by the enzyme (or enzymes) involved is essential.

Unfortunately, until comparatively recently, enzyme kinetics were often regarded as being a highly specialised field which was most conveniently left well alone by the general biochemist. In fact many papers on enzyme kinetics appear to have been written with the aim of preserving a mystique around the subject rather than informing the general reader. A number of kineticists have, however, written papers on kinetics that are more or less readily intelligible to the 'biochemist in the street' and some of these are given in the bibliography at the end of this chapter. This chapter, which is largely devoted to outlining the use of kinetic studies for investigating enzymes which have more than one substrate, is intended to indicate the logic underlying the interpretation of enzyme kinetics. The systems considered will be relatively simple, in the belief that an understanding of the principles of kinetic analysis is more important than a catalogue of different kinetic effects and anomalies.

Because the emphasis of this chapter is on logic rather than on the detailed algebraic manipulation of rate constants no description of the derivation of steady-state equations is included. However simple methods for deriving these equations have been detailed in several readily available sources (see e.g. Mahler and Cordes 1971; Dixon and Webb 1964).

2 Some preliminaries

2.1 Single substrate reactions

The kinetics of single-substrate reactions are discussed in some detail in the basic textbooks of biochemistry (see e.g. Mahler and Cordes 1971; White *et al.* 1964; Lehninger 1970). For a reaction which can be represented as

$$E + S \underset{k_{-1}}{\overset{k_{+1}}{\rightleftharpoons}} ES \xrightarrow{k_{+2}} E + P \tag{1}$$

the rectangular hyperbola (the Michaelis curve) which is obtained when one plots the initial velocity (v) of the reaction against the substrate concentration ([S]) will be described by an equation of the form

$$v = \frac{V[S]}{K_m + [S]} = \frac{V}{1 + K_m/[S]} \tag{2}$$

This equation, which is known as the Michaelis–Menten, or more simply as the Michaelis equation, contains two constants. V is the maximum velocity that will be obtained when the substrate concentration is sufficiently high for essentially all the enzyme to be in the form of the *ES* complex under which conditions

$$v = V = k_{+2}e \tag{3}$$

where e represents the total enzyme concentration. The other constant is K_m or the Michaelis constant. If the substrate concentration is assumed to be equal to K_m substitution of [S] for K_m in equation (2) yields

$$v = \frac{V}{[S]/[S] + 1} = \frac{V}{2} \tag{4}$$

and thus K_m may be defined as the substrate concentration which gives half-maximum velocity.
Under steady-state conditions the Michaelis constant in equation (2) is given by:

$$K_m = \frac{k_{-1} + k_{+2}}{k_{+1}} \tag{5}$$

In some cases the rate of breakdown of the ES complex to give products may be so slow that this complex remains essentially at equilibrium with the free enzyme and substrate (i.e. $k_{+2} \ll k_{-1}$). In this case equation (5) will simplify to

$$K_m = \frac{k_{-1}}{k_{+1}} = K_s \tag{6}$$

in which K_m is equal to the dissociation constant for the ES complex (K_s) and is therefore a measure of the affinity of the enzyme for its substrate; the lower the K_s the higher the affinity.

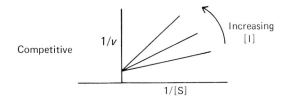

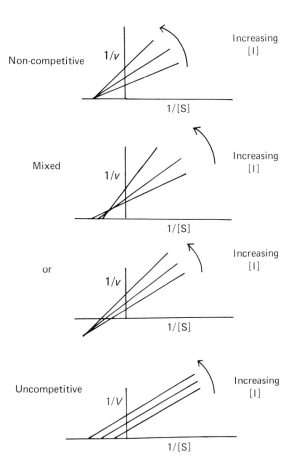

Note: Some writers make no distinction between non-competitive and mixed inhibition and refer to both types as non-competitive (see e.g., Cleland 1963).

Fig. 6.1.

The form of the Michaelis curve under such equilibrium conditions will however be no different from that which would be obtained if steady-state conditions applied, and any discussion of K_m values in terms of affinities is valueless unless equilibrium conditions have been shown to apply.

The values of the two constants in the Michaelis equation may be determined by a graphical procedure known as the reciprocal or Lineweaver–Burk plot. Taking reciprocals of equation (2) gives

$$\frac{1}{v} = \frac{K_m}{V} \cdot \frac{1}{[S]} + \frac{1}{V} \tag{7}$$

and thus a graph of $1/v$ against $1/[S]$ will give a straight line of slope K_m/V which will intersect the $1/v$ axis at a value corresponding to $1/V$ and, if extrapolated, will cut the $-1/[S]$ axis at a point corresponding to $-1/K_m$. The reciprocal plot is not the only method of obtaining accurate values of these constants and several other graphical and statistical methods are available (see e.g. Dixon and Webb 1964; Cleland 1967) which may be more accurate. However the reciprocal plot is, at present, by far the most commonly used method and will thus be used throughout this discussion.

2.2 Reversible inhibition

The different types of reversible inhibition that can arise in single-substrate reactions are discussed in most textbooks of biochemistry. They can be distinguished by the characteristic patterns of reciprocal plots which are obtained in the presence of a series of fixed concentrations of the inhibitor and these patterns are shown in Fig. 6.1. It should be noted that a specific type of inhibition is defined by the form of the reciprocal plot given and not by the type of interaction which takes place. Even in the simple single-substrate reaction it is possible for a given type of inhibition to arise in more than one way and detailed analyses of reversible inhibition (and activation) of single substrate enzyme reactions have been given by Frieden (1964) and Dixon and Webb (1964).

3 Two-substrate reactions: the general kinetic equation

Although a great deal has been written on the kinetics of enzyme reactions which involve only one substrate, the vast majority of enzyme-catalysed reactions involve two or more substrates. Many two-substrate systems are group transfer reactions of the type:

$$Ax + B \rightleftharpoons A + Bx \tag{8}$$

where Ax could be ATP, NADH, acetyl-CoA, etc.

An enzyme-catalysed reaction of this type could proceed by a pathway or mechanism in which the enzyme bound its two substrates in a fixed order or one in which either of the substrates could bind to the enzyme before the other. This first type of interaction is known as a compulsory order mechanism and may be represented as:

$$E \underset{}{\overset{Ax}{\rightleftharpoons}} EAx \underset{}{\overset{B}{\rightleftharpoons}} EAxB \rightleftharpoons EA|Bx \underset{}{\overset{Bx}{\rightleftharpoons}} EA \underset{}{\overset{A}{\rightleftharpoons}} E \tag{9}$$

The second type of reaction is known as a random order mechanism and may be represented as:

$$
\begin{array}{c}
\text{Ax} \nearrow \text{EAx} \searrow \text{B} \qquad\qquad \text{Bx} \nearrow \text{EA} \searrow \text{A} \\
E \underset{B \searrow \text{EB} \nearrow \text{Ax}}{\overset{}{\rightleftarrows}} \text{EAxB} \rightleftharpoons \text{EABx} \underset{A \searrow \text{EBx} \nearrow \text{Bx}}{\overset{}{\rightleftarrows}} E
\end{array}
\tag{10}
$$

In the case of enzymes that obey Michaelis kinetics, i.e. reactions in which a plot of v against $[Ax]$ (or $[B]$) at a fixed concentration of the other substrate will give rise to a rectangular hyperbola, the kinetic equation for the reaction must take a form similar to that of equation (2). For the majority of two-substrate reactions that obey Michaelis kinetics this equation takes the form:

$$
v = \cfrac{V}{1 + \cfrac{K_m^{Ax}}{[Ax]} + \cfrac{K_m^B}{[B]} + \cfrac{K_s^{Ax} . K_m^B}{[Ax][B]}}
\tag{11}
$$

This equation contains a Michaelis constant for each substrate (K_m^{Ax} and K_m^B) together with a combined constant term ($K_s^{Ax} . K_m^B$). At any fixed concentration of one of the substrates, e.g. B, the equation may be rearranged to give

$$
v = \cfrac{\cfrac{V[B]}{K_m^B + [B]}}{1 + \left(\cfrac{K_s^{Ax} . K_m^B + K_m^{Ax}[B]}{K_m^B + [B]} \right) \cfrac{1}{[Ax]}}
\tag{12}
$$

Thus variation of the concentration of Ax will give a Michaelis curve (or a linear reciprocal plot) in which the apparent K_m for Ax will be dependent on the concentration of B. At very large concentrations of B equation (11) will simplify the following form:

$$
v = \cfrac{V}{1 + \cfrac{K_m^{Ax}}{[Ax]}}
\tag{13}
$$

and a similar type of equation will be obtained at saturating concentrations of Ax. Equation (13) is, of course, identical to the simple Michaelis equation, and hence K_m^{Ax} can be defined as the concentration of Ax which will give half maximum velocity at saturating concentrations of B and similarly K_m^B can be defined as the concentration of B which will give half maximum velocity when Ax is saturating. Thus these two constants are true Michaelis constants.

The combined constant term $K_s^{Ax} . K_m^B$ can be shown to contain an apparent dissociation constant (K_s^{Ax}) rather than two Michaelis constants by the following reasoning. Consider a compulsory-order reaction sequence of the type:

$$
E + Ax \underset{k_{-1}}{\overset{k_{+1}}{\rightleftarrows}} EAx \underset{k_{-2}}{\overset{Bk_{+2}}{\rightleftarrows}} EAxB \ldots \text{etc.}
\tag{14}
$$

The forward rate of the step in which B is bound will depend on $k_{+2}[B]$. Thus if the concentration of B is very low this step will become very slow and the combination of Ax with the enzyme will tend to an equilibrium situation, true equilibrium for Ax binding being approached as $[B]$ tends to zero. Thus the apparent K_m for Ax will tend to the dissociation constant of the EAx complex (K_s^{Ax}) as [B] tends to zero. In addition it can be seen that as [B] tends to zero equation (12) will simplify to

$$v = \frac{\dfrac{V[B]}{K_m^B}}{1 + \dfrac{K_s^{Ax}}{[Ax]}} \tag{15}$$

and thus the apparent Michaelis constant for Ax will tend to K_s^{Ax} (the constant in the combined term) as [B] tends to zero. In the compulsory order case, therefore, the combined constants term will be the product of the dissociation constant for the first substrate to be bound and the Michaelis constant for the second. In the case of an enzyme which obeys a random-order mechanism such as that shown in equation (9) it can be easily seen that as, for example, the concentration of B falls to zero the above reasoning will also hold.

It should be emphasised that this apparent dissociation constant (K_s^{Ax}) does not necessarily represent the true dissociation constant for the reaction E + Ax = EAx since if the EAx complex undergoes any further transformations before the binding of B, e.g.

$$\text{E} + \text{Ax} \rightleftharpoons \text{EAx} \rightleftharpoons \text{ExA} \rightleftharpoons \text{ExAB} \ldots \text{etc.} \tag{16}$$

all the steps which occur before B is bound will be included in the apparent dissociation constant. If the reaction with Ax includes an irreversible step before B is bound, e.g.:

$$\text{E} + \text{Ax} \rightleftharpoons \text{EAx} \rightarrow \text{ExA} \rightleftharpoons \text{ExAB}, \ldots, \text{etc.} \tag{17}$$

the value for the apparent dissociation constant for Ax will obviously be zero and hence the term $K_s^{Ax} \cdot K_m^B / [Ax] [B]$ will vanish and equation (11) will reduce to

$$v = \frac{V}{1 + \dfrac{K_m^{Ax}}{[Ax]} + \dfrac{K_m^B}{[B]}} \tag{18}$$

The constants in equation (11) may be determined by a reciprocal method analogous to the Lineweaver–Burk method. If reciprocals are taken the equation can be rearranged to give

$$\frac{1}{v} = \left(\frac{K_m^{Ax} + \dfrac{K_s^{Ax} \cdot K_m^B}{[B]}}{V} \right) \frac{1}{[Ax]} + \left(\frac{1 + \dfrac{K_m^B}{[B]}}{V} \right) \tag{19}$$

A reciprocal plot of $1/v$ against $1/[Ax]$ at a series of concentrations of B will give a series of straight lines which may intersect above, on or below the $-1/[Ax]$ axis (such plots are sometimes called primary plots). The algebraic values of the slopes and intercepts of these plots are shown in Table 6.1. The intersection point of this set of lines can be solved in terms of its position above or below the $-1/[Ax]$ axis by taking the $1/v$ values to be equal for any two concentrations of B in equation (19) giving:

$$\left(K_m^{Ax} + \frac{K_s^{Ax} \cdot K_m^B}{[B_1]} \right) \frac{1}{[Ax]} \cdot \frac{1}{V} + \left(1 + \frac{K_m^B}{[B_1]} \right) \frac{1}{V} = \left(K_m^{Ax} + \frac{K_s^{Ax} \cdot K_m^B}{[B_2]} \right) \frac{1}{[Ax]} \cdot \frac{1}{V}$$

$$+ \left(1 + \frac{K_m^B}{[B_2]} \right) \frac{1}{V} \tag{20}$$

Rearranging yields

$$\left[\left(K_m^{Ax}+\frac{K_s^{Ax}.K_m^B}{[B_1]}\right)-\left(K_m^{Ax}+\frac{K_s^{Ax}.K_m^B}{[B_2]}\right)\right]\frac{1}{[Ax]}=-K_m^B\left(\frac{1}{[B_1]}-\frac{1}{[B_2]}\right)\tag{21}$$

and

$$K_s^{Ax}.K_m^B\left(\frac{1}{[B_1]}-\frac{1}{[B_2]}\right)\frac{1}{[Ax]}=-K_m^B\left(\frac{1}{[B_1]}-\frac{1}{[B_2]}\right)\tag{22}$$

therefore

$$\frac{1}{[Ax]}=-\frac{K_m^B}{K_s^{Ax}.K_m^B}=-\frac{1}{K_s^{Ax}}\tag{23}$$

Thus when $1/[Ax]$ (where Ax is the first substrate to bind in a compulsory order mechanism) is being varied at a series of fixed concentrations of B, the lines intersect above, on or below the $1/[Ax]$ axis at a value of $-1/K_s^{Ax}$. If $1/[B]$ is being plotted, the intersection point can be shown, in a similar way, to correspond to $-K_m^{Ax}/K_s^{Ax}.K_m^B$. Thus for compulsory order and random-order equilibrium (see later) reactions the apparent dissociation constant for one of the substrates can be directly determined from kinetic studies provided that, in the former case, it is known which one binds to the enzyme first.

From Table 6.1 it can be seen that the lines of the primary plots when Ax is varied will intersect the $1/v$ axis at a value corresponding to $\left(1+\dfrac{K_m^B}{[B]}\right)\Big/V$. These values represent the

Table 6.1   Graphical Determination of Constants from Equation (19)

Plot	Slope	Intercept on $1/v$ axis	Intercept on $-1/[S]$ axis
Primary plot			
$1/v$ vs $1/[Ax]$	$\left(K_m^{Ax}+\dfrac{K_s^{Ax}.K_m^B}{[B]}\right)\dfrac{1}{V}$	$\left(1+\dfrac{K_m^B}{[B]}\right)\dfrac{1}{V}$	$\dfrac{1+K_m^B/[B]}{K_m^{Ax}+K_s^{Ax}.K_m^B/[B]}$
Secondary plots			
(i) $1/v$ intercept vs $1/[B]$	$\dfrac{K_m^B}{V}$	$\dfrac{1}{V}$	$\dfrac{1}{K_m^B}$
(ii) Slope vs $1/[B]$	$\dfrac{K_s^{Ax}.K_m^B}{V}$	$\dfrac{K_m^{Ax}}{V}$	$\dfrac{K_m^{Ax}}{K_s^{Ax}.K_m^B}$
Primary plot			
$1/v$ vs $1/[B]$	$\left(K_m^B+\dfrac{K_s^{Ax}.K_m^B}{[Ax]}\right)\dfrac{1}{V}$	$\left(1+\dfrac{K_m^{Ax}}{[Ax]}\right)\dfrac{1}{V}$	$\dfrac{1+K_m^{Ax}/[Ax]}{K_m^B+K_s^{Ax}.K_m^B/[Ax]}$
Secondary plots			
(i) $1/v$ intercept vs $1/[Ax]$	$\dfrac{K_m^{Ax}}{V}$	$\dfrac{1}{V}$	$\dfrac{1}{K_m^{Ax}}$
(ii) Slope vs $1/[Ax]$	$\dfrac{K_s^{Ax}.K_m^B}{V}$	$\dfrac{K_m^B}{V}$	$\dfrac{1}{K_s^{Ax}}$

reciprocals of the maximum velocities that can be obtained at the fixed concentration of the second substrate (B). If a series of such intercept values is plotted against $1/[B]$ (a secondary plot) a straight line will be obtained which will cut the base line at a value of $-1/[B] = 1/K_m^B$. Another secondary plot can be made of the slopes of the individual primary plot lines against $1/[B]$. The slopes and intercept values of primary and secondary plots for systems that obey equation (19) are shown in Table 6.1, from which it can be seen that these plots can be used to obtain values for all the constants in this equation.

4 Types of reaction mechanism

As already mentioned there are several possible reaction pathways or mechanisms by which two-substrate reactions can proceed and these will be considered in turn. Examples of enzymes which obey each of the different mechanisms are listed in the bibliography at the end of this chapter.

4.1 Compulsory-order mechanisms involving the formation of a ternary complex

Such a system has been represented by the mechanism shown in equation (9) where there are two ternary complexes (i.e. complexes between three species: EAxB and EABx). Derivation of the steady-state rate equation for such a system gives an equation which is identical to equation (11). The system represented by equation (9) need not necessarily involve steady-state kinetics under all conditions. The binding of both substrates could represent a simple equilibrium process or only the binding of the first substrate could be in simple equilibrium. If the binding of Ax to the free enzyme were in equilibrium this would mean the second step, that is the binding of B, would have to be relatively slow. Since the rate of this step depends on the concentration of B one would expect that at very high values of B equilibrium conditions for binding of Ax would not continue to hold. However a number of enzymes are known in which the binding of the first substrate appears to obey equilibrium kinetics in the range of concentrations of the second substrate normally used. Under these conditions the value of K_m^{Ax} in equation (11), will become vanishingly small since at saturating concentrations of B the equilibrium

$$E + Ax \rightleftharpoons EAx \tag{24}$$

will be pulled over completely to the right provided only that there is as much Ax present as there is enzyme. Hence K_m^{Ax} will tend to a value of half the enzyme concentration and, provided that the enzyme concentration is very small, as is usually the case, the term $K_m^{Ax}/[Ax]$ will be negligible in the presence of reasonable concentrations of Ax and equation (11) will become

$$v = \frac{V}{1 + \dfrac{K_m^B}{[B]} + \dfrac{K_s^{Ax} \cdot K_m^B}{[Ax][B]}} \tag{25}$$

In reciprocal form this equation gives

$$\frac{1}{v} = \left(\frac{K_s^{Ax} \cdot K_m^B}{[B]} \right) \frac{1}{[Ax]} \cdot \frac{1}{V} + \left(1 + \frac{K_m^B}{[B]} \right) \frac{1}{V} \tag{26}$$

Thus a reciprocal plot of $1/v$ against $1/[Ax]$ at a series of concentrations of B will give a family of lines intersecting above the $1/[Ax]$ axis which will each have a slope of

$$\frac{K_s^{Ax} . K_m^B}{[B]} \frac{1}{V}$$

and hence a secondary plot of these slopes against $1/[B]$ will give a straight line that passes through the origin. In addition, rearrangement of equation (26) for a primary plot against $1/[B]$ gives

$$\frac{1}{v} = \left(K_m^B + \frac{K_s^{Ax} . K_m^B}{[Ax]} \right) \frac{1}{[B]} \cdot \frac{1}{V} + \frac{1}{V} \tag{27}$$

At a series of concentrations of Ax this primary plot will give a family of lines which all intersect on the $1/v$ axis. Such a plot should not be taken to imply that, at saturating $[B]$, the enzyme can reach maximum velocity in the absence of Ax but merely that under equilibrium conditions for the binding of Ax, the concentration of this substrate required is negligible at very high B concentrations. If equilibrium conditions are assumed to apply for the binding of both substrates an equation identical to (25) will be obtained. It is therefore possible to determine the order in which the substrates are bound to the enzyme from the form of the reciprocal plots in this special case.

4.2 Compulsory-order mechanisms in which a ternary complex is not formed

4.2.1 The Theorell–Chance mechanism

This mechanism, which was originally suggested for liver alcohol dehydrogenase by Theorell and Chance (1951), can be represented by the equation:

$$\text{E} \xrightleftharpoons{\text{Ax}} \text{EAx} \xrightleftharpoons[\substack{\text{B} \quad \text{Bx}}]{} \text{EA} \xrightleftharpoons{\text{A}} \text{E} \tag{28}$$

This mechanism should not be taken to imply that the two substrates (Ax and B) do not meet on the surface of the enzyme; it merely indicates that the rate of formation of the ternary EAxB complex is so slow and the rate of its breakdown is so fast that the concentration of this complex is too low to be kinetically significant. The rate equation for this mechanism is identical to equation (11).

4.2.2 The double-displacement or ping-pong mechanism

In a double-displacement mechanism the two substrates never meet on the surface of the enzyme but the reaction proceeds in two halves involving the formation of a free modified form of the enzyme:

$$\text{E} \xrightleftharpoons{\text{Ax}} \text{EAx} \xrightleftharpoons{} \text{ExA} \xrightleftharpoons{\text{A}} \text{Ex}$$
$$\text{Ex} \xrightleftharpoons{\text{B}} \text{ExB} \xrightleftharpoons{} \text{EBx} \xrightleftharpoons{\text{Bx}} \text{E} \tag{29}$$

For the initial rate measurements in such a system the concentration of the products (A and Bx) will, of course, be essentially zero and hence the step in which A is released from the enzyme will be irreversible. This is, therefore, a case in which there is an irreversible step following the binding of Ax but before B is bound; thus the apparent dissociation constant for the EAx complex will be zero and this mechanism will obey equation (18). Taking reciprocals of equation (18) will give

$$\frac{1}{v} = \frac{K_m^{Ax}}{V} \cdot \frac{1}{[Ax]} + \left(1 + \frac{K_m^B}{[B]} \right) \frac{1}{V} \tag{30}$$

Primary plots of $1/v$ against $1/[Ax]$ at a series of concentrations of B will thus give a family of parallel lines and similar patterns will be obtained if $1/[B]$ is plotted. Values for the two Michaelis constants can be obtained from secondary plots of the intercept values. If the initial rate determinations are made in the presence of a fixed amount of one of the products of the reaction the release of A from the enzyme will no longer be an irreversible step and hence K_s^{Ax} will have a definite value. Under these conditions equation (11) will be obeyed and intersecting primary plots will be obtained.

Primary plots which give rise to families of parallel lines are diagnostic for a double-displacement mechanism, but it is not always easy to decide whether the lines are really parallel or whether they will intersect at a point far distant from the $1/v$ axis (which would be the case if K_s^{Ax} were finite but very low). There are several methods for confirming that a double-displacement mechanism really is operative and one of the simplest involves varying both substrates simultaneously. If Ax and B are mixed together such that $[Ax] = y[B]$ and a series of initial rates are determined at different concentrations of this mixture, the equation for a reciprocal plot may be obtained by substituting $y[B]$ for $[Ax]$ in equation (30) and rearranging:

$$\frac{1}{v} = \left(\frac{K_m^{Ax}}{y} + K_m^B \right) \frac{1}{V[B]} + \frac{1}{V} \tag{31}$$

Under such conditions a double-displacement mechanism will therefore give rise to a linear reciprocal plot. If, however, the mechanism obeys equation (11) substitution and taking reciprocals gives

$$\frac{1}{v} = \left(\frac{K_m^{Ax}}{y} + \frac{K_s^{Ax} . K_m^B}{y[B]} + K_m^B \right) \frac{1}{V[B]} + \frac{1}{V} \tag{32}$$

from which it can be seen that a reciprocal plot will depend on $[B]^2$ and will hence be non-linear.

There is another simple test for the operation of a double-displacement mechanism: since the reaction proceeds in two distinct halves it should be possible to demonstrate the formation of the product A in amounts stoichiometric to the amount of enzyme present when the enzyme is incubated with excess Ax in the absence of B.

4.3 Random-order mechanism

A random-order mechanism can be written as shown in equation (10). Such a mechanism may obey either equilibrium or steady-state kinetics. If equilibrium kinetics are obeyed the rate of interconversion of the ternary complexes must be very slow relative to the rate of product binding so that the ternary complex EAxB remains in equilibrium with the free enzyme and substrates. Under these conditions only the first half of the reaction need be considered for the determination of the initial rate equation in the forward direction.

$$\tag{33}$$

In this mechanism the four constants shown are dissociation constants. We can also write an equation for the initial velocity of the reaction,

$$v = k[\text{EAxB}] \tag{34}$$

and an equation indicating how the total amount of enzyme (e) is distributed among the various complexes

$$e = ([\text{E}] + [\text{EAx}] + [\text{EB}] + [\text{EAxB}]) \tag{35}$$

Taking our various dissociation constants we can rearrange them to give expressions for the concentrations of the different enzyme forms in terms of EAxB. Thus:

$$K_m^{\text{Ax}} = \frac{[\text{EB}] \cdot [\text{Ax}]}{[\text{EAxB}]} \quad \therefore [\text{EB}] = \frac{K_m^{\text{Ax}}}{[\text{Ax}]}[\text{EAxB}] \tag{36}$$

$$K_m^{\text{B}} = \frac{[\text{EAx}] \cdot [\text{B}]}{[\text{EAxB}]} \quad \therefore [\text{EAx}] = \frac{K_m^{\text{B}}}{[\text{B}]}[\text{EAxB}] \tag{37}$$

$$K_s^{\text{Ax}} = \frac{[\text{E}] \cdot [\text{Ax}]}{[\text{EAx}]} \quad \therefore [\text{E}] = \frac{K_s^{\text{Ax}}}{[\text{Ax}]}[\text{EAx}] \tag{38}$$

Combining (38) with (37)

$$[\text{E}] = \frac{K_s^{\text{Ax}} \cdot K_m^{\text{B}}}{[\text{Ax}][\text{B}]}[\text{EAxB}]$$

If the relationships derived above are substituted into (35) we get

$$e = \left(1 + \frac{K_m^{\text{Ax}}}{[\text{Ax}]} + \frac{K_m^{\text{B}}}{[\text{B}]} + \frac{K_s^{\text{Ax}} \cdot K_m^{\text{B}}}{[\text{Ax}][\text{B}]}\right)[\text{EAxB}] \tag{39}$$

which can be substituted into (34) to give

$$v = \frac{ke}{1 + \dfrac{K_m^{\text{Ax}}}{[\text{Ax}]} + \dfrac{K_m^{\text{B}}}{[\text{B}]} + \dfrac{K_s^{\text{Ax}} \cdot K_m^{\text{B}}}{[\text{Ax}][\text{B}]}} \tag{40}$$

Since the maximum velocity (V) will be reached when essentially all the enzyme is in the form of the EAxB complex, V may be substituted for ke in equation (40) which is thus identical to equation (11). It can be seen that although there are four dissociation constants in the pathway (33) only three of them occur in the final equation. This is because the fourth is redundant since it is defined by the other three, by the equation:

$$K_s^{\text{Ax}} \cdot K_m^{\text{B}} = K_m^{\text{Ax}} \cdot K_s^{\text{B}} \tag{41}$$

If we have a case where the binding of one substrate has no effect on the affinity of the enzyme for the other, we have a situation in which $K_s^{\text{Ax}} = K_m^{\text{Ax}}$ and $K_s^{\text{B}} = K_m^{\text{B}}$ and thus equation (40) simplifies to

$$v = \frac{V}{1 + \dfrac{K_s^{\text{Ax}}}{[\text{Ax}]} + \dfrac{K_s^{\text{B}}}{[\text{B}]} + \dfrac{K_s^{\text{Ax}} \cdot K_s^{\text{B}}}{[\text{Ax}][\text{B}]}} \tag{42}$$

In this case the primary reciprocal plots determined at a series of concentrations of the other substrate will intersect on the horizontal (i.e. 1/[substrate]) axis and the dissociation constants may be determined simply by varying one substrate at a fixed concentration of the other.

If steady-state conditions apply the kinetic equation given for the initial rate of mechanism (10) is extremely complicated, taking the general form

$$v = \frac{(C_1[Ax][B] + C_2[Ax]^2[B] + C_3[Ax][B]^2)e}{C_4 + C_5[Ax] + C_6[B] + C_7[Ax][B] + C_8[Ax]^2 + C_9[B]^2 + C_{10}[Ax]^2[B] + C_{11}[Ax][B]^2}$$

(43)

Where $C_1, \ldots, C_{11}$ are complex combinations of rate constants. The appearance of squared terms in both substrate concentrations in this equation arises because there are two alternative routes by which each substrate can bind to the enzyme to form the ternary complex (see (10)).

Equation (43) is of little practical use and will give rise to non-linear reciprocal plots. The occurrence of non-linear reciprocal plots has been frequently regarded as being diagnostic of random-order mechanisms in which steady-state conditions apply. The degree of curvature will, however, depend on the relative values of the different constants and the absolute values of the substrate concentrations, and a number of workers have pointed out that in many cases the curvature of the reciprocal may be unnoticeable and this mechanism may then be difficult to distinguish from the equilibrium case by the methods outlined in this chapter (see, e.g. Cleland 1971).

Another system that will give rise to non-linear reciprocal plots is one in which the enzyme catalyses a reaction between two identical substrate molecules, e.g. the enzyme adenylate kinase (myokinase) catalyses

$$ADP + ADP = AMP + ATP \tag{44}$$

In such cases an enzyme that follows any mechanism which gives rise to an equation of the form of (11) would, writing [A] for the concentration of the substrate, give

$$v = \frac{V}{1 + \dfrac{K_m^A}{[A]} + \dfrac{K_{m'}^A}{[A]} + \dfrac{K_s^A \cdot K_{m'}^A}{[A]^2}}$$

(45)

where K_m^A and $K_{m'}^A$ are the Michaelis constants for the 'first' and 'second' molecules of A to bind. Equation (45) would give rise to upwardly curved reciprocal plots and the Michaelis curve would be sigmoid. The only circumstances in which such a system would be expected to give rise to a linear reciprocal plot would be if the binding of the two substrate molecules were separated by an irreversible step, so that K_s^A would be zero. Such a situation would occur if a double-displacement mechanism were followed in which case equation (45) would become

$$v = \frac{V}{1 + \dfrac{K_m^A}{[A]} + \dfrac{K_{m'}^A}{[A]}}$$

(46)

5 Distinction between kinetic mechanisms

5.1 Product inhibition

All the mechanisms discussed which give rise to linear reciprocal plots give similar patterns of primary plots, except the double-displacement mechanism and the compulsory-order mechanism in which the first substrate is bound in an equilibrium. It is however frequently possible to distinguish between kinetic mechanisms by determining the types of inhibition given by the products of the reaction. In order to understand how this is done it is necessary to consider what effects the modification of individual kinetic constants will have on the reciprocal plots obtained. Two situations should be considered for systems that obey

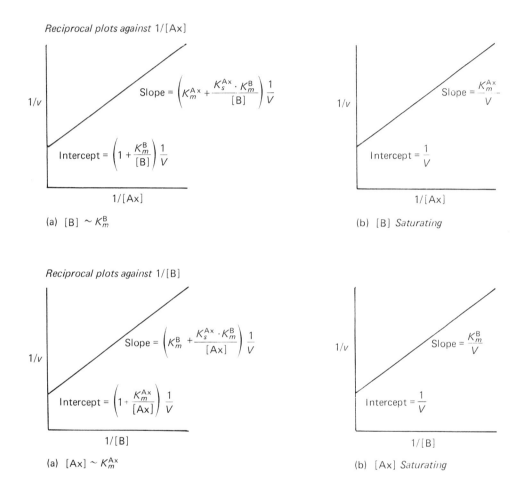

Fig. 6.2.

equation (11): (a) when the concentration of the second substrate is of the order of its K_m and (b) when the concentration of the second substrate, e.g. B, is so high that the terms $K_s^{Ax} \cdot K_m^B/[Ax][B]$ and $K_m^B/[B]$ are negligible (the second substrate is said to be saturating under these conditions). The slopes and intercepts of reciprocal plots under these conditions are shown in Fig. 6.2.

If the effect of an inhibitor is to increase one of the constants that appears in the slope of the primary plot without affecting the intercept term inhibition will appear to be competitive. Alternatively if the inhibitor alters the constants in both the slope and intercept terms the inhibition will be mixed. Finally if the effect of the inhibitor is to alter a constant which appears in the intercept term without affecting the slope the inhibition will be uncompetitive.

A number of possible effects in terms of equation (11) are shown in Table 6.2. The possibilities given in this table are by no means exhaustive but they cover those most commonly encountered in two-substrate enzyme systems. We can now consider the ways in which the alterations in the constants listed in Table 6.2 can come about.

Table 6.2   Some possible effects of inhibitors of two-substrate reactions

Effect of Inhibitor	*Type of Inhibition Observed*			
	With respect to Ax		*With respect to B*	
	B Not saturating	*B Saturating*	*Ax Not saturating*	*Ax Saturation*
(a) Increase in K_m^{Ax} and K_s^{Ax}	Competitive	Competitive	Mixed	None
(b) Increase in K_m^B	Mixed	None	Competitive	Competitive
(c) Equal decrease in V, K_m^{Ax} and K_m^B	Uncompetitive	Uncompetitive	Uncompetitive	Uncompetitive
(d) Equal decrease in V, K_m^{Ax} and K_m^B *plus* increase in K_m^B	Mixed	Uncompetitive	Mixed	Mixed

The effects of inhibitors on the individual kinetic constants in a two-substrate system are similar to the effects of an inhibitor on a single-substrate reaction. Thus an inhibitor which binds to the same form of the enzyme and at the same site as Ax will be competitive with respect to it and will thus increase the amount of Ax required to half-saturate the enzyme. The effect of such an inhibitor will therefore be to increase both K_m^{Ax} and K_s^{Ax} by the same factor $(1 + i/K_i)$, where i represents the concentration of the inhibitor and K_i is the apparent dissociation constant of the enzyme–inhibitor complex (see references in sections 2.1 and 2.2). As can be seen from Table 6.2 (case (a)) such an inhibitor will appear competitive with respect to Ax at all concentrations of B, and would be mixed with respect to B except if Ax were saturating when no inhibition would be observed. This last observation is not surprising since, if the inhibitor is competitive with respect to Ax, saturation with this substrate will drive all the inhibitor from the enzyme. The case in which the inhibitor binds to the same form of the enzyme and at the same site as B will give rise to the exact parallel of the inhibition patterns described above, as shown in case (b) of Table 6.2.

In a single-substrate reaction an inhibitor that only binds to the enzyme–substrate complex exerts a dual effect of decreasing V and of decreasing K_m by the same factor by pulling the reaction $E + S \rightleftharpoons ES$ in the direction of ES. (See, e.g. Mahler and Cordes 1971; Gutfreund 1965). This gives rise to an equation:

$$v = \frac{\dfrac{V}{(1 + i/K_i)}}{1 + \dfrac{K_m}{(1 + i/K_i)}\dfrac{1}{[S]}} = \frac{V}{(1 + i/K_i) + K_m/[S]} \tag{47}$$

which describes the uncompetitive inhibition pattern. In the case of a two-substrate reaction an inhibitor that binds to an enzyme–substrate intermediate after both substrates have been bound will act in the same way, decreasing V and the Michaelis constants. Thus K_m^{Ax}, K_m^B and V will all be decreased by a factor of $(1 + i/K_i)$. K_s^{Ax} will not be affected since this represents the limiting value of K_m^{Ax} as B tends to zero and hence cannot be affected by an inhibitor which only binds to the enzyme after B has bound. An inhibitor that binds in this way will therefore give rise to an equation of the form

$$v = \frac{V/(1 + i/K_i)}{1 + \dfrac{K_m^{Ax}}{(1 + i/K_i)}\dfrac{1}{[Ax]} + \dfrac{K_m^{B}}{(1 + i/K_i)}\dfrac{1}{[B]} + \dfrac{K_s^{Ax}.K_m^{B}}{(1 + i/K_i)}\dfrac{1}{[Ax][B]}}$$

$$= \frac{V}{1 + \dfrac{i}{K_i} + \dfrac{K_m^{Ax}}{[Ax]} + \dfrac{K_m^{B}}{[B]} + \dfrac{K_s^{Ax}.K_m^{B}}{[Ax][B]}} \tag{48}$$

which, as seen from Table 6.2, will give rise to uncompetitive inhibition with respect to both substrates regardless of the concentration of the second substrate.

Where inhibition of enzyme reactions occurs by binding of one of the products of the reaction in a ternary complex it might also perhaps be expected to reverse the reaction to some extent. For example in the compulsory-order mechanism:

$$E \xrightleftharpoons{Ax} EAx \xrightleftharpoons{B} EAxB \rightleftharpoons EABx \xrightleftharpoons{Bx} EA \xrightleftharpoons{A} E \tag{49}$$

Product A can bind to the free enzyme and presumably at the same site as Ax and hence one would expect this product to give rise to the inhibition pattern described by (a) in Table 6.2. Product Bx however cannot bind to the free enzyme but only to EA to form a ternary complex. Thus one might at first sight expect it to give rise to the uncompetitive inhibition pattern shown by (c) in Table 6.2. However since Bx is a product (and therefore a substrate for the reverse reaction) it would be expected that high concentrations of Bx could reverse the reaction and displace B from the enzyme. This will of course also give rise to an increase in K_m^{B} which will result in an inhibition pattern which is a combination of cases (b) and (c) of Table 6.2 as shown in (d) where inhibition will be mixed with respect to all substrates except when B is saturating, under which conditions no B will be displaced from the enzyme and hence we get simple uncompetitive inhibition with respect to Ax.

These lines of reasoning are summarised in Table 6.3 which shows the inhibition patterns to be expected from the products of two-substrate reactions. In the case of the compulsory-order mechanism with ternary complexes the inhibition pattern expected has already been discussed

Table 6.3   Product inhibition patterns in two-substrate reactions

| | | With respect to Ax | | With respect to B | |
		B Not saturating	B Saturating	Ax Not saturating	Ax Saturating
Compulsory-order with ternary complexes					
(equation (9))	A	Competitive	Competitive	Mixed	None
	Bx	Mixed	Uncompetitive	Mixed	Mixed
Theorell–Chance	A	Competitive	Competitive	Mixed	None
(equation (28))	Bx	Mixed	None	Competitive	Competitive
Double-displacement	A	Mixed	None	Competitive	Competitive
(Equation (29))	Bx	Competitive	Competitive	Mixed	None

above. In the Theorell–Chance mechanism products A and Bx behave like cases (a) and (b) respectively of Table 6.2. The inhibition patterns given by the double-displacement mechanism are straightforward provided one remembers that in this case A binds to the same form of the enzyme (Ex) as B (see equation (29)) and hence will be competitive with respect to B rather than Ax. It should also be remembered that in the presence of a product the term $K_s^{Ax}.\ K_m^B/[Ax][B]$ will be present in the final kinetic equation and hence the inhibition pattern will be as expected from mechanisms obeying equation (11).

Product inhibition studies can therefore allow one to decide which type of mechanism is operative and it should be noted that they may also enable the order in which the products are released to be distinguished. In all the ordered mechanisms considered here it has been assumed that Bx is released from the enzyme before A, but the Theorell–Chance mechanism, for example, could proceed like this

$$
E \rightleftharpoons \overset{\text{Ax}}{} \text{EAx} \underset{\text{B}}{\overset{}{\rightleftharpoons}} \overset{}{\underset{\text{A}}{}} \text{EBx} \overset{\text{Bx}}{\rightleftharpoons} E \tag{50}
$$

In this case the product inhibition pattern would be the converse of that shown for this mechanism in Table 6.3; it would, in fact, be identical to that given by the double-displacement mechanism. Thus it is theoretically possible to determine the order of release of the products of a reaction provided that one knows, perhaps from direct binding studies, the order in which the substrates bind to the enzyme.

In the case of the random-order equilibrium mechanism prediction of the inhibition patterns is complicated by the number of possible forms that interaction with product may take. Thus the binding of product A will not necessarily prevent substrate B from binding, giving rise to an abortive EAB complex:

$$
\begin{array}{c}
\text{EAx} \\
\nearrow \quad \searrow \\
\text{E} \quad\quad \text{EAxB} \ldots \text{etc.} \\
\nearrow \searrow \quad \nearrow \\
\text{EA} \quad \text{EB} \\
\searrow \quad \nearrow \\
\text{EAB}
\end{array} \tag{51}
$$

The type of inhibition pattern obtained will depend on the relative affinities of B for E and for EA. A will obviously be competitive with respect to Ax and in the special case (equation (42)) in which the binding of Ax to the enzyme has no effect on the binding of B one would, perhaps, expect the presence of A to have no effect on the binding of B and hence, since equilibrium conditions hold, inhibition will be non-competitive with respect to B. If however the presence of bound A does affect but not prevent the binding of B the inhibition will be mixed with respect to that substrate. An extreme case can occur in which one of the products, e.g. Bx, occupies sufficient of both substrate binding sites to be able to displace either of them from the enzyme. In this case Bx would be competitive with respect to both substrates.

From the above discussion it can be seen that in some cases it is not possible to distinguish the Theorell–Chance mechanism from a random-order equilibrium mechanism by simple product inhibition studies of this type. However a slightly different use of product inhibition will allow such a distinction to be made. In the case of the Theorell–Chance mechanism shown in equation (28) the K_i value for product A acting as a competitive inhibitor with respect to Ax represents the dissociation constant of the EA complex and will therefore be independent of the concentration of B in the assay medium. However, in the case of a random-order equilibrium mechanism in which an abortive complex is formed, inhibition by A is due to its combination with two forms of the enzyme (E and EB) and hence in this case, if the K_i values for the EA and the EAB complexes are different (which is likely to be the case when the product inhibition patterns are indistinguishable from those given by the Theorell–Chance mechanism) the K_i value for A as a competitive inhibitor with respect to Ax will depend on the concentration of B.

5.2 'Dead end' inhibitors

The use of reversible inhibitors that are neither substrates nor products of an enzyme can, in some cases, assist in distinguishing between the possible kinetic mechanisms. These compounds will combine with the enzyme to form a 'dead-end' complex which unlike a true product of the reaction is incapable of reacting further. The inhibition patterns given by non-reactive product analogues will, of course, be similar to those given by the products themselves in most cases, except where the product causes inhibition by reversing the reaction. Thus in a compulsory-order mechanism involving ternary complexes (as shown in equation (9)) an analogue of Bx will give rise to the uncompetitive pattern shown in Table 6.2, case (c).

The inhibition patterns given by product analogues in a double-displacement mechanism are of particular interest since, if these are unreactive they will be incapable of reversing the product release steps and so the constant K_s^{Ax} will still be zero. Thus the kinetic equation in the presence of an unreactive analogue of A (A') which will bind to Ex, and hence compete with B, will become

$$v = \frac{V}{1 + \dfrac{K_m^{Ax}}{[Ax]} + \dfrac{K_m^{B}}{[B]}\left(1 + \dfrac{[A']}{K_i}\right)} \tag{52}$$

or in reciprocal form

$$\frac{1}{v} = \frac{K_m^{Ax}}{V}\frac{1}{[Ax]} + \left[1 + \frac{K_m^{B}}{[B]}\left(1 + \frac{[A']}{K_i}\right)\right]\frac{1}{V} \tag{53}$$

Hence A' will be competitive with respect to B at all concentrations of Ax and will be uncompetitive with respect to Ax unless B is saturating when no inhibition will be observed.

In the case of 'dead-end' inhibitors which are analogues of the substrates of the reaction, the relationship between the inhibition patterns will in the main be straightforward to predict. A particularly interesting form of inhibition can occur in compulsory-order systems such as that shown in (9). In this case an analogue of Ax (Ax') which is not a substrate for the enzyme will bind to the free enzyme and compete with Ax; however, in addition, since Ax' resembles Ax it may be possible for substrate B to bind to the EAx' complex.

$$
\begin{array}{c}
\text{EAx}' \xrightleftharpoons{K_{iB}} \text{EAx}'\text{B} \\[4pt]
{\scriptstyle K_i}\Big\Updownarrow \\[4pt]
\text{E} \rightleftharpoons \text{EAx} \rightleftharpoons \text{EAxB} \dots \text{etc.}
\end{array} \tag{54}
$$

If this situation applies the substrate B will, in addition to its normal role, actually act to increase the inhibition given by the substrate analogue by combining to form an abortive EAx'B complex. The equation for such a system can be written

$$v = \frac{V}{1 + \dfrac{K_m^{Ax}}{[Ax]}\left(1 + \dfrac{[B][Ax']}{K_i K_{iB}} + \dfrac{[Ax']}{\tilde{K}_i}\right) + \dfrac{K_m^{B}}{[B]} + \dfrac{K_s^{Ax} \cdot K_m^{B}}{[Ax][B]}\left(1 + \dfrac{[Ax']}{K_i}\right)} \tag{55}$$

in which K_i and K_{iB} are the dissociation constants of Ax' from EAx' and B from the EAx'B complexes respectively and $\tilde{K}_i$ is a more complex constant. This equation predicts that the reciprocal plot against 1/[B] may become non-linear in the presence of Ax' while the plot against 1/[Ax] will be linear. In contrast an analogue of B (B') will bind to the EAx complex but no binding of any substrate to the EAxB' complex would be expected and thus reciprocal plots would be linear. Random-order equilibrium mechanisms involve the possibility of either substrate binding to a complex of the enzyme with an analogue of the other substrate, however in this case, since equilibrium conditions hold, the existence of alternative species to which B can bind will not result in non-linearity, e.g. for the system shown below, in which K_i, K_{iB} and K_{Bi} are dissociation constants:

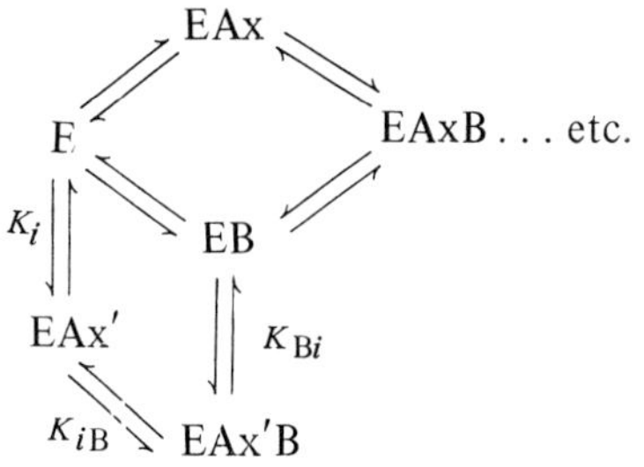

$$\tag{56}$$

derivation of the rate equation by the method given on p. 11 will give:

$$v = \frac{V}{1 + \dfrac{K_m^{Ax}}{[Ax]}\left(1 + \dfrac{[Ax']}{K_{Bi}}\right) + \dfrac{K_m^{B}}{[B]} + \dfrac{K_s^{Ax} \cdot K_m^{B}}{[Ax][B]}\left(1 + \dfrac{[Ax']}{K_i}\right)} \tag{57}$$

which will give rise to linear reciprocal plots.

Thus the appearance of curved reciprocal plots in the presence of a substrate analogue can distinguish between random- and compulsory-order mechanisms and can also indicate which substrate binds first. The method will not necessarily distinguish between compulsory-order mechanisms involving a ternary complex and Theorell–Chance mechanisms, since although in the latter mechanism the concentration of EAxB is negligible this need not apply to a complex of the form EAx'B which represents a dead-end. It is, of course, also possible that product (or product analogue) binding to the free enzyme in a compulsory-order system could be followed by the binding of the second substrate and the formation of an abortive complex. The compulsory-order mechanism shown below

$$E \rightleftharpoons EAx \rightleftharpoons EAxB \rightleftharpoons EABx \rightleftharpoons EA \rightleftharpoons E \tag{58}$$
$$\updownarrow$$
$$EAB$$

would in the presence of A, give rise to an equation of a form similar to (55) when inhibition by A is being studied, although in this case the inhibition is further complicated by the fact that B also acts as an uncompetitive inhibitor of the reaction.

5.3 Haldane relationships

Haldane first showed that the equilibrium constant (K_{eq}) for a reversible single-substrate reaction of the type

$$E + S \rightleftharpoons ES \rightleftharpoons E + P \tag{59}$$

could be related to the Michaelis constants for the reaction in the forward and reverse directions by the expression

$$K_{eq} = \frac{V^f K_m^b}{V^b K_m^f} \tag{60}$$

where the superscripts f and b denote the forward and reverse reactions respectively (see Dixon and Webb 1964 for the derivation of this relationship).

In the case of reversible two-substrate reactions obeying equation (8) the Haldane relationships can sometimes be diagnostic of the reaction mechanism prevailing and some of

Table 6.4   Some Haldane relationships for two-substrate reactions

Mechanism	Haldane relationship
Compulsory-order (equation (9))	$K_{eq} = \dfrac{V^f . K_s^A . K_m^{Bx}}{V^b . K_s^{Ax} . K_m^B}$
Theorell-Chance (equation (28))	$K_{eq} = \dfrac{V^f . K_s^A . K_m^{Bx}}{V^b . K_s^{Ax} . K_m^B} = \left(\dfrac{V^f}{V^b}\right)^3 . \dfrac{K_m^A . K_m^{Bx}}{K_m^{Ax} . K_m^B}$
Double-displacement (equation (29))	$K_{eq} = \left(\dfrac{V^f}{V^b}\right)^2 . \dfrac{K_m^A . K_m^{Bx}}{K_m^{Ax} . K_m^B}$
Random-order equilibrium (equation (10))	$K_{eq} = \dfrac{V^f . K_s^A . K_m^{Bx}}{V^b . K_s^{Ax} . K_m^B}$
	$= \left(\dfrac{V^f . K_m^A . K_s^{Bx}}{V^b . K_s^{Ax} . K_m^B} = \dfrac{V^f . K_s^A . K_m^{Bx}}{V^b . K_m^{Ax} . K_s^B} = \dfrac{V^f . K_m^A . K_s^{Bx}}{V^b . K_m^{Ax} . K_s^B}\right)$

these relationships (adapted from Dalziel 1957) are listed in Table 6.4. The Haldane relationships for three-substrate reactions have also been listed by Cleland (1963) and Dalziel (1969).

6 Three-substrate reactions

Space does not allow any detailed discussion of reactions with more than two substrates, but the kinetic analysis is, in many ways, similar to the two-substrate systems. There are many possible variations on mechanism for three-substrate reactions. As well as completely random and completely ordered systems there can be 'hybrid' mechanisms in which, e.g. two of the substrates bind randomly and the third binds in a compulsory order, or two of the substrates are involved in a double-displacement type of mechanism. The general kinetic equation obeyed by three-substrate reactions can be written as:

$$v = \frac{V}{1 + \dfrac{K_m^A}{[A]} + \dfrac{K_m^B}{[B]} + \dfrac{K_m^C}{[C]} + \dfrac{K^{AB}}{[A][B]} + \dfrac{K^{AC}}{[A][C]} + \dfrac{K^{BC}}{[B][C]} + \dfrac{K^{ABC}}{[A][B][C]}} \tag{61}$$

unless random order binding of two (or three) substrates occurs under steady-state conditions. In this equation A, B and C represent the three substrates, K_m^A, K_m^B and K_m^C represent the true Michaelis constants for these substrates and K^{AB}, etc., represent multiples of constants. If one of the substrates, e.g. A, is present at saturating concentrations the terms containing that substrate concentration will become negligible and equation (61) becomes

$$v = \frac{V}{1 + \dfrac{K_m^B}{[B]} + \dfrac{K_m^C}{[C]} + \dfrac{K^{BC}}{[B][C]}} \tag{62}$$

which is identical in form to equation (11), and thus three-substrate reactions can be treated in the same way as the two-substrate case if one of the substrates is held at saturating concentrations.

A particularly interesting case occurs with the compulsory-order system such as

$$E \xrightarrow{\ A\ } EA \xrightarrow{\ B\ } EAB \xrightarrow{\ C\ } EABC \ldots \text{etc.}$$

The equation given by this system will be similar to (61) except that the constant K^{AC} will be absent,

$$v = \frac{V}{1 + \dfrac{K_m^A}{[A]} + \dfrac{K_m^B}{[B]} + \dfrac{K_m^C}{[C]} + \dfrac{K_s^A . K_m^B}{[A][B]} + \dfrac{K_s^B . K_m^C}{[B][C]} + \dfrac{K_s^A . K_s^B . K_m^C}{[A][B][C]}} \tag{64}$$

In this equation K_s^B represents the dissociation constant of B from the EAB complex. If the middle substrate to add (B) is present in saturation amounts the equation becomes

$$v = \frac{V}{1 + \dfrac{K_m^A}{[A]} + \dfrac{K_m^C}{[C]}} \tag{65}$$

which is identical to the equation given by the two-substrate double-displacement mechanism (equation (18)) and thus reciprocal plots will yield families of parallel lines. Saturation with one of the other substrates will not yield an equation of this simplicity. For example if A is saturating we get

$$v = \frac{V}{1 + \dfrac{K_m^B}{[B]} + \dfrac{K_m^C}{[C]} + \dfrac{K_s^B . K_m^C}{[B][C]}} \tag{66}$$

Which will not, of course, yield parallel reciprocal plots. Thus in compulsory-order three-substrate reactions it is possible to ascertain which of the substrates adds in the middle since saturation with that substrate will give parallel reciprocal plots. It will also be noted that if the products are released in an ordered fashion, the middle product to be released will be an uncompetitive inhibitor with respect to all three substrates since it binds to a complex formed after all three substrates have bound but will not be able to reverse the reaction to displace any substrate from the enzyme in the absence of the first product to be released. A detailed account of the kinetics of three-substrate reactions has been presented by Dalziel (1969) and an account of the use of 'dead-end' inhibitors with such systems has been given by Fromm (1967).

7 Different kinetic notations

One of the principal problems of understanding published work on enzyme kinetics is that different authors frequently express the same kinetic equation in very different ways. This means that the reader is frequently required to 'translate' published material into his favourite nomenclature before he can understand what is going on. Unfortunately it is not easy to say that one specific formulation is clearly the best and so there is no general agreement on nomenclature. The way of writing equations in this paper is adapted from that of Alberty (1953) and the kinetic constants are as recommended by the Enzyme Commission of The International Union of Biochemistry, but in addition to this method there are other systems in wide use and it is hoped that the brief glossary (sections 7.1, 7.2) of the constants used in the two most widely used of these methods will aid in 'translation' of published papers on kinetics.

7.1 The formulation of Dalziel

Dalziel (1957) would write equation (11) in the form

$$\frac{e}{v} = \phi_0 + \frac{\phi_1}{[Ax]} + \frac{\phi_2}{[B]} + \frac{\phi_{12}}{[Ax][B]} \tag{67}$$

This is obviously a reciprocal expression (cf. equation (19)) and values for the ϕ's, which are referred to as kinetic coefficients, can be determined graphically from primary and secondary plots. If the reaction is reversible the coefficients for the backwards reaction are denoted by primes, e.g. ϕ_0', ϕ_1', etc. Definitions of these constants in terms of the constants used in this paper are shown in Table 6.5.

Table 6.5   Relationships between kinetic
constants in different formulations

This paper	Dalziel	Cleland
V (or V^f)	e/ϕ_0	V_1
K_m^{Ax}	ϕ_1/ϕ_0	Ka
K_m^B	ϕ_2/ϕ_0	Kb
K_s^{Ax}	ϕ_{12}/ϕ_2	Kia

7.2 The formulation of Cleland

Cleland (1963) would write equation (11) in the form

$$v = \frac{V_1 Ax . B}{KiaKb + KbA + KaB + AB} \tag{68}$$

an equation which can be converted to the same form as equation (11) by dividing all through by $Ax . B$. The maximum velocity for the reaction in the backwards direction would be denoted by V_2. The constants used in this equation are defined in terms of equation (11) in Table 6.5.

Cleland has also devised a convenient method for writing kinetic equations which can be illustrated with reference to the mechanisms discussed here. A compulsory-order reaction such as that shown in equation (9) would be written as

$$
\begin{array}{cccccc}
Ax & B & & Bx & A & \\
\downarrow & \downarrow & & \uparrow & \uparrow & \\
\hline
E & EAx & (EAxB - EABx) & EA & E &
\end{array}
\tag{69}
$$

and would be termed an ordered bi bi (two substrates bind and two products are then released) mechanism.

The random-order mechanism (equation (10)) would be written

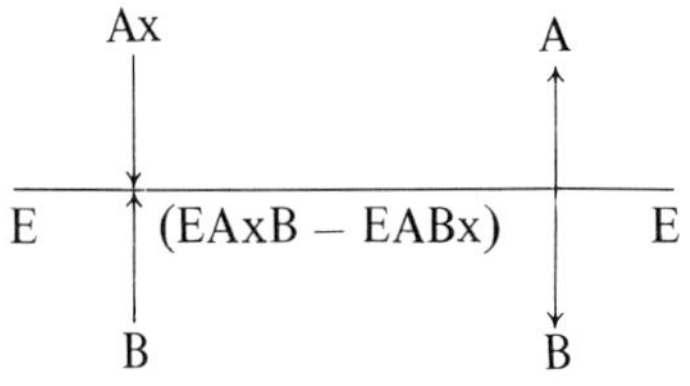

$$
\begin{array}{ccc}
Ax & & A \\
\downarrow & & \uparrow \\
\hline
E \quad \uparrow & (EAxB - EABx) & \downarrow \quad E \\
\downarrow & & \downarrow \\
B & & B
\end{array}
\tag{70}
$$

while the double-displacement mechanism (equation (29)) would be written

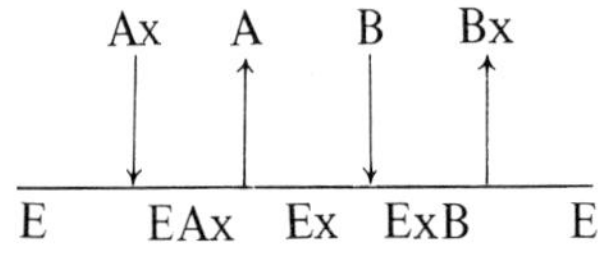

$$
\begin{array}{ccccc}
Ax & A & B & Bx & \\
\downarrow & \uparrow & \downarrow & \uparrow & \\
\hline
E & EAx & Ex & ExB & E
\end{array}
\tag{71}
$$

and would be termed a ping-pong bi bi reaction.

8 Snags and limitations

The foregoing discussion may have, I hope, implied that a distinction between different kinetic mechanisms is quite straightforward and in many cases this is indeed true; sometimes, however, it is considerably more difficult to make this distinction. For example some reactions are, for all practical purposes, irreversible and in such cases the release of one of the products from the enzyme may represent an irreversible step. Product inhibition patterns are usually unable to distinguish between kinetic mechanisms if only one of the products of the reaction is inhibitory. If 'dead-end' inhibitors for the enzyme are available these may, of course, be used and in addition kineticists have devised an armoury of 'back-up' methods which can be used in such cases and also to confirm decisions reached from initial rate and product inhibition studies. Space does not permit discussion of these methods but good accounts of the use of isotope exchange methods are given by Wong and Hanes (1964) and Cleland (1971). The use of alternative substrates is described by Fromm (1964) and Rudolph and Fromm (1970) and alternative product inhibition is discussed by Cleland (1971). The complications that can arise if one of the substrates is inhibitory at high concentrations have been discussed by Dalziel (1957) and Cleland (1971).

This discussion has so far made no mention of allosteric enzymes which are outside the limited scope of this review. It should, however, be mentioned that the 'major' theories of allosteric interactions are equilibrium theories in that the sigmoid Michaelis curve represents a true binding curve for the substrate. Thus the step occurring after substrate binding is, often tacitly, assumed to be rate limiting and the equations developed are solely for substrate binding. This approach is certainly valid in a number of cases since substrate or effector binding has been shown to give allosteric kinetics in the absence of any overall reaction. However several steady-state systems which could give rise to sigmoid kinetics have been suggested and these have been reviewed in detail by Whitehead (1970).

It is all too easy to get the subject of enzyme kinetics out of proportion by assuming either that it can on its own provide important answers to biochemical problems or, on the other hand, that the information it yields is trivial. In fact, as has already been emphasised, the answer probably lies between these extremes. Kinetics represent an important part of any detailed study of an enzyme or enzyme system. However it is important to realise some of the limitations that are inherent in kinetic studies. For example, studies of the type outlined above cannot tell us how many intermediates are involved in a reaction. In the mechanisms for two-substrate reactions which involve ternary complexes we have assumed here that there are two such complexes (EAxB and EABx); however there could be only one or even a hundred without altering the form of the kinetic equations.

It is impossible to prove absolutely that a specific kinetic mechanism is obeyed by a given enzyme because it is usually possible to work out alternative mechanisms of more complexity which would also fit the observed results. Thus one has to rely on the principle of Occam's razor; that is to choose the simplest mechanism that is consistent with the results whilst being prepared to increase the complexity if further results require this and remembering that 'God does not always shave with Occam's razor' (G. D. Greville, unpublished observation).

Finally it should be remembered that there is nothing hard and fast about the kinetic mechanism obeyed by a given enzyme. For example the difference between different mechanisms only represents differences in the values of individual rate constants in the mechanism. Thus one should not necessarily assume that an enzyme will obey the same mechanism regardless of the conditions. Specific enzymes have been shown to change the kinetic mechanism that is followed with changes in the substrate structure, pH, ionic strength and temperature. These possibilities are obviously of considerable importance in any attempts to apply results from steady-state kinetic experiments to the understanding of the behaviour of enzyme systems in the intact cell.

9 Bibliography

9.1 General references

References marked with an asterisk have not been cited in the text but contain useful accounts of kinetic theory for further reading.

ALBERTY, R. A. (1953). 'The relationship between Michaelis constant, maximum velocities and the equilibrium constant for an enzyme-catalysed reaction.' *J. Amer. Chem. Soc.,* **75**, 1928–32.

* BLOOMFIELD, V., PELLER, L. and ALBERTY, R. A. (1962). 'Multiple intermediates in steady-state enzyme kinetics II and III.' *J. Amer. Chem. Soc.,* **84**, 4367–81.

250

CLELAND, W. W. (1963). 'The kinetics of enzyme-catalysed reactions with two or more substrates or products. I. Nomenclature and rate equations.' *Biochim. Biophys. Acta,* **67**, 104-37.

CLELAND, W. W. (1967). 'Enzyme kinetics.' *Ann. Rev. Biochem.,* **36**, 77-112.

CLELAND, W. W. (1971). 'Steady-state kinetics.' In *The Enzymes,* vol. 1 pp. 1-65. Ed. Boyer, P. D. New York and London. Academic Press.

DALZIEL, K. (1957). 'Initial steady-state velocities in the evaluation of enzyme-coenzyme-substrate reaction mechanisms.' *Acta Chem. Scand.,* **11**, 1706-23.

DALZIEL, K. (1969). 'The interpretation of kinetic data for enzyme-catalysed reactions involving three substrates.' *Biochem. J.,* **114**, 547-56.

DIXON, M. and WEBB, E. C. (1964). *Enzymes,* 2nd ed. pp. 54-166. London. Longman.

FRIEDEN, C. (1964). 'Treatment of enzyme kinetic data. I. The effect of modifiers on the kinetic parameters of single-substrate enzymes. *J. Biol. Chem.,* **239**, 3522-31.

FROMM, H. J. (1964). 'The use of alternative substrates in studying enzymic mechanisms involving two substrates.' *Biochim. Biophys. Acta* **81**, 413-17.

FROMM, H. J. (1967). 'The use of competitive inhibitors in studying the mechanism of action of some enzyme systems utilizing three substrates.' *Biochim. Biophys. Acta,* **139**, 221-30.

GUTFREUND, H. (1965). *An Introduction to the Study of Enzymes,* pp. 72-116. Oxford. Blackwell.

LEHNINGER, A. L. (1970). *Biochemistry,* pp. 147-167. New York. Worth Publishers.

MAHLER, H. R. and CORDES, E. H. (1971). *Biological Chemistry,* 2nd ed., pp. 267-324. New York. Harper and Row.

* PLOWMAN, K. M. (1972). *Enzyme Kinetics.* New York. McGraw-Hill Book Co.

* REINER, J. M. (1959). *Behaviour of Enzyme Systems.* Minneapolis. Burgess Publishing Co.

RUDOLPH, F. B. and FROMM, H. J. (1970). 'Use of isotope competition and alternative substrates for studying the kinetic mechanism of enzyme action. I. Experiments with hexo-kinase and alcohol dehydrogenase.' *Biochemistry,* **9**, 4660-65.

THEORELL, H. and CHANCE, B. (1951). 'Studies on liver alcohol dehydrogenase. II. The kinetics of the compound of horse liver alcohol dehydrogenase and reduced diphospho-pyridine nucleotide.' *Acta Chem. Scand.,* **5**, 1127-44.

* WESTLEY, J. (1969). *Enzymic Catalysis.* New York. Harper and Row.

WHITE, A., HANDLER, P. and SMITH, E. L. (1964). *Principles of Biochemistry.* 3rd ed. pp. 204-62. New York. McGraw-Hill Book Co.

WHITEHEAD, E. (1970). 'The regulation of enzyme activity and allosteric transition.' *Prog. Biophys.,* **21**, 321-97.

* WONG, J. T. F. and HANES, C. S. (1962). 'Kinetic formulations for enzymic reactions involving two substrates.' *Canad. J. Biochem.,* **40**, 763-804.

WONG, J. F. T. and HANES, C. S. (1964). 'Isotopic exchange at equilibrium as a criterion of enzymatic mechanisms.' *Nature, Lond.,* **203**, 492-94.

9.2 Examples of enzymes obeying specific kinetic mechanisms

The following brief list of examples has been chosen to illustrate the different approaches which have been applied in kinetic studies as well as to provide examples of enzymes which follow the different types of kinetic mechanisms.

(*a*) *Compulsory-order mechanisms involving a ternary complex*

MORRISON, J. F. and EBNER, K. E. (1971). 'Studies on galactosyl transferase. Kinetic investigations with N-acetylglucosamine as the galactosyl group acceptor.' *J. Biol. Chem.*, **246**, 3977–84. (In this paper Mg^{++} is shown to be the first substrate to bind and its binding has been shown to be in simple equilibrium.)

ORSI, B. A. and CLELAND, W. W. (1972). 'Inhibition and kinetic mechanism of rabbit muscle glyceraldehyde 3-phosphate dehydrogenase.' *Biochemistry*, **11**, 102–9.

RAVAL, D. N. and WOOLFE, R. G. (1962). 'Malic dehydrogenase. II. Kinetic studies of the reaction mechanism, and III. Kinetic studies of the reaction mechanism by product inhibition.' *Biochemistry*, **1**, 263–69 and 1112–17.

(*b*) *Theorell–Chance mechanisms*

DALZIEL, K. (1962). 'Kinetic studies of liver alcohol dehydrogenase.' *Biochem. J.*, **84**, 244–54.

DUNCAN, R. J. S. and TIPTON, K. F. (1971) 'The kinetics of pig brain aldehyde dehydrogenase.' *Europ. J. Biochem.*, **22**, 538–43.

ZEWE, V. and FROMM, H. J. (1962). 'Kinetic studies of rabbit muscle lactate dehydrogenase.' *J. Biol. Chem.*, **237**, 1668–75.

(*c*) *Double-displacement (ping-pong) mechanisms*

MIDDLETON, B. (1972). 'The kinetic mechanism of 3-hydroxy-3-methylglutaryl-coenzyme A synthetase from baker's yeast.' *Biochem. J.*, **126**, 35–47.

UHR, M. L., MARCUS, F. and MORRISON, J. F. (1966). 'Studies on adenosine triphosphate: arginine phosphotransferase. Purification and reaction mechanism.' *J. Biol. Chem.*, **241**, 5428–35.

TIPTON, K. F. (1968). 'The reaction pathway of pig brain mitochondrial monoamine oxidase.' *Europ. J. Biochem.*, **5**, 316–20.

(*d*) *Random-order mechanisms*

   (*i*) *Equilibrium conditions*

CHASE, J. F. A. and TUBBS, P. K. (1966). 'Some kinetic studies on the mechanisms of action of carnitine acetyltransferase.' *Biochem. J.*, **99**, 32–40.

MORRISON, J. F. and JAMES, E. (1965). 'The mechanism of the reaction catalysed by adenosine triphosphate-creatine phosphotransferase.' *Biochem. J.*, **97**, 37–52.

(ii) *Steady-state conditions*

FROMM, H. J., SILVERSTEIN, E. and BOYER, P. D. (1964). 'Equilibrium and net reaction rates in relation to the mechanism of yeast hexokinase.' *J. Biol. Chem.*, **239**, 3645–52.

GULBINSKY, J. S. and CLELAND, W. W. (1968). 'Kinetic Studies of *Escherichia coli* galactokinase.' *Biochemistry*, **7**, 566–74.

7
Structure and Function of Nucleic Acids

J. Hindley

Department of Biochemistry, University of Bristol

1 Introduction

In a discussion of the relation between structure and function in nucleic acids it needs to be recognised that both DNA and RNA may contain two different types of primary structure; the potentially coding (or translatable sequences) and the non-coding sequences which appear to have a variety of different functions. In some instances these functional differences may be clearly separated as, for example, in the case of the non-translated ribosomal and tRNA sequences and the largely (but not entirely) translatable sequences of m-RNAs. In other instances, however, the distinction may be blurred and polycistronic m-RNAs appear to contain coding sequences (cistrons) separated by non-translatable 'spacer' sequences. Furthermore there is increasing evidence that particular sequences in the small RNA phage messengers have evolved which have both coding and non-coding functions.

In even the simplest DNA molecules a complex arrangement occurs of transcribed and non-transcribed sequences which, in addition to coding for polypeptide chains, also serve as signals controlling the ordered duplication and expression of the genome. Any attempt therefore to relate DNA structure to function at the molecular level must rely heavily on methods for identifying and isolating particular regions of chromosomal DNA. Progress in this direction is discussed in the first part of this chapter.

Very much more is known, however, about RNA primary structure. The three main classes of RNA — tRNA, ribosomal RNA and mRNA — have each been extensively studied and the complete sequences of many tRNAs and also the 5S ribosomal RNAs from several different species have been determined. Much information is also available about the primary structure of the polycistronic messengers derived from the small RNA coliphages (MS2, Rl7, $Q\beta$) and a rather convincing account can now be given at a molecular level of some aspects of the translation and transcription of these messengers and their relation to RNA structure. This will be described in the second section of this chapter.

2 DNA Structure and Function

The last five years have witnessed a remarkable advance in our understanding of the mechanisms and controlling elements involved in the highly ordered transcription of several phage DNAs, e.g. T_2, T_4 and T_7 λ, ϕX174. Refined genetic analysis has defined the order, position and products of many of the phage genes and an overall picture of how these interact and also modify the host transcriptional and translational machinery has emerged.

At the molecular level, however, difficulties in devising methods for DNA sequence analysis have until very recently hampered detailed studies of these interactions. Even the smallest DNA molecules contain several thousand nucleotides and for that reason the main objectives of sequence analysis have been directed to smaller sequences within the molecules that are involved in some specific function.

2.1 In phages and bacteria

2.1.1 The "sticky" ends of phage λ DNA

It is sometimes found that a DNA molecule (e.g. phage λ DNA) which exists in the phage particle in a linear double-stranded form can assume a closed circular configuration in the phage-infected cell. The interconversion of these two species is due to short single stranded segments of DNA at each end of the linear molecule which are strictly complementary to each other and can therefore form base pairs to give a closed circle (Fig. 7.1). An enzyme, DNA ligase, then joins the broken ends (shown by the dotted arrows) yielding a covalently bonded circular duplex. The important point is that in the interconversion of these species, the 'makes' and 'breaks' do not occur randomly but at one specific region in the DNA molecule. In the reverse reaction breakage of the intact circle is brought about by an endonuclease of remarkable specificity; in the whole molecule of about 5×10^4 base pairs only two breaks are made, one in each strand and separated from each other by 12 base pairs. The only bonds now holding the molecule in the circular form are the H-bonds between the two 'nicked' strands. Thermal agitation alone is sufficient to rupture these and consequently regenerate the linear double-stranded DNA. The specificity of the endonuclease attack is demonstrated by the unique sequence of nucleotides found at each end of the linear molecule since random fragmentation would generate random sequences. This must clearly have some important

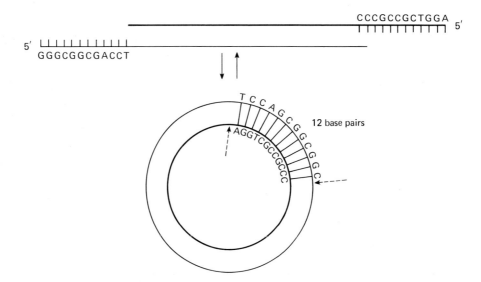

Fig. 7.1 Interconversion of linear and circular forms of λ DNA. (The dotted arrows show where the rings are sealed with DNA-ligase and also the sites of cleavage by endonuclease.)

function in the cell and it is now thought that it provides a controlling mechanism for DNA replication. Since only nicked DNAs (i.e. those possessing a free 3'-hydroxyl end) can be replicated by DNA polymerase, the covalently closed circles are inactive as templates for replication. The observation that circular DNAs have a widespread occurrence (the entire *E. coli* chromosome and yeast mitochondrial DNA can exist as closed circles) suggests a general role for this mechanism in limiting and controlling chromosome duplication.

2.1.2 Modification and restriction

Probably the best studied examples of the specificity of particular DNA endonucleases are encountered in the phenomenon of host-induced DNA modification and restriction. Many strains of *E. coli* can recognise and degrade 'foreign' DNA from other strains. This obviously has an important function in maintaining the genetic uniqueness of the cell against possible interference by extraneous biologically active DNAs. Whether or not a foreign DNA is degraded depends on certain non-heritable properties (or host induced modifications) introduced into the DNA molecule by the host cell. These modifications probably constitute a pattern of specific methylation generated in particular nucleotide sequences of the DNA molecule by a modification enzyme. A classic example is the ability of phage λ, after growth in different *E. coli* host strains, to reinfect fresh *E. coli* cells successfully. Phages grown in *E. coli* K strains possessing the modification character m_K will multiply with high efficiency in fresh K cells, whereas phage previously grown in strains lacking the m_K character are very poorly infective. Instead, their DNA is rapidly broken down into several large pieces ('restricted') on entering the K strain cells. The simplest hypothesis to account for this observation is that the host cells contain a restriction character which is a gene directing the formation of a nuclease directed against DNA lacking the corresponding modification. A restriction enzyme, called R.K., has been isolated and purified from *E. coli* strain K.

It is apparent that a restriction enzyme must possess a remarkable degree of specificity to avoid cleavage of host cell DNA and the simplest hypothesis here is that it must recognise a region of unique base sequence not found in the host DNA. Consistent with this, the restriction enzyme, endonuclease R, from *H. influenzae* has no nucleolytic activity against *H. influenzae* DNA but will degrade T_7 DNA, into pieces of average size about 1000 nucleotide base pairs. In other words, only about 0.1 per cent of the phosphodiester bonds are broken and this suggests that the enzyme recognises only a relatively small number of sites within the foreign DNA molecule. A further feature of the enzyme is that it appears to bring about only double-stranded breaks, yielding products terminating in 5'-phosphoryl:3'-hydroxyl groups. No single-stranded breaks are found in the products. Analysis of the sequences around the 5'-phosphoryl ends (using polynucleotide kinase) and the 3'-hydroxyl ends (using micrococcal nuclease digestion) suggested that the restriction enzyme recognises the following specific sequence:

$$5' \ldots \text{pGpTpPy} \uparrow \text{pPupApCp} \ldots 3'$$

$$3' \ldots \text{pCpApPup} \downarrow \text{PypTpGp} \ldots 5'$$

The arrows indicate the points of cleavage by endonuclease R (Kelly and Smith 1970); Pu = purine and Py = pyrimidine base.

The most striking feature of the recognition site is its symmetry: when read with the same polarity the sequence of bases in one strand is the same as that in the other. This may reflect an underlying symmetry in the enzyme, since it brings about the cleavage at equivalent points in two DNA sequences of opposite polarity. Modification is brought about by the methylation of particular adenine residues in the DNA molecule (to yield 6-methyladenine) and there is increasing evidence that this occurs at those sites which, in the unmodified form, are susceptible to cleavage by the restriction enzyme.

2.1.3 Recognition of promoter and operator sites

Two biologically important instances of specific recognition and interaction between proteins and DNA sequences are the binding of RNA polymerase to the promotor sites in DNA and repressor binding to a site in the operator locus.

Initiation of transcription by DNA-dependent RNA polymerase takes place by the initial binding of the enzyme to particular regions (promotor sites) on the DNA template. In the model of Jacob and Monod (1961) the chromosomal DNA is pictured as being divided up into a series of separate 'obligatory units of transcription' or operons. The transcription of any operon yields a messenger RNA, which can be either monocistronic or polycistronic in nature (i.e. it can code for one or several proteins). At any given time not all operons are active, and an elaborate switch gear exists in the cell which can turn on or turn off different operons in response to the requirements of the cell for the synthesis of particular enzymes. This control is exerted in Jacob and Monod's terminology by a separate locus, the 'operator', which immediately precedes the genes in the operon that code for messenger RNA. The operator locus can also be regarded as containing two different elements. One element recognises and responds to signals — the repressor molecules — which govern whether or not a particular operon is transcribed, whilst the other, the promotor site, specifies the point of attachment of the RNA polymerase which is responsible for catalysing transcription of the operon. Any elementary text in molecular biology develops this theory in detail (e.g. Watson, 1970).

Both the binding of RNA polymerase to the promotor and the attachment of repressor to its unique operator site involve the specific recognition of a particular DNA sequence by proteins and this again provides, in principle, a means of isolating a particular region of DNA and of relating its sequence to a defined function. Preliminary attempts to isolate and characterise a promotor site sequence have been made by deoxyribonuclease digestion of complexes of RNA polymerase and DNA prepared in the absence of the nucleoside triphosphate precursors. The assumption here is that the fragments of DNA which are resistant to hydrolysis and attached to the RNA polymerase, represent a part of the promotor binding site. A 2·7 S DNA fragment, (which would correspond to 38 nucleotide pairs) was found to reassociate with RNA polymerase. There is no good evidence, however, that this represents a unique fragment nor has any sequence analysis been reported.

The evidence for specific recognition of DNA sequences by repressor proteins *in vitro* is much stronger. The λ-phage repressor is coded for by the C_1 gene of *E. coli* phage λ. *In vivo*, this repressor switches off the other phage genes, which would result in vegetative multiplication of the phage and cell lysis and is therefore responsible for maintaining the phage chromosome in the lysogenic (prophage) state. The phage repressor into which radioactive amino acids have been incorporated has been highly purified. *In vitro*, the repressor binds tightly and specifically to DNA containing the appropriate operator locus giving a complex readily identified by sucrose or glycerol gradient centrifugation. The *lac* repressor, isolated from *E. coli* cells, has also been extensively purified and shown to be a tetramer of mol. wt. 150 000. It binds specifically to the *lac* operator region of the *E. coli* chromosome and also to a variety of inducers, in general β-galactosides. A detailed molecular model for this interaction in which a sequence of approx. 50 amino acids at the N-terminus if the *lac* repressor is thought to bind directly to the *lac* operator DNA has been proposed by Adler *et al.* (1972).

From studies of the equilibrium binding of repressor (R) to the DNA operator site (O):

$$R + O \; \underset{k_r}{\overset{k_f}{\rightleftharpoons}} \; RO$$

the equilibrium constant

$$K_d = \frac{k_r}{k_f} = 10^{-13}\text{M}$$

indicative of very tight binding.

The selectivity of these associations clearly open the way for their analysis at the molecular and sequence level. X-Ray diffraction analysis of the tertiary and quaternary structure of the repressor, coupled with sequence analysis of the DNA binding site, would be expected to give valuable information about how a protein can recognise a unique DNA sequence. Such information is likely to be basic to our understanding of many genetic control processes.

These examples show clearly how small, defined regions of macromolecular DNA with particular functions can be selected for detailed study, and illustrate how the ability of proteins to recognise specific regions of DNA sequence is an important aspect of the relation between structure and function in DNA.

2.1.4 Ribosome binding sites in phage DNA

A different approach to the problem of picking out particular DNA sequences of biological interest, recently exploited by Robertson et al. (1973), was based on the observation that under the conditions for initiation of protein synthesis, ribosomes would bind specifically to a particular region of the single-stranded DNA from phage ϕX which is then resistant to nuclease digestion. This is analogous to the approach used in studying the ribosome binding sites of the RNA phages (see section 3.7). Since the phage DNA strand contains the same sequences as the corresponding messenger RNA (except that the uridine residues in the RNA are replaced by thymine residues in the DNA), ribosomes would be expected to attach to a DNA sequence corresponding to the start of one (or more) of the phage cistrons. This indeed was the case and a fragment containing 51 deoxyribonucleotides, corresponding to a region around the start of the phage coat protein cistron was isolated after nuclease digestion and its sequence determined.

This work shows that direct analysis of extended DNA sequences is now possible, and the implications of this in the following discussion will be obvious.

2.2 Repetitive DNA in higher organisms

It might be thought that the sheer size and complexity of DNA in eukaryote chromosomes would effectively exclude any meaningful analysis of gene arrangement and structure. For example, the haploid genome of a mammalian cell contains approx. 5×10^9 base pairs, or about 1000 times the amount found in the E. coli chromosome. If only one copy of each gene were present per haploid chromosome then about two million different proteins could theoretically be specified. Such hypothetical deductions, however, can be highly misleading since it can be shown that many repeated copies of certain genes occur and, fortunately for our purpose, the reiterated DNA segments corresponding to these genes differ sufficiently in density from the bulk of the chromosomal DNA to allow their separation by centrifugation in dense salt gradients.

This important finding stemmed from the discovery that the band pattern of calf thymus DNA, after equilibrium sedimentation in a neutral caesium chloride gradient, had a pronounced skew distribution towards the heavier side. Subsequently both lighter and heavier DNA components were found in other species, the extreme example of this being the discovery of a DNA fraction from the crab which was composed almost entirely of a repeating dAT sequence. These minor DNA components are called 'satellite' DNAs. The other feature which distinguishes satellite DNAs is their ability to reassociate rapidly after denaturation suggesting that they are composed of largely reiterated sequences strung together. For example, from its reassociation rate mouse satellite DNA ($\sim 10^5$ base pairs) appears to be composed of hundreds of repeated sequences of about 200–300 base pairs. Although more recent work has emphasised that the quantitative aspects of results based on rates of reassociation need to be interpreted with caution, the widespread occurrence of repetitive sequences in a variety of different DNAs show them to be an important feature of chromosome structure in higher organisms.

Analysis of the smaller RNA sequences, such as the transfer and 5 S RNAs, shows that a characteristic feature is the small number of defined oligonucleotides produced after degradation by enzymes specific for one (ribonuclease T_1) or two (pancreatic ribonuclease) bases. Although there are no corresponding base-specific DNases, DNA can be degraded by the diphenylamine–formic acid reaction to yield oligopyrimidines of the form p-(pyr-p)$_n$ the number of such different tracts depending on the complexity of the sequence. Thus crab dAT-rich satellite DNA gives mainly pTp by this reaction whereas more complex sequences, like ϕX174 DNA, give a very large number of oligopyrimidines.

In the α-component of guinea-pig satellite DNA, as in the corresponding mouse satellite, the two strands produced by denaturation differ sufficiently in their GT content to permit their separation by density gradient centrifugation in alkaline CsCl solution. In this way Southern (1970) separated the heavy (H) (i.e. GT rich), and light (L) strands from guinea-pig α-satellite DNA. Analysis of their polypyrimidine sequences gave remarkable results: the H strand yielded only three main pyrimidine tracts namely p(Tp)$_2$, pTp and p(Tp)$_3$. The L strand also gave a highly non-random pattern of polypyrimidines, the most common sequence being pCpCpCpTp. Further analysis of all the polypyrimidines and calculation of the average repeat length for each oligomer suggested an extremely simple basic repeating sequence, namely:

$$5'\text{-}(CCCTAA)_n\text{-}3' \qquad \text{L strand}$$
$$3'\text{-}(GGGATT)_n\text{-}5' \qquad \text{H strand}$$

Thus satellite DNA contains a simple sequence, repeated many times in which random base substitutions have accumulated during evolution. It may be significant that all possible base substitutions do not appear to have occurred with equal frequency. In the L-strand, the change from CCCT to TCCT is ten times more frequent than that to CCCC. The evolutionary implications of these results have been excellently reviewed by Walker (1971). A primary structure composed of a basic repeating sequence seems to rule out the possibility that the satellite codes for a protein; on transcription it would give either alternating nonsense codons or code for repeating dipeptides, depending on the reading frame. The function and origin of these sequences poses an important and intriguing problem.

Several lines of investigation have shown that the satellite DNAs are concentrated mainly in the heterochromatin (condensed) fraction of isolated interphase chromatin. When metaphase chromosomes are extracted with 2M NaCl about 70 per cent of the DNA is extracted into the supernatant, but all the satellite material is in the remaining 30 per cent. The technique of *in situ* hybridisation shows that a large fraction of the satellite DNA is located near the centromere in metaphase chromosomes. The weight of evidence suggests that the satellite sequences are not transcribed into RNA; so what, then, is their function? One suggestion is that the untranscribed DNA has a 'housekeeping' function and is involved in controlling the specific folding and condensation of chromosomes or that it is somehow involved in the 'recognition' of the centromeric regions of homologous chromatids. On the other hand, as discussed by Walker (1971), there can be extreme differences in sequence and amount of satellite DNAs between closely related species. Three species of *Drosophila*, that can interbreed, and show chromosome pairing along their entire length, nevertheless have three heavy satellites of different densities and one species has in addition, two light satellites. It is difficult to envisage how any common recognition function involving specific DNA-protein interaction could be accommodated within such marked variations in DNA structure.

2.3 Repetitive DNA cistrons coding for the ribosomal RNAs

A particular type of repetitive DNA is found in the multiple DNA cistrons coding for ribosomal RNA. The extent of gene redundancy (i.e. the multiplicity of occurrence of a particular gene) is determined by hybridising the DNA with an excess of the homologous ribosomal RNA (rRNA). These experiments show that the number of ribosomal cistrons is roughly proportional to the overall DNA content of the genome. For example, bacteria, with a DNA content of $2-3 \times 10^9$ daltons have $5-10$ ribosomal cistrons, plants have several thousand and in amphibia this figure reaches 10^4 or more. In the amphibian oocyte, presumably in anticipation of oocyte's need for intense rRNA and protein synthesis, the number of ribosomal cistrons is increased to several million by a process of gene amplification. A further feature, relevant to the following discussion, is that the larger ($23-28$ S) and smaller ($16-18$ S) ribosomal RNA species do not hybridise to the same DNA sites, and in addition the 28 S and 18 S cistrons occur in a 1 : 1 ratio. This whole topic has been reviewed by Birnstiel *et al.* (1970).

Two important observations, which permit both the isolation and detailed analysis of the ribosomal cistrons by hybridisation, come from buoyancy studies in CsCl gradients. In the first place, for many eukaryotes the DNA containing the ribosomal cistrons ('ribosomal DNA') bands at a density greater than that of the bulk of the DNA. In *Xenopus*, all the ribosomal cistrons are found in a DNA satellite with a buoyant density of 1·723, whereas the main bulk of the DNA bands at a density of 1·699. The second observation is that ribosomal DNA, when hybridised to rRNA, shows a marked increase in buoyant density. Analysis on gradients therefore gives a clear separation between the denser rRNA : DNA hybrids and the lighter, major DNA band. DNA from an anucleolate mutant of *Xenopus* tadpoles does not give the characteristic rDNA satellite nor does it hybridise with rRNA. These findings thus gave powerful support to the idea that the nucleolar organiser region, which occupies a specific site on the chromosomes and is deleted in the anucleolate mutants, in fact contains the ribosomal cistrons. Since a single deletion mutation which brings about the loss of a single chromosome locus also results in the loss of about 800 ribosomal cistrons, it must be assumed that the ribosomal cistrons are clustered together at one site corresponding to the nucleolar organiser region. *In situ* hybridisation of radioactive rRNA to a single chromosome locus also supports this interpretation.

2.4 Mapping the ribosomal cistrons

The discovery that both 28 S and 18 S rRNA of *Xenopus* hybridise to a single ribosomal DNA satellite showed that the separate cistrons must be intermingled in the chromosomal DNA. Moreover, the hybridisation data showed that the two cistrons were present in a 1 : 1 ratio. Since the 28 S and 18 S rRNAs differ widely in their (G + C) content, the corresponding rDNAs would be expected to band at widely different densities in CsCl. If all the 800 or so 28 S cistrons were clustered together in one block (equivalent to a mol. wt. of 2×10^9) and all the 18 S cistrons in another block (mol. wt. 10^9) this would give rise to two distinct rDNA satellites banding at their corresponding densities. However, even after shearing the DNA to an average mol. wt. of about 10^7, only a single satellite is found containing equal proportions of the 28 S and 18 S cistrons. Hybridisation experiments using even shorter single-stranded DNA segments of mol. wt. 3×10^6 showed that these fragments contained on average a single 28 S cistron ($1 \cdot 5 \times 10^6$) and a single 18 S cistron ($0 \cdot 7 \times 10^6$). Further analysis confirmed these findings and also showed the presence in the rDNA satellite of a heavier DNA segment distinct

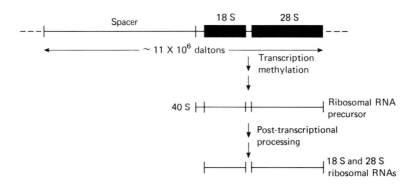

Fig. 7.2 One model for repeating unit of rDNA of *Xenopus laevis*. It is not known which end of the 40 S precursor is synthesised first or from which end the excess RNA is lost.

from the ribosomal DNA cistrons. Hybridisation experiments showed that about 40 per cent of the DNA in the satellite band was complementary to rRNA sequences indicating that about 60 per cent of the DNA corresponded to the heavier non-ribosomal DNA. The combined mol. wt. of a single molecule each of 28 S and 18 S rRNA is $2 \cdot 2 \times 10^6$ daltons and the corresponding double-stranded DNA cistrons would contain $4 \cdot 4 \times 10^6$ daltons. If this corresponds to 40 per cent of the basic repeating unit the heavier non-ribosomal 'spacer' DNA comprises a further $6 \cdot 6 \times 10^6$ daltons.

It is known that the 28 S and 18 S rDNA cistrons are not transcribed separately but that the two RNA species arise from the post-transcriptional processing of a single precursor RNA. This is a general phenomenon in all organisms examined. In the case of *Xenopus* the precursor RNA sediments at 40 S and has a mol. wt. of about $2 \cdot 5 \times 10^6$, about $0 \cdot 3 \times 10^6$ in excess of a 28 S + 18 S tandem. The following model can therefore be deduced for rDNA in *Xenopus* (Fig. 7.2): $4 \cdot 4 \times 10^6$ daltons code for the 28 S and 18 S rRNAS; $0 \cdot 6 \times 10^6$ daltons code for other transcribed, but non-conserved, sequences in the RNA precursor; and almost 6×10^6 daltons of non-transcribed spacer DNA alternates with the ribosomal transcriptional unit. It should be emphasised that all the quantitative data in this model derives from hybridisation experiments, and while the general picture has received strong support from direct electron micrographs of oocyte nucleoli engaged in RNA synthesis (Miller and Beatty 1969) the precise amounts of spacer and transcribed DNA in the rDNA satellite may be subject to revision. As yet no function can be ascribed to the DNA spacer segments, nor is there any information relating to their sequence.

2.5 Heterogeneous nuclear RNA

An important difference which distinguishes eukaryotic organisms from prokaryotes is the physical separation of the genetic material in the cell nucleus from the translational machinery which occurs in the cytoplasm. The messenger RNA synthesised in the nucleus therefore has to be exported through the nuclear membrane to the cytoplasmic ribosomes for subsequent translation. When cultures of mammalian cells, e.g. HeLa cells, are grown in the presence of a radioactive RNA precursor, such as ^{3}H-uridine, much of the newly synthesised RNA is found in the nucleus. After careful extraction this RNA has a sedimentation constant of 70-100 S, corresponding to a minimum mol. wt. of about 3×10^6. In contrast, the messenger RNA which can be extracted from the cytoplasmic polyribosomes, has a mol. wt. of only about one-tenth of this and further experiments showed that, surprisingly, the bulk of the labelled RNA never leaves the nucleus and is subsequently degraded to low mol. wt. material.

The relationship between this heterogeneous nuclear RNA (Hn RNA) and the cytoplasmic mRNA has been studied in many laboratories and it now seems clear that the mRNA sequences are selected from the HnRNA by a post-transcriptional processing. For example, both the HnRNA and mRNA from HeLa cells contain long polyadenylic acid (poly(A)) sequences and if addition of poly (A) to the HnRNA is prevented this also blocks the appearance of most of the mRNA in the polyribosomes. These findings raise the problem of why so much mRNA precursor is made only, apparently, to be broken down again and why a considerable proportion of the nuclear DNA is given over to making RNA which is not translated into protein nor, as far as we can tell, has any ascertainable function. At present we have no answer to these problems. It may be that the untranslated RNA sequences reflect the need for an elaborate switchgear − perhaps analogous to the promoter and operator controls of bacteria − for controlling the activity of the different genes. Alternatively it is possible that much of the untranslated RNA arises from evolutionarily 'old' DNA sequences which, in the present stage of evolution of the organism, have been superseded and are no longer translated into protein, These problems are among the most fascinating in modern biology and the answers are likely to be basic to our understanding of the organisation of the chromosome and the control of gene function. These ideas are developed and discussed by Britten and Davidson (1969), Crick (1971), and Paul (1972).

2.6 Summary

The aim of this discussion has been to highlight some of those areas of research in which a beginning has been made in the task of relating DNA structure to function. Definitive analysis of DNA sequences involved in particular functions (e.g. RNA polymerase binding, modification and restriction, repressor binding) is probably possible with the techniques at hand. The isolation and sequence analysis of longer DNA segments from the small single-stranded DNA phages such as fd, fl and ϕX174 is under active study (Ziff, Sedat and Galibert 1973) as is the search for new enzymes and chemical techniques for the specific fragmentation of DNA chains. The outstanding problems which may yield to this approach are the nature of the recognition and control sequences involved in replication and transcription. The wealth of genetic data to hand on the structure and control processes in, for example, phage λ provide a more distant goal for interpretation in terms of molecular structure.

In considering eukaryote chromosome structure one of the most puzzling and significant facts is the general occurrence of highly repetitious DNA sequences which, in some cases, are estimated to account for as much as 80–90 per cent of the total DNA. The complexity of the repetitive sequences can cover a wide range varying from a repeat unit of $\sim 10^4$ base pairs in the ribosomal DNA cistrons to almost homopolymer-like simplicity. A 9 S DNA segment, coding for histone messenger in the sea urchin, is about 400-fold repetitive and the evidence implies that the repetitive sequences are closely clustered and therefore potentially separable from the bulk of nuclear DNA. This opens up the possibility of isolating such genes from mammalian cells. While at present the function of the non-translated repetitive DNAs is not known, any model relating chromosome structure and function must clearly take into account their existence and the remarkable evolutionary divergence found among the DNA satellites in closely related species.

3 Structure and function of phage RNA

3.1 Introduction

During the last few years our understanding of the mechanisms whereby the small RNA phages (R17, Qβ, MS2, f2) parasitise and replicate in the *E. coli* host has made great progress and it is now possible to describe, at least in general terms, the molecular basis for a number of these processes. These advances stemmed largely from the convergence of two separate approaches. In the first instance, the development of radiochemical methods for determining RNA sequences by Sanger *et al.* (1967) has permitted long nucleotide sequences, corresponding to particular regions of the phage genome, to be delineated. The second breakthrough was the discovery and development of purified *in vitro* systems (Billeter *et al.* 1969) which allowed the accurate synthesis of viral RNA. Moreover, the RNA from these phages have been shown to code for only three (or in the case of Qβ, four) recognisable proteins. Thus the ease with which the viral RNAs (which serve directly as the RNA messenger) can be purified and used to programme *in vitro* protein synthesis, affords a unique system in which their messenger function can be related to a knowledge of their primary nucleotide sequence.

3.2 Structure and classification of RNA phages

The RNA coliphages are regular polyhedrons about 200 Å in diameter with a mol. wt. between $3 \cdot 6 \times 10^6$ (f2 and R17) and $4 \cdot 2 \times 10^6$ (Qβ). They contain about 30 per cent by weight RNA, the remainder being protein.

The RNA, encapsuled in the phage, is a single strand 3300–4500 nucleotides long. It is clear, therefore, that it must exist as a compact folded structure in the phage; studies on the extent of internal sequence complementarity and hyperchromicity, indicate about 70 per cent secondary structure even for isolated RNA.

The intact phage particles have been classified serologically into three groups. Group I includes f2, R17, MS2, M12 and R23 and these phages have been used for most of the work on the translation of mRNA. Phage Qβ belongs to a different antigenic group (group III) and differs in several other respects from those of group I. Qβ has a slightly higher particle weight and has an RNA chain of approx. 4000 nucleotides; group I phage RNAs are about 3300 residues long. Hybridisation studies have shown that Qβ RNA has no long nucleotide sequences in common with group I phage RNAs; the phage coat proteins differ considerably and the MS2 and Qβ induced replicase enzymes are specific for their respective templates.

Group I phages consist of a single-stranded RNA, about 180 molecules of coat protein and one of A (or maturation) protein, a component necessary for the functional integrity of the virus particle. The major protein component of the phage is therefore the coat protein and this has a mol. wt. of about 14 000. The sequences of the coat proteins of f2, R17, MS2, and fr and Qβ have been determined. An important difference found in Qβ is that, in addition to the coat protein, two minor protein components (A_1, mol. wt. 38 000 and A_2, mol. wt. 44 000) are encountered in the intact particle. A_2 corresponds to the A (maturation) protein of R17.

3.3 Replication of phage RNA

The single-stranded RNA genome of the small RNA phages fulfils three functions: (1) on infection of the host cell it serves as a messenger coding for three (or four) phage proteins; (2) it is specifically recognised and replicated by the phage-induced RNA replicase; and (3) it forms an integral part of the mature phage particle.

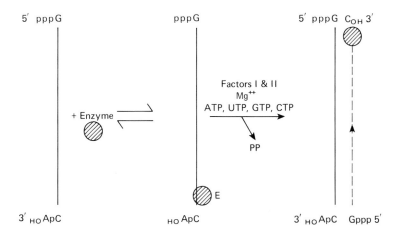

Fig. 7.3 Synthesis of Qβ minus strand on a Qβ template.

It is clear that for the phage RNA to be efficiently replicated in the infected cell some selective mechanism is required specifically to recognise and replicate the phage RNA and neglect the (very large) excess of possible host RNA templates. To do this, the viral RNA codes for the β-subunit of a new enzyme complex (the replicase or RNA-dependent RNA-polymerase), the three other polypeptides being coded for by host DNA. Thus the Qβ induced replicase is specific, and will transcribe only intact Qβ RNA (the 'plus' strand) and the strand complementary to Qβ RNA (the 'minus' strand). Figures 7.3, 7.4 illustrate the two step process by which the phage RNA is duplicated via the intermediate formation of the complementary 'minus' strand. Transcription of the minus strand will regenerate a sequence identical to the original plus strand. strand will regenerate a sequence identical to the original 'plus' strand.

One feature of both these steps is that the replicase begins synthesis at the penultimate C residue rather than at the 3'-terminal A (see fig 7.3), and directs the incorporation of a pppG residue as the first nucleotide as the 5'-end of both plus and minus strands. After completion of the strand, the enzyme adds a final A residue to the 3'-C, though this is not specified by the template, to reform the ubiquitous . . . $CCCA_{OH}$ terminal sequence.

Immediately after infection the phage RNA plus strand must serve as a messenger, for until the replicase cistron has been translated the phage RNA cannot be replicated. It is likely that

264

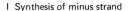

I Synthesis of minus strand

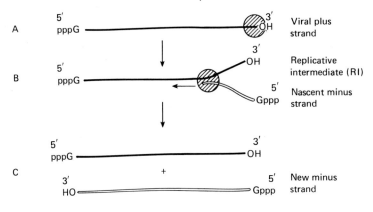

II Synthesis of plus strand

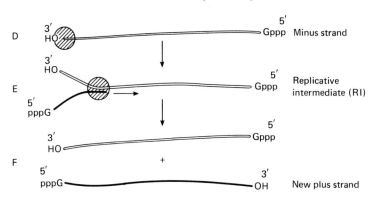

Fig. 7.4 Mechanism of synthesis of phage RNA.
A.   Replicase plus host factors represented by ⬤ interacts with its recognition site at the 3'-end of the viral RNA.
B.   The enzyme synthesises a complementary strand which grows in the 5'- to 3'-direction.
C.   Completion of synthesis of complementary minus strand.
D–E. Replicase, without host factors, interacts with Qβ minus strand and catalyses synthesis of a new plus strand which grows in 5'- to 3'-direciton.
F.   Completion of one round of synthesis generating a new plus strand.
(Taken from 'Progress in Biophysics and Molecular Biology' (1973). Vol. 26, chapter 6.)

partial translation of the coat protein cistron is required before initiation of translation of the replicase gene can take place and this is supported by the observation that *in vitro*, ribosomes bind predominantly to the start of the coat protein cistron (see section 3.8.2). One of the early events in infection is therefore the synthesis and accumulation of the viral coat protein and the replicase β-subunit. Shortly after the phage-specific replicase is assembled, the enzyme attaches to the 3'-terminus of the input RNA which it uses as a template for the synthesis of the complementary minus RNA strand. This association of the RNA with the replicase to form the initiating complex has been demonstrated both by zone centrifugation and retention of the complex by nitrocellulose filters. There seems to be only one molecule of enzyme bound per RNA.

In addition to enzyme, RNA, Mg^{2+} and the ribonucleoside triphosphates, two protein factors, termed Factors I and II, are required for the synthesis of the minus strand. Both these factors are present in extracts of uninfected as well as infected cells. These factor requirements, however, are mandatory only for the synthesis of minus strands on a plus template. The reverse process, i.e. the synthesis of plus strands on a minus template, has no requirement for either factor.

The scheme for synthesis of Qβ minus strands is shown in Fig. 7.3 and 7.4. After one round of synthesis the minus strand is released from the template. The same enzyme (with no host factors needed) can then attach to the 3'-end of the minus strand which is now used as a template for the synthesis of a new viral plus strand. Both strands are therefore synthesised in a $5' \rightarrow 3'$ direction (also the direction of RNA synthesis on a DNA template, catalysed by DNA-dependent RNA polymerase), Under optimal conditions the rate of elongation of both plus and minus strands is about 35 nucleotides per second at 37°C.

3.4 Specificity of phage-induced replicases

As described above, the phage-induced replicases exhibit a remarkable specificity for the replication of their homologous plus and minus templates. Since the basis of this specificity must presumably lie in the recognition by the enzyme of specific regions (primary sequences or secondary structures) in the RNA, important information relating to this problem might be expected to emerge from a knowledge of the primary structure of different phage RNAs. It would be a reasonable guess that the simplest way for the replicase to identify specifically the plus and minus strands would be by recognition of some feature which they have in common. Since synthesis of nascent strands begins at the 3'-end of both templates, the recognition site might be a sequence of nucleotides near the end of the template. As shown in Fig. 7.5 a

$$\text{3'-end of plus strand} \quad . \quad . \quad . \quad G\ U\ U\ A\ \boxed{C\ C\ A\ C\ C\ C}\ A_{OH}$$

$$\text{3'-end of minus strand} \quad . \quad . \quad . \quad G\ G\ U\ C\ \boxed{C\ C\ A\ C\ C\ C}_{OH}$$

Fig. 7.5 Nucleotide sequences at the 3'-end of the R17 plus strand and the 3'-end of the minus strand (this latter sequence was deduced from a knowledge of the sequence of the 5'-end of the complementary plus strand.

common sequence of six nucleotides —CCACCC— is present at the 3'-end of both plus and minus strands of phage R17 RNA. It can be argued that this is very unlikely to occur by chance and that these terminal sequences are part of the recognition site for the R17 RNA replicase. A comparison of the terminal regions of Qβ plus and minus strands shows that the correspondence between sequences extends to only four nucleotides ($-CCCA_{OH}$); also, the 3'-terminal A may be removed without affecting template activity. Since several other small viral RNAs, such as MS2, R17 and TMV RNA, which also terminate with $-CCCA_{OH}$ at the 3'-end, are inactive as templates, it can be concluded that the presence of a —CCC— sequence may be a necessary, but is certainly not a sufficient, condition for template activity.

Since it has been shown that 61 and at least 32 nucleotides, respectively, at the termini of Qβ RNA do not lie within cistrons, it is possible that these, and perhaps, internal sequences too, are involved in the recognition function, possibly by providing some characteristic secondary or tertiary structure (such as internal loops, Fig. 7.6). Thus plus and minus strands may possess similar secondary structural features despite their largely different nucleotide sequences. It may

266

Possible secondary structure of $Q\beta$ minus strand

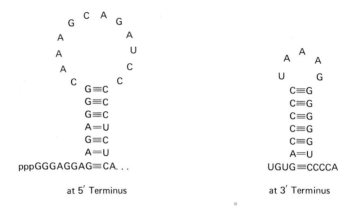

at 5′ Terminus   at 3′ Terminus

Possible secondary structure of $Q\beta$ plus strand

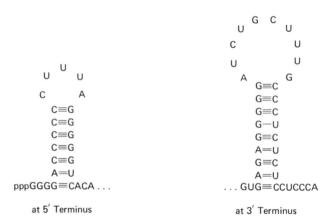

at 5′ Terminus   at 3′ Terminus

Fig. 7.6 Possible secondary structure at the 5′- and 3′-termini of $Q\beta$ plus and minus strands. (From Goodman, *Proc. Nat. Acad. Sci.* (1970), 67, 921.)

be, therefore, that $Q\beta$, replicase possesses two recognition sites, one of which is dependent on the presence of the host specified factor(s). Further evidence regarding the $Q\beta$ replicase recognition site could come from sequence analysis of the small $Q\beta$ 'variants' (Mills *et al.* 1967) which arise when $Q\beta$ RNA is transcribed by partially purified $Q\beta$ replicase under certain conditions. They are themselves replicated very rapidly by the enzyme and must therefore contain the replicase recognition site. This whole topic has been reviewed by Stavis and August (1970).

3.5 Translation of viral RNAs

During the past few years our understanding of the processes involved in the initiation of polypeptide synthesis in cell-free systems has increased enormously and the outlines of the overall process are fairly clear (see Chapter 1). First, a 30 S ribosomal subunit forms a complex with messenger RNA and formylmethionyl-tRNA ($tRNA_f^{Met}$) in the presence of GTP and initiation factors. Subsequently a 50 S ribosomal subunit is attached to the complex and a second tRNA, directed by the codon next to the initiator codon enters the 'A' site on the 70 S complex; the first peptide bond is then forged. In every case so far studied, the codons AUG and GUG specify the initiator tRNA (formylmethionyl-tRNA$_f$) resulting in the insertion of N-formyl methionine at the amino terminus of the nascent polypeptide. Viral messenger RNAs however contain many AUG and GUG triplets which are not involved in polypeptide chain initiation from which it is clear that the machinery for initiation can recognise only those triplets at the beginning of a cistron. It was originally thought that this distinction might be due to the proximity of the initiation triplet to the 5′-end of the mRNA but this idea now appears untenable. In the case of Qβ RNA, no AUG or GUG occurs within 61 residues of the 5′-end Billeter *et al.* (1969) and in R17 the first AUG triplet occurs 46 residues from the 5′-end. The novel finding that initiation of polypeptide synthesis can take place directly on circular phage fd DNA also supports the idea that initiation does not require a free 5′-end. Probably the most telling evidence against the idea of a single initiation site near the 5′-end is the observation that initiation of translation of the different cistrons can occur independently. Thus, for example, Skogerson *et al.* (1971) have shown that using a highly purified cell-free system from *E. coli*, in which translocation of the viral message is prevented, Qβ RNA directs the synthesis of two fMet-containing dipeptides. One of these, fMet-Ala corresponds to the N-terminal dipeptide of the coat protein and the other, fMet-Ser, to the N-terminal dipeptide of the replicase. Under these conditions, therefore, the translation machinery has independently selected two of the possible three initiation triplets. Using phage f2 RNA as messenger each of the three cistrons is capable of independent translation, the ratio of initiation of coat protein: polymerase: protein being 100 : 30 : 5·5. It is reasonable to assume, and look for, structural and/or sequence characteristics which could determine these differences.

In vivo studies of the translation of phage RNA are usually complicated by the large background of host cell protein synthesis, and most experiments have relied on the use of specific inhibitors of host messenger RNA synthesis (and therefore host protein synthesis) to isolate the effect of the phage messenger RNA. Actinomycin D and rifampicin are both potent inhibitors of host DNA-dependent RNA synthesis but have little or no effect *in vitro* on the activity of the phage specific (RNA-dependent) polymerase. *In vivo*, however, both phage yield and RNA synthesis are inhibited to varying degrees and the results from such experiments need to be interpreted with caution.

A very important reason for using phage RNAs as a model for studying the relationship between structure and function in a polycistronic messenger is the relative ease with which the proteins synthesised *in vitro* can be identified and shown to correspond to the natural proteins. When synthesised *in vitro*, each of the phage specific proteins contains N-formylmethionine as the N-terminal amino acid. Using phage f2 RNA as messenger, the amino terminal sequence of the coat protein is found to be N-formyl met-ala- - -, that of the maturation 'A' protein N-formyl-met-arg ... and that of the polymerase β-subunit N-formyl-met-ser ... Amber mutants of phages f2 and R17 have been isolated in which the glutamine codon CAG has been substituted by the amber codon UAG in a position corresponding to the sixth amino acid of the coat protein. When the RNA is used as messenger in the *in vitro* system the mutation is found to result in premature termination of the nascent coat protein yielding a peptide of sequence fMet-Ala-Ser-Asn-Phe-Thr. This agrees with the N termini of the complete sequences of the coat proteins of these two phages and inspires confidence that the correct processes of initiation, decoding and termination occur *in vitro*.

In this discussion only an outline has been presented of our present knowledge regarding the *in vivo* and *in vitro* translation of the phage RNAs and for further description the excellent review of Kozak and Nathans (1972) should be consulted. The important point is that translation of a phage polycistronic messenger *in vivo* is closely regulated and geared to the requirements for the synthesis and assembly of progeny phage, and that aspects of this control are manifested *in vitro*. Our understanding of the molecular basis of these controls and their relation to the structure of the phage genome will next be considered.

3.6 Organisation of the phage genome

Genetic analysis has revealed three complementation groups (genes) in both Group I and Group III phages. These correspond to the three phage-specified proteins, coat protein (mol. wt. 14 000), maturation (A or A_2) protein (mol. wt. 42 000–44 000) and replicase β-subunit (mol. wt. 63 000–67 000). Phage $Q\beta$, however, contains a fourth virus-specified protein, designated A_1 (mol. wt. 38 000) which occurs as a minor component of the intact phage particle. Amino-acid sequence analysis revealed that the N-terminal sequence of A_1 was identical to that of the coat protein and also that suppression (with a UGA suppressor) markedly increased the production of A_1. The interpretation of these findings is that the A_1 protein is initiated at the start of the coat cistron but that the message is read through the normal coat termination site (presumably a UGA codon) to a later termination signal. Thus termination at the first site yields coat protein, while suppression at the first and termination at the second site gives rise to A_1 protein.

Since recombination has not been found in the RNA phages, the problem of gene order had to be resolved by direct biochemical analysis. Three different approaches were used, but the principle behind each was characterisation of RNA sequences derived from particular regions of the genome in terms of their specific coding potential for the known phage proteins. Each method gave the gene order in all RNA phages examined as:

$$5'-A(A_2)\text{protein} - \text{Coat} - \text{Replicase } \beta\text{-subunit} - 3'$$

3.7 Ribosome binding sites of phage RNAs.

If we assume that some distinguishing feature of the mRNA, presumably in the vicinity of the true initiator codons must direct the ribosomes (or initiation factors) to specifically attach at these sites, nucleotide sequence analysis of these regions should provide definite information about the molecular basis of this recognition.

Under conditions in which the translocation of the RNA messenger was prevented, highly labelled phage RNA was attached to ribosomes and the 70 S initiation complex isolated by surcrose gradient centrifugation. Treatment of this complex with RNase degraded all the unprotected RNA sequences leaving intact those regions bound to the ribosome in the 70 S complex. These binding site segments were subsequently isolated and their sequences determined. As an example, Table 7.1 shows the sequence of 26 nucleotides deduced for the ribosome binding site of the coat protein cistron of phage $Q\beta$ RNA. If AUG specifies fMet-tRNA$_f$, a sequence of six amino acids can be proposed from the following nucleotide sequence and this corresponds precisely to the N-terminal amino acid sequence of the $Q\beta$ coat protein. Using similar methods the binding site sequences corresponding to all the R17 and $Q\beta$ cistrons have been determined (Table 7.1). Detailed discussion of these results is not possible here but it is clear that marked similarities in sequence characterise certain binding sequences.

If largely intact high-molecular-weight phage RNAs are used for these binding experiments, the ribosomes attach almost exclusively to the start of the coat protein cistron. On the other hand, fragmented or formaldehyde-denatured RNA, in which the secondary structure has been largely destroyed, allows ribosomes to attach equally well to all three cistrons. This implies that

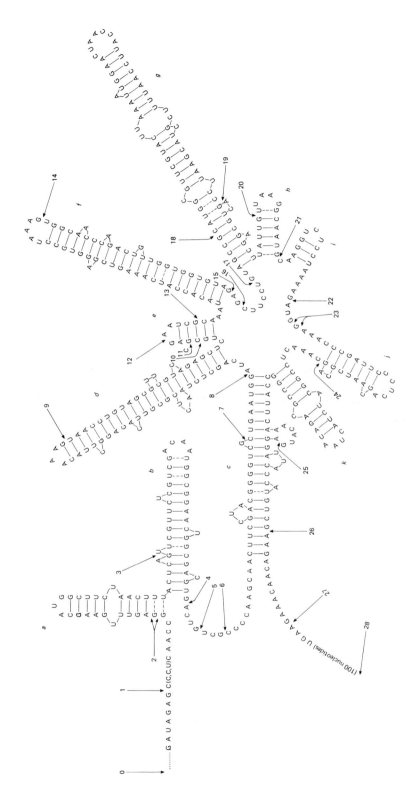

Fig. 7.7 A model for the secondary structure of phage MS-2 coat protein gene (the 'Flower' model). The arrows indicate the splitting points in partial T$_1$-ribonuclease digests and the base paired regions are listed a–k (Min-Jou *et al.* 1972).

Table 7.1   Ribosome binding sites of phage RNAs.   (Potential nonsense codons are underlined - - -)

Qβ coat protein
...pyAAUUUGAUCAUG . GCA . AAA . UUA . GAG . AC...
fmet . ala . lys . leu . glu . thr

R17 coat protein
...pyAGAGCCUAACCGGGGUUUGAAGCAUG . GCU . UCU . AAC . UUU...
fmet . ala . ser . asn . phe

f2 coat protein
...pyAAAAUAGAGCCUAACCG(G,A)GUUUGAAGCAUG . GCU . UCC . AAC . UUU . ACU . CAG
fmet . ala . ser . asn . phe

Qβ polymerase
...(AG)UAACUAAGGAUGAAAUGCAUG . UCU . AAG . ACA . GC...
fmet . ser . lys . thr . ala

R17 polymerase
...pyAAAACAUGAGGAUUACCCAUG . UCG . AAG . ACA . ACA . AAG...
fmet . ser . lys . thr . thr . lys

Qβ maturation protein (A$_2$)
...pyGAGUAUAAGAGGACAUAUG . CCU . AAA . UUA . CCG . CGU . G
fmet . pro . lys . leu . pro . arg

R17 maturation protein
...pyCCUAGGAGGUUUGACCUAUG . CGA . GCU . UUU . AGU . G
fmet . arg . ala . phe . ser

the secondary structure of the native RNA molecules strongly influences the sites at which protein synthesis can start and, as discussed in the next section, this provides the basis of an important translational control mechanism. The source of ribosomes is also important in determining which cistrons are translated. When ribosomes from *B. stearothermophilus* are used for the translation of f2 RNA, only the A protein is synthesised. Similarly, when ribosomes from this source are used to prepare the 70 S complex with R17 RNA, it is found that essentially only the initiation sequence of the A-protein cistron is protected. In other words, the ribosomes from *B. stearothermophilus* are able to distinguish between the different initiation signals in the RNA messenger.

3.8 Interactions controlling translation and transcription

Evidence from many different directions, using both *in vivo* and *in vitro* systems, has shown that transcription and translation of the phage messenger is closely regulated and geared to the requirements for the synthesis and assembly of progeny phage, and a molecular basis for at least three of these control elements can now be proposed.

3.8.1 Translational repression by phage coat protein

It has been shown that amber mutants in the coat protein cistron result in overproduction of replicase, and also that the phage coat protein may function as a repressor of translation of the replicase cistron. This idea has received strong support from *in vitro* experiments in which a complex of phage RNA with the homologous coat protein as messenger gave a markedly decreased translation of the replicase cistron.

Recent experiments have shown that, in the case of R17, coat protein binds specifically to an RNA segment which includes the start of the replicase cistron thereby effectively blocking all translation of this gene. This provides a clear example of translational repression and explains the observation that, in the late phase of infection, when the demand for coat protein is maximal, replicase synthesis is greatly reduced.

3.8.2 RNA secondary structure

As described previously, however, there is good evidence that, using native RNA as template, polypeptide synthesis cannot be initiated at the replicase cistron unless at least part of the coat cistron is first translated. To explain this it was suggested that an interaction between the replicase cistron initiation site and some region of the coat protein cistron may prevent ribosome attachment to the former. During translation of the coat cistron this interaction is reversed thus allowing ribosome attachment and translation of the replicase gene. Min-Jou *et al.* (1972) have pointed out that a complementary relationship exists between an RNA segment early in the coat protein cistron and the initiation site of the replicase cistron. If this complementarity results in hydrogen-bonding between these two regions, this provides a molecular basis for this observation and implies that sequences widely separated at the primary level may associate closely at the secondary level in the control of translation. The importance of secondary structure is also manifested in ribosome binding-site experiments. The A_2-protein cistron is available for ribosome binding only in short 5′-terminal RNA segments; in segments longer than about 1300 nucleotides the A_2-cistron binding site is masked and ribosome attachment occurs exclusively at the start of the coat protein cistron.

3.8.3 Qβ replicase as repressor

Qβ replicase binds strongly and specifically to Qβ RNA and this binding affords a marked degree of protection to a segment of the RNA against nuclease digestion (Weber *et al.* 1972). By analogy with the ribosome binding site experiments, it seemed likely that the replicase-protected sequence would constitute at least a part of the enzyme recognition sequence, presumably located at or near the 3′-end of the RNA strand. An unexpected finding was that the protected RNA sequence of 100 nucleotides included the AUG of the coat protein cistron. In other words, Qβ replicase binds specifically to a site immediately preceding the coat cistron and overlaps the ribosome binding site of that cistron.

In the phage-infected cell, the phage plus-strand can function either as the messenger for phage protein synthesis or as template for the synthesis of complementary minus strands. In the first instance, the phage RNA exists in a polysome with ribosomes travelling in the 5′- to 3′-direction. During replication, the replicase attaches to the the 3′-terminus and advances towards the 5′-end, i.e. on a collision course with the translating ribosomes. With an eye to resolving this dilemma, Kolakofsky and Weissmann (1971) showed (i) that replicase cannot dislodge ribosomes bound to Qβ RNA and (ii) that Qβ replicase strongly inhibits the binding of ribosomes to Qβ while at the same time allowing translating ribosomes to complete protein synthesis and become detached. Since initiation at the replicase cistron is dependent on some prior translation of the coat cistron, and since initiation of A_2-synthesis does not normally occur in the mature Qβ strand, it is evident that replicase attachment will ultimately lead to a total stripping of ribosomes from the RNA. The RNA is then free to function as a template and the attached replicase can initiate synthesis of minus strands.

These findings provide a molecular explanation for the translational repressor activity of the Qβ replicase. The replicase has evolved to have a remarkable dual function: as well as catalysing the transcription of plus and minus strands, it recognises specifically a different region of the phage genome. In the infected cell this provides a mechanism for making available RNA for replication in face of competition by the ribosomes which utilise the RNA as a messenger. At present it is not clear whether the binding of replicase to the initiation region of the coat cistron is an essential part of the template recognition process, or whether this interaction has evolved solely for conversion of Qβ RNA from a message in a polysome to a template in a replicating complex.

3.9 Structure of phage RNA

From the preceding discussion it is evident that much of our understanding of the properties and functions of the phage RNA rests heavily on our knowledge of RNA sequences. Moreover, the deeper we dig in attempting to explain this plethora of functions at the molecular level, the more urgent becomes our need for sequence analysis of large areas of the molecule. Equally important, however, may be the role played by secondary structure in determining the

```
    82Trp   Arg    Ser    Tyr    Leu    Asn    Met  Glu    Leu
5′...U G G.C G U.U C G.U A C.U U A.A A U.A U C.G A A.U U A.
       ||| |||   ||| || || ||| ||| || ||       || || ||     || ||| ||| || || ||   A
3′...G.C C U.C A A.G C A.U C G.C U U.U U A.A C C.U U A.U C Thr
      99Ser   Asn    Thr    Ala    Phe    Ile    Pro    Ile
```

Fig. 7.8 Secondary structure of a fragment from the coat protein cistron of R17 RNA (Jeppesen *et al.* 1970). (The phasing of codons is shown by dots and the corresponding amino-acid sequence is written alongside. Vertical lines indicate hydrogen bonds between base pairs.)

specificity of the varied interactions controlling the processes of translation and replication. Finally it should be recognised that the conformation of an RNA molecule may be different in different functional states, and we need to envisage a dynamic variable structure rather than a static minimum free energy structure.

Hydrodynamic studies on phage RNA indicate a compact structure and a sedimentation coefficient of about 27 S. Extensive hydrogen bonding is also implicated by thermal denaturation studies; and from the hyperchromicity on treatment with formaldehyde a helical (double-stranded) content of about 70 per cent was deduced. A specific secondary structure for phage RNA (at neutral pH and moderate ionic strength) is also suggested by the formation of specific cleavage products on digestion with ribonucleases at low temperature. Further, extensive stretches of internal complementarity in nucleotide sequences, giving rise to hairpin-type structures, have been proposed for large regions of the coat protein cistron in phage R17 and MS2 RNA (cf. Fig. 7.7). There seems little doubt, therefore, that these phage RNAs are characterised by a large degree of helical self-complementary internal structure and it is likely that this configuration has evolved, at least in part, as a device to simplify the compact packaging of the RNA in the phage particle. In addition to this constraint, the RNA must have evolved to specify the different biologically active phage proteins, and it is clear that this need for a highly self-complementary structure can limit the codon sequences, and therefore the amino acids, which can be used to specify the different proteins.

An example of a self-complementary sequence, present in the coat cistron of R17, is shown in Fig. 7.8. From the phasing of the codons it is seen that the third positions do not lie opposite one another. (A feature of the genetic code is that codons which differ only in the third position often code for the same amino acid and this allows for considerable variation in the nucleotide sequence and secondary structure without altering the amino acid sequence.) An absolute requirement for a localised area of secondary structure imposes severe restrictions on the amino acid sequence. To take a simple example: assuming that the dipeptide sequence Met-Glu, occupying positions 88–89 in the R17 coat protein is essential for its biological function and must be conserved, the corresponding nucleotide sequence (1) and the possible complementary sequences (2) with the phasing indicated in Fig. 7.8 are as shown:

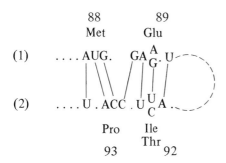

Thus the sequence at position 88–89, given a fully base-paired structure, will limit the dipeptide sequence at 92–93 to only two possibilities out of the 400 possible dipeptides that could be specified in a completely unpaired structure. It seems highly unlikely that selection pressures which generate self-complementary RNA sequences will inherently tend to produce biologically active amino-acid sequences, and we have to assume that the relation between secondary structure and amino-acid sequence is the best compromise achieved to date in the evolutionary history of the virus.

For R17 and MS2, where extended sequences of the coat cistron and also the amino-acid sequence of the coat are known, it is possible to check the codon assignments proposed by Nirenberg, Khorana and Ochoa, which were based mainly on synthetic messengers (see Chapter 1). As expected, the codon assignments have been completely vindicated, but the available evidence suggests that not all possible codons are used. For example, the codons AUU and AUC for isoleucine are each used five times, but AUA not at all. Tyrosine is coded for four times

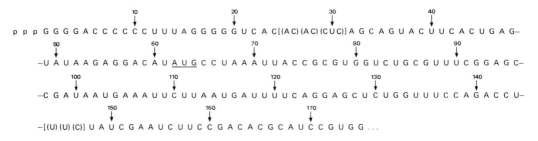

p p p G G G G A C C C C C C U U U A G G G G G U C A C[(AC)(AC)(CUC)] A G C A G U A C U U C A C U G A G–

–U A U A A G A G G A C A U A$\underline{UG}$C C U A A A U U A C C G C G U G G U C U G C G U U U C G G A G C–

–C G A U A A U G A A A U U C U U A A U G A U U U U C A G G A G C U C U G G U U U C C A G A C C U–

–[(U)(U)(C)] U A U C G A A U C U U C C G A C A C G C A U C C G U G G . . .

Fig. 7.9 Nucleotide sequence at the 5′-terminus of Qβ RNA synthesised *in vitro* (from Billeter *et al.* 1969). The AUG sequence underlined (62–64) corresponds to the start of the A$_2$-cistron. (The secondary structure at the 5′-end is shown in Fig. 7.6.)

by UAC but never by UAU, and glutamine five times by CAG and never by CAA. On the basis of these findings a modulation-type control of translation of the coat cistron involving specific codons cannot be excluded. For example, ACA for threonine, AGU for serine, CAA for glutamine are possible rate-limiting markers.

Sequence analysis has also demonstrated the presence of extended non-translated nucleotide sequences at the 5′-terminus (62 nucleotides in Qβ (Fig. 7.9), 129 in the MS2-R17 group) and at the 3′ end (at least 32 in Qβ, and at least 10 in the MS2-R17 group). In addition, non-coding sequences are also found between the different cistrons; for example, there are 36 nucleotides between the coat and replicase cistrons of R17. Regions of the intercistronic non-coding regions are known to be included in the ribosome binding sites and, in one case (Qβ), in the replicase binding site. It is reasonable to suppose that these 'spacers' are involved in control mechanisms regulating the independent translation of the different cistrons and the balance between the messenger and template functions of the genome. The non-coding sequences at both ends of the genome may well include specific features signalling replicase recognition, and may also be capable of hydrogen-bonding to other regions of the molecule to provide a particular secondary structure. A comparison of the sequences found in the closely related R17 and MS2 RNAs has shown that, although the coat protein sequences differ in only one amino acid, there are at least nine nucleotide changes distinguishing the coat cistrons. By contrast, the non-coding regions show no differences in their nucleotide sequence. The translatable regions thus show genetic variation, within the framework of a virtually unaltered secondary structure, whereas the extracistronic regions appear inviolate.

3.10 Concluding remarks

Over the past five years a number of remarkably sophisticated control mechanisms governing both translation and replication of the phage RNA have been uncovered. Vital to our understanding of the molecular basis of these controls have been nucleotide sequence analysis of several areas of the phage genome and, in the case of Qβ, the development of methods for isolating the replicase in pure form thus allowing detailed *in vitro* studies of the mechanism of replication and the factors controlling it. The small RNA phages, with sufficient genetic information for only 3-4 cistrons and relatively short extracistronic regions, have evolved a dual role for several of the phage products. For instance, the phage RNA needs to be efficiently and specifically replicated in an environment containing a large excess of other, potentially replicable, RNA species. This demand is met by the specification of a new protein, the β-subunit of the replicase which, in combination with host-specified components, produces a new enzyme with a remarkable template specificity. In addition, the replicase has evolved the property of interacting specifically with a further area of the genome where it appears to function as a translational repressor thus permitting, during the early stages of infection, the essential built up of an adequate supply of templates for subsequent translation. Evolutionary pressures have also developed a dual role for the virus coat protein. As well as fulfilling a vital role as the main structural component of the viral envelope, the coat protein interacts with the genome messenger to repress specifically translation of the replicase cistron and thus promote the almost exclusive synthesis of coat protein in the later stages of infection when the viral progeny are being assembled.

Within the framework of these adaptations, the genome has also evolved a structure capable of being compactly folded and encapsulated in a symmetrical particle. This has produced an RNA molecule containing up to 70 per cent of self-complementary base-paired sequences, with consequent severe limitation on the variety of amino-acid sequences which can be utilised.

It is the unravelling of this array of molecular processes which provides the challenge for those engaged in this work.

Bibliography

ADLER, K., BAYREUTHER, K., FARNING, E., GEISLER, N., GRONENBORN, B., KLEMM, A., MÜLLER-HILL, B., PFAHL, M. and SCHMITZ, A. (1972). 'How *lac* repressor binds to DNA', *Nature, Lond., 237*, 322-27.

BILLETER, M. A., DAHLBERG, J. E., GOODMAN, H. M., HINDLEY, J. and WEISSMANN, C. (1969). 'Sequence of the first 175 nucleotides from the 5'-terminus of Qβ RNA synthesized *in vitro', Nature, Lond., 224*, 1083-86.

BIRNSTIEL, M. L., CHIPCHASE, M. and SPIERS, J. (1970). 'The ribosomal RNA cistrons'. In: *Prog. Nucleic Acid Res. and Molec. Biol.*, Vol. II, pp. 351-89. (eds. J. N. Davidson and W. E. Cohn) New York, Academic Press.

BRITTEN, R. J. and DAVIDSON, E. H. (1969). 'Gene regulation for higher cells: A theory', *Science, 165*, 349-57.

CRICK, F. H. C. (1971). 'General model for the chromosomes of higher organisms', *Nature, Lond., 234*, 25-7.

JACOB, F. and MONOD, J. (1961). 'Genetic regulatory mechanisms in the synthesis of protein', *J. Molec. Biol., 3*, 318-56.

JEPPESEN, P. G. N., STEITZ, J. A., GESTELAND, R. F. and SPAHR, P. F. (1970). 'Gene order in the bacteriophage R17 RNA: 5'-A protein-coat protein-synthetase-3'', *Nature, Lond., 226*, 230-37.

KELLY, T. J. Jr. and SMITH, H. O. (1970). 'A restriction enzyme from *H. influenzae*. II. Base sequence of the recognition site', *J. Molec. Biol., 51*, 393-409.

KOLAKOFSKY, D. and WEISSMANN, C. (1971). 'Possible mechanism for transition of viral RNA from polysome to replication complex', *Nature New Biol., 23*, 42-6.

KOZAK, M. and NATHANS, D. (1972) 'Translation of the genome of a ribonucleic acid bacteriophage', *Bact. Rev., 36*, 109-34.

MILLER, O. L. Jr. and BEATTY, B. R. (1969). 'Extrachromosomal nucleolar genes in amphibian oocytes', *Genetics,* **61**, Suppl. 1, 133–43.

MILLS, D., PETERSON, R. C. and SPIEGELMAN, S. (1967). 'An extracellular Darwinian experiment with a self-duplicating nucleic acid molecule', *Proc. Natn. Acad. Sci. U.S.A.,* **58**, 217–24.

MIN-JOU, W., HAEGEMAN, G., YSEBAERT, M. and FIERS, W. (1972). 'Nucleotide sequence of the gene coding for the bacteriophage MS2 coat protein', *Nature, Lond.,* **237**, 82–8.

PAUL, J. (1972). 'General theory of chromosome structure and gene activation in eukaryotes', *Nature, Lond.,* **238**, 444–48.

ROBERTSON, H. D., BARRELL, B. G., WEITH, H. L. and DONELSON, J. E. (1973). 'Isolation and sequence analysis of a ribosome-protected fragment from bacteriophage ϕX174 DNA',*Nature New Biol.,* **241** 38–40.

SANGER, F. and BROWNLEE, G. G. (1967). 'A two-dimensional fractionation method for radioactive nucleotides', *Meth. Enzym.,* **12**, Part A, 361–81.

SKOGERSON, L., ROUFA, D. and LEDER, P. (1971). 'Characterization of the initial peptide of Qβ RNA polymerase and control of its synthesis', *Proc. Natn. Acad. Sci., U.S.A.,* **68**, 276–79.

SOUTHERN, E. M. (1970). 'Base sequence and evolution of guinea pig α-satellite DNA', *Nature, Lond.,* **227**, 794–98.

STAVIS, R. L. and AUGUST, J. T. (1970). 'The biochemistry of RNA bacteriophage and replication', *A. Rev. Biochem.,* **39**, 527–60.

WALKER, P. M. B. (1971). 'Repetitive DNA in higher organisms'. In: *Prog. Biophys. and Molec. Biol.,* **23**, 145–90. (eds. J. A. V. Butler and D. Noble).

WATSON, J. D. (1970). *Molecular Biology of the Gene* (2nd edn). New York, Benjamin.

WEBER, H., BILLETER, M. A., KAHANE, S., WEISSMANN, C., HINDLEY, J. and PORTER, A. G. (1972). 'Molecular basis for repressor activity of Qβ replicase', *Nature New Biol.,* **237**, 166–70.

ZIFF, E. B., SEDAT, J. W. and GALIBERT, F. (1973). 'Determination of the nucleotide sequence of a fragment of bacteriophage ϕX174 DNA', *Nature New Biol.,* **241**, 34–7.

8
Viruses

A. A. Newton
Department of Biochemistry, University of Cambridge

1 Introduction

Viruses have been of interest to man for thousands of years; this is due to the dramatic and frequently fatal nature of viral diseases in man himself, as well as in his domestic animals and his crop plants. The search for a means of curing these diseases has continued from the time of the ancient Chinese (1000 B.C.) to the present day. So far this search has been largely unsuccessful and its continuation is one of the reasons why so much current research is directed towards an elucidation of the fundamental aspects of virus growth. The hope is that a complete understanding of this at a biochemical level will reveal processes unique to the virus, which may therefore be susceptible to chemotherapeutic agents.

Viruses however also provide a tool that the biochemist can use to investigate problems of cellular and molecular biology. It is probably true to say that the science of Molecular Biology owes more to the study of viruses than to that of any other organism, even *Escherichia coli*. In the cell-free, infectious state the virus is a metabolically inert 'super molecule'; in this form it provides a comparatively simple system for a study of assembly and interactions of macromolecules. The intracellular phase of virus growth, however, involves subversion and modification of cellular machinery in order to replicate the comparatively simple molecules of which viruses are composed. The way in which this is achieved in various cells can tell one much about the functions of a normal cell.

Viruses have now been discovered in cells of most of the main groups of organisms. Over the last 20 years it has become evident that the more types of cells that are examined, and the greater the detail in which they are examined, the more viruses will be recognised. Thus in recent years viruses have been discovered in blue-green algae, in fungi, in fish and amphibians (see Table 8.1). Viruses of protozoa and of yeasts have not, however, been demonstrated conclusively.

In order to be sure that a newly discovered entity is indeed a virus it is necessary to have a clear definition of a virus. This in itself is not easy. Ideas about the nature of a virus have fluctuated between the extremes of a crystallisable, self-replicating macromolecule to that of a slightly degenerate, small bacterium. The most satisfactory definition should indicate the special nature of the virus as: 'A virus is a virus is a virus' (Lwoff), but should also serve to distinguish between viruses and all other organisms and subcellular organelles. Examination of the following definition, also by Lwoff, reveals something of the nature of viruses.

'Viruses are submicroscopic entities capable of being introduced into specific living cells and of reproducing inside such cells only. They possess only one type of nucleic acid, multiply in the form of their genetic material, are unable to grow by binary fission and possess no system of enzymes for energy production.'

The following criteria must therefore be met, to show that an entity is a virus.

a   Viruses are submicroscopic entities. Any definition of an organism by size is necessarily an arbitrary one and indeed viruses exhibit a great range in size; the smallest have a diameter of about 10 nm. However, since the largest viruses, such as pox viruses, are visible in the light microscope (resolving power 250 nm) only when devices such as ultra-violet illumination are used to increase the resolving power, this arbitrary criterion is also a practical one.

b   Viruses are capable of being introduced into specific living cells. This part of the definition implies that a virus is capable of infecting some cell. Since this is the only part of the definition serving to distinguish viruses from some intracellular particles it is essential that this criterion is met in order to characterise a newly discovered particle as a virus. This is frequently difficult to achieve, since it entails having a susceptible cell and also being capable of recognising that the cell has indeed been infected. Where a virus causes overt cell damage and death, identification of infection may not present too great a problem. Many viruses, however, will grow only in whole organisms and then cause only subtle changes in a few cells of that organism such as the induction of malignancy after very long latent periods. Moreover, ethical considerations may restrict the performance of experiments necessary to prove that a particle found in human tissues is a virus capable of infecting another human.

c   Viruses are capable of reproducing inside living cells only. This emphasises that viruses are obligate intracellular parasites, which distinguishes them from most other organisms—except for a few types that lie on the borderline between bacteria and viruses such as rickettsiae and psittacosis-like organisms. Unlike true viruses, however, these organisms multiply by binary fission.

d   Viruses contain one type of nucleic acid only. This serves to distinguish viruses from all other types of organism but not from some subcellular particles, e.g. ribosomes. (In fact, one group of RNA-containing viruses—the oncorna viruses—contain also small quantities of DNA, but this is thought to be a host component and is not essential for infectivity.)

e   Viruses multiply in the form of their genetic material. A unique property of viruses that is essential to an understanding of their special position amongst living organisms is the fact that they multiply as the nucleic acid only. That they represent the essential replication mechanisms of more complex organisms reduced to a minimum accounts for the fascination that viral growth processes hold for molecular biologists and biochemists.

That they can replicate as the nucleic acid means that the nucleic acid alone may be sufficient to initiate infection. In many cases infectious nucleic acid can be prepared and is capable of producing all the features of a normal growth cycle. Recent studies have shown that the infectious entity of potato spindle tuber disease is naked nucleic acid. Such a particle, representing the very minimum requirement for a virus, has been named a viroid. It seems possible that the causative agents of such diseases of the nervous system as scrapie in sheep, and kuru, a disease restricted to certain tribes in New Guinea (whose ritual involves cannibalistic devouring of dead relatives) are similar pieces of infectious nucleic acid: it is certain that the agents of these diseases are very small and have proved extremely elusive.

The life cycle of viruses falls into two distinct phases, the extracellular phase when a virus is infectious but metabolically inert, and an intracellular phase when viral components are replicated but where a virus may not be recognisable as an entity.

2 The structure of extracellular virus

In the extracellular form viruses are packages of genetic information that must be delivered safely. A successful virus particle, therefore, must fulfil certain functions. The nucleic acid component must contain sufficient information to specify the production of new complete virus. This nucleic acid must be protected from harmful environmental conditions, such as nuclease attack, by the protein material enclosing it. The external layers of the virus should also provide some mechanism for recognition of and attachment to susceptible cells. Since viruses are usually comparatively simple in structure they provide an unique opportunity for studying the relationship between structure and function and the chemical and physical properties of the components necessary for assembly of the structures. Moreover, since viruses may be recognisable as entities only in the extracellular phase, the shape and composition of the particle is used as a basis for classification of viruses (see Table 8.1).

According to present nomenclature the complete virus particle is known as a virion. The symmetrical assembly of nucleic acid and protein is known as the nucleocapsid; the nucleocapsid may or may not be surrounded by a membranous envelope.

2.1 Viral nucleic acid

Viruses contain either RNA or DNA, and the nucleic acid may be either single-stranded or double-stranded. In one case, the minute virus of mice, particles contain single-stranded DNA, but the material when isolated is double stranded; it has been suggested therefore that some particles contain + strands and others the complementary − strand. In some viruses the DNA component is present as a covalently closed circle which may show supercoiling. In a few bacteriophages single-stranded self-complementary ends are present in the DNA so that it readily assumes a circular form; it is not known whether this is the form present in the particle. No circular RNA molecules have been described. Very little is known of the way in which nucleic acid is packaged inside the spherical or complex particles. It seems fairly certain that in most isometric viruses the nucleic acid is highly condensed, perhaps in association with internal proteins or polyamines, and is surrounded by the coat-protein. Little is known also about the forces between nucleic acid and protein but nucleic acid does not seem to be covalently bound to protein in any particle.

The amount of information in viral nucleic acid varies considerably. The smallest independent viruses, such as MS2, have RNA coding for three proteins only. These proteins are the viral polymerase, the coat-protein and the A protein which is a minor component of the virion. A few viruses have less nucleic acid than this, but this is insufficient information to allow for replication of viral nucleic acid (e.g. tobacco necrosis satellite virus has only 1200 nucleotides, which is just sufficient to code for its coat-protein of about 400 amino acids); such viruses can multiply only in the presence of another virus that supplies the necessary enzymes. At the other extreme, T-even phages and pox viruses have nucleic acid of molecular weight 120×10^6 and 160×10^6 daltons respectively. These contain sufficient information to code for 200–300 average size proteins.

Although most viruses possess only one molecule of nucleic acid, the genome of a few RNA viruses has been shown to consist of separable and distinct molecules. Influenza virus genome is present in 5–7 pieces, and the double-stranded RNA of Reovirus is present in ten pieces each of which is distinct from the others. Some plant viruses, e.g. tobacco rattle virus, are multi-component, and their full complement of genetic information is distributed into two or more distinct virions, all of which must infect a cell if new infectious virus is to be produced.

Table 8.1(A) Examples of RNA-containing viruses

Type of nucleic acid (double stranded = ds) (single stranded = ss)	Symmetry helical (H) or isometric (I)	Presence of membranes	Group name	Virus name	Host	Size of nucleic acid (daltons × 10^6)	Virion enzymes	Comments
ssRNA	H	–	Protovirus	Tobacco mosaic	Plants	2	–	
	H	–	Pachyvirus	Tobacco rattle	Plants	2 and 0·7	–	Genome divided between two particles
	I	–	Androphage	Qβ.MS2.R17	E. coli	1·2	–	Only three cistrons
	I	–	Picornavirus	Polio	Primates	2·4	–	
			"	Foot and mouth disease	Cattle			
			"	Bee paralysis	Insects		–	
			"	Turnip yellow mosaic	Plants	2	–	
			"	Tomato bushy stunt	Plants	2	–	
			"	Tobacco necrosis satellite	Plants	0·36	–	Dependent on another virus
	H	+	Myxovirus	Influenza	Mammals	4–5	+	Genome in seven pieces
			"	Fowl plague	Poultry		+	" "
	H	+	Paramyxovirus	Sendai	Mammals	7	+	Viral nucleic acid not messenger
			"	Mumps	Mammals	7	+	
			"	Newcastle disease	Poultry	7	+	
	H	+	Oncornavirus	Rous sarcoma	Chickens	10	+	Multiplies as DNA provirus
	H?	+	Arbovirus	Yellow fever	Mammals + insects	3	–	Two host systems
	H	+	Rhabdovirus	Vesicular stomatitis	Cattle	4	+	Viral nucleic acid not messenger
			"	Lettuce necrotic yellows	Plants + insects		+	Two host systems
dsRNA	I	–	Reovirus	Reo	Animals + insects	10	+	Genome in ten pieces

Table 8.1(B) Examples of DNA-containing viruses

Type of nucleic acid (double stranded = ds) (single stranded = ss)	Symmetry helical (H) or isometric (I)	Presence of membranes	Group name	Virus name	Host	Size of nucleic acid (daltons x 10^6)	Virion enzymes	Comments
ssDNA	H	−	Ionophage	fd.	E. coli	2	−	Circular DNA
	I	−	Picodnavirus	φX174	E. coli	1·7	−	,, ,,
	I	−	Parvovirus	Minute mouse virus	Mice		−	Complementary strands in separate particles
		?	,,	Adeno associated	Mammals	3	?	Dependent on another virus
dsDNA	H?	−	Polyhedrosis	Polyhedrosis	Insects		−	
	Complex	−	Bacteriophage	T phages	E. coli	120	−	
	,,	−	Cyanophage	LPP1	Blue-green algae		−	
	?	+	—	PM2	E. coli		?	Only enveloped phage known
	I	−	Papovavirus	Polyoma	Mammals	3	−	
	I	−	Adenovirus	Adeno	Mammals	23	−	
	I	+	Iridovirus	Tipula	Insects	160		
		+	,,	Lymphocystosis	Fish	130		
		+	,,	Cauliflower mosaic	Plants	4		
	I	+	Herpes virus	Herpes simplex	Man	100	−	
		−	,,	Marek's disease	Poultry		−	
		−	,,	Lucke carcinoma	Frogs		−	
	? Brick shaped	+	Poxvirus	Smallpox	Man	160	+	
	shaped	+	Vagoiavirus	Spindle disease	Insects		+	

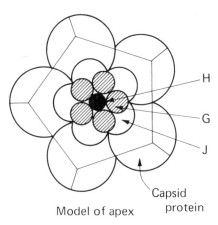

Model of apex

Capsid protein

Fig. 8.1 Properties of ϕX174 virus.
Isometric symmetry, envelope absent. Diameter 25 nm. Characteristics of nucleic acid: single-stranded circular DNA, molecular weight 1.7×10^6 daltons. Number of proteins = 5–6.

1. Major capsid protein. Molecular weight 49 000. Sixty capsid subunits present as trimers giving 20 faces.
2. Minor capsid protein. Molecular weight 23 000. Only 1–2 copies present/particle. Possibly acts as maturation factor.
3. Spike proteins present at 12 apices, can be subdivided into
   3a. H protein – actual spike, involved in adsorption. One copy/apex. Molecular weight 37 000.
   3b. G protein. Five copies/apex. Attached to H. Molecular weight 19 000.
   3c. J protein. Five copies/apex. Underlies G protein, attached to capsid protein. Molecular weight $5-9 \times 10^3$.

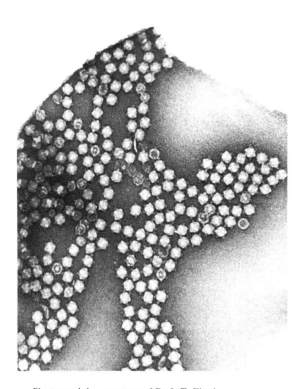

Photograph by courtesy of Dr J. T. Finch. Diagram from Burgess (1969). *Proc. Natn. Acad. Sci. U.S.A.*, **64**, 613.

Infectious nucleic acid is readily prepared from the simple viruses but it is clear that this is an extremely difficult task for viruses containing multi-component genomes. Furthermore certain viruses require the presence of enzymes present in the virus particle (see section 3.4) in order to initiate infection. It is not surprising that it is impossible to prepare infectious nucleic acid from these viruses.

Since viruses replicate in the form of their nucleic acid, the nucleic acid component of the virus provides a fundamental basis for classification. Viruses of similar nucleic acid content but differing form show similar patterns of replication, even in widely differing host cell systems.

2.2 Viral proteins

The form of virus seen in the electron microscope is characteristic and therefore an easy criterion for identification. The appearance of the nucleocapsid depends upon the way in which the protein coat of the virus is arranged. In all simple viruses the protein shell surrounding the nucleic acid is composed of relatively few types of protein molecule, and these molecules are

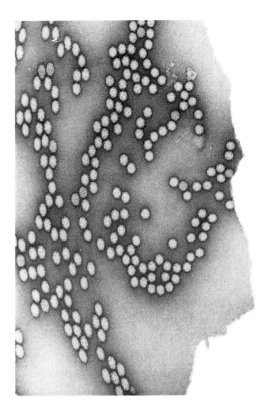

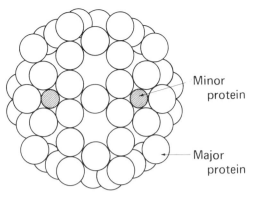

Minor
protein

Major
protein

Fig. 8.2 Properties of tomato bushy stunt virus. Isometric symmetry, no envelope. Diameter = 30 nm. Particle weight 9×10^6 daltons. Characteristics of nucleic acid: single-stranded RNA, molecular weight 2×10^6. Number of proteins = 2.

1. Major protein. Molecular weight $38-42 \times 10^3$ daltons. 180 structure units clustered into 90 dimers that represent structure units.
2. Minor component. Molecular weight 28×10^3 daltons. Twelve subunits located at apices.

Photograph by courtesy of Dr J. T. Finch. Diagram adapted from Finch, Flug and Leberman (1970). *J. Mol. Biol.*, **50**, 215.

arranged to give either a rod shape or a quasi spherical conformation. Caspar and Klug proposed that the many copies of a protein molecule present should all occupy equivalent or quasi equivalent places in the structure. The simplest form generated in this way is a helix, and this type of arrangement of identical protein molecules is found in tobacco mosaic virus (TMV). Packing identical asymmetric subunits to form a nearly spherical shell can only be achieved in a limited number of ways; in fact all 'spherical' viruses are icosahedral in shape. An icosahedron is a regular figure with 20 faces and 12 vertices, showing two-, three- and five-fold symmetry. The morphological subunits (capsomeres) seen in the electron microscope may be formed from several polypeptide chains, which may or may not be identical; thus the small bacteriophage ϕX174 has 12 apical capsomeres each containing at least 11 polypeptides (see Fig. 8.1). The morphological subunits may show a clustering which gives the characteristic surface features to the virus particle seen by negative staining and may themselves not be identical (see Fig. 8.2).

Thus in the assembly of a large icosahedron such as adenovirus, the capsomeres are arranged so that they are surrounded by either five or six nearest neighbours; the 240 hexons, or capsomeres forming the centre of six-membered rings, form the 20 equilateral triangles that are the faces of the icosahedron, while the pentons found at the vertices of this structure, have only five neighbours (see Fig. 8.3).

The protein shell of the virus serves to protect the nucleic acid, and is frequently found to be resistant to proteolytic digestion and to be stable over a considerable range of pH values and ionic strengths. However, the nucleic acid must be released from the particle with relative ease when the particle reaches a susceptible cell. In some cases the nucleic acid of one virus may be coated with the protein of another virus (phenotypic mixing). This has been achieved *in vitro* with plant viruses; for example tobacco mosaic virus RNA has been enclosed in the shell of

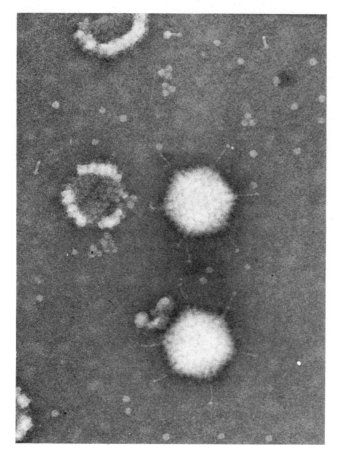

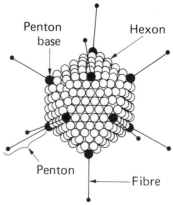

Fig. 8.3 Properties of adenovirus.
Isometric symmetry, envelope absent. Diameter 80 nm. Particle weight 175×10^6. Characteristics of nucleic acid: double-stranded DNA, molecular weight 23×10^6 daltons. Number of proteins in particle, at least five. Number of capsomeres 252.
Characteristics of proteins determined so far:

1. Hexon in 240 capsomeres.
2. Penton in 12 capsomeres situated at apices, can be subdivided into
   2a. Penton base and
   2b. Fibre.
3. Two-core proteins, arginine rich.

Photograph from the collection of the late Dr R. Valentine, kindly supplied by Dr W. Russell. Diagram from Russell (1971). Wolstenholme *et al.* (eds.), 'Strategy of the viral genome', *CIBA Foundation Symp.*

cowpea chlorotic mottle virus, a spherical virus. Other viruses, such as Rous sarcoma virus, do not normally produce a coat-protein in the infected cell, and rely on another virus to code for coat-protein material.

The protein portion of the virus is involved in attachment to the cell surface (see later) and therefore the coat-protein may contain sequences involved in combination with specific regions on the cell surface. The 'spike' proteins present in the capsid of ϕX174 are thought to be involved in attachment to the cell surface, as is the 'A' protein present as a single molecule in the coat of MS2 phage.

The haemagglutinin of myxoviruses is another well characterised protein involved in cell attachment, but this is present in the outer membrane and not the nucleocapsid (see Fig. 8.4).

Bacteriophages may possess specialised tails that are involved in attachment to the cell surface, and in injection of the nucleic acid into the host cell. These may be very complicated and contain many different proteins. Phage T4 contains at least 40 different proteins and its structure and symmetry are very complex.

Many viruses also contain internal proteins. Some have been shown to be basic arginine-rich proteins whose function is to combine with and neutralise the charge on the nucleic acid. Viruses of eukaryotic cells also contain enzymes which are internal proteins, frequently closely

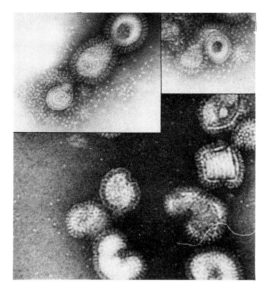

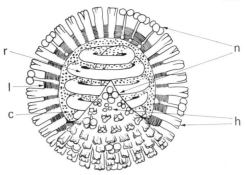

Fig. 8.4 Properties of influenza virus.
Helical symmetry, envelope present. Diameter 80–200 nm. Particle weight 150×10^6 daltons. Characteristics of nucleic acid: single-stranded RNA, present in up to seven pieces. Molecular weight 4–5×10^6 daltons.
Characteristics of proteins known to date:

1. Neuraminidase (n) glycoprotein in surface spikes.
2. Haemagglutinin (h) glycoprotein in surface spikes.
3. Capsid protein (c). Associated with RNA, arginine rich, probably many copies of the same protein form helical ribonucleoprotein (r), diameter 10 nm.
4. RNA polymerase. Internal protein.

Other components. Lipid (1) derived from host membrane. Composition varies with host. Carbohydrate in two glycoproteins n and h. Composition varies with host.
Photograph and diagram by courtesy of Dr I. Griffith.

associated with the nucleocapsid. It used to be thought that viruses contained only those enzymes involved in cell attachment and penetration, but this view has been modified by the finding that many viruses contain enzymes of vital importance for the replication of their nucleic acid (see Table 8.2). These enzymes are enclosed within the outer envelope or shell, and are almost certainly inactive in the extracellular particle. The virus is stripped of its outer layers on entering the cell and the enzyme is then able to function. As will be described later there are many types of virus for which no host cell enzyme is available for an essential first step in replication, and it is in these cases that the enzyme is imported with the virion. Some enzymes have been described whose precise function is unclear to date. Since the major protein portion of bacterial viruses is left behind on entering the cell, it is unlikely that many bacterial viruses will be found that contain enzymes involved in replication, although a recent report suggests that the unique enveloped phage PM2 does contain a polymerase within the virion.

2.3 Viral envelopes

Many viruses also possess an outer lipid-containing envelope that surrounds the inner symmetrical nucleocapsid, possibly obscuring the underlying symmetry. In viruses of eukaryotes these membranous envelopes are usually acquired as the virus 'buds' through a membrane of the cell in which it has been replicated. The viral envelope thus contains some components that originated in the host cell, and some of the characteristics will be determined by the cell and not by the virus. In all the cases that have been investigated so far the viral envelope contains no host proteins but only proteins specified by the virus. At least one of these envelope proteins is always a glycoprotein. The lipids of the membrane however may

Table 8.2 Enzymes present in virions

Virus	Enzyme	Function
T$_2$	Lysozyme	? Release from cell
	ATPase	ATP hydrolysis in tail contraction
Influenza	Neuraminidase	Release from cell
Reovirus	dsRNA $\longrightarrow$ RNA polymerase	Synthesis of viral mRNA
	ATPase	
Newcastle disease virus	ssRNA $\longrightarrow$ RNA polymerase	Synthesis of viral mRNA
Pox virus	DNA $\longrightarrow$ RNA polymerase	Synthesis of viral mRNA
	DNAase	?
	ATPase	
RNA tumour viruses	RNA $\longrightarrow$ DNA polymerase	Synthesis of c-DNA
	DNA $\longrightarrow$ DNA polymerase	Synthesis of dsDNA
	DNA ligase	Integration of DNA
	Exo- and endo-DNAase	,,
	RNAase H	
Avian myeloblastosis	ATPase	
	Protein kinase	

originate in the host cell. One enveloped bacterial virus, PM2, has also been described but this is not formed by budding through the bacterial membrane and the characteristics of the envelope are thus determined by the virus. The envelopes of the viruses of eukaryotic cells are involved in attachment to the plasma membrane of the host cell, and may be involved in entry of the virus into the cell. Viral proteins, such as influenza haemagglutinin, that are involved in adsorption, protrude through the lipid layer of the envelope.

3 Intracellular replication of viruses

The intracellular replication of all viruses follows a similar sequence of events. This particular pattern was first established for T-even phages but has since been shown to be generally true for all types of virus in which it has been investigated. The growth pattern shows the following features (see Fig. 8.5(a)).

Virus attaches to a susceptible cell, which it then enters with loss of the protecting protein coat; at this stage the virus loses its identity as an infectious entity and is said to be eclipsed. The exposed nucleic acid is then able to provide information for the synthesis of 'early proteins'. These usually consist of enzymes involved in nucleic acid replication. Synthesis of viral nucleic acid and 'late' (capsid) proteins proceed in parallel and mature virions are assembled from these preformed components. Virus particles may not be recognised as such until the final stage of assembly when they are released into the extracellular environment.

Killing of the host cell does not always accompany this type of replication. Some phages such as T2 kill cells on entry whereas others, such as fd, allow cell multiplication to occur at almost normal rates. Some cells are lysed when virus is released but in many systems the virus is released slowly over long periods of time without apparent damage to the cell.

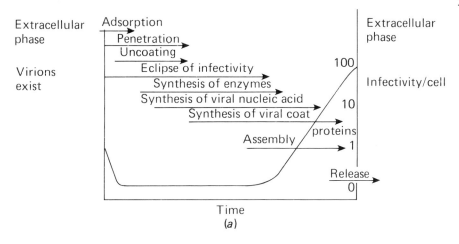

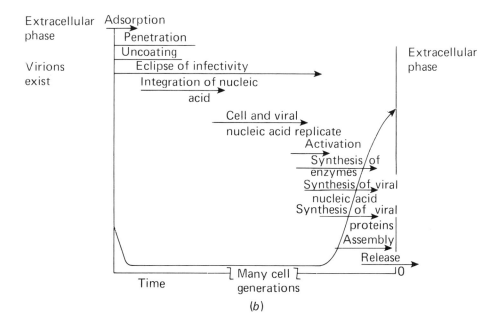

Fig. 8.5 Time course of viral replication. (*a*) Virulent viruses. (*b*) Lysogenic viruses.

In lysogenic systems, however, the sequence of events described above does not follow immediately as a consequence of infection. After uncoating of the nucleic acid, and transcription of some viral information, the viral nucleic acid is integrated with the host genome and is replicated with it; cells multiply normally and viral information is distributed to daughter cells together with the host genome. A little, or in a few cases, much of the viral information may be transcribed into viral mRNA during the integrated phase. Some stimulus interrupts the sequence and then viral replication may proceed as in the virulent cycle (Fig. 8.5(*b*)).

Some of the biochemical processes involved in virus replication will now be considered in greater detail.

3.1 Adsorption

Viruses are not motile and therefore viruses in suspension can rely only on diffusion to reach the surface of cells. The efficiency of infection will obviously depend on the ability of the virus to adsorb to the cell surface once it has collided with it. Clearly the most efficient viruses will be those that adsorb only to cells that can be profitably infected. In fact, the host specificity typical of many viruses depends on this initial interaction between the host cell and the virus. Such specificity may be very great; thus different strains of salmonella may be identified by their ability to support the growth of various phages. Little is known of the mechanism of this specific recognition but it is assumed that it results from an interaction between specific regions on the host cell surface and areas in the viral surface in a manner similar to enzyme–substrate or antibody–antigen combination. In a few cases some details of the nature of cell receptors and of viral recognition sites are known. Thus the proteins responsible for binding and attachment of the T-even phages are present in the phage tail fibres; different T-phages bind to different receptors in the bacterial surface. Some of these have been identified and partially purified, e.g. receptors for T2 and T6 are in the outermost lipoprotein layer, while sites for T3, T4 and T7 are in the rigid mucopeptide layer beneath. In the isolated form the receptors will react with the phage and neutralise its infectivity.

The small RNA phages will adsorb only to pili of male bacteria, and are unable to infect bacteria lacking this structure: although details of this reaction have not been studied it is known that the A protein, a minor component of the virion, is necessary for attachment to occur.

Viruses of eukaryotes probably possess many recognition sites on their surfaces since many molecules of antibody are necessary to neutralise infectivity. Many animal viruses exhibit the property of haemagglutination, where the virus agglutinates red cells by adsorbing to more than one cell. This indicates that the virus contains at least two combining sites; however influenza virus is known to contain at least 200 haemagglutinin molecules per particle. The properties of the haemagglutinin of the myxovirus group (of which influenza is one) have been studied in great detail, and it is known that it combines with neuraminyl residues of glycoproteins and glycolipids present in the cell surface of many types of cell including erythrocyte membranes.

Polio virus will infect only cells of primate origin and *in vivo* only those of the digestive tract and nervous system. It has been shown that these cells possess a receptor material in the cell membrane that is absent from other cell types; partially purified preparations of this material contain lipid and glycoprotein. Since infectious RNA prepared from polio virus can infect a large number of cell types from different species, it is clear that specificity of infection results from some interaction between the capsid protein and the cell receptor material, but the nature of this reaction is not known. In the case of morphologically similar plant viruses, however, similar experiments show that the cell specificity does not reside in the protein coat, but in the RNA itself.

Many viruses of eukaryotes possess membranous envelopes, and the initial interaction between virus and cell surface involves an alignment of the two membranes. Again the nature of this interaction is unknown, but the fact that most viral envelopes originate from the host cell and that membrane systems of normal eukaryotic cells frequently exhibit this type of alignment suggests that some normal membrane behaviour is being exploited by the virus for attachment.

In all cases so far studied, the initial interaction is reversible and can occur at low temperatures. It can be prevented by antibodies directed against either cell receptors or the virus. Raising the temperature to the normal one for virus growth results in some change in the

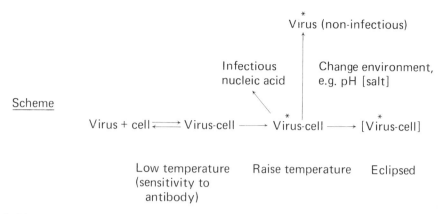

Fig. 8.6 Scheme illustrating attachment of virus to an animal cell.

system so that virus is irreversibly bound, and the infectivity is no longer sensitive to antibody inactivation. In some cases several stages are discernible in this irreversible reaction. Thus there may be an irreversible change in the structure of the virion. This may be detected as a change in reaction with specific antibodies, or as a loss of some of the virion protein. In some cases the capsid protein shows increased sensitivity to digestion by proteolytic enzymes, and the nucleic acid may no longer be protected from nuclease attack. Infectious nucleic acid may, however, still be recovered from these altered particles, at least in the earlier phases of the process. Little is known about the mechanism of this process but it is possible that the interaction of the capsid protein with the hydrophobic material of cell membranes may cause a change in stability and so of conformation of the viral capsid assembly. (See figure 8.6.)

3.2 Entry into the cell

In order to initiate infection the nucleic acid of the virus has to enter the cell, traversing not only the cell membrane but also any outer protective layers of the cell surface. Many bacterial viruses possess a specialized tail in order to achieve this. The detailed mechanism of attachment and the injection process of phage T4 have been elucidated (Fig. 8.7). The attachment of the long tail fibres to the specific receptor sites on the bacterial surface triggers a change in conformation of the fibre proteins so that the short tail spikes are brought into close contact with the cell surface. These attach firmly to the bacterial cell wall; at the same time the base plate to which they are attached changes conformation and this in turn initiates the contraction of the proteins comprising the tail sheath. This protein resembles actomyosin in many of its properties and the contraction of the tail is accompanied by the hydrolysis of bound ATP molecules. The contraction of the tail sheath combined with the firm attachment of base plate fibres to the surface results in the tail tube being forced through the outer layers of the bacterial cell surface. It is not known whether the tail tube also penetrates the cell membrane, or, if it does not, how the viral DNA enters the cell. As the tail tube penetrates the bacterial cell wall, the DNA is ejected from the head, through the tail tube and into the bacterial cytoplasm. How a molecule of the length and rigidity of DNA can be propelled through a tube of diameter only slightly greater than that of the DNA remains a matter for speculation. In the case of T5, protein synthesis in the infected host cell is necessary to effect complete transfer, but for other T-phages metabolic cooperation by the host cell is unnecessary. Only a few internal proteins of the phage are injected along with the DNA, the rest remaining outside the cell.

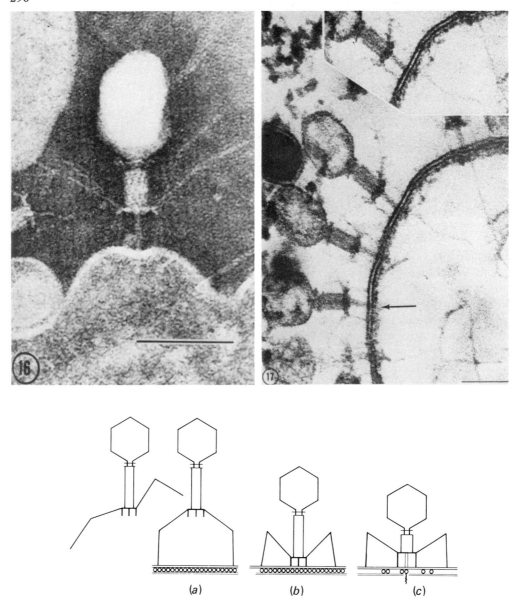

Fig. 8.7 Attachment and penetration of the cell wall by T_2 and T_4 bacteriophages. (*a*) Attachment of tail fibres to specific receptor sites. (*b*) Change in conformation of the long tail fibres, bringing short tail-plate pins into contact with the cell wall. (*c*) Contraction of the tail sheath. The inner tail tube penetrates the lipoprotein, lipopolysaccharide and mucopeptide layers but may not penetrate the membrane (Simon and Anderson (1967). Copyright © Academic Press, London and New York).

Many bacterial viruses do not possess a tail, and the mechanism by which their nucleic acid gains entry is unknown. The spikes of ϕX174 are involved in attachment and the rest of the coat protein adsorbs and remains attached to the cell membrane while the DNA penetrates the cell. After attachment of small RNA phages to the sides of the sex pili there is some change in the conformation of the virion so that the RNA becomes transiently sensitive to ribonuclease;

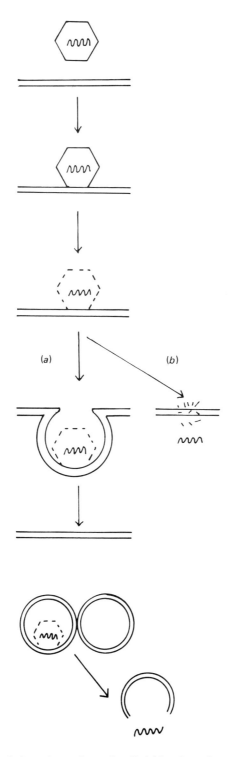

Fig. 8.8 Entry of non-enveloped viruses into eukaryotic cells (a) by viropexis or (b) by passage through the membrane.

subsequently the RNA can be isolated from the pili while the protein remains outside. Recent work suggests that the A protein may enter the cell with the RNA. Filamentous DNA phages attach to the ends of the pili and although DNA may enter the cell via a possible internal channel in the pili, it is also possible that the pilus contracts pulling the attached virus level with the bacterial surface. This is the probable function of the pili in bringing bacteria into close proximity for conjugation to occur.

Little is known about the mechanism of nucleic acid transport across the bacterial membrane, either for viral nucleic acid or indeed for any nucleic acid. In all cases the coat-proteins of bacterial viruses remain outside the cell, although some internal proteins may enter with the nucleic acid.

There are three theories for the mechanism of penetration of viruses into animal cells. The first theory suggests that after adsorption the virus is taken into the cell cytoplasm by the normal process of endocytosis (Fig. 8.8(a)) in this process, which has been named viropexis by Dales, the area of cell membrane to which the virus has become adsorbed forms an invagination which is then pinched off to form a vesicle enclosing the attached virion. As in the normal process of endocytosis these vesicles become associated with lysosomes and it is thought that lysosomal enzymes are involved in the digestion and removal of outer layers of the virus. Intact Reovirus can be isolated from the lysosomal fraction of cells shortly after infection, and as time proceeds the outer protein is gradually removed. The outer capsid proteins of Reo can also be digested by lysosomal proteolytic enzymes *in vitro*. The viral cores released by these enzymes are probably liberated into the cytoplasm when the vesicles break down.

Another theory proposed for entry of viruses into cells suggests that simple viruses are transported directly across the membrane (Fig. 8.8(b)). No mechanism for this process has been suggested, but there is some evidence that some capsid proteins of adenovirus and poliovirus remain attached to cell membranes while the core material passes through. Plant virus proteins may dissociate from the nucleic acid following attachment to the cell membrane, with only nucleic acid entering the cell.

There is considerable evidence to suggest that enveloped viruses possess a special mechanism for gaining entry to the cell cytoplasm (Fig. 8.9). Following alignment of cell membrane and viral envelope the membranes fuse, and then break down creating a channel between virus and cytoplasm through which the nucleocapsid can pass (Fig. 8.10). Several enveloped viruses possess the ability to fuse adjacent cells together when added in high concentration. Thus Sendai virus (a paramyxovirus) is now widely used in the technique of cell fusion, first employed by Henry Harris and his group and now of general use in studies of molecular biology and cellular genetics. In this case it is known that a virus particle fuses with two cells, causing the formation of a bridge between them; many such bridges allow a complete breakdown of cell boundaries in this region (Fig. 8.11). This process may be mimicked by the addition of lysolecithin to cells, but there is some doubt whether virus induced fusion is due to the presence of lysolecithin in the virus envelope.

The way in which viruses enter plant cells is not known; most plant viruses rely on damage caused to the tough cell wall either by an insect vector or mechanically. It is quite probable that once having passed the barrier of the cell wall the virus crosses the cell membrane by viropexis. Certainly TMV particles are taken up by plant cell protoplasts in this way and may initiate infection.

In multicellular organisms virus can pass from cell to cell through cell bridges. These bridges may be formed as a result of virus-induced alteration of cell membranes, or they may occur naturally in the host. Thus plant viruses have been seen passing through plasmadesmata in

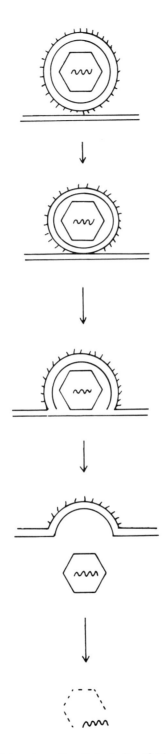

Fig. 8.9 Entry of enveloped viruses into eukaryotic cells by fusion of cell and viral membranes.

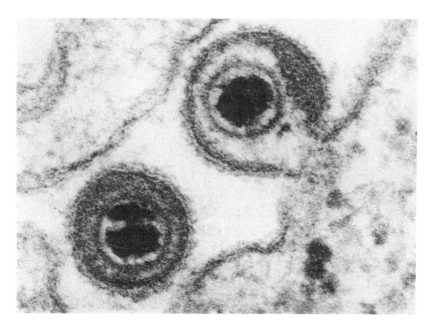

Fig. 8.10 Herpes simplex virus attached to the membrane. Note the breakdown of cell and viral membranes in area of contact. From Morgan, Rose and Mednis (1968). *J. Virol.*, **2**, 507.

sections of infected plants. It is also probable that immature forms of the virus containing the nucleoprotein can infect adjacent cells.

Although viruses of eukaryotes may enter cells by different mechanisms it is clear that each process allows for much of the virion protein to be imported into the cell; this includes enzymes enclosed in the viral particle. This represents an extremely significant difference between viruses of bacteria and those infecting higher organisms, and allows for a greater variety in virus type.

In many instances the intact virion seems to enter the cytoplasm. There is thus a problem of removing protective protein so that the nucleic acid can be exposed for transcription and translation, and so that the enclosed enzymes may be activated. This process may, as we have seen, involve lysosomal enzymes. It is of interest in this connection that many virions are resistant to proteolytic degradation outside the cell; however the change in conformation that occurs on attachment to the cell surface may render them susceptible to proteolytic cleavage, and thus allow uncoating to occur. Some enteroviruses have been shown to be associated with a

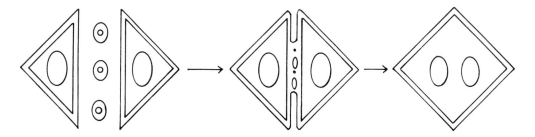

Fig. 8.11 Diagrammatic representation of cell fusion by an enveloped virus, e.g. Sendai.

proteolytic enzyme, although it is not clear whether this is an integral part of the virus, and this may have a function in uncoating the virus.

Removal of the outer proteins may result in activation of the enzymes contained. The RNA polymerase of Reovirus can be activated *in vitro* by treatment of the virus particles with chymotrypsin; trypsin digestion, on the other hand, inactivates the enzyme. This suggests that the proteins of the virus must be 'tailor-made' to meet the specificity requirements of cellular enzymes if infection is to be productive.

Uncoating of large viruses such as pox viruses is a more complex, two-stage process. The first stage is dependent on host enzymes and is similar to that already described. This process activates the virion RNA polymerase and mRNA is then transcribed from the viral genome. This codes for enzymes that complete the uncoating of the nucleic acid; only after this second stage can viral DNA be replicated.

3.3 Synthesis of viral nucleic acid

Once the genetic information is present unmasked, inside the cell, replication of viral material can begin. Although synthesis of new enzymes is the first stage in replication of a few viruses, most begin with synthesis of virus coded nucleic acid. Since more types of nucleic acid are

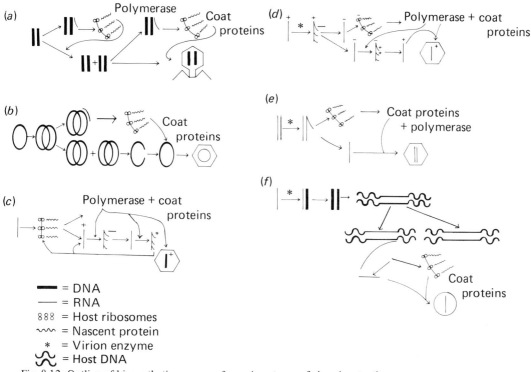

Fig. 8.12 Outline of biosynthetic processes for various types of virus (see text).

(a) Double-stranded DNA virus, e.g. phages T_2 and T_4.
(b) Single-stranded DNA virus, e.g. ϕX174.
(c) Single-stranded RNA virus, e.g. $Q\beta$. polio.
(d) Single-stranded RNA virus, e.g. NDV. VSV.
(e) Double-stranded RNA virus, e.g. Reo.
(f) Single-stranded RNA virus–DNA provirus, e.g. Rous sarcoma.

found in viruses than in cells, it is clear that enzymes other than those present in normal cells will be involved in synthesis of viral nucleic acid (see Table 8.3). These systems will now be described (Fig. 8.12).

Table 8.3 Enzyme systems for nucleic acid synthesis

Present in normal cell systems	*Present only in virus infected cells*
dsDNA⟶dsDNA	dsDNA ⟶ ssDNA
(ssDNA⟶dsDNA)	ssRNA ⟶ ssRNA
DNA ligase	dsRNA ⟶ ssRNA
	ssRNA ⟶ dsRNA
dsDNA⟶ssRNA	ssRNA ⟶ ssDNA ?

⟶ represents information transfer.

3.3.1 Viruses containing double-stranded DNA

Synthesis of nucleic acid in cells infected by DNA-containing viruses follows a similar pattern to that in normal cells. Thus double-stranded DNA is replicated in a semi-conservative fashion and mRNA is transcribed from one strand of the duplex DNA. Host cell enzymes may be used for these processes, though new DNA polymerases are frequently found in these systems. T4-induced DNA polymerase is known to be involved in phage replication since mutants lacking this enzyme cannot replicate viral DNA. Since DNA polymerases examined *in vitro* are not template-specific, the reason for synthesis of new enzyme in these cases is not understood. It is possible that the host replicating system demonstrates *in vivo* a template specificity, or is subject to controls, that make it incapable of replicating viral DNA. These problems may be solved when host enzymes known to be involved in DNA replication have been more fully characterised

Messenger RNA is transcribed from viral DNA by DNA-directed RNA polymerase. In the case of T4, λ and SPO1 there is good evidence that the host enzyme is used for this purpose throughout the growth cycle. This enzyme is inhibited by rifamycin, which combines with a subunit (β') of the enzyme. In cells sensitive to rifamycin, viral replication is also sensitive to the drug at all stages of the growth cycle, whereas virus growth is resistant to rifamycin when grown in cells that are resistant. However although it is clear that at least the β' subunit of the enzyme is used for transcription of viral RNA, the enzyme is altered during the course of viral replication. It becomes less able to transcribe host cell DNA, and more efficient at using T4 DNA as template. This change has been ascribed to a change in the affinity for host σ-factor, and possibly to the synthesis of new σ-like proteins following infection. In the presence of host σ, only very early mRNA is transcribed from T4 *in vitro* but the presence of virus-specified proteins allows synthesis of late mRNA (see section 3.6). The host enzyme is also modified by adenylation of the α subunits.

The growth of T7, in contrast, is unaffected by rifamycin except in the very early stages of infection. It has been shown that this virus codes for the synthesis of a new enzyme unaffected by rifamycin (see Fig. 8.12(*a*)).

3.3.2 Viruses containing single-stranded DNA

When the viral genome is single-stranded DNA, as in φX174, the strand complementary to the incoming DNA is synthesised using host DNA polymerase. The double-stranded replicative form first becomes attached to some host 'site' and then serves as template for the synthesis of further closed circular duplex molecules. These in turn act as template for the synthesis of single-stranded progeny DNA. These latter processes require virus-coded proteins. The replicative form of φX174 also serves as template for the synthesis of viral mRNA which is complementary to the − strand (i.e. it has a base sequence identical with that of virion DNA, except that U replaces T) (see Fig. 8.12(b)).

3.3.3 Viruses containing single-stranded RNA

If the virus contains an RNA genome the information transfer is normally from RNA to RNA, a process that probably does not occur in normal cells. There has been considerable discussion as to the mechanism of synthesis of viral RNA in these systems. Both double-stranded RNA (replicative form, RF) containing the viral RNA and its complementary strand, and replicative intermediate (RI) which contains not only double-stranded RNA but also some single-stranded RNA of molecular weight less than the viral RNA (Fig. 8.13), have been isolated from infected cells. Many schemes have been proposed to account for this, but there is considerable agreement that the following is most likely to account for the observations:

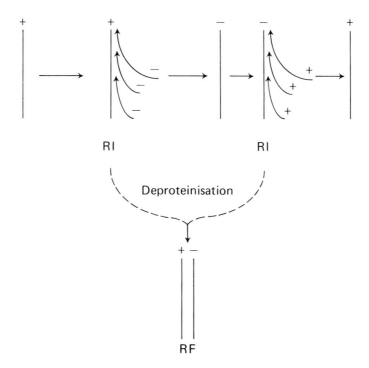

Fig. 8.13 Scheme showing replication of single-stranded RNA and possible relationship between replicative intermediate (RI) and replicative form (RF). Arrowheads indicate direction of polymerase movement.

The incoming + strand serves as template for the production of complementary − strands. Several of these strands would be synthesised at any one time, and this structure would represent replicative intermediate. The newly formed − strands released from this structure would serve in turn as templates for new + strands in a similar manner. The double-stranded RF is probably an artefact caused by annealing of complementary + and − strands. In the case of phage Qβ (which has been investigated in greater detail than most other RNA viruses) a newly formed enzyme Qβ replicase (Chapter 7) catalyses synthesis of both + and − strands. This enzyme has four subunits, and only one of these is coded by the virus; the remaining three are proteins present in the infected cell (including T_s and T_u (see Chapter 1)). The enzyme, unlike most polymerases, shows a high degree of template specificity. Host cell proteins, in addition to these, are needed for copying of + strand to give − strand, but not for the complementary reaction. The RNA polymerase responsible for synthesis of another small RNA phage, R17, contains two subunits sequestered from the normal host DNA-directed RNA polymerase. Although only the RNA phage enzymes have been studied in detail, most viruses containing single-stranded RNA have been shown to be replicated by the process described. Furthermore, in many of these cases also new RNA-directed RNA-polymerases have been isolated from the infected cell (see Fig. 8.12(c)).

3.3.4 Viruses containing double-stranded RNA

In viruses having double-stranded RNA genomes, the single-stranded mRNA is copied from the double-stranded RNA (although the details are still unclear). An entirely new enzyme imported with the virus is used to catalyse this reaction, and the enzyme seems to be firmly associated with the RNA template within the viral core. The single-stranded RNA product is used not only as mRNA but also in turn as template for the synthesis of complementary strands, in the formation of double-stranded progeny RNA. This last process occurs within a particle which is probably a newly formed virion; the reaction has not been studied in detail (Fig. 8.12(e)).

3.3.5 RNA-containing tumour viruses

The RNA tumour viruses present the most extreme departure from the usually accepted scheme of information transfer. These viruses possess single-stranded RNA as their genome, but as Temin and others have shown there is much evidence that information for viral replication is integrated into the host cell genome in the form of double-stranded DNA. In cells infected by these viruses, single-stranded RNA is copied by an RNA-directed DNA polymerase into complementary single-stranded DNA. When this enzyme was discovered it was thought to be a unique property of this virus group; however recent results suggest that it may be present in the differentiating cells of embryonic tissue and that it may have some role in gene amplification. The single-stranded DNA product of the viral enzyme is copied into double-stranded DNA by a DNA-directed DNA polymerase, and the product of this is then inserted into the host chromosome by a process probably similar to that thought to be involved in recombination, i.e. nicking of intact double-stranded DNA by endonuclease, partial digestion by exonuclease to form 'sticky ends', annealing of sticky ends of host and viral DNA, and finally complete integration by the formation of covalent bonds catalysed by a ligase. Similar processes are involved in the integration of lysogenic phage DNA such as λ into bacterial DNA. In virally transformed cells integrated DNA may then be transcribed into mRNA by host DNA-directed RNA polymerase. Evidence from cells transformed by adenovirus suggest that integrated viral DNA and adjacent regions of host DNA are transcribed sequentially in one operation, so that very large RNA molecules containing both viral and host information are formed. Transcription of integrated λ DNA is controlled by the presence of λ repressor; only the small region specifying the repressor is transcribed in the lysogenic state. Inactivation of repressor allows transcription of other regions of λ, one of which codes for a protein allowing modification of the host-specified DNA-directed RNA polymerase, which in turn transcribes the rest of the λ genome (see Fig. 8.12(f)).

3.4 New enzymes present in infected cells

Synthesis of new viral components is the main result of metabolic activity in a cell infected by an efficient virus. It used to be thought that only the metabolic systems of the host cell were used for this, but the discovery by Cohen that a whole new system of enzymes involved in synthesis of hydroxymethylcytosine appeared in cells infected by T-even phages changed these ideas. Since then many new enzymes have been found in cells infected by viruses. There are various reasons for the necessity for synthesis of new enzymes following infection:

a Synthesis of viral components may require a process not normally occurring in the cell, e.g. replication of double-stranded RNA.

b The virus may contain a component not normally present or synthesised in the cell, e.g. hydroxymethylcytosine in phage DNA.

c The host cell enzyme may be localised at a site inaccessible to replicating virus, e.g. pox viruses replicate in the cell cytoplasm and cannot react with the host RNA polymerase located in the nucleus.

d The host cell enzyme may be controlled so that it is specific for regions of macromolecules not present in the virus, e.g. σ mediated binding of E. coli RNA polymerase to promoter regions of the DNA.

e The host cell enzyme may be repressed or have only low activity in the cell before infection, e.g. thymidine kinase has very low activity in non-dividing animal cells; many DNA-containing viruses code for the synthesis of this enzyme, and the virus-coded enzymes have a lower K_m than the host, to the advantage of the virus.

Synthesis of new enzymes is not, however, the only way in which these problems may be overcome. Thus the activity of existing host enzymes may be changed either by modification of amino acids or by addition of extra virus-coded subunits. Both of these changes occur in the RNA polymerases present in E. coli infected by T4 phage, and result in a change of specificity of the enzyme so that it uses phage DNA as template in preference to host DNA. Derepression of host enzymes may also occur and such a process probably accounts for the stimulation of host enzymes involved in DNA synthesis following infection of slowly growing cells by polyoma virus. If new enzymes or subunits of enzymes are necessary for replication of viral nucleic acid it is clear that synthesis of these enzymes is an obligatory early event in virus replication. Synthesis of viral mRNA must precede synthesis of virus-coded proteins. The larger DNA-containing viruses most often make use of host RNA polymerase to transcribe mRNA required for synthesis of early enzymes The nucleic acid of many of the small RNA viruses of both pro- and eukaryotic cells is itself the mRNA. Infecting RNA can therefore attach to the host ribosomes and code for the synthesis of enzymes required for its own multiplication.

However some of these small RNA viruses (e.g. Newcastle disease virus, vesicular stomatitis virus) contain a genome which is complementary to the mRNA. Since copying of the incoming + strand into complementary − strand requires the presence of an RNA-directed RNA polymerase, an enzyme not present in cells, these viruses necessarily have to import the enzyme in order to synthesise mRNA. A similar argument applies in the case of viruses containing a double-stranded RNA genome. Single-stranded RNA species copied from this template are the mRNA molecules and an enzyme capable of carrying out this reaction is not found in normal cells. In this case also the enzyme is present in the particle and starts to synthesise mRNA as soon as it is activated by removal of virus coat protein (see Fig. 8.12(*d*)).

3.5 Protein synthesis

The classical experiments of Jacob, Monod and Brenner demonstrated that no new ribosomes were made in cells infected by T-phages, and that viral protein synthesis used the protein synthesising machinery of the cell directed by viral mRNA. It is now clear that this is a general phenomenon. No virus has yet been discovered that contains information for ribosome synthesis, and this seems unlikely on the grounds of size of genome required (i.e. $10 - 20 \times 10^6$ daltons of DNA). Ribosomes have occasionally, but not consistently, been found enclosed within the membranes of some of the larger viruses; since these are formed by budding from the cell surface the ribosomes are thought to have been trapped by chance.

It is clear, however, that the machinery of protein synthesis may not remain entirely unmodified in the infected cell. Thus new species of tRNA may be synthesised, e.g. up to 15 have been reported in cells infected by T4 and T5 phages, and new activating enzymes corresponding to these tRNAs are presumably then also required. Subak Sharpe has suggested that if the viral mRNA uses codons for a particular amino acid that differ from those used normally by the host cell, virus protein could only be synthesised following the synthesis of the appropriate tRNA species. In many infected cells, synthesis of host proteins ceases shortly after infection. Polyribosomes characteristic of the host cell break down, and may be replaced by polysomes containing viral mRNA. The mechanism of this is not understood but it may be due to an alteration in the various factors involved in initiation and translocation. Thus superinfection by T4 of a cell infected by MS2 phage inhibits replication of the RNA phage. Ribosomes from normal *E. coli* cells are able to support protein synthesis directed by MS2 RNA *in vitro*, but ribosomes from T4 infected cells allow only inefficient translation from MS2 or *E. coli* mRNA whereas T4 mRNA is translated normally. This difference is thought to be due to a change in F3 initiation factor. If such a change increases the specificity of the system so that viral mRNA is translated preferentially, this is clearly to the advantage of the virus. Similar changes may well take place in eukaryotic systems, but until recently no satisfactory *in vitro* systems have been available to test these ideas.

Interferon, which is a protein produced in animal cells following treatment by a variety of agents including inactivated viruses and synthetic double-stranded RNA, inhibits multiplication of both RNA and DNA viruses. It has been shown that it acts by preventing translation of viral mRNA without affecting translation of cellular mRNA; this suggests that there is some difference between viral mRNA and cellular RNA. The development of good *in vitro* systems should allow these ideas to be tested. Many, but not all, viral mRNA species in eukaryotic cells contain poly-A sequences which are thought to be characteristic of mRNA.

Bacterial systems are capable of translating polycistronic messenger into several individual proteins. Indeed most of our knowledge of the mechanism and control of these processes comes from such systems programmed with viral mRNA. In eukaryotes, however, the position seems to be different. This was first suggested by work with polio virus, an RNA virus containing information for 8-10 proteins. Polysomes from cells infected by polio virus are larger than polysomes from normal cells: RNA released from these polysomes is the same length as intact viral RNA. Baltimore and his co-workers found that about 14 new proteins were synthesised in polio-infected cells instead of the expected 8-10, and that some of the proteins made were of very high molecular weight. Further work demonstrated that the large proteins were in fact precursors of the smaller ones and the results suggest that the entire genome of polio virus is transcribed in one unit, and that the product is then cleaved by proteolytic enzymes to give the required individual proteins. 'Initiation' and 'termination' signals must thus be present in the amino-acid sequence of the large protein and must be recognised by proteolytic enzymes. The origin of the proteolytic enzymes involved is unknown, but they are probably host enzymes. Inhibition of the proteolytic action by tosylphenylalanine chloromethyl ketone or by incorporation of amino-acid analogues into viral proteins leads to an increased yield of precursor proteins and inhibition of virus. This type of post-translational cleavage is seen in several other viruses of eukaryotic cells, and also in uninfected cells, as for example in the conversion of pro-insulin to insulin. However the process is by no means universal. Some RNA viruses such as influenza or Reo have fragmented genomes, and the mRNA copied from these is present in small pieces that probably represent monocistronic message.

Glycoproteins are components of many viruses grown in eukaryotic cells; all enveloped viruses contain at least one glycoprotein in the envelope. In uninfected cells glycoproteins are formed by the addition of monosaccharides to specific amino-acid residues in the completed protein, this process probably taking place in the Golgi apparatus. The structure and composition of the carbohydrate moiety is determined by the specificity of the glycosylating enzymes in the particular cell. Although this process has not yet been investigated in detail it seems likely that viral glycoproteins are processed in the same way; indeed it has been shown that the enzymes that glycosylate host proteins continue to add the same sugars in the same order to viral proteins. Thus the glycoproteins of influenza virus carry the same carbohydrate portion as those of the cells in which it has been grown, although the amino-acid sequence is determined entirely by the viral genome. Since the specificity of the glycosylating enzymes varies in different cell types, the carbohydrate groups added to the virus will vary with the type of cell in which the virus grows. Moreover since the carbohydrate moiety of glycoproteins is strongly antigenic, the presence of host-specified carbohydrate material accounts for the fact that enveloped viruses react with antisera prepared against uninfected host cells.

3.6 Control of synthesis of macromolecules

Virus replication is frequently a highly efficient process, in which many thousands of mature virions are produced in a comparatively short time. A cell infected by RNA phage produces up to 100 000 new particles in 20 min; this is a mass equivalent to that of the host cell. Thus the processes of macromolecular synthesis must be carefully controlled so that products appear at the correct time and in balanced quantities. The RNA synthesised during the life cycle of a small RNA virus has to serve multiple functions, as mRNA for the synthesis of viral proteins, as template for the synthesis of complementary strands and as the genetic material for assembly into progeny virions. Early in infection RNA is used mainly for the first two of these processes, whereas RNA synthesised towards the end of infection is almost all incorporated in virions. MS2 coat protein combines with viral RNA and prevents its translation into viral RNA polymerase (see Chapter 7).

More complex viruses show intricate patterns of synthesis of macromolecules. During synthesis of T-phages, 'early' proteins include the enzymes required for viral nucleic acid synthesis whereas late proteins comprise virion proteins, and enzymes such as lysozyme which are involved in release of virus from the cell. The mRNA synthesised at each time is characteristic, and classes of 'pre-early' 'immediate early', 'delayed early' and late mRNA are recognised; moreover late RNA can only be transcribed from DNA that has been replicated. By the use of specific inhibitors, it has been easy to show that the production of macromolecules at any stage is dependent on the correct sequence of events having preceded it.

This is fertile ground for investigation of control mechanisms, which may be at the level either of transcription or translation. Much information has already been gained from these investigations and is discussed elsewhere in this book (Chapters 1 and 7).

3.7 Role of membranes in viral biogenesis

The replication of many viruses occurs in association with one of the membrane systems of the cell. Thus parental phage DNA becomes associated with the bacterial membrane shortly after infection. The parental replicative form of ϕX174 DNA is firmly bound to a 'site' in the bacterial membrane throughout the growth cycle; there is only a limited number of such sites which are thought to provide some essential function in DNA replication. Viruses that multiply in the cytoplasm of eukaryotic cells are often assembled in membrane-bound 'factories'. These can be isolated in the large particle fraction by conventional cell fractionation techniques and are found to contain all the enzymes involved in viral biosynthesis as well as virions in various stages of completion. Two distinct types of membrane-bound body are found in polio-infected cells; one, associated with smooth endoplasmic reticulum contains the RNA synthesising system, while the other enclosed by rough endoplasmic reticulum is the site of virus directed protein synthesis. Enclosure of cytoplasmic 'factories' by membranes may serve to isolate viral replication from harmful lytic enzymes in the cytoplasm as well as to limit and so concentrate the reactants.

Many viruses of eukaryotic cells are completed by budding through a cell membrane. They may be released from the cell by budding outwards through the plasma membrane, or into vesicles that subsequently fuse with the cell membrane. During this process the nucleocapsid becomes enclosed in an envelope by pinching off part of the membrane. Pre-labelling of cells before infection shows that the lipids of the viral envelope may have been synthesised by the host cell before infection. Moreover the lipid composition of viral envelopes varies according to the type of cell in which the virus was grown. However no host proteins are present in the envelopes of myxoviruses or arboviruses, and the proteins that are found are synthesised only after infection. Electron microscopy of cells infected by these viruses shows that the viral nucleoprotein accumulates in certain regions of the cell, just under the cell membrane (Fig. 8.14). The membrane in these areas appears thicker and labelling with ferritin-labelled antibody directed against the virus shows that these regions of the plasma membrane contain viral proteins. It thus appears that the virus directs the synthesis of proteins that are then inserted into the host membrane, possibly displacing the host proteins already present. These virus-specified proteins contain a strongly hydrophobic region which is embedded in the membrane, and are usually glycoproteins with host determined carbohydrate moieties (see earlier). It is not clear why virus proteins are present only in certain regions of the membrane, nor whether their presence modifies the properties of the membrane in any way. Nor is it known whether the presence of viral nucleoprotein directs the modification at specific sites, or if it is attracted to regions that have been modified by the presence of viral protein; the latter seems more likely since modification of membranes occurs before ribonucleoprotein synthesis. The net result of the process is that viral ribonucleoprotein passes out of the cell enclosed by an envelope composed of a mosaic of host and viral material. Viruses of the herpes group are assembled in the nucleus and they acquire their envelope by budding through the nuclear membrane into the perinuclear space; recent evidence suggests that both the lipid and the protein composition of the nuclear membrane is altered in some regions.

Viruses that bud from the membranes in this way seldom cause cell lysis, but are released gradually from the cell surface. The alterations caused by the presence of viral proteins in the plasmalemma may promote fusion of cells one with another and in this way allow passage of virus into adjacent cells.

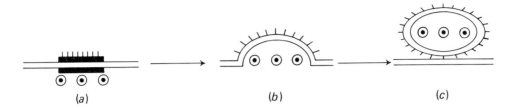

Fig. 8.14 Release of paramyxo virus SV5 from the infected cell.
  (a) Accumulation of ribonucleoprotein under modified membrane.
  (b) Membrane buds off, enclosing ribonucleoprotein.
  (c) Released virus.

When vertebrate cells have been transformed by tumour viruses the growth properties of the cell are changed. Normal cells in culture cease multiplying when they have formed a uniform layer one cell thick; they cannot grow or move over each other and are said to be 'contact inhibited'. Transformed cells, in contrast, do not exhibit this property and can grow into very dense cultures with cells piled up above each other to give layers several cells deep. This change is associated with many changes in the properties of the cell membrane, including permeability changes. At the same time changes are observed in the composition of the transformed cell membrane, with an alteration in the type and concentration of glycoproteins and glycolipids. It is not clear whether these changes are the cause of the altered behaviour of the cells, or are the result of an increased rate of cell division in the cultures, following transformation; normal cells in mitosis show cell membrane changes comparable with those present at all stages in the transformed cell. There is a strong possibility, however, that information transcribed from the genome of the transforming virus codes for new enzymes which in turn cause a change in the composition of the cell membrane.

3.8 Assembly of viruses

Examination of the assembly of virus particles both *in vivo* and *in vitro* provides an excellent system for the study of the morphogenesis of comparatively simple objects. Many of the simple viruses that contain only nucleic acid and protein have now been assembled *in vitro*. Most of the detailed studies have been carried out using tobacco mosaic virus which indeed was reassembled from its separated components as long ago as 1955. These studies have been facilitated by the fact that TMV contains only one species of protein. TMV protein alone aggregates to form stacked discs, or if incubated at low pH values gives rods resembling intact virus. The presence of viral RNA is necessary for the formation of particles of the correct length and structure at physiological values of pH and ionic strength. Recent evidence suggests that the virus is assembled from discs of 34 subunits, composed of two rings of 17 subunits closer together at the outside edges than the inner. When viral RNA is added to a suspension of these discs they begin binding to the 5' end, and completed virus particles are formed within 5 min. Many of the smaller icosahedral RNA viruses can also be assembled *in vitro*. The protein of many plant viruses can form virus-like particles enclosing nucleic acid; the source of the nucleic acid is not particularly important and may include rRNA or even DNA; its function seems to be that of a nucleation centre for the protein.

Virus-like particles may also be assembled from a mixture of MS2 protein and its RNA, but these particles are of very low infectivity. The naturally produced virions contain one molecule of an additional protein, the A protein, that is necessary for adsorption to the host cell. Addition of a purified preparation of A protein to the *in vitro* assembly system results in a several hundredfold increase in infectivity. *In vivo* a few coat proteins are thought to combine with viral RNA to form an initiation complex which can then bind further subunits: the A protein is thought to assist in the early stages of this process.

More complex viruses containing several proteins have not yet been reassembled *in vitro* from their purified components. However several of these viruses have been formed by incubation of infected cell extracts containing partially assembled precursors of viral structures.

Polio virus capsids contain four species of protein (VP1, VP2, VP3 and VP4) which are assembled to give 60 structure units. Empty capsids (73 S) are also found in extracts of infected cells; these also contain VP1 and VP3, but lack VP2 and VP4 having in their place another protein VP0 not found in mature virus. Labelling of the cells shows that VP0 is a precursor of VP2 and VP4. Infected cells also contain 5 S particles, equal in size to the structure units, and 14 S particles, and labelling experiments suggest that these as well as 73 S procapsids are precursors of mature virions. The following scheme, Fig. 8.15, has been proposed to account for these observations:

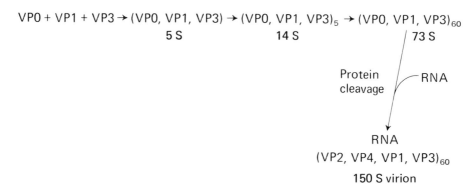

Fig. 8.15 Scheme for assembly of polio virus from precursor protein and RNA.

The final stage, in which RNA is added to the procapsid, is accompanied by a cleavage of VP0 to give the two smaller proteins VP2 and VP4. Extracts of infected, but not control, cells catalyse the conversion of 14 S to 73 S particles and the properties of these extracts indicate that a membrane component is important for this process. This is not surprising since synthesis of the components of polio virus is known to take place in membrane bound 'factories'.

The very complex structures of bacteriophages T2 and T4 have also been assembled by incubation of extracts of infected cells. Edgar and Wood developed the 'extract complementation method' to investigate the assembly of these phages. Cells were infected with phage mutants defective in the assembly process; extracts of these cells were mixed with extracts of cells infected with a second mutant also deficient in this process. If the mutants were able to complement each other mature viruses were formed *in vitro*. Testing of many mutants in this way revealed the existence of many different steps in assembly, and the nature of the components accumulating in mutant extracts indicated the nature of many of these processes. These studies also indicate that viral structural proteins are not synthesised sequentially but simultaneously. The accumulated proteins then combine together in an orderly series of reactions in which the product of one reaction serves as substrate for the next. Most of these reactions are, however, probably not enzymically catalysed but are true self-assembly processes. There is some evidence that host factors such as membrane structures have a role in this assembly.

3.9 Inhibition of cellular metabolism

Little has been said so far about the effect of viruses on cells. This varies very greatly both with the virus and with the cell type. At one extreme the T-even viruses cause an immediate inhibition of the synthesis of host macromolecules; although the mechanism by which these viruses kill cells is not known, it is thought to result from the initial disturbance of the bacterial cell membrane that occurs during the interaction of the phage with the bacterial surface. At the other extreme the RNA tumour viruses stimulate cell growth and division, although transformed cells may be producing infectious virus at all times. Many viruses kill cells only towards the end of the viral growth cycle, and this may just be an incidental effect resulting from disruption of cellular organisation caused by the accumulation of viral products. The most successful viruses are those that have little effect on the cells, thus providing a continuing environment for viral replication. Thus the filamentous DNA phages like fd are continually released from the surface of infected cells, whose growth rate is only slightly diminished by infection. The myxovirus SV5 grows to a high titre in monkey kidney cells, and the cells show no visible effect of infection: in contrast the same virus kills hamster cells with only a poor yield of infectious particles. This difference has been related to a difference in fragility of the host cell membranes.

4 Chemotherapy of viruses

One of the most urgent problems facing a virologist is the development of successful chemotherapeutic agents. Since viruses share many metabolic processes with the host cell it is difficult to find a selective inhibitor of viral biosynthesis. However recent discoveries of new enzymes synthesised in infected cells, and present in viral particles, raises hopes of new developments in this field. For example the reverse transcriptase present in RNA tumour virions is inhibited by derivatives of antibiotics such as rifamycin. A few compounds are now being used successfully to combat viral diseases. Analogues of nucleic acid bases, of which iododeoxyuridine is the most useful, are used to inhibit the growth of herpes viruses; the base analogue is incorporated into viral nucleic acid rendering it noninfectious. Since these compounds are also incorporated into host nucleic acid they are too toxic for internal administration. However viruses of the herpes group largely cause surface lesions (e.g. chicken pox, shingles, eye lesions) where the host cells are not dividing, and so the agents may be used for topical application. The other highly successful drug is isatin-β-thiosemicarbazide, which can be used to combat smallpox infections both as a prophylactic and as a curative treatment. This compound interferes with the assembly of pox viruses, thereby preventing formation of infectious virus.

306

Summary

Viruses are a unique type of organism. They possess only one type of nucleic acid; this may be single stranded or double stranded and either RNA or DNA. The nucleic acid is surrounded by protein which is assembled in a symmetrical arrangement. This nucleocapsid may be surrounded by an envelope, part of which contains host membrane material. Enzymes may be enclosed within the virion.

Bacterial viruses possess special mechanisms for entering the host cell, and the coat proteins remain outside. Viruses of eukaryotic cells enter either by a process similar to phagocytosis or by fusion of viral envelopes with cell membranes. Most of the viral protein enters the cell.

Since different systems for nucleic acid synthesis are present in infected cells, new enzymes to catalyse these processes are frequently required. These are either synthesised after infection or imported with the viral nucleic acid. Protein synthesis uses machinery already present in the cell, but this may be modified to enable viral mRNA to be translated more efficiently.

The growth of many viruses results in a modification of cell membranes. These may play some role in virus synthesis and assembly.

Bibliography

Books for general reading

COHEN, S. S. (1968) *Virus-Induced Enzymes*, New York, Columbia University Press.

CRAWFORD, L. V. and STOKER, M. G. P. (eds.) (1968) *The Molecular Biology of Viruses,* 18th Symposium of Society for General Microbiology.

FRAENKEL-CONRAT, H. (ed.) (1968) *The Molecular Basis of Virology,* Reinhold, New York.

HERSHEY, A. D. (ed.) (1971) *The Bacteriophage* λ, Cold Spring Harbor Conference.

LEVY, H. B. (ed.) (1969) *Biochemistry of Viruses*, Marcel Dekker, New York and London.

LURIA, S. E. and DARNELL, J. E. (1968) *General Virology,* 2nd edn. John Wiley, New York.

WOLSTENHOLME, G. E. W. and O'CONNOR, M. (eds.) (1971) *Strategy of the Viral Genome,* CIBA Foundation Symposium, Churchill Livingstone, Edinburgh and London.

Review articles dealing with special aspects

BALTIMORE, D. (1971) 'Expression of animal virus genomes'. *Bact. Rev.* **35**, 235 – 41.

CASPAR, D. L. D. and KLUG, A. (1962) 'Physical principles in the construction of regular viruses'. *Cold Spring Harb. Symp. Quant. Biol.,* **27**, 1.

DALES, S. (1965) 'Penetration of animal viruses into cells'. *Prog. Med. Virol.* **7**, 1.

EISERLING, F. A. and DICKSON, R. C. (1972) 'Assembly of viruses'. *Ann. Rev. Biochem.*

LWOFF, A. (1957) 'The concept of virus'. *J. Gen. Microbiol.,* **17**, 239 – 53.

NEWTON, A. A. (1970) 'The requirements of a virus', in *20th Symposium of Society for General Microbiology,* pp. 323 – 58.

SUGIYAMA, T., KORANT, B. D. and LONBERG-HOLM, K. K. (1972) 'RNA virus gene expression and its control'. *Ann. Rev. Microbiol.* **26**, 467 – 502.

TEMIN, H. M. (1971) 'Mechanism of cell transformation by RNA tumour viruses'. *Ann. Rev. Microbiol.,* **25**, 609 – 48.

9
Polysaccharide Structure and Function

Peter J. Winterburn

Department of Biochemistry, University College, Cardiff

Introduction

The major advances in the chemistry and biochemistry of nucleic acids and proteins over the past two decades pushed the polysaccharides out of the limelight. However, especially over the last five years, the technology required for investigating polysaccharides has been refined and this is now leading to a greater understanding of the relationship between the properties of these ubiquitous natural polymers and their chemical formulae.

There is a great structural diversity within the polysaccharides which is reflected in the variety of functions they perform. An important aspect of the roles played by polysaccharides concerns their location. Apart from the storage polysaccharides virtually all of these carbohydrate polymers are synthesised intracellularly to perform their function in the extracellular environment. Thus the well known fibrous types such as cellulose are embedded in an extracellular gel formed by other polysaccharides. The gels demonstrate another fascinating property of these polymers because they are able to modify the external environment of the cell. For example, 0·1 per cent w/v agarose in water forms a gel which effectively 'solidifies' the second component of the system, the 99·9 per cent water. Such a gel creates a hydroskeleton which augments the tensional resistance provided by the fibrous polysaccharides. These polymers are implicated in other extracellular functions such as recognition phenomena (e.g. the antigenic capsular polysaccharides of bacteria, see Chapter 10) and they also have protective roles (e.g. the gum exudates of plants). The intracellular polysaccharides are storage reserves, examples of these being amylose, amylopectin and glycogen.

Many polysaccharides have had a commercial importance for centuries and subtle aspects of their properties were known long before the advent of molecular predictions. Those of prime importance are cellulose for fibre and paper manufacture, and pectins and alginates which are used extensively in the food and other industries as gelling agents and emulsifiers. This chapter is an attempt to relate composition, conformation and interactions which hopefully may lead the topic away from a recipe of ingredients towards an understanding of the biological systems.

1 Conformations and nomenclature

1.1 Monosaccharides

The majority of carbohydrates exist in solution in a ring form which is either five-membered (furanose) or six-membered (pyranose). The configuration of the hydroxyl group attached to the asymmetric carbon atom farthest from the aldehyde or ketone function in the straight chain representation dictates whether a sugar is classified as a member of the D or L series. Thus, for a hexose such as glucose this is C_5 while for a pentose it is C_4. On formation of a

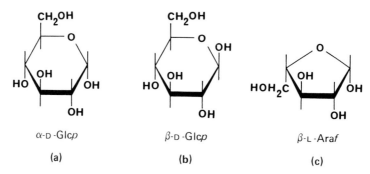

α-D -Glc*p*   β-D -Glc*p*   β-L -Ara*f*

(a)   (b)   (c)

Fig. 9.1 Haworth structural formulae of (a) α-D-glucopyranose, (b) β-D-glucopyranose and (c) β-L-arabinofuranose.

pyranose ring another asymmetric carbon atom is created at C_1; this is termed the anomeric carbon atom and the orientation of the attached hydroxyl is designated by α or β. The formulae of α-D-glucopyranose, β-D-glucopyranose and β-L-arabinofuranose are shown in Fig. 9.1. Note that the ring hydrogens are not shown and that in this figure the shorthand notation has been used to describe the sugars where *p* = pyranose and *f* = furanose.

These planar representations (Haworth structures) bear little resemblance to the true shape of the molecule since they take no account of the normal valence angles. To be free from strain the ring adopts non-planar conformations of which there are three types – chair, boat and skew. The non-bonded interactions in the latter two, for instance repulsion of prow and stern groups in the boat form, render them energetically unfavourable and they occur only as possible reaction intermediates or covalently locked bicyclic derivatives. The chair forms may be considered as derivatives of cyclohexane (Fig. 9.2(a)) in which one $-CH_2-$ group has been replaced by the geometrically similar oxygen atom and substitutions effected at various other positions around the ring. For example, α-D-glucopyranose (Fig. 9.2(b)) can be formed from the cyclohexane structure shown in Fig. 9.2(a), if C_6 is replaced by $-O-$, the hydrogens at the positions labelled A, B, C and D replaced by hydroxyl groups and $-CH_2OH$ substituted at E. Cyclohexane can exist in an alternative chair form (Fig. 9.2(c)) which normally is indistinguishable from the first since all $-CH_2-$ groups are equivalent. However, these positions become non-equivalent when either a ring atom or a substituent is changed. Also the presence of the substituents affects the free energies of the two forms and therefore influences the rapid interconversion which exists between the forms of cyclohexane. By performing the substitutions described above but on the structure Fig. 9.2(c) a form of α-D-glucopyranose (Fig. 9.2(d)) is generated in which the relationship of the groups to the plane of the ring has changed. To distinguish between the two chair conformations the Reeves convention is adopted – that shown in Fig. 9.2(b) is termed C1 while that in Fig. 9.2(d) is 1C. Inspection of the two structures reveals that for each ring carbon atom one of the two bonds carrying substituents is approximately in the plane of the ring while the other is almost perpendicular to this plane.

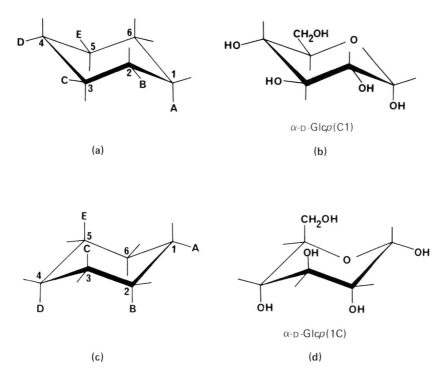

Fig. 9.2 Chair forms of carbohydrates. Two different chair conformations of cyclohexane (a and c) via the substitutions detailed in the text lead to (b) α-D-glucopyranose (C1) and (d) α-D-glucopyranose (1C).

This leads to another method of describing the configuration of the substituents: those in the plane of the ring are equatorial (e) while those perpendicular are axial (a). Thus, using α-D-glucopyranose (C1) as in Fig. 9.2(b), the hydroxyls at C_{1-4} are a, e, e and e while in the 1C form (Fig. 9.2(d)) they are e, a, a and a, respectively. This demonstrates another principle, namely that a change between C1 and 1C changes a substituent from either axial to equatorial or *vice versa*.

The C1 and 1C forms are in equilibrium in aqueous solution although in the case of D-glucose, as for most common monosaccharides, the equilibrium favours the C1 species. The factors which affect the equilibrium are mainly steric. Substituents (particularly if they are bulky) contribute a lower free energy to the structure in equatorial than in axial positions. The steric repulsion of two axial substituents on the same side of the ring (for example, in Fig. 9.2(d), the C_3–OH and –CH_2OH on C_5) would distort the ring and raise the free energy. There is also an electrostatic interaction between the C_1–O_1 and C_5–O_5 bonds termed the anomeric effect. These rules have been quantitated (Durette and Horton 1971) and enable the carbohydrate chemist to calculate the total free energy of various possible forms and hence to predict for each monosaccharide which conformation is the most stable. It has been calculated that the free energy change between the C1 and 1C forms of β-D-glucopyranose is 6 kcal/mole in favour of the C1 species. Therefore, in solution the percentage of 1C will be small. For other sugars the free energy difference may be much smaller and appreciable amounts of the 1C form will be present.

1.2 Oligosaccharides and polysaccharides

The glycosidic linkage is almost invariably formed between the anomeric carbon atom C_1 of one monosaccharide and one of the several hydroxyl groups available on another residue. For a hexopyranose there are five such groups available, thus there are $1 \rightarrow 1$, $1 \rightarrow 2$, $1 \rightarrow 3$, $1 \rightarrow 4$ or $1 \rightarrow 6$ linkages of which the commonest are the latter three. Consideration of a generalised hexopyranose in C1 conformation reveals that the number of possibilities is greater than this since the hydroxyl groups at C_{1-4} and the $-CH_2OH$ group at C_5 may be either axial or

D-Glc$p$-(1e → 4e)-D-Glc$p$

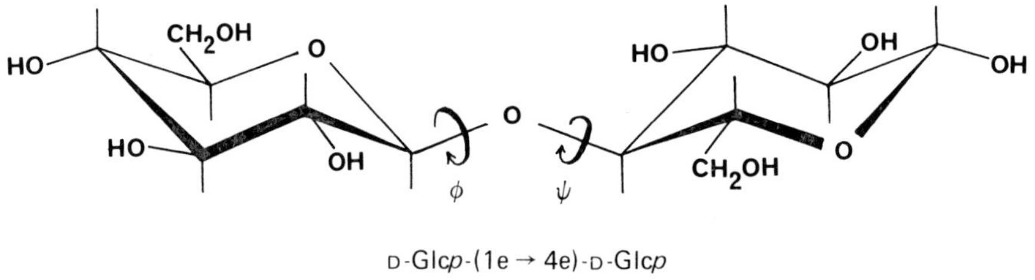

Fig. 9.3 β-D-Glucopyranosyl-(1 → 4)-D-glucopyranose, trivially known as cellobiose.

equatorial. Therefore, there are twenty combinations of two hexopyranoses (C1) in forming a disaccharide when the anomeric carbon of one unit is engaged in the linkage. Typical is cellobiose (Fig. 9.3) which is a β(1 → 4) or (1e → 4e) linked disaccharide of glucose derived from acid hydrolysates of cellulose. Note that in the figure the right-hand glucose unit has been inverted for reasons which will be described later.

Without considering such modifications as methylation, sulphation or acetylation, there are about twenty different monosaccharides used in polysaccharides. Only a limited number of these are present in any one polymer, usually one or two but up to a maximum of about six. Another feature characteristic of carbohydrate polymers is branching. This is possible because each of the several hydroxyl groups on a monosaccharide can accept another unit. The properties and hence biological function of a polysaccharide are derived by selection of the appropriate units, their sequence and the linkages between them.

The homoglycans are polysaccharides which contain only one type of monosaccharide and these may be either linear or branched. They are named according to the type of sugar; for example a homoglycan of entirely glucose units is a glucan while one of galacturonic acid residues is a galacturonan. The exceptions within this system are those that have traditional names, for example cellulose and chitin. Those polysaccharides composed of two or more different units are the heteroglycans. A branched glycan with a main chain or backbone of xylose units to which are attached arabinose side chains is an arabinoxylan while a linear heteroglycan containing both glucose and mannose units in the same chain with the latter type predominating is termed a glucomannan. Many of the linear heteroglycans possess regular repeating units, for example the disaccharide unit of hyaluronic acid, although these regular regions may be interrupted occasionally by the insertion of a different type of residue.

2 Methods for structural investigation

To a large extent the advance of a subject is related to the development of its techniques. This is true of polysaccharide chemistry which has benefited from the improvements in chromatographic procedures used in the purification of the polymers and the characterisation of monosaccharides and their derivatives. X-ray diffraction and spectroscopic techniques such as NMR, ORD and CD have been powerful adjuncts to the more conventional physico-chemical techniques in the investigation of the shape and properties of polysaccharides. This second section is devoted to an outline description of the methods employed in structural studies.

2.1 Primary structure

A well-characterised covalent structure is essential for worthwhile interpretation of the structure of the polymer in three dimensions. The chemical procedures used in determining the monosaccharide sequence and linkages (primary structure) of glycans are described in the book by Aspinall (1970) and are not duplicated in this article. The techniques such as partial acid hydrolysis, methylation or periodate oxidation are used individually or in combination to degrade the polysaccharide. After careful isolation and identification of the fragments it is often possible to reconstruct a picture of the original polymer.

2.2 Higher structural orders

2.2.1 X-ray diffraction

Some polysaccharides occur in a highly crystalline fibrous form while others can be induced to form fibres or oriented films. The oriented films are produced by drying a concentrated solution of the polysaccharide on a glass slide. On stretching the dried film the linear chain molecules orient themselves along the strain lines improving the alignment and packing in the fibre. The procedure as employed for the preparation of hyaluronate fibres is described in greater detail by Atkins and Sheehan (1973). Within the finished fibre there are local areas of regular structure or crystallites separated by amorphous regions where the chains have not packed efficiently. These crystallites have one axis in common which is parallel to the fibre axis but the other two axes are randomly oriented perpendicular to the fibre axis. A collimated monochromatic X-ray beam passing through such a fibre is scattered as though the structure is a fibre of a single crystal rotating about the fibre axis. The result of the destructive interference or reinforcement of these rays is recorded on a photographic film as a series of spots arranged along the layer lines (Fig. 9.4(a)). In practice the crystallites are not accurately aligned; therefore arcs are obtained instead of sharp spots. The amorphous regions between the crystallites cause random scattering which gives rise to a diffuse darkening on the photographic film. An example of an X-ray photograph is shown in Fig. 9.4(b).

The diffraction of X-rays by polysaccharide fibres has been described by Marchessault and Sarko (1967); therefore only the points relevant to the purposes of this chapter will be mentioned. The first task is to calculate the dimensions of the unit cell, which is defined as the smallest portion of the structure that by continuous repetition yields the entire structure. The linear glycan chains within the crystallites are packed in a regular array across the fibre and so scatter X-rays perpendicular to the fibre axis. These are termed equatorial reflections (Fig. 9.4(a)). Applying the law of reciprocal space, the dimensions of the base of the unit cell are deduced from the spacings of the equatorial reflections and hence the interchain distances are found.

Vertical displacements, termed meridional reflections, are caused by repeat structures along the chain axis, for example the rise per residue, while regularities along a helix axis, such as rise

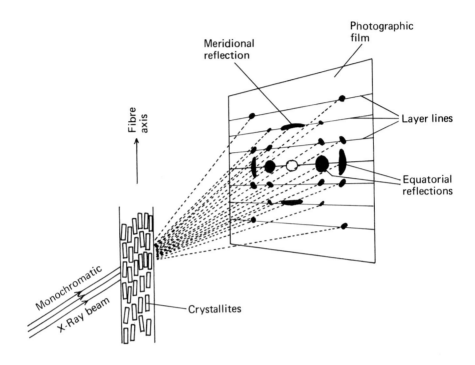

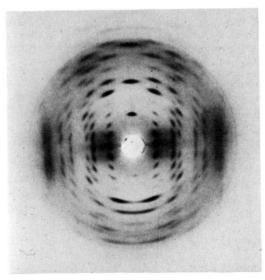

Fig. 9.4 (a) X-Ray diffraction by polysaccharide fibres.
(b) X-Ray fibre diffraction photograph of the sodium salt of chondroitin 4-sulphate (kindly supplied by Dr E. D. T. Atkins).

per turn, are revealed as diagonal displacements (Fig. 9.5). (Note that, as shown in the figure, the chain and helix axes need not be identical.) If there are n identical units per turn, where n is an integer, then the first meridional reflection occurs on the nth layer line and this is the unit cell periodicity along the fibre axis, otherwise known as the fibre repeat. This parameter is

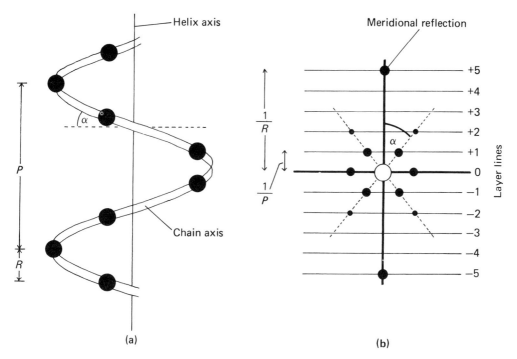

Fig. 9.5 (a) A five-fold helix. This is characterised by a pitch (P) also called the rise per turn or fibre repeat, the angle α and the rise per repeat unit in the primary structure (R).
(b) The fibre diffraction pattern obtained from the helix shown in (a); only a few of the spots are shown. The fibre repeat and rise per repeat unit are calculated from the layer line and meridional reflection spacings.

particularly sensitive to the conformation of the monosaccharide units and linkages. Thus the fibre repeat distance, coupled with the limitations imposed by the nature of the primary structure, yields information on the polysaccharide conformation.

Fibres aligned in one dimension generate far fewer reflections than a crystal of a low molecular weight carbohydrate. Consequently a Fourier synthesis using the spot intensities to deduce the spatial organisation of the components of the unit cell is not possible. However, the intensities may be used in an inverse sense. Starting with a hypothetical model of the structure the spot positions and intensities are predicted; comparison with the experimental observations verifies or repudiates the model. This could be an extremely time-consuming hit or miss process unless the model is a sterically reasonable one. The building of these models is described in the next section.

2.2.2 Conformational analysis

The manual construction of a molecular model is tedious even when the particular conformation is known. To extend this process to polysaccharides where initially it would appear that the number of possible conformations is very large would be virtually impossible.

314

Computers are ideal for tackling this type of problem and calculate which structures are sterically feasible and which conformation is energetically the most favourable. The accuracy is naturally dependent on the assumptions within the programme used. Assuming that the pyranose ring is virtually rigid and that the glycosidic angle can be varied by only a few degrees from 117°, the shape of the polysaccharide backbone is dependent on the rotation around the two bonds attached to the glycosidic oxygen (the angles marked ϕ and ψ in Fig. 9.3). The 1,6-linked polymers are more complex because there is the additional rotation, ω, of the

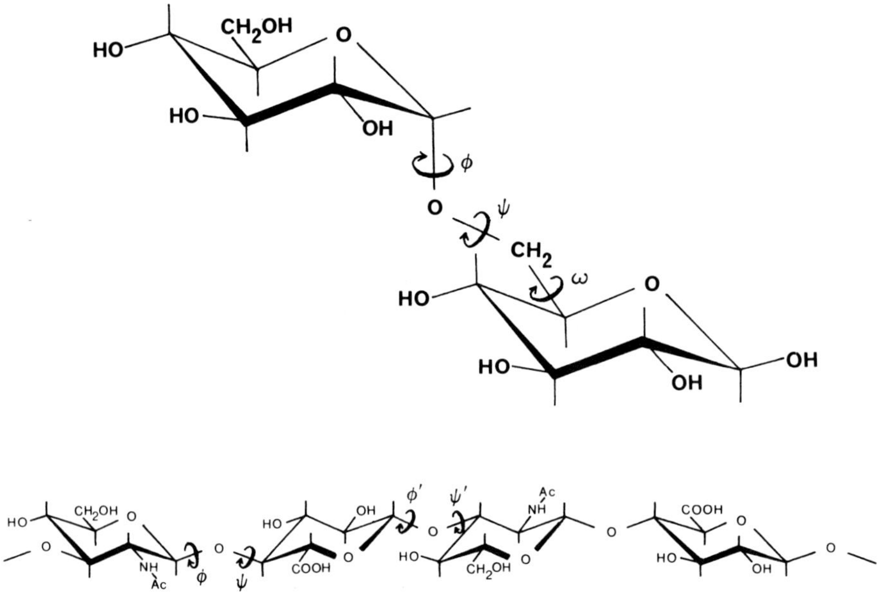

Fig. 9.6 (a) α-D-Glucopyranosyl-(1 → 6)-D-glucopyranose, or isomaltose, showing the three torsion angles in the inter-residue linkage.
(b) Hyaluronic acid, an example of a repeating disaccharide polymer, showing the four torsion angles which control the conformation of the chain.

C_5–C_6 bond to be considered (Fig. 9.6(a)). A polysaccharide based on a repeating disaccharide unit has two glycosidic linkages around which rotation can occur and hence four torsion angles (Fig. 9.6(b)). The bond angles and lengths from crystallographic data on simple sugars are used to calculate the three-dimensional coordinates of the atoms in each monosaccharide. The computer uses these coordinates and the glycosidic angle to calculate the new coordinates when the torsion angles ϕ and ψ are varied. On the basis of these coordinates and the van der Waals atomic radii the values of ϕ and ψ which yield permissible conformations are determined. Application of this procedure to cellobiose using 10° angular increments is shown as a map in Fig. 9.7. Only twenty-four of the 1 296 conformations are free of steric clash and twenty-one more have only slight compression of atoms, i.e. 96 per cent of the conformations are excluded. The values of ϕ and ψ determined by X-ray crystallography on cellobiose lie within the allowed region of the map. More refined treatments calculate interaction energies between all the atoms taking into account van der Waals attraction and repulsion, electrostatic forces and hydrogen bonding. The sum of these forces is presented as an energy contour map from which is seen the position of minimum potential energy. For cellobiose the predicted values of ϕ and ψ were within 2° of those determined experimentally — an indication of the precision that can be achieved. When the probable torsion angles of the basic repeat unit have been determined the entire polymer is constructed and used to calculate relevant dimensions such as pitch of helix, number of residues per turn, etc.

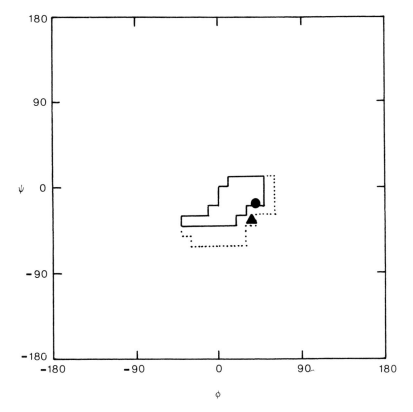

Fig. 9.7 Computer analysis of permitted values of ϕ and ψ (see Fig. 9.3) for the (1e → 4e) linkage of cellobiose. The combinations which yield structures free of steric clash lie within the region bounded by the solid line and those having slight atomic compression within the dotted line. The experimentally determined conformations of cellobiose and cellulose have ϕ, ψ values represented by ● and ▲ respectively.

The nature and configuration of the substituents on the carbon atoms adjacent to the linkage dictate the values of ϕ and ψ and the limits of their rotation. These parameters decide the shape of the polymer and its flexibility, respectively. If the area of allowed rotations is large the polymer is flexible because it assumes many conformations of equal free energy; this is observed especially in the 1,6-linked glycans where the greater separation of the rings leads to fewer steric clashes. Restricted rotation, which is revealed as a small area, means that the chain is stiff, for example the cellobiose results are consistent with the rigidity of cellulose chains, and that there is a constant geometrical relationship of adjacent residues giving a symmetrical polymer.

At present conformational analysis considers the polysaccharide in isolation and makes no allowance for solvation. Therefore the information has to be correlated with the properties of the hydrated polymer as revealed by other techniques.

2.2.3 Spectroscopic techniques

The infra-red region of the electromagnetic spectrum encompasses many of the bond frequencies pertinent to the study of polysaccharides. Stretching and bending frequencies of bonds are sensitive to their environment and their relationship with other atoms. This provides information on anomeric configurations, positions of ester groups and whether a substituent is axial or equatorial. The absorption maximum of a primary hydroxyl group is at a frequency of $3\,642$ cm^{-1} while that of a secondary hydroxyl is lower ($3\,629$ cm^{-1}). When hydroxyls participate in hydrogen bond formation the stretching frequency is decreased to a value of about $3\,500$ cm^{-1} and the magnitude of this displacement is related to the strength of the bond. Unfortunately this useful information is obscured by the strong absorption by water in this region of the spectrum. This problem may be overcome by exposing the sample to D_2O vapour or working in D_2O solutions. The hydroxyl groups involved in solvent interaction exchange with the deuterium atoms while those shielded from the solvent by inter- or intramolecular hydrogen bonding do not exchange. The stretching frequency of the O–D bond ($2\,630$ cm^{-1}) and the solvent absorption ($2\,500$ cm^{-1}) are displaced so permitting the identification of the hydrogen bond which remains at about $3\,500$ cm^{-1}. The maximum absorption of a stretching frequency is observed when the radiation is polarised parallel to the bond direction. Therefore if the sample is crystalline or otherwise oriented the alignment of a particular bond with respect to a crystal axis can be deduced from the angular dependence of absorption.

Although the study of specific optical rotation as a function of wavelength yields useful information on polymer conformation, its application to the study of polysaccharides is limited because the only chromophore in the accessible wavelength region is the carboxylate function present in the uronic acid-containing glycans. Nevertheless the optical rotation at a single wavelength is also dependent on the ring conformation and substituents. Rees (1972) has discussed the relationship between the optical rotation and the torsion angles, ϕ and ψ, of the glycosidic linkage and was able to predict quite accurately the specific rotations of many di- and oligosaccharides. The ability to relate specific rotation to conformation is extremely important because, unlike most other techniques, measurement of optical rotation yields information on the shape of the polysaccharide in solution. The values of the glycosidic torsion angles derived from the optical rotation are correlated with those obtained for the solid state by X-ray or conformational analysis. A good agreement would suggest that the polymer assumes similar shapes in both states.

In this chapter the approach adopted for the discussion of the various structures is a classification firstly by function with subsequent sub-divisions according to the composition of the main chain. However, whatever the basis of classification, divisions are artificial because in the natural context the polysaccharides occur within complex mixtures evolved so that a particular function is provided by the combined properties.

3 Fibrous polysaccharides

Generally the fibrous polysaccharides are composed of bundles of efficiently packed ribbon-like linear glycan chains. The close packing essential to prevent solvent penetration and retain insolubility would be disturbed by branching although the occasional short side chain can be accommodated without distortion. The greatest strength of the structure is provided along the fibre axis by the covalent linkages within the chains, whereas the weaker van der Waals forces and hydrogen bonds hold adjacent chains together. In plant cell walls where these fibrous polysaccharides are of commonest occurrence the fibrils are cemented together with gel-forming polysaccharides (hemicelluloses and pectins) to form a material of high tensile strength akin to a carbon-fibre reinforced resin. By adjusting the orientation of these fibrils during development, the cell can adapt its framework to provide flexibility during primary growth or rigidity during secondary growth (see Chapter 11).

3.1 Cellulose

Cellulose, a $\beta(1 \rightarrow 4)$ glucan of up to 10 000 glucose units, is the principal cell wall component of land plants and many algae. It accounts for 50 per cent of the carbon on this planet. The first X-ray analysis of the virtually crystalline fibres was performed in 1937 by Meyer and Misch and led to calculation of the unit cell dimensions and the packing arrangement of the chains.

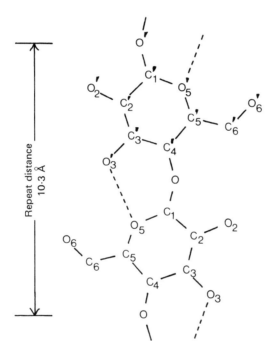

Fig. 9.8 Conformation of the cellulose chain showing the O'_3–O_5 intramolecular hydrogen bond between adjacent glucose residues.

The repeat distance derived from the first layer line spacing is 10·3 Å and because the first meridional reflection is on the second layer line the helix has two glucose units per turn. The 180° rotation between adjacent residues (Fig. 9.3) characteristic of a two-fold helix is confirmed by conformational analysis of cellobiose (section 2.2.2). The disaccharide repeat of 10·3 Å is the maximum extension of a (1e → 4e) linked disaccharide. Therefore the helix must be very narrow with the fibre axis passing through the glucose units creating a ribbon structure. In addition to the van der Waals repulsions limiting the rotation between residues, extra stability is provided by intramolecular hydrogen bonding parallel to the chain axis between O'_3 and O_5 of

318

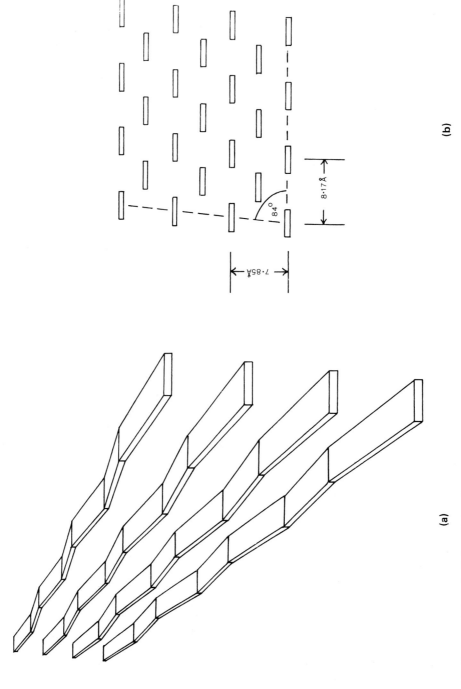

Fig. 9.9 The association of cellulose chains. Several ribbon-like chains form a vertical pile (a) which associates laterally with other stacks creating a micro-fibril. The arrangement is shown in cross-section in (b).

(a)

(b)

7·85Å

8·17Å

84°

the next residue (Fig. 9.8). The presence of this bond and its orientation have been confirmed by polarised infra-red spectroscopy. All the substituents of this glucan are equatorial which creates apolar surfaces of H_{1-5} and O_5 above and below the plane of the chain. The hydrophobic surfaces of the cellulose chains associate with those of others to form a stack of parallel chains (Fig. 9.9(a)), reminiscent of the way in which the folding of a polypeptide minimises the aqueous interaction of its hydrophobic groups. The stacks associate to form microfibrils by hydrophobic contacts and intermolecular hydrogen bonding between O_6 and O'_1 (the glycosidic oxygen joining two glucose units in the adjacent pile of chains). This packing arrangement of parallel chains is seen in cross-section in Fig. 9.9(b). Native microfibrils have a minimum diameter of 35 Å corresponding to about thirty-six chains although fibrils up to 100 Å diameter have been reported. Analysis of reconstituted cellulose, i.e. cellulose fibres obtained by precipitation from solution, reveals that the adjacent stacks pack in an antiparallel fashion. Such differences in the direction of packing are encountered frequently in polymer chemistry and reflect the relationship between the polymerisation and crystallisation stages of fibril formation. The crystallisation of completed polymers from solution is initiated at nucleation sites where either two chains come into contact (intermolecular) or two regions of the same chain associate (intramolecular). Generally the latter situation predominates creating chain folding. However if crystallisation immediately follows polymerisation the free segments of the chains are too short for intramolecular associations to develop before they are packed into the fibril. Thus the directional control during biosynthesis produces the parallel arrangement in the native fibres (see also Chapter 11). For a fuller discussion of the theories on fibril structure consult Shafizadeh and McGinnis (1971).

3.2 Chitin and peptidoglycan

Chitin which is found in bacteria, fungi and invertebrates serves a role similar to that of cellulose in plants. It closely resembles cellulose in structure and properties but is even more insoluble, inert and rigid. Structurally chitin is a (1e → 4e) linked glucan but with an acetamido group in place of the equatorial C_2 hydroxyl. This bulky substituent further limits the rotation around the glycosidic linkage so making the chain stiffer than cellulose. The X-ray pattern of a deproteinised sample of chitin is similar to that of cellulose and shows that the chains have the same two-fold screw axis and a disaccharide repeat of 10·4 Å. Likewise the flexibility is further reduced by the O'_3–O_5 intramolecular hydrogen bond between neighbouring N-acetylglucosamine residues. As with cellulose the chains stack in parallel fashion on top of each other using hydrophobic contacts, but these are supplemented by interchain hydrogen bonding via the amide groups. This bonding, which has been confirmed using polarised infra-red spectroscopy, is shown in Fig. 9.10 where the chains are drawn perpendicular to the page. In nature the stacks of chains may be packed either antiparallel (α-chitin) or parallel (β-chitin) as in reconstituted and native cellulose respectively. The α-form is believed to result from chain folding and gives rise to chitin sheets while the β-form is fibrous.

Peptidoglycan like chitin is a β(1 → 4) polymer of N-acetylglucosamine but alternate residues carry lactyl groups on O_3 (see also Chapter 10). Individually the chains probably have a rigidity equivalent to chitin and a similar conformation. The main difference is in the association of the chains. Whereas cellulose and chitin utilise lattice forces (van der Waals and hydrogen bonding), the adjacent chains of peptidoglycan are covalently linked via short peptides attached to the lactyl groups. The network so formed possesses considerable strength along the main chain axis but permits expansion during growth until the peptides are maximally extended.

320

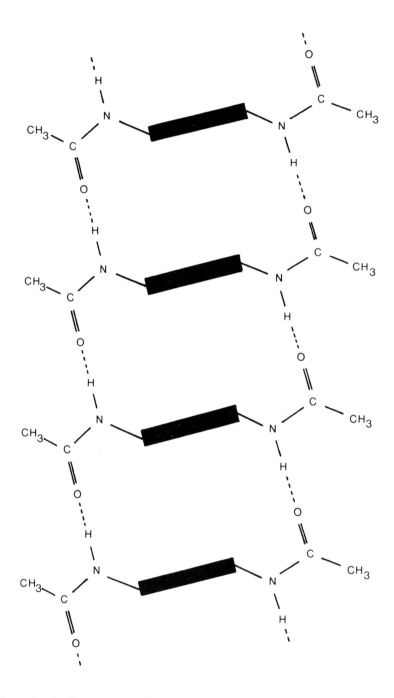

Fig. 9.10 Hydrogen bonding between acetamido groups within a stack of chitin chains. The chain axes shown as black bars are perpendicular to the page.

3.3 Mannans

The mannans are members of the group of plant polysaccharides termed hemicelluloses because unlike cellulose they can be extracted from plant cell walls with alkali. Pure $\beta(1 \to 4)$ mannans are rare but there are many heteroglycans, especially in the gymnosperms, which utilise the simple linear mannan backbone to produce polymers which can replace or associate with cellulose. Pure mannan as obtained from the ivory nut is a slightly more flexible form of cellulose. This is borne out by the X-ray and conformational analyses. The mannan appears as the typical ribbon-like polymer but with less restriction on the rotation of the $(1e \to 4e)$ linkage because the equatorial group at C_2 is the small hydrogen atom. This conformational similarity of mannan to cellulose enables the chains to pack together and possibly accounts for the close association of these polymers in the plant cell wall. However the packing of mannans in fibres is less efficient than that of cellulose because of the decreased rigidity and this leads to the slight increase in solubility. This feature is enhanced in the galactomannans of leguminous seeds such as clover and locust bean. Single D-galactopyranose units are linked $\alpha(1 \to 6)$ to the mannan backbone and further hinder the association of chains. In fact, as the degree of substitution with galactose residues increases, the function of the galactomannans probably changes from structural to that of a reserve carbohydrate.

3.4 Xylans

There are two types of xylan characterised by the nature of the interunit linkage. Firstly the $\beta(1 \to 4)$ xylan will be considered in order to examine further the influence of the substituents on the conformation of $(1e \to 4e)$ glycans and this will then be contrasted with the $\beta(1 \to 3)$ xylan, a $(1e \to 3e)$ linkage, to show the profound difference in shape induced by the change in linkage.

Like mannans the $\beta(1 \to 4)$ xylans are hemicelluloses often associated in plant cell walls with cellulose. They are particularly common in hardwoods during secondary growth (cell wall thickening) and in grasses where they constitute up to 30 per cent of the dry weight. These xylans usually occur in a modified form which may be due to partial acetylation at either O_2 or O_3, or substitution with single L-arabinose or 4-O-methyl-D-glucuronic acid residues. Generally the hardwood xylans have a lower frequency of side chains (one 4-O-methyl-D-glucuronic acid per ten xylose units) than the conifers. The xylan chain can be considered as cellulose devoid of the bulky equatorial $-CH_2OH$ groups at C_5. In comparison with the observations on mannans one would expect even greater flexibility about the glucosidic linkage. X-ray diffraction pictures of white birch xylan reveal a fibre repeat distance of 15 Å and meridional reflections on the third and sixth layer lines from which it is deduced that the helix is three-fold, i.e. three xylose residues per turn. The rise per residue along the helix axis in cellulose, chitin and mannan, which have two-fold helices, is $10\cdot4/2$ Å $= 5\cdot2$ Å. In this xylan, although the helix is three-fold the rise per residue is virtually the same, $15/3$ Å $= 5\cdot0$ Å. This is close to the maximum extension of the chain. Therefore as observed with the other $(1e \to 4e)$ glycans the chain and helix axes must be approximately the same. The minimum potential energy position by conformational analysis agrees with the X-ray data and favours a left-handed helix rather than a right-handed one, because the former can establish an O'_3-O_5 intramolecular hydrogen bond (Fig. 9.11). This has been verified spectroscopically.

The three-fold helix adopted by the $\beta(1 \to 4)$ xylan with the $120°$ rotation between residues converts the ribbon chain into a cylinder and distributes the two remaining hydroxyl groups more evenly around the helix than they are in cellulose, so increasing the solubility. As already

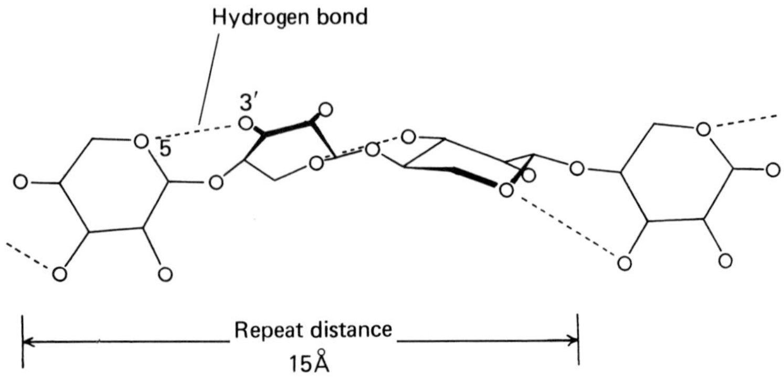

Fig. 9.11 Diagrammatic representation of the three-fold helix of the $(1e \rightarrow 4e)$ linked $\beta(1 \rightarrow 4)$ xylan showing the $O_3'-O_5$ hydrogen bond.

mentioned native xylans are partially acetylated. Nieduszynski and Marchessault (1972) studied artificially acetylated xylan and observed that the three-fold helix was converted to a two-fold helix typical of cellulose. Thus, the incorporation of large equatorial substituents at C_2 and/or C_3 restores the 180° rotation around the $(1e \rightarrow 4e)$ linkage. The possible biological implications are fascinating. The acetylated regions of the xylan possess a cellulose-like structure and could co-crystallise with the microfibrils, while the more soluble non-acetylated segments could either provide a hydrophilic coating for the cellulose or act as crosslinks between microfibrils.

For skeletal polysaccharides it is frequently observed that extensive branching hinders the establishment of the interchain associations which are so important for conferring strength. In cereal flour the xylan backbone carries numerous long arabinan side chains which abolish the structural characteristics of the main chain and convert the glycan into a reserve material.

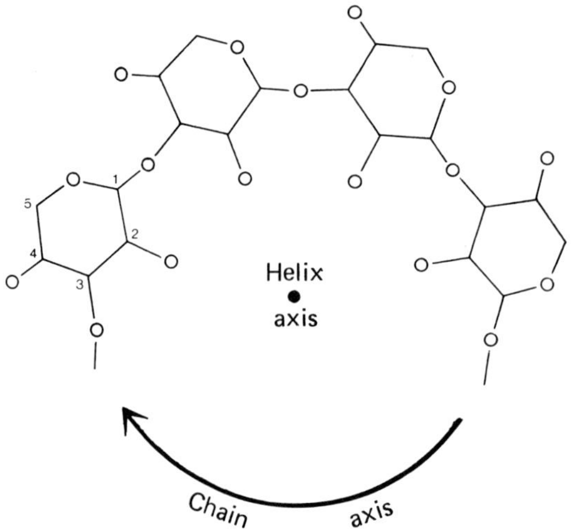

Fig. 9.12 The inter-residue relationship in a $(1e \rightarrow 3e)$ linked xylan. Note that there is no rotation between neighbouring residues, and that H_5 is positioned on the outside of the helix.

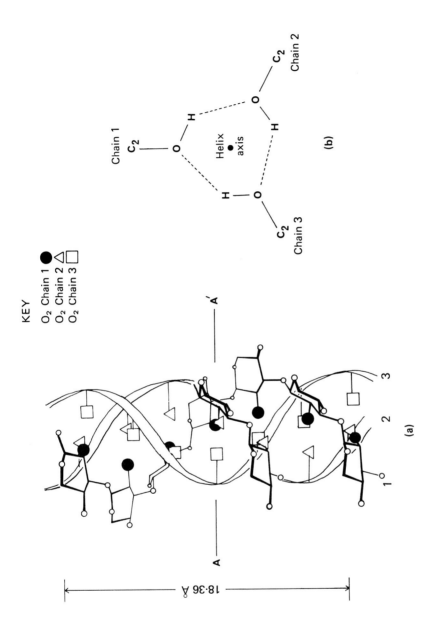

Fig. 9.13 (a) $\beta(1 \rightarrow 3)$ Xylan triple helix. Chain 1 is shown complete, chain 2 partially and chain 3 as a continuous ribbon. The positions of the C_2 hydroxyl groups are represented by symbols to emphasise their relationship. (Adapted from Atkins *et al.*, *Proc. Roy. Soc. B* (1969).)
(b) AA' cross-section of the triple helix in (a) showing the cyclic triad system of hydrogen bonds between the C_2 hydroxyl groups.

The skeletal polysaccharide of the green algae is the fibrous $\beta(1 \rightarrow 3)$ xylan. Unlike the $(1e \rightarrow 4e)$ linkage which creates an essentially linear chain, the $(1e \rightarrow 3e)$ type rotates the chain axis about the helix axis with no rotation between adjacent residues (Fig. 9.12). The open helix produced is like a wire spring. This polysaccharide was studied in detail by Atkins *et al.* (1969). They deduced that the open helix which is not compatible with the formation of a strong fibre achieves the necessary stability by intertwining with two other chains in a triple helix. Each strand of the right-handed helix has six xylose residues per turn in a fibre repeat of 18.36 Å. The structure is shown in Fig. 9.13(a) which is adapted from the original paper. The association of the three chains is stabilised by a cyclic triad system of hydrogen bonds formed from three C_2-hydroxyl groups, one donated from each chain. This can be seen by viewing the cross-section AA' of Fig. 9.13(a), as shown in Fig. 9.13(b). Examination of the xylan fibres by infra-red spectroscopy revealed an absorption at $3\,477$ cm^{-1} which was approximately perpendicular to the fibre axis and resistant to deuteration. Each strand has six residues per turn; therefore the helix is stabilised by a total of eighteen hydrogen bonds per turn. Because the triad system is buried in the centre of the helix and is shielded from the weakening influence of the aqueous medium, the fibres are extremely stable as befits their skeletal role.

4 Reserve polysaccharides

The differences in function between a reserve and a skeletal polysaccharide are reflected in certain features of their structures. These do not result from changing the nature of the component monosaccharide units but by selection for the position of interunit attachment and the linkage configuration. These two factors influence the steric relationship of adjacent residues and control the flexibility of the glycosidic linkage. Storage polysaccharides are generally more flexible than fibrous polymers due to less steric limitation on the torsional rotations. Further it may be significant that within this group the extremely flexible $1 \rightarrow 6$ linkage (section 2.2.2) is common. Since these chains have many conformations of similar free energy and equal probability they do not form the strong, regular, intermolecular associations characteristic of fibrous polymers. The weak interchain bonding can be readily disrupted to facilitate access to catabolic hydrolases which liberate the stored material. Another common feature of the reserve polysaccharides is extensive branching. The side chains which are long and themselves branched further hinder the packing of the chains into a regular structure and confer possible functional advantages of compactness and provision of more end groups for exo-enzymes.

4.1 Amylose

Starch, present in the cytoplasmic granules of many plant cells particularly seeds, is a mixture of predominantly two glucans, amylose and amylopectin. Although the proportions vary according to the source amylose is usually the minor component and represents 15–30 per cent of the total polysaccharide. Amylose is a linear $\alpha(1 \rightarrow 4)$ glucan of various chain lengths up to a maximum of about 3 000 residues. The analysis of the glucan $(1a \rightarrow 4e)$ linkage shows that its conformation is completely different from the $(1e \rightarrow 4e)$ seen in cellulose. Using maltose crystals and oriented films of amylose, crystallographic studies demonstrate that the inversion of alternate glucose units, a characteristic of cellulose, does not occur and in the crystal state the conformation is stabilised by an intramolecular hydrogen bond between O'_3 and O_2 (Fig. 9.14(a)). Computerised model building confirms this analysis of the $(1a \rightarrow 4e)$ linkage. When

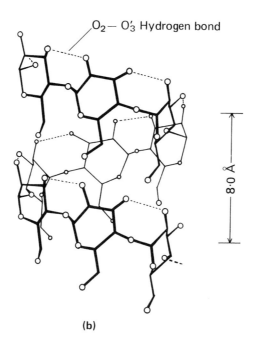

(a)

(b)

Fig. 9.14 (a) Conformation of maltose in the crystalline state.
(b) Six-fold helical structure proposed for amylose (V form).

amylose is precipitated from solution it assumes one of several related helical forms in which the torsion angles of the glycosidic linkage are close to those of its disaccharide, maltose. The units comprising these helices are oriented so that the hydrophilic hydroxyls project outwards and are solvated but the C-H point inwards creating a hydrophobic core (Fig. 9.14(b)). The information gathered by viscometry and light scattering measurements suggests that these helices break down in aqueous solution to give random coils. However, a small polar organic molecule such as 1-butanol can be inserted into the hydrophobic centre of the helix so stabilising it. Similarly the well-known blue iodine–amylose complex is a result of the helix accommodating one iodine molecule per turn. The number of glucose residues per turn depends on the size of the molecule fitted within the helix; thus the iodine complex has about six per turn whereas the larger 1-butanol molecule expands the helix to seven per turn. This adjustment in diameter demonstrates the flexibility of the wide helices characteristic of reserve polysaccharides. If the complexing agent is removed from an oriented amylose film or the humidity increased, the chains assume a different helical form. The fibre repeat is longer but there are still six residues per turn from which it follows that the diameter is decreased – further exemplifying the flexibility of the (1a → 4e) glucan chain. The X-ray diffraction patterns given by this form of amylose are similar to those obtained from tuber starch grains. It is tempting to speculate that in starch grains there is a requirement for the helical form to provide efficient packing but that solvation destroys these associations forming random coils suitable for rapid degradation.

4.2 Amylopectin and glycogen

On the basis of primary structure these two glucans are closely related. Amylopectin occurs in plants and glycogen is the animal kingdom equivalent. The $\alpha(1 \to 4)$ linked chains carry further chains by means of $\alpha(1 \to 6)$ branch points and these chains may in turn bear others to produce a compact multibranched structure. Generally the glycogens are more highly branched than the amylopectins, a factor that is revealed by a shorter average chain length between branch points – 19–26 glucose units in amylopectins and 12 in glycogens. However the main differences reside in the properties that reflect higher orders of structure. In solution the amylopectins have high intrinsic viscosities and behave as spherical molecules with molecular weights up to 10^6. The glycogens usually have higher molecular weights but are more compact because of the increased frequency of branching, and therefore exhibit lower viscosities. The two polysaccharides may also be distinguished by their reactivity to iodine.

Various branched structures have been proposed for amylopectin and glycogen, of which probably the best known is the tree-like model described by Meyer and Bernfeld in 1940 (Fig. 9.15(a)). This is compatible with most experimental evidence although recently Whelan (1971) has obtained evidence which he has interpreted in terms of a different structure for amylopectin, shown in Fig. 9.15(b).

These reserve polysaccharides enable the cell to store considerable quantities of simple carbohydrates in an inert but readily available form. Each glycogen molecule represents a high concentration of glucose yet it exerts an osmotic pressure which is only a minute fraction of that of its 30 000 glucose units if they were free. The numerous branches provide the sites for the attachment of the exo-enzymes and, whichever model is assumed, about 50 per cent of the glucose can be liberated without the enzymes' encountering branch points.

4.3 Other reserve polysaccharides

Dextran is a high molecular weight (10^7) microbial storage polysaccharide, well known as a starting material in the manufacture of some chromatographic media. It is an $\alpha(1 \to 6)$ glucan which has a small number of branches attached to either the 3 or 4 positions. In solution these polymers have indefinite shapes because of the high proportion of the very flexible 1,6 linkage.

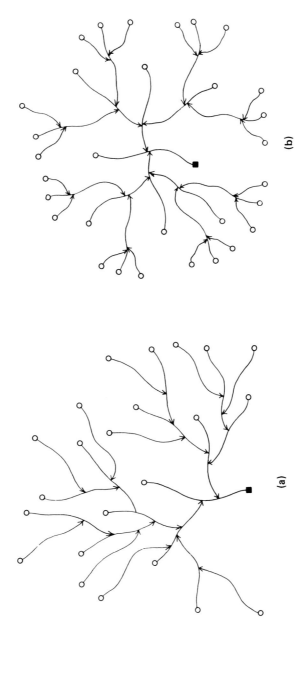

KEY

Free reducing end, C_1 ■
Non-reducing end, C_4 ○
$1 \rightarrow 6$ Branch points ↓

(a)

(b)

Fig. 9.15 Two models proposed for the structure of amylopectin according to (a) Meyer and Bernfeld, (b) Whelan.

Paramylon is a reserve glucan in which the residues are linked $\beta(1 \rightarrow 3)$. As seen with the xylans (section 3.4), the (1e → 3e) linkage creates open helical structures and since H_5 is on the outside of the xylan helix, the CH_2OH substituent in paramylon should not disrupt this flexible conformation. Many cereal grains possess a linear glucan which has both $\beta(1 \rightarrow 3)$ and $\beta(1 \rightarrow 4)$ linkages. This polysaccharide probably serves a dual function: the (1e → 4e) linked portions resembling cellulose may provide structural support during dormancy while on germination the polymer may act as a reserve polysaccharide by virtue of the helical (1e → 3e) regions.

5 Gelling polysaccharides

Certain polysaccharides occur in nature as hydrated viscoelastic gels. The different polymers within this group exhibit gradations in the viscoelastic properties which range from viscous, where the solution deforms slowly under stress and is therefore a good lubricant, to elastic when the gel is a rubbery material resistant to shock and which returns to its original shape when the stress is removed. The gel properties are produced by cross-linking the glycan chains into a continuous network through non-covalent bonds. The properties of a particular gel depend on the extent of the cross-linking – the greater the number of junctions between chains the more rigid the gel and hence the more elastic. Similarly, by producing weaker or fewer tie-points the viscous properties are enhanced. On a weight basis the polysaccharide may constitute <1 per cent of the gel, but this dominates the properties of the bulk of the material which is water. Therefore although both fibrous and gelling polysaccharides serve structural functions, the former utilise covalent forces along the fibre axis while the latter immobilise water and use the high compression resistance of the solvent in a hydroskeleton. This affinity for water also enables gels to be used as anti-desiccants.

5.1 Agars

The agars are a family of algal polysaccharides which range from a neutral species, agarose, to the highly charged, sulphated carrageenans. Although the carrageenans are used extensively in the food industry as emulsifiers and stabilisers their biological function in the intercellular matrix of red seaweeds is probably to prevent desiccation at low tide and provide mechanical support. The gels formed by agars melt when warmed at a temperature characteristic of the particular polysaccharide, and reset on cooling. The entire process is readily reversible and reproducible. The interesting aspect is that the melting and setting temperatures are not the same and may differ by as much as 50°C. Therefore the model for the gel structure must offer an explanation for this phenomenon as well as for network formation.

The primary sequence of the entire family is based on a repeating disaccharide unit of two galactose derivatives. These residues, labelled A and B, are linked:

$$-A-(1e \rightarrow 4e)-B-(1e \rightarrow 3e)-A-(1e \rightarrow 4e)-B-(1e \rightarrow 3e)-$$

Some of the AB combinations are listed in Table 9.1 and the conformations of κ-carrageenan and agarose are shown in Fig. 9.16. Normally a D-galactose would have an axial hydroxyl at C_4 because the C1 ring conformation is more stable than 1C. However the equatorial 4-position in the B residue is achieved by using either the 3,6-anhydro bridge to stabilise the 1C conformation, as in κ-carrageenan, or L-galactose fixed in a C1 conformation (L-galactose prefers 1C) by the same type of bridge. The two linkages present in this polymer have been discussed in earlier sections. The ribbon-generating (1e → 4e) alternates with the helix-forming (1e → 3e) to give an effect which is helical. Another feature of their primary structures which is

Table 9.1 Components of the repeating $-(A-B)_n-$ unit of some agars

A gar	A residue	B residue
κ-Carrageenan	D-Gal$p$ 4-sulphate (C1)	3,6-anhydro-D-Gal$p$ (1C)
ι-Carrageenan	D-Gal$p$ 4-sulphate (C1)	3,6-anhydro-D-Gal$p$ 2-sulphate (1C)
Agarose	D-Gal$p$ (C1)	3,6-anhydro-L-Gal$p$ (C1)

important in gel formation is the presence of occasional changes in the B residue. For example, in ι-carrageenan about every tenth B unit is replaced by D-galactose 2,6-disulphate or less frequently D-galactose 6-sulphate. These irregularities interfere with the preparation of good fibres for X-ray analysis. Their removal by treating the ι-carrageenan with sodium borohydride eliminates the 6-sulphate and forms the bicyclic 3,6-anhydro-D-galactose 2-sulphate, the normal B unit. The X-ray picture from these regularised chains shows the first meridional reflection on the third layer line consistent with three disaccharides per turn (Rees 1972). The density of the sample is twice that predicted for a unit cell containing single chains, leading to the proposal that the agars form double helices. Computer model-building for ι-carrageenan requires the systematic analysis of four torsion angles at $10°$ intervals, i.e. $1·68 \times 10^6$ conformations. Forty of these combinations give helical structures, of which only seventeen can accommodate a second helix wrapped around the first and only one, a right-handed double helix with parallel chains, is free from constraint. The double helix is stabilised by a hydrogen bond between O_6 of an A unit on one chain and O_2 of an A unit on the other chain. This interchain bond which is buried in the centre of the molecule is resistant to exchange in D_2O, and its orientation to the fibre axis is confirmed by polarised infra-red spectroscopy. The polar sulphate groups are attached to hydroxyl groups on the surface of the structure where they interact with the solvent. It is interesting that agarose which has the B residue in the L

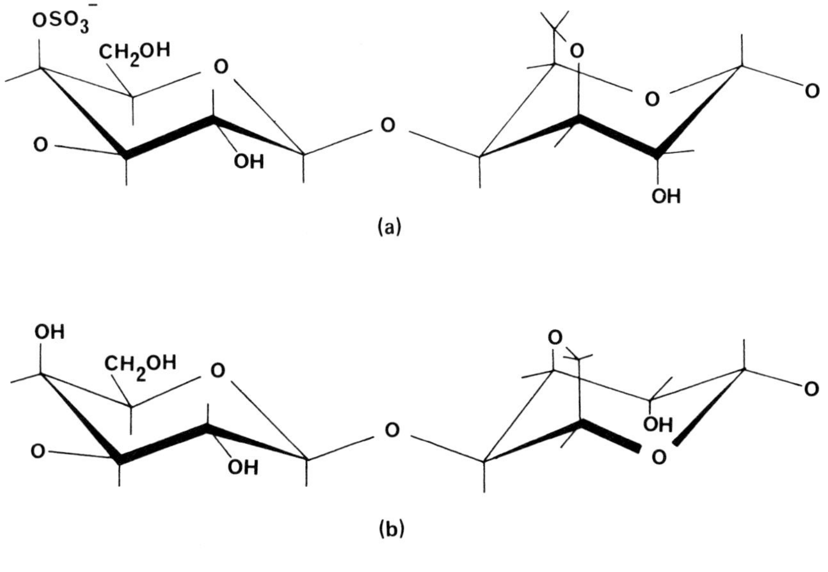

(a)

(b)

Fig. 9.16 Disaccharide repeat units of (a) κ-carrageenan and (b) agarose.

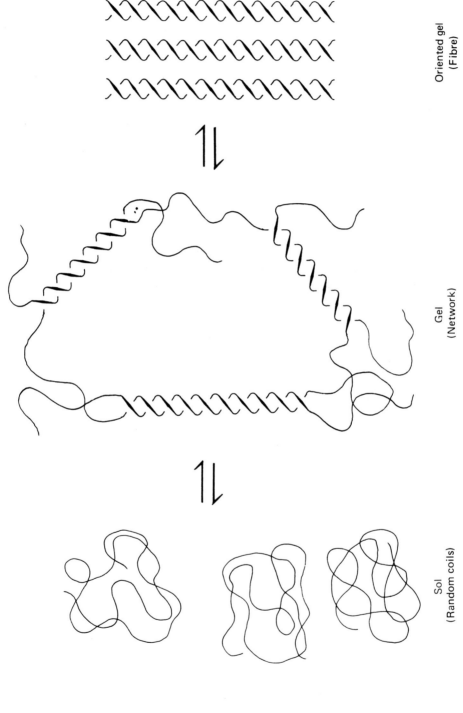

Oriented gel
(Fibre)

Gel
(Network)

Sol
(Random coils)

Fig. 9.17 Scheme for the gelling behaviour of agars. The randomly coiled chains of the sol state associate when the gel sets to create a network held by double helical tie-points. An oriented gel as used for X-ray fibre diffraction is produced by removing the kinking residues.

configuration is believed to form left-handed helices; the inversion from D to L sugar changes the screw sense of the helix.

Rees (1972) proposed that, on cooling, the random coils of the sol state intertwine in double helices which are the tie-points of the network (Fig. 9.17). Since these helical chains alone would form long junctions and not a network, he stressed the importance of the irregularities in the primary sequence which had been smoothed out in preparing 'good' fibres for X-ray analysis. The replacement B residues have a different conformation to the 3,6-anhydro-D-galactose. For example, the substitutions in ι-carrageenan are residues with a C1 conformation instead of the normal 1C unit; these galactose derivatives have no anhydro bridge to anchor the 1C state and so revert to C1. These residues cannot participate in the formation of the carrageenan double helix. (A similar effect is observed in polypeptide conformations where proline breaks an α-helix.) The minor components provide kinks in the chain which disrupt helix formation and make the chain find another partner. In this way the three-dimensional network is created. The distances between these kinks dictate the length of the helical regions and therefore the strength, porosity and elasticity of the gel. The optical rotation changes accompanying the sol–gel transition are close to those predicted from the torsion angles of the helical and random coil states and so provide experimental evidence supporting this model (Rees 1972).

Algal extracts possess an enzyme activity which catalyses the conversion of galactose 6-sulphate residues (the kinks) within the polymer to the normal B unit. This enzyme, trivially known as dekinkase, eliminates the 6-sulphate and forms the anhydro bridge which changes the ring from C1 to 1C. Since this modification is performed post-synthetically the plant is potentially able to increase the rigidity of the existing gel networks by decreasing the frequency of kinks. This may relate to observed increases in the anhydro-galactose content of some red seaweeds which are subject to battering by waves on exposed parts of the shore.

The model also offers an explanation for the difference in melting and setting temperatures – an hysteresis effect. Reactions in which a gel state is more stable as the temperature is decreased are characterised by the gellation process being initiated at nucleation sites. Other examples of similar systems are the association of DNA into its double helix, or triple helix formation by collagen. In the sol state the chains are flexible because the B residue, being 1C, does not have large equatorial substituents restricting rotation around the (1e $\rightarrow$ 4e) linkage. To construct the first turn of the helix, two chains must come together in the correct orientation and wind round each other. This is the nucleation site. The assembly of the second and subsequent turns is easier because the chains are in contact. Each turn contributes six hydrogen bonds to the structure ensuring that as it grows the cooperative action of these small forces increases the helix stability. Once the first turn is complete the rest of the helix lengthens rapidly. When the gel is warmed and sufficient thermal energy is imparted to the helices to break the first few bonds, the melting process goes quickly to completion as the large number of weak bonds are ruptured. In theory if the sol state is cooled to just below the melting point the helices would be thermodynamically stable but because of the difficulty in starting the first turn they do not form until the motions of the chains are slowed down by further cooling. This is the basis of the difference in melting and setting temperatures.

The agars also participate in the setting of mixed component gels under conditions where the individual polysaccharides do not gel. Agarose at concentrations below 0·1 per cent does not gel but when mixed with a galactomannan, a non-gelling polysaccharide, a gel is formed which exhibits the melting and setting behaviour of an agar gel. The optical rotation evidence (Rees 1972) is consistent with the agarose forming double helices but the concentration is so low that there is insufficient material to create the network. These few helices interact with galactoman-

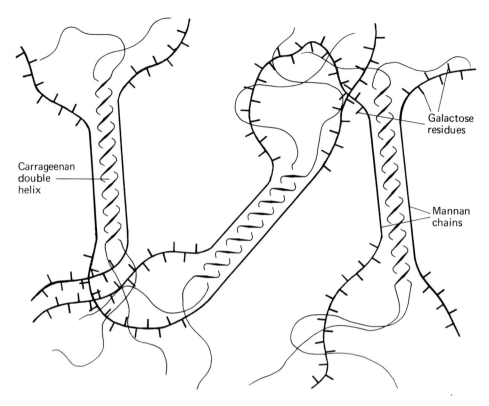

Fig. 9.18 The proposed model for the two component gel formed between a galactomannan and an agar.

nan chains and so provide the junctions for a galactomannan network. In the galactomannans (section 3.3) the galactose residues are distributed unevenly along the mannan backbone and tend to be clustered in blocks. The gelling ability of the galactomannans is inhibited as the galactose content increases which suggests that the unsubstituted ribbon-like mannan regions associate with the agarose or carrageenan helices (Fig. 9.18), although the nature of the non-covalent forces is unknown. The gel properties resemble those of the single component agars since the stability of both systems depends on the presence of double helices. If the gel is warmed the helices break, the junctions between galactomannan chains are destroyed and the gel melts. In nature galactomannans and agars do not coexist; consequently this is a non-biological interaction. However the study of this system provides an insight into the possibilities of natural associations among fibrous polysaccharides and the hemicelluloses in the organisation of plant cell walls.

5.2 Alginates

Up to 40 per cent dry weight of brown seaweeds are alginates. These polysaccharides form gels which, unlike the agars, are not formed or melted by temperature changes but by a change in the counter-ion. Setting is induced by the addition of Ca^{2+} ions whereas removal of these ions or replacement with an alkali-metal ion such as Na^+ 'melts' the gel.

The alginates are linear glycuronans of $1 \rightarrow 4$ linked D-mannuronic and L-guluronic acid units. (Note L-guluronic acid is the 5-epimer of D-mannuronic acid.) The proportions of the two components depend on species and on the region of the plant from which the alginate is extracted, and also show seasonal variations. The residues are arranged in blocks of either entirely mannuronic acid (M) or guluronic acid (G) units interspersed by regions of irregular alternation:

-M-M-M-M-M-M-G-M-G-G-M-G-G-G-G-G-G-G-G-G-G-M-M-G-M-M-M-M-M-

The association of like regular regions from several chains yields microcrystallites which can be examined by X-ray diffraction. Although the $\beta(1 \rightarrow 4)$ mannuronan in the free acid form has the same structure as mannan, the steric and electrostatic repulsions of the carboxylate anion in

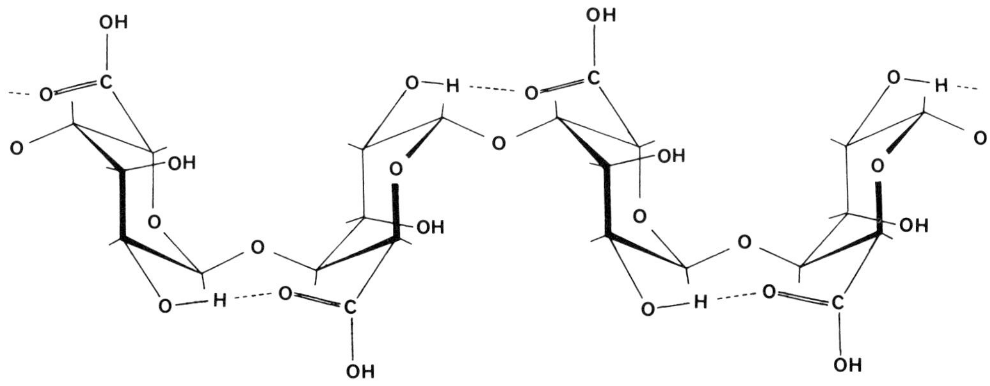

Fig. 9.19 Guluronan chain showing the two-fold symmetry and the intramolecular hydrogen bonds between neighbouring L-guluronic acid residues.

the salt form of the polymer convert the two-fold helix into a three-fold one – like $\beta(1 \rightarrow 4)$ xylan – with the carboxylates distributed evenly around the axis. The NMR spectrum of the $\alpha(1 \rightarrow 4)$ guluronan segments shows the L-guluronic acid residues adopting a 1C conformation which keeps the large carboxylate group in an equatorial position. This, together with X-ray evidence and computer model building, suggests that the $(1a \rightarrow 4a)$ linked polymer forms a two-fold helix stabilised by an intramolecular hydrogen bond (Fig. 9.19). These rod-like chains have a high affinity for Ca^{2+} which is probably a property of the geometrical arrangement of the carboxylate group, the ring oxygen and the equatorial O_2' of the adjacent residue (Rees 1972). The Ca^{2+} ions replace the hydrogen bonds to stiffen the chains which then aggregate and become the junctions in the network. The mannuronan or alternating segments cannot participate in this type of complex and constitute the flexible part of the network. Since the segments have different affinities for the gelling agent, the gel consistency depends on the relative proportions of mannuronic and guluronic acids, as expected. A high guluronan content leads to strong but brittle gels while an increase in the mannuronan element creates more flexible gels. This observation correlates with the biological occurrence of these polymers. The mannuronan content of alginates is higher in the tissues undergoing growth and expansion whereas the support tissues at the plant base have a greater proportion of the rigid guluronan. Some bacteria possess a 5-epimerase which converts polymer-bound D-mannuronic acid into L-guluronic acid but this has not as yet been detected in algae. The brown seaweeds may adjust the physical properties of the alginates by chemical modification after synthesis (c.f. the agars).

5.3 Pectins

The name pectins encompasses a group of polysaccharides based on an $\alpha(1 \to 4)$ galacturonan backbone which occur in the intercellular spaces of plants, especially in young tissues such as primary cell wall. Sugar beet and citrus fruit peel are the main sources of these commercially important polysaccharides. The gelling types are used in jams and jellies while those with enhanced viscous properties are added to baby foods as lubricants. Although D-galacturonic acid is the principal component, often a proportion of the carboxylates are blocked as methyl esters and usually the galacturonan main chain carries side chains, the length and composition of which depend on the source of the material. The galacturonans have a high affinity for water and this may account for their occurrence in biological situations which require water retention or the uptake of water, for example in germinating seeds.

Whether esterified or not the shape of the linear galacturonan backbone is a right-handed extended three-fold helix which conformational analysis reveals is stiffer than cellulose because of the steric limitations imposed by the $(1a \to 4a)$ linkage. As with the alginates the gellation of the non-esterified galacturonans is pH sensitive and cation dependent. In fact there is evidence (Rees 1972) that galacturonans can form a Ca^{2+} complex similar to that proposed for the guluronan segments in alginates. As the methyl ester content increases the ionic and viscous properties are suppressed and gels can form in the absence of multivalent cations. On this evidence it is probable that the junctions in the gel network are a non-covalent association of Ca^{2+}-complexed or methylated galacturonan regions. Microcrystallites which could correspond to this type of aggregate have been observed in celery stem tissue. The pectins like the carrageenans possess kinking residues, in this case L-rhamnose units, inserted in the galacturonan chain. These terminate the junctions and cause the chains to find other partners.

As noted with the alginates, there is an apparent correlation of structure with biological location. The rhamnose content and the incidence of side chains are higher in the walls of rapidly growing plant cells; both of these effects would be expected to hinder chain packing and lead to viscous solutions rather than gels, so permitting expansion in cell volume. The galacturonans with the highest branching frequencies are the gums, e.g. tragacanthic acid, which are exuded as viscous fluids at a site of injury, and the mucilages, e.g. from slippery elm bark, which probably function as anti-desiccants in seeds and bark. In summary, although the principal factor in the properties of pectins is the conformation of the galacturonan backbone, the viscoelasticity, porosity and ion-exchange behaviour can be controlled by methylation, insertions and branching.

5.4 Glycosaminoglycans

Within the intercellular matrix of animal connective tissues is a family of anionic polysaccharides formerly referred to as mucopolysaccharides but more recently as glycosaminoglycans. In conjunction with collagen they serve structural roles in tissues such as skin, bone and cartilage. One cannot help noting the similarity in construction of the animal connective tissues and plant cell walls — both possess a gelling polysaccharide, glycosaminoglycan and hemicellulose respectively, and a fibrous element which in plants is cellulose but in animals is replaced by a protein, collagen. Other functions which have been ascribed to the glycosaminoglycans include (a) joint lubrication by synovial fluid, (b) regulation of the flow of cations between cells by using the ion-exchange properties, and (c) control of the rate of diffusion of solutes, particularly macromolecular ones, by adjusting the porosity of the gel.

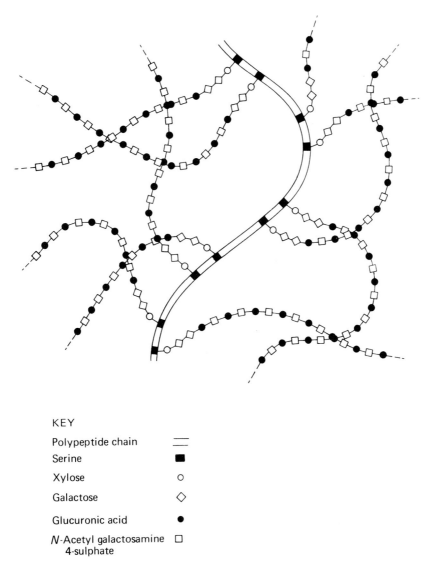

KEY

Polypeptide chain	=
Serine	■
Xylose	○
Galactose	◇
Glucuronic acid	●
N-Acetyl galactosamine 4-sulphate	□

Fig. 9.20 Part of the chondroitin 4-sulphate proteoglycan complex.

Each member of this group is a linear heteroglycan based on a repeating disaccharide unit with the same general formula and linkages as the agars (section 5.1) but as the name suggests one of the components is an hexosamine. The nature of the A and B residues in some of the glycosaminoglycans is shown in Table 9.2. All the evidence favours the sugars' being in the C1 conformation; even L-iduronic acid within dermatan sulphate chains is apparently also C1 in the solid state although energy calculations predict it to be 1C. As can be seen from the list of primary structures keratan sulphate differs from the rest in two respects. It does not have an uronic acid and the alternation of the residues is reversed, with the hexosamine occurring as the B unit not the A. As with the other gelling polysaccharides described in previous sections there are occasional irregularities in the sequence although these have not been so well characterised. Branching within the chains is believed to be absent but because the polysaccharide chains are

Table 9.2 Components of the repeating $-(A-B)_n-$ unit of some glycosaminoglycans

Glycosaminoglycan	A residue	B residue
Hyaluronic acid	D-GlcNAc	D-GlcA
Chondroitin 4-sulphate	D-GalNAc 4-sulphate	D-GlcA
Chondroitin 6-sulphate	D-GalNAc 6-sulphate	D-GlcA
Dermatan sulphate	D-GalNAc 4-sulphate	L-IdoA
Keratan sulphate	D-Gal or D-Gal 6-sulphate	D-GlcNAc or D-GlcNAc 6-sulphate

usually covalently linked to a polypeptide backbone in a proteoglycan complex the result is a branched macromolecule. The chondroitin and dermatan sulphate chains are linked to the polypeptide via a trisaccharide glycosidically joined to the hydroxyl group of a serine residue in the polypeptide chain:

$$-A-B-A-B-A-D\text{-}GlcpA-\beta(1 \to 3)-D\text{-}Galp-\beta(1 \to 3)-D\text{-}Galp-\beta(1 \to 4)-D\text{-}Xylp-Ser-$$

The proteoglycan complex has the form shown in Fig. 9.20. Keratan sulphate chains may be linked to the polypeptide in the same way or via an aspartamido linkage between an N-acetylhexosamine and the amide nitrogen of an asparagine residue. This type of linkage is

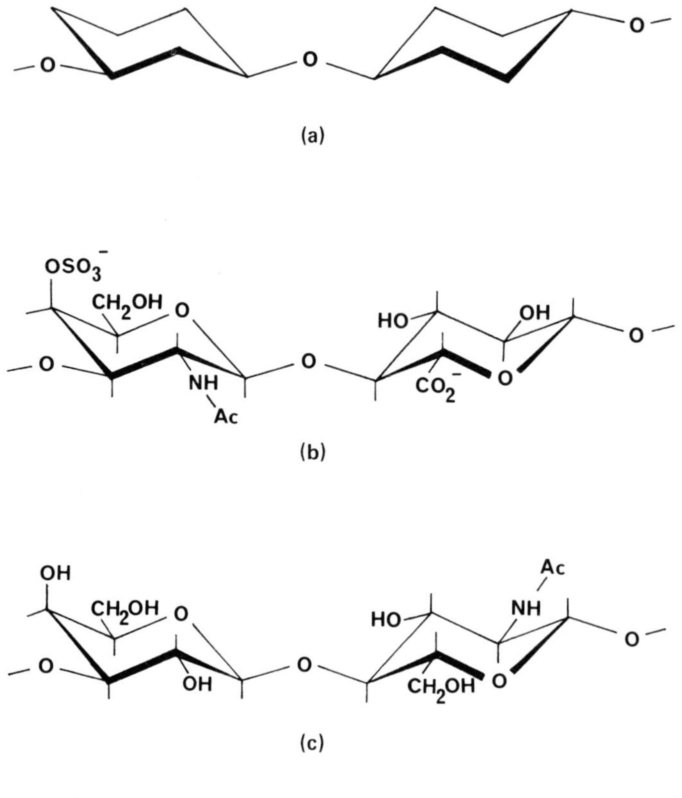

Fig. 9.21 (a) Basic repeating unit of the agars and glycosaminoglycans. Substitutions and additions to this structure yield the disaccharide repeating units of chrondroitin 4-sulphate (b) and keratan sulphate (c).

frequently encountered in the plasma glycoproteins. The exception to this branched complex is hyaluronic acid. Almost certainly this polysaccharide is a single long glycan chain which may be attached to a small amount of protein (less than 2 per cent) by an as yet uncertain linkage.

The chain conformations are very similar to those of the agars. All possess the same geometrical relationship of the rings and the entry and exit points of the linkages are identical. It can be considered that the agars and glycosaminoglycans are derived from a common structure (Fig. 9.21(a)) which, by the insertion of heteroatoms and the addition of substituents, generates the individual members of the group (Figs. 9.6(b), 9.16 and 9.21(b) and (c)). The final conformation and flexibility of the polymer is decided by the four torsion angles of the two glycosidic linkages which are in turn governed by the nature and size of the equatorial substituents. In general the animal polysaccharides are stiffer because of greater restriction about the (1e → 4e) linkage.

The recent advances in applying X-ray analysis to oriented fibres have been achieved only with deproteinised fractions. As yet the intact proteoglycans do not yield usable samples. Because the structure of the polypeptide and the way it influences the polysaccharide conformation are largely unknown, the ensuing discussion will concentrate on the glycan part of the proteoglycan complex. The elucidation of the role of the polypeptide is awaited with interest.

5.4.1 Hyaluronic acid

Chemically the single linear chain of this glycosaminoglycan makes it the simplest of the group and it is also the most studied. It is found in all connective tissues although it is particularly abundant in umbilical cord, vitreous humour and synovial fluid. In dilute solution the long chains of up to 25 000 disaccharide units (molecular weight 10^7 daltons) assume a random coil conformation. These coils occupy large molecular domains, most of which is solvent, and form highly viscous solutions. It has been calculated, using random coil dimensions, that at a concentration of only 0·2 mg/ml the hyaluronate molecules occupy the whole volume of the solution without the chains overlapping. As the concentration is increased the chains progressively become more entangled forming a continuous network. Under these conditions the hyaluronate solution is a viscoelastic gel. As expected of a polyanion these properties are pH dependent and the maximum of gel strength is at pH 2·5. A 1 mg/ml solution at this pH behaves like siloxane 'bouncing' putty in that when cut surfaces are pressed together they heal and under shock it does not deform but it will flow slowly when a long-acting stress is applied. These are properties expected of non-covalently cross-linked gels.

Dea *et al.* (1973) have used the pH 2·5 putties to prepare highly crystalline oriented fibres which have been analysed by X-ray diffraction. The results, taken in conjunction with other evidence, lead these authors to argue that the hyaluronate chains are intertwined in double helical structures (Fig. 9.22). Each chain has four disaccharide units per turn and is coiled in a left-handed helix which accommodates a second chain in a similar conformation but travelling in the opposite direction. Atkins *et al* (1972) and Atkins and Sheehan (1973) using X-ray diffraction have observed different conformations of hyaluronic acid according to the conditions of fibre preparation. Some of their samples contained double helices while others were single helices. The free acid form of hyaluronic acid in which the carboxylates are fully protonated occurs as an extended ribbon in a single helix having two-fold symmetry shown schematically in Fig. 9.23(a). As observed with the mannuronan segments of the alginates, the addition of a bulky cation cannot be accommodated in the two-fold helix so the (1e → 4e) linkage rotates by 60° yielding a three-fold helix. By adjusting the tension, humidity and

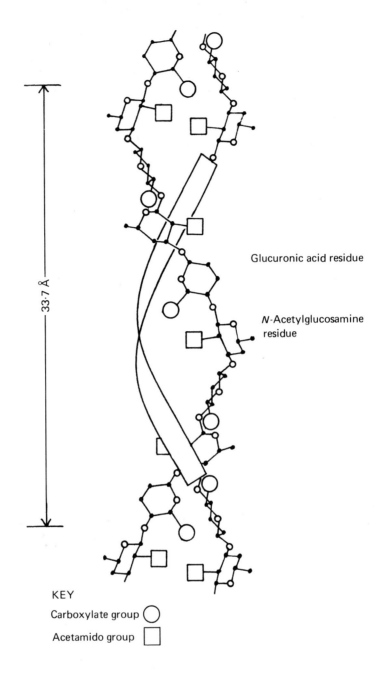

33·7 Å

Glucuronic acid residue

N-Acetylglucosamine residue

KEY

Carboxylate group ◯

Acetamido group ▢

Fig. 9.22 Structure of the proposed double helical form of hyaluronate. Each chain has four disaccharide units per turn.

temperature Atkins and Sheehan (1973) were able to convert double to single helices within the fibres and Dea *et al.* (1973) reported X-ray pictures of states containing both types of helix. This ability of hyaluronate to exist in so many interconvertible conformations in the solid state fibres suggests that the double helix is not as stable as the agar type, and that in solution there is an equilibrium of coil and double helical forms with the double helices providing the junctions of the gel network. The viscous element of the viscoelastic properties depends on the coil content, and elasticity is a function of the number of junctions, which is proportional to the amount of helix. As with agar gels the viscoelastic behaviour of hyaluronate solutions is

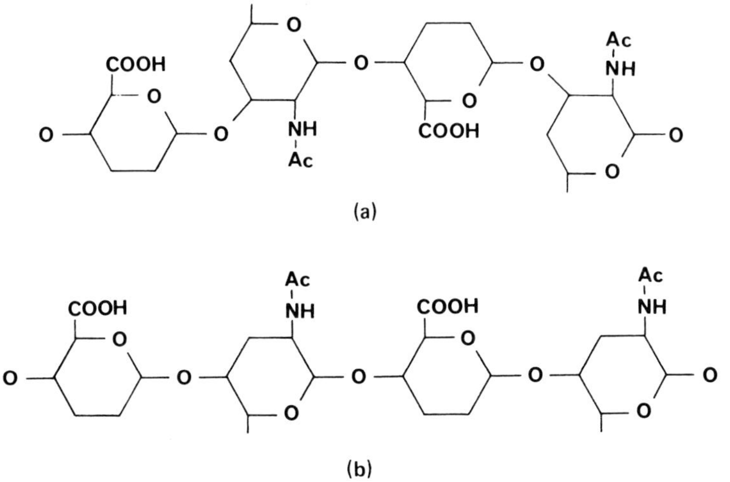

(a)

(b)

Fig. 9.23 (a) Diagrammatic two-fold hyaluronic acid chain showing the distribution of the functional groups on either side of the chain axis.
(b) Hypothetical hyaluronic acid-type of polymer in which the (1e → 3e) linkage is replaced by a (1e → 4e).

temperature dependent, with the sol-gel transition occurring between 30 and 40°C. Since this range has physiological significance there is possibly an appreciable contribution *in vivo* from the double helical species.

This model can be used as the molecular basis for theorising on the origin of the many anomalous properties of hyaluronate solutions. It has been known for many years that the sedimentation and diffusion coefficients and viscosity are markedly concentration dependent. This may now be interpreted as an increase in helical content as the concentration rises so making the gel firmer and by virtue of the cross-linking behave as though the molecules were larger.

An interesting question to pose concerns the advantage of an alternating (1e → 3e : 1e → 4e) linkage arrangement. If all the linkages were (1e → 4e) there would be 180° inversion between each residue, e.g. as in cellulose, and all the carboxylate and acetamido groups would be on the same side of the chain (Fig. 9.23(b)). A (1e → 3e) polymer, e.g. $\beta(1 \to 3)$ xylan, has no inversion. Therefore, by combining these properties the functional groups are positioned on each side of the two-fold helix (Fig. 9.23(a)). (Note that single residue repeating polymers achieve the same effect employing only the (1e → 4e) linkage.) In the three-fold helix the distribution of these groups around the helix axis is further evened out.

The hyaluronate gels exercise considerable control over the transport of materials through intercellular spaces. Apart from their role in ionic regulation, the large hydrated polyanions retard the movement of water and stabilise the water distribution in the tissue. This hydrodynamic damping is of particular importance in articular cartilage to resist the high compression forces tending to squeeze the water out of the tissue. The gel has a filtering effect on the passage of proteins. As in gel chromatography the permeability to solutes depends on the degree of cross-linking or porosity, which in the case of hyaluronates is adjustable by altering the concentration, and on the size of the penetrant macromolecule. This aspect has important implications for the movement of asymmetric molecules like collagen.

5.4.2 Other glycosaminoglycans

There is considerably less known at present concerning the intra- and intermolecular associations of the other glycosaminoglycans. However several structural aspects are worthy of note. Although the individual polysaccharide chains are much shorter than those of hyaluronic acid, (lengths of 25–40 disaccharide units from present evidence), the proteoglycan complexes have particle weights in excess of 10^6 daltons. Single chains of chondroitin 4- and 6-sulphates have been induced to form oriented fibres which exhibit two-fold and three-fold helical symmetry in the free acid and salt forms respectively, comparable with the single helical conformations of hyaluronate. Preliminary work on dermatan sulphate fibres gives X-ray diffraction pictures consistent with this polymer's possessing a conformation different from that of other glycosaminoglycans. This is all the more exciting since this polymer has occasional D-glucuronic acid replacements of the normal L-iduronic acid residues which may represent kinking units.

From a biological viewpoint chondroitin and dermatan sulphates and heparin exhibit two very interesting intermolecular associations. Firstly, light scattering studies on solutions of monomeric tropocollagen fibrils show that both chondroitin and dermatan sulphates increase the interaction of the fibrils and so may function in regulating the extracellular assembly and alignment of the fibres (Öbrink 1973). Second, when glycosaminoglycan chains are covalently linked to agarose beads using the cyanogen bromide technique, it is observed that dermatan sulphate and heparin chains but not other types of chain bind light density lipoprotein and very light density lipoprotein under ionic conditions which approximate to physiological (Iverius 1972). These two lipoproteins have been implicated in atherosclerosis and it may be significant that the arterial content of dermatan sulphate increases with age.

6 Summary

Over the past few years there has been a rapid expansion of interest in polysaccharides. If a reason for this had to be found it would probably be that there is a better understanding of the relationship between primary structure and the physical and biological properties.

Polysaccharides differ from other biological macromolecules in possessing long regions composed of identical repeating units. These are responsible for the numerous cooperative interactions which are the basis of the tertiary and quaternary structure. In these cohesive forces between like and unlike chains resides the key to the architecture of the heterogeneous systems which are the biological tissues. The systems surrounding cells are constructed from gelling and fibrous polymers. It is fascinating that although the two types of component have strikingly different properties, in terms of primary and secondary structure they are very similar. Both comprise linear chains which inherently tend to associate. The profound difference in properties arise from the ability of the gelling polysaccharides to limit the lengths of these aggregated segments by the insertion of terminating residues. In the fibrous polysaccharides such punctuation marks are absent and in consequence the associations continue over greater lengths. The reserve polysaccharides have evolved to a state in which the tertiary forces are not strong enough in solution to offset the high conformational entropy of these flexible polymers. Interchain associations are unstable and the polysaccharide assumes a random coil conformation. Another device employed to reduce tertiary interactions is irregular or multiple branching. This is encountered in the storage polysaccharides and the extremely viscous gums and mucilages.

References Cited

ASPINALL, G. O. (1970). *Polysaccharides.* Pergamon Press. Oxford. (A useful reference book especially on primary structures.)

ATKINS, E. D. T., PARKER, K. D. and PRESTON, R. D. (1969). 'The helical structure of the β-1,3-linked xylan in some siphoneous green algae. *Proc. Roy. Soc. B,* **173**, 209-21.

ATKINS, E. D. T., PHELPS, C. F. and SHEEHAN, J. K. (1972). 'The conformation of the mucopolysaccharides: hyaluronates.' *Biochem. J.,* **128**, 1255-63.

ATKINS, E. D. T. and SHEEHAN, J. K. (1973). 'Hyaluronates : relation between molecular conformations.' *Science,* **179**, 562-64.

DEA, I. C. M., MOORHOUSE, R., REES, D. A., ARNOTT, S., GUSS, J. M. and BALAZS, E. A. (1973). 'Hyaluronic acid: a novel, double helical molecule.' *Science,* **179**, 560-62.

DURETTE, P. L. and HORTON, D. (1971). Conformational analysis of sugars and their derivatives.' *Adv. Carb. Chem. & Biochem.,* **26**, 49-125.

IVERIUS, P-H. (1972). 'The interaction between human plasma lipoproteins and connective tissue glycosaminoglycans.' *J. Biol. Chem.,* **247**, 2607-13.

MARCHESSAULT, R. H. and SARKO, A. (1967). 'X-Ray structure of polysaccharides.' *Adv. Carb. Chem.,* **22**, 421-82. Interpretation of fibre X-ray diagrams and details of the crystallography and structure of cellulose, chitin, amylose and other polysaccharides.

NIEDUSZYNSKI, I. A. and MARCHESSAULT, R. H. (1972). 'Structure of β,D(1 → 4')-xylan hydrate.' *Biopolymers,* **11**, 1335-44.

ÖBRINK, B. (1973). 'The influence of glycosaminoglycans on the formation of fibres from monomeric tropocollagen *in vitro.' Europ. J. Biochem.,* **34**, 129-37.

REES, D. A. (1972). 'Shapely polysaccharides.' *Biochem. J.,* **126**, 257-73. (Colworth Medal lecture, mainly concerning gelling polysaccharides.)

SHAFIZADEH, F. and McGINNIS, G. D. (1971). 'Morphology and biosynthesis of cellulose and plant cell-walls.' *Adv. Carb. Chem. & Biochem.,* **26**, 297-349.

WHELAN, W. J. (1971). 'Enzymic explorations of the structures of starch and glycogen.' *Biochem. J.,* **122**, 609-22.

General References

REES, D. A. (1967). *The Shapes of Molecules.* Oliver & Boyd, Edinburgh. (Relationship of biological function to conformation for polysaccharides.)

STODDART, J. F. (1971). *Stereochemistry of Carbohydrates.* John Wiley, New York. (Conformations, mainly of monosaccharides and their derivatives.)

10
Structure and Synthesis of Bacterial Walls

Pauline M. Meadow
Department of Biochemistry, University College, London

1 Introduction

Bacterial envelopes, comprising both walls and cytoplasmic membranes, can be isolated from disrupted bacteria by sedimentation at about 17 000 *g*. They are composed of a number of discrete polymers. Their relationship one to another can be seen by electron microscopy, but the linkages which make the complete shape of the bacteria are by no means fully understood. Research in the last 10–15 years has provided interesting information of the ways in which insoluble complex molecules can be assembled outside what is normally regarded as the metabolic limit of the bacterium. In this essay I shall discuss the structure and biosynthesis of the major polymers on which most of this work has been done but first I will try to summarise as briefly as possible the present state of our knowledge of bacterial walls and membranes.

2 The bacterial envelope

2.1 General properties

The outer layers of a bacterium determine its response to the Gram stain, its antigenic behaviour and its susceptibility to bacteriophages. The host defence mechanisms are directed against them and they play a major part in the response of the organism to antibacterial compounds. The individual components of the envelope can be differentiated both structurally and functionally. The innermost layer, the cytoplasmic membrane, is the permeability barrier across which substrates must be transported, often by active transport; it is the site of oxidative phosphorylation, of septum formation and of chromosome attachment; it may be the site of attachment of some bacteriophages and it is the part responsible for the biosynthesis of wall and membrane polymers. The mucopeptide (also known as glycopeptide, peptidoglycan or murein) has the same shape as the bacterium and is responsible also for its mechanical strength. In Gram-positive bacteria the teichoic acids are the antigenic determinants, provide some of the sites of attachment for bacteriophages and may play a role in ion binding and transport. In Gram-negative bacteria the lipopolysaccharides have these functions.

2.2 Gram-positive and Gram-negative bacterial envelopes

The envelopes isolated from Gram-positive and Gram-negative bacteria are easily distinguishable by chemical analysis while electron microscopy shows a quite different organisation in the two classes of bacteria (Fig. 10.1). Both have the same basic structure of an inner cytoplasmic

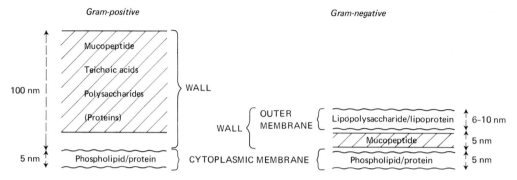

Fig. 10.1 Diagrammatic representation of bacterial envelopes

membrane surrounded by a wall containing the mucopeptide, but the additional components and their arrangement in the walls are different. The walls of Gram-positive bacteria are thicker than those of Gram-negative species and hence supply a higher proportion of the dry weight of the bacterium. They are separated clearly from the cytoplasmic membrane which can be isolated free of wall components either by differential centrifugation of disrupted bacteria or from lysed protoplasts (osmotically-fragile forms made by dissolving the wall layers). In Gram-negative bacteria the more complex structure of the wall layers makes it very difficult to remove them without damaging the underlying cytoplasmic membrane. True protoplasts are not formed and walls cannot be separated from membranes by differential centrifugation.

The mucopeptide is a major part of the relatively unstructured Gram-positive wall. The residue is made up of a variety of negatively charged polysaccharides which include the teichoic acids and in some species there are proteins. The double membrane structure of the Gram-negative bacterial envelope shows up clearly in thin sections. The inner membrane apparently corresponds to the cytoplasmic membrane of the Gram-positive organisms whereas the outer membrane contains the antigenic lipopolysaccharide and lipoprotein. The two are separated by a periplasmic region within which is a thin mucopeptide layer visible as an electron-dense band. In some species the mucopeptide is linked covalently to the inner surface of the outer membrane through a lipoprotein, but in at least one organism, a marine pseudomonad, it appears to be associated rather more closely with the outer surface of the cytoplasmic membrane.

Bacterial walls, therefore, are composed of a number of chemically distinct and often physically separable polymers. In the complete envelope they must be integrated and the study of their biosynthesis has involved finding, firstly, ways of establishing the structure of the individual components and subsequently trying to separate the synthesis of one polymer from another. The structures of walls from a wide variety of bacterial species are now known and have been well reviewed (Salton, 1964; Rogers and Perkins, 1968; Ghuysen, Strominger and Tipper, 1968; Glauert and Thornley, 1969; Reaveley and Bruge, 1972). I shall limit this account to the three polymeric components whose synthesis has been studied most thoroughly — the mucopeptide, the lipopolysaccharide and the teichoic acids. I shall start with the mucopeptide because this is common to both Gram-positive and Gram-negative bacteria.

3 The mucopeptide

The work of Salton and of Cummins and Harris in the 1950s established that the mucopeptide of Gram-positive bacterial walls had a relatively simple composition. Weidel and his collaborators showed that Gram-negative bacteria contained similar material. In all bacteria it seems to be the component mainly responsible for shape and certainly to a major extent for mechanical strength. Its biosynthesis is particularly interesting in that it could lead to the study of morphopoietic enzymes. It was soon established that the amount of mucopeptide in any species might vary with growth conditions but its major chemical composition was a stable species characteristic. It has a unique composition both in terms of its constituent molecules and in their organisation to form a polymer. It is the site of action of some antibiotics. For this reason attention has been focused on its synthesis, but it is also of purely biochemical interest since we know so little about the formation of any kind of insoluble structural polymer.

3.1 Structure

In all species the mucopeptide consists of a backbone of alternating units of the amino sugars N-acetylglucosamine (GlcNAc) and its O-lactyl ether, N-acetylmuramic acid (MurNAc). They are $\beta(1 \rightarrow 4)$-linked like the glucose units in cellulose. Projecting from the amino sugar backbone are short peptide chains in a specific sequence. The most common of these consists of the four amino acids L-alanine, D-glutamic acid, a diamino acid and finally D-alanine. In staphylococci the diamino acid is L-lysine while in *Escherichia coli* and all Gram-negative bacteria it is *meso*-diaminopimelic acid. In some species the γ-carboxyl group of glutamic acid is amidated or substituted with an additional amino acid. The peptide chains are cross-linked from the carboxyl group of the terminal D-alanine to the free amino group of the diamino acid either directly or by a short peptide bridge. The type of cross-linking is mainly a species characteristic.

Staphylococcus aureus

—GlcNAc—MurNAc—GlcNAc—MurNAc—GlcNAc—

Escherichia coli

—GlcNAc—MurNAc—GlcNAc—MurNAc—GlcNAc—

Key: GlcNAc; N-acetylglucosamine; MurNAc, N-acetylmuramic acid; DAP, diaminopimelic acid

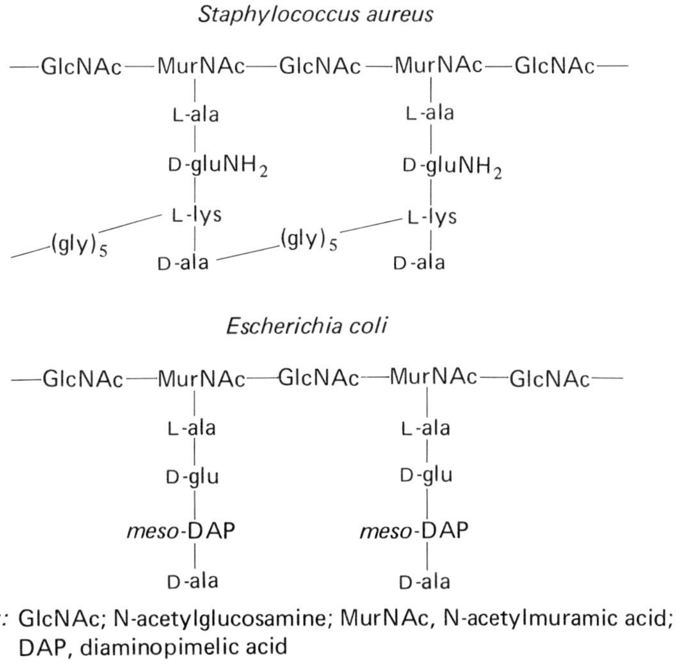

Fig. 10.2 Mucopeptide structures

The structures have been determined by careful degradative studies using specific enzymes (see Ghuysen, 1968).

The two mucopeptides whose synthesis I shall discuss, those of *Staphylococcus aureus* and *Escherichia coli* are shown in Fig. 10.2. Although the peptide bridges are drawn linking adjacent peptide side chains, three-dimensional cross-linking could be achieved by bridges between peptides attached to different amino sugar backbones. The amount of cross-linking and the length of the polymer varies with growth conditions.

Since the mucopeptide component of Gram-positive bacteria is a major part of the wall, it was logical to select these organisms for biosynthetic work. The chemical composition of the mucopeptide of *Staphylococcus aureus* was deduced from analyses of isolated walls from which the teichoic acids had been extracted with trichloracetic acid. Before its synthesis could be studied it was necessary to know how its constituent molecules were linked. The first clues came from work which had been going on since the early 1950s to find the effects of penicillin on bacteria. Park had shown that the addition of sublethal concentrations of penicillin to logarithmically growing cultures of *S. aureus* strain Copenhagen caused the accumulation of uridine nucleotides which contained alanine, glutamic acid, lysine and an unidentified amino sugar. As the analyses of bacterial walls gained momentum it became clear that the amino acids were those in the mucopeptide and that the amino sugar was the O-lactyl ether of glucosamine which had been called muramic acid. In 1957, Park and Strominger determined the structure of the accumulated uridine nucleotides and postulated that they were precursors of a mucopeptide whose synthesis was inhibited by penicillin.

3.2 Synthesis

The biosynthetic sequence can be divided conveniently into three stages. The first involves the synthesis of the uridine nucleotide peptides which accumulate in the presence of penicillin. In the second stage a disaccharide of N-acetylglucosamine and N-acetylmuramyl peptide is formed and transferred to an existing mucopeptide acceptor. Finally the mucopeptide is completed by cross-linking the peptide chains of the polymer. Although the mucopeptide appears to be covalently attached to teichoic acids in Gram-positive bacteria and to lipoprotein in Gram-negative bacteria nothing is yet known of the mechanism.

3.2.1 Synthesis of UDP-muramyl pentapeptide

As a result of work largely by Park, Strominger and their collaborators the sequence of events shown in Fig. 10.3 has been established in *Staphylococcus aureus*. Most of the enzymes require Mg^{2+} or Mn^{2+} and ATP and some of them have been purified in this and other species. The first few stages of this cycle leading to UDP-N-acetylmuramic acid are typical of the reactions to be expected in forming nucleotide precursors of carbohydrate polymers, but the synthesis of the pentapeptide chain shows some unusual features. It is quite different from protein synthesis. The amino acids are added sequentially to the nucleotide precursor, the order being determined not by a messenger RNA template but by the specificity of each enzyme for the amino acid and its nucleotide acceptor. For example, in the mucopeptide of organisms such as *Escherichia coli* and *Corynebacterium xerosis,* lysine is replaced by its carboxylated form diaminopimelic acid. Extracts from these two organisms will transfer diaminopimelic acid but not lysine. Only UDP-MurNAc-ala-glu can be used as the acceptor for the transfer. Although the synthesis of the peptide chain extends as in protein synthesis from the N-terminal to the C-terminal end, there is no requirement for RNA or ribosomes and the amino acids are not activated before addition.

The first three amino acids are added singly in the sequence shown but the final addition to

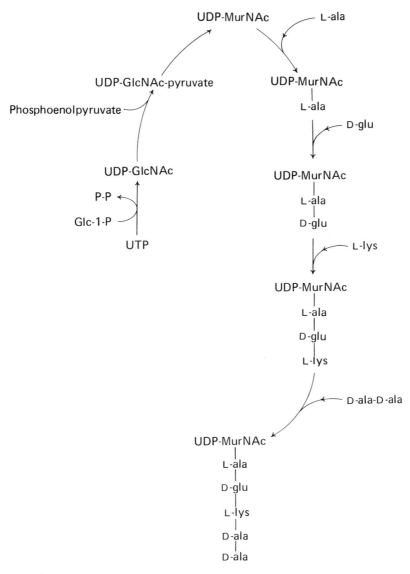

Fig. 10.3 Biosynthesis of UDP-N-acetylmuramylpentapeptide in *Staphylococcus aureus*

make the pentapeptide is of a preformed D-alanyl dipeptide. The synthesis of this dipeptide is particularly interesting because it is inhibited by the antibiotics D-cycloserine and O-carbamyl-D-serine. The first indication of this came from the discovery that cyloserine caused lysis of sensitive bacteria and the accumulation of UDP-MurNAc-ala-glu-lys. The inhibitory effect can be reversed by D-alanine suggesting direct competition between the two molecules which are structurally similar. The mechanism of inhibition has been the subject of some elegant experiments on the individual enzymes shown in Fig. 10.4 (Neuhaus, 1967). Although both the racemase and the D-alanine-D-alanine ligase are inhibited by D-cycloserine it is clear from the concentrations required and from the Lineweaver–Burke plots that the primary site of action of D-cycloserine is the ligase. It competes with D-alanine for the first stage of the reaction in which two molecules of D-alanine are separately bound to the enzyme. Like cycloserine

O-carbamyl-D-serine inhibits the D-alanine racemase possibly by competing with D-alanine. Unlike cycloserine, however, it is utilised by the ligase to form a D-alanine-O-carbamyl-D-serine

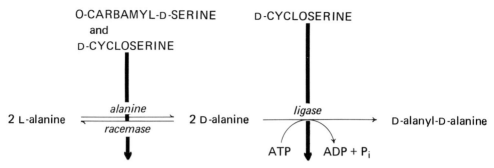

Fig. 10.4 Synthesis of D-alanyl-D-alanine

dipeptide which is inactive. It is interesting that D-cycloserine is taken into the bacteria by the transport system which normally carries glycine and D-alanine and which is quite separate from the L-alanine transport system.

3.2.2 Synthesis and polymerisation of disaccharide peptide

All the enzymes which synthesise the UDP-muramyl pentapeptide are soluble. The completed mucopeptide is an insoluble polymer outside the cytoplasmic membrane so naturally it was assumed that membrane particles would be involved in later stages of the synthesis. Almost simultaneously Park's and Strominger's laboratories reported that particulate preparations used the pentapeptide to form a polymer. The early results were confused largely because of the difficulties of synthesising a substance whose detailed structure was unknown using radioactive precursors which were available only by isolation from staphylococci. The scarcity of the substrates necessitated the use of microtechniques and the use of particulate preparations both as enzyme and acceptor added to the difficulties of identifying the product formed. Subsequent experiments confirmed the original conclusions but showed an unexpected complication. This arose from the observation that incubation of particulate preparations with UD^{32}P-MurNAc-pentapeptide and UDP-GlcNAc released UM^{32}P and ^{32}P$_i$. When UD^{32}P-GlcNAc was used with cold UDP-MurNAc-pentapeptide, the radioactive product was UD^{32}P. This suggested that the two nucleotide substrates were probably involved in separate processes. Eventually the scheme represented in Fig. 10.5 was established (Strominger *et al.*, 1967). Synthesis of the polymer starts with transfer of muramyl pentapeptide from its UDP derivative to a lipid acceptor in the particulate membrane fraction. A pyrophosphate bond is formed and UMP released thus explaining in part at least the ^{32}P experiments. N-Acetylglucosamine is then added from its UDP derivative to form a disaccharide pentapeptide still attached to the lipid. In some organisms this disaccharide peptide is transferred to the mucopeptide acceptor without further modification resulting in a polymer of alternating units of GlcNAc and MurNAc-pentapeptide. In *Staphylococcus aureus*, however, the mucopeptide contains a peptide bridge of five molecules of glycine and its glutamic acid is amidated (see Fig. 10.2). Two modifying reactions are needed before polymerisation occurs. Glycine is added from its tRNA and ammonia is added to the glutamic acid. There are some points of special interest in this cycle.

1. The lipid carrier. This was thought at first to be a phospholipid. It has now been shown to be the phosphate ester of a C$_{55}$ isoprenoid alcohol (undecaprenol) whose structure is shown in Fig. 10.6. It is soluble in lipid solvents and is extracted with phospholipids from isolated

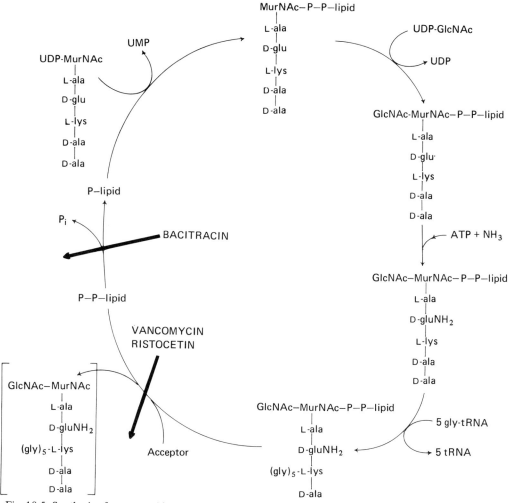

Fig. 10.5 Synthesis of mucopeptide polymer in *Staphylococcus aureus*

$$CH_3-\overset{\overset{\displaystyle CH_3}{|}}{C}{=}CH-CH_2\left(CH_2-\overset{\overset{\displaystyle CH_3}{|}}{C}{=}CH-CH_2\right)_9 CH_2-\overset{\overset{\displaystyle CH_3}{|}}{C}{=}CH-CH_2-O-\overset{\overset{\displaystyle OH}{|}}{\underset{\underset{\displaystyle O}{||}}{P}}-OH$$

Fig. 10.6 Structure of the lipid carrier undecaprenol phosphate

membranes. It has been suggested that its role is to solubilise mucopeptide precursors in the membrane and transport them to the exterior for the final assembly of the wall. Several isoprenoid lipids have been isolated recently from plants, animals and bacteria and this particular one functions similarly in the synthesis of O-antigenic polysaccharides and teichoic acids. It may be considered as the lipid alternative to a nucleotide activator.

2. Antibiotics inhibiting polymer formation. Ristocetin and vancomycin at growth inhibitory concentrations inhibit the transfer of the disaccharide pentapeptide to the existing acceptor in the wall. The disaccharide lipid intermediate is formed normally and accumulates because polymer synthesis is inhibited. These two antibiotics apparently act by binding specifically to the D-alanine dipeptide end of the chain since a peptide analogue (diacetyl-L-diaminobutyryl-D-ala-D-ala) complexes with vancomycin and prevents its inhibition (Neito, Perkins and Reynolds, 1972). Three other antibiotics enduracidin, moenomycin and prasinomycin also inhibit the synthesis of mucopeptide polymer but whether their mechanisms of action are the same remains to be seen.

Bacitracin interferes with the reaction cycle at another stage by inhibiting the phosphatase which dephosphorylates the lipid pyrophosphate. Consequently the lipid acceptor is trapped as its pyrophosphate and further disaccharide synthesis is inhibited until more undecaprenol phosphate is made. Strominger and his colleagues have studied an enzyme from the membranes of *Staphylococcus aureus* which transfers phosphate from ATP to undecaprenol. It is unusal in that it is extracted from membranes by butanol at pH 4·2, is soluble in methanol and ethanol but insoluble in water. It has now been separated into protein and phospholipid components, the latter conferring lipid solubility on an otherwise insoluble protein. Since the enzyme occurs in a lipid environment in the cytoplasmic membrane this phospholipid/protein molecule presumably has been evolved specifically to do so.

3. Formation of glycine peptide bridge. The peptide bridges are apparently synthesised from glycine molecules activated by attachment to their tRNA by stepwise addition since no tRNA-linked peptides have been found. Despite the requirement for tRNA there is no evidence that any of the other components of protein synthesis are needed. Indeed this part of the sequence, unlike the pentapeptide chain, grows from the C-terminal to the N-terminal end in the reverse direction from proteins. Strominger and his colleagues have studied the mechanism of formation of the interpeptide bridge in a number of species. They found the tRNA derivative to be required whenever the bridge amino acids were of the L-configuration. In the lactobacilli, however, the interpeptide bridge often contains D-aspartate. There is presumably no tRNA which can recognise it and the activated intermediate is β-D-aspartyl phosphate. In the two species studied in detail, *Staphylococcus aureus* and *S. epidermidis,* there are four types of glycyl tRNA (tRNA^{gly}), all of them function in mucopeptide synthesis but only three are active in protein synthesis. It is possible, therefore, that one of the tRNA^{gly} has evolved specifically for wall synthesis. It is present as 40 per cent of the total tRNA^{gly}, as would be expected from the relative amounts of mucopeptide and protein in these organisms (Stewart, Roberts and Strominger, 1971).

3.2.3 Cross-linking of the mucopeptide polymer

The preceding sequence of reactions would form polymers of alternating units of GlcNAc and MurNAc peptides (Fig. 10.7). They are very similar to those found in the walls of *Staphylococcus aureus* and *Escherichia coli* but there are two major differences. The product of the biosynthetic sequence contains an extra D-alanine per unit and the peptide chains are not cross-linked. It seemed likely that the last step in synthesis would release D-alanine using its bond energy to carry out a transpeptidation reaction.

The final stage has proved exceptionally difficult to study in cell-free systems. This is not surprising when one realises the degree of spatial organisation which must be achieved to make a three-dimensional network outside the metabolic activities of the bacterium. It is clear now that the reaction does occur and that it is inhibited by penicillin (Fig. 10.7). In *Staphylococcus aureus* penicillin causes the formation of walls with increased free amino- and carboxyl-groups

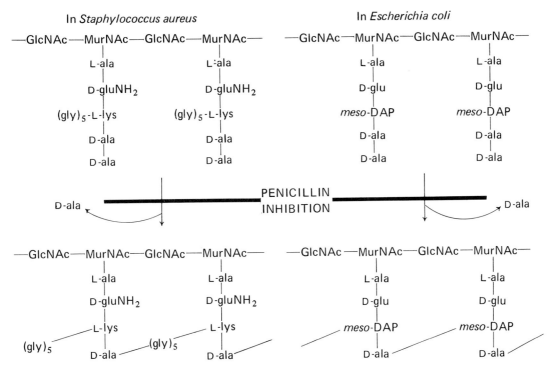

Fig. 10.7 Proposed transpeptidation reaction to cross-link the mucopeptide

as well as excess D-alanine. Furthermore, the product formed *in vitro* in the presence of penicillin is more soluble than that produced in its absence. However the staphylococcal transpeptidase has not yet been purified and the first clear demonstration of the action of penicillin and its derivatives came from *Escherichia coli.* Two enzymes, a D-alanine carboxy-peptidase and a transpeptidase were both sensitive to penicillin derivatives. The carboxy-peptidase was competitively inhibited by concentrations much lower than those required to inhibit growth. This suggests either that the enzyme is not important for the bacterium or that it is not normally accessible to the antibiotic. On the other hand the amounts of penicillin derivatives inhibiting the transpeptidase correlated well with the concentrations required to inhibit growth. The only exception was Penicillin G to which *E. coli* is resistant. It inhibited the transpeptidase at only a tenth of the growth inhibitory concentration. It appears that the lack of activity of Penicillin G against *E. coli* may reflect its inability to penetrate to its site of action (see Strominger, 1969).

Penicillin-sensitive transpeptidases and D-alanine carboxypeptidases have been studied in a number of organisms, particularly the bacilli. Their properties are consistent with the hypo-thesis that at least one of the targets of penicillin inhibition is a membrane component which catalyses transpeptidation. Some of the experiments are described by Strominger in a Royal Society Discussion held to commemorate thirty years of penicillin therapy (Strominger *et al.,* 1971). More recently Hartmann, Holtje and Schwarz (1972) have presented evidence of a second target for penicillin action, an endopeptidase which hydrolyses preformed mucopeptide in the walls of *Escherichia coli.* Such hydrolases are almost certainly necessary to provide acceptor sites for newly synthesised disaccharide peptide units. The enzymes allowing elongation might differ from those leading to septation and division and could vary in their sensitivity to penicillin. It will be interesting to see the outcome of further experiments.

Despite all this work it is still not possible to define the inhibitory activity of penicillin in molecular terms. The penicillins are cyclic dipeptides of L-cysteine and D-valine and one suggestion is that they behave as analogues of the D-alanyl dipeptide end of the peptide chain. They might form penicilloyl-enzyme derivatives and hence inactivate the transpeptidase, but until the enzyme can be purified it is impossible to prove this hypothesis.

Summary

The mucopeptide of both Gram-positive and Gram-negative bacterial walls is important in providing mechanical strength and shape. It consists of a $\beta(1 \rightarrow 4)$-linked amino-sugar backbone from which project short peptide chains. These may be cross-linked to form a three-dimensional network. Synthesis can be divided into three stages. (1) Synthesis of UDP-N-acetylmuramyl pentapeptide by soluble enzymes activated by Mg^{2+} or Mn^{2+}. Their specificity for the nucleotide acceptor as well as the amino-acid substrate governs the sequence. (2) Synthesis by membrane-bound enzymes of a lipid-linked disaccharide peptide which is then transferred to the growing mucopeptide chain. Modifications to the peptide chain such as amidation or addition of interpeptide bridges are made before transfer. The antibiotics bacitracin, ristocetin and vancomycin inhibit the cyclic process. (3) Cross-linking by a membrane-bound transpeptidase to form the final three-dimensional network. The terminal D-alanine is split off in the process. Penicillin G and its derivatives bind to the transpeptidase and inactivate it.

4 The lipopolysaccharide

In Gram-positive bacteria the major component of the wall is the mucopeptide which although important is only 5-20 per cent of the total in Gram-negative bacteria. The walls of these latter bacteria contain a phenol-extractable lipopolysaccharide (LPS) which together with lipoprotein makes up the outer membrane. Interest was focused on the LPS when it was discovered that the O-antigenic behaviour and hence the identification of the Enterobacteriaceae which include the pathogenic *Salmonella* and *Shigella* species, could be related to the composition of their lipopolysaccharides. Mutation from smooth to rough colonial appearance was correlated with changes in the polysaccharide components of the LPS and in sensitivity to bacteriophages as well as loss of pathogenicity.

Interest in the structure and biosynthesis of lipopolysaccharides has been limited until very recently to the Enterobacteriaceae. In this group at least there is a basic structure with relatively minor modifications between species. It is these modifications which give the lipopolysaccharides their bacteriophage and antigenic specificity. The chemistry of the structure has been worked out mainly in LPS-defective mutants by Luderitz, Westphal and their collaborators and has been the subject of recent reviews (Luderitz, Jann and Wheat, 1968; Luderitz *et al.*, 1971). This section will discuss the biosynthesis and structure of the lipopolysaccharide of only one species, but it is probable that the sequence and style of synthesis is common to all enterobacteria.

4.1 Structure

The lipopolysaccharide on which the most detailed biochemical studies have been done is that of *Salmonella typhimurium*. Its heat-stable O-antigens are part of the LPS. Like all the lipopolysaccharides so far studied in detail its structure can be divided into three, perhaps four, distinct regions (Fig. 10.8). The lipid part (lipid A) can be split from the polysaccharide by mild

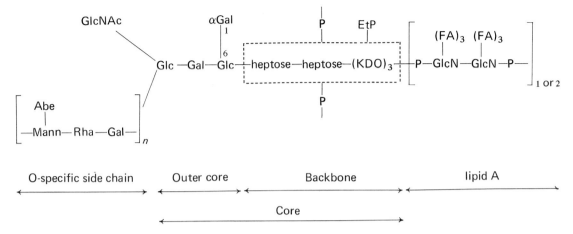

O-specific side chain   Outer core   Backbone   lipid A

Core

Key: Abe, Abequose; Mann, mannose; Rha, rhamnose; GlcNAc, N-acetylglucosamine; Glc, glucose
   Gal, galactose; P, phosphate; KDO, 2-keto-3-deoxyoctonate; EtP, ethanolamine phosphate;
   FA, fatty acid; GlcN, glucosamine.

Fig. 10.8 Lipopolysaccharide of *Salmonella typhimurium*

acid hydrolysis. It consists of a glucosamine disaccharide substituted with long chain fatty acids both O-ester- and amide-linked. All enterobacterial lipid A's contain β-hydroxymyristic acid (3-OH, 14:0) but in some other species such as pseudomonads this is replaced by hydroxylauric and hydroxydecanoic acids. The lipid A is responsible for the toxicity of the LPS. It is joined to the polysaccharide part of the molecule by 2-keto-3-deoxyoctulosonate (KDO) which is linked to a heptose disaccharide. Ethanolamine phosphate is attached to the heptose-KDO part of the molecule and there are phosphodiester bridges cross-linking adjacent chains into large polymers. This part of the molecule is called the backbone since it may be envisaged as stretching the length of the bacterium with the lipid A and the rest of the polysaccharide projecting from it. The sugars making up the core of the polysaccharide vary slightly from one species to another but are common to most Salmonellae. The polysaccharide is completed by a chain of about seven or eight molecules of a tetrasaccharide. This repeating side-chain is responsible for O-antigenic specificity and its loss causes smooth strains to appear rough. In this part of the molecule many species contain unusual sugars such as deoxy sugars or the L-isomers of the more common D-sugars. *Salmonella typhimurium* has two such components in its antigenic side chain, abequose (3,6-dideoxy-D-galactose) and L-rhamnose (6-deoxy-L-mannose).

4.2 Biosynthesis

Before the synthesis of a complex molecule can be studied *in vitro* it is essential not only to know what constituent molecules are involved but in what sequence they are joined. In a complex insoluble polymer this is particularly difficult. However, a combination of chemical, immunological and genetic information gave the biochemists a very good idea of the structure and the biosynthetic work has continued with much useful interdisciplinary collaboration. The sequence of events has been determined mainly by Nikaido, Osborn, Robbins, Rothfield and

354

their collaborators and has recently been reviewed (Nikaido, 1968; Osborn and Rothfield, 1971; Robbins and Wright, 1971; Rothfield and Romeo, 1971).

There is almost no information of the way in which lipid A is synthesised and none at all about how it is linked to the polysaccharide part of the molecule. This is because its structure has only just been determined and no one has succeeded in isolating lipid A-defective mutants. It is possible that they would not be viable. The synthesis of the backbone also remains largely unknown. In *Escherichia coli*, KDO can be transferred from CMP-KDO on to a degraded lipid A fraction from an LPS-defective mutant but so far nothing is known of its incorporation in other organisms. Mutants deficient in heptose and phosphate have recently been isolated and a start has been made on the study of their incorporation into the backbone but not enough is known at this stage to warrant detailed discussion.

On the other hand studies of the biosynthesis of the outer core and antigenic side-chains of the lipopolysaccharides have provided some interesting answers to the problem of synthesising complex insoluble polysaccharides. By analogy with the mucopeptide component one might expect that polysaccharide would be synthesised by soluble enzymes catalysing the stepwise addition of carbohydrates from their nucleotide derivatives and subsequent polymerisation from lipid intermediates. This does occur but there are some interesting variations which use phospholipid-activated enzymes as well as lipid-linked intermediates. The biosynthesis of the outer core region and the O-antigenic side-chains are quite different.

4.2.1 Synthesis of the outer core

Both the structure and the mode of synthesis of the outer core region have been established using lipopolysaccharide-defective mutants. Such mutants were originally isolated because they had lost their normal smooth appearance on agar plates and grew as 'rough' colonies. Analysis of their isolated lipopolysaccharides showed that they had lost the O-antigenic side-chains and in some cases varying amounts of the outer core components as well. Some of these mutants occur spontaneously but bacteriophage selection can also be used since many of the rough mutants show altered sensitivity patterns. Some of the rough mutants were defective in

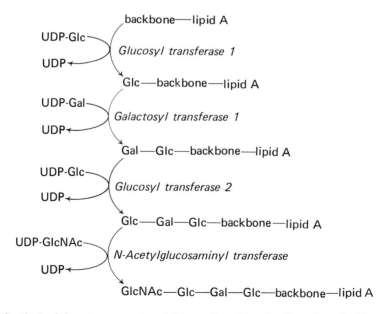

Fig. 10.9 Synthesis of the outer core region of *Salmonella typhimurium* lipopolysaccharide

UDP-galactose-4-epimerase or phosphoglucoisomerase and formed lipopolysaccharides lacking most of the outer core region as well as the antigenic side-chains. Addition of galactose or glucose to the growth medium produced normal LPS and a smooth appearance. This suggested that the outer core might be synthesised by the stepwise addition of sugars from their nucleotide precursors to vacant sites on the lipopolysaccharide core. One of the first reactions to be studied used a rough mutant deficient in the synthesis of UDP-glucose. Its LPS contained none of the outer core or O-specific side-chain components but a particulate envelope fraction of the mutant catalysed the transfer of glucose from UDP-glucose to the non-reducing end of its defective lipopolysaccharide. This could then accept galactose from UDP-galactose. Envelopes from wild type organisms did not act as acceptors for the sugars. It seemed therefore that the addition of the sugars from their nucleotide precursors required both the requisite nucleotide sugar and a suitable acceptor site on the lipopolysaccharide as well as a transferase enzyme. In the defective mutant the envelope fractions were providing both the enzyme and the acceptor.

The sequence of events shown in Fig. 10.9 has now been established. There is genetic evidence that the addition of 1–6 linked galactose to the first glucose residue is probably catalysed by a separate transferase but whether this acts before or after the main additions to the chain is not known nor has the enzyme been demonstrated in a cell-free system. All the transferases are present in crude envelope fractions which also contain membrane particles, lipopolysaccharide acceptors and some soluble proteins. The enzymes themselves appear to be membrane bound and are only solubilised by prolonged sonication. The solubilised enzymes, however, could not transfer the sugars to purified lipopolysaccharide and it soon became clear that a phospholipid fraction present in the envelope was also needed for activity. So far the phospholipid requirement has been established only for glucosyl transferase 1 and galactosyl transferase 1 but it seems at least possible that the five enzymes involved in the sequence are similar. I shall discuss the evidence for galactosyl transferase 1 because this enzyme has been studied most thoroughly.

Requirement for phospholipid in core transferases. A substantial part of the enzyme UDP-galactosyl transferase 1 is solubilised by prolonged sonication. The solubilised protein catalysed the transfer of galactose from UDP-Gal to heat-treated envelopes derived from galactose-deficient bacteria but not to similar preparations from other mutants or the parent strain. However, this protein did not transfer galactose when isolated lipopolysaccharide was used as acceptor. Heat-treated envelopes lost their activity as acceptors after extraction with lipid solvents and Rothfield and Horecker showed that the component extracted was a phospholipid. The most active of the phospholipids tested was phosphatidyl ethanolamine in which the two acyl groups were either unsaturated- or cyclopropane-fatty acids. Phosphatidyl choline was inactive. The phosphatidyl ethanolamine (PE) seems to be required to bring the lipopoly-saccharide, transferase enzyme and the nucleotide sugar into contact by a three-stage process:

    i. Galactose-deficient LPS + PE → LPS . PE
    ii. LPS . PE + transferase → transferase LPS . PE
    iii. Transferase LPS . PE + UDP-Gal → Gal-LPS + UDP + transferase

Whether the PE remains bound to the LPS for the next stages in the reaction is not known. Its role seems to be a purely physical one. It is not covalently bound in the complexes which have been isolated and characterised: rather it becomes annealed to the LPS when the two molecules are heated and allowed to cool together. It has been suggested that they form a mixed bilayer fixing the LPS in the membrane while the transferase enzymes, also bound in the membrane, are extending it. A more detailed account of this process can be found in the reviews by Rothfield and Romeo (1971) and by Osborn and Rothfield (1971).

4.2.2 Synthesis of the O-antigenic side-chains

None of the mutants with single enzyme defects which was used for the studies of core biosynthesis contained any of the components specific to the side-chain region (abequose, mannose or rhamnose). Even mutants known to be defective only in phosphomannose isomerase lacked the entire side-chain. Thus they were quite different from the mutants defective in core biosynthesis which synthesised incomplete cores. On the other hand, some mutants with complete core structures but lacking antigenic side-chains accumulated polymeric antigenic material in the medium. It seemed likely that the side-chains might be synthesised as tetramers and possibly polymerised as well before addition to the existing core. These reactions, like those leading to mucopeptide synthesis might involve a lipid carrier. The first indication of this came from experiments with a galactose-deficient strain which incorporated galactose from UDP-Gal into its LPS core. When extracts were incubated with UDP-Gal in the presence of TDP-rhamnose some of the galactose incorporated was found in a chloroform–methanol soluble fraction. Subsequent work mostly by Robbins, Osborn and their collaborators has shown the existence of a cycle rather like that found in mucopeptide synthesis. The same undecaprenol phosphate is the carrier lipid (Fig. 10.10). The cycle can be considered in three stages; firstly, synthesis of a lipid-linked tetrasaccharide; secondly, polymerisation of the tetramer and finally transfer of the polymeric tetrasaccharide to the core.

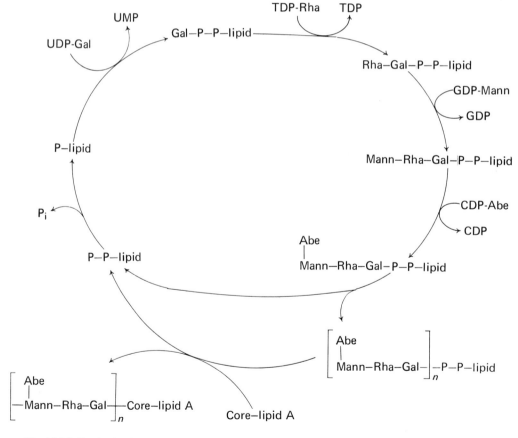

Fig. 10.10 Synthesis of the antigenic side-chains of the lipopolysaccharide of *Salmonella typhimurium*

1. Synthesis of lipid-linked tetrasaccharide. This part of the sequence is analogous to the synthesis of the disaccharide pentapeptide in mucopeptide biosynthesis. The sugars are added sequentially from their nucleotide derivatives to the non-reducing end of the molecule. A phosphate residue is transferred with the first sugar forming a pyrophosphate bond with the lipid acceptor and releasing UMP. The next three sugars, rhamnose, mannose and abequose, are added sequentially from their nucleotide derivatives to form a lipid-linked tetrasaccharide. All these enzymes are membrane-bound but can be solubilised by detergents. The only one so far studied in detail is the first enzyme (galactose-1-phosphate transferase) which like the first enzyme in mucopeptide synthesis (phospho-N-acetylmuramyl pentapeptide translocase) is reversible. Once solubilised it requires for activity both the antigen carrier lipid and cell membranes as well as its substrate UDP-galactose. It seems probable that, like the enzymes of core biosynthesis, it needs to be directionally orientated in the membrane for activity.

2. Polymerisation of tetramer. The next stage in the reaction sequence is quite different from that of mucopeptide synthesis where the disaccharide with its peptide side-chain is transferred from its lipid acceptor to the growing wall. In O-antigen synthesis, polymerisation apparently occurs while the tetrasaccharide is still attached to the carrier lipid in the membrane. The early experiments were rather confusing because both monomeric and polymeric tetrasaccharides were transferred to the complete core in some species. Some of these results may have been artefacts. Lipid-linked dimers of the tetrasaccharide unit have now been isolated and so polymerisation probably occurs before transfer to the core although this has not yet been proven. In the reaction a tetrasaccharide unit must be transferred from its carrier lipid to another polymer also linked to a carrier lipid. Robbins suggested that this could take place by two different mechanisms:

a The new tetrasaccharide could be transferred to the non-reducing end (the end not attached to the lipid) of the growing polymer as in the synthesis of most polysaccharides.
b The polymer could be transferred to the non-reducing end of the newly formed tetrasaccharide. This would be analogous to fatty-acid synthesis where the growing acyl chain is transferred from its acyl carrier protein to the new acetyl group, or to protein synthesis where the growing polypeptide chain is transferred from its own tRNA to the new amino acid located on the ribosomes through its tRNA. Pulse labelling experiments support this hypothesis (Fig. 10.11). In mechanistic terms it is probably easier to understand than the

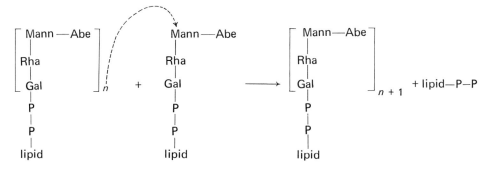

Fig. 10.11 Elongation of O-antigenic side-chains

alternative suggestion. Both the polymerase enzyme and the lipid carrier are membrane-bound so that the growing polysaccharide chain can move along it extending by one tetrasaccharide unit at a time.

With each turn of the cycle a tetrasaccharide unit is transferred to the lipid-linked polymer leaving a pyrophosphorylated carrier lipid molecule. In order to start the cycle again this must be dephosphorylated by the phosphatase which is inhibited by bacitracin.

3. Transfer of complete O-antigen to the core — ligase reaction. This is a most unusual reaction since it must involve the covalent linking of two macromolecules. It has been studied using a pair of mutants. One of these was defective in core biosynthesis but synthesised lipid-linked O-antigen, the other produced a complete core but made no side-chains. The LPS from the latter was used as acceptor, the first mutant providing the ligase enzyme preparation and the lipid-linked O-antigen. The reaction required low concentrations of non-ionic detergents possibly to achieve the close interrelationship necessary for reaction between two partially hydrophobic substrates.

Summary

The lipopolysaccharides of the Enterobacteriaceae consist of lipid A joined to a polysaccharide which can be divided into backbone outer core and side-chains. Little is known of the synthesis of the lipid or backbone.

The core polysaccharide is common to many Gram-negative bacteria. It is synthesised by the stepwise addition of sugars from their nucleotide derivatives. The sequence is determined by the specificity of the enzymes for both substrate and acceptor. The enzymes are probably arranged as a multienzyme complex in the membrane, but can be solubilised and require phosphatidyl ethanolamine for activity. The side-chains are synthesised by the addition of sugars from their nucleotide derivatives to an isoprenoid carrier lipid in the membrane. Polymerisation takes place while the oligosaccharides are linked to the lipid. Synthesis is inhibited by bacitracin.

The ligase which transfers the polymeric side-chain to the core is again membrane-bound and *in vitro* requires non-ionic detergents.

5 Teichoic acids

The teichoic acids, first described by Baddiley in 1959, are soluble polymers containing ribitol- and/or glycerol-phosphate units joined by phosphodiester linkages. They occur in all Gram-positive bacteria and may represent as much as 10 per cent of their dry weight. Both glycerol- and ribitol-teichoic acids can be extracted from isolated walls of Gram-positive bacteria by 10 per cent trichloracetic acid in the cold. They probably lie somewhere in the unstructured region towards the outside of the wall because they are the major antigenic determinants in some species, they protect against autolysis and they are involved in bacteriophage absorption. It is likely that they are attached covalently to the mucopeptide since teichoic acid chains linked to one or more mucopeptide units are released from walls by lysozyme. Hydrolysis of these chains yields muramic acid phosphate, but its linkage to the teichoic acid backbone is unknown.

In addition to the teichoic acids in the wall itself, glycerol teichoic acids occur somewhere between the membrane and the wall, probably covalently attached to phospholipid in the membrane. Unlike wall teichoic acids they are extracted by hot water or aqueous phenol. Originally they were called membrane teichoic acids, but since those which have been investigated more fully have been shown to contain lipid, they have been renamed lipoteichoic acids. They occur in all species so far examined even if there are no teichoic acids in the wall. The relationship between the amounts of the two polymers found under different growth conditions provides a clue as to their role. When magnesium is growth-limiting, for instance, the amount of teichoic acids in the walls is highest, while when phosphate is limiting they are almost completely replaced by different anionic carbohydrate polymers called teichuronic acids (Ellwood and Tempest, 1972). These contain no phosphate and their anionic nature is caused by acidic sugars. The main function of the teichoic acids appears to be to concentrate magnesium in or near the cytoplasmic membrane though this has been demonstrated unequivocally in only a few species (Hughes *et al.*, 1971).

5.1 Structure

Since their initial discovery a wide variety of teichoic acid polymers has been isolated from different species largely by Baddiley and his group. The details of their structures are described in a number of excellent review articles (Archibald, Baddiley and Blumson, 1968; Baddiley,

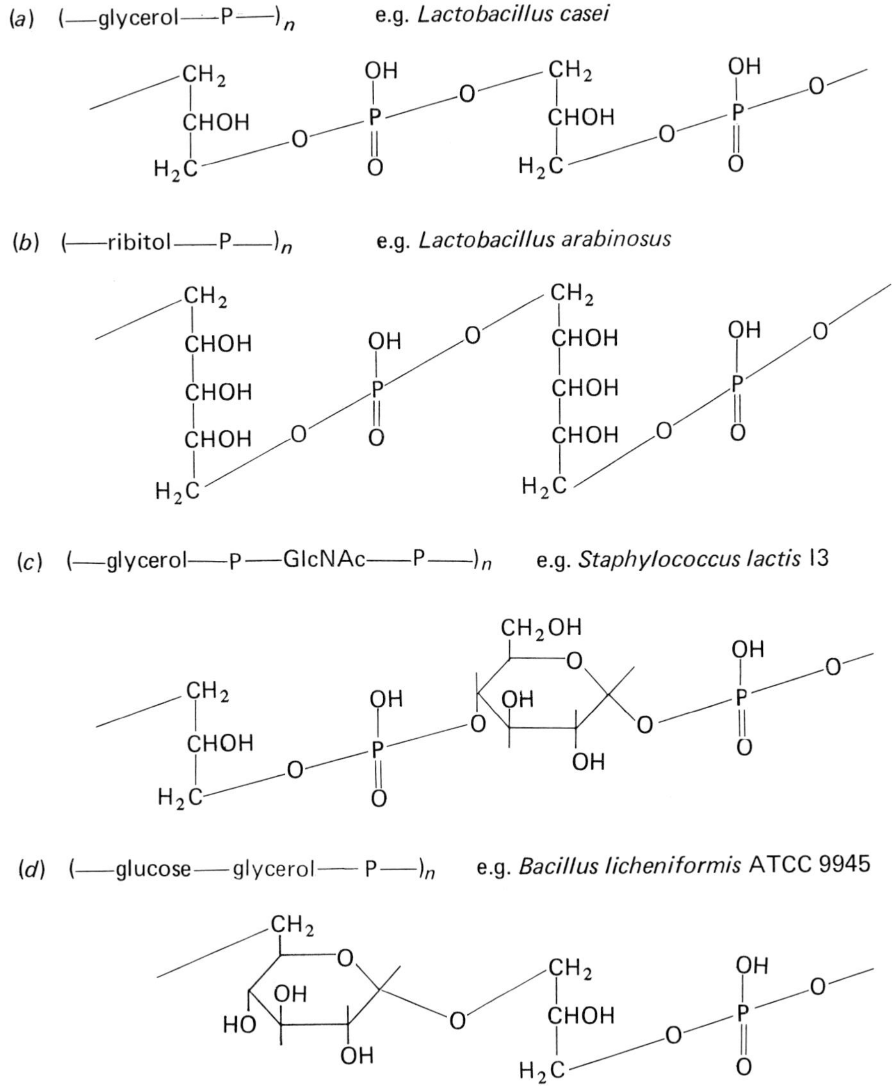

(a) $(—\text{glycerol}—P—)_n$ e.g. *Lactobacillus casei*

(b) $(—\text{ribitol}—P—)_n$ e.g. *Lactobacillus arabinosus*

(c) $(—\text{glycerol}—P—\text{GlcNAc}—P—)_n$ e.g. *Staphylococcus lactis* 13

(d) $(—\text{glucose}—\text{glycerol}—P—)_n$ e.g. *Bacillus licheniformis* ATCC 9945

Fig. 10.12 Some teichoic acid structures. (D-Alanyl and glycosyl substituents on the backbone have been omitted.)

1968, 1970, 1972). The teichoic acids all contain one or both of ribitol- or glycerol-phosphate; they have D-alanine and sugar substituents ester-linked to the hydroxyl groups of the polyols, though the amount of substitution varies. Some of them also contain sugar residues directly linked in the polyolphosphate backbone. A few of the structures with the alanine and sugar residues omitted for simplicity are shown in Fig. 10.12.

5.2 Synthesis

Teichoic acids are unusual among wall polymers in that we knew something of their synthesis before they were known to exist. The presence of CDP-glycerol and CDP-ribitol in bacterial extracts lead Baddiley to postulate that they must be precursors of glycerol- or ribitol-phosphate polymers. Subsequent work has confirmed this idea. In *Bacillus subtilis*, for instance, the teichoic acid is a polyglycerophosphate which is synthesised from CDP-glycerol by particulate preparations of lysed protoplast membranes while in *Lactobacillus plantarum* the ribitol teichoic acid is synthesised from CDP-ribitol. The membrane-bound enzymes cannot be solubilised by sonication or treatment with detergents. They require high concentrations of magnesium or calcium ions suggesting the need for some degree of organisation in the membrane fragments. Synthesis is inhibited by the antibiotic novobiocin though the molecular basis of the effect is by no means clear. The direction of polymer growth has been determined in a few species. It occurs by addition to the glycerol or sugar terminal end of the chain. This is characteristic of polysaccharide synthesis and different from that which is used to extend the polysaccharide chains in the lipopolysaccharide. The sugar residues esterified to the polyol backbone appear to be added soon after if not with backbone synthesis. The addition of alanine is thought to require ATP and may be one of the final stages in biosynthesis but at present little is known of the details.

The discovery of teichoic acids with sugars as well as glycerol in the polymer backbone has allowed more detailed study of biosynthesis because of the possibility of isolating intermediates by limiting one of the substrates. It now seems probable that as in mucopeptide and lipopolysaccharide synthesis part of the biosynthesis of teichoic acids takes place while bound to an isoprenoid lipid carrier in the membrane. The lipid intermediates have not been characterised chemically but indirect evidence suggests that the isoprenoid carrier lipid is shared between the mucopeptide and teichoic acid-synthesising systems as we shall see later.

Teichoic acid synthesis appears to proceed similarly whatever the backbone. I shall discuss the synthesis of two of the more complex ones. The wall teichoic acid of *Staphylococcus lactis* I3 consists of alternating units of glycerol and N-acetylglucosamine joined by phosphodiester bonds while that of *Bacillus licheniformis* ATCC 9945 contains glucose glycosidically-linked to glycerol-phosphate (see Fig. 10.12). Particulate preparations of both organisms use CDP-glycerol as one of the substrates for teichoic acid synthesis, the other substrate being UDP-GlcNAc for *S. lactis* and UDP-Glc for *B. licheniformis*. By using radioactive substrates and pulse-labelling, these precursors were shown to be incorporated sequentially into butanol-soluble lipid intermediates whose properties are consistent with their identification as isoprenoid compounds. The reaction sequences shown in Fig. 10.13 have been proposed. The first step is the transfer of sugar or sugar-phosphate from its nucleotide derivative to the membrane-bound lipid; the addition of glycerol phosphate from CDP-glycerol being the second stage. The complete phosphodiester-linked unit is then transferred from the lipid carrier to the growing teichoic acid chain.

Two different types of phosphate transfer occur in these organisms. In *Staphylococcus lactis* a pyrophosphate bond is formed when N-acetylglucosamine phosphate is transferred to the

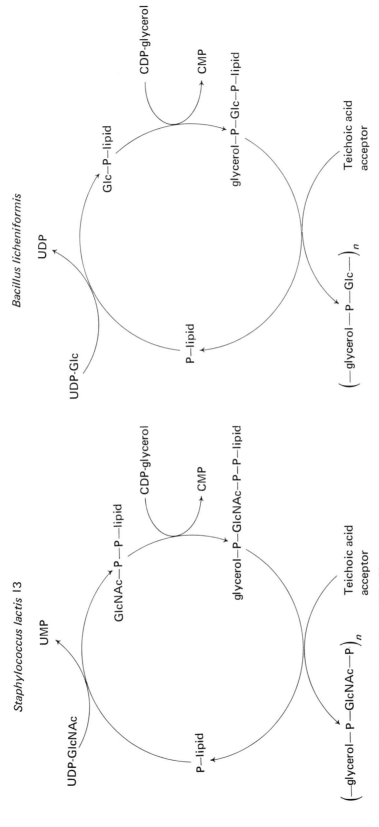

Fig. 10.13 Synthesis of the teichoic acid backbone.

lipid acceptor but it is broken when the complete unit is added to the growing polymer. The lipid monophosphate acceptor is thus released and is ready to restart the cycle without further dephosphorylation. In *Bacillus licheniformis,* where the teichoic acid contains only one phosphate per glucosyl-glycerol unit, the initial step is the transfer of glucose from its nucleotide derivative to form an ester with the phosphate of the lipid carrier. Glycerol-phosphate is then added, but the transfer to the growing chain is a transglycosylation and not a transphosphorylation leaving lipid monophosphate in the membrane. Neither of the cycles involves a step like that in mucopeptide or lipopolysaccharide synthesis in which transfer of the precursor to the polymer leaves a lipid pyrophosphate which must be dephosphorylated before it can start a new cycle.

It is because there is no lipid pyrophosphate-phosphatase in teichoic acid synthesis that it has been possible to show without isolating it, that the same acceptor is used for both mucopeptide and teichoic acid synthesis. If such a phosphatase were involved in the cycle it should be inhibited by bacitracin as is mucopeptide synthesis. Teichoic acid synthesis is not normally inhibited by bacitracin nor by vancomycin which inhibits the transfer of disaccharide-peptide units to the mucopeptide chain. If, however, the same lipid acceptor were shared between mucopeptide and teichoic acid synthesis then any compound preventing release of lipid monophosphate should inhibit teichoic acid synthesis. The same particulate preparations contain the enzymes for both functions. In the presence of the substrates required for the mucopeptide, UDP-GlcNAc and UDP-MurNAc-peptide, teichoic acid synthesis is inhibited by bacitracin and vancomycin. Under these circumstances all the available lipid acceptor is bound as the disaccharide peptide pyrophosphate, trapped by vancomycin, or as the lipid pyrophosphate trapped by bacitracin. The most probable explanation is that the lipid acceptor is shared between the two systems and has a higher affinity for the enzymes of mucopeptide synthesis than for those which synthesise teichoic acids (Hussey and Baddiley, 1972; Hancock and Baddiley, 1972).

Summary

Teichoic acids are synthesised by the stepwise addition of the backbone components from nucleotide derivatives to a lipid in the membrane before transfer to the growing chain. The transferase enzymes are membrane-bound and are not readily solubilised. Synthesis occurs by addition to the end away from the phosphate. It is inhibited by novobiocin, but only indirectly by bacitracin and vancomycin. Glycosylation of the backbone and addition of D-alanine probably occur after the main chain is synthesised.

6 General

One of the interesting aspects of the synthesis of these wall polymers is the way in which biosynthetic pathways can be modified to allow a complex polymer of specific sequence to be assembled outside the cytoplasmic membrane. The sequence of all three polymers is determined inside the membrane by the specificities of the transferase enzymes both for their nucleotide or amino-acid substrates, and for the acceptors. Transfer to an isoprenoid lipid in the membrane is a necessary part of synthesis and in the case of the O-antigen polymerisation occurs at this stage. Even before polymerisation some of the biosynthetic enzymes appear to require phospholipid membrane components. As more of these enzymes are solubilised we may see more phospholipid involvement, not necessarily as covalently-linked carriers, but as activators used to bring the water-soluble proteins and polymer precursors into closer orientation around the membrane. In several stages of the synthesis the need for some kind of physical organisation is apparent as discussed by Ellar (1970). Rothfield and Romeo (1971) have put forward a neat hypothesis for the synthesis of the lipopolysaccharide on enzymes fixed in the phospholipid membrane matrix and Anderson, Hussey and Baddiley (1972) have suggested a similar system for teichoic acid synthesis. The idea of fixed enzymes passing growing polymer chains along a biosynthetic pathway is an attractive one but remains to be proven.

The other facet of this work which is so far unexplained is the determination of shape interestingly discussed by Rogers (1970). When the mucopeptide component was first isolated and shown to have the shape of the bacterium from which it was derived, it seemed obvious that it was the component which determined that shape. Removal of mucopeptide by whatever means resulted in the conversion of rod forms to round ones. But although in both Gram-positive and Gram-negative bacteria the mucopeptide may give mechanical support, it is difficult to see how it could *define* the shape. There are mutants which grow as round forms in some growth conditions and as rod forms in others. Preliminary evidence appeared to relate these changes in morphology to alterations in chain length of the mucopeptide or the amount of teichoic acids present. However, in at least two species, *Escherichia coli* and *Bacillus subtilis*, there is no detectable difference between the walls of the rod and round forms. Rogers, McConnell and Hughes (1971) suggest that the changes in *B. subtilis* may be related to the presence of mesosomes, while Henning *et al.* (1972) have implicated the cytoplasmic membrane. Whatever the final conclusions a great deal of work will be necessary to determine the interrelationship between the biosynthetic and degradative enzymes of wall-polymer synthesis and the roles played by membranes and mesosomes in the morphogenetic processes of bacterial elongation and septation.

References

ANDERSON, R. G., HUSSEY, H. and BADDILEY, J. (1972). 'The mechanism of wall synthesis in bacteria. The organisation of enzymes and isoprenoid phosphates in the membrane', *Biochem. J., 127,* 11-25.

ARCHIBALD, A. R., BADDILEY, J. and BLUMSON, N. L. (1968). 'The teichoic acids', *Adv. Enzymol. Relat. Areas Mol. Biol.,* **30,** 223-53.

BADDILEY, J. (1968). Leeuwenhoek Lecture. 'Teichoic acids and the molecular structure of bacterial walls', *Proc. Roy. Soc.* Ser. B., **170,** 331-48.

BADDILEY, J. (1970). 'Structure, biosynthesis and function of teichoic acids', *Accounts Chem. Res.,* **3,** 98-105.

BADDILEY, J. (1972). 'Teichoic acids in cell walls and membranes of bacteria', in *Essays in Biochemistry* (P. N. Campbell and F. Dickens eds.) Vol. 8, 35-77. Academic Press, London & New York.

ELLAR, D. J. (1970). 'The biosynthesis of protective surface structures of prokaryotic and eukaryotic cells', in *Organisation and Control of Prokaryotic and Eukaryotic cells* (H. P. Charles and B. C. J. G. Knight eds.) *Symp. Soc. Gen. Microbiol.,* **20,** 167-202.

ELLWOOD, D. C. and TEMPEST, D. W. (1972). 'Effects of environment on bacterial wall content and composition', in *Advances in Microbial Physiology* (A. H. Rose and D. W. Tempest eds.) **7,** 83-117. Academic Press, London & New York.

GHUYSEN, J. M. (1968). 'Use of bacteriolytic enzymes in determination of wall structure and their role in cell metabolism', *Bacteriol. Rev.,* **32,** 425-64.

GHUYSEN, J. M., STROMINGER, J. L. and TIPPER, D. J. 'Bacterial cell walls', in *Comprehensive*

Biochemistry (M. Florkin and E. H. Stotz eds.) **26A**, 53–104. Elsevier, Amsterdam.

GLAUERT, A. M. and THORNLEY, M. J. (1969). 'The topography of the bacterial cell wall', *Ann. Rev. Biochem.*, **23**, 159–98.

HANCOCK, I. C. and BADDILEY, J. (1972). 'Biosynthesis of the wall teichoic acid of *Bacillus licheniformis*', *Biochem. J.*, **127**, 27–37.

HARTMAN, R., HOLTJE, J.-V. and SCHWARZ, U. (1972). 'Targets of penicillin action in *Escherichia coli.*', *Nature, Lond.*, **235**, 426–29.

HENNING, U., REHN, K., BRAUN, V. and HOHN, B. (1972). 'Cell envelope and shape of *Escherichia coli* K 12. Properties of a temperature-sensitive *rod* mutant', *Eur. J. Biochem.*, **26**, 570–86.

HUGHES, A. H., STOW, M., HANCOCK, I. C. and BADDILEY, J. (1971). 'Function of teichoic acids and effect of novobiocin on control of Mg^{2+} at the bacterial membrane', *Nature New Biol.*, **229**, 53–55.

HUSSEY, H. & BADDILEY, J. (1972). 'Lipid intermediates in the biosynthesis of the wall teichoic acid in *Staphylococcus lactis* I3, *Biochem. J.*, **127**, 39–50.

LUDERITZ, O., JANN, J. and WHEAT, R. (1968). 'Somatic and capsular antigens of Gram-negative bacteria', in *Comprehensive Biochemistry* (M. Florkin and E. H. Stotz eds.) **26A**, 105–228. Elsevier, Amsterdam.

LUDERITZ, O., WESTPHAL, O., STAUB, A. N. and NIKAIDO, H. (1971). Isolation and chemical and immunological characterization of bacterial lipopolysaccharides', in *Microbial Toxins* (G. Weinbaum, S. Kadis and S. J. Ajl eds.) **4**, 145–232. Academic Press, New York & London.

NEITO, M., PERKINS, H. R. and REYNOLDS, P. E. (1972). 'Reversal by a specific peptide (diacetyl-L-diaminobutyryl-D-alanine-D-alanine) of vancomycin inhibition in intact bacteria and cell-free preparations', *Biochem. J.*, **126**, 139–49.

NEUHAUS, F. C. (1967). 'D-Cycloserine and O-carbamyl-D-serine', in *Antibiotics Mechanism of action* (D. Gottlieb and P. D. Shaw eds.) **1**, 40–83. Springer-Verlag, Berlin.

NIKAIDO, H. (1968). 'Biosynthesis of cell-wall lipopolysaccharide in Gram-negative enteric bacteria', *Adv. Enzymol. Relat. Areas Mol. Biol.*, **31**, 77–124.

OSBORN, M. J. and ROTHFIELD, L. I. (1971). 'Biosynthesis of the core region of lipopoly-saccharide', in *Microbial Toxins* (G. Weinbaum, S. Kadis and S. J. Ajl eds.) **4**, 331–60. Academic Press, New York & London.

REAVELEY, D. A. and BURGE, R. E. (1972). 'Walls and membranes in bacteria', in *Advances in Microbial Physiology* (A. H. Rose and D. M. Tempest eds.) **7**, 1–81. Academic Press, New York and London (pp. 30–46 and 47–68 are particularly relevant).

ROBBINS, P. W. and WRIGHT, A. (1971). 'Biosynthesis of O-antigens', in *Microbial Toxins* (G. Weinbaum, S. Kadis and S. J. Ajl eds.) **4**, 351–68. Academic Press, New York & London.

ROGERS, H. J. (1970). 'Bacterial growth and the cell envelope', *Bacteriol. Rev.*, **34**, 194–214.

ROGERS, H. J., McCONNELL, M. and HUGHES, R. C. (1971). 'The chemistry of the cell walls of *rod* mutants of *Bacillus subtilis*', *J. Gen. Microbiol.*, **66**, 297–308.

ROGERS, H. J. and PERKINS, H. R. (1968). *Cell Walls and Membranes*. E. & F. Spon Ltd., London. (This book covers plants and animals as well as bacteria.)

ROTHFIELD, L. and ROMEO, D. (1971). 'Role of lipids in the biosynthesis of the bacterial cell envelope', *Bacteriol. Rev.*, **35**, 14–38.

SALTON, M. R. J. (1964). *The Bacterial Cell Wall*. Elsevier, Amsterdam. (A comprehensive review of structure and synthesis up to 1964.)

STEWART, T. S., ROBERTS, R. J. and STROMINGER, J. L. (1971). 'Novel species of tRNA', *Nature, Lond.*, **230**, 36–38.

STROMINGER, J. L. (1969). 'Penicillin-sensitive enzymatic reactions in bacterial cell wall synthesis', *The Harvey Lectures*, **64**, 179-213.

STROMINGER, J. L., BLUMBERG, P. M., SUGINAKA, H., UMBREIT, J. and WICKUS, G. G. (1971). 'How penicillin kills bacteria: progress and problems', *Proc. Roy. Soc.* Ser. B **179**, 369-83.

STROMINGER, J. L., IZAKI, K., MATSUHASHI, M. and TIPPER, D. J. (1967). 'Peptidoglycan transpeptidase and D-alanine carboxypeptidase: penicillin-sensitive enzymatic reactions', *Fed. Proc. Fedn. Am. Socs Exp. Biol.*, **26**, 9-22. (A concise review including the experimental results from which the conclusions are drawn.)

11
The Structural Development
of Plant Cells

Keith Roberts,

John Innes Institute, Colney Lane, Norwich

1 Introduction

1.1 General remarks

Within the scale of this chapter it would be impossible to offer a comprehensive summary of the basic facts and ideas relevant to the development of plant cells, and considerable selection (and exclusion) has had to be exercised. In the first place, the ideas covered in the subsequent sections refer by and large to higher plants only, and in particular to the angiosperms. In choosing the higher plant cell as a starting point, it is further assumed that the reader has a general understanding of both the structure and functions of the major cell organelles, which leaves more room to discuss here the relevance of those structures peculiar to plant cells (e.g. cell walls) and those plant cell structures not usually discussed at length (e.g. microtubules and the membrane system in relation to cell development). The general plan then is to describe the main ultrastructural basis of the growth and development of the plant cell together with some biochemical data where applicable.

1.2 Plant growth

1.2.1 Meristems and primary growth

In contrast to animals, the production of new cells within a plant is confined to a few permanently embryonic regions called meristems. These perpetually young tissues, concerned primarily with growth by cell division, occur at the two ends of the main axis of the plant, i.e. at the shoot and root apex, where they are called apical meristems. Derivatives of these may become new apices and hence give rise to the branching of the shoot or root system. The cellular derivatives of divisions within a meristem follow different paths of development. Some cells remain and constitute the permanent initiating cells of the meristem, while others, following expansion and elongation, develop into the various tissue cell types that contribute to the overall growth of the plant. This growth, by cell division and elongation which can be clearly seen in the cells derived from a root apical meristem, is called primary growth, and it has both a quantitative and a qualitative component. Quantitatively the cell, derived for example from a root apical meristem, may undergo perhaps an hundredfold increase in volume. This is accompanied by extensive vacuolation, resulting in a reduction of the cytoplasm to a thin peripheral layer, coupled with an enormously increased water uptake by the cell. To accommodate this volume change the cell wall must also increase in area and this is correlated not only with an increase in cell-wall plasticity, but with active cell-wall deposition to maintain its thickness. Although the cell is now more vacuolate, there is also active synthesis of new cytoplasm during the phase of cell elongation. The qualitative changes in a plant cell during growth usually involve both a radical restructuring of the cell wall, with the deposition of new components, and an alteration of cytoplasmic components, usually associated with the acquisition of a specific cell function. Such changes constitute the process of cell differentiation. In a shoot for example, cells derived from the meristematic initials at the shoot apex, eventually elongate and may eventually differentiate to form primary tissue cells, for example those concerned with transport. Two main transport systems exist in the plant, the phloem and the xylem, and both arise by elaborate differentiation of the vacuolate elongated cells comprising the procambium. Phloem, and here we are only considering angiosperms, consists of a continuous conducting system of living cells leading from a source of carbohydrate to a place in the plant where it can be used or stored. The main substance translocated is sucrose, in concentrations as high as 30 per cent, and at speeds in the region of 1 m/h. Phloem comprises four main cell types, the sieve elements, which are the principal conducting cells, companion cells, unspecialised parenchyma cells and fibres. Xylem on the other hand is the main water conducting tissue of the plant and is composed of numerous tracheary elements having characteristic annular or spiral thickenings in their cell walls. When mature they are basically dead cells and the events which lead to the wall thickenings and the dissolution of the cell contents involve an elaborate planned sequence of cell organelle movements.

1.2.2 Secondary growth

The vascular tissue just described is usually, in the stem, arranged in discrete bundles of cells

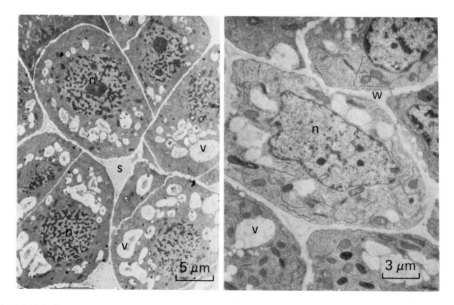

Fig. 11.1(a) Cells from a transverse section of a wheat root meristem. The cells have dense cytoplasm, thin walls and large nuclei (n). The vacuole (v) is represented by a few large vesicles. Intercellular spaces (s) may be seen.

(b) Similar cells from a pea root meristem. The thin walls (w), large nucleus (n) and small vacuole (v) are clearly visible. (Micrograph, courtesy of B. Wells.)

with the phloem towards the outside and xylem towards the centre. For any large scale increase in stem width new cells will be produced by a process of secondary growth. Secondary tissues, largely xylem and phloem, are produced from cells derived by divisions in a newly formed secondary meristem, the vascular cambium. This is a system of thin-walled, elongate, vacuolate cells forming a cylinder about three or four cells deep within the stem, separating the xylem and phloem. Divisions occur tangentially to the stem axis and cell derivatives cut off on the outside will differentiate to form phloem cells, while those on the inner side will form xylem cells.

1.3 The plant cell

The various tissue systems in the intact plant are formed by the specific differentiation of cells at particular locations, and it is to the cellular level that we must go in order to understand the structural basis of such development. It is convenient to regard a cell in the apical meristem as a basic cell type and to discuss any subsequent modifications of it through cell differentiation in relation to this basic type. What then are the important structural features of a young meristematic cell? The normal cell, as found for example in the apex of a root tip, is a roughly isodiametric entity (10–20 μm in diameter) surrounded by a thin primary cell wall (Fig. 11.1). Several thousand will together constitute the apical meristem. The primary cell-wall is a semi-rigid structure laid down outside the cell's bounding membrane, or plasmalemma, and is usually about 0·1–0·5 μm in thickness. It consists of an amorphous matrix of pectin and hemicellulose within which are laid down orientated bands of cellulose microfibrils, in much the same way as glass fibres are found in a sheet of fibreglass. The cell wall is not a continuous barrier between adjacent cells, however, and nearly all the cells of a plant are interconnected by a system of slender cytoplasmic strands crossing the cell wall through structures known as plasmodesmata. Several thousand such plasmodesmata can be found for example in the end wall of a root tip cell, although fewer are usually found on the lateral walls. They presumably provide for the intercellular movement of certain large and small molecules, although how such transport is controlled is not known. Structurally they are formed from a tube through the cell wall, about 50 nm in diameter, lined by the plasmalemma membrane. The membrane systems of adjacent cells are therefore in physical continuity. Occasionally branched plasmodesmata are found. Through the centre of a plasmodesmata is usually found a central dense core, of unknown structure, which appears to be derived from endoplasmic reticulum associated with either side of the pore (Fig. 11.2).

Within the cytoplasm of all plant cells is found a system of single membrane bound cavities, which in many cases may occupy at least 90 per cent of the cell volume, and which is called the vacuole. In the meristematic cell the vacuole is usually small, and consists of a diffuse network of cavities and vesicles. As the cell elongates, the vacuole enlarges by the fusion of these smaller vesicles accompanied by a massive uptake of water by osmosis, and results eventually in a large, central, fluid filled cavity with the cytoplasm reduced to a thin layer adjacent to the cell wall. Cytoplamic strands usually exist, crossing the vacuole, and concentrated predominantly on the nucleus. The membrane surrounding the vacuole is called the tonoplast and it possesses very specialised permeability properties compatible with the various functions of the vacuole. Of these functions three main ones may be defined. First is the requirement for the maintenance of the cell's turgor pressure by the transport of water across the tonoplast. Second, in many cases smaller vacuoles may act as storage organelles. Products of metabolism may accumulate selectively within the vacuole and may even crystallise there. A parallel with animal cell microbodies may be made and, in fact, several classes of single membrane bound bodies within the plant cell have been designated microbodies (Frederick *et al.* 1968). Third, the vacuole represents an extracytoplasmic compartment which can cope with waste products. Hydrolases and acid phosphatases have been detected there and encourage an analogy with the animal cell lysosomes. Non-metabolisable waste products may also accumulate in the vacuole, e.g. tannins

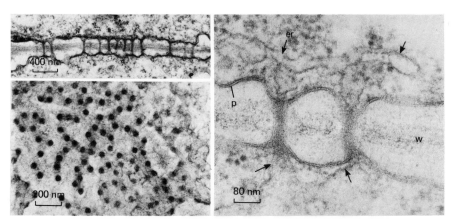

Fig. 11.2 (*a*) A vertical section through the cell wall of the male fern *Dryopteris filix-mas*. Numerous plasmodesmata can be seen crossing the wall.

    (*b*) A glancing section of a wall (w) from the same material showing the plasmodesmata in cross-section.

    (*c*) A section of a cell wall from *Dryopteris*. Two plasmodesmata are seen at high magnification lined by the plasmalemma membrane (p). The close association of profiles of endoplasmic reticulum (er) with either side of the pore may be seen. (Micrographs taken by Dr J. Burgess.)

and anthocyanins, and may crystallise there, e.g. calcium oxalate. Material may enter the vacuole either by transport across the membrane or by the pinching off of regions of cytoplasm by a reverse pinocytotic mechanism through the tonoplast. It must be emphasised that not only do the size and function of the vacuoles change from cell to cell, and even within one cell, but that a single vacuole may change its nature during the life cycle of a cell. For example the osmotic pressure of the vacuolar sap varies with the age of the cell. The cytoplasm of the plant cell is not a static entity but exhibits a constant motion, usually seen as the phenomenon of cytoplasmic streaming. Both the ground cytoplasm and the organelles within it can show motility, and in most higher plant cells the rate of this streaming is within the range of 1-10 μm/s. The whole cytoplasm may show a general cyclical motion or the organelles within it may show organised directed motion within small microchannels. Streaming also occurs in the strands of cytoplasm which cross the vacuole. The exact function of streaming is unknown but presumably the fast organised transport of material around the cell and from one cell to another is facilitated by it. Two specific cell organelles have been implicated with the propagation and direction of streaming, the microtubules and the microfilaments. Microtubules are slender protein tubules, about 24 nm in diameter, found in almost all plant cells, and in many cases they have been shown to be aligned parallel to the direction of streaming. In other cell types bundles of microfilaments, protein rods about 5-10 nm in diameter, have been demonstrated. Whether either or both are the causative agents in streaming, or whether they merely provide directionality, is as yet uncertain.

    One of the key problems, in structural terms, of the growth and development of the plant cell is the organised synthesis and transport of new materials, both within the cell and outside the cell, to specific locations. Several pathways of transport may be seen in the cell, concerned with the movement of different entities, and one of these is the continuous membrane system. The cytoplasmic membranes constitute the selectively permeable boundaries of a series of cellular compartments, and also provide an enormous surface area on which numerous specific biochemical reactions can occur. A lot of evidence has accumulated now to suggest that we should also regard the cells membrane system as bounding a continuous lumen involved in transport. The predominant direction of flow through this system is from the nuclear envelope to the endoplasmic reticulum, to the golgi apparatus, and via vesicles to the plasmalemma.

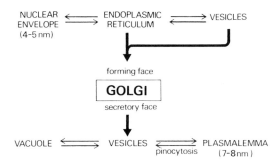

Fig. 11.3 A diagrammatic representation of the probable membrane flow in plant cells, which emphasises the crucial regulatory position held by the golgi apparatus as a 'one-way valve'.

Although either end of this pathway is to some extent reversible (e.g. uptake of material by pinocytosis through the plasmalemma), the flow to and from the golgi apparatus is one way, and this highlights the organelle as a one-way valve in the system and hence as an important control point. This flow pattern of both membrane and contained materials is shown diagrammatically in Fig. 11.3. The endoplasmic reticulum and the nuclear envelope are in direct continuity with each other and both are composed of unit membranes about 4-5 nm across. The relationship between the golgi apparatus and the endoplasmic reticulum is more indirect and may involve either a direct incorporation of sheets of endoplasmic reticulum into one face of a golgi body, or an indirect incorporation via a vesicular shunt. Important in this context is the concept of the golgi body as a polarised cell organelle, with a structural and chemical gradient existing from one face to the other. The generally accepted model for golgi function envisages one face as being an immature or forming face, where membrane material is accumulated, followed by a movement and modification of the individual golgi cisternae across the stack, being secreted in the form of vesicles at the other face, the mature or secretory face. Modifications to both the membranes themselves and the cisternal contents occur across the golgi as a whole. For example the membrane thickness in sectioned material is much larger at the secretory face and resembles that of the plasmalemma (7-8 nm). The golgi apparatus therefore is a key organelle in the mediation of membrane synthesis and modification, but also of prime importance is the material contained within the lumen of the golgi apparatus and its associated vesicles (Fig. 11.4). The golgi apparatus is the site of both the synthesis and modification of complex polysaccharides and glycoproteins, in particular those destined for transport to the cell wall. The type and amount of polysaccharide formed depends on the metabolic state of the cell, its age and hormonal and nutritional status, as well as where the cell is within the whole plant. This may be seen, for example, in the root tip. The golgi apparatus in the maize root cap synthesises a mucopolysaccharide slime which is excreted from the cell, and this is characterised by fucose comprising 30 per cent of its neutral sugar residues. However,

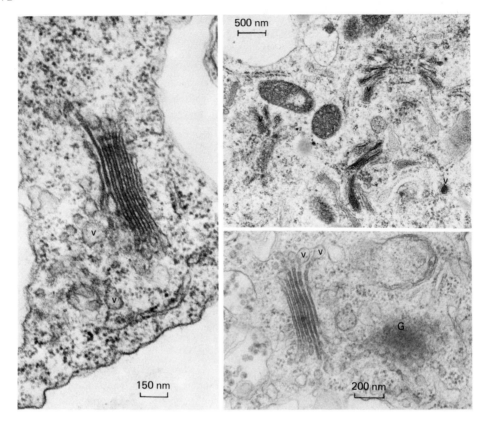

Fig. 11.4 (*a*) A golgi body from a sycamore cell in vertical section, showing the cisternae from which it is made and vesicles (v) budding off from their edges.

(*b*) Golgi bodies from the root cap of maize demonstrating the specialisation of golgi function. Hyper-trophied cisternae are budding off large dense vesicles (v).

(*c*) Golgi bodies from a sycamore tissue culture cell, showing both a vertical section and a horizontal section (g). Both are budding off vesicles (v).

only a few cells away, in the root apical meristem, the golgi apparatus is producing polysaccharides which have no fucose present at all (Harris and Northcote 1970). This specialisation of golgi function can also be seen at cell division when the golgi apparatus produced numerous specialised vesicles which contribute to the matrix materials for the cell plate and the new plasmalemma (e.g. Hepler and Newcomb 1967). Glycoproteins are also made within the golgi apparatus, or rather the glycosylation steps are carried out there. Proteins, made on the ribosomes of the rough endoplasmic reticulum, are thought to be transported within this membrane system, via the smooth endoplasmic reticulum and vesicles to the golgi apparatus, where the appropriate glycosyltransferase can attach the oligosaccharide side-chains. (For speculations on the significance of glycosylation for the fate of a protein molecule, see Winterburn and Phelps 1972.) The evidence from which the above functions of the golgi apparatus have been deduced is largely obtained from labelling studies combined with high resolution autoradiography, but unfortunately far more experiments have been done with animal-cell systems than with plants. (For a comprehensive review of the origin and functions of the golgi apparatus see Morré, Mollenhauer and Bracker (1971). For a detailed discussion of its role in the determination of cell surface and extracellular materials see Whaley, Dauwalder and Kephart (1972), and for a discussion of synthesis and membrane flow in plants, see Northcote (1971).)

The lumen of the cell's membrane system is not the only transport route of importance in the cell. Central to any idea of cell development is the sequential expression of the information encoded in the nuclear genome, and from this follows the importance of controls on the transport of such information between the nucleus and cytoplasm. Depending on the material to be transported there are three routes by which material can cross the nuclear envelope. First,

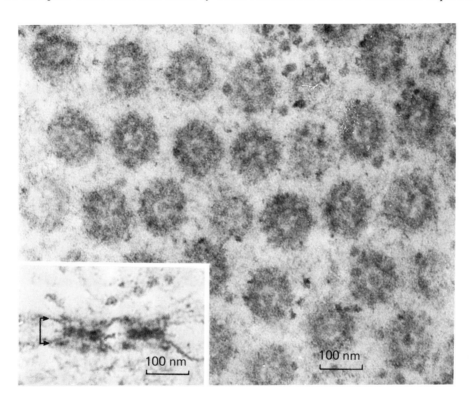

Fig. 11.5 (*a*) A horizontal section through the nuclear envelope of a pollen mother cell from *Fritillaria lanceolata*. Numerous nuclear pores may be seen possessing a central granule surrounded by a ring of electron dense sub-units.

(*b*) A vertical section through the nuclear envelope (arrowed) of a similar cell showing two nuclear pores from the side. The granules which act as a structural barrier within the pore may be seen. (Both micrographs by courtesy of L. F. La Cour of the John Innes Institute, Norwich.

and probably not of great import, is the formation of vesicles by the outer membrane, or both membranes, of the nuclear envelope. Second, is the direct diffusion of material across the membranes, and third, is the passage of material through the specialised regions of the nuclear envelope called nuclear pores. The nuclear pore is a circular gap in the envelope about 85 nm in diameter, the rim of which supports three stacked annuli each composed of about eight granules. The overall pore diameter is about 115 nm. In the centre of most pores is another granule of uncertain structure (Northcote 1971). The pore may be seen, therefore, as an elaborately structured organelle, which does not present a completely free passage for diffusion (Fig. 11.5). Experiments using colloidal gold particles have shown that the largest particles that can be taken into the nucleus are about 15 nm in diameter. The granules within the pore still present problems of interpretation, and although they probably contain ribonucleoprotein their exact role in nucleocytoplasmic transport is unclear.

Changes in cells during the process of growth and differentiation always involve changes and redistributions in the organelles of those cells. These changes can be of various kinds. There may be changes in the type of organelle present, for example, in the shoot apex and leaf, differentiating cells will develop large numbers of chloroplasts, whereas before only proplastids and a few chloroplasts were present. The numbers of a particular organelle may also change during cell development. In the root cap the outermost cells contain large numbers of highly specialised golgi bodies which are synthesising slime, but the root cap initials possess fewer less specialised golgi. The localised distribution of organelles may also change as for example in the complicated redistributions of endoplasmic reticulum and microtubules during cell division and xylem formation (see sections 2 and 3). Coupled with this is the fact that for all these variations a strict timetable operates, such that the right organelles are in the right place at the right time, and about such overall problems of cellular dynamics we have very little information.

1.4   Plant hormones

Both the rate of growth and the pattern of differentiation of any particular plant cell, and hence of the whole plant, are in part under the control of chemicals which are translocated from other regions of the plant. These growth regulating substances have generally been called plant growth hormones, and broadly speaking they fall into three main classes; the auxins; the gibberellins; and the cytokinins. Some information about the structure and function of each of these classes is considered in turn.

1.4.1   Auxins

These were the first discovered plant hormones, and although several related molecules have auxin-like activity, the main one is indole-3-acetic acid (IAA) (Fig. 11.6(*a*)). This is thought to be primarily synthesised in stem apices and leaves, and in shoots is transported basipetally only (from apex to base) i.e. shows polar transport. As with all hormones, the precise concentration at the site of its action is of crucial importance, and this will depend on the rate of synthesis, the rate of degradation (by both photoxidation and IAA oxidase) and its inhibition by various endogenous inhibitors. The details of its synthetic pathway are still unclear. The effects that auxin produces vary with its site of action and also on the presence or absence of other hormones. In general, in stems it is essential for cell elongation to occur, and for cambial cell divisions to take place. It is also involved in fruit growth and apical dominance. The exact mechanism of auxin action is not understood but some clues are emerging from recent work. It was originally demonstrated by several workers that IAA can influence RNA synthesis in plant cells. Using a purified system from the chromatin of coconut nuclei it has been further shown that RNA synthesis is only stimulated if both IAA and an acceptor protein are present. Under these conditions, RNA/DNA hybridisation studies revealed that new RNA species were being synthesised (e.g. Mondal et al. 1972). It must be emphasised, however, that we are far from understanding the complete mechanism of even one of IAA's many effects.

1.4.2   Gibberellins

There are many naturally occurring gibberellins, all based on the gibbane carbon skeleton (Fig. 11.6(*b*)) with various side-chain substituents. The most common one worked on is gibberelic acid (Fig. 11.6(*c*)). This is synthesised in the plant using five-carbon isoprene units to build various terpinoid intermediates. Unlike auxin it shows non-polar movements within the plant, but like auxin it appears to be involved in the extension growth of stem tissues. There is a close synergistic interaction between these two hormones, both often being necessary for a full tissue response. As with auxin no definite general mechanism of gibberelin action has been elucidated.

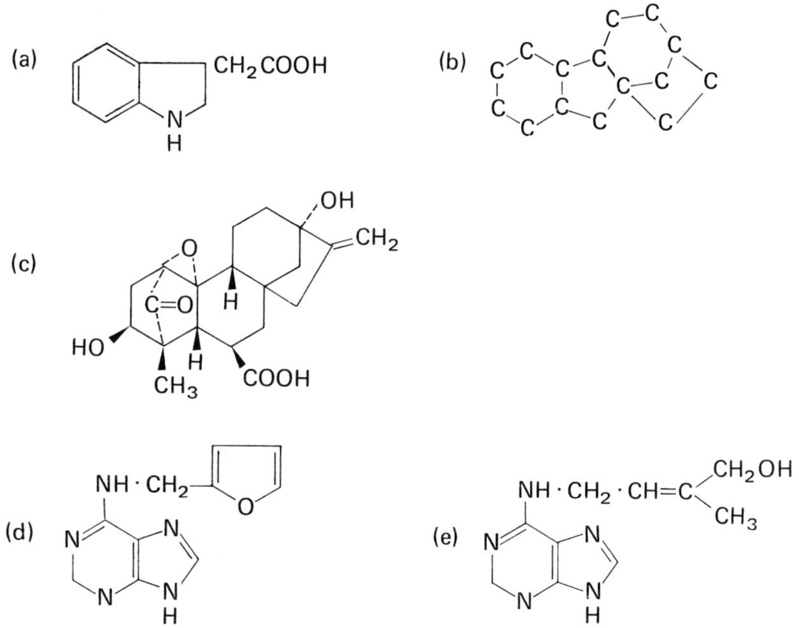

Fig. 11.6 The structural formulae of some plant hormones: (a) Auxin (indole acetic acid) (b) Gibbane carbon skeleton, (c) Gibberellic acid (GA$_3$), (d) Kinetin [6-furfurylaminopurine], (e) Zeatin [6-(4-hydroxy-3-methyl-but-2-enyl) amino purine].

A system which has been used to study the details of gibberelin action is the aleurone tissue of barley. This homogeneous population of cells responds to as little as 10^{-10}M gibberelic acid by producing the starch digesting enzyme α-amylase. RNA synthesis is essential for the hormone to elicit this response and it is probable that the hormone binds to an acceptor protein and that this modifies a post-transcriptional step in the expression of α-amylase mRNA (Carlson 1972).

1.4.3   Cytokinins

These are all purine derivatives based on adenine which are essential, together with auxin (with which they act synergistically) for the promotion of active cell division. Variable amounts of cytokinin and auxin also appear to be necessary for controlled cell differentiation. Kinetin, a synthetic cytokinin is 6-furfurylamino purine (Fig. 11.6(d)) but naturally occurring cytokinins are probably different, such as zeatin or 6-(4-hydroxy-3-methylbut-2-enyl)amino purine (Fig. 11.6(e)). In all cases the adenine moiety is essential for hormonal activity. In nature, cytokinins probably exist as the nucleoside or nucleotide and many are even found bound as components of transfer RNA adjacent to the anticodon loop. The significance of this fact for hormone action is at present unclear.

Although the exact mechanisms of action of any of the above classes of hormone are unknown there has been speculation that, as in many animal cells, cyclic AMP may act as an intermediary. For example, in the barley aleurone tissue mentioned above, it has been shown

376

that cyclic AMP can replace gibberellic acid in the induction of α-amylase production and it has been suggested that the cyclic nucleotide induces DNA synthesis which results in gibberellin biosynthesis and that this in turn activates the synthesis of α-amylase (Kessler and Kaplan 1972).

1.5 Tissue culture

Many different plant tissues, for example the storage tissue from a carrot, can be maintained and grown in sterile culture. When such sterile tissue cultures are induced to undergo cell division by the administration of growth hormones, such as auxin, to the medium, the resultant growth lacks the organisation of the parent tissue, and a relatively homogeneous and

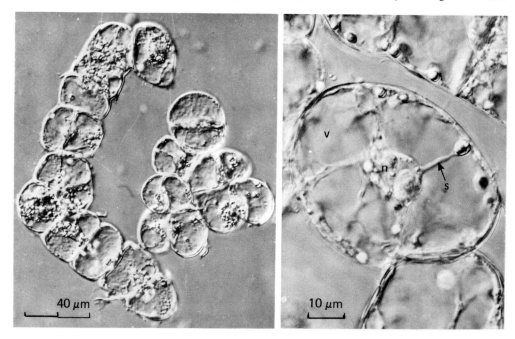

Fig. 11.7 (a) A low power light microscope view of a clump of suspension cultured callus cells derived from a sycamore tree (*Acer pseudoplatanus*). Cells have divided to form pairs, rows and groups.

(b) A higher power view of a single isolated plant cell in culture (*Acer pseudoplatanus*) showing clearly the nucleus (n) suspended in the very large vacuole (v) by cytoplasmic strands (s).

undifferentiated mass of cells results. This callus, as it is called, can be removed from the tissue and grown indefinitely in isolation on a suitable sterile medium. If grown in an agitated liquid medium containing a potent source of growth factors, such as coconut milk (a liquid endosperm tissue) a callus will tend to break up and divide more rapidly. This will yield eventually a liquid suspension culture of plant cells composed of both single cells and small cell clumps (Fig. 11.7). It has been shown, particularly using cultures derived from carrot tissue, that a whole plant can be regenerated from a single isolated cell (Steward, Kent and Mapes 1966). This demonstration of the totipotency of many plant cells, while impressive, has still to be achieved with many other plant types. The component cells in both callus and liquid suspension cultures all look similar, having large vacuoles, active cytoplasmic streaming and rapid cell division (Fig. 11.7). Callus cultures have been used in numerous experiments in which roots and buds typical of the parent plant have been induced to regenerate. With tissue derived from tobacco pith for example it has further been demonstrated that the nature of the regenerated meristem is dependent (among other factors) on the balance of plant hormones used in the induction. High auxin and low cytokinin levels favour the production of root primordia and the opposite combination favours shoot apical meristems.

1.6 Cell differentiation

The central problems of cell differentiation are, first, the elucidation of the factors which decide that a particular plant cell will go to form, for example, a xylem cell or a phloem cell, and second, how that decision is effected. We do not have very precise answers to such key questions but what evidence we do have comes from two main sources; experiments on whole plants, and work on isolated sterile tissue and cells. When it is remembered that among the factors which may be involved are hormone concentrations, sucrose concentrations, the position of the cells, the ploidy of the cells and the precise previous history of the cells in relation to such things as micronutrients, it can be seen that there are great experimental difficulties involved in isolating any one particular effect. Using isolated tobacco pith, and examining the effects of plant hormones on this relatively undifferentiated tissue type, it has been shown that a combination of cytokinin and auxin can initiate cells divisions, and subsequently cell differentiation and organisation into new meristem regions. Not only the specific hormones and their levels, but the exact balance between the hormones, are important in determining the patterns of differentiation. If a block of sterile callus cells, derived from a bean plant, is placed in contact with a source of auxin, it has been demonstrated that, at specific locations within the callus, regions of differentiation are induced (Wetmore and Rier 1963). Furthermore, it has been shown that the auxin diffuses into the callus forming a gradient, and that differentiation only occurs at a specific point along this auxin gradient (Jeffs and Northcote 1966). The type of cell formed here can be regulated by introducing sucrose in addition to auxin. This auxin (0·1 mg/l IAA) together with low sucrose concentrations (1 per cent or below) induce xylem cell formation, but if the sucrose is raised to 2 per cent or above both xylem and phloem are formed. Differentiation occurs in small nodules, in which actively dividing cells (as in the cambium) are located between developing phloem and xylem cells, as in the intact plant. Such nodules induced in sterile cultures are of great importance in both biochemical and structural studies of differentiation. A biochemical measure of the degree of differentiation present can be made by examining the sugars of the cell wall as, in general, specialisation in a plant cell means specialisation in the cell-wall structure and composition. One such parameter is the ratio of xylose to arabinose, which is low in primary walls and high in secondary walls. In secondary thickening there is also a repression of pectin synthesis and a rapid rise in lignin content. Such changes are reflections of underlying changes in the patterns of polysaccharide synthesis within the cytoplasmic membrane system (see section 1.3). (For a good general discussion of plant growth and differentiation, see Wareing and Phillips (1970), chs. 1–3.)

2 Cell division

2.1 General remarks

During plant growth and development the overall shape and size of the plant organs are largely the result of cell divisions followed by cell elongation and differentiation. The tissue patterns observed in these organs are dependent on the orientation of the plane of cell division. It can be appreciated then that one of the crucial stages in the life of the cell is the process of cell division, and in order to begin to understand development in general it is essential to attempt an explanation of the various events that constitute cell division. On the level of descriptive cytology, both within the light and the electron microscope range, we posses an enormous amount of factual data from a very wide range of plant types and tissues. Unfortunately, however, when it comes to explanations, at a molecular level, of any aspect of cell division, from the control of orientation to the mechanics of chromosome separation, we are woefully ignorant, there being very few facts and a great deal of speculation. In this section therefore, the emphasis will be on a description of the physical and structural events during division.

Cell division, in both plants and animals, shows three main sequential phases, which are however separable both in nature and artificially (see Chapter 12, Pasternak). There are, first, the reproductive events, in which the genetic complement of the cell is doubled. This DNA replicative phase (usually designated the S phase) may occur at any time during the interphase period between consecutive cell divisions. This S phase may occur without a subsequent cell division and in this case polyploid cells will result. Second, there is the phase of nuclear separation, which in all higher plants constitutes the mitotic or meiotic events in which the chromosomes are partitioned through the agency of a spindle. Third is the phase of cell partition. Usually, following nuclear separation, the contents of the cell are separated into two by the formation of a new cell wall, in a process called cytokinesis. In this section it is with the last two events that we shall be concerned, that is mitosis and cytokinesis. The material that has been used in studies of cell division is diverse but has centred on either root tip meristems, in which artificial synchrony can be induced, endosperm cells, which show a natural synchrony of cell division and can be examined *in vivo*, or free suspension culture cells, in which divisions can be examined *in vivo* and in which the effects of hormones, for example, are easily studied. In general it can be said that, for sustained cell division in isolated plant tissues, the interaction of two plant hormones is essential. These are the auxins (e.g. indole acetic acid) and the cytokinins (e.g. 6-furfurylamino purine or kinetin). The position is complicated, however, as very different balances of these are required both by different plants and by different plant organs such as root and shoot.

2.2 Mitosis

Mitosis, in higher plants, usually occupies between 2 and 12 hours of the whole cell cycle. It is usually subdivided into four phases which are convenient for the purposes of structural description.

The observations from which the following description is drawn have been made on electron micrographs of thin sections from a wide variety of tissues. These have been supplemented by a detailed frame by frame analysis on time lapse cine-films of living cells, notably the endosperm cells of *Haemanthus katherineae* (e.g. Bajer and Mole-Bajer 1969).

2.2.1 Prophase

Prophase is defined as that portion of mitosis starting with the first observable signs of approaching division and ending with the establishment of the mitotic spindle. It is during this phase that the position and orientation of the mitotic spindle is determined. Unlike animal cells, plant cells have a rigid cell wall and hence can maintain an exact orientation with respect to the whole plant. It is probable that the influence of its adjacent cells is very great in the determination of a cell's spindle polarity; but the exact physical causes of this remain obscure.

One hint at a possible feature that may play an important role, however, is the presence in many dividing cells, at very early prophase, of an organised band of microtubules, encircling the cell adjacent to the cell wall (Fig. 11.8). (Microtubules are intimately involved at all phases of mitosis and may be seen in electron microscope sections at a series of locations within the cell at each phase. The problems of microtubule assembly and function are discussed further in section 2.4.) This band may contain up to two or three hundred microtubules and often lies in a position where the eventual cell plate will fuse with the mother cell wall. However, this is not

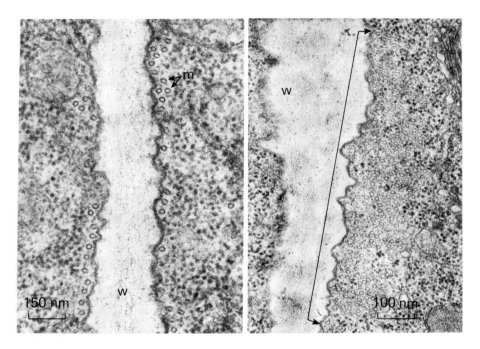

Fig. 11.8 (a) A section through a longitudinal wall (w) in a root meristem from sycamore. Adjacent to the plasmalemma can be seen numerous microtubules (m), usually one deep, in transverse section.

   (b) A similar wall (w) from a root tip cell of timothy grass (*Phleum pratense*) which has just entered the prophase stage of mitosis. The wall microtubules are now in a concentrated band of several hundred tubules about eight deep (between the arrows). The rest of the cell wall has no microtubules adjacent to it. (Micrograph by Dr J. Burgess.)

always the case, and it seems possible that its primary role is one of positioning the future spindle in the correct orientation at right angles to it (Burgess and Northcote 1967). Before the appearance of this band, the cytoplasmic microtubules which lie, during interphase, just below the plasmalemma disappear, and it is possible that they are broken down and reassembled to form the band. As the band disappears during prophase microtubules are found near the nucleus, running along the outside of the nuclear envelope, in the direction of the future spindle, in a zone from which cell organelles appear to be absent. Most of the organelles such as plastids and mitochondria have been transported to regions corresponding to the two future spindle poles. At this stage, while the chromosomes are undergoing the last stages of condensation, the nuclear membrane begins to break down. This starts at the poles, and the envelope fragments become indistinguishable from elements of the endoplasmic reticulum, the nuclear pores being no longer visible. Microtubules now appear in the nuclear region itself, and will eventually constitute the full spindle apparatus. This is composed of microtubules which originate at the polar regions and invade the spindle area. Higher plant cells contain no centrioles, unlike animal cells, and the polar regions are rather vaguely defined cytologically. Chromosomal spindle fibres on the other hand originate from a discrete, apparently amorphous region on the chromosome called the kinetochore. The final result of these events is the mitotic spindle apparatus, a barrel shaped structure composed of numerous microtubules and attached chromosomes, which eventually align themselves in an organised way in the equatorial plane.

2.2.2 Metaphase

In metaphase the chromosomes are stationary in the equatorial plane, in a dynamic equilibrium, before their separation at anaphase (Fig. 11.9). The electron microscope has revealed only one major component of the spindle which could correspond to the spindle fibres seen in the light

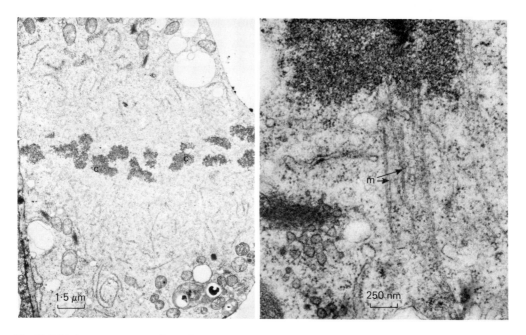

Fig. 11.9 (a) A sycamore suspension culture cell in mitosis. The metaphase arrangement of chromosomes (c) may be seen. The major cell organelles are grouped at the two spindle poles and the spindle area can be seen invaded from the poles by profiles of endoplasmic reticulum.
   (b) A metaphase chromosome (c) from a similar cell to show the kinetochore region (k) where the spindle fibre is attached. The spindle fibre can be seen to be composed of numerous microtubules (m).

microscope and that is the microtubule. The number of these varies with the source, but is between a few hundred and several thousand. In general they tend to lie parallel with each other and in some cases have been seen to be connected by cross bridges. In the light microscope, the spindle apparatus shows birefringence using polarising optics, and it is thought that the microtubules are the main contributing factor towards this.

2.2.3 Anaphase

The problem of the poleward movement of the chromosomes during cell division is a very old one, and one to which unfortunately we still do not have unambiguous answers. Certain general statements can be made, based on observations of living animal and plant cells. Chromosome motion is parallel to the main pole/pole axis of the spindle apparatus, and the chromosomes are attached to this apparatus by chromosomal spindle fibres (microtubules). The two poles of the spindle are defined by the ends of the interpolar spindle fibres. Poleward forces exist at anaphase which act on the kinetochore region of each chromosome. Anaphase movement involves both these poleward forces and a certain degree of spindle elongation. The actual chromosome velocity, and the force required for it, is very low when compared to most other cellular movements. Beyond such general statements, we enter the realms of speculation. Numerous theories on the mechanism of movement have been advanced, the most attractive of these being based partly on the known properties and distribution of microtubules, and partly on the presence of cross bridges between adjacent microtubules. Such bridges have been proposed as active mechanochemical force transducers (in much the same way as actin and myosin might interact). However all theories have so far encountered difficulties, and much remains to be clarified. Again, we have no adequate direct evidence for any hypothesis of force production. (For an example of such theories see McIntosh, Hepler and Van Wie (1969) and for detailed criticism see Nicklas (1971).)

2.2.4 Telophase

At telophase two main events occur, the uncoiling of the chromosomes and the reconstitution of the nuclear envelope. Elements of the endoplasmic reticulum, which have been located at the two poles and at metaphase even invade the spindle region, now appear to fuse to form the new nuclear envelope. Small pieces attach to the chromosomes and these eventually link up. The nuclear pores make their reappearance at about this time. The two half-spindles have now moved apart, resulting in numerous microtubules with their ends interdigitated in the equatorial region between the two daughter nuclei. This forms the region where the initial cell plate will be laid down.

2.3 Cytokinesis

The two processes of nuclear and cytoplasmic division, because they tend in the usual course of events to occur consecutively, have come to be regarded as one more or less continuous event. That this is not always the case may be seen from two examples. In many tissues, continued nuclear division may occur naturally to give rise to a large multinucleated syncytium which may at a later stage undergo cytoplasmic cleavage, e.g. in endosperm tissue. The two events may also be separated in plants experimentally by chemical treatment. Cells treated with caffeine, for example, undergo nuclear division but cytokinesis is blocked, whereas treatment with ethidium bromide blocks nuclear division but allows a certain amount of cytokinesis. It is better, therefore, to regard mitosis and cytokinesis as two separately controlled processes that normally closely follow one another. Within the process of cytokinesis itself we may distinguish three main phases, the relative contributions of which depend on the plant cell type examined. These are the formation of the initial cell plate, the extension of the plate and furrowing at the cell periphery. (The latter occurs commonly in many lower plants but has been largely replaced in higher plants by the cell plate.) The initial phase of cell-plate formation depends on both the production of cell-wall precursors in the form of vesicles, and the presence of the microtubules left between the two daughter nuclei at telophase (Fig. 11.10). These microtubules appear to extend from the new nuclei to the equatorial region where there is a plane of limited overlap or interdigitation. It is in this plane that vesicles accumulate and eventually fuse to form the cell plate. The region marked out by the microtubules and in which no other major cell organelles appear has been termed the phragmoplast. In many cells examined it has been demonstrated that the vesicles, usually about 100 nm in diameter, appear to be associated with the phragmoplast microtubules, and in conjunction with observations on living plant cells, the idea has evolved that in some unspecified way the microtubules guide, and possibly actively transport, the vesicles to the equatorial plane. This is a good example of one of the limitations of electron microscopy, in that static images only are examined, and it is very difficult to infer from these the nature of the actual dynamic events. For the cell plate to extend and reach the

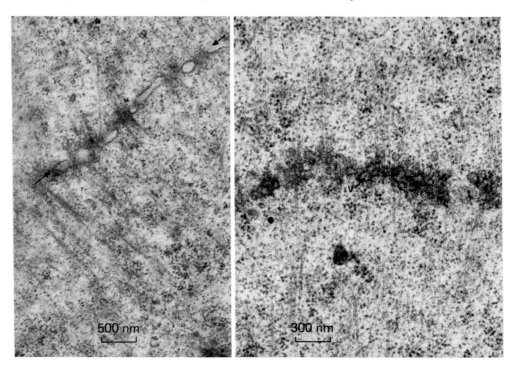

Fig. 11.10 (a) Cytokinesis in a cell from a maize root tip. The phragmoplast region (→) shows numerous microtubules ending in amorphous electron dense regions in the plane of the cell plate. Limited regions of new cell plate can be seen (see stage 2 in Fig. 11.11).

(b) Initial stage of cytokinesis in a root tip cell. Numerous spindle remnant microtubules interdigitate in the equatorial region and vesicles (v) accumulate there prior to fusion forming the cell plate. (Corresponds to stage 1 in Fig. 11.11).

mother cell wall it is obvious that new microtubules must be assembled at the leading edge of the cell plate to align further vesicles. As the original spindle remnant microtubules disappear at this stage, it seems reasonable to infer that as the tubules break down at regions of cell-plate consolidation, the sub-units are transferred laterally and are reassembled to form the requisite new tubules at the leading edge (Fig. 11.10). The process of cell-plate formation is summarised in Fig. 11.11. The origin and exact nature of the vesicles that contribute to the cell plate has been somewhat controversial. Some workers (Porter and Machado 1960) have shown that elements of the endoplasmic reticulum give rise to the vesicles, while others (Hepler and Newcomb 1967) have implicated vesicles derived from the golgi bodies, which appear numerous around the phragmoplast. The truth probably lies in between; vesicles derived from both sources contributing to the final cell plate. Nor is this surprising when it is considered that the endoplasmic reticulum and the golgi apparatus are both parts of a functionally integrated membrane system (see section 1.3). On the question of what these initial cell-plate vesicles contain, opinions are more united. It is believed that during the very early stages the predominant material laid down is pectin, together with some hemicelluloses, i.e. matrix materials. As the plate is consolidated cellulose microfibrils are eventually woven into the texture of this matrix material. Finally, the extending cell plate reaches the mother cell wall and fuses with it, thus completing cell separation. Separation of plant cells is seldom total, however, as cytoplasmic continuity is often maintained via the plasmodesmata; and cell-plate formation is one stage at which such discontinuities in the wall may be established.

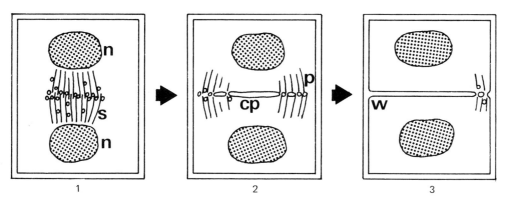

Fig. 11.11 A diagrammatic representation of the sequence of events in cytokinesis. Stage 1 shows the remnants of the mitotic spindle (s), and the vesicles accumulating at the equator between the two daughter nuclei (n). Stage 2 shows the cell-plate consolidated at the centre (cp) and growing outwards by a peripheral phragmoplast region (p) containing microtubules. Stage 3 shows the plate fusing with the mother cell wall (w), often at one side first.

2.4 Microtubules

All the phases of cell division described above show a remarkable dependence on the presence of precise arrays of cytoplasmic microtubules and it is now proposed to describe some of the more important properties of these cell organelles. Microtubules were first described as general features of plant cells (Ledbetter and Porter 1963) following the introduction of glutaraldehyde as a fixative in electron microscopy (previous fixatives such as potassium permanganate did not preserve microtubules), and they have now been found in all classes of eukaryotic cell, forming elaborate structures in such material as nerve cells, flagella, cilia, protozoa and mitotic spindles. In both sectioned and negatively stained material, microtubules appear as slender, unbranched, tubular structures having a diameter between 20 and 25 nm. The wall of the tubule is made up of twelve or thirteen longitudinally arranged filaments, each composed of globular sub-units about 4 nm across. These sub-units are arranged in pairs, each dimer having a sedimentation coefficient of 6S and a molecular weight of 110 000. Although little work has been done on microtubules isolated from plant material, in those from animal cell sources each dimer is made of two closely related proteins of MW about 55 000 (which show only a few differences of amino acid residues) and is therefore more correctly called a heterodimer. The two component proteins of the heterodimer can be separated by careful polyacrylamide gel electrophoresis (Bryan and Wilson 1971). Each heterodimer can bind two molecules of nucleotide (GTP and GDP), which appear to be very important in the assembly process, and one molecule of colchicine, a drug that is used experimentally to disrupt microtubules (e.g. Berry and Shelanski 1972). It has been concluded, from observations that microtubules are readily disrupted by such other treatments as low temperature, pH changes and high hydrostatic pressure, and are stabilised by heavy water (D_2O), that they exist in the cell in a

384

state of equilibrium with a pool of reusable sub-units. The balance between the sub-units in this pool and the polymerised microtubules is very delicate, and even slight changes in the cytoplasmic environment are capable of tipping it one way or the other. Some classes of microtubules are, however, less readily dissociable than others; flagellar microtubules, for example, are far more stable than spindle microtubules. Assuming that they are built of much the same sub-units, this apparent variation in their stability could be explained in two ways. Either the binding of small molecules to the tubule sub-units may increase stability, or else the existence of small cross bridges between adjacent tubules may help to preserve their integrity.

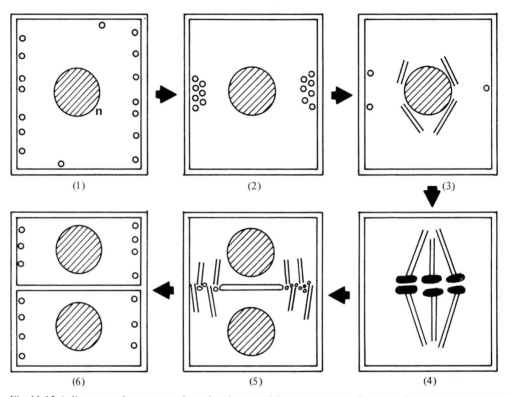

Fig. 11.12 A diagram to demonstrate the ordered sequential rearrangement of groups of microtubules through-out the cell cycle. (1) shows the situation in a root tip cell at interphase. Microtubules are arranged along the longitudinal wall (see Fig. 11.8 (a)). At prophase (2) they are reformed into a band around the centre of the cell (see Fig. 11.8 (b)). They gradually disappear from this location and reappear around the nucleus (3), and as the nuclear membrane breaks down they form the mitotic spindle (4). At cytokinesis (5) they are used to construct the phragmoplast region (see Figs. 11.10 and 11.11), until finally the interphase location is assumed again (6).

The experimental evidence for both is at the moment only slight (see Tucker 1971). If microtubules are constructed by a self-assembly process, which it appears that they are, then the cell must have very precise mechanisms for determining when and where they are made. In general, microtubules are initiated at precise cell locations, or nucleating sites, usually regarded as being local concentrations of sub-unit material. Examples of such nucleating sites may be seen in the kinetochore region of the chromosome to which the microtubules of the spindle fibres are attached, and in the equatorial region of the cell plate (Figs. 11.9 and 11.10). These rather amorphous electron dense areas are probably more ordered than their electron microscope appearance suggests. The question of the control of cell differentiation very often involves control of the distribution of microtubules within the cell, and since the nucleating sites not only initiate polymerisation, but also influence the direction of growth, the nature of the polymer and its polarity, the factors which control the distribution of these nucleating sites is of the greatest importance. The cell cycle in a dividing cell from a root meristem highlights the successive arrangements and recycling of microtubules that occurs (Fig. 11.12), starting with the position at interphase when tubules are found predominantly scattered against the longitudinal walls, at right angles to the root axis (Fig. 11.7). At each stage microtubules are being broken down in one part of the cell and reassembled in another part, and at each step in this recycling process a close association with profiles of endoplasmic reticulum is found. The function of the membrane system here could lie in two directions. One is with the directed transport of the microtubule sub-units about the cell within its lumen, and the other is with the localised binding of small molecules concerned in the assembly and disassembly of micro-tubules. From observations on a wide variety of cell types the general conclusion is that microtubules have two main functions in the cell, depending on their particular location. In many cells they serve a skeletal function in the initiation and maintenance of cell shape, and in particular in the establishment of asymmetries within cells. In other locations, for example in flagellar and spindle fibres, they have a functional role in various aspects of motility, helping to transport cellular material and organelles in specific directions within the cell. This highlights their structural importance in the process of plant cell division and differentiation. (For a general discussion on the structure and function of microtubules see Porter (1966) and for their importance in plant cells see Newcomb (1969).)

2.5 Polarised cell divisions

The overall structural and physiological polarity of a plant organ such as a root is a direct consequence of a polarity within its component cells, and this is clearly visible when the cells divide. It is the plane of cell division and the position of the new cell wall, which, together with the processes of cell development, will determine the precise morphology of the resultant organ. For example, cytokinesis at right angles to the axis will produce an increase in length, and cytokinesis parallel to the axis increase in width. It is apparent therefore that the plant must have mechanisms for controlling the plane of cell division to within very fine limits. In a callus culture such controls appear to have broken down and an amorphous mass of cells is the result. Many cell divisions are not only polarised but also give rise to unequal daughter cells. It seems to be generally true that any structural differentiation in plant cells is preceded by a polarised cell division, and this may be seen most clearly when such divisions are unequal. In

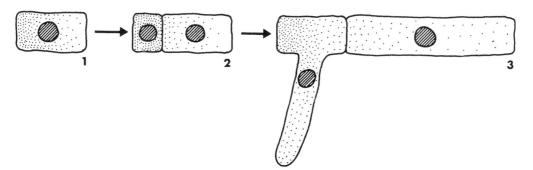

Fig. 11.13 A drawing to show the origin of a root hair cell in *Phleum pratense*. A cell becomes polarised, being more dense at one end than the other and an asymmetric division occurs. The smaller, denser, daughter cell eventually develops into the root hair.

grasses, root hairs arise from the products of cell division in the outer epidermis. A cell, in which the contents are polarised in their distribution, divides to give a smaller dense cell (towards the root apex) and a larger less dense cell. The small cell subsequently develops to form the root hair (Fig. 11.13). Another example is found in the development of the pollen grain, in which the nucleus becomes asymmetrically placed before mitosis. A gradient of cytoplasmic components is formed within the cell and division results in a small dense cell, the generative cell, and a larger less dense one, the vegetative cell. Similar cases of unequal cell divisions conferring a developmental polarity on the two products may be observed in the initial divisions of the egg cell which give rise to the young embryo, in the development of stomatal guard cells, and in the apical cells of such algae as *Chara*. The structural basis of the initial polarisation of the dividing cell is at present not understood but it would seem likely that the ordered movement of endoplasmic reticulum, coupled with the initiation of oriented groups of microtubules, will be found to play a part.

3 Cell walls

3.1 General remarks

The plant cell wall, a semi-rigid envelope surrounding the protoplast, supplies the compressional and tensile strength of the cell and ultimately of the whole plant. One of the most noticeable changes during plant cell development or differentiation is the sequential modification and enlargement of this wall. In fact the changing patterns in deposition of polysaccharides and other wall materials during cell growth provide us with a convenient measure of that growth. Plant cell walls are usually described as consisting of two main components, the primary wall and the secondary wall. The primary cell wall constitutes that structure which is laid down between cytokinesis and the end of the phase of cell elongation. The secondary cell wall, on the other hand, is formed from materials laid down after this period and remains a static structure until cell death. The primary wall is composed of an amorphous matrix of pectins and hemicelluloses, within which are laid down organised groups of cellulose microfibrils.

3.2 The primary cell wall

3.2.1 Pectic substances

Pectins together with hemicelluloses, make up the cell wall matrix materials, and they are defined as those substances that are extracted from the wall in water or solutions of chelating agents. Hemicelluloses, on the other hand, are extracted, after pectins, by treatment with aqueous alkali. Further fractionation of the two classes may be achieved by gel filtration, electrophoresis or fractional precipitation. The residue after alkali extraction represents the crude cellulose fraction. The pectic substances are amorphous and highly hydrophilic and may play a role in controlling the plasticity of the cell wall. They consist of two interrelated polysaccharide types, the ratio of which varies with the age of the cell wall. In very young walls an acidic fraction is laid down first composed primarily of an $\alpha(1 \rightarrow 4)$ linked poly D-galacturonic acid, the carboxyl groups of which are methylated to varying degrees. As the primary wall establishes itself blocks of neutral sugar components, containing predominantly galactose and arabinose, are added to the polygalacturonic acid backbone. Many of the acidic groups are also methylated thus reducing the overall acidity of the pectin. Free neutral arabinogalactans are also present and these may represent a pool of source material for the neutral blocks.

3.2.2 Hemicelluloses

There is a certain degree of overlap between this class of alkali soluble polysaccharides and the pectins, but a few general features help to distinguish them. They are amorphous or paracrystalline polymers of D-xylose, L-arabinose, D-mannose, D-glucuronic acid and D-galacturonic acid. Two broad overall types occur, the xylans and the gluco- and galactoglucomannans. The degree and type of branching, the degree of polymerisation, and the extent of methylation varies with both cell type and age.

3.2.3 Cellulose

Characterised by rigidity and great tensile strength, cellulose provides the structural framework material of the cell wall. It is represented empirically by the alkali insoluble residue obtained from primary cell walls. The bulk of this fraction is accounted for by α-cellulose, a $\beta(1 \rightarrow 4)$

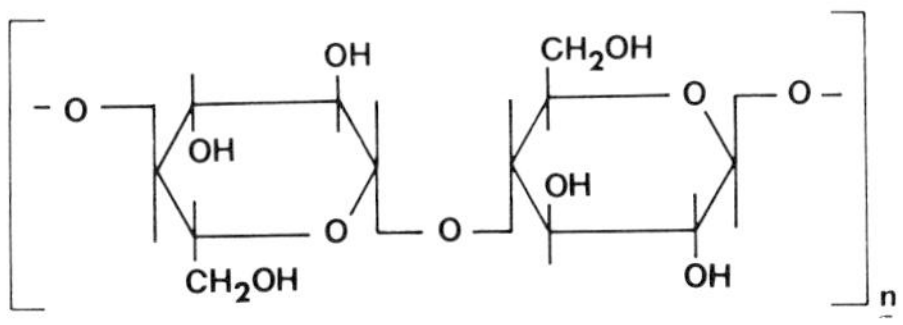

Fig. 11.14 The repeating unit of cellulose ($\beta(1\rightarrow4)$ glucan).

linked glucose polymer. (Fig. 11.14). Several such glucan chains are arranged in a paracrystalline structure, the microfibril (5-15 nm in diameter) which constitutes the main building block of the wall. Numerous models exist to account for the packing of individual polymers in a microfibril, varying from clustered crystalline groups of parallel chains to folded chains themselves helically wound. However there is no clear cut evidence yet to favour any particular model.

3.2.4 The protein component

Several different protein species are found in the primary cell wall, most of which are glycosylated to some extent. Apart from a few enzymes that are found in the wall, such as certain glycosyltransferases, it is a protein, or proteins, containing the unusual amino acid *trans*-4-hydroxyl-L-proline that has been the centre of interest over the last decade (Lamport 1970). This amino acid, not found elsewhere except in animal collagen, has been the focus of much debate, and only recently has the consensus of research confirmed its specific location in a wall bound form, not only in higher plants but also in algae. Much of the research in this field has been carried out using callus cells derived from such plants as sycamore. The protein, in most cases, appears to be linked through an O-glycosidic bond from the hydroxyproline to the arabinose moiety of a polysaccharide which also contains galactose. This glycoprotein is usually tightly bound to the cellulose wall fraction and is difficult to extract. In higher plants a theory has been proposed for the role of the hydroxyproline containing glycoprotein in cross-linking some of the wall polysaccharides. The protein-glycan network so formed could constitute a point of control for cell-wall extensibility (Lamport 1970). Both the synthesis of the glycoprotein itself, and the hydroxylation of the proline, occur in a membrane-bound cytoplasmic fraction before its export across the plasmalemma to the cell wall. Another important linkage within the glycoprotein appears to be through the hydroxyl group of serine to a galactan. The full significance of this for wall integrity is not yet known.

3.3 The secondary cell wall

After cell expansion has ceased, secondary thickening of the wall becomes a major feature of cell development. This transition from primary to secondary thickening is related to changes in the pattern of polysaccharides synthesised and incorporated into the cell wall. The matrix material for the cellulose microfibrils is now entirely composed of hemicelluloses, pectin synthesis having been turned off. The ratio of cellulose to hemicellulose also increases, and

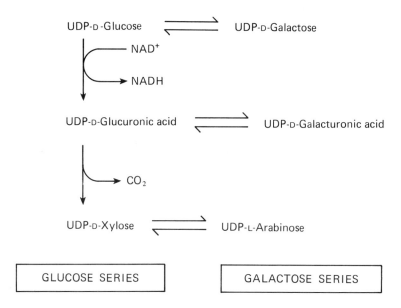

Fig. 11.15 Polysaccharide synthesis depends on the transfer of sugars from nucleoside diphosphate-sugars. The nucleotide derivatives of the glucose and the galactose series of sugars are interconvertible by a series of specific epimerases as shown in the diagram.

various encrusting substances, e.g. lignin, cutin, suberin, callose and waxes, may also be laid down. The matrix material polysaccharides show a distinct shift in their chemical composition during secondary wall formation, from the series of sugars based on galactose (e.g. arabinose and galacturonic acid) to the series based on glucose (e.g. glucuronic acid and xylose). The balance between these two sets of sugar precursors, present as the sugar nucleotides, is controlled in part by a series of epimerases (Fig. 11.15) and these enzymes may well be the control point in the synthetic shift.

The cellulose microfibrils laid down during secondary thickening are static, unlike those deposited during primary cell-wall growth which may change their orientation as the cell expands. The microfibrils may show a marked layered arrangement, each layer having microfibrils orientated in a different direction from those adjacent to it. In many secondarily thickened cells, e.g. xylem, lignin is laid down at particular sites in the cell wall. Lignin is a chemically inert aromatic high molecular weight material derived by the dehydrogenation and subsequent polymerisation of coumaryl, coniferyl and sinapyl alcohols, the proportions of which vary with the type of plant.

3.4 Sites of synthesis

3.4.1 Polysaccharide synthesis

There is now general agreement that two distinct mechanisms contribute to the total synthesis of plant cell polysaccharides. One is the actual formation of the glycosidic bonds by the addition to an acceptor molecule (e.g. an oligo or polysaccharide) of a sugar from a nucleoside diphosphate-sugar compound. The other involves transglycosylation steps to modify the initial polysaccharides. Such steps may produce branched structures, new polysaccharides, or transfer

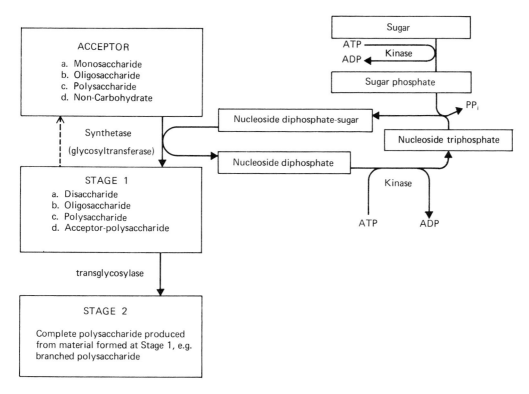

Fig. 11.16 A diagram to show the essential steps in polysaccaride biosynthesis, involving glycosyltransferase and transglycosylase steps.

modifying material on to other polysaccharide chains. These two mechanisms of synthesis are summarised in Fig. 11.16 which shows the primary synthetic reaction as the transfer of sugars from nucleoside diphosphate-sugars, which apart from cellulose synthesis are usually based on uridine diphosphate (UDP). (For a review of polysaccharide synthesis, see Hassid 1969.) It is essential now for an understanding of how the various processes of cell development in plants are carried out and controlled, to discuss the places within the cell where the various wall components are synthesised, and how they reach their final destination.

3.4.2 Pectin and hemicellulose

The basic polygalacturonic acid backbone of pectin is probably assembled by successive transfer of the uronic acid from UDP-D-galacturonic acid to the growing chain. Particulate enzyme preparations have been isolated from plants which will catalyse this reaction and in much the same way reactions, involving the appropriate nucleoside diphosphate-sugar, have been identified in cell preparations producing hemicellulose like polysaccharides such as xylans, glucomannans and glucuronans. It has been suggested that glycosyl lipids may also be important in the transfer of sugar residues during wall polysaccharide synthesis.

It has become increasingly clear recently that the major portion of the pectin and hemicellulose cell-wall matrix polysaccharides are both synthesised and transported by the golgi apparatus and its associated vesicles. The experimental evidence for this conclusion is drawn from three main approaches to the question.

1. Structural. By electron microscopy it has been shown in numerous plant tissues that during the early stages of cytokinesis the cell plate is assembled from vesicles derived in part from the golgi apparatus, and this early cell plate is thought to contain predominantly pectin and hemicellulose.

2. Radioactive labelling. It has been found in several plant tissues, notably root tips, that by feeding the cells with radioactively labelled sugar precursors, e.g. ^{3}H-D-glucose, it is possible to locate the label very rapidly using high resolution autoradiography, in the golgi bodies. This bound label can be 'chased' with cold glucose from the golgi into the cell wall. The labelled material when isolated has been shown to resemble pectic polysaccharides. It has been demonstrated that active golgi in the root cap cells secrete a polysaccharide slime which lubricates the root cap. This too can be labelled in the golgi apparatus and 'chased' in the way described above.

3. Cell fractionation. A number of workers have fed plant cells with radioactive sugar precursors and have then used centrifugation techniques to obtain various cell fractions. It has been discovered that the fraction rich in golgi bodies contained labelled polysaccharides resembling the pectin and hemicellulose found in the cell wall. The clear structural identification of particulate cell fractions containing polysaccharide synthetases has been very difficult, but these have been shown to be membrane-bound and thus support the role of the golgi apparatus in synthesis.

3.4.3 Cellulose

Considering that cellulose is the most abundant polysaccharide in nature, it is rather surprising that in fact we know very little about how and where it is synthesised. Experiments to determine how it is synthesised have centred on the preparation of particulate enzymes from various plant sources (e.g. mung bean) and the incubation of these with labelled precursors. No pure enzyme has been isolated. Incorporation of label into cellulose like polysaccharides has been achieved with both UDP-D-glucose-^{14}C and GDP-D-glucose-^{14}C. However, other polysaccharides in addition to cellulose seem to be formed using UDP-D-glucose, and overall it seems likely that GDP-D-glucose is the true *in vivo* precursor for the $\alpha(1\rightarrow4)$ linked chain of cellulose. It has been suggested by some workers that a glucosyl-lipid precursor may actually be formed inside the cell and that this carries the glucose extracellularly across the membrane where it is polymerised to cellulose; but this has yet to be confirmed.

Just as doubt still exists on which precursor is used to build cellulose, so there is still uncertainty as to exactly where and how outside the cell the chains are synthesised. The picture is complicated by the fact that at least two major kinds of cellulose are present in cell walls, that in primary walls being of a much shorter chain length (2000) than that in secondary walls (14 000). The evidence from freeze-etched cells is conflicting. Small, 8 nm diameter, bodies, which have been interpreted as synthetic particles attached to the outside of the plasmalemma, have also been seen embedded in the matrix of the wall, although autoradiographical evidence suggests that synthesis proceeds at the plasmalemma surface. If we accept the idea of a synthetic particle, possibly membrane attached, the problem still remains of how it reaches the outside of the cell. In higher plants evidence from autoradiography tends to suggest that neither the endoplasmic reticulum nor the golgi apparatus is involved in synthesis. On the other hand, in some algae, cellulose has been clearly shown to be formed in the golgi apparatus. Two possibilities exist for higher plants. One is that the synthetic particles are formed attached to the membrane within the golgi vesicles which by reverse pinocytosis become extracellular. Here, the particles may then engage in chain formation, either attached to the plasmalemma or after detachment. The other possibility, perhaps less likely, is that the particles either pass across the membrane direct or are an integral part of the membrane. At present the overall picture is far from clear.

3.4.4   Cell-wall protein

The hydroxyproline containing cell-wall proteins have been shown by labelling studies to be synthesised in the cytoplasm. The peptide bound proline becomes hydroxylated and the protein is then exported, and there are suggestions that this occurs within a membrane-bound particle. (Reviews of the material covered in section 3.4 may be found in Lamport (1970) and Northcote (1969).)

3.5 The wall as a dynamic cell organelle

Once upon a time the cell wall was regarded as a dead structure, a static box within which the cell led an active existence. Times have changed however and facts have accumulated which have transformed our view of the cell wall into a more dynamic one. We know for example that many components are not stable after their incorporation into the cell wall. The type and amount of polysaccharide laid down in the wall changes in response to the stage of development of the cell, and hence to the environmental situation, hormonal, nutritional and positional, within the whole plant. Transglycosylation reactions are in part responsible for some of these changes. Others, such as the switch in synthesis from primary to secondary wall polysaccharides, reflect programmed changes in the golgi apparatus. Enzymes of various sorts are found within the cell wall, and in certain cases autolysis and specific wall degradation may occur. All these factors can have an important influence on the changing chemical and mechanical properties of the wall. During the phase of cell elongation, two processes are at work. One is a large, and usually polarised, stretching of the wall, while the other is the actual synthesis of new wall material. The wall thus both stretches and increases in thickness, and these two processes are distinct and experimentally separable. The irreversible plastic stretching of the cell wall involved in elongation has been shown to be very greatly enhanced by the plant hormone auxin (idole-acetic acid). As the cellulose microfibrils themselves are non-extensible and the cellulose is in very close association with the matrix polysaccharides, it seems clear that wall plasticity must depend on bond breaking within the wall matrix materials themselves. How auxin effects such wall loosening and how new material is inserted into the expanding wall, is however still unclear.

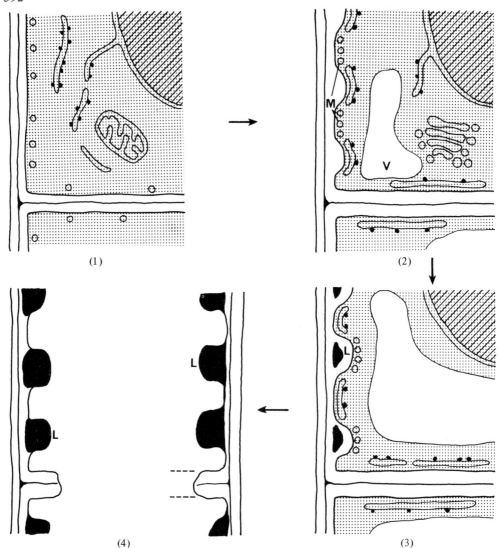

Fig. 11.17 A drawing summarising the main steps in the formation of a xylem vessel (4) from a cambial initial (1). Stages 2 and 3 show successive rearrangements of endoplasmic reticulum and microtubules (m) and the enlargement of the vacuole. Stages 3 and 4 show increasing lignification of the wall thickenings (1).

3.6 Cell differentiation and wall specialisation

3.6.1 Xylem formation

One of the most obvious facets of the differentiation process in plant cells is the elaboration and modification of the cell wall. In a mature xylcm cell the events which lead to the dissolution of the cell contents and the building of the wall thickenings involve an elaborate planned sequence of cell organelle movements. The future xylem cell arises as a procambial cell which initially undergoes both elongation and vacuolation. At this stage the endoplasmic reticulum is randomly distributed throughout the cell and the microtubules are found parallel to the cell wall, but with no preferential distribution. The golgi bodies now show considerable activity, budding off numerous vesicles. Thickened areas appear on the cell walls and these are associated with realignments of other cell organelles. Microtubules, now no longer randomly distributed, are found concentrated against the wall between the thickened areas, and elements of endoplasmic reticulum appear close to the wall over the thickenings. A change in the cell-wall components being laid down now becomes apparent. Pectin seems to be preferentially removed from the wall and the ratio of cellulose to hemicellulose rises rapidly. Before vacuolation is complete (usually about three or four cells from the cambium) lignification of the thickened wall areas begins, and this is associated with a reversal in the positions of the microtubules and endoplasmic reticulum. The former are now found over the lignified thickenings and the latter close to the wall between the thickenings. The golgi apparatus activity drops off at about this stage. The final stages of the process involve the complete lignification of the thickenings and the breakdown, by partial hydrolysis, of the end walls of the cells, resulting in a continuous tube. The cytoplasmic components and the nucleus undergo final dissolution and removal resulting in the mature xylem cell. This complex series of events is illustrated in the simplified diagram Fig. 11.17.

3.6.2 Phloem formation

Phloem comprises four main cell types, the sieve elements, companion cells, unspecialised parenchyma cells and fibres. Only the first two will concern us here. The sieve elements are cells whose end walls are perforated to form a sieve plate, and together they go to make continuous conducting strands called sieve tubes. Each sieve element is associated, in most cases, with a companion cell which has dense cytoplasm and numerous pores in the wall connecting it to its sieve element. Circumstantial evidence suggests that the companion cell forms the major route of sucrose transport between the sieve tubes and the surrounding tissue. Each companion cell and sieve element develop from the division products of a single cambial cell, and they illustrate perfectly one of the central problems of differentiation. What is it that specifies that one of the two daughters, initially identical, genetically and structurally, shall become a sieve element and the other a companion cell? What determines which specific programmed sequence of structural changes they will follow? All we can do at present unfortunately is to describe the changes.

One of the first developments is that the walls of both cells become secondarily thickened, a process associated with the large-scale production of golgi vesicles. Each future sieve element is joined by plasmodesmata to other sieve elements and to its companion cell. The companion cell retains its nucleus and keeps both numerous ribosomes and active cell organelles during its development, whereas the sieve element usually gradually loses its nucleus. The plasmodesmata between companion cell and sieve tube develop elaborate branchings on the companion cell side in close association with elements of the endoplasmic reticulum. Callose thickenings are deposited on the sieve tube side around each pore (callose is a $\beta(1 \rightarrow 3)$ linked glucose polymer). Elements of endoplasmic reticulum also come to lie in either side of the pores which perforate the end walls of the sieve element. Callose is laid down between the wall and plasmalemma surrounding the pore, and gradually the wall is eroded at the level of the middle lamella, leaving two inverted cones of callose at either side of the pore. The callose is then also eroded to some extent, leaving a pore about 50 times the area of the original plasmodesmata and lined with a thin layer of callose. The end walls, full of these large pores, are called the sieve plates. While the pores are being formed aggregates of protein fibrils have been accumulating in the cytoplasm and when the pores are completed these aggregates break up and the constituent 10 nm P-protein fibrils become dispersed throughout the lumen of the cell. (It has been proposed that the P-protein fibrils may play some role in the translocation mechanism.) A simplified diagram of phloem development is shown in Fig. 11.18. Two main problems remain about the

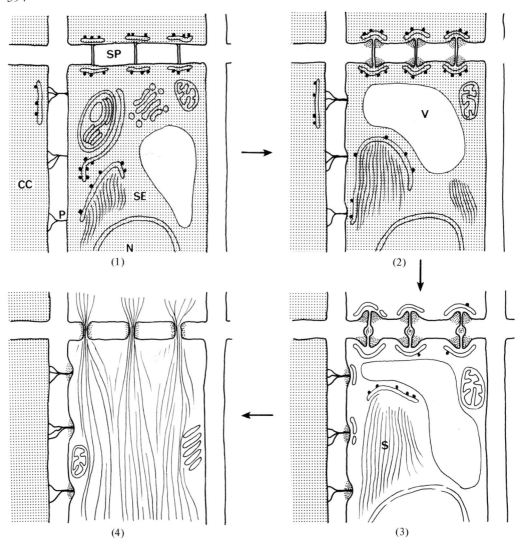

Fig. 11.18 A diagrammatic summary of phloem cell differentiation. Stage 1 shows a future sieve element (se) next to its companion cell (cc) and connected by branched plasmodesmata (p). The sieve plate (sp) is perforated by pores associated with endoplasmic reticulum. Stages 2 and 3 show the increase in the degree of vacuolation (v), the elaboration of the sieve plate pores, and the development of the slime bodies (s). At stage 4 the nucleus (n) and cytoplasmic contents are gone, leaving a few organelles and transcellular slime strands.

structure of the mature sieve element which highlight the difficulty of trying to interpret dynamic functioning systems on the basis of evidence from static fixed thin sections. The first is to what extent the sieve plate pores are free communication channels in the intact living plant. It has been shown that, as a response to any sort of injury, callose can be deposited around the pore within seconds, and as it is very difficult to avoid either cutting or fixation injury to a sealed phloem system which may be under osmotic pressures of many atmospheres the extent of pore occlusion is hard to assess. Rapidly frozen and fixed cells would suggest that *in vivo* the pore is open and only has a very thin layer of callose lining it. The second problem concerns the distribution of the P-protein fibrils *in vivo*, and the two main opposing ideas have been proposed. In one theory the fibrils are evenly distributed throughout the lumen of the sieve tube and very few are associated with the open sieve pore. The flow of material would be through the pores and the cell lumen, perhaps facilitated in some way by the fibrils. In the other theory material is translocated solely within strands, consisting of membrane-bound tubules lined with P-protein fibrils, that cross the sieve pores. It seems clear that the experimental difficulties involved have so far prevented any clear cut choice between these ideas. (For a more detailed discussion of both the structural development and function of phloem see Wooding (1971), and Clowes and Juniper (1968).)

References

The references listed below are in no sense to be regarded as complete and enough recent general and review articles have been included to enable the reader to locate many of the other important original papers on which this chapter is based.

BAJER, A. and MOLE-BAJER, J. (1969). 'Formation of spindle fibres, kinetochore orientation and, behaviour of the nuclear envelope during mitosis in Endosperm, *Chromosoma*, **27**, 448–84.
Much work on plant cell division has been done on endosperm tissue. This paper is a good example of one particular aspect.

BERRY, R. W. and SHELANSKI, M. L. (1972). Interactions of tubulin with vinblastin and guanosine triphosphate', *J. Mol. Biol.* **71**, 71–80.
One of many papers from Shelanski and his group describing aspects of the biochemistry of the microtubule protein tubulin.

BRYAN, J. and WILSON, L. (1972). 'Are cytoplasmic microtubules heteropolymers?' *Proc. Nat. Acad. Sci. USA* **68**, 1762–6.
One of the original papers describing the separation of two species of microtubule protein.

BURGESS, J. and NORTHCOTE, D. H. (1967). 'A function of the preprophase band of microtubules in *Phleum pratense*', *Planta*, **75**, 319–26.
This short paper discusses the role of microtubules during the early stages of division in the root tip grass.

CARLSON, R. S. (1972). 'Notes on the mechanism of action of Gibberellic acid', *Nature New Biology (Lond.)* **237**, 39–41.
Describes the role of gibberellic acid in the induction of α-amylase in barley.

CLOWES, F. A. L. and JUNIPER, B. E. (1968). *Plant Cells*, Blackwell, Oxford.
Although a few years old, this book is an excellent comprehensive survey of the structure and development of plant cells. Well illustrated.

FREDERICK, S. E., NEWCOMB, E. H., VIGIL, E. L. and WERGIN, W. P. (1968). 'Fine structural characterisation of plant microbodies', *Planta,* **81**, 229–52.
A well illustrated survey that unifies our concepts of many of the single membrane bound bodies found in plant cells.

HARRIS, P. J. and NORTHCOTE, D. H. (1970). 'Patterns of polysaccharide biosynthesis in differentiating cells of maize root tips', *Biochem J.*, **120**, 479–91.

HASSID, W. Z. (1969). 'Biosynthesis of oligosaccharides and polysaccharides in plants', *Science*, **165**, 137-44.
A concise review of progress in the biochemical analysis of polysaccharide biosynthesis in plants. Well referenced.

HEPLER, P. J. and NEWCOMB, E. H. (1967). 'Fine structure of cell-plate formation in the apical meristem of *Phaseolus* roots', *J. Ultrastruct. Res.,* **19**, 498-513.
One example from many papers on this aspect of cell division.

JEFFS, R. A. and NORTHCOTE, D. H. (1966). 'Experimental induction of vascular tissue in an undifferentiated plant callus', *Biochem. J.,* **101**, 146-52.

KESSLER, B. and KAPLAN, B. (1972). 'Cyclic purine mononucleotides: Induction of gibberellin biosynthesis in barley endosperm', *Physiol. Plant* **27**, 424-31.
One of several recent papers linking cyclic AMP with plant hormone actions.

LAMPORT, D. T. A. (1970). 'Cell wall metabolism', *A. Rev. Pl. Physiol.,* **21**, 235-70.
Well-written review that emphasises the problems still remaining. Long section on cell-wall proteins.

LEDBETTER, M. and PORTER, K. R. (1963). 'A 'microtubule' in plant cell fine structure', *J. Cell Biol.,* **19**, 239-50.
One of the key papers recognising the general occurrence of microtubules following gluta-raldehyde fixation.

McINTOSH, J. R., HEPLER, P. K. and VAN WIE, D. G. (1969). 'Model for mitosis', *Nature, Lond.,* **224**, 659-63.
A speculative paper on the role of microtubules in chromosome movement.

MONDAL, H., MANDAL, R. K. and BISWAS, B. B. (1972). The effect of indole acetic acid on RNA polymerase in vitro', *Biochem. and Biophys. Research Comm.* **49**, 306-11.

MORRÉ, D. J., MOLLENHAUER, H. H. and BRACKER, C. E. (1971). 'Origin and continuity of golgi apparatus', in *Origin and Continuity of Cell Organelles* (J. REINERT and H. URSPRUNG eds.). pp. 82-118, Springer-Verlag, Berlin,
Although predominantly about animal cells this review has much relevance to the role of the golgi in the plant cell.

NEWCOMB, K. R. (1969). 'Plant microtubules', *A. Rev. Pl. Physiol.,* **20**, 253-88.
A good well illustrated review of the structure and functions of microtubules in plants.

NICKLAS, R. B. (1971). 'Mitosis', in *Advances in Cell Biology II*, pp. 225-98. Appleton Century-Crofts, New York.
Although mainly dealing with animal cells, most of the recent theories of the mechanism of mitosis are critically reviewed here.

NORTHCOTE, D. H. (1969). 'The synthesis and metabolic control of polysaccharides and lignin during the differentiation of plant cells', *Essays in Biochemistry*, **5**, 89-138.

NORTHCOTE, D. H. (1971). 'Organization of structure, synthesis and transport within the plant during cell division and growth', *Symp. Soc. Exp. Biol.,* **25**, 51-69.

NORTHCOTE, D. H. (1972). 'Chemistry of the plant cell wall', *A. Rev. Pl. Physiol.*, **23**, 113-32.
The above three articles comprehensively assess what is known about plant cell walls and how they are made.

PORTER, K. R. (1966). 'Cytoplasmic microtubules and their functions' in *Principles of Biomolecular Organization*, CIBA symposium, G. E. W. WOLSTENHOLME and M. O'CONNOR eds., pp. 308-56. Churchill, London.

PORTER, K. R. and MACHADO, R. D. (1961). 'Studies on the endoplastic reticulum IV its form and distribution during mitosis in cells of onion root tip', *J. Biophys.* and *Biochem. Cytol.*, **7**, 167-80.
Although the photographs now seem very dated the essential information is still sound.

STEWARD, F. C., KENT, A. E. and MAPES, M. O. (1966). 'The culture of free plant cells and its significance for embryology and morphogenesis', in *Current Topics in Developmental Biology* (A. MONROY and A. A. MOSCONA eds). Vol. 1. Academic Press, New York.
A review by one of the pioneers of free plant cell culture.

TUCKER, J. B. (1971). 'Spatial discrimination in the cytoplasm during microtubule morphogenesis', *Nature, Lond.* **232**, 387-89.
This deals with microtubule movements in a ciliated protozooan, but the statement of the problems is of general relevance.

WAREING, P. F. and PHILLIPS, I. D. J. (1970). *The Control of Growth and Differentiation in Plants*, Pergamon Press, Oxford.
A sound general text on plant cell growth and development. Of particular note are the sections on control and the role of the plant hormones.

WETMORE, R. H. and RIER, J. P. (1963). 'Experimental induction of vascular tissue in callus of angiosperms', *Am. J. Bot.*, **50**, 418-30.

WHALEY, W. G., DAUWALDER, M. and KEPHART, J. E. (1972). 'Golgi apparatus: influence on cell surfaces', *Science*, **175**, 596-99.
Summarises recent work on the nature of the cell surface and the implication of the golgi on this. Much of the work described is on animal cells.

WINTERBURN, P. J. and PHELPS, C. F. (1972). 'The significance of glycosylated proteins', *Nature, Lond.* **236**, 147-51.
Polemical discussion of the nature and function of glycoproteins in both animal and plant cells.

WOODING, F. B. P. (1971). *Phloem* Oxford Biology Readers, **15** (J. J. HEAD and O. E. LOWEINSTEIN eds), Oxford University Press, Oxford.
Short but excellent booklet on the structure and function of phloem cells critically reviews the theories of transport.

12
Biochemical Aspects
of the Cell Cycle

C. A. Pasternak
Department of Biochemistry, University of Oxford

1 Introduction

As cells pass from one generation to the next, they undergo a periodic sequence of events known as the cell cycle (Fig. 12.1); the evidence on which the division of the cycle into S (synthesis of DNA) and G (gap) periods is based is discussed in section 3.1. In many microbes and cultured cells of higher organisms, an exact doubling in size and in content of constituent

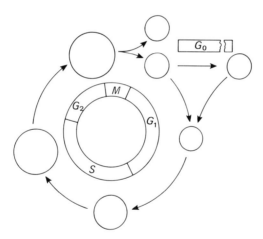

Fig. 12.1 The Cell Cycle. The abbreviations used are M, mitosis; G_0, G_1 and G_2, gap periods; S, (DNA) synthetic period. The entire G_1–G_2 period is often referred to as interphase. The concept of G_0 is discussed in section 5. Measurement of the length of G_1, S and G_2 is discussed in detail in Chapter 4 of Mitchison (1971).

molecules takes place prior to cell division. In many egg cells following fertilisation, on the other hand, there is no increase in size and few molecules other than nuclear components such as DNA double in amount during the initial cell divisions, known as cleavage. Most other cells inside an animal or plant fall into an intermediate category in that there is generally a doubling in size, but without an exact duplication of all molecules. It is just the selective synthesis of particular proteins at successive cell divisions that is the biochemical basis of cellular differentiation (Pasternak 1970).

399

Three inter-related questions are of importance with regard to the cell cycle.

1. At what stages are the constituents of cells synthesised? Is DNA made before RNA, for example; are mitochondria synthesised before plasma membrane? Or are all constituents synthesised continuously, the rate being adjusted so as to ensure the right amount of synthesis before each cell division?
2. How is the synthesis of each molecular species switched on and off (assuming that the last-mentioned mechanism of continuous synthesis does not apply in every case)?
3. How is the overall rate of the cycle set? In other words, what makes intestinal epithelial cells divide every day or so, but liver cells only once a month? By what mechanism is the rate of cell proliferation set so as to exceed during growth, or to equal in an adult, that of cell destruction? Related to this question is that of the *extent* of growth. This is of particular medical interest, since it is failure to restrict the growth of cells that is one of the underlying faults in cancer.

In order to answer these questions, suitable experimental systems must first be found. Ideally, single cells provide the best material. However culture and analysis of individual cells is extremely tricky and only limited information has so far been obtained (Chapter 2 of Mitchison 1971). An alternative approach is to use synchronised cells, in which all cells are at the same stage of the cell cycle. Several microbial and animal cell types in culture have proved amenable to synchronisation techniques and these are accordingly the systems of choice for studying the cell cycle. In addition naturally synchronous cells, such as animal embryos immediately following fertilisation, or plant cells at certain stages of development, have been used.

The next section will describe various methods for obtaining synchronous cells. It is followed by a summary (sections 3–5) of current knowledge regarding the above three questions. Microbes will be treated alongside cells of higher organisms, since many of the basic processes have proved to be the same. It will become apparent that section 3 is largely factual, whereas sections 4 and 5 contain much material that is as yet essentially speculative.

2 Methods for obtaining synchronous cells

The various methods that have been used fall in two categories, (1) induction techniques and (2) selection techniques. In the first, cells are *induced* to enter a particular phase of the cell cycle, while in the second, cells that *happen* to be at a particular phase are separated from all others by physical means. Theoretically, the maximum yield by the first method is 100 per cent, whereas in the second it is determined by the relative length of the particular phase under study; separation of mitotic from non-mitotic cells, for example, is rarely able to yield more than 5 per cent of the population. Both methods have recently been reviewed (Chapter 3 of Mitchison 1971.)

2.1 Induction techniques

Chemical induction is most widely used, and depends on the selective action of an inhibitor. Inhibitors of DNA synthesis such as fluorodeoxyuridine and hydroxyurea or high concentrations of thymidine (2 mM), for example, block cells in S, while inhibitors of mitosis, such as colchicine, colcemid or vinblastine, block cells in metaphase. Following removal of the inhibitor, the accumulated cells proceed around the cycle in synchrony. Since the S period is generally rather long, only a partial synchrony is achieved by this technique. It is improved by repetition of the blockade as illustrated in Fig. 12.2. Induction synchrony depends on effective removal of the inhibitor and on cells remaining 'static' during the blockade. The first criterion is generally satisfied, but the second often not. Thus cells arrested in S continue to synthesise protein and other molecules for a period of time ('unbalanced growth'), so that when the block is lifted, these cells are physiologically distinct from cells that are passing through S naturally. A consequence of this is that subsequent cell periods such as G_2 and G_1 are shorter than they would otherwise have been.

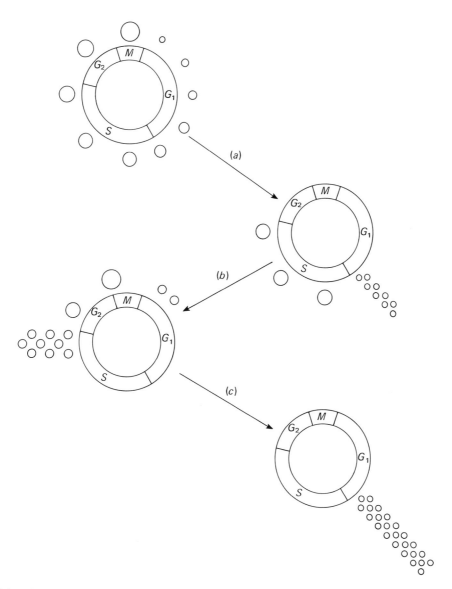

Fig. 12.2 Induction of synchronous cells by double thymidine blockade. Exposure to excess thymidine for a period of time equal to $G_2 + M + G_1$ results in the situation depicted in (a). Removal of thymidine by washing and exposure to thymidine-free medium for a period equal to S results in (b). Re-exposure for a further period equal to $G_2 + M + G_1$ results in (c). In practice, the exposure periods are slightly longer than those mentioned, to ensure complete transition by all cells.

Nevertheless the advantage of the method with regard to yield is obvious and in several instances induction and selection synchrony have led to the same result.

Another way of inducing synchrony is by starving cells, in which case they often stop growing in G_1; restoration of favourable conditions leads to partial synchrony. This method works better with microbes than with animal cells. Temperature shock is another technique that has been used for synchronising microbial and animal cells (see section 5.1). Since exposure to the unfavourable temperature is often short compared with the length of the cycle (e.g. one hour at 4°C for cultured human cells having a generation time of 17–18 hr) it is not always clear how the method works; again it has proved more useful for microbial than for animal cells.

2.2 Selection techniques

These comprise selection according to differences (1) in size and density, (2) in adhesion to glass or membrane filters, and (3) in sensitivity to certain toxic agents.

1. Cells in G_2 are clearly larger than cells in G_1, and provided this increase is fairly gradual throughout the cell cycle, it is possible to separate cells by gradient centrifugation (Fig. 12.3). The method, originally devised for microbial cells (p. 49 of Mitchison 1971) works

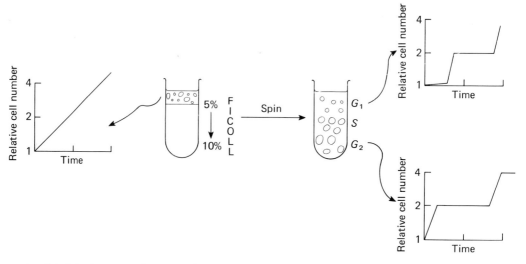

Fig. 12.3 Selection of synchronous cells by gradient centrifugation. The growth patterns of asynchronous, G_1 and G_2 cells are shown.

well with mammalian cells also (Pasternak 1973). Rate, rather than isopycnic sedimentation is used and successful separation is generally due to the fact that cell density remains constant during most of the cycle. Mitotic cells, however, are sometimes lighter than interphase cells and may be separated by isopycnic sedimentation. The differences in size have also been exploited by filtration methods (in the case of microbes), by electronic sorting and by counter-current centrifugation. Selection by size has the advantage of providing cells at various stages of the cell cycle simultaneously, but is subject to the drawback that synchrony is not very sharp, especially if increase in size does not occupy the entire cycle.

2. Animal cells that attach to glass or plastic in culture can be separated into mitotic and interphase cells simply by gentle shaking. Cells in mitosis detach, whereas the rest remain attached. Although the yield, as mentioned above, is necessarily low, the method is a gentle and very effective one. The yield may be increased by temporary exposure to colcemid (see section 2.1), but the drawbacks of induction synchrony then apply. A slightly different technique has been used for bacteria. A cell suspension is applied to a membrane filter and washed so that the filter is saturated with attached cells. As these begin to divide, half the daughter cells are detached (because there is no room for them) and provide a relatively pure sample of post-mitotic cells.

3. Certain animal cells in *S* can be selectively killed by incorporation of high specific activity [^{3}H]thymidine or by treatment with hydroxyurea (which can also be used to *induce* synchrony, see above). The surviving cells are then partially synchronised, but the degree of synchrony is obviously small and the presence of dead cells complicates this procedure considerably.

2.3 Summary

Cultured cells may be synchronised by *induction* or by *selection*. Induction of synchrony by inhibitors of DNA synthesis or of mitosis results in a high yield of cells, but possible side effects detract from the general applicability of the method. Selection of cells according to size or to adhesive properties is a preferred technique, but the yields are necessarily lower.

3 Biosynthetic activities during the cell cycle

The synthesis of cellular constituents during the cell cycle may be either continuous or step-wise. In the latter case, the period of synthesis may be long or short, and may occur at any stage in the cycle. Superimposed on synthesis may be degradation, that is, turnover, which may coincide with synthesis or be temporarily distinct so that a series of steps, or oscillations, ensues. In this section the pattern by which the major chemical species, as well as some specific molecules, are synthesised during the cell cycle will be examined. First, some points of methodology will be discussed.

With synchronously-growing cells in culture, samples are taken at intervals and analysed. Since the degree of synchrony deteriorates progressively, measurements taken at the start of synchrony are more accurate than those taken subsequently. Hence a selection technique such as separation according to size, in which cells from all parts of the cycle are analysed simultaneously, has distinct advantages. Moreover, if biosynthetic potential is measured by incorporation of radioactive precursor, cells may by exposed to the precursor prior to separation, thus ensuring that all cells are in an identical milieu.

The use of radioactive precursors allows one to study one aspect of biosynthetic activity even in an asynchronous population including intact organisms. This concerns the *extent* of biosynthesis; i.e. whether synthesis is continuous or step-wise, and if step-wise for how long. The technique is to expose the culture or organism to a pulse of precursor and then, by autoradiographic microscopy, assess the number of cells that have become labelled. If all cells are labelled, synthesis is continuous. If only some cells are labelled, synthesis is discontinuous and the percentage of cells that are labelled is related to the duration of the synthetic activity being studied. This method cannot give an indication of *when* in the cycle synthesis occurs; in the case of DNA synthesis (Fig. 12.1) for example, the length of S but not of G_1 or G_2 may be ascertained. In so far as cells of higher plants are generally not so easy to grow in culture as some animal cells, much of our knowledge on the cell cycle in plants is restricted to this kind of information.

3.1 DNA synthesis

The earliest experiments (1953) were of the type just described and showed that in certain animal cells, DNA synthesis is discontinuous and is separated from mitosis by two gap periods (determined by measurements with synchronous cells, e.g. Fig. 12.4)–called G_1 and G_2 (Figs. 12.1, 12.4 and 12.5). This has proved to be an almost universal situation for all higher

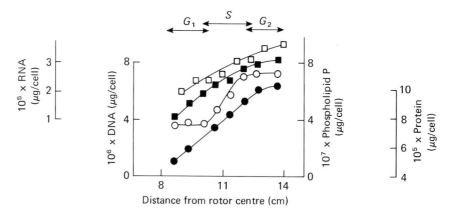

Fig. 12.4 Total content of cellular constituents during the cell cycle. P815Y mouse mastocytoma cells were separated by gradient centrifugation and the content of RNA □——□, phospholipid ■——■, DNA ○——○ and protein ●——●, analysed. Approximate allocation of cells to G_1, S and G_2 periods is indicated. This result, taken from Warmsley, A. M. H. and Pasternak, C. A. (1970), *Biochem. J.*, **119**, 493–99, is typical of many cell types.

organisms, including plants. The duration of S in adult cells is fairly constant and the increase in span of the cell cycle as cells become progressively more differentiated is generally due to a lengthening of G_1; the concept of a G_0 period (Fig. 12.1) has been introduced to describe the position of non-growing or rarely-dividing cells. In very rapidly-dividing cells, such as amphibian and sea-urchin embryos following fertilisation, there is virtually no G_1 or G_2 period after the first cell cycle and even S is very much shorter than in adult cells. In adult cells DNA synthesis is initiated at many points on a single chromosome; in the early embryo it is likely that most of these are used simultaneously, or that there are even more initiation points, since the rate of synthesising nucleotides is probably already maximal in adults. The factors that control the length of the cell cycle and the entry into G_0 are discussed more fully in section 5.

   The synthesis of *mitochrondrial* DNA has been studied in cultured cells. It does not occur concurrently with nuclear DNA synthesis, but often takes place only in late S-G_2. To what extent the synthesis of mitrochondrial DNA reflects the biogenesis of intact mitochondria (section 3.5) is not yet clear.

   In microbes the situation is similar to that in animals, in that the presence of G_1 and G_2 periods is determined by the rate of cell division. Only in very slowly-growing cultures is DNA synthesis found to be step-wise; generally S occupies the entire cell cycle. In such situations, there is often only one initiation site per chromosome, so that the sequence of genes on the chromosome may be assessed by determining the appearance of the gene-products (e.g. enzymes) in the cytoplasm. This is referred to again in section 3.3 and illustrated in Fig. 12.7. In very rapidly-growing cultures, the S period becomes necessarily shortened and the initiation of DNA synthesis now occurs at more than one point, as on animal chromosomes.

3.2 RNA synthesis

In contrast to DNA synthesis, RNA synthesis appears to be continuous throughout the cell cycle, whether in animals, plants or microbes. All cells, except mitotic ones, in a culture exposed to a pulse of radioactive uridine, for example, incorporate radioactivity and the total content of RNA increases gradually throughout interphase (Fig. 12.4); the difference between synthesis of DNA and RNA is clear, and makes it unlikely that duplication and transcription restricted by identical factors (see section 4.2).

The rate of uridine incorporation, indicative of the rate of RNA synthesis (or rather of the rate-limiting step of RNA synthesis, often *presumed* to be the polymerisation step), often increases more than two-fold during the cell cycle (Fig. 12.5): hence there is a decrease in G_2 to

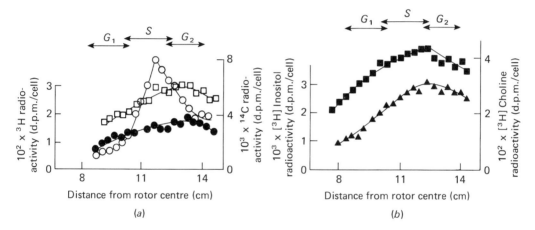

Fig. 12.5 Incorporation of precursors during the cell cycle. P815Y mouse mastocytoma cells prelabelled with ^{14}C-thymidine ○——○, ^{3}H-proline ●——●, ^{3}H-uridine □——□, ^{3}H-choline ■——■ or ^{3}H-inositol ▲——▲, were separated by gradient centrifugation as in Fig. 12.4, and incorporated radioactivity assayed. Approximate allocation of cells to G_1, S and G_2 periods is indicated. This result, taken from Warmsley, A. M. H. and Pasternak, C. A. (1970), *Biochem. J.*, **119**, 493–99, is typical of many cell types.

prevent over-production of RNA. The rate decreases even further prior to mitosis than is apparent from Fig. 12.5, and is probably zero in mitosis, when the DNA template – at least in the nucleus – is in the condensed, heterochromatin form. In other systems an exact doubling in synthetic rate is observed, and since this coincides with the onset of S, it is possible that control by gene-dosage (section 4) operates. Since the exact specific activity of intracellular metabolites in such experiments is generally not known, it is difficult to correlate the synthetic rate with the net increase in the number of molecules formed (i.e. Fig. 12.5 versus Fig. 12.4), but in most situations extensive turnover is superimposed on synthesis. It should be noted that whereas the bulk of cellular RNA is ribosomal, it is the smaller amounts of DNA-like* or messenger-RNA that have the fastest turnover.

---

* By DNA-like RNA is meant that species that most resembles DNA in its base ratio and hence shows the greatest degree of hybridisation to DNA.

The synthesis of different RNA species has been relatively little studied. Ribosomal and transfer RNA appear to be synthesised continuously in interphase, and the same is true of DNA–like RNA. Attempts to detect fluctuation in base ratio, or in hybridisation capacity, during the cell cycle have been more successful in microbes than in animal cells.

Under certain developmental conditions such as the cleavage of amphibian embryos or the germination of seeds, ribosomal, transfer and DNA-like RNA synthesis are initiated at distinct times, and such patterns of synthesis are superimposed upon any fluctuation due to the cell cycle alone.

3.3 Enzymes and other proteins

In most cells, bulk protein synthesis follows that of RNA. In other words synthesis is continuous throughout the cell cycle (Fig. 12.4) and the rate of precursor incorporation (Fig. 12.5) often exceeds the rate of net synthesis, indicating turnover of certain species.

The relative ease of assaying individual proteins (compared with their respective RNA templates) has led to the accumulation of much data regarding changes in enzyme activity during the cell cycle. Where total activity of a particular enzyme just doubles during the cycle of an established cell line in culture, there is good reason to suppose that enzyme activity is a measure of total enzyme content. Confirmation may be achieved by titration with specific antibody raised against the purified enzyme. Where the activity increases more than two-fold, or even undergoes a series of oscillations, it is clear that either the activity is being modulated by tightly-bound effectors or that actual enzyme molecules are being synthesised and degraded sequentially. Simultaneous degradation and resynthesis, i.e. turnover in its usual sense, can be assessed only by measuring changes in the labelling pattern of highly-purified proteins. The two situations are well illustrated (Fig. 12.6) by comparing the activity of lactate dehydrogenase in

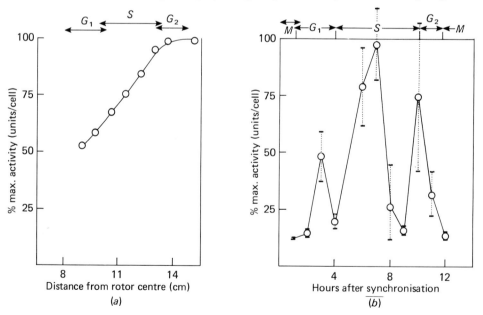

Fig. 12.6 Patterns of enzyme activity during the cell cycle. Lactate dehydrogenase in P815Y mastocytoma cells separated by gradient centrifugation (a) and in Don C chinese hamster cells synchronised by treatment with colcemid (b). Taken from Warmsley, A. M. H., Friedrichs, B. and Pasternak, C. A. (1970), *Biochem. J.*, **120**, 683–88 for (a) and from Klevecz, R. R. and Ruddle, F. H. (1968), *Science*, **159**, 634–36 for (b).

mouse P815Y cells and in Don C chinese hamster cells (albeit synchronised by different techniques). A third situation is one in which an enzyme doubles in step-wise manner, like DNA (Fig. 12.4). Although many different enzymes have now been examined, in microbes as well as in animals, no general pattern appears to emerge and no generalisations are therefore possible.

In bacteria in which duplication of the haploid chromosomes begins at one site only, step-wise appearance of some enzymes has been correlated with the position of their respective genes on the chromosome (Fig. 12.7). Note, however, that the activity of other enzymes is modulated in such a way as to give a continuous increase in activity.

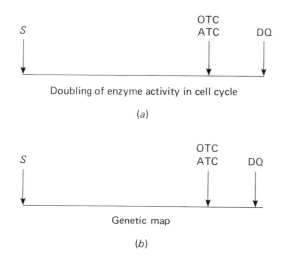

Fig. 12.7 Sequential doubling of some 'step' enzymes during the cell cycle. *Bacillus subtilis* was synchronised by starvation, and the doubling in activity or ornithine transcarbamylase (OTC), aspartate transcarbamylase (ATC) and dehydroquinase (DQ) during one cell cycle measured (indicated by the arrows in (*a*)). The similarity to the sequence of the relevant genes (indicated by the arrows in (*b*)) is clear. The potentiality for synthesising a 'continuous' enzyme, sucrase (S), has been aligned with the position of the sucrase gene on the chromosome. Taken from Masters, M. and Pardee, A. B. (1965), *Proc. Nat. Acad. Sci.,* **54**, 64–70.

Although the difficulties of assaying specific RNA templates has been referred to, an indirect way of examining the behaviour of messenger RNA during the cycle has been attempted. This is to determine the period when *induction* of a particular enzyme is possible. In most cases inducibility is limited to a particular phase of the cell cycle, even in situations where the activity of the uninduced enzyme increases continuously. Interpretation of such results in terms of messenger RNA has therefore to be made with caution; factors other than availability of RNA templates may prove to be rate limiting.

Turning to some non-enzymic proteins, the most widely studied in lower and higher eukaryotes are the histones (prokaryotes do not appear to contain histones). Histone synthesis seems to be step-wise and to take place in *S*. The concomitant synthesis of histones and DNA has been observed in another situation also. In sea urchin embryos in early cleavage, in which there is generally rather little net protein synthesis accompanying DNA synthesis, half of the total proteins that are synthesised are histones. Whether acidic proteins (section 4.2) and other nuclear components are synthesised at the same time remains to be seen; *some* acidic proteins (Baserga and Stein, 1971) appear to be synthesised rather early on in the cell cycle (section 4.1). Certainly nuclear size and dry weight seem to increase later on in the cycle (section 3.5).

Immunoglobulin production in cultured lymphoid and myeloma cells is likewise restricted to a part of the cell cycle, in this case late G_1 and most of *S*. Microtubular protein, assayed by its binding affinity for colchicine, on the other hand, is synthesised continuously throughout the cell cycle. The most obvious function of microtubular proteins is during mitosis, when they aggregate to form the mitotic spindle responsible for the alignment of chromosomes. During interphase microtubules may by involved in the movement of other subcellular components. Other non-enzymic proteins which are relatively easy to recognise and to measure, and would therefore be profitable for study, are haemoglobin, collagen, myosin and actin.

3.4 Other cellular constituents

Since phospholipids are major constituents of membranes and occur intracellularly only in conjunction with membranes, their synthesis may be taken as a measure of membrane biosynthesis (section 3.5). In animal cells phospholipid synthesis is continuous and follows bulk protein and RNA synthesis (Figs. 12.4 and 12.5). The pattern for total phospholipid synthesis (Fig. 12.4) is compatible with that for choline- and inositol-containing species (Fig. 12.5) and therefore probably applies to the other classes also. Phospholipid turnover is continuous throughout the cycle, maximal degradation accompanying maximal synthesis in S.

In bacteria, on the other hand, phospholipid synthesis appears to be maximal during mitosis. Since bacterial plasma membranes continue to grow at the time of cell division, whereas in animal cells plasma and other membranes appear to be assembled only throughout interphase (section 3.5), these two observations are not as contradictory as they appear at first sight.

The synthesis of low molecular weight constituents during the cell cycle follows no set pattern. Amino acids for example, show individual oscillations in *Chlorella* but vary rather less in animal cells. ATP probably follows cellular dry weight (i.e. the pattern shown by protein, RNA and phospholipid in Fig. 12.4), whereas thymidine nucleotides appear to decrease in S, when their utilisation is maximal. In the case of compounds which are derived from the medium rather than synthesised *de novo*, effects of altered permeability must be taken into account. Changes in intracellular sodium and potassium ions, for example, may reflect the change in plasma membrane ATPase that has been observed (section 3.5).

3.5 Subcellular organelles

The assembly of structures such as nuclei or plasma membranes might be thought to be a matter merely of microscopic observation. But since the nucleus as such disappears entirely in mitosis, and since increase in cell size could initially be due either to stretching of the plasma membrane or to net synthesis of its components, this is not so. Hence it is important to measure the components of organelles in addition to observing their gross disposition during the cell cycle.

Many of the characteristic constituents of subcellular membranes are synthesised on the endoplasmic reticulum. Since no accumulation there is detected at any point in the cell cycle, it appears that insertion into the respective membranes follows the synthesis of membrane components without much delay.

3.5.1 Nucleus

As referred to above (sections 3.1 and 3.3) DNA and histones double during S, and this is reflected in the assembly of whole daughter chromosomes at this time. Nuclear dry weight and volume, on the other hand, increase only later on in the cell cycle. Whether nuclear membrane is stretched or replenished at this time is not known. It is not even clear whether the membrane components which reform around the two daughter nuclei are those that surrounded the parent nucleus, or whether dissolution and complete resynthesis takes place. The nuclear membrane has a particular importance in that chromosomes may become attached to it during their replication, just as bacterial DNA is duplicated from a point of attachment on the protoplast membrane. The behaviour of the nucleolus during the cell cycle has been little investigated, though its participation in RNA and protein synthesis is well documented.

3.5.2 Cell surface

Current evidence favours the view that in animal cells the major components of the plasma membrane are replenished throughout interphase. Protein, phospholipids and some marker enzymes such as 5'-nucleotidase, for example, increase at the same rate as bulk cellular protein. Cytokinesis, the splitting of the cell into two halves, may thus be essentially a physical 'pinching' process, with any deficit in membrane density being made good in the next G_1 period. Minor components may behave in a different manner. Sodium/potassium-activated ATPase, for example, appears to reach a peak in late S-G_2, and the insertion of some glycoproteins and glycolipids may likewise take place in step-wise fashion, around the time of cell division.

It must be remembered that certain membrane components such as phospholipid are metabolically rather labile and that their continual turnover (section 3.4) probably reflects physical movements within the membrane. Certainly pieces of membrane become inserted or removed during phagocytosis and pinocytosis, some types of protein secretion or the 'budding' of membrane-coated viruses; more subtle modifications accompany other surface phenomena like the serum-stimulated resumption of cell division in static cultures. Such changes in the architecture of the cell surface are particularly apparent if one measures the expression of some antigenic markers; this is at a minimum at the very time that their synthesis is likely to be maximal (Cikes and Friberg 1971; Pasternak *et al.* 1971). Although the reasons for this apparent discrepancy are not yet clear, some kind of re-orientation of membrane components is probably involved.

In bacteria the growth of the plasma membrane may be somewhat different. The increase in size during interphase is probably accompanied, as in higher cells, by the continuous insertion of phospholipid and protein components. During cytokinesis, however, the septum that is formed appears to arise by net synthesis (section 3.4) rather than by pinching of the existing membrane. The growth of the outer wall of bacteria likewise takes place by synthesis and insertion of new units, their size in general depending on the species concerned.

3.5.3 Mitochondria

The biogenesis of mitochondria during the cell cycle is complicated by the fact that it is not definitively known whether new organelles are assembled *de novo* or by growth and fission of existing structures. An accurate assessment of the number and relative size of mitochondria throughout the cell cycle might well resolve this point. Whatever mechanism proves to be correct, it is clear that most of the components such as phospholipids and many proteins are synthesised outside the mitochondria. Such markers as have been assayed appear to be synthesised and inserted continuously. As with plasma membranes, however, minor components may be added at a discrete time. Mitochondrial DNA, for example, appears to be synthesised in late S-G_2, (section 3.1).

3.5.4 Endoplasmic reticulum and other organelles

Various enzyme markers have been measured, but it is premature to draw definitive conclusions. One study reveals a continuous increase in microsomal enzymes, while another (using a different method of synchrony and a different cell line) indicates a periodic increase in some lysozomal enzymes. The exact programming of intracellular assembly in higher organisms is clearly a matter that requires further experimentation. One organelle that has been rather carefully studied is the centriole and this does appear to be assembled step-wise rather than continuously (Chapter 9 of Mitchison 1972).

3.6 Summary

Some biosynthetic activities, such as DNA synthesis and associated events concerned with chromosome replication, occur at a discrete point in the cell cycle of higher organisms. In microbes or in very rapidly dividing animal cells, this period generally occupies the whole of interphase. The synthesis of most other bulk constituents such as RNA, protein or phospholipid is continuous throughout interphase. The rate of synthesis generally exceeds the rate of net accumulation, since turnover is appreciable. Some individual molecular species are synthesised step-wise, others are synthesised continuously; no general pattern is apparent with regard to enzymes, for example. Cellular organelles such as plasma membrane or mitochrondria are probabliy assembled by continual synthesis and insertion of constituent molecules, but nuclei undergo rather more sharply defined changes during the cell cycle.

4 Control of biosynthetic activity

The question of how the biosynthetic activities discussed in section 3 are initiated and terminated must now be considered.

4.1 DNA

The initiation of DNA synthesis in eukaryotic cells is prevented if protein synthesis is inhibited prior to S. Once DNA synthesis has begun, it is less sensitive to inhibition of protein synthesis; in bacteria also, initiation but not completion of DNA synthesis requires the synthesis of certain proteins. Inhibitors of RNA synthesis have a rather similar effect in eukaryotes, and it may be concluded that a protein(s), synthesised on an unstable template, has to be synthesised before DNA synthesis can commence. Synthesis of the unstable template may be triggered by those proteins synthesised early in the cell cycle (section 3.3). In amphibian embryos, on the other hand, DNA synthesis can be initiated in the virtual absence of protein synthesis. This may mean that fertilisation or activation overcomes the need for protein synthesis or that the relevant protein is already present (having a more stable template in egg cells than in somatic cells; since the stability of RNA is a characteristic feature of egg cells, this would not be too surprising). The nature of such initiating proteins is not known; it is unlikely to be DNA polymerase itself.

Another approach has been to determine the site of stimulation. From experiments in which an *Amoeba* nucleus at one stage of the cell cycle is transplanted into an enucleate cell at another, it appears that the state of the cytoplasm, not the nucleus, is decisive. Thus G_2 nuclei in S cytoplasm do not synthesise DNA, whereas S nuclei in G_2 cytoplasm do. One might try to correlate this result with the previous one by postulating that extranuclear protein synthesis is the cytoplasmic stimulant necessary for initiating DNA synthesis. However proteins are much more likely to come into contact with chromosomes during mitosis than during interphase, and other results (reviewed in Chapter 7 of Pasternak 1971) show that the cytoplasmic signals are quite general (e.g. human cytoplasm can activate hen chromosomes), and therfore less likely to be proteins. In the case of bacteria, it has been suggested that DNA synthesis is initiated whenever the concentration of a postulated inhibitor falls below a critical level as a result of an increase in volume during growth. Such a mechanism can clearly not be operative in controlling DNA synthesis in early embryogenesis, since cell size remains constant throughout each cell cycle. Nevertheless a related mechanism, that of critical mass, has been proposed to trigger DNA synthesis in animal cells.

4.2 RNA

In the case of RNA synthesis it has been proposed that 'unmasking' of the relevant sites on the DNA template – normally covered by proteins or other molecules – is the trigger for transcription. Histones, modified histones, acidic proteins and RNA have all been postulated to be involved in the degree of masking; acidic proteins are the most likely to be *specifically* involved. It should be pointed out that the same mechanism of unmasking has been suggested to play a part in the initiation of DNA synthesis also. Since the entire chromosome is duplicated, whereas only very limited regions are transcribed, it seems unlikely that the processes are identical. In addition to the availability of free templates and the presence of RNA polymerase, certain protein initiation factors have been shown to be necessary for RNA synthesis.

Turning from the *specificity* of transcription to the general increase in rate during S, it has been suggested that if template availability is indeed rate limiting, a doubling of templates in S would result in the observed increase. However, such 'gene-dosage' obviously cannot account for the decrease in G_2 (Fig. 12.5).

4.3 Proteins, other molecules and organelles

Gene-dosage has been postulated to control protein synthesis also and in the case of certain bacterial enzymes (Fig. 12.7) this is clearly so. It cannot of course be operative in the case of proteins whose synthesis begins prior to S or decrease in G_2 (Fig. 12.5). Whether 'messenger RNA-dosage' controls translation under such conditions is not known. Post-transcriptional factors seem to be involved in some types of induced enzyme synthesis (section 3.3).

As regards other molecules, 'enzyme-dosage' may play a part where synthesis closely follows synthesis of the relevant enzymes. In general, however, it is becoming clear that it is the *activity* rather than the *amount* of enzyme that is rate-limiting in many biosynthetic sequences.

The nature of the signals that control the size and number of cellular organelles is even less well understood. Quite apart from all the uncertainties just mentioned, there is the further question of how some proteins (and other molecules) synthesised on the endoplasmic reticulum are specifically channelled toward either mitochondia, plasma membrane or other structure. It seems likely that sophisticated *in vitro* systems will have to be developed before such questions can be satisfactorily answered.

4.4 Summary

Protein synthesis appears to be necessary to trigger DNA synthesis in microbes and higher organisms. Despite considerable knowledge of factors involved in the biosynthesis of most other cellular constituents, it has not yet proved possible to pin-point the rate-limiting steps that initiate their synthesis during the cell cycle.

5 Control of cell division

The rate at which the cell cycle operates is the rate of cell division. Factors that trigger mitosis must therefore be considered. Such factors, as will become apparent, probably act *after* DNA synthesis has been completed. But the difference between rapidly-growing and slowly-growing cells lies in the length of G_1 and more specifically in the entry into G_0 (section 1), so that additional factors must be involved. In other words, entry into mitosis and entry into S are under separate types of control. The first ensures that cells that have duplicated their genome go on to divide. The second is concerned with whether a cell is going to enter the cycle at all, or 'opt-out' in G_0.

5.1 Entry into mitosis

This problem has been approached in a fashion similar to that discussed with respect to biosynthetic activity (section 4). That is, the effect of inhibitors and other treatments that interrupt the normal course of events is assessed. It must be stressed, however, that any interpretation of such results is complicated by the need to distinguish between *general* and *specific* effects; this is, of course, true of all inhibitor studies. To give a simplified example, if the temperature of a cell culture is raised $10°C$ and it is noted that cell division stops, the interpretation that a heat-sensitive step is involved in triggering division is hardly justified. But if cells early on in the cycle prove to be more sensitive to increasing the temperature than cells later on, such an interpretation becomes more realistic. With such reservations in mind, the evidence for the existence of a mitotic factor acting in G_2 may be considered.

If a culture of *Tetrahymena pyriformis*, a large ciliated protozoan which grows optimally at $29°C$, is raised to $34°C$, cell division is prevented. If the temperature is reduced back to $29°C$, division takes place, but only after some delay. By applying a series of such temperature shocks a synchronously-growing culture can eventually be induced. The reason for this is that the period of delay varies with the stage in the cell cycle at which a cell happens to be. Immediately after cell division there is little delay, but the period of delay builds up gradually until it is maximal (e.g. 60 min in cells dividing every 120 min) in late G_2; thereafter it falls to zero again just before mitosis. A somewhat similar situation is obtained if protein synthesis is inhibited for brief periods throughout the cell cycle. One may conclude (a) that a heat labile, metabolically unstable protein(s) is required to trigger cell division and (b) that there is a transition point in late G_2 at which cells become committed to division.

Other cells, bacteria as well as mammalian cell lines, have been exposed to similar treatment and it appears that the concept of a mitotic protein factor acting in G_2, or at any rate after DNA synthesis has begun, is generally valid. That the role of such a factor is to ensure that cells which have duplicated their DNA go on to divide, is conjecture. Inded, in certain cases, such as heat shock in *Tetrahymena*, DNA synthesis and cell division are obviously not coupled; in others, such as the assembly of polytene chromosomes in Dipteran salivary glands, a natural uncoupling occurs. But it is a striking fact that, except at particular stages of plant or animal development, all cells contain a fixed amount of DNA. Since cell division and the next round of DNA synthesis in daughter cells is clearly *not* tightly coupled (section 5.2), a mechanism whereby initiation of DNA synthesis is accompanied by the build-up of a protein designed to trigger cell division once DNA synthesis has been completed is an attractive one.

5.2 Entry into S

The concept of G_0 (Fig. 12.1) has been introduced to describe the situation of non-dividing cells. Such cells require an external stimulus before cell division is resumed. The reason for placing G_0 between M and S is based on the fact that when G_0 cells are stimulated to divide, a period of DNA synthesis invariably precedes cell division. Of course not *all* non-dividing cells can re-enter the cell cycle. Terminally-differentiated cells such as erythrocytes or polymorphonuclear leucocytes never resume cell division, but many 'quiescent' cells, such as liver, kidney or skin do so when stimulated by hepatectomy, nephrectomy or wounding, respectively. The induction of malignancy in such cells probably involves the same progression from G_0 into the cycle; the difference lies in the fact that when normal liver, kidney or skin cells have completed a period of cell cycles necessary to repair the damage, they return to G_0, whereas malignant cells have somehow lost the ability to do so. What, then, are the biochemical triggers that (a) release a cell from G_0 and (b) commit a cell to G_0?

The fact that when animal cells in culture reach saturation density and stop dividing, they generally enter G_0, provides a useful experimental model. Many factors can cause the resumption of cell division. Increases in serum concentration, the addition of proteolytic and other enzymes and of certain hormones, transformation by RNA and DNA viruses, and even wounding by scratching across the surface of a cell layer, have been shown to initiate DNA synthesis within a matter of hours; mitosis follows some time later. Protein and RNA synthesis, as might be expected from what has been said previously, are initiated prior to DNA synthesis; phospholipid turnover also increases early on. One of the earliest events is an increase in the permeability of certain metabolites and this, coupled with other evidence, has focused attention on the plasma membrane. As a result the concept has arisen that much of the growth control of normal and cancer cells lies on the cell surface – a phenomenon termed 'topoinhibition'. Just how an alteration of the surface membrane leads to a resumption of the G_1 activities necessary for entering S is at present unknown.

Some of the triggers that cause cells already in G_1 to proceed to S have been discussed in connection with the control of DNA synthesis (section 4.1).

The mechanism that commits dividing cells to G_0 is likewise unclear. Certainly the stimulus is likely to act well before M, perhaps at the time that specific repressors (section 4.2) are made; indeed the signal may well be initiated some divisions prior to the final one. In any case the transition from rapid cell division to virtual cessation is generally a gradual one, and entry into G_0 is often preceded by a gradual lengthening of G_1. The implications of such signals to the differentiation of cells lies in the fact that the onset of specific protein synthesis, such as that of myosin in developing myoblasts, is generally accompanied by a decrease in the rate of cell division. At this point a note of caution regarding the relative length of G_1 should be made. In cell culture, and certainly in a tissue growing *in vivo*, the alternatives of (a) all cells having a long G_1 period and (b) some cells having a short G_1 period and some being in G_0, are not easy to assess. In other words, it may be that the G_1 period is relatively constant, and that what varies is the proportion of cells that enter G_0.

5.3 Summary

The overall rate of the cell cycle appears to be regulated at several points. A protein(s) that triggers cell division begins to be built up after DNA synthesis has been initiated and probably controls the length of G_2. Cells in G_0 or G_1 are primed to enter S by factors that are often extracellular in origin; the cell surface appears to be the initial site of stimulation. Commitment to G_0, which is an important factor in cellular differentiation, probably occurs early on in the previous cell cycle(s).

414

6 Conclusions

In this chapter an attempt has been made to sketch the sequence of biochemical events that take place between one cell division and the next. It will probably have become obvious to the reader that topics such as the control of biosynthetic activity (section 4) or the rate of cell division with respect to differentiation and cancer (section 5) constitute some of the most fundamental problems in biology today. In that case he will appreciate that a detailed study, using some of the techniques described in section 2, of certain aspects of the cell cycle is likely to yield rich rewards. That is why the biochemistry of the cell cycle is being so actively investigated in many laboratories.

Acknowledgements

I am grateful to Dr. C. F. Graham for his critical appraisal of this chapter, to Drs. Klevecz and Pardee and the Editors of the *Biochemical Journal, Science* and the *Proceedings of the National Academy of Sciences,* for permission to reproduce original material and to the Medical Research Council for financial support.

References

Most of the key papers and reviews on this topic are referred to in the excellent book by Mitchison (1971). Only articles *not* mentioned in that text are accordingly listed below.

BASERGA, R., ed. (1971). *The Cell Cycle and Cancer.* Marcel Dekker, Inc., New York. Several useful articles, especially with respect to section 5.

BASERGA, R. and STEIN, G. (1971). 'Nuclear acidic proteins and cell proliferation.' *Fed. Proc.*, **30**, 1752–59. Part of a symposium on *Neoplasia and the Cell Cycle*; references for further reading with respect to sections 3.3, 4.2 and 5.2.

CIKES, M. and FRIBERG, S. (1971). 'Expression of H-2 and Moloney leukaemia virus-determined cell-surface antigens in synchronised cultures of a mouse cell line.' *Proc. Nat. Acad. Sci.*, **68**, 566–69. Antigenic expression (section 3.5) measured by immunofluorescence.

MITCHISON, J. M. (1971). *The Biology of the Cell Cycle.* Cambridge University Press, Cambridge. A very comprehensive account, with full references, of all the matters discussed in sections 2–4.

PASTERNAK, C. A. (1970). *Biochemistry of Differentiation.* Wiley-Interscience, London. Includes discussion of events in naturally synchronised systems (section 3), of control of biosynthetic activity (section 4) and of development and cancer (section 5).

PASTERNAK, C. A. (1973). 'Synchronisation of mammalian cells by size separation.' In: *Methods in Molecular Biology* (eds. A. I. Laskin and J. A. Last.) Volume 3. Marcel Dekker, Inc. New York. In Press. A technical review of this method of achieving synchrony (section 2.2).

PASTERNAK, C. A., WARMSLEY, A. M. H. and THOMAS, D. B. (1971). 'Structural alterations in the surface membrane during the cell cycle.' *J. Cell Biol.*, **50**, 562–64. Antigen expression (section 3.5) measured by immune cytolysis.

13
Microbial Growth

A. T. Bull

Biological Laboratory, University of Kent at Canterbury

1. Introduction

> The study of the growth of bacterial cultures does not constitute a
> specialised subject or a branch of research; it is the basic method of
> microbiology
>
> *Jacques Monod* (1949)

Despite the veracity of Monod's assertion and indeed its extension to a major portion of
biochemistry, the principles of microbial growth all too often have not been fully appreciated
or have been neglected entirely by experimental biologists. Microorganisms respond to
physico-chemical changes in their environment by processes of metabolic adaptation and the
corollary, that a microorganism is able to exist in a broad spectrum of physiological states, must
be recognised when microbial activity and behaviour is investigated in laboratory cultures.
Unfortunately the critical experimental requirements essential for elucidating phenotypic
variations in microbial structure, chemistry and biochemistry are not always met. Consequently, a failure to observe the basic principles of microbial growth inevitably leads to
experiments which are largely invalidated or at best have substantial limitations. Although the
development of continuous-flow cultures has been considerable over the past two decades,
microbiologists and biochemists – with only a few, albiet notable, exceptions – have been
tardy in exploiting the great advantages of the technique. However, our innate conservatism (or
suspicion) in this context is disappearing rapidly and in the discussion which follows I shall
examine some aspects of the theory and applications of continuous-flow culture. In short, the
intent of this essay is to focus attention on the necessity for careful consideration of growth
conditions in biochemical and microbiological work and to suggest that regard for the
techniques also can further our understanding of the nature of microbial growth itself.

2 Closed and open growth systems

Microbial cultures can be distinguished conveniently as closed or open systems based on their
mode of operation. Closed cultures are those used traditionally in the laboratory and by the
process biochemist and microbiologist in industry, and they are characterised by the absence of
an input and output of materials once a batch of growth medium has been inoculated with the
desired organism. Thus, the *batch culture* is a discontinuous process in which the environment

and the organism are changing continually; in other words only non steady or transient states occur in such cultures. In contrast, open-culture systems are capable of attaining steady-state conditions in which the average values of all culture and organism properties remain constant. The open culture has a balanced input of growth substrates and output of organisms, metabolic products and unused nutrients: such a system is termed a *continuous-flow culture* and, at least in theory, can be operated indefinitely. Ideal steady states are not attainable in practice due to problems such as non-perfect mixing of organisms and nutrients and the selection of mutants during the extensive culture period, nevertheless, as we shall discover, continuous-flow cultures approximate quite closely to steady-state systems and their observed behaviour can be described very adequately in terms of steady-state theory.

The peculiar advantages of continuous-flow cultures over batch cultures may be summarised as follows:

1. The growth rate can be controlled readily and maintained at a predetermined value for long periods.
2. The growth rate can be set and held constant and subsequently any physical or chemical parameter can be varied systematically; only such a system enables the effects of these parameters on microbial activity to be determined unequivocally.
3. Organism concentration also can be set and maintained independently of the growth rate.
4. It allows substrate-limited growth (see section 3.1) conditions to be established; the implications and experimental advantages of different substrate limitations will become apparent later in this chapter.
5. Data obtained from continuous-flow cultures are more reliable and reproducible than those obtained from batch cultures with the result that fewer confirmatory experiments are necessary.
6. Organisms can be grown for longer periods under constant conditions.
7. Higher productivity per unit volume per unit time (productivity here refers to the output of cellular material).

The great benefit that continuous-flow culture offers to the experimenter is the extremely fine control which can be exercised over growing microorganisms. For example, the growth rate of an organism can be varied in batch systems but only by quite major alterations of the physical conditions or by changing the quality of the growth medium; 'shift' experiments of this sort are discussed below (section 5.2). In contrast, a much more delicate control over growth rate can be made in continuous-flow systems simply by altering the concentration of the limiting substrate; in practice this is achieved by varying the dilution rate (see section 3.2).

It is common to find microbial growth referred to as 'balanced', an epithet simply implying that all extensive properties (e.g. total nucleic acid and protein contents, mass of organisms) of the growing culture are increasing at a common rate. The rate of increase, k_a, of a molecular species is given by $k_a = \frac{1}{a} \cdot \frac{da}{dt}$ where a is the concentration of the particular molecule. If k_a, k_b, k_c to the n th molecular species are equal to each other and to μ, the specific growth rate, then the population is in a state of balanced growth. Unbalanced growth describes the situation in which the major cell components are being synthesised at dissimilar rates, a condition which may prove to be lethal. Two points must be stressed regarding balanced growth: (*a*) the term is apposite to populations but not to individuals because different molecular species are synthesised at different rates during the cell cycle; (*b*) use of the word balanced to qualify growth does not signify here a normal or ideal condition nor convey information about the relative quantities of the different molecules which are present.

At this juncture kinetic analysis of microbial growth in closed and open systems is necessary both to substantiate the claims made above and to define the population dynamics with which we are concerned.

3 Kinetic considerations

3.1 Growth of batch cultures

The growth of a microbial population can be viewed as an autocatalytic event which, under ideal conditions, proceeds at an exponential rate. (Note that growth of individual organisms is approximately linear.) During exponential growth the time interval for the population to double is its *generation or doubling time* (t_d). Thus, if an initial population mass x_0, doubles in time, t_d, the final population x after time t will be $x_0 2^n$ where n is the number of generations through which the population has passed. Rearranging, we have $x = x_0 2^{t/t_d}$, the natural logarithmic form of which is

$$\ln x = \ln x_0 + \frac{\ln 2}{t_d} t \tag{1}$$

where $\ln$ is the abbreviation for logarithm to the base e. If $\ln x$ is plotted against time, the slope of the curve $(\ln 2/t_d)$ gives us the *specific growth rate* (μ) of the organism under the given experimental conditions; μ has the units of reciprocal time (h^{-1}). Substituting μ in (1) gives

$$\ln x = \ln x_0 + \mu t \tag{2}$$

which is the *basic growth equation* and describes exponential growth. This relationship can be approached also by considering the autocatalytic nature of growth. Now $dx/dt = \mu x$, or, the amount of organism produced (dx) in a small time interval dt is proportional to that amount present initially (x). Integration of this equation gives $x = x_0 e^{\mu t}$ which, on taking logarithms, leads to our basic growth relationship (2) from which discussions of microbial growth kinetics usually stem. It will be evident that a logarithmic plot of x against t produces a straight line when growth is exponential. Although exponential growth is a characteristic of microorganisms, it persists for a relatively short time in laboratory batch cultures. Following the introduction of microorganisms into a culture medium nutrients are assimilated, growth ensues and metabolic products accumulate extracellularly. Clearly the act of growth causes profound changes in the environment which lead eventually to a cessation of growth. The usual causes of growth arrest are depletion of nutrients, development of an unfavourable pH in the medium and, where aerobic organisms are being studied, oxygen limitation. Failure to compensate for the increasing oxygen demand of a growing culture by ensuring adequate aeration is a frequent cause of poor growth and is a trap into which the unmindful worker can readily fall.

Microorganisms grown in batch culture pass through a familiar and characteristic succession of changes as shown, in an idealised form, in Fig. 13.1. Most textbooks of microbiology treat this so-called 'batch culture cycle' in detail and it will suffice here to make only a few brief comments. First, batch cultures obviously are not cyclic processes in the sense that they represent some intrinsic property of organisms. A second misnomer is the 'stationary' phase of growth; at this point in the batch culture although the net rate of growth has fallen to zero, the population remains metabolically active. It is much more appropriate to term this state of growth the maximum population phase. We might digress briefly at this point to consider the

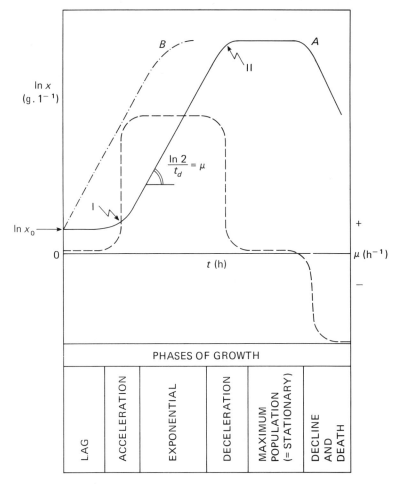

Fig. 13.1 Course of microbial growth in a batch culture. The changes in organism concentration, ln x(—— and ·—·—·) and specific growth rate, μ(– – – –) are shown as functions of time.

concept and reality of a maximum population. Taking a value of 2·5 μm^3 as the volume of an average eubacterium, in one millilitre of medium there can be a maximum of $10^{12}/2\cdot5 =$ 4×10^{11} organisms. Further assuming that 1 mg dry weight is roughly equivalent to 2×10^9 bacteria, the theoretical maximum population density will be 200 mg ml^{-1}. In practice the highest organism densities in batch cultures are about one-tenth of this theoretical value. A 'biological space' requirement − a sort of microbial lebensraum − has been advanced as a reason for lower population maxima but the hypothesis has attracted little experimental support. The limit to population size is due in most cases to autotoxicity coupled with oxygen starvation in aerobic cultures. The specific growth rate has a constant positive value for only a small proportion of the total culture time and eventually it becomed negative as the population dies. Finally, batch cultures may not pass through a lag phase (this represents a lag in cell division). If a new culture is established by adding an already exponentially growing inoculum to medium of similar composition, the lag phase can be alleviated (Fig. 13.1, B) but if spores or maximum population phase cells are used, or, changes are made in the nature of the growth conditions, a lag phase ensues (Fig. 13.1, A).

The value of μ is specific for each microorganism and the physico-chemical conditions under which it is growing, i.e. it is a function of a genotype and the environment. The specific growth rate is affected particularly by the concentration of nutrients in the medium. If we consider a medium in which all nutrients are present in excess save one, the *growth-limiting substrate*, then, at low concentrations μ is dependent on the growth-limiting substrate concentration s. During his classic studies of microbial growth Monod showed that the variation of μ with s was satisfactorily expressed by the hyperbolic relation:

$$\mu = \mu_{max} \cdot \frac{s}{K_s + s} \tag{3}$$

in which μ is the specific growth rate corresponding to limiting substrate concentrations, μ_{max} is the maximum specific growth rate which occurs when s is large and K_s is the *substrate saturation constant* (numerically equivalent to s at $\mu_{max}/2$ and is characteristic of the organism and the substrate).

Thus, we can write an expression for the rate of growth such that

$$\frac{dx}{dt} = \mu_{max} x \frac{s}{K_s + s} \tag{4}$$

Accurate determinations of μ_{max} and K_s can be made from double reciprocal plots of μ against s (Fig. 13.2). Values of K_s generally are very low: for bacteria they are of the order of mg l^{-1} (carbohydrate substrates) or μg l^{-1} (amino acids). Consequently, because the carbon substrate of most growth media is of the order of g l^{-1}, a K_s of say 10–100 mg l^{-1} reduces μ only a few per cent with respect to μ_{max} in batch cultures.

Monod investigated also the efficiency with which microorganisms utilise substrates and defined a *yield* factor (Y) as being equal to the weight of cells produced divided by the weight of substrate utilised, i.e. $Y = -dx/ds$. Substituting in (4) gives us the equation of the rate of substrate utilisation:

$$-\frac{ds}{dt} = \mu_{max} \frac{x}{Y} \frac{s}{K_s + s} \tag{5}$$

Although at first Y was thought to be a constant, we know now that it is a function of growth rate and the environment and we shall take up this point later. To conclude, if we know the values of K_s, μ and Y we can make a thorough quantitative analysis of any batch culture system. Moreover, having derived the relations between these growth parameters it is but a small step to the kinetic analysis of simple continuous-flow systems.

3.2 Simple theory of the chemostat

Exponential growth can be maintained indefinitely in a batch system by a regular withdrawal of part of the culture and it replenishment with fresh medium. As the number of withdrawals-replenishments per unit time approaches infinity, a continuous-flow system becomes established. Under these conditions of unlimited nutrient supply, growth rate is limited by the inherent properties of the organisms, i.e. growth is under *internal control*. Continuous-flow cultures operated in this mode are known as *turbidostats* and a constant predetermined organism density is maintained via a photo-detector system which controls the inflow of fresh medium into the fermenter vessel. Here the population grows at or very close to μ_{max}. Alternatively, continuous-flow cultures may be subjected to *external control* when growth rate

420

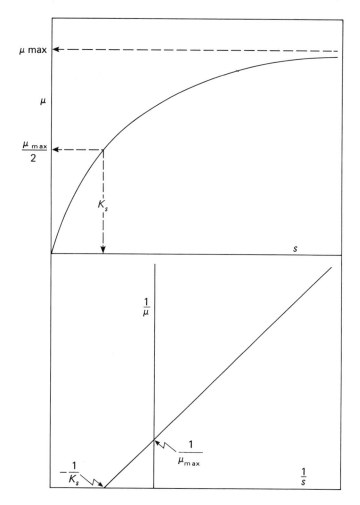

Fig. 13.2 The Monod relationship between growth-limiting substrate concentration and specific growth rate. The substrate saturation constant K_s is that concentration of s which produces half the maximum specific growth rate μ_{max}. Accurate determination of K_s and μ_{max} is made from double reciprocal plots of the Lineweaver-Burk type.

is dependent on the rate of supply of a limiting substrate; such cultures are termed *chemostats*. In the latter system μ can be varied over a wide range independently of organism concentration but μ_{max} is not attainable. Figure 13.3 summarises the relationship between the two control systems.

At present, the chemostat is used most widely in laboratory studies and our discussions will be concentrated upon it. A comprehensive discussion of the turbidostat has been made recently by Watson (1972). For details of chemostat construction and operation the reader similarly is directed to the excellent article of Evans, Herbert and Tempest (1970). Although the relatedness of the two control systems is such that kinetic analyses of turbidostats and chemostats are very similar, the mathematical description which follows relates specifically to a simple (i.e. single stage without feedback) chemostat.

A continuous-flow culture is characterised by its fractional rate of medium replacement or its *dilution rate (D*; having units of h^{-1}). The dilution rate is defined as

$$D = F/V \qquad (6)$$

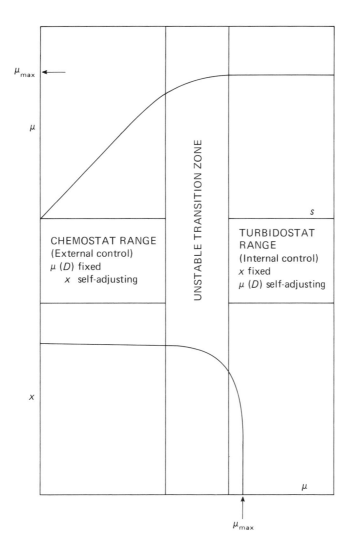

Fig. 13.3 Control characteristics and operational ranges of continuous-flow cultures.

where $F(1\ h^{-1})$ is the flow through the fermenter vessel and $V(1)$ is the culture volume. The reciprocal of the dilution rate is the *mean residence time* (θ), or the average time an organism remains in the vessel and it is essential to distinguish it from the organism's doubling time t_d. The two parameters are related thus, $t_d = \theta \ln 2$ and the chance of an organism remaining in a continuous-flow fermenter until cell division occurs is e^{-Dt_d}; in other words, the longer becomes t_d, the greater chance of the organism being washed out before it divides. Returning to the culture *per se* and disregarding the fate of individual organisms, we will find it convenient to analyse chemostat theory in terms of balance equations for x and s. Thus, the net change in x with time is the sum of growth and output of cells:

increase = growth − output

$$dx/dt = \mu x - Dx \tag{7}$$

and when $\mu = D$ and dx/dt is zero, a steady state will be established. Substituting in equation (4) we can write:

$$\frac{dx}{dt} = x\left[\mu_{max}\left(\frac{s}{K_s + s}\right) - D\right] \tag{8}$$

Similarly the balance equation for s is:

increase = input − output − consumption

$$ds/dt = DS_R - Ds - \mu x/Y \tag{9}$$

where S_R is the growth-limiting substrate concentration as it flows into the fermenter from the medium reservoir. Again if we substitute for $\mu(=D)$ from equation (3):

$$\frac{ds}{dt} = D(S_R - s) - \frac{\mu_{max}x}{Y}\left(\frac{s}{K_s + s}\right) \tag{10}$$

These simultaneous equations (8 and 10) can be solved for the steady-state conditions $dx/dt = 0$ and $ds/dt = 0$, such that

$$\bar{x} = Y(S_R - \bar{s}) \tag{11}$$

and

$$\bar{s} = K_s\left(\frac{D}{\mu_{max} - D}\right) \tag{12}$$

The bar terms $\bar{x}$ and $\bar{s}$ define steady-state values. Finally, we can substitute for $\bar{s}$ in equation (11) and derive:

$$\bar{x} = Y\left[S_R - K_s\left(\frac{D}{\mu_{max} - D}\right)\right] \tag{13}$$

These equations enable us (a) to predict the steady-state organism and substrate concentrations in a chemostat operating with known values of S_R and D; and (b) to understand the self-adjusting capacities of a chemostat because the dilution rate determines s (equation (12)) which itself determines μ (equation (3)) and when $D = \mu$ steady-state conditions are restored. The theoretical steady-state conditions for given values of S_R, K_s, Y and μ_{max} are presented graphically in Fig. 13.4. Clearly D cannot be made greater than μ_{max} and as a result there is an upper limit for D (nearly equal to μ_{max}) known as the *critical dilution rate* (D_{crit}). When $D > D_{crit}$ non-steady conditions obtain and an exponential washout of organisms from the fermenter occurs. Figure 13.4 also illustrates that the productivity ($D\bar{x}$) of a chemostat has a maximum value.

In passing it may be noted that washout data can be used for the accurate determination of μ_{max}. Equation (2) can be rearranged to give

$$x = x_0 e^{(\mu - D)t}$$

and solved for μ thus:

$$\mu = D + \frac{1}{t} \ln \frac{x}{x_0}$$

Similarly the washout rate $-A$ (h^{-1}) can be written

$$-A = \frac{1}{t} \ln \frac{x}{x_0}$$

and hence the maximum specific growth rate will be given by $D + (-A)$.

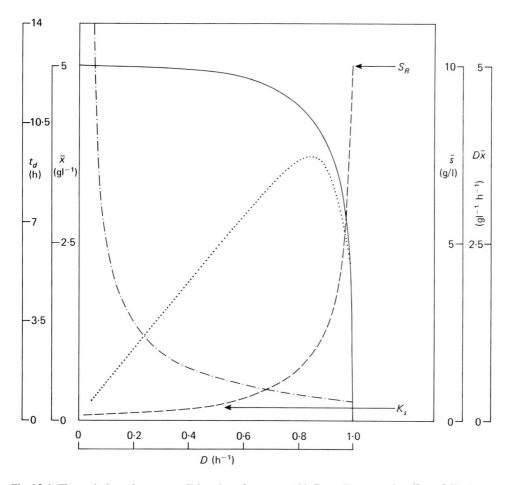

Fig. 13.4 Theoretical steady-state conditions in a chemostat. This figure illustrates the effect of dilution rate D on steady-state organism concentration $\bar{x}$ (——), steady-state substrate concentration $\bar{s}$(– – – –), doubling time t_d(·—·) and organism productivity $D\bar{x}$(······). The following values were used to construct the curves: S_R 10 g l^{-1}, K_s 0·2 g l^{-1}, Y 0·5 μ_{max} 1·0 h^{-1}.

3.3 Validity of the chemostat theory

In certain cases the theory agrees very well with the observed behaviour of chemostat cultures. However, exact agreement tends to be infrequent. This disparity between theory and practice should not come as a surprise: the simple theory makes few assumptions about the stoichiometry of growth and Monod's relationship (equation (3)) can be taken only as an approximation. Furthermore, organism and substrate concentration effects appear to produce significant departures from the simple theory. Among the most important and commonly encountered reasons for non-ideal chemostat behaviour are the following:

1. The yield factor does not have a constant value.

a. In carbon-limited chemostats operating at a low value of D, organism concentrations often are lower than predicted (Fig. 13.5, A). This deviation is explained by a requirement for substrate to maintain cell integrity and viability, or a *maintenance energy* requirement (m; units of g substrate (g organism)$^{-1}$ h^{-1}). Pirt has derived an expression for m as follows. In the production of Δx amount of organism, suppose that $(\Delta S)_G$ amount of substrate is used for biosyntheses and $(\Delta S)_M$ amount for maintenance. The observed yield, Y, is

$$Y = \frac{\Delta x}{(\Delta S)_G + (\Delta S)_M} \tag{14}$$

and when $m = 0$,

$$Y_G = \frac{\Delta x}{(\Delta S)_G}$$

where Y_G is the *true growth yield*. The substrate balance can be written as

$$ds/dt = (ds/dt)_M + (ds/dt)_G \tag{15}$$

substituting in (15) we have

$$-\mu x/Y = -mx + (-\mu x/Y_G)$$

which reduces to

$$1/Y = m/\mu + 1/Y_G \tag{16}$$

Given that m and Y_G are constants it is a simple matter to determine m from the slope of $1/Y$ against $1/\mu$. Some typical values of m for organisms grown aerobically under carbon-limited conditions are: 22 (*Penicillium chrysogenum*), 29 (*Aspergillus nidulans*), 94 (*Klebsiella aerogenes*) and under anaerobic conditions 473 (*K. aerogenes*), all expressed as mg substrate (g dry weight cells)$^{-1}$ h^{-1}.

b Conversion of substrate into intracellular reserves such as glycogen, polyhydroxybutyrate and volutin (Fig. 13.5, B) a phenomenon which is most pronounced when growth is other than carbon-limited.

c. Variable yields may reflect shifts in metabolism with changing dilution rate. For example, a switch to a more fermentative metabolism has been observed in both yeasts and moulds when D is increased beyond a certain value and the subsequent less energetically efficient metabolism results in a reduced yield (Fig. 13.5, C).

2. The organism population is not completely viable. Clearly if a proportion of the population is non-viable, the remaining viable organisms have to grow at a rate greater than D. The doubling time of the viable organisms (t_d^v) equals $\ln 2 + \ln \alpha/D$, where α is the *viability index* [$\alpha = (V + 1)/2$, where V is the percentage of viable organisms]. The magnitude of this effect was demonstrated by the Porton group who found that *K. aerogenes* was about 37 per cent viable when grown under glycerol-limitation at $D = 0.004$ h^{-1}; consequently the values of t_d and t_d^v were 173 and 78 h respectively (Tempest *et al.* 1967).

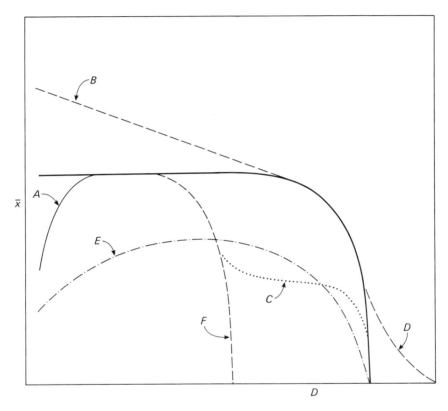

Fig. 13.5 A schematic representation of experimental departures from the simple chemostat theory. The solid line indicates the theoretical variation of the steady-state organism concentration $\bar{x}$ with dilution rate D. The curves A to F indicate various observed non-ideal behaviour the details of which are discussed in the text.

3. Imperfect mixing in the fermenter; fewer organisms than predicted are washed out, consequently D values apparently greater than μ_{max} are observed (Fig. 13.5, D). This is a problem particularly with filamentous microbes which tend to form mycelial aggregations and accretions on fermenter surfaces and are not easily washed out from the fermenter. Indeed the effect of growth on the fermenter walls may be to extend the operational range of a continuous-flow culture well beyond D_{crit}. Variations of this type also may reflect the selection and establishment of faster growing mutants.

4. Substrate concentration effects

a. Monod's relationship (equation (3)) describes s as asymptotic to μ_{max} while in fact at high substrate concentrations μ may be reduced. Some limiting substrates may be toxic even at low levels; Fig. 13.5, E illustrates the growth of a yeast on phenol and depicts both a high maintenance requirement at low D values and the toxicity of the substrate at higher D values.

b. D_{crit} may be much smaller than the μ_{max} measured in batch cultures. The explanation may reside in a partial requirement by the organism for a vitamin, sufficient of which is supplied in the inoculum in batch cultures but which leads to washout in continuous-flow cultures unless D is low (Fig. 13.5, F). Such observations have been made in chemostat cultures of moulds.

5. Organism concentration effects. Deviations from the simple chemostat theory will occur when organisms synthesise substances that promote or inhibit utilisation of the growth limiting substrate, and such deviations will be population dependent. It has been shown, for example, that in Mg^{2+}-limited chemostats an increase in *Bacillus subtilis* concentration was paralleled by an enhanced assimilation of Mg^{2+} ions. This enhancement was attributed to the excretion of an uptake-promoting substance whose identity has yet to be determined but whose effect on μ_{max} can be written as

$$\mu = \mu_{max} \left(\frac{1 + \lambda p}{1 + p} \right) \left(\frac{s}{K_s + s} \right) \tag{17}$$

where p is the concentration of the growth promoting compound, λ is a constant and μ_{max} is the maximum specific growth rate when $p = 0$. Thus, at saturating levels of p, D approaches $\mu_{max} \lambda$ which is greater than μ_{max}. This phenomenon may be referred to as *hypertrophic growth*.

3.4 Efficiency of growth: the yield concept

The quantification of growth efficiency may be examined from both thermodynamic and biochemical standpoints but in this brief discussion an examination of the biochemical approach is the most appropriate. We have noted previously Monod's expression of growth efficiency in terms of a yield factor (Y). Subsequently a much more useful concept of growth yield has been formulated, namely the *molar growth yield*, which we will designate as Y_S^M, and has the units of grammes dry weight of organisms produced per mole of substrate consumed. Such an expression allows easy comparisons of growth efficiencies on different substrates. Alternatively growth yields can be related to the amount of ATP derived from the catabolism of the substrate such that the coefficient Y_{ATP} defines the grammes dry weight of organisms produced per mole of ATP generated. Clearly a knowledge of Y_{ATP} values permits accurate comparisons to be made of the growth efficiencies of different species and of catabolic routes on the basis of the energy available for cell synthesis. It is now well established that the average Y_{ATP} value for organisms growing anaerobically is approximately 10·5. This value is much less than that predicted by assuming that the ATP generated is coupled solely to anabolic reactions; values of 28–33 g mole^{-1} have been estimated on the basis of the latter assumption. Conversely, quite frequently it is found that Y_{ATP} levels are abnormally low, i.e. less than 10·5 g mole^{-1}. Two main explanations can be offered for these observations: (a) uncoupling of catabolic and anabolic reactions. For example. Progressive uncoupling occurs in *Aspergillus nidulans* grown in carbon-limited chemostats when the dilution rate is increased over the range 0·03–0·07 h^{-1}; (b) maintenance energy requirements. A useful application of Y_{ATP} is the estimation of ATP yields from unknown fermentation pathways. Simply by determining the molar growth yield and using the average value of 10·5 for Y_{ATP} an approximate figure for the ATP yield can be obtained.

The discussion of growth yields which we have considered so far applies to microorganisms growing in complex media and under anaerobic conditions. Similar efforts have been made to determine aerobic growth yields but, whereas all the ATP produced via fermentative routes is derived from substrate phosphorylation, during aerobic growth ATP derives mainly from oxidative phosphorylation, the efficiency of which in bacteria has yet to be established unequivocally. Because oxygen is consumed during aerobic growth, yield coefficients have been defined accordingly. Thus, Y_{O_2} (g mole O_2^{-1}) and Y_O (g atom O^{-1}) have been proposed, the latter having the advantage that when divided by Y_{ATP} is equivalent to the P/O ratio which is used as the conventional measure of oxidative phosphorylation efficiency. However, values of P/O ratios for aerobically growing bacteria have ranged from 0·5 to 3·0 when the calculations have been based on an average Y_{ATP} of 10 g mole^{-1}. We may conclude that, in general, the yields of aerobically grown microorganisms are lower than those anaerobically grown microorganisms although they may attain the value of the latter under certain circumstances. In mechanistic terms the efficiency of coupling is believed to be lower in oxidative phosphorylation than it is in substrate phosphorylation with the result that factors such as the dissolved oxygen tension of the medium and the rate of catabolism will exert a profound effect on the efficiency of growth.

4 Intermission: the growth of filamentous fungi

Most general discussions of microbial growth tend to be limited to unicellular species such as bacteria and yeasts and in contrast relatively few critical analyses have been made of the growth of filamentous fungi. Moreover, many workers have claimed that the growth of such fungi in submerged liquid culture does not conform to the basic growth equation (2) discussed above. Because this latter view remains widespread and perpetuated in the literature it requires an emphatic refutal. Substantial evidence now exists for the applicability of the exponential growth and yield relationships (equations (4) and (5)) to filamentous fungi and several detailed chemostat studies have shown that their behaviour in continuous-flow culture is similar to that of unicellular organisms. It is important when working with fungi to operate homogeneous cultures, homogeneous in the sense that the growth morphology should be completely filamentous. Many experiments have been made with cultures in which the organism grows with a pellet morphology; under these conditions the kinetic descriptions of growth considered in the previous section cannot be expected to apply. Filamentous fungi grow by forming new cells at the apices of their hyphae. Assuming that the age of the cells is zero when they are first formed and that the age of the mother cells becomes progressively greater, the mean age of a filamentous organism at time t is the cumulative age of daughter and mother cells divided by the total number of cells comprising that organism. In a small time interval Δt an increment of growth equivalent to $\mu x \Delta t$ will be formed (x is the organism concentration at time t). The age of this mycelial increment will be less than or equal to t. Thus, a *mean cumulative age* ($\bar{\Lambda}$) can be defined by summing the ages of all such increments and dividing by the sum of the amounts of increments having each particular age. The Japanese workers Aiba and Hara (1965) have shown that the mean cumulative age of an exponentially growing batch culture is described by:

$$\bar{\Lambda} = \frac{1}{\mu}(1 - e^{-\mu t}) \tag{18}$$

provided $\Delta t \to 0$; while for a continuous flow system

$$\overline{\Lambda} = \frac{1}{\mu} [1 - e^{-\mu(t-t_0)} + \overline{\Lambda}^0 e^{-\mu(t-t_0)}] \tag{19}$$

Equation (19) shows that $\overline{\Lambda}$ approaches $1/\mu$ (i.e. t_d/in 2) as t becomes large, i.e. with an increase in the number of generations. Thus, the mean cumulative age of a filamentous organism is always greater than t_d. On the other hand for organisms dividing by binary fission, $\overline{\Lambda}$ will be less than t_d. The significance of the mean cumulative age concept comes in deciding whether or not rate relationships in multicellular mycelial populations can be treated in terms of average kinetics. What little data there is with which to test this hypothesis supports the view that the distribution of relative cell ages is not critical in determining the activities of mould cultures. Furthermore, isotopic labelling experiments have shown a uniform distribution of biosynthetic activity among cells of a mycelial population. Once again, however, it must be emphasised that these conclusions do not hold for fungal cultures growing in pellet morphologies.

5 Strictures on the rate of growth

5.1 The reality of finite maximum and minimum growth rates

Throughout our discussions so far we have spoken of rates of microbial growth without reflecting on what limitations there are, if any, to such rates. The observed maximum growth rates of prokaryotic microbes may be an order of magnitude greater than those of eukaryotic types (the record appears to be held by the marine bacterium *Beneckea nitriegens* for which a t_d of 9-10 min has been claimed) while the growth rates of metazoan cells may be yet another order slower. Several inter-related questions need answering among which are: what is the upper limit to growth rate? and why are certain groups of organisms capable of much faster growth than others? In attempting to answer these questions I have relied much on the ideas of Maynard Smith (1969) whose thoughtful essay should be read for further details.

For a bacterium growing in a nutritionally rich medium, growth rate appears to be limited by the requirements of protein synthesis. The rate of protein synthesis in a confined cell volume is controlled by the fact that the reactants (ribosomes and tRNA-amino acid complexes) are large molecules and that, above a limiting concentration a further increase in their concentrations would reduce the rate of synthesis. Maynard Smith has viewed this rate limitation in terms of a queuing problem during protein synthesis. Assuming that T_c, T_s and T_b represent respectively the 'collision time' (time taken for a collision between a codon on the mRNA and a correctly oriented anticodon on a tRNA molecule), the 'shunting time' (time interval during which mRNA is shifted relative to the ribosome and exposes a new codon) and the 'backing out time' (time taken for tRNA molecule to move sufficiently away for the codon to be re-exposed), it can be expected that T_s will be independent of tRNA concentration, that T_c will decrease with increasing tRNA concentration and that T_b will increase with tRNA concentration. Consequently with increasing concentrations of tRNA, the rate of increase in T_b will be greater than the rate of decrease in T_c. An explanation for the slower growth rates of eukaryotic organisms may be found in the following facts:

a. A much smaller ratio of rRNA to protein than typical of bacterial cells;

b. A temporal sequence of cellular development involving a number of biochemical states; transition from one state of differentiation to the next might be limited by the time required to degrade redundant mRNA species and synthesise new ones;

c. Morphological complexity; it is conceivable that increases in cell size and structural differentiation lead to longer times for the operation of control circuits.

It is equally interesting to explore the possibility of a minimum growth rate for microorganisms and in recent years sufficient data has accumulated to support the thesis of a finite minimum growth rate. This most convincing and extensive evidence for a μ_{min} in bacteria results from work on *Klebsiella aerogenes* (Tempest, Herbert and Phipps 1967). Results from both glycerol- and ammonia-limited chemostats demonstrate a μ_{min} value of 0.009 h^{-1} ($t_d = 77$ h) but, because the estimates of organism viability may have been in error (Pirt 1972) a more realistic value for μ_{min} may be 0.06 h^{-1} ($t_d = 11.5$ h). This latter value is equivalent to about 7.5 per cent of μ_{max}. Also there is evidence to believe that μ_{min} for *Escherichia coli* corresponds to about 6 per cent of μ_{max}. The few relevant studies of fungal growth suggest that a minimum growth rate again is definable and is about 5 per cent of μ_{max} for *Aspergillus nidulans* and *Penicillium chysogenum*. To summarise, we can claim, with some confidence, that a finite minimum growth rate equivalent to about 6 per cent of the maximum occurs in bacteria and fungi and below μ_{min} the population begins to differentiate progressively and enters a non-steady state characteristic on a non-growing system. However, it must be emphasised that this conclusion is based on few data and may need to be revised in the light of new findings. Recently it has been observed that very little loss in viability of the soil bacterium *Arthrobacter globiformis* occurs in carbon-limited chemostats over the dilution rate range $0.01-0.35$ h^{-1} (B. M. Luscombe 1972; personal communication). The maximum specific growth rate of this species was found to be 0.37 h^{-1}, thus, μ_{min} is less than 3 per cent of μ_{max}.

5.2 Experimental control of growth rate

It will be obvious from what has gone before that the most satisfactory means of varying the growth rate of a microbial culture is by changing the dilution rate in a chemostat. In this apparatus the experimenter may study microbial behaviour over the range μ_{min} to D_{crit}, in other words from about 5 to 95 per cent of μ_{max}. If we substitute for μ in equation (3), the range of growth-limiting substrate concentrations s, corresponding to the range of growth rates is 0.05 to 19 K_s. In batch cultures the substrate concentration will frequently be 100–1000 times K_s and as a result the organisms are saturated with substrate for most of the growth period (see section 3.1 above). Therefore, the two important features of the chemostat in this context are: it allows for a continuous variation in μ and provides steady-state conditions of growth where the organism is not saturated with substrate.

However, batch cultures may be modified so that the organisms are not saturated with substrate. This is achieved by feeding the growth-limiting substrate into the culture and if different rates are available, different values of μ will be obtained. Although this type system usually is established by pumping the substrate into the culture, Pirt has developed an alternative method involving a diffusion capsule. The growth-limiting substrate is added to the culture within the diffusion capsule and, depending on the thickness of the retaining membrane, the diameter of the membrane orifice and the initial concentration of the substrate, the growth rate can be varied as required. Both pump and capsule feeding of substrate enable a continuous variation in μ to be made. Finally, the growth rate of batch cultures can be

430

controlled by changing the nature of the substrate or some other environmental variable such as temperature or pH. Such changes in cultivation conditions are termed shifts: if the change engenders an increase in μ we call it a *shift-up*, conversely a decrease in μ is called a *shift-down*.

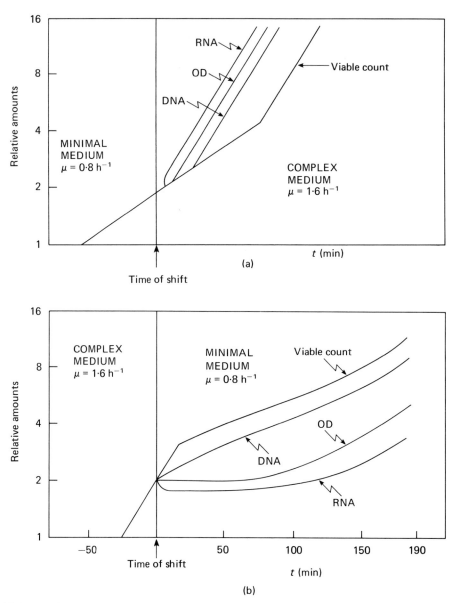

Fig. 13.6 Rate changes following nutritional 'shifts'. Graph (a) illustrates a shift-up experiment in which *Salmonella typhimurium* was transferred from a minimal synthetic to a complex medium with a corresponding increase in specific growth rate from 0·8 to 1·6 h^{-1}. The size of the bacteria, their chemical composition and the growth characteristic of the rich medium is attained about 70 min after the shift.

Graph (b) illustrates a shift of comparable magnitude but in the reverse direction. Following the shift to the poor medium DNA synthesis proceeds for a short time and small, uni-nucleoid bacteria are formed which have low RNA contents. Net protein and RNA synthesis cease for about 2 h during which time turnover of these macromolecules occurs and enables the organism to differentiate and adapt to the new medium. (After Maaløe and Kjeldgaard 1966).

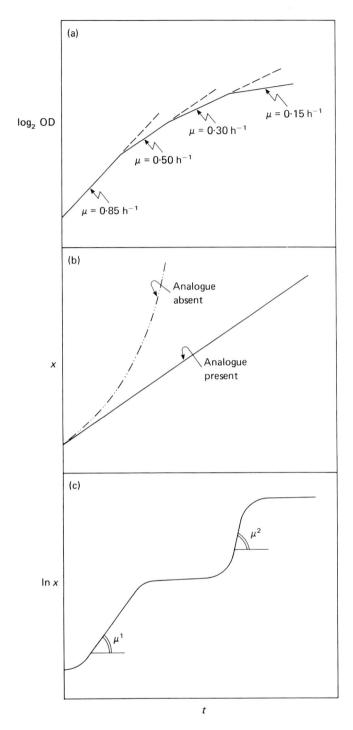

Fig. 13.7 Complex patterns of batch culture growth. (a) Multiple exponential growth phases in a culture of *Escherichia coli* due to progressive exhaustion of endogenous metabolites (from Monod 1949). (b) Linear growth resulting from the incorporation of an amino acid analogue into cell proteins. (c) Diauxic growth: two exponential growth phases separated by an intervening lag.

Extensive experiments of this type have been made by Maaløe and his collaborators in Copenhagen and at this point it is useful to examine the adaptive responses which micro-organisms make when subjected to shifts in their environment. Following a shift-up, the rate of RNA synthesis increases almost instantaneously while comparable increases in the rates of protein and DNA synthesis occur within a few minutes. The cell division rate, however, continues at the pre-shift level for 70 min by which time the cells have the composition characteristic of the new medium. Figure 13.6(a) illustrates these rate changes following a shift-up of *Salmonella typhimurium* growth in a medium giving $\mu = 0.8$ h^{-1} to one giving $\mu = 1.6$ h^{-1}. When bacteria are shifted-down from nutritionally rich to poor media, DNA synthesis and cell division continue for quite a long time but RNA and protein synthesis stop abruptly. These initial effects are succeeded by slow rate changes that lead eventually to a new balanced growth being established (see Fig. 13.6(b) for a shift-down of comparable magnitude to that shown for shift-up in the preceding figure). The reader is directed to Maaløe and Kjeldgaard's monograph (1966) for further details and discussions of shift experiments. Of course it is not necessary to manipulate shift experiments *per se* in order to alter μ in batch cultures. Instead culture media can be judiciously formulated so that they support growth at a very wide range of rates. But it will be equally obvious that experimental systems of this type will only permit stepwise changes in the growth rate.

6 Additional features of batch growth

6.1 Complex batch kinetics

A number of departures from the simple kinetics of batch cultures which were described above (section 3.1) may be evident. Sometimes a succession of exponential growth phases may be observed; these successive phases are not separated by intervening lags. Kinetics of this type may reflect exhaustion of an endogenous reserve. An example of this is the growth of *Escherichia coli* under suboptimal levels of CO_2 and the progressive fall off in growth rate reflects the exhaustion of several endogenous metabolites each synthesised anaplerotically from CO_2 (Fig. 13.7(a)). Occasionally the kinetics of batch growth are linear rather than expo-nential. Explanations of this behaviour usually are based on physiological imbalances brought about by growing the organism in a medium lacking a required growth factor. Under these conditions growth cannot proceed autocatalytically and, instead, becomes directly proportional to time. Similar kinetics are produced when the organism is grown in a medium containing amino acid analogues. The 'false' proteins resulting from the incorporation of the analogue lack catalytic properties and consequently the catalytic capacity of the culture is frozen so that linear growth ensues (Fig. 13.7(b)).

A third and familiar type of variation is introduced when the growth medium contains more than one carbon source; here multiple batch growth patterns may be observed due to the preferential utilisation of one substrate before a second one is metabolised (Fig. 13.7(c)). Monod first studied this phenomenon in two-substrate media and called the behaviour *diauxie*. Diauxic growth will occur either when: (*a*) adaptation to the less preferred carbon source is prevented by the presence of the preferred substrate, an effect manifest usually by catabolite repression; or (*b*) adaptation to the less preferred substrate occurs under non-gratuitous conditions. This latter situation pertains if the less preferable substrate permease is essential for generating intracellular inducer, or, if the induced enzymes are needed to metabolise the carbon source. The operation of chemostats on mixed substrates is germane to this discussion. The few investigations made to date all demonstrate interference of utilisation of a secondary carbon source by a primary carbon source when the dilution rate is high; at lower values of D both carbon sources are likely to be completely utilised. Figure 13.8 illustrates a very complex situation where the yeast *Saccharomyces fragilis* is growing in a multiple carbon substrate chemostat. Fructose and glucose both are utilised over nearly the whole dilution rate range; sucrose is utilised only when $D < 0.2$ h^{-1} (when $D > 0.2$ h^{-1} small, steady-state levels of glucose repress invertase synthesis); while all the sugars repress sorbitol metabolism and the

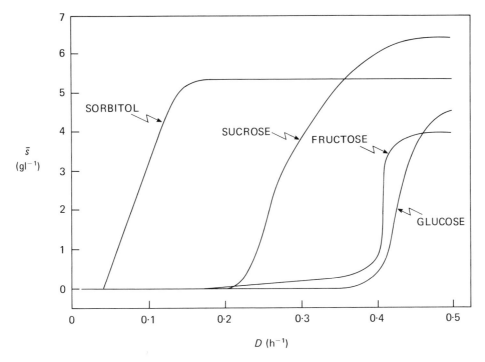

Fig. 13.8 Substrate utilisation curves for *Saccharomyces fragilis* growing on a mixed feed of glucose, fructose, sucrose and sorbitol. Sorbitol and sucrose utilisation are subject to a distinct 'on-off' effect presumably controlled by very low levels of glucose and/or fructose in the medium; sucrose utilisation is known to be regulated by catabolite repression of invertase synthesis. Fructose utilisation is somewhat poorer than that of glucose at growth rates about 0.2 h^{-1} but a sharp 'on-off' control of its metabolism is not apparent and information on enzyme levels will be required to substantiate catabolite repression as the cause of this utilisation pattern. (M. E. Smith and A. T. Bull 1972; unpublished expts.).

polyol is utilised only when $D < 0.1$ h^{-1}. Because commercial culture media are usually complex in terms of their carbon nutrients, considerations of this type are crucial in the optimisation of continuous-flow fermentation processes such as the production of microbial protein and of beer, and also to the design of waste-treatment processes.

6.2 Influence of the energy source

One example will indicate the significance of substrate selection in studies of microbial growth and this concerns the effect on the yield of aerobic organisms when the oxygen demand exceeds the rate of oxygen supply. *Escherichia coli* continues to grow in a batch culture in a lactate medium even after oxygen supply falls below demand, whereas in a glucose medium further growth is prevented under such conditions. Oxygen-limited growth on glucose enhances organic acid production which in turn inhibits further growth. With lactate as the energy source, however, oxygen-limited growth becomes linear, the rate being directly proportional to the aeration efficiency. The practical lesson here is that high growth yields may be possible without the expense and trouble of high aeration systems.

6.3 Differential exhaustion of substrates

We have remarked that the termination of exponential growth in a batch culture commonly is due to substrate exhaustion. This statement now requires qualifying inasmuch as the further development of the culture is very dependent on *which* substrate is first to become depleted. One of the most thorough studies of differential substrate exhaustion was made some years ago by Borrow and his colleagues (1961) during an investigation of the gibberellic acid fermentation by the mould *Gibberella fujikuroi*. Experiments were set up in which the four medium constituents C, N, Mg and P were exhausted in all possible sequences, the first nutrient being exhausted at a constant organism concentration. From the enormous amount of data which was collected from these experiments the following illustrate the consequence of different patterns of substrate exhaustion:

Nutrient first exhausted	Consequent effects on mould development
C	growth arrested immediately or after the endogenous carbon reserves had been utilised.
N	growth arrested; small increases in fat and carbohydrate contents; synthesis of gibberellin.
P	endogenous metaphosphate utilised; growth continued until C or N depleted; uptake of Mg and K curtailed.
Mg	growth continued until C or N depleted; carbohydrate and fat increase.

It is clear that the sequence of substrate exhaustion greatly influences the chemical composition of the mould and its synthesising capability. The implications of these findings in the study of microbial biochemistry and in the design of fermentations are obvious and are of a general applicability.

6.4 Relationships between batch and continuous-flow cultures

It is worthwhile at this point to consider precisely what relationships exist between batch and continuous-flow culture growth. It is tempting to suppose that the various transient states characteristic of batch growth can be simulated and maintained indefinitely in a continuous-flow culture simply by selecting the appropriate dilution rate. From a batch culture curve we can recognise two points (I and II. Fig. 13.1) at which the growth rates μ_I and μ_{II} are equal. Moreover, a chemostat can be operated at a particular dilution rate such that $D = \mu_I = \mu_{II}$. The problem is which of the two batch states, I or II, is analogous to the chemostat state? The answer, unfortunately for our initial supposition, is neither: in state I nutrients are in great excess and usually the cell size is much larger than that typical of exponential growth, in state II nutrients are being depleted, i.e. substrate-limited growth may have developed, but the organisms are not growing under steady-state conditions. In short, batch cultured cells have a history — the constant change in the environment is such that the behaviour of organisms at any particular time is influenced by the history of their recent ancestors which have developed under different environmental conditions. The state of affairs in a chemostat is very different — because of the constancy of the culture conditions the organisms, once they are growing in a steady state, can be viewed as not having a history.

On the other hand, properties of the other type of continuous-flow culture to which we drew attention, the turbidostat, resemble much more closely certain features of batch cultures. During at least the late exponential growth phase of the batch culture a balanced maximum rate of growth undoubtedly occurs and a similar state can be provided for in a turbidostat.

7 The research potential of continuous-flow cultures

Development of the continuous-flow culture system represents one of the most important conceptual and technical advances in microbiology for very many years. Now what is significant is the coming of age of continuous-flow culture: we are passing from a phase during which the theory and apparatus were developed to one where the application to research problems is rapidly occurring. Far from continuous-flow cultures merely providing a convenient means of mass producing microorganisms, they furnish a unique and often indispensable means of investigating microbial chemistry and behaviour. Indeed, continuous-flow cultivation is fast becoming the method of choice in many areas of microbiology and biochemistry. We have emphasised repeatedly that the establishment of steady-state conditions is a major attraction of open flow systems; this enables the experimenter to make precise and systematic analyses of the effects of all environmental parameters on any organism property, i.e. it provides the only unequivocal means of studying phenotypic variation. Although at first sight it may appear paradoxical, a second pre-eminence of this form of culture system is in the study of transient state behaviour. Thus, regulatory processes may be revealed and defined when a steady-state culture is perturbed by say a stepwise change in dilution rate, and adapts to the new steady-state condition. I hope that the following discussions will fully exemplify these claims for continuous-flow cultures and, again, most reference will be made to the chemostat.

7.1 Macromolecular constitution as a function of environment

Herbert (1961) classified the major macromolecular constituents of microorganisms as *storage* materials (polysaccharides, polyphosphates, fats) which may be dispensable and *basal* materials (DNA, RNA, proteins) which are essential. The environment affects the synthesis of these two groups of substances in significantly different ways. Carbon storage products tend to be synthesised when a nutrient other than the carbon source is growth limiting and their synthesis frequently occurs independently of growth. In contrast, the cellular levels of basal materials are largely growth dependent and are not related to the nature of the growth-limiting substrate. Typical data from Herbert's laboratory which illustrate these differences are shown in Fig. 13.9.

Some years ago Maaløe and his colleagues, on the evidence of experiments with their 'shift' culture technique, concluded that the rate of protein synthesis was approximately proportional to the number of ribosomes present in a bacterium. This led to the concept of a *constant ribosomal efficiency* hypothesis. Subsequently evidence in support of and against this notion was obtained and it is only recently that the application of chemostat studies has resolved the conflict. Koch (1971) has presented an excellent review of his own and other people's work on

436

this problem and only the salient points in the argument will be included here. The invalidity of the constant efficiency hypothesis was proved finally by the finding that the rate of protein synthesis of carbon-limited cultures of *Escherichia coli* switched from $D = 0.06$ h^{-1} to $D = 1.03$ h^{-1} increased seven-fold before an appreciable change in RNA occurred. Thus, there was 'extra' RNA in the slowly growing bacteria which, although it was not being utilised, was present in a rapidly utilisable form. That some of these cells were not active in protein synthesis or that large-scale protein turnover was occurring were not responsible for these observations.

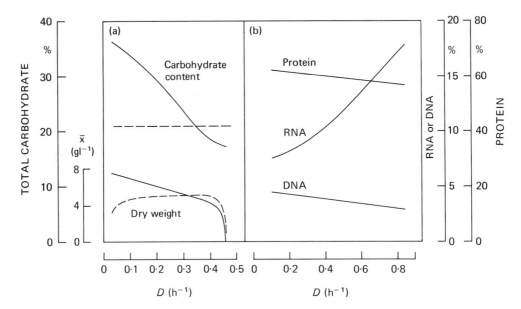

Fig. 13.9 Effect of culture conditions on macromolecular composition of microorganisms. (a) Carbohydrate content of the yeast *Torula utilis* as a function of growth rate and nutrient limitation (– – – –, carbon-limited growth; ——— nitrogen-limited growth). The changes shown reflect variations in glycogen synthesis, i.e. in the formation of storage materials. (b) Effect of growth rate on the synthesis of basal materials by *Klebsiella aerogenes*. The data are from carbon-limited chemostat cultures but almost identical changes occur when growth is limited by nitrogen. (Redrawn from Herbert 1961.)

Therefore, we conclude that carbon-limited, slowly growing cells synthesise rRNA, tRNA and ribosomal protein that are not actually used to full efficiency. Teleologically, Koch considers that this 'extra' RNA could reflect a response of the bacterium in the expectation that environmental conditions will improve. Moreover, Koch has shown that there is considerable selective advantage to the organism having reserve protein-synthesising capacity, for example it permits growth to proceed rapidly and without a lag when the appropriate conditions arise.

Turning from a general appraisal of how growth conditions influence macromolecular composition, it will be instructive to examine one particular cell structure in detail, namely the bacterial wall. These studies, made mainly by D. W. Tempest and D. C. Ellwood, show the full value of the chemostat in exploring environmental effects on cell chemistry and, thereby, cell functioning. In chemical and structural terms the walls of bacteria are very complex (see chapter 10 of this volume) but in the present context emphasis' will be placed on the anionic polysaccharides (teichoic acids and teichuronic acids) found only in Gram-positive species.

The teichoic acids are composed of polyribitol – or polyglycerol – phosphates variously substituted with sugars and alanine; the anionic character of the teichuronic acids, however, is due to the presence of uronic acids. Various combinations of these anionic polysaccharides can be found in bacterial walls the types and quantities of which are influenced profoundly by the growth conditions. Most striking is the effect of phosphate nutrition: in all Gram-positive bacteria analysed, the wall teichoic acid is replaced by a teichuronic acid when growth is made in PO_4-limited chemostats. Ellwood and Tempest provided convincing evidence that this changeover was a truly phenotypic response and not due to mutant selection. Thus, when a Mg-limited chemostat was switched to PO_4-limitation, the teichoic acid content fell much more quickly than the predicted rate and, conversely, the teichuronic acid content rose faster than the predicted rate. Clearly, wall turnover in addition to *de novo* synthesis had occurred under these conditions. The invariable presence of an anionic polysaccharide in walls of these bacteria has suggested a role for them in the assimilation of cations such as magnesium. Magnesium- or potassium-limited *Bacillus subtilis* var. *niger* bind Mg^{2+} much more voraciously than PO_4-limited organisms and Mg^{2+} binding to the wall is competitively inhibited by Na^+. Thus, if Mg^{2+} binding is an essential prerequisite to Mg^{2+} assimilation, subjection of the bacterium to high levels of Na^+ in a PO_4-limited chemostat should, ultimately, lead to growth inhibition and washout. The results from such an experiment, therefore, are surprising; the cells responded by replacing progressively the teichuronic acid as the NaCl concentration was raised until, at 6 per cent NaCl, these PO_4-limited organisms had teichoic acid as their sole anionic polysaccharide. This elegant illustration of bacterial adaptation was made possible only by the use of chemostat cultures. Similarly, teichuronic acids appear to be poor binders of Mg^{2+} at low pH values, so much so that even in PO_4-limited chemostats teichoic acids displace teichuronic acids in the wall in response to a shift of pH from 8 to 5.

The lipopolysaccharides characteristic of Gram-negative bacterial walls also have received some preliminary study with respect to their phenotypic variability. For example, the nature of the growth-limiting substrate and the growth rate exert an influence both on lipopolysaccharide content and chemistry in *Klebsiella aerogenes*. These results have wide implications because the wall lipopolysaccharides of Gram-negative bacteria possess the specific determinants of O-antigenicity and possess toxigenic and pyrogenic properties and the relevance to vaccine production is unmistakable. Evidence that the major structural component of bacterial walls — the peptidoglycan — also is subject to phenotypic variation is beginning to accumulate. The reader is directed to Ellwood and Tempest's (1972) comprehensive review for an up-to-date account of environmental effects on bacterial walls.

The microbiological and biochemical literatures contain numerous statements of the type 'the content/composition of X in microorganism Y is such-and-such a value'. It should be obvious now that such statements are of little meaning unless account has been taken of phenotypic variation and the precise conditions of growth.

7.2 Enzyme and metabolic regulation

The special advantages of chemostats in the study of enzyme regulation have been endorsed by many workers and the sizeable body of information on this subject has been reviewed recently by Clarke and Lilly (1969), Bull (1972) and Dean (1972). Critical quantitative changes in enzyme levels can be made only with organisms growing under steady-state conditions, whereas batch cultures can be used quite adequately to follow qualitative variations. In addition, the chemostat is an ideal system with which to explore control mechanisms and offers considerable scope for the selection of strains which are constitutive and hyperproductive for certain enzymes. Let us consider a few examples. Tempest (1970) provides a persuasive illustration of the first point by reference to C_2 metabolism. Acetate utilisation in *Pseudomonas ovalis* occurs

438

via the glyoxylate cycle and in batch cultures the key enzyme isocitrate lyase is completely repressed by succinate. However, it became apparent from acetate-limited chemostat studies that isocitrate lyase was unaffected or only partially repressed by succinate *per se* although a change to ammonia-limitation did make succinate additions severely repressive. The clue to the interpretation of these results lies probably in the 'pool' concentration of succinate or succinate-derived repressor substances. Thus, despite the extremely small steady-state levels of succinate in NH_4-limited cultures (< 1 mM), the 'pool' level would be much greater than in acetate-limited cultures, with the result that the lyase becomes repressed.

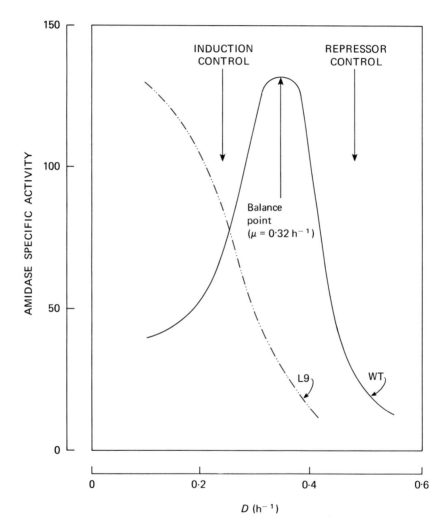

Fig. 13.10 Dual control of enzyme synthesis. Aliphatic amidase synthesis by wild type (———) and mutant (··—··) stains of *Pseudomonas aeruginosa* growing in carbon-limited chemostats on acetamide (20 mM) plus succinate (10 mM). The wild type strain shows induction — repression control of the enzyme and the specific activity passes through a peak at $D = 0.32$ h^{-1}. The explanation of these results probably involves inducer control of amidase synthesis as D is increased from low values and catabolite repression at high dilution rates. The mutant strain illustrated appears to have a double mutation which has affected both inducibility (it is constitutive mutant) and catabolite repression (the activity–dilution rate curve is displaced to the left indicating increased repressibility). (From Clarke and Lilly 1969.)

We have seen previously that the dual control (induction and repression) of an enzyme synthesis often is manifest as diauxic growth in a batch culture. However, the balance between induction and repression is far more obvious when examined in chemostats. A good example of this phenomenon is aliphatic amidase synthesis by *Ps. aeruginosa*. Patricia Clarke and her colleagues followed amidase induction when the carbon status of the inflowing medium was changed from succinate to succinate + acetamide and found that the duration of the induction lag was growth-rate dependent. Likewise it was possible to demonstrate the balance between induction and repression in steady-state cultures growing on acetamide (Fig. 13.10). Analogous control mechanisms have been revealed for the synthesis of other catabolic enzymes including β-galactosidase, pullulanase and β-1,3 glucanase. Constitutive and hyperproductive mutants for inducible enzymes frequently are selected when a wild-type organism is grown at very low substrate inducer concentrations: at high concentrations an excess of enzyme in the hyper strain is superfluous and growth of the wild-type organism is favoured; at low concentrations, however, cells producing above normal levels of enzyme have a selective advantage and outgrow the wild-type. Finally, it is noteworthy that chemostat studies have revealed that glucose enzymes are controlled by induction by glucose. Dawes' group have demonstrated this type of regulation for enzymes of the Entner–Doudoroff pathway in *Ps. aeruginosa*, while a similar regulation of Embden–Meyerhof and hexose monophosphate shunt enzymes in the mould *Aspergillus nidulans* has been reported from our laboratory.

I wish to conclude this section by considering the applicability of chemostasis to metabolic studies and to cite recent work on glutamate synthesis (reviewed by Brown and Stanley 1972) in support. In many bacteria, yeasts and microfungi glutamate is synthesised from 2-oxoglutarate via glutamate dehydrogenase (GDH). When grown under carbon- or phosphate-limited conditions *Klebsiella aerogenes* has such a synthetic sequence but when growth is ammonia-limited very little GDH is produced and certainly insufficient to meet the biosynthetic needs of the organism. Under these latter conditions other known routes of glutamate synthesis are inoperative. However, by pulsing ammonia-limited chemostats with extra ammonia and monitoring the transient fluxes in pool constituents the following route of glutamate synthesis was elucidated:

$$
\begin{array}{ccc}
NH_3 & Glutamate & Glutamate \\
ATP & & NADP \\
 & GS \quad GOGAT & \\
ADP & & NADPH + H^+ \\
+ & Glutamine & 2\text{-}Oxoglutarate \\
P_i & &
\end{array}
$$

The synthesis occurs in two steps, first the formation of glutamine (via glutamine synthetase, GS) and then the amidation of 2-oxoglutarate via the newly discovered glutamine (amide)-2-oxoglutarate amido-transferase (GOGAT). The respective K_M values for ammonia of GDH and GS are the vital clues to the synthesis of glutamate under different growth conditions. The K_M for GDH is 10 mM and much higher than the NH_4 content of NH_4-limited cells (c. 1·0 mM); the K_M for GS is 0·5 mM and consequently the enzyme is able to scavenge successfully for NH_4 in NH_4-limited cultures. Subsequently the GS-GOGAT system has been detected in a variety of other bacteria but it is important to remember that its discovery was a direct result of chemostat investigations.

The reader is directed finally to the very elegant application of continuous-flow cultures to bioenergetic research. Systematic modification of the chemical environment has been used to induce phenotypic changes in the respiratory chain of *Torulopsis utilis* with the aim of characterising as yet unidentified components. Recently Light (1972) has appraised the success of this approach in the definition of energy conservation and inhibitor sensitive sites in this yeast. She has shown that, under conditions of iron-limited growth, the cytochrome content of intact cells may be reduced by as much as 70 per cent and that accompanying changes in the mitochondria include

a. loss of energy conservation site 1 between NADH and the cytochromes;
b. loss of sensitivity to the inhibitors rotenone and piericidin A; and
c. loss of the electron paramagnetic resonance signal at $g = 1 \cdot 94$.

The e.p.r. signal is attributable to the non-haem iron of the NADH dehydrogenase system and its absence reflects the loss of energy conservation at site 1. Light also made the interesting discovery that, by manipulating the ratio of iron to glycerol (the carbon and energy source) in the medium, mitochondria could be produced which possessed a site 1 energy conservation but were no longer sensitive to rotenone or piericidin A; these properties formerly were claimed to be inseparable. It is important to emphasise again that these changes in mitochondrial constitution were entirely phenotypic.

7.3 Other applications

The foregoing discussions have dealt mainly with the benefits of using continuous-flow cultures in microbial biochemistry. The scope of these culture systems in biological research is much wider, however, and I wish to conclude this chapter by drawing attention, albeit briefly, to their utility in microbial genetics and in microbial ecology.

Continuous-flow culture is a powerful tool for the study of mutation. In batch cultures the possibility of 'jackpot' mutations may produce serious over-estimates of mutation rates and also obscure selection history when this occurs. (A 'jackpot' effect simply refers to the situation where a mutation occurs early in the course of a batch culture so that the population comes to contain a high proportion of mutant organisms; usually mutational events are less probable during the first few cell generations.) Chemostat cultures are much to be preferred for such studies. The rate of mutant increase is given by

$$\frac{\mathrm{d}m}{\mathrm{d}t} = K^{\mathrm{f}}(N - M) - K^{\mathrm{b}}M + (\mu^M - \mu)M \tag{20}$$

where M and N are the mutant and total organism populations, μ^M and μ are their respective growth rates and K^{f} and K^{b} are the rates of forward and back mutation. Assuming that $K^{\mathrm{f}} \simeq K^{\mathrm{b}}$ and $M \ll N$, equation (20) reduces to $\mathrm{d}m/\mathrm{d}t = K^{\mathrm{f}} + (\mu^M - \mu)M$. This equation can be used to determine the progress of mutant accumulation under conditions of favourable selection ($\mu^M > \mu$), no selection ($\mu^M = \mu$) or adverse selection ($\mu^M < \mu$). The majority of chemostat experiments have considered mutations in the absence of a selective pressure, e.g. resistance to bacteriophage. Under these circumstances M increases linearly and hence the rate of mutant accumulation is easily and accurately measured. Clearly the possibility of long term experiments is a great boon to the microbial geneticist interested in mutation. Additionally, study of the action of mutagens and anti-mutagens is facilitated by chemostasis because toxic compounds can be supplied at sublethal levels over long periods. The chemostat can be used to direct the evolution of enzymes and metabolism and a recent interesting application of this thinking is found in the work of Francis and Hansche (1972) on yeast acid phosphatase.

Chemostats were operated under conditions of phosphate-limitation (β-glycerophosphate) at a pH of 6; the latter reduces acid phosphatase activity by 70 per cent. Thus, selection favoured mutants in which (a) phosphatase had increased activity at the adverse pH and a reduced K_M, and (b) there was a more efficient metabolism of orthophosphate. The system was run for about 1 000 generations (c. 8 months; $D = 0\cdot15 \text{ h}^{-1}$, $t_d = 4\cdot6$ h) during which time selected mutants were analysed in terms of their biochemical adaptability. The first adaptive event was a mutation ($M1$) detected after about 180 generations: increase in μ of 25 per cent at the value of s when it arose; increase in Y of 30 per cent. After about 400 generations mutant $M2$ was detected: 60 per cent increase in enzyme activity at pH 6 (pH optimum shifted from 4·2 to 4·8). $M2$ resulted from a single mutation in the acid phosphatase structural gene. A third but distinct adaptive event (mutant $M3$) occurred after about 800 generations. Cells of $M3$ had a greater propensity to clump and hence a greater settling rate. Consequently, the clumped cells had a lower washout rate than the free cells and clumping may be viewed as an adaptation to the equipment. The significance of these results is twofold; chemostasis enables a rational experimental approach to be made to evolutionary processes; and, chemostasis offers the opportunity of modifying enzyme properties by specific manipulations of the environment.

Natural microbial populations are best described as a complex of different *open* systems each regulated by its own limiting factor and influenced by environmental changes. Implicit also are the facts that populations will be mixed and that substrate concentrations generally are low. In the laboratory most investigators use *closed* cultures in studies of microbial interaction. Such systems are unsatisfactory on several grounds, in particular, the substrate concentrations almost invariably are much higher than those of natural ecosystems and, consequently, μ_{max} becomes the prime growth parameter determining selective ability. The chemostat is a superior model for ecological work. It provides high population densities at low substrate concentrations and as such approaches much more closely natural populations. Moreover, K_s has been shown to be a decisive growth parameter in selection (cf. discussion of mutant selection above) in mixed populations and its significance cannot be gauged from closed culture studies. Finally, the chemostat permits the reproducible enrichment of organisms with low substrate specificities; in contrast, replicate 'non-selective' enrichments in closed systems produce very variable data. To conclude, although the chemostat is only a model system and does not reproduce natural ecosystems, it does enable the microbial ecologist to identify underlying ecological principles. For further details and examples of the chemostat applied to mixed culture studies the reader is directed to the stimulating review of Veldkamp and Jannasch (1972).

8 References

DEAN, A. C. R., PIRT, S. J. and TEMPEST, D. W. (Eds.) (1972). *Environmental Control of Cell Synthesis and Function.* London. Academic Press. This book contains the proceedings of the 5th International Symposium on Continuous Culture of Microorganisms held in Oxford in 1971 and is the most recent and comprehensive discussion of the role of continuous culture in microbiological and biochemical research. The papers referred to in this chapter are:
BROWN, C. M. and STANLEY, S. O. ('Regulation of glutamate synthesis in bacteria'); BULL, A. T. ('Synthesis of exocellular macromolecules, including enzymes'); DEAN, A. C. R. (Control of enzyme synthesis); LIGHT, P. A. (Mitochondrial function); PIRT, S. J. (Current developments in continuous culture); VELDKAMP, H. and JANNASCH, H. W. (Mixed cultures, microbial ecology); WATSON, T. G. (Turbidostats).

AIBA, S. and HARA, M. (1965). 'Studies in continuous fermentation. Part 1. The concept of the mean cumulative age of microorganisms.' *J. Gen. Appl. Microbiol.,* **11**, 25–40.

BORROW, A. *et al.* (1961). 'The metabolism of *Gibberella fujikuroi* in stirred culture.' *Canad. J. Microbiol.,* 7, 227–76. (Extensive studies of differential substrate exhaustion and its influence on culture development.)

CLARKE, P. H. and LILLY, M. D. (1969). 'The regulation of enzyme synthesis during growth,' pp. 113–59. In *Microbial Growth,* Eds: P. M. Meadow and S. J. Pirt. Cambridge University Press.

ELLWOOD, D. C. and TEMPEST, D. W. (1972). 'Effects of environment on bacterial wall content and composition.' *Adv. Microbial Physiol.,* 7, 83–117. (An excellent essay by two of the pioneers of chemostasis as a tool in analysing cell chemistry.)

EVANS, C. G. T., HERBERT, D. and TEMPEST, D. W. (1970). 'The continuous cultivation of microorganisms. 2. Construction of a chemostat.' In *Methods in Microbiology* Vol. 2, eds: J. R. Norris and D. W. Ribbons. London, Academic Press.

FRANCIS, J. C. and HANSCHE, P. E. (1972). 'Directed evolution of metabolic pathways in microbial populations.' *Genetics,* **70**, 59–73. (Use of the chemostat technique to modify the acid phosphatase of yeast.)

HERBERT, D. (1961). 'The chemical composition of microorganisms as a function of their environment', pp. 391–416. In *Microbial Reaction to Environment.* Ed: G. G. Meynell and H. Gooder. Cambridge University Press. (An important essay which has been instrumental in reorienting views on microbial cell composition.)

KOCH, A. L. (1971). 'The adaptive responses of *Escherichia coli* to a feast and famine existence.' *Adv. Microbial Physiol.,* **6**, 147–217. (A fascinating analysis of ecological success discussed in biochemical terms.)

MAALØE, O. and KJELDGAARD, N. O. (1966). *Control of Macromolecules Synthesis. A Study of DNA, RNA and Protein Synthesis in Bacteria.* New York and Amsterdam. W. A. Benjamin Inc. (This monograph is a major contribution to our understanding of cell synthesis and molecular biology; written by the chief exponents of the 'shift' culture technique.)

MAYNARD SMITH, J. (1969). 'Limitations on growth rate', pp. 1–13 In *Microbial Growth.* Eds: P. M. Meadow & S. J. Pirt. Cambridge University Press.

MONOD, J. (1950). 'The technique of continuous culture. Theory and applications.' *Ann. Inst. Pasteur,* **79**, 390–410. (One of the classic papers on microbial growth; has laid the foundation to the theory and development of continuous culture.)

TEMPEST, D. W. (1970). 'The place of continuous culture in microbiological research.' *Adv. Microbial. Physiol.* 4, 223–50. (The case for open-culture systems as analytical tools in research.)

TEMPEST, D. W., HERBERT, D. and PHIPPS, P. J. (1967). 'Studies on the growth of *Aerobacter aerogenes* at low dilution rates in a chemostat.' In *Microbial Physiology and Continuous Culture.* Eds: E. O. Powell *et al.* (One of the few studies of growth at very low dilution rates; detailed analysis of viability.)

14
Applications of Prokaryotic Genetics in Biochemical Studies

J. F. Collins

Department of Molecular Biology, University of Edinburgh

1 Introduction

The explosive growth in our understanding of the structure and function of genetic material, particularly from studies with the prokaryotic bacterial and bacteriophage systems, has been a two-way process. Biochemical studies have explored the molecular basis of many of the processes involved in the replication and expression of the genetic material, while genetic studies have identified elements and properties of biochemical systems or reactions for further biochemical analysis. Genetic manipulations also allow tests of gene function in different combinations of genetic markers in a single chromosome or in diploid situations. The genetic markers also help to define the patterns of inheritance and to demonstrate the transfer of functional genetic material between different genetic structures or different organisms. For detailed analyses of the fundamental events in prokaryotes and the molecular interactions involved, both biochemical and genetic studies need to be combined for maximum efficiency and adequacy.

In this chapter, it is not proposed to lay the foundation for a thorough understanding of the science of genetics, but the intention is to illustrate the way in which genetic events and situations can be exploited to aid and extend biochemical studies. While sophisticated genetic analysis may be required for the ultimate resolution of some systems, even the simplest genetic studies may provide powerful tools in almost any field of biochemical study.

Genetic differences between organisms form the starting point for investigation, but the genetic differences between independently isolated strains, no matter how closely they appear to be related, cannot be assumed to be limited to the differences that may be observed between the strains. For this reason, such strains are usually unsuitable for detailed comparative study. The simplest study free from this objection is made with strains that are isogenic (that is, genetically identical) in all respects save the character under analysis. Such pairs of strains are easy to obtain, for mutations can be made in a wide range of genetic characters to give stable derivatives of the parental organism. The different forms of a gene (the determinant of a basic unit character in the genetic material) are called alleles; the original strain isolated (or wild type) carries wild-type alleles for each character, the derivative (or mutant) strain has a mutant allele for one character but the rest are unchanged. There are many possible mutant alleles for each wild-type allele, but with any one mutant allele it may be quick to establish the significance of that allele by comparative studies of the mutant and the wild type strains.

Why are these changes so useful to the biochemist? Consider the way in which living systems are analysed. Living organisms are rather like mysterious 'black boxes' at first sight; they take in various raw materials and modify them in unknown ways to produce new organisms and assorted waste products. Analysis starts with the whole organism, establishing the gross structure and composition, and also identifying intermediates in the production of the new organism from the growth substrates. The enzymes that catalyse some of the chemical conversions are found, and a picture of the metabolism emerges. How accurate or complete is this picture? And how does the picture change when the organism responds to some new environment? The technique used to investigate such a 'black box' is common to many fields of enquiry – prod it and see what happens. The classical flowering of biochemistry employed this method to great effect. Add to an organism, for instance, some suspected intermediate or inhibitor, and watch for changes in the levels of the different intermediates that are metabolically linked to the test compound, changes that spread through the system like ripples spreading on the surface of a pond. Intermediates close to the test compound respond first; others, more remote, follow later with less extreme changes. In a number of cases, the occurrence and coordination of reactions into metabolic pathways can be demonstrated, provided the compound tested is on or very close to the pathway. Relatively few intermediates can be tested this way, however, because entry into the cell is limited to compounds that can be transported across the cellular permeability barrier. Many interesting compounds are therefore excluded from study. Moreover, quantitative measurements needed to assess the role played by the different pathways in intact and undisturbed cells cannot be made.

The introduction of radioisotopes provided an extremely powerful tool for the detection of metabolic relationships in the intact organism, subject, of course, to the limitations that the compound supplied must pass into the cell and it must be provided with a suitable radioisotope label. The use of a radioisotopic compound by a cell does not disturb the flow through the metabolic pathways, at least to a first approximation. But there is a disturbance – to the isotopic content of the intermediates and cellular components derived from the compound supplied. While the isotopic label is spreading throughout the cell, the radioactivity of any compound can give information about its metabolic relation to the test substance, even identifying the atoms in its structure derived from particular atoms in the compound supplied. In some cases, the rates of the different reactions can be assessed, together with the stability of the labelled intermediates, by using a short pulse of radioactive material, and measuring the passage of the radioisotope into and out of the intermediate under study. Thus the use of radioisotopes adds a new method of investigating normal metabolic patterns, while providing at the same time a probe of the metabolism of cells during any period of interest – during adaptation or response to stress, for example.

In a similar manner to the extension of earlier methods by the use of isotopes, the introduction of genetic variants adds a new dimension as well as providing new material to which all the earlier methods of analysis can be applied. The metabolic state of a mutant organism will differ from that of the parental organism under similar growth conditions if the expression of the altered gene is different in the slightest way; whether this change is qualitative or quantitative in nature, the metabolism of the mutant settles to a steady state during growth. Biochemical analysis of this state can be very revealing, for though it may be possible to force the parental organism into a similar condition (at least superficially) by manipulating the environment (e.g. by inducing or repressing the formation of a particular enzyme), the response of the organism in the latter case originated from interactions between components in the medium and the surface of the organism, and has been transmitted through an unknown host of intermediates to modify expression of the gene concerned. In the mutant, on the other hand, the new metabolic state has sprung from the modified expression of the gene involved, under normal growth conditions, within the organism, and all changes should be causally connected back to this event.

The first part of this chapter is a brief review of the nature of the genetic material and of mutations. Next, the basic effects of mutations are discussed, followed by methods of inducing mutations and of isolating mutants. Some general properties of genetic transfer systems and their consequences are described next. A number of more detailed cases of genetic and biochemical analysis are then given, with some aspects of the more complex systems in which mutants are now making substantial contributions as aids to biochemical studies.

2 Heredity and the genetic material

The capacity of living things to reproduce themselves with remarkable accuracy over many generations, and the complex cycle of operations that go to accomplish this process, are aspects of the process of heredity. Though this arose historically as the science of genetics from a study of higher organisms, today bacteria and their viruses are the systems best understood at the molecular level. A remarkably detailed picture has been built up of many of the processes involved in the growth cycle, and of the way in which bacteria and viruses interact together. Much remains to be discovered, and further analysis will continue to exploit these systems, by building on the broad basis of knowledge that exists. However, bacteria are prokaryotic organisms, and differ from higher organisms (eukaryotes) in important aspects; there is no substitute for the eukaryotic cell when it is desired to study the way in which eukaryotic genetic information is expressed, or the way this expression depends on the presence of a nuclear membrane, or the manner in which the multiple chromosomes behave during the meiotic and mitotic cell cycles.

Bacteria are haploid organisms, containing one molecule of DNA with the whole of the genetic information required by the organism; this DNA molecule is replicated to give two genomes during the growth cycle. By contrast, eukaryotes often contain duplicate sets of chromosomes, one set derived from each parental gamete, and thus may be diploid; during the mitotic cell cycle, all the chromosomes are replicated and each daughter cell receives a complete duplicate set of chromosomes. This genetic situation is extremely complex compared to that in bacteria. The simplicity of the bacterial systems would be of limited advantage if it were impossible to study the interactions of genomes or to produce new combinations of genes in a single genome. However, genetic transfer systems have been found in a limited (but increasing) number of bacteria, and these organisms have been extensively studied. The outstanding example is *Escherichia coli,* and many of the studies quoted later relate to this organism.

The genome, this single circular DNA molecule in bacteria, has to fulfil two roles in the cell cycle; it has to provide the information that is selectively transcribed into messenger and other RNA molecules, and it has to be replicated, accurately. In some organisms there are other DNA elements outside the chromosome which are also replicated and inherited; these are called plasmids. Plasmids have a variety of genetic markers in them, but are not essential to the growth of the organism. These plasmids are often unstable in the organism and occasionally do not replicate or partition into the daughter cells properly, though they are dependent on the cellular metabolism for the timing of their replication.

2.1 Genetic Information and Mutations

The information content of the genome is the basis of the inherited characteristics of an organism. These characteristics include not just the form and content of the existing organism, but the form and changes that are potentially possible if the organism adapts to a new situation. Only part of the total genetic information is expressed in the growing organism at any time, and the qualitative and quantitative aspects of the regulatory processes involved are the central areas of biochemical research. The genetic information, carried in the DNA in the sequence of base pairs, can be divided into functionally distinct categories; the effects of mutations depend on the types of change in the genetic material and on the function of the region in which they occur.

Consider replication. All the DNA has to be replicated. However, it is not all equivalent, for replication is believed to start at a specific point, and then to proceed in both directions from this origin. It is not known whether the two replicating complexes meet at a specific termination point on the circular genome, or whether they meet at a random point. The specificity of the origin of replication must reside in the base sequence of the DNA in this region. Mutants modified in their ability to initiate DNA replication could include those with altered initiation sequences, though many other mutant components might have the same effect.

Consider transcription. Some regions are transcribed into specific RNA molecules; these regions are characterised by a binding site for an RNA polymerase at one end, at which the polymerase can initiate synthesis. The direction the polymerase will travel may be uniquely determined by the sequence at the binding site. Close to these sites, other binding sites for regulatory proteins may be found, and these may have either positive or negative effects on the expression of the gene, depending on whether they stimulate or inhibit the polymerase initiating RNA synthesis.

At the end of the region transcribed, there must be a sequence at which the polymerase stops; this region has not been identified, but might be formed in practice by a region binding some protein which formed a block to the further progress of the polymerase.

Regions that are never transcribed may serve functions as binding sites for different types of regulating elements or polymerase, or they may have no function. Any functional region should be capable of modification by mutation; are mutational events in non-functional regions significant? They would not be detected by most criteria, but these regions, and mutations occurring in them, could be significant if they are the sites of embryonic genes or are regions where pre-existing gene fragments evolve.

Consider further the region transcribed into RNA. The RNA can be the stable RNA species in a cell, such as transfer RNA's and ribosomal RNA, or it can be the rather unstable messenger RNA. Some messenger RNA molecules, termed polycistronic, contain the information for several distinct polypeptides to be synthesised. Hence in the RNA there are regions at which the ribosome will initiate polypeptide synthesis, separated by the regions coding for the amino acids and regions that terminate polypeptide synthesis. The information for these regions is coded in the DNA; mutations affecting these control properties of the messenger RNA molecule will be detectable phenotypically, as are mutations that alter the coding for an amino acid in a polypeptide.

At this point, it is worth distinguishing two things: the genotype and the phenotype. The form an organism adopts and the characteristics it shows under specific circumstances are called the phenotype. The phenotype depends on the expression of the genetic information under those particular growth conditions, and, as expression of the genetic information is variable, the phenotype of the organism is variable. The information stored in the genetic material with all the potential to express different genes in specific ways represents the genotype of the organism and is fixed.

The phenotypic changes in a mutant stem from the mutation in DNA which can have a direct effect only on the expression of a small region of the entire genome. These direct effects may be magnified in the entire organism by secondary metabolic consequences, but however bizarre the phenotype may appear, mutations are fundamentally simple events. Since they consist of changes to the DNA base pair sequence, they are also easy to classify. They can involve the substitution of a purine by another purine or a pyrimidine by another pyrimidine (transition), or the replacement of a purine by a pyrimidine and vice versa (transversion). The addition of extra base pairs into the sequence, and the deletion of base pairs from the DNA, form the only other basic changes possible in the DNA. It is not always easy to identify the type of change that has occurred in any particular mutant, but fortunately it is not always necessary to do so. Some mutants turn out on analysis to contain more complex changes in the DNA, but however these have arisen, it is possible to visualise them as combinations of the three basic possible modifications.

2.2 Effects of Mutations

What mutations can be found? Those mutations producing phenotypic changes in the whole organism should be obvious. 'Silent' mutations, however, causing no phenotypic change, do exist, and are not simply mutations occurring in non-functional regions of the genome. At least two types of silent mutation are known; in one, the mutational change has failed to change the function of the genetic information when it is expressed in the organism because, though the mutation may alter an amino acid in some polypeptide, the functional properties of the polypeptide do not noticeably change. In the second case, the mutation does not even alter the amino acid sequence of the polypeptide coded in the gene with the mutation, for though the change has altered a codon, both the original and mutated sequence give codons in the messenger RNA that are translated as the same amino acid (possible because of the redundancy in the genetic code).

The common mutations produce the following types of effects (though these are not listed to suggest either the frequency of occurrence or the ease of detection):

(i) changes in the properties of any binding site in the DNA for polymerases or regulator proteins;

(ii) changes in the structural genes resulting in amino acid changes in the polypeptide coded by the gene;

(iii) changes in the initiation signal in the messenger RNA that reduce or eliminate the start of translation by the ribosomes;

(iv) changes in the termination sequence for translation of messenger RNA that result in additional amino acids being added to the polypeptide at the carboxy terminal;

(v) changes in the structural gene that create a termination sequence in the messenger RNA and prematurely stop translation.

The first type is characterised by changes in the level of gene expression, which may be accompanied by changes in the regulation of the genes controlled through the binding site. Addition or deletion mutations are believed to be particularly likely to abolish the binding properties of the site, though even modifications due to single base changes can be detected in some cases.

The remaining types are concerned with the expression of genetic information in the form of polypeptides. The substitution of one amino acid by another in a polypeptide will have phenotypic effects on the properties of the polypeptide that depend on the degree of disturbance to the tertiary structure and to the activity of any key groups involved in enzymic activity or interactions with cellular components. These missense mutants are usually spotted from the phenotypic changes that follow major changes in properties of the polypeptide in question. These changes may be obvious as modified activity, though the modified properties may also be revealed only under specific conditions; e.g. the change may result in altered stability to heat, and the defective phenotype will only be apparent on growth at an elevated temperature (known as a restrictive temperature if growth of the organism is grossly affected).

The mutants that have modified initiation or termination sequences may fail to make any functional product from the messenger RNA from that gene. Where the messenger RNA is polycistronic, it is normal for all the polypeptides to be produced by ribosomes that have initiated at one end and travel the length of the messenger, terminating one polypeptide and then initiating the next. Messenger RNA molecules appear to be broken by endonucleases in the cells unless they are occupied by ribosomes actively translating the structural information. Failure to initiate correctly at any of the control points may therefore affect the synthesis of the polypeptides coded later on the same messenger, as the messenger may be degraded prematurely. These effects, known as polar effects, can also be caused by the presence of a false termination sequence in a structural gene which results in premature termination of the synthesis of the polypeptide product by the ribosome, giving a shortened and usually defective product. In the case of a polycistronic messenger, these mutants show polar effects which vary in severity depending on the distance between the mutation and the next initiation sequence; the ribosomes wander along the RNA after terminating polypeptide synthesis but are spread out and probably have a tendency to detach from the messenger, allowing endonuclease attack and messenger breakdown to occur. The shorter the distance between the termination sequence and the initiation sequence the less chance there is of this happening, and the less polar the effect, even though in each case the product from the mutated gene is defective.

False termination sequences can arise in two distinct ways: first, by the modification of a base pair to give one of the three termination codons in the RNA, the triplets UAG, UAA or UGA (found in 'amber', 'ochre' and what are occasionally referred to as 'opal' mutants respectively); second, the generation of a termination codon by an addition or deletion mutation. The simple gain or loss of a single base pair in the structural gene for a polypeptide is called a frameshift mutation, for the information in the gene, when transcribed into RNA for translation by ribosomes, has to be read in groups of three bases, or codons; the sequence of bases has different coding properties depending on the phase in which the bases are read. The effect of the frameshift mutation is to alter the phase in which bases after the mutations are read by the ribosome; the amino acid sequence is unrelated to the original, and the polypeptide will be extended until some new termination sequence is reached. As this may lie beyond the existing end of the polypeptide in the RNA, new regions of the messenger will be interpreted as polypeptide, and further initiation will of course be dependent on encounter with the next initiating codon; this may be a novel one, found by reading the messenger still out of the normal phase, or the ribosome may reinitiate at a normal initiating codon in the original and correct phase again.

2.3 Production of Mutations

The accuracy of DNA replication is normally extremely high; nevertheless, it is possible to examine so many bacteria that every type of mutation can be shown to arise spontaneously.

Many agents have been found that will raise the mutation rate several orders of magnitude above the spontaneous rate, and some act by specific interactions with the target DNA. Base analogues (such as 2-amino-purine or 5-bromouracil) can be activated by the bacterial cell and incorporated into the newly synthesised strand of DNA in place of the natural bases (adenine or thymine, respectively). These bases do not have the same hydrogen bonding properties as the normal bases, so that on further replication, an incorrect base (cytosine or guanine, respectively) can occasionally be inserted into the new DNA strand, which can now be stabilised into a mutant DNA sequence. Therefore, these mutagens have some specificity about the type of base change they induce in the DNA. Other mutagens modify the DNA chemically (e.g. hydroxylamine, nitrous acid, ethylmethanesulphonate, diethylsulphate and nitrogen mustard derivatives) or induce damage photochemically or through ions or radicals (e.g. ultra-violet light or gamma irradiation). The sequence of events varies with the mutagen used, but basically these agents produce DNA structures which cannot be replicated by the normal mechanisms. Bacteria have evolved systems to deal with damage to the DNA of this type, and are able to detect and cut out from a damaged strand a short region that includes the modified bases. The repair process continues with a stage of DNA synthesis of material complementary to the intact strand opposite this gap, and completes the operation by joining the ends of the newly synthesised strand and the original strand together. The process of repair is not so accurate as that of the primary DNA replication for mutations are introduced at the same time. Since the introduction of any of the normal bases into the repaired structure incorrectly means that the normal base-paired structure of the DNA is not restored immediately, the repair mechanisms can remove the offending region on the other strand, and produce perfect but now mutant DNA. This process is very important, for it accounts for the appearance of mutants in a single step without replication of the entire chromosome to stabilise the new DNA.

One chemical agent, N-methyl-N'-nitro-N-nitrosoguanidine, is believed to damage only single-stranded regions of the DNA; as these are characteristic of the regions about to be replicated (and of the gaps between Okazaki fragments (see section 4.6) before they are closed), this agent produces mutations principally in replicating cells, and can produce multiple mutations in the same region of DNA for this reason. A practical application has been to use this property to identify the region being replicated, and in the case of the *Escherichia coli* chromosome, for which about 450 genetic markers are mapped, the order and the timing of the replication process can be detected by following the number and types of mutations occurring on synchronous replication of the chromosome.

Another set of interesting chemical agents is related to the nitrogen mustard agents, but carry only one reactive group attached to an acridine derivative. The presence of the heterocyclic ring, which bears some resemblance to the planar structure of a hydrogen-bonded base pair in normal DNA, disturbs the repair process in such a way that base addition or deletion mutants are formed almost exclusively. The molecular basis for this phenomenon is not yet well-defined.

2.4 Detection and isolation of mutants

After mutagenesis, the population of cells will contain potential mutants, many dead cells and unchanged parental cells. A period of growth is particularly useful to help stabilise any potential mutants by allowing DNA replication and also it allows the segregation of mutant genomes if the cells tend to grow in small clusters thereby including normal genomes with the mutants. A further reason for a period of growth is that the phenotype which will be selected may not appear immediately the DNA has mutated; for instance, resistance to a bacterial virus or bacteriophage depends on the loss of the attachment site on the surface of the cells. As these are not destroyed, resistant cells will arise only after a sufficient number of cell cycles have occurred to dilute these sites, and give progeny cells free from them. Similarly, resistance to streptomycin can arise by modification of the ribosomes; an organism is phenotypically sensitive to streptomycin while there are sensitive ribosomes present, and growth must take place to allow the sensitive ribosomes to be replaced by the new mutant ribosomes before streptomycin selection is used to find the mutants. This phenotypic lag can also occur during the construction of strains by genetic transfer between organisms.

Isolation of mutants is usually related to the detection of a phenotypic difference between parent and mutant; it must be remembered that the same phenotypic difference may arise from mutations in a variety of genes in some cases (e.g. in strains constitutive for histidine biosynthesis in *Salmonella*, see p. 459). Detection methods should therefore be made as specific and sensitive as possible; for instance, by using conditions which allow the growth of the mutant but not of the parent. Mutants resistant to antibiotics or toxic analogues of normal metabolites can be readily isolated by incorporating the toxic agent into solid media and picking the colonies able to grow. Another type of mutant is one which has recovered a vital function; e.g. in a strain mutated to be auxotrophic for some metabolite, or temperature-sensitive, growth in the absence of the metabolite, or at the high temperature, would not be possible, and mutants restored to the wild-type state of prototrophy (for this compound) or temperature insensitivity would be the only survivors. Reversion studies of this type can be used to characterise the mutant (which mutagenic agents can revert it?) or the mutagen (does the mutagen revert mutants already characterised as base transitions, transversions or frameshifts?). Some revertants do not reverse the original damage exactly, but introduce compensating changes either in the same gene, or in some other gene which provides an alternative method of restoring the defective function; e.g. an alternative metabolic path, or modification of a tRNA to provide inaccurate translation of particular codons (principally nonsense codons) and allow partial expression of the defective gene (suppressor mutations).

A second method of mutant detection is the use of media on which the mutants have a distinctive appearance, either in the form of colonial differences (mutants in surface characteristics or pigment formation) or differences which can be seen when a suitable reagent is poured onto the colonies. Chromogenic substrates are known for many enzymes, and mutants that have appreciably more or less than the parent organism of such enzymes as galactosidase, phosphatase, glucosidase, amylase and many others can be detected on plates. Staining characteristics of colonies also vary with the pH round the colony, and this can be turned into the classical method for detecting the fermentation of different sugars, when acid is also produced.

Where the mutant phenotype is an inability to grow under conditions which the parent organism can tolerate, a more complex approach is needed. This problem was solved by Lederberg in his replica pad technique. The initial stage is the growth of the mutagenised culture as single separated colonies on a solid medium where all cells, wild-type and mutant, survive. The plate is then used as a master, and an impression of the colonies is transferred onto the second selective plate by a sterile velvet pad. Each colony is now exposed to the conditions which kill the mutant – e.g. lack of a specific nutrient, or high temperature. After incubation, colonies on the second plate correspond to normal organisms, but any mutant organism will fail to grow. Comparison of the plates allows the mutant colonies to be identified on the first plate. This approach has been very fruitful, though the use of mutagens to raise the proportion of mutants present in the populations examined has been as vital a factor, for a single plate may contain only 100–300 colonies, and screening 10 000 colonies represents a considerable effort. If the frequency of the desired mutation is low, or if the frequency is not known because the desired type of mutant has never been isolated before, it may be necessary to use an enrichment technique before screening the individual colonies. Once the population has been mutagenised and the mutants have been stabilised, the entire culture may be treated to kill the parental organisms without killing the mutants.

Penicillins, and other antibiotics that interfere with normal cell wall or membrane function, kill cells that are actively growing. By transferring the culture to a medium in which the mutants cannot grow, and then adding the antibiotic, the parental organisms will succumb to the lethal effects of the antibiotic while the mutants remain quiescent. Later, the drug is removed, and the growth conditions restored so that the surviving organisms can grow. This new population can be enriched for mutants by a factor of 10 or 100, and the entire process can be repeated till the original screening technique can be used successfully. Temperature-sensitive mutants which do not themselves die at the restrictive temperature can be enriched by this method.

Another version of the enrichment technique exploits any differences in growth rate that may exist between the parental organisms and the mutants. If the mutant can grow even slightly faster than the parental organism — if, for instance, it is better adapted to utilise a particular growth substrate — then continuous growth of the mixed population under these conditions should allow the mutant to increase with respect to the normal organisms present. Even if the difference is only 5 per cent, continuous culture for a period of a week (not unusual for such types of culture) can raise the proportion of the mutant by a factor of 1000; from an initial frequency of 10^{-6}, it could rise to 10^{-3}, when the standard plate methods of screening can be applied.

Much ingenuity goes into the design of effective and efficient methods of detecting and scoring mutants, particularly on plates. Nevertheless, even the simple method of screening single isolates on a random basis for the particular mutant property sought has had some spectacular successes, and will continue to do so in the future.

3 Genetic Manipulations

While the initial study of mutants is essentially physiological, it is often desirable and sometimes necessary to know the location of the mutation and the behaviour of the mutation in combination with other mutations in a single chromosome or in diploids. Three methods are known for the transfer of genetic material between different organisms, but they are not universally available, and the method used depends on the properties of the strain under investigation.

3.1 Conjugation

DNA can be transferred on cell-to-cell contact between strains of *E. coli* that act as a donor and recipient respectively. This polarised form of transfer is termed conjugation; though it resembles, and indeed is termed, mating, as in higher organisms, it is fundamentally distinct, and prokaryotes do not have a true sexual cycle. The donor strains owe this property to an F (for fertility) episome which genetically determines transfer functions; these strains, known as males, transfer their DNA in an ordered manner, starting from a particular genetic marker and continuing round the chromosome till the mating process is interrupted naturally or artificially. The DNA transferred into the recipient or female strain must align physically along the regions of the chromosome containing similar genetic sequences (this may happen as part of the process of transfer) and the two DNA molecules undergo a sequence of steps involving strand breakage, unwinding and re-pairing of homologous base sequences, followed by closure of gaps by repair enzymes. The consequence of these operations, repeated along the chromosome, is the physical incorporation of stretches of single-stranded DNA from the donor DNA into the recipient chromosome. Where a sequence is not identical, the modified allele is now part of a new genome, and the organism produced is termed a recombinant, detected through the new combination of genetic characters it carries, and differs from either parental organism.

Genetic material can also be transferred on cell contact when one of a group of plasmids (the autonomous genetic elements that can replicate independently from the chromosome) which has transfer characteristics is present in the donor. These plasmids may also act as carriers for segments of the main chromosome, and thus effect transfer.

3.2 Transduction

Bacteriophage are infective viral parasites which in the course of their life cycle normally kill the host cell (see p. 462). The progeny particles contain the newly synthesised infective viral genomes; however, occasionally (1 in 10^6) DNA from the host cell chromosome is packaged in the virus instead. When this is injected into another cell, no viral functions are expressed, but the DNA can pair with a suitable homologous region of the chromosome and undergo recombination. In the case where any piece of the chromosome can be transferred, it is termed generalised transduction. However, when the viral genome has some specific region in the chromosome where it can pair and perhaps be linked by recombination, faulty replication of the viral DNA still attached to a segment of chromosome gives a high proportion of defective phage particles carrying one specific region of host cell DNA. This is termed specialised transduction; in both forms of transduction, relatively small pieces of DNA (perhaps 1 per cent of the total genome) are carried and co-transduction of two genetic markers establishes a physical closeness on the genome.

3.3 Transformation

The third transfer method is transformation; isolated DNA fragments are absorbed into cells of certain strains by a process which allows recombination of the DNA fragment with the genome of the cell. The DNA has to be of reasonable size (M.W. 10^6 or higher) and probably suffers degradation during the uptake process and before recombination is completed. Markers generally show lower linkage than in transduction.

Transformation is efficient in some bacteria (up to 10 per cent of all cells transformed for *Pneumococcus* or *Bacillus subtilis*) but in others it has not yet been demonstrated. It can be used to transfer any part of the chromosome but a good selection method is desirable to find the transformants.

3.4 Mapping

The measurement of linkage between markers can be extended to demonstrate that there is a unique linear order of genetic characters, consistent with the linear structure of the DNA. The ability to separate different markers in recombinants decreases as the markers are closer together, and this can be interpreted as a linkage 'distance', to form an ordered and scaled map. Recombinants can be obtained from mutations at adjacent base pairs, though extremely infrequently. The ordering and linkage determination are based on the relative proportions of

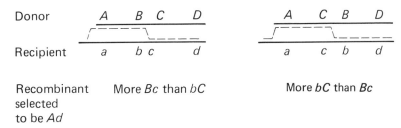

Fig. 14.1. Evidence of genetic marker order from cross. Markers *A* and *D* would lie in neighbouring markers, while *B* and *C* represent different mutations in the gene under study; *a, b, c* and *d* represent the parental alleles.

different recombinant classes obtained (Fig. 14.1). Ordering of genes in the gene clusters found in many systems in *E. coli* and *Salmonella* is easily done by the transduction process, particularly if a specialised transducing phage can be used to give a high proportion of crosses involving the region concerned.

3.5 Diploids

When a phage genome has picked up a region of the bacterial chromosome, it has usually lost some of its original genes and becomes non-virulent. The non-lethal transducing phage can then enter into recombination with the chromosome in two ways: either through homology of the bacterial DNA it carries with the corresponding region in the chromosome, or through homology of some region of the phage genome with a region in the chromosome (a specific

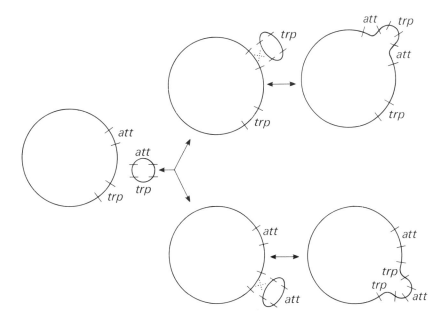

Fig. 14.2. Production of partial diploids involving an episome carrying a chromosomal region. Initial state, and alternative states if the episome can integrate at a specific attachment site (*att*) or at the site of the homologous chromosomal region (the region coding for the tryptophan biosynthetic enzymes, or *trp*, used as an example).

attachment site). In either case, the genome then contains two allelic regions for a particular set of bacterial genetic markers (Fig. 14.2). A cell that acquires a plasmid carrying bacterial markers also establishes a diploid state for the common region of bacterial DNA. Such diploids are fairly stable, enabling interactions of different alleles to be studied, measured by the expression of the genes concerned.

3.6 Complementation

In any diploid, whether fairly stable, as described above, or more transient — say, when a cell is infected by different viruses at the same time, or acquires DNA in conjugation or transformation — the duplicated regions can be expressed independently but interact to determine the phenotype of the cell. If two different sets of genes contribute all the functions for, say, growth (in two similar auxotrophic mutants) or phage multiplication (in a cell infected by two different mutant phage), when neither alone could suffice, they are said to complement each other. For instance, two strains defective in tryptophan synthesis because of mutations in different genes of the *trp* operon could give *trp* diploids able to make all the enzymes for tryptophan synthesis; the diploid could grow without tryptophan supplements. In general, mutations in different genes that produce products used in the cell complement one another; mutations in the same gene rarely complement. Hence this test may screen quickly for mutants of a certain gene based on this latter property, though polar mutations can complicate the analysis.

A mutation in a region that is not transcribed — such as a binding site for regulation or enzyme attachment — cannot have its function restored. On the other hand, the active products from this mutant can continue to complement missing functions on the other set of genes, and phenotypically complement in the diploid. If the expression of each set can be determined separately (by direct measurements of the different products produced) this situation can be distinguished experimentally. It is of considerable significance to determine whether or not a gene makes a product in the cell, or whether it only influences the expression of neighbouring genes in the same DNA molecule.

4 Examples of combined genetic and biochemical analysis

Four examples have been chosen of bacterial metabolic pathways which have been subjected to extensive analysis, and one detailed summary is given of a viral system. While the information has become exceedingly detailed, the ordered nature of the processes emerges as a striking feature that runs through each case. It must be pointed out that these represent some of the simplest systems in the cell, however; they involve the interaction of soluble components with simple receptors, and can be reproduced (in the case of the *lac* system) with considerable success in the test tube from isolated and purified components. The impact of genetics on more complex systems is increasing; but the main contribution in the complex cases (which will be outlined) is the identification of the components involved in processes such as transport properties of membranes, cell growth and division, DNA replication, etc. The biochemical task is to understand the functional relationships of the components, and this is now proving one of the areas of most rapid advance.

Table 14.1 Properties of β-galactosides and analogues used in analysis and selection of *lac* phenotypes in *Escherichia coli*

Compound	Induction properties	Hydrolysis by β-galactosidase	Comment
Lactose	poor	fair	Growth substrate; strong fermentation can be detected by dyes in agar plate.
Phenyl-β-D-galactoside	v. poor	good	Selects constitutive mutants, since it does not induce.
(β-Phenylethyl)-β-D-galactoside	–	fair	Releases toxic alcohol on hydrolysis; selects non-fermenters.
o-Nitrophenyl-β-D-galactoside	nil	fair	Releases nitrophenol, used in assay for enzyme and stain for high level colonies.
Isopropyl-β-D-thiogalactoside	good	nil	Best inducer.
Methyl-β-D-thiogalactoside	good	nil	Good inducer.
Phenyl-β-D-thiogalactoside	nil	nil	–
5-Bromo-4-chloro-3-indolyl-β-D-galactoside	nil	fair	Stains any *lac*⁺ colony blue.
Melibiose	v. poor	fair	Can be used for growth by different enzymes and transport system; but transport system fails at 42°C; can be transported by lactose system; hence selects for expression of lactose transport.
Raffinose	nil	nil	Growth requires use of active lactose transport system to get raffinose into cell.

456

4.1 The lactose system in *Escherichia coli*

Strains of *E. coli* able to utilise lactose for growth do not attack the sugar until the cells adapt to the presence of lactose and only if they have no other carbon source more readily used. The ability to transport lactose into the cell and the ability to hydrolyse it to glucose and galactose both appear within three minutes of the addition of an inducer such as lactose or certain closely related compounds. These two functions are stimulated specifically and, ultimately, about 1000-fold — indeed, the enzyme β-galactosidase forms over 2 per cent of the total protein in the cell.

The genes controlling these functions were the first system subjected successfully to combined genetic and biochemical analysis. The *lac* system remains the best documented system, and it was during this investigation that many basic facts of bacterial gene expression and regulation were first established. The concepts of operon, operator, repressor, effector, promoter and messenger RNA are now so familiar that they are assumed to be true (possibly too readily in some cases) during the interpretation of data for other regulated systems. These concepts were defined to account for the experimental observations in the *lac* system and amply confirmed by many workers.

Physiological studies showed that the *lac* enzymes could deal with galactosides other than lactose, but not all β-galactosides were transported into the cells, could be split by β-galactosidase, or would act as inducers for the system. Of those entering the cells, the best substrates were found to be poor inducers, and the best inducers (thio analogues of β-galactosides) were non-substrates. Exploiting the compounds shown in Table 14.1, many *lac* mutants were found, genetically mapped and characterised physiologically both in the haploid state and in a variety of diploids constructed principally with the F'*lac* episome (which carries the *lac* region of the chromosome).

The genetic map of the *lac* region (about 0·1 per cent of the *E. coli* chromosome) is shown in Fig. 14.3, and contains a group of three closely linked structural genes coding for β-galactosidase, the M protein, which forms a part of the galactoside transport system, and the enzyme thiogalactoside transacetylase (this enzyme does not appear to be significant for lactose utilisation). Expression of these three genes is coordinated, and nonsense mutations in the β-galactosidase gene (*Z*) lower expression of the other two genes; together, these genes form an operon, and are transcribed as a group into messenger RNA. The *I* gene exerts a negative

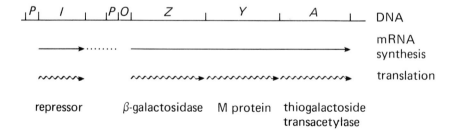

Fig. 14.3. Organisation of the *lac* region in the genome of *Escherichia coli*. Functions related to transcription:
  P = promoter (concerned with RNA polymerase binding)
  O = operator (binding site for repressor)
Structural genes:
  I = repressor (a tetramer)
  Z = β-galactosidase (a tetramer)
  Y = M protein, involved in galactoside transport
  A = thiogalactoside transacetylase (dimer)

regulatory effect; its product is the genetic repressor, produced steadily during growth through continuous transcription of the gene. The active form of the repressor is in fact a tetramer of the I gene polypeptide, and it blocks transcription of the *lac* operon by binding to the operator region (O). The RNA polymerase attaches at the promoter (P), and must cross the O region to transcribe the structural genes. The affinity of the repressor for the operator is markedly lowered when an inducer molecule interacts with the protein, and presumably causes a conformational change in it. The inducer thus acts to free the operator and allow transcription to start. However, for good levels of expression, the polymerase should contain a sigma factor and there should be a catabolite gene activator protein (CAP) bound to a site in the promoter region and finally cyclic AMP bound to the CAP protein. Parenthetically, mutants lacking the CAP factor or adenyl cyclase fail to express a number of sugar fermentation characters, so this control mechanism is common to a set of operons; in addition, compounds such as glucose are known to lower the cyclic AMP levels in the cells drastically, and may block formation of β-galactosidase even in the presence of a good inducer for this reason.

Direct measurements of messenger RNA from the *lac* genes, using hybridisation techniques with *lac* DNA carried on an episome, confirm that cells making high levels of the *lac* enzymes contain considerably more *lac* messenger RNA than repressed cells; *lac* regulation is through control of transcription, not translation or activation of some inactive precursors. Translation of the messenger starts while it is being synthesised; on induction new β-galactosidase can be detected several minutes before transacetylase appears; similarly, when the inducer is removed, β-galactosidase synthesis stops a minute or so before transacetylase synthesis. This reflects the time needed to translate the individual messenger molecules.

The main elements of this detailed model were determined from the properties of many mutants; only a few can be mentioned, but those omitted serve only to confirm the model. High level constitutive producers of galactosidase (I^- mutants) show no control by galactosides, though they can still be catabolite repressed. When they are tested in diploids with wild-type *lac* alleles, both sets of *lac* genes are inducible. The normal I allele has supplied the component that was inoperative in the mutant, and this must repress gene expression and must also be diffusible within the cell. The repressor has been isolated on a large scale by using mutants which overproduce it. These were derived from a mutant in which repressor synthesis is temperature sensitive; these strains are phenotypically inducible at normal growth temperatures but constitutive at high temperatures. Among new mutants which have lost the expression of the *lac* genes at the high temperature, there are mutants which have a better promoter for the I gene and make ten times the normal amount of repressor.

In vitro studies of the repressor have confirmed that it has a strong binding affinity for the operator region, and that this affinity is lowered in the presence of an inducer. I^- mutants either make no product or a defective product which does not bind to the operator; another class of I mutants expected and found produces a repressor that binds to the operator, but is not modified by the inducer. This super-repressed (I^s) mutant cannot be induced, and in diploids it renders the expression of both I^+ or I^- alleles low and uninducible.

Since the affinity of the repressor depends on the base sequence in the operator region, mutations in the operator lower the effectiveness of the repressor; these O^c mutations have a higher level of expression of the *lac* genes in the absence of inducer, but can still be stimulated to the full induced level by the inducer. In diploids, the level of expression stays high whatever the second *lac* contains; the O^c mutation has no effect on the expression of the second *lac* region in the diploid since it produces no product from this region.

The promoter is also concerned only in the expression of the *lac* region it occurs in; it was first identified as an essential function whose loss virtually eliminated *lac* expression. It can be

458

lost by deletion; some *lac* expression still occurs, which depends on polymerases attaching to the promoter of the I gene and travelling into the *lac* genes. Regulation in these strains depends on whether the operator is functional or has been lost in the deletion. This provides confirmation of the orientations shown in the diagram for the transcription of the *lac* genes (Fig. 14.3).

4.2 The arabinose system in *Escherichia coli*

The *ara* genes are a group of regulator and structural genes concerned with the conversion of L-arabinose to D-xylulose-5-phosphate (Fig. 14.4) for utilisation in the pentose cycle. In this case, the evidence for the organisation and regulation of the system has come primarily from genetic evidence, and biochemical verification of the model is still in progress. L-Arabinose induces the formation of the three enzymes coordinately, and the structural genes form an operon. The *ara C* gene (C^+ in the wild type) produces a regulating protein which determines the inducibility characteristics, and C gene mutants lacking inducibility can have either high (C^c) or low (C^-) levels of expression of the *ara* enzymes. The behaviour of these mutants in diploids distinguishes them from similar mutants in the *lac* system, for C^-/C^+ diploids are inducible, and C^-/C^c diploids are constitutive, for both the alleles present.

The C gene product is needed to activate expression of the operon, and, while the C^c mutant produces a product able to activate expression at all times, the C^+ wild type needs the inducer to render the C gene product stimulatory. The site at which the C gene product activates the operon is termed the I region, close to the end of the operon. Work with two deletion mutants has shown that the O region is a second site at which the C gene product acts, but only to limit expression of the *ara* genes. This complex situation is proposed to arise from the interconversion of the C gene product between two states; in one form it binds to the O region and inhibits expression, and in the other form it binds to the I region and stimulates expression. The inducer acts to shift the equilibrium of the C gene product from the predominantly inhibitory form to the stimulatory form, presumably through a conformational change

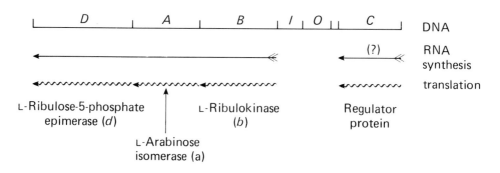

Fig. 14.4 Organisation of the *ara* genes in *Escherichia coli*, and the pathway concerned. *A, B, C* and *D* are structural genes for the proteins; *I* is the region where the regulator protein activates operon expression when stimulated by inducer; *O* is the region where the regulator protein inhibits operon expression in the absence of inducer.

associated with complexing of the inducer and the C gene product. A strain carrying a deletion of the O and C gene regions but with an intact I region is stimulated in a diploid with a C^+ *ara* region, and can be induced to normal levels. Constitutive mutants can also be obtained from this deletion which are mutated in the I region, and insensitive to C gene product in a complementation test. These I^c strains do not modify the expression of *ara* genes in the trans position in diploids. The promoter may lie in this region with I and O, and the regulation would again arise from control of transcription of the operon. The nature of the positive stimulus to transcription is not known, but may come from modifications to the secondary structure of the DNA, on combination with the stimulatory form of the C gene product, which leads to better polymerase binding or easier initiation of RNA synthesis.

Most *ara* mutants can be isolated on the basis of high or low levels of the *ara* enzymes, and good or bad utilisation and fermentation of L-arabinose. In some cases, secondary properties of the mutants provide selective characteristics – for instance, *ara* D^- strains are inhibited by L-arabinose, even in a complete medium, probably because L-ribulose-5-phosphate accumulates in the cells (the accumulation of high levels of phosphorylated compounds in bacterial cells is toxic for unknown reasons). Mutants of the *ara* D^- strains that are able to grow in the presence of L-arabinose, do not accumulate L-ribulose-5-phosphate; the new mutations are found in the *ara* A, B or C genes, or in the form of deletions in this region. Transport-defective strains may also be isolated – the genes determining transport of L-arabinose are unlinked to this group of *ara* genes, though they are also regulated by the C gene product.

In another example, constitutive mutants may be selected as strains able to grow on L-arabinose in the presence of D-fucose. D-Fucose is transported into cells by the arabinose transport system, but it is not metabolised; it acts as a competitive inhibitor of arabinose transport. The mutants are derepressed for arabinose transport, and are of the C^c class.

4.3 Histidine biosynthesis in *Salmonella typhimurium*

Histidine biosynthesis in this organism is controlled by inhibition of the first enzyme in the pathway, phosphoribosyl-ATP synthetase, by histidine, and by repression of the formation of the enzymes. The production of all the enzymes is repressed when adequate histidine is available to the cell. Mutations that result in loss of the ability to make histidine map in a single cluster of genes coding for the enzymes in the pathway (Fig. 14.5). These genes form a single operon, being transcribed as a unit from studies of polar mutants, though curiously there is evidence of two regions in this cluster that can act as low level promoters, restoring in such cases low degrees of expression of late genes, though of little physiological significance.

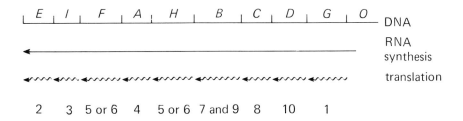

Fig. 14.5. The *his* operon in *Salmonella typhimurium*. Genes A to I code the histidine synthetic enzymes; O is the operator region. The genes associated with production of the repressor, which may be a complex of histidyl tRNA with its synthetase, are scattered elsewhere on the chromosome. The numbers indicate the order in which the enzymes are involved in the pathway.

The regulation seemed analogous to that proposed for the *lac* system by Jacob and Monod, in which expression of an operon is limited by a repressor — in this case, the level of available histidine would determine the affinity of the repressor for the operator, to limit histidine biosynthesis in the presence of sufficiently high levels of histidine. Regulator mutants were obtained by selecting mutants resistant to a variety of compounds that act as analogues of histidine or of its precursors. Since these compounds limit histidine synthesis by supplying false signals leading to repression of the pathway, the resistant mutants are derepressed for the expression of the operon. Six genes have been found which mutate to give this constitutive phenotype, but none of them resembles the expected gene coding for the repressor protein. There is a set of cis-dominant *his O* mutants located at the end of the operon, and these appear to define the operator. The other five genes are concerned not with histidine itself but with the production of histidyl tRNA in the cell, and the true effector regulating the expression of the *his* operon is now believed to be histidyl tRNA or a close derivative. The *his S* mutants have defective histidyl tRNA synthetase; all these mutants have some synthetase activity, however, perhaps because there is only a single synthetase for this essential product in the cell. Some of them have synthetases with lowered affinity for histidine, and high levels of histidine added to the cells stimulate the production of histidyl tRNA and repress the *his* operon. *His R* mutants contain about half the amount of tRNA^{his} found in the wild type; *his U* and *his W* are believed to be altered in the maturation process during the biosynthesis of tRNA^{his} though they are physiologically distinct. *His T* mutants may also be concerned with the reactions of histidyl tRNA, though this gene appears to influence the expression of other biosynthetic pathways, such as the synthetic pathways for valine and tyrosine.

The way to study a repressor system has been shown in the work on the *lac* system, but what can be done if the repressor gene cannot be identified? One possibility is that the repressor product is itself vital to the cell in some other capacity, and total loss mutants cannot be recovered. The implication of histidyl tRNA in negative regulation could mean that the free or loaded form of this molecule acted directly on the operator; both forms will be found in any cell. On the other hand, both the free and aminoacylated forms of the tRNA are found complexed with the synthetase, and it has been suggested that the complex of histidyl tRNA synthetase with histidyl tRNA may be the effective agent. It is interesting to note how this approach has provided a way of finding genes concerned with the synthesis, maturation and function of a specific tRNA species — an unexpected spin-off from the original investigation.

4.4 The tryptophan system in *Escherichia coli*

In *Escherichia coli* and in *Salmonella typhimurium*, the genes coding for the enzymes catalysing the conversion of chorismate to tryptophan form a single operon. In *E. coli*, the regulation is through negative control by a repressor coded by the *trp R* gene, which is located about one-quarter of the genome distant from the *trp* operon. The regulator protein is activated by the tryptophan itself, and the system is therefore much simpler to analyse than the *his* system in *Salmonella*. Genetic evidence has accumulated for a promoter/operator region at one end of the set of genes, and the system seems well defined (Fig. 14.6). Detailed studies of gene expression have, however, shown that inside the second structural gene there is a region which has low but detectable activity as a promoter, and genes *3, 4,* and *5* can be expressed at low levels even when the main promoter is fully repressed. This may not have much physiological significance, but sequence determination of the polypeptide coded by this region would give a partial nucleotide sequence of the weak promoter in the DNA, which would be of considerable interest.

Genetic analysis set the outlines of this system, and polar mutants in the first gene were interpreted to mean that a single operon was present. Mutants with false termination sequences in the second gene were unexpectedly found to lack the activity of the first enzyme in the pathway as well as the second, and this 'anti-polar' effect was extremely puzzling until the first

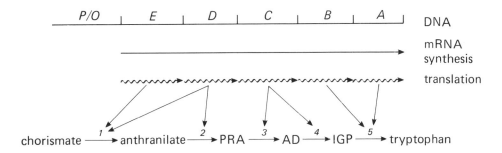

Fig. 14.6. Tryptophan synthesis in *E. coli;* the pathway and the operon. PRA = phosphoribosylanthranilate; AD = anthranilic deoxyribulotide; IGP = indoleglycerol-phosphate; enzyme 1 = anthranilate synthetase; 2 = phosphoribosyl transferase; 3 = PRA isomerase; 4 = IGP synthetase; 5 = tryptophan synthetase.

enzyme was studied more carefully and found to be composed of the polypeptides coded by both genes *1* and *2*. Gene *1* product by itself was inactive, though the gene *2* product was independently active as the second enzyme. The third and fourth enzymes are also interesting, for they are coded by a single gene and are present in a single polypeptide; the activity of each can be lost independently by mutation, and the mutations which affect each activity cluster into groups at opposite ends of the gene. The last two genes code for two polypeptides, the β and the α components of the last enzyme, tryptophan synthetase. The intact protein consists of two α and two β subunits, and catalyses reaction (1). Subunit α catalyses a different reaction, (2), and subunit β (as a dimer) catalyses reaction (3). While the sum of these two reactions gives the overall reaction catalysed, the complete tryptophan synthetase molecule acts without releasing any free indole, and the activity of the two components is intimately linked.

$$\text{Indoleglycerol-phosphate + serine} \longrightarrow \text{tryptophan + glyceraldehyde-3-phosphate} \qquad (1)$$

$$\text{Indoleglycerol-phosphate} \rightleftharpoons \text{Indole + glyceraldehyde-3-phosphate} \qquad (2)$$

$$\text{Indole + serine} \longrightarrow \text{tryptophan} \qquad (3)$$

This system has been of great interest to enzymologists, for it is possible to purify the α and β subunits separately, and reconstitute the complete enzyme simply by allowing the subunits to reaggregate. When either of the subunits has been prepared from a mutant, it is possible to construct *in vitro* aggregates from different strains and assess the effects of the subunit interaction on the catalytic behaviour. Active subunits are each stimulated 30–100-fold when the complementary subunit is added, even when the reaction measured is only that catalysed by the separate subunit. This is also true when the extra subunit added is itself inactive, and demonstrates the importance of subunit interactions in this system. It is interesting to note that in *Neurospora*, the tryptophan synthetase consists of a single protein unit which appears to combine the functions of the α and β subunits, as each of the partial reactions can also be demonstrated.

4.5 Viral systems

Viral systems have been intensively studied because they represent the simplest natural self-reproducing entities (even though they rely on the hospitality of the host cell for a variety of enzymes and building materials).

A bacteriophage particle attaches to the host cell, and injects the nucleic acid through a tube that penetrates the bacterial surface structures. The production of the progeny particles depends solely on the nucleic acid, on the information present in it that can be expressed by the bacterial systems (initially, at least).

The successful infection of a cell by a phage depends on a number of properties, all subject to modification by mutation, namely:

> attachment,
> penetration of cell,
> injection of DNA (or RNA in certain RNA viruses),
> formation of circular nucleic acid structure from linear form by closure of homologous single-stranded ends, or by recombination of sequences repeated at each end of the genome (terminally redundant),
> replication,
> controlled expression of genes,
> assembly of new particles,
> lysis of cell and release of progeny.

The pattern of behaviour may be modified if the host cell chromosome can recombine with the viral genome, or if the cell can degrade, or repress the expression of, the viral genome.

Viral mutants are screened by testing single particles on solid media covered with a lawn of susceptible bacteria. The initial cell bursts after perhaps an hour, releasing a considerable number (50–100) of infective particles, and these in turn attach to the neighbouring cells and repeat the lytic cycle, forming a plaque or clear patch on the plate. The appearance of the plaques is characteristic for different mutants, if they have altered lytic activities or leave some cells unlysed (giving turbid plaques, for instance). The host cells can themselves carry determinants which permit only certain types of virus to grow on them, and, for instance, they may be immune to one virus because they already carry a quiescent phage of the same immunity properties in the chromosome. This state is called lysogeny, and a viral or phage chromosome entering this association with the cell chromosome must recombine into the genome, and regulate its genes so that no deleterious functions are expressed (it will replicate when the chromosome replicates because it is truly incorporated into the genome as an integral part of the DNA). It retains the expression of certain genes, notably those that provide immunity by excluding, inhibiting or destroying any challenging virus of similar properties.

In the case of bacteriophage λ, the regulation of its life cycle is controlled temporally by a combination of positive and negative control systems (see Fig. 14.7). In a genome entering the lytic growth cycle, only two sections of the DNA, one of which includes the N gene, are initially transcribed. The N gene product acts as a positive control agent at three different sites, but it is not certain whether it acts to promote initiation of transcription at these regions, or whether it has a permissive role, removing a block to transcription already in progress. The regions transcribed are concerned with recombination functions, with the synthesis of phage DNA, and with further control of late function, exercised through the Q gene product. When enough Q product has been synthesised, transcription of the so-called late genes occurs, and the components required for the assembly of the heads and tails of the new particles are formed. The regulatory region also carries a gene CI, which, when transcribed produces a repressor of

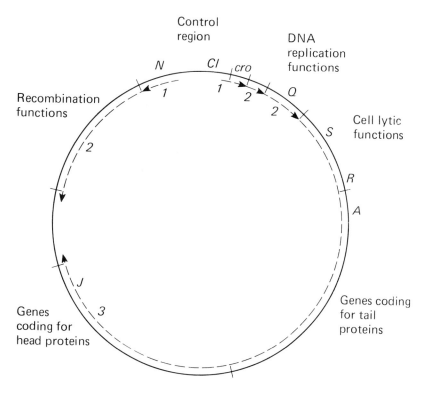

Fig. 14.7. Covalently closed DNA circle of phage λ genome. Sequence of gene expression during lytic growth cycle.
1   Initial transcription.
2   Second stages: continuation of transcription if *N* gene product has been formed.
3   Third stage: transcription if *Q* gene product formed.

both the 1st stage transcription regions, acting through operator regions near the promoters for these regions. This *CI* product acts to inhibit expression of all the other viral functions, directly or indirectly, and the viral genome in a cell goes into a dormant state. If it is integrated into the chromosome, the λ DNA survives in each cell, but never multiplies or produces other viral components until some disturbance interrupts repressor action — this is the lysogenic state. Since repressor is made continuously, a superinfecting λ phage is repressed and does not survive in the cell.

More complex interactions of the regulatory elements amount to control circuits, e.g. turning on transcription at one point in the life cycle, and turning it off again later. The recombination region and *N* gene stop being transcribed 10 min. after infection because the product of the *cro* gene, produced during the first stage of transcription, exerts a negative repressor action on the operators involved; it is not known whether these operators are precisely those at which the *CI* repressor works.

A technique extensively used in phage genetics is the use of mutants able to plaque on a mutant bacterial strain that carries a nonsense suppressor mutation, but unable to plaque on suppressor-free strains. These phage mutants are therefore nonsense mutants in some essential gene coding for a polypeptide. They can be examined to see what products formed are recognisable when they infect suppressor free strains: under the electron microscope a fascinating collection of phage parts (for instance, empty heads, filled heads, detached extended

464

and contracted tails, tail fibres and base plates) can be seen. The missing pieces in each case suggest the nature of the defective function.

The origin of the notation C for the repressor gene arises from the fact that in normal λ plate tests, all plaques are slightly turbid because lysogenic cells are formed which are resistant to λ attack. A mutant in the CI gene unable to make active repressor cannot establish lysogeny, and all cells infected are lysed, giving clear plaques – hence, C for clear plaque mutant.

This system probably has more surprises in store; biochemical studies of the molecular interactions and events during phage development are still very productive. Since the phage shows temporal regulation of its genetic material, could it be considered a primitive form of differentiation?

4.6 Complex functions involving DNA

From the importance of DNA in the living cell, it is reasonable to assume that the processes of replication, repair and recombination are the most significant, since they are involved in the continued growth, survival and evolution of the organism. Repair and recombination have been studied biochemically and genetically in *E. coli,* and partial understanding of the systems has been reached. There are two mechanisms for repair; one involves detection, removal and replacement of the damaged region in a single strand of DNA, the information in the undamaged strand determining the bases inserted into the gap. The other method of repair relies on a recombination event between sister strands following replication, when there may be a gap opposite the damaged bases in the strand used as template, which cannot be repaired by the excision mechanism first described. A recombination with the sister duplex can generate recombinants where the perfect strands in the good duplex have been separated into the different DNA chains, and the perfect chain opposite the gap can act as a template for the

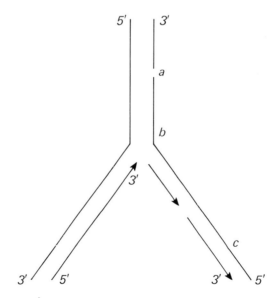

Fig. 14.8. Some aspects of DNA replication at a replication fork.
a – potentially lethal nick.
b – region of DNA unwinding and synthesis of new strand using left hand strand as template.
c – DNA synthesis on right-hand strand as template, producing fragmented DNA (Okazaki fragments) which require gap closure.

correction of the defect. If, instead of transferring the damaged DNA strand at the same time, a gap is left, then the second DNA duplex can be resynthesised and the defect is repaired. Mutants that fail to nick the damaged strand or to excise the damaged region thereafter are known to be excision repair defective, and are phenotypically ultra-violet sensitive. Mutants defective in recombination map in three genes, two of which specify the activity of an exonuclease, but the function of the third is not known. Mutants defective in both excision repair and recombination are extraordinarily sensitive to ultra-violet radiation, and the formation of a single thymine dimer by irradiation is a lethal event. Mutants lacking other steps in the process are not known, so these processes may be essential to the cell, and may be involved in the replication process.

The difficulty with the genetic analysis of the replication process is precisely that it is vital to the cell, and the only mutants that can be sought are conditional ones – temperature-sensitive mutants that fail to replicate DNA at the restrictive temperature, or nonsense mutants generated in the presence of, and dependent on, an active suppressor gene. In addition, there are polymerases that are involved in repair and recombination also present, making direct assay of DNA synthesis no measure of replicase activity.

If DNA synthesis proceeds by extension of the 3' end of the chain (and no other activity has yet been found) replication of both strands of DNA with antiparallel chains must involve a number of distinct operations in addition to polymerisation and these might well be common to the processes of repair and recombination (Fig. 14.8). For instance, all gaps must be closed before replication; but the synthesis of the right-hand strand in the figure cannot proceed as a single operation, but must be discontinuous, and generates gaps that need to be closed. So gaps must always be repaired even in normal replication, and those functions must be essential to the cell.

Seven groups of DNA replication mutants have been characterised in *E. coli*. The groups are scattered in the chromosome, and have been partially analysed biochemically and genetically. Groups *A* and *C* are mutants that fail to initiate rounds of DNA replication, they are recessive in diploids, so there is an active product normally produced from these genes. Groups *B* and *G* fail to elongate the DNA chain, and yet they can synthesise a strand of DNA during the process of conjugation. It is possible that these mutants fail to perform one of the operations necessary to join the Okazaki fragments, or to initiate synthesis of the DNA on this strand (see Fig. 14.8). Group *D* mutants are probably similar to group *C* mutants; group *F* mutants lack ribonucleotide reductase, and so fail to produce DNA precursors. The last group, *E* mutants, are particularly interesting, for they determine the activity of polymerase III, and this can perform semiconservative replication of DNA. More suggestive, *E* mutants include some with unusually high rates of spontaneous mutation, and this property has been proposed as one the true replicase mutant should possess, since a loss of fidelity in replicating the DNA is the obvious way in which such a phenotype could be explained.

4.7 Restriction and modification of DNA

A simpler system involving DNA was found in the restriction and modification properties of different bacterial strains. Some bacteria destroy foreign DNA that penetrates into the cell, but DNA from related strains survives. This discrimination has been traced to the (protective?) mechanism for which the cell employs two enzymes; the first enzyme binds to specific regions in DNA and modifies (often by methylation) one or two of the groups present. The second enzyme recognises the same sequence, but produces a lethal double-stranded break unless the sequence contains the modified group(s). DNA is thus distinguished by the modified groups,

466

and this explains the difference in behaviour of the cell to DNA from modifying and non-modifying strains. The two enzymes are genetically determined, of course, and mutants are readily obtained using phages to test the cells' properties; a phage genome usually contains sites which have to be modified for survival in the restricting strain, and both the phage DNA (and

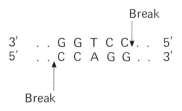

Fig. 14.9 DNA base sequence (reported by Drs Bigger, K. Murray and N. Murray) at which the restriction enzyme carried on the RII plasmid in *E. coli* produces a staggered double-stranded break. Note the symmetry surrounding the central base pair.
G = guanine; C = cytosine; A = adenine; T = thymine.

the host chromosome) are modified as they are synthesised to protect them from the restriction enzyme. Strains can be tested for restriction activity by their ability to grow phages from modifying and non-modifying hosts, and for their modifying activity by testing phage propagated on them on a restricting strain. While mutants in both markers are known, a strain cannot be defective for modification but active for restriction, since this would be a lethal combination.

The restriction enzymes have been characterised for specificity by identifying the base sequence at the point of breakage. These often show a bilateral symmetry (Fig. 14.9) and are potential tools in nucleic acid sequence studies because they produce a defined set of fragments from a pure DNA preparation (e.g. phage DNA) which can be separated and ultimately sequenced chemically.

4.8 Membrane Properties

The membrane is the seat of many reactions as well as acting as the partition between cytoplasm and environment. The study of membrane properties is difficult because of their complexity and the background of normal metabolic activity that may influence membrane behaviour. The preparation of sealed vesicles by Kaback means that bacterial membrane activity can be studied in the absence of the cytoplasmic contents, and direct comparisons can be made of normal and mutant membrane function. Bacterial membranes can be disaggregated into 20–30 major protein components and more minor components, but have no component that could be called a structural protein. Two kinds of membrane mutants arise indirectly. Glycerol-dependent strains cannot make lipid or phospholipid in the absence of glycerol but continue to grow by spreading the available lipids ever more thinly. The membrane proteins continue to be produced and inserted into the membrane, but can be found to some degree in the cytoplasm. Other membrane mutants have defective fatty acid synthesis; by supplying different fatty acids in the medium, the type of lipid synthesised can be dictated, and the effects of variations in lipid composition can be related to function. The major properties remain the same, including the main classes of transport, but there may be physical change related to the different micellar structures formed, reflected in the mobility of membrane components at different temperatures.

4.9 Transport Studies

It is somewhat surprising to realise that the transport properties of isolated mitochondria and chloroplasts are better documented than those of bacteria in some respects at the moment. This situation will probably be reversed soon, as the use of mutants has opened the way for detailed study of bacterial systems. Passive diffusion across the largely lipid membrane is not very important; it is non-selective, and for compounds insoluble in the lipid material, particularly slow. Speed and selectivity can be provided when a specific carrier is incorporated into the membrane, which can complex with a particular substrate, act as a carrier for it into the membrane structure, and release it at the inner surface of the membrane. This process is called facilitated diffusion; like passive diffusion, it can only equalise the concentrations of the compound carried on each side of the membrane, and it will do this whether the compound is initially more concentrated inside or outside the cell. The process is not spatially oriented, therefore, and most suited to components found in the medium which are rapidly utilised inside the cell – in other words, during active growth, the compound would be continuously passing into the cell. Some compounds are accumulated by cells to a concentration often far higher than that in the medium, and this process is selective, energy dependent, and oriented across the membrane. The interest in such processes of active transport has been related to the mechanism by which it occurs and to their importance in the growing cell. Many different transport mutants are known, particularly for amino acids, sugars, peptides and ions.

The active transport systems are defined by a carrier for the component moved, and some device for coupling the carrier to the energy derived from terminal oxidation in the membrane. Some cases of defective transport – e.g. an inability to concentrate potassium to the normal high internal levels – seem to be associated with the loss of the coupling factor. This coupling factor is not that lost in another mutant, in which oxidative phosphorylation is uncoupled. This mutant lacks membrane-bound ATP'ase activity; another shows defective terminal respiration because it lacks coenzyme Q.

The main active transport systems are those for amino acid and sugar transport. Amino acid transport mutants can be found in a number of ways – resistant mutants to amino acid analogues include strains which fail to transport the analogue into the cell. These can be detected since the cells fail to transport the natural amino acid as well, and the number of amino acids transported by any specific system can be determined. There are five or six systems known for the amino acids; some share common carriers, such as leucine, valine and isoleucine, but most are specific.

Since many microorganisms grow on proteins and do not excrete the enzymes for their breakdown to free amino acids but only to peptides, the uptake of peptides has been examined. The utilisation of peptides can be demonstrated by feeding an auxotrophic organism that requires histidine on a peptide containing histidine; this proves to be an effective way of supplying the cell with its histidine requirement, as intracellular peptidases liberate the free amino acid in the cell once the peptide has been transported into the cell. Mutants unable to grow on a peptide that contains an amino acid required for growth lack either the transport mechanism for the peptide or a suitable intracellular peptidase. Similarly, mutants resistant to the toxic effects of a dipeptide containing a toxic analogue of an amino acid such as norleucine or norvaline either fail to accumulate the peptide or fail to split it. If the resistant mutant can utilise a tripeptide for growth, then the peptidases are working in the mutant but the dipeptide permease has been lost. There are separate peptide permeation systems for dipeptides and oligopeptides; the latter is a single character, and loss mutants can be selected in the same manner with a toxic tripeptide. The dipeptide transport system may also be a single character, though this is not established. The identification of these mutants has allowed characterisation of the properties and specificity of these transport systems; for instance, the oligopeptide transport system seems to cope with tri- and tetrapeptides, but not with larger peptides.

The transport of these compounds seems not to be related to any reactions they may undergo inside the cell. This is not true for many other compounds, particularly some sugars. The components of the sugar transport system which release the sugar as a phosphorylated derivative inside the cell have been characterised biochemically and genetically. The system starts with a small protein, called HPr, which is phosphorylated in the cytoplasm of the cell by a specific enzyme I.

$$\text{HPr} + \text{phosphoenolpyruvate} \xrightarrow{\text{I}} \text{Hpr-phosphate} + \text{pyruvate}$$

This phosphorylated protein undergoes a complex reaction with two membrane proteins, of which one component seems to be common to many sugars that are transported this way, and the other determines the specificity of the particular complex.

$$\text{HPr-phosphate} + \text{sugar} \longrightarrow \text{Sugar-phosphate} + \text{HPr}$$

In *E. coli,* where this transport system metabolises glucose, mannose, fructose, mannitol and α-glucosides, there is a specific protein determinant for each sugar, and the transport system for each sugar can be lost independently. On the other hand, mutants lacking HPr or enzyme I fail to grow on a range of sugars. The exact connection between phosphorylation and translocation of the sugar across the membrane needs biochemical clarification. In some cases, transport and phosphorylation are coincidental but unlinked. Thiamine is transported into *E. coli* and phosphorylated by a membrane-bound ATP-dependent kinase. Mutants can be found which cannot utilise thiamine but can utilise thiamine pyrophosphate; these mutants are defective in the phosphorylation of thiamine, but they continue to accumulate thiamine in the cell; so in this case transport is not dependent on phosphorylation.

The accumulation of phosphorylated derivatives of a sugar does not necessarily prove that the scheme described above applies in all cases. The best documented exception is glycerol; accumulating in the cell as glycerol phosphate, the compound is transported across the membrane by a carrier that is not actively energised (i.e. by facilitated diffusion) and trapped in the cell as the phosphate to allow concentration of the metabolite. The carrier and the soluble glycerol kinase are coded by two genes that form an operon and are coordinately expressed.

A further example from *E. coli* of a possible linked system for uptake and derivatisation is that for fatty acid transport. Mutants defective in acyl coenzyme A synthetase cannot use fatty acids; they also fail to transport them into the cell. Since fatty acids are never found free in the cell, this would suggest a strong link between the transport and the acylation reaction.

The partial constituents for many transport systems have been found as proteins with specific binding properties for individual compounds or groups of compounds, such as ions, amino acids, sugars and vitamins. If these carriers can bind a toxic agent and transport it into the cell, a general method exists to select transport loss mutants. The use of toxic analogues of amino acids has been mentioned; another interesting case is in *E. coli,* where the toxic ion arsenate is transported by the phosphate transport system. Arsenate-resistant mutants are defective in phosphate transport, but not completely in the first instance. There seem to be at least two transport systems for phosphate, only one of which transports arsenate as well. From the arsenate-resistant strains, now dependent on a single phosphate transport system, mutants can be found that are unable to use phosphate but are dependent on organic phosphate compounds.

4.10 The cell cycle and division

The growth of a cell usually appears as a continuous process, but the presence of timing mechanisms that determine the initiation of DNA replication and the division of the daughter cells is well recognised. The molecular basis of the temporal control during the cell cycle should prove amenable to analysis if suitable mutants can extend the observations possible with wild type strains. Mutants that fail to divide can be selected among the larger cells in any population, using a filter of such porosity that most normal cells pass through. These mutants may be conditional (usually temperature-sensitive) and open up possible avenues of biochemical investigation. One mutant type found in *E. coli* has the unusual characteristic that it fails to divide at internal sites (it is a rod-shaped organism that grows by increasing its length but with a constant diameter), but can divide at an end (i.e. where some previous division had occurred) and form a mini-cell, which is small, as its name implies, lacks DNA and so has no genetic material, and rather limited expectations. The population length distribution has been shown to fit with that predicted if the cell, at division, has a certain finite number of division processes it can undertake, but unlike the wild-type strains, the ends are active sites competing for this 'division potential'. This implies a unitary character for the division potential, as well as a diffusible nature for it, since the commitment to divide at one site negates the possible division at another site.

Other studies have produced mutants which form long filaments unless they are supplied with specific polyamines, such as putrescine. These division mutants generally fail to form a septum, and the filament is one compartment with many genomes in it. Other mutants form long filaments, but these are chains of properly divided cells which have failed to separate, and lack (and can be phenotypically repaired by) the lytic enzymes normally cleaving the wall between cells.

Some bacteria have a resting stage, when they form heat stable spores quite different from the vegetative cells. Sporulation is a simple process of differentiation, therefore, and the study of the sporulation process has been greatly helped by the mutants which are unable to sporulate, and can be examined to find which stage of the process has failed, and what changes are characteristic and essential for successful spore formation.

4.11 Cell surfaces

The wall structures of bacteria have been characterised for many different species, and the rigid supporting polymer is a peptidoglycan. Because of the essential need for this layer (except under conditions that osmotically support the cell from bursting) mutants have only recently been characterised in specific steps of peptidoglycan synthesis; these are temperature-sensitive mutants, and are screened initially by looking for colonies in replica testing that not only do not grow at a restrictive temperature, but also die by lysis. This exploits the known involvement of lytic enzymes in these layers which are needed to cope with the problems of growth and expansion of the cell inside a rigid support; if these enzymes continue to degrade the existing material while the cell continues to expand, the wall becomes too fragile to hold the membrane intact. Temperature-sensitive strains that do not die under restrictive conditions probably have weakened walls but fail to exert sufficient pressure to burst because they have stopped some other major class of synthesis, particularly DNA, RNA or protein synthesis.

A particularly interesting case of genetic and biochemical analysis has been the examination of the lipopolysaccharide layer on the surface of organisms such as *Salmonella*. Originally interesting because they determine the antigenic properties of the organism which were used in

470

classification methods of medical bacteriology, they showed a bewildering variety of antigenic determinants. What was so complex can now be resolved, by analysis of mutants which lack enzymes capable of performing one or other step in the complex sequence of reactions needed to assemble the complete polymer (see Fig. 14.10). Chemical analysis of such a structure is

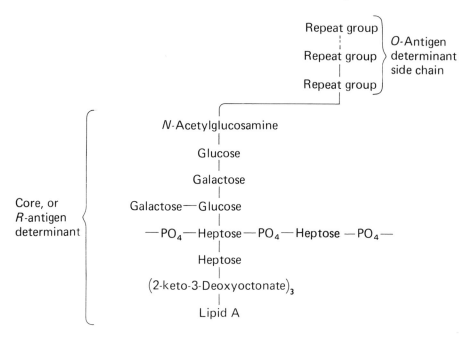

Fig. 14.10. Sequence of residues in the Lipopolysaccharide found on the surface of *Salmonella*. The repeat group contains four or five residues characteristic of each strain, but is determined in part by the lysogenic phage present in the strain.

daunting, but as colonial morphology mutants were obtained with modified lipopolysaccharide antigens, it became possible to see the mutants reflecting the ordered assembly of the complete molecule, as well as providing material for analysis with discrete changes in structure.

Starting at the outer part, the repeat groups contain about four or five sugars variously linked, and which are modified in wild-type strains by the acquisition of specific phages in the lysogenic state. The phage appears to produce an enzyme modifying the structure which may result in the modified recognition of the cell by other phage.

The colonial variants that have been recognised show loss of some (D mutants), all but one (C mutants) or all (B mutants) of the repeated groups. D mutants appear to have patchy attachment of the repeat groups they contain, since the antigenic determinant of the core is also detectable in this strain. Strains which only make the core or some part of it have a characteristic rough colonial appearance, and the genetic determinants are called *rou* genes, for rough. The formation of the core requires certain sugar derivatives and specific transferases. If these are lacking at any point, a truncated product is again found, and the ultimate (E) mutants contain no sugar after 2-keto-3-deoxyoctonate.

This is an extreme case of genetic simplification of a complex chemical problem, but such suitable cases for this antigenic analysis are not often found.

Other surface properties of cells such as flagella or hair-like structures called pili or fimbriae are probably protein in nature, and failure to complete the subunit structure probably destroys all vestige of the structure. Flagella are associated with motility of the organisms, and mutants can easily be selected for loss of motility; these have either defective or no flagella, or paralysed flagella which are not coupled properly to the cellular energy source. Mutants with no flagella sometimes contain a protein similar to the normal flagellin subunits of the flagellum, and may lack some assembly function.

5. Concerning antibiotics

The use of antibiotics as tools to study biological processes has proved extremely fruitful. Many areas of action of antibiotics overlap with areas where genetic studies are in progress. In other cases, the antibiotic can be used as pressure from which mutants can escape. In this manner, genetic determinants for enzymes that detoxify the antibiotic, for loss of transport for the antibiotic, and for modification of the sensitive site are found. Natural antibiotics seem to have a wide spectrum of sites of action; bacterial wall synthesis is affected by penicillin, novobiocin, cephalosporin, bacitracin, cycloserine and vancomycin, for example; protein synthesis (in prokaryotes) by agents such as chloramphenicol, streptomycin, kanamycin, erythromycin and puromycin; RNA synthesis by rifamycin or rifampicin (which inhibit the initiation of RNA polymerase action); DNA synthesis is inhibited by nalidixic acid.

One piece of information derived from streptomycin-resistant mutants is that resistance can arise through modification of a ribosomal protein; when this is present in the ribosomes some degree of misreading of the messenger RNA occurs which can be detected by the phenotypic restoration of activity from genes that are already mutant, particularly when the mutant is a nonsense mutation. These mutants are called ribosomal ambiguity mutations (*ram* for short). This effect ties in with another property of streptomycin and similar antibiotics, where the same phenotypic correction of a defect can occur if the cells are grown in the presence of the antibiotic; when it binds to the ribosome, it causes misreading, which may be the origin of its toxic effect.

The other area of interest with regard to antibiotics is the occurrence of plasmids that confer drug resistance on the strain carrying them. Hospitals tend to have problems with infections that are recalcitrant to drug therapy, which arise from these resistance transfer factors. Unfortunately, the number of resistance markers that can be carried is large; up to six have been reported on a single factor, and the drugs affected are the most useful in clinical practice – penicillin (ampicillin and cephalosporin), chloramphenicol, kanamycin, erythromycin, sulphonamides and streptomycin are all affected. The epidemiology of these resistance factors is unusual, for the factors can pass between pathogenic strains to organisms that are not regarded as pathogens normally, and these act as reservoirs from which the pathogens can pick up resistance properties all too rapidly.

6. Bacteria as marked strains

Problems of natural ecology can be studied at the microbiological level by the use of suitably marked strains. These should be recognisable by some genetic character such as an antibiotic resistance marker, rather than an auxotrophic marker, so that growth of the test organism is not handicapped during the study. Even antibiotic resistant strains do not always grow as well as the wild-type organisms from which they were derived, and marking through some plasmid or lysogenic phage may have advantages.

Problems could be related to the natural determination of population level and composition, or to specific problems such as the movement of material under natural conditions — the spread of organisms through natural events or through man-made disturbances like water treatment or disposal systems. An ingenious application, not specifically involving genetics (but which could of course become involved), is the tracing of water movement through the spread of mu phage. This can be prepared in enormous numbers (10^{13} particles) but detected as single particles in small water samples (1 in 0·1 ml) though there is some problem of background contamination at this level. By using the cells to adsorb the phage in liquid samples (say, 10 ml) and marking the phage so that it has some conditional property (say, temperature sensitivity) that would enable it to be distinguished from any natural phage, the sensitivity of this method could be raised several orders of magnitude.

7 Practical contributions from mutants

The use of mutants of bacteria, and of the eukaryotes traditionally used in the fermentation industry (yeast and fungi) have formed the basis of the commercial production of enzymes, amino acids, vitamins and antibiotics. The fungal systems have been exploited most, producing the major antibiotics, but bacterial systems have become of increasing potential. Many amino acids, such as glutamic acid, lysine, methionine and valine, are produced in the pure L-form through fermentation, and have value in specific problems of nutrition. The yields are extremely high — glutamic acid at 60 g/l, and lysine at 42 g/l, for example. Vitamins such as riboflavin and B12 were once laboratory curiosities; now they are produced in fermentations at a concentration of 5 g/l and 30 mg/l respectively.

ATP is now produced through fermentation very cheaply. In each case, the strains used have been improved by repeated mutagenesis and selection of better strains, though a knowledge of the biochemical steps involved has been of advantage and has stimulated research into ways of improving the strains still further by rational attempts at modification of specific features of regulation that may still limit production.

The use of enzymes is increasing steadily; there are requirements for large quantities of carbohydrate hydrolytic enzymes, while the mixed enzyme preparations used in biological detergents have had a well-publicised history. The proteolytic enzymes used in such applications would be of little value if they failed to work under standard washing conditions, and mutants that retained proteolytic activity under alkaline conditions and the high temperatures used in washing operations were selected very simply on solid media. With the improvement of strains, some extraordinary levels can be attained — an *E. coli* diploid has been described that makes 25 per cent of its protein as β-galactosidase! Even the strains used to produce the *lac* repressor protein made 1000 times the amount found in the wild-type strain, though this is still much less than the previous example.

8 Conclusions

The molecular analysis of some systems has reached a degree of sophistication from which we may feel that the general properties of all biological systems are within our grasp. But it should be plain from the examples given that the more complex the system, and the more it is integrated into total cellular functions, the less we can say about it. Since the prokaryotes, and their viruses, are also the simplest complete living systems for analysis, there is a long way to go before we should feel confident that the major problems have been solved. What is also apparent is that total analysis is the outcome of biochemical and genetic work so integrated that to separate them in teaching or research is now artificial, and possibly restricting to future biologists.

For those who will follow this path, two quotations:

'There are no straight roads in the world; we must be prepared to follow a road that twists and turns . . . ' (Mao Tsetung)
'First catch your hare.' (Mrs Glasse).

Bibliography

ACHTMAN, M. (1973) 'Genetics of the F sex factor in enterobacteriaceae', *Current Topics in Microbiology and Immunology*, **60**, 79 – 124.

BECKWITH, J. R. and ZIPSER, D. (eds.) (1970) *The Lactose Operon*. Cold Spring Harbor Laboratory, New York.

BRENNER, M. and AMES, B. N. (1971) 'The histidine operon and its regulation', in *Metabolic Pathways*, 3rd edn., H. J. Vogel (ed.), Vol. 5, pp. 350 – 88. Academic Press. New York and London.

DEMAIN, A. L. (1972) 'Riboflavin oversynthesis', *A. Rev. Microbiol.*, **26**, 369 – 88.

EISERLING, F. A. and DICKSON, R. C. (1972) 'Assembly of viruses', *A. Rev. Biochem.*, **41**, 467 – 502.

ENGLESBERG, E. (1971) 'Regulation in the L-arabinose system', in *Metabolic Pathways*, 3rd edn., H. J. Vogel (ed.), Vol. 5, pp. 257 – 96.

GROSS, J. (1972) 'DNA replication in bacteria', *Current Topics in Microbiology and Immunology*, **57**, 39 – 74.

HALL, R. M. and BRAMMAR, W. J. (1973) 'Increased spontaneous mutation rates in mutants of *E. coli* with altered polymerase III', *Molec. gen. Genet.*, **121**, 271 – 76.

HAROLD, F. M. (1972) 'Conservation and transformation of energy by bacterial membranes', *Bact. Rev.*, **36**, 172 – 230.

HAYES, W. (1968) *The Genetics of Bacteria and their Viruses*, 2nd edn. Blackwell Scientific Publications. Oxford and Edinburgh.

HERSHEY, A. D. (ed.) (1971) *The Bacteriophage Lambda*. Cold Spring Harbor Laboratory, New York.

KLEIN, A. and BONHOEFFER, F. (1972) 'DNA replication', *A. Rev. Biochem.*, **41**, 301 – 32.

MARGOLIN, P. (1971) 'Regulation of tryptophan synthesis', in *Metabolic Pathways*, 3rd edn., H. J. Vogel (ed.), Vol. 5, pp. 389 – 415.

MESELSON, N., YUAN, R. and HEYWOOD, J. (1972) 'Restriction and modification of DNA', *A. Rev. Biochem.*, **41**, 447 – 66.

MINDICH, L. (1970) 'Membrane synthesis in *Bacillus subtilis*', *J. molec. Biol.*, **49**, 433 – 39.

OXENDER, D. L. (1972) 'Membrane transport', *A. Rev. Biochem.*, **41**, 777 – 814.

PATO, M. L. (1972) 'Regulation of chromosome replication and the bacterial cell cycle', *A. Rev. Microbiol.*, **26**, 347 – 68.

STENT, G. S. (1971) *Molecular Genetics*. Freeman. San Francisco.

474

SUSSMAN, A. J. and GILVARG, C. (1971) 'Peptide transport and metabolism in bacteria', *A. Rev. Biochem.*, **40**, 397 – 408.

WATSON, J. D. (1970) *Molecular Biology of the Gene,* 2nd ed. Benjamin. New York.

ZUBAY, G. and CHAMBERS, D. A. (1971) 'Regulating the *lac* operon', in *Metabolic Pathways,* 3rd ed., H. J. Vogel (ed.), Vol. 5, pp. 297 – 349.

15
Biochemistry of Microbial Pathogenicity

H. Smith

Department of Microbiology, University of Birmingham

1 Introduction

Perhaps no discoveries have had such a dramatic impact on human affairs as the proof that micro-organisms cause disease and the subsequent recognition of the species involved. The resulting public health measures have controlled in many countries the worst effects of infectious disease, especially death from bacterial disease. This has occurred without sophisticated research on the biochemistry of infectious processes. Even vaccines and drugs, which played a significant but smaller part in controlling infection have been developed largely by empirical methods requiring little or no biochemical knowledge of the microbial or host determinants of pathogenicity. Is it surprising then that the biochemistry of these determinants is still obscure? What is known is confined almost entirely to bacterial diseases and even here information is scanty. While the biochemistry of the toxins responsible for the classical toxaemias (e.g. tetanus, diphtheria) has been studied in detail and the chemical nature of some bacterial substances which inhibit ingestion by phagocytes is known, many of the mechanisms of bacterial pathogenicity are still not clear. The bacterial products responsible for the harmful effects of some diseases (e.g. pneumonia) remain obscure and knowledge is only just emerging about the biochemical bases for communicability (i.e. spread of disease), for survival and growth of certain bacteria (e.g. those causing brucellosis and tuberculosis) within phagocytes, for the specific attack of certain hosts and tissues, and for long term bacterial survival in chronic diseases. And, of course, these phenomena are not confined to bacterial diseases, they occur in the even less understood infections caused by viruses, mycoplasmas, fungi and protozoa.

A student might ask, what is the point of investigating the biochemistry of microbial pathogenicity if infectious disease can be controlled? Well, despite the advances of the past century, infectious disease remains a major problem in human and veterinary medicine with the economically important nuisance and chronic aspects gaining prominence as the fatal consequences become less frequent. Effective chemotherapy of troublesome virus diseases is still lacking. Drug resistance of bacteria, protozoa and fungi is increasing. Many vaccines remain unsatisfactory, either due to producing incomplete protection or to the hazard of injecting live organisms. The alarming increase in gonorrhoea is a current reminder of the ability of a once-regulated disease to rebound with changing social conditions. New methods of attacking infectious disease are needed. Undoubtedly, as in the past, empirical screening procedures will provide some new methods. However, after nearly 50 years, such procedures may be reaching the limit of their usefulness; so far they have failed to provide an effective therapy for virus diseases. There appears good reason for an inclination towards a more rational approach to the problem, namely that of attempting to recognise and then to neutralise the biochemical determinants of pathogenicity.

This chapter summarises our present knowledge of the biochemistry of microbial pathogenicity and emphasises the gaps in this knowledge. Only broad principles can be covered in the space available but relevant examples will be cited. Bacteria form the main subjects. Then the almost unknown mechanisms of pathogenicity of viruses, mycoplasmas, fungi and protozoa are discussed briefly in the context of concepts used for bacterial pathogenicity. The terms pathogenicity and virulence are nearly synonymous. They mean the capacity to produce disease; and the former is used with respect to species and the latter with respect to degrees of pathogenicity of strains within species. In each section below a description of the methods and difficulties of research in this field forms a prelude to a discussion of five main aspects of pathogenicity:

1. Entry to the host usually by surviving on and penetrating mucous membranes;
2. Multiplication *in vivo*;
3. Inhibition or non-stimulation of host defence mechanisms;
4. Damaging the host;
5. The reasons for tissue and host specificity.

Most of the references for the work described will be found in several books and recent reviews that are cited for further reading; only references that are not contained in these sources are specifically mentioned here.

2. Bacterial pathogenicity

Pathogenic bacteria are peculiarities. The great majority of bacteria are harmless and often beneficial. Obviously pathogenic bacteria have a chemical armoury which enables them to invade a host and produce disease. The problem is to identify the weapons in this armoury, their chemical nature, mode of action and order of importance.

2.1 Methods and difficulties of studying biochemical determinants of pathogenicity

The fact that pathogenicity is usually determined by a number of products and not a single powerful toxin as in tetanus and diphtheria provides one difficulty in biochemical studies. But the main factor contributing to difficulty in this field is that virulence, the disease-producing capacity of a microbial propulation, can be measured only *in vivo* (by determining the minimum number of organisms needed to produce a certain pathological effect, e.g. those needed to kill half of a group of animals—L.D.$_{50}$), and it is markedly influenced by changes in growth conditions due to selection of types and to phenotypic change. Under different growth conditions new enzymes will be induced and others repressed. Bacterial virulence is at its maximum in bacteria obtained directly from infected animals. Usually it is reduced by subculture *in vitro* because under laboratory conditions bacteria lose the capacity to form one or more of the full complement of virulence attributes manifested in infected animals. Also *in vitro*, apparent virulence factors might be produced which are not formed and therefore have no relevance *in vivo*. There is now abundant evidence for many bacterial species including staphylococci, streptococci, gonococci, plague bacilli, anthrax bacilli and tuberculosis bacilli that organisms grown in infected animals are different chemically and biologically from those grown *in vitro*. Thus, although most studies are made on bacteria grown *in vitro*, these bacteria can be incomplete or misleading as regards the possession of virulence attributes. This is the essence of the difficulties encountered in studies of pathogenicity.

How then can the various factors involved in the pathogenicity be identified? Obviously these factors can be produced in laboratory cultures if the correct nutritional conditions can be found. This has already happened in studies of the classical bacterial toxins and some antiphagocytic substances. However, for problems of pathogenicity which have defied solution by conventional procedures using cultures *in vitro*, the above discussion suggests that one approach would be to study bacterial behaviour *in vivo*. There are no vitalistic leanings behind this suggestion, merely a realistic assessment of an approach which might reveal aspects of pathogenicity which later could be reproduced *in vitro* by appropriate changes in cultural conditions.

This chapter describes several examples of virulence factors that were recognised by this approach. Information on bacterial behaviour *in vivo* can be gained in several ways. Bacteria and their products can be separated directly from the diseased host for biological examination and for chemical and serological study *in vitro*. The behaviour of organisms growing *in vivo* and their repercussion on the host can be examined either in the whole animal or in tissues. The largest gaps in our knowledge of pathogenicity occur here; detailed experimental pathology and precise biochemical determinations are not easily and safely accomplished during infection and only in a few cases have such detailed studies supplemented the clinical pictures. Yet this information is vital, if mechanisms of pathogenicity are to be understood and relevant biological tests for potential virulence factors designed. Not the least of the difficulties is the lack of suitable laboratory animals in which natural infections of man (e.g. gonorrhoea) can be truly simulated. Some light can be shed on particular phases of microbial behaviour *in vivo* by making observations in tissue or organ culture. Finally, tests *in vitro* can be made more relevant to microbial behaviour *in vivo*.

The classical method of studying bacterial virulence is to compare the properties of virulent and avirulent strains. Properly used, this method of bacteriology is a powerful tool which could be applied to studies of pathogenicity of other micro-organisms (see later). The techniques of microbial genetics have increased the scope of the method; and studies *in vitro* on enzymes, metabolic characteristics and antigens of different bacterial strains have indicated many virulence *markers*, i.e. factors associated with virulence. However, only relatively few of these factors have been shown to be virulence *determinants*, i.e. produced during infection and having biological activities directly connected with virulence, e.g. the power to inhibit or destroy phagocytes. Studies on virulence can benefit from a comparison of the behaviour and properties of virulent and avirulent strains and from examinations of the influence of the products of a virulent strain on the behaviour of an avirulent strain, provided tests are carried out *in vivo* or in simulant conditions *in vitro*. These methods have been used to good effect in bacteriology to identify cell-wall and capsular materials which inhibit humoral and cellular defence mechanisms (see below).

2.2 Entry: survival on and penetration of mucous membranes

Although some bacteria enter the host directly by trauma or vector bite, most bacterial infections start on the mucous membranes of the respiratory, alimentary and urogenital tracts. Hence pathogenic bacteria must at least survive and better multiply on these membranes from a small inoculum; and this process must occur, not in pure culture but in competition with commensals, the non-pathogenic microbes abundantly present on mucous surfaces. Subsequently the bacteria must penetrate into the tissues. Together with other factors, a differential ability of bacterial species for accomplishing these early stages of infection might explain why some diseases (e.g. brucellosis) are more communicable than others (e.g. anthrax). Precise information on early bacterial attack of mucous membranes is scanty. There is much current work *in vivo* and *in vitro* on mixed culture of gut bacteria, both commensals and pathogens, showing that growth of one organism has a profound biochemical effect on another. Apart from complex antibacterial factors such as antibiotics which might be produced by commensals simple products of metabolism may be important. For example, short chain fatty acids produced by fusiform gut commensals have an inhibitory action in a reducing environment on pathogens such as dysentery bacilli and salmonellae. But the biochemical mechanisms which allow these pathogens to overcome this inhibitory action in dysentery, food poisoning and typhoid fever are unknown. Furthermore, although there has been some electron microscopy of the early attack of gut epithelium by these pathogens the biochemical mechanisms involved are obscure.

2.3 Multiplication *in vivo*

Virulent bacteria must multiply in the host tissues in order to produce their disease syndrome either by increasing locally or by spreading throughout the host. Two qualities are needed for multiplication. First, an inherent ability to multiply in the biochemical conditions of the host tissue and second a propensity to inactivate or not to stimulate host defence mechanisms which would otherwise kill or remove them (see later). The effects of these two qualities *in vivo* are not easy to separate and often it is difficult to assess their relative importance, in the increase of a single infecting population, in the differential behaviour of virulent and attenuated strains and in the different susceptibilities of tissues or hosts to infection. In this section the first quality – ability to multiply – is discussed.

Avirulence can arise from inability to grow and divide in the environment *in vivo*. Thus, nutritionally deficient mutants of pathogenic species were shown to be avirulent unless injected with their required nutrients; this was first accomplished with purine requiring mutants of typhoid bacilli. However, for most bacteria the tissues and body fluids probably contain sufficient nutrients to support some growth. Few naturally occurring bacterial strains will be avirulent due solely to inability to grow in the host. Nutritional considerations will, however, affect rate of growth *in vivo*. The more rapid it is, the more the chance of establishing the infection against the activity of the host defence mechanisms. What do we know of multiplication rate *in vivo*? The numbers of viable bacteria in the tissues of an infected host can be counted at any time after inoculation. But these numbers are only the resultants of multiplication and destruction or removal. Only recently has a method been evolved for measuring true bacterial division rates *in vivo*. Pathogenic bacteria, genetically labelled with biochemical markers retained by a known proportion of the progeny at each division, were used. Both division rates and death rates were determined by measuring the proportion of organisms with the marker at various times after inoculation into animals. Remarkable results were obtained; in the spleens of mice *Salmonella typhimurium* divided at only 5–10 per cent of the maximum rate *in vitro* and the death rate was extremely small. This type of approach may be used for measuring true division and death rates of other microbes *in vivo*. We shall see later how variation of the biochemical conditions for bacterial multiplication can explain some examples of tissue preference of pathogenic bacteria.

2.4 Inhibition of host defence mechanisms

To increase within the host tissues metabolic ability to multiply in the nutritional environment is not enough. Pathogenic bacteria must also be able to inhibit host defence mechanisms which otherwise would destroy them. The bacterial compounds which inhibit these mechanisms are called 'aggressins' an old term which well describes their biological role.

Aggressins act in the decisive, primary lodgement period of infection, that is, during the first few hours when the few invading bacteria are most vulnerable to the protective reactions of the host. At this early stage, aggressins must inhibit non-specific bactericidal mechanisms; not only those already existing in tissues but also those agencies, especially phagocytic cells (amoeboid cells which ingest and destroy bacteria) that are mobilised by inflammatory processes soon after the tissues are irritated. If some bacteria survive the primary lodgement and grow, spread of infection is opposed by the fixed phagocytes of the reticulo-endothelial system (lymph nodes, spleen, liver); and again, to make headway, bacteria need aggressins, possibly different from those operating during the early lodgement phase. The clinical outcome of the disease depends on the interplay of these defensive reactions of the bacteria and of the host, and varies from complete subjugation of the host to complete destruction of the bacteria including near stalemate in chronic infections. Various types of bacterial aggressins are described below.

2.4.1 Inhibitors of blood and tissue bactericidins

Body fluids and tissues contain a variety of bactericidal factors such as basic polypeptides, lysozyme, complement (acting with antibody or possibly non-specific substances), and possibly a system involving the iron-binding protein transferrin. Since clearly there are several different types of bactericidins, virulent bacteria must produce different types of aggressins to inhibit them. Resistance to these bactericidins has been associated with virulence in strains of many bacterial species such as enteric pathogens, staphylococci, anthrax bacilli, brucellosis bacilli and, recently, gonococci isolated directly from patients and tested without subculture. But only rarely have the aggressins been chemically identified. Those from anthrax bacilli are capsular poly-D-glutamic acid and the three-component anthrax toxin (see later). Resistance of *Brucella abortus* to the bactericidal action of bovine serum is due to a cell-wall component containing protein, carbohydrate, formyl residues and much (35–41 per cent) lipid.

2.4.2 Inhibitors of the action of phagocytes

Once a microbe has penetrated into the tissues the phagocytic activity of the wandering and fixed cells of the reticulo-endothelial system forms the main protective mechanism of the body; a mechanism which acts non-specifically but which is greatly enhanced by immunisation. Phagocytes vary in origin, morphology, constituents and bactericidal function. There are two main types each having two subdivisions: polymorphonuclear (neutrophils and eosinophils) and mononuclear (blood monocytes and tissue macrophages) phagocytes. Polymorphonuclear phagocytes are end-cells with a short life derived from different stem cells from the long-lived mononuclear phagocytes. Inflammatory exudates contain cells of all types, the polymorphonuclear cells predominating initially but later dying to leave the mononuclear phagocytes ascendent. Macrophages form the fixed phagocytic system in the lymph-nodes, spleen and liver.

Phagocytosis of bacteria involves three stages, contact, ingestion and intracellular killing and digestion. Virulent bacteria may produce aggressins which inhibit any of these stages.

2.4.3 Inhibitors of contact

Contact with bacteria is effected by random hits, by trapping on uneven surfaces in confined tissue spaces, by filtration systems in lymph-nodes, spleen and liver and by chemotaxis. Bacterial products could hardly interfere with the mechanical processes, but they can inhibit chemotaxis. For example, it has been demonstrated that certain fractions from tubercule bacilli and a cell wall material from staphylococci inhibit leucocyte migration.

2.4.4 Inhibitors of ingestion

Ingestion of bacteria involves engulfment within a phagocytic vacuole – the phagosome – which is derived by inversion of the phagocyte membrane. Specific and non-specific opsonins (serum factors) enhance this process of ingestion. Once inside phagocytes, many bacteria (e.g. pneumoccoci, *Bacillis anthracis* and *Salmonella typhi*) are rapidly killed and digested. Resistance to ingestion thus avoiding intracellular bactericidins, is therefore essential for the survival of virulent strains of these species. Aggressins that inhibit ingestion have been recognised but the chemical basis for their activity is unknown. They fall into two main types.

First, there are surface and capsular products which do not harm phagocytes. Examples are the capsular polysaccharides of pneumococci, the cell-wall M protein and capsular hyaluronic acid of streptococci, the capsular poly-D-glutamic acid of *B. anthracis*, the O somatic antigens of some Gram-negative organisms, the Vi antigen (poly-*N*-acetyl-D-galactosaminuronic acid) of *Salm. typhi* and the protein carbohydrate envelope substance of plague bacilli. Further investigation of the detailed chemistry of these aggressins may reveal common structural features. Although acidic components occur in most of the above compounds including the various pneumococcal polysaccharides, they are absent from some active compounds (e.g. type IV and XII pneumococcal polysaccharides) and hence do not appear to determine aggressive activity. There is recent exciting work correlating the chemical nature of the polysaccharide side chains of the O antigens of *Escherichia coli* and *Salmonella* with antiphagocytic activity. At present, however, there is no obvious connection between structure and aggressive activity. Equally the modes of action of aggressins are not clear. Interference with ingestion may be purely mechanical but other mechanisms can be involved such as inhibition of adsorption of serum opsonin or possibly the aggressin is so 'host like' that the host cannot recognise the bacterial surface as foreign. Although several aggressins are capsular in origin, a common impression, possibly arising from the classical work on the pneumococcal polysaccharides, that all capsulated bacteria resist phagocytic ingestion and are virulent, is not true. The chemical nature of the surface material determines virulence not the presence of a capsule *per se.*

Excreted bacterial products which have a toxic action on phagocytes form the second type of aggressins interfering with ingestion. The leucocidins of the staphylococci are good examples.

2.4.5 Inhibitors of intracellular bactericidins; promotion of intracellular growth

When bacteria are phagocytosed, granules from the cytoplasm of the phagocyte discharge into the phagosome (ingestion vacuole). The granules contain the bactericidins which normally kill and digest the phagocytosed organism. However, virulent strains of some bacteria such as tubercule bacilli, leprosy bacilli and brucellae resist the internal phagocytic bactericidins which destroy other micro-organisms and grow intracellularly. This property is the most important aspect of the pathogenicity of these bacteria; it is seen in infected animals and in cell maintenance culture *in vitro*. Obviously, virulent strains possess aggressins which interfere with the internal bactericidal mechanisms but only very recently have we begun to catch glimpses of the nature of these aggressins and their mode of action. By making observations first on brucellae isolated directly from infected animals, a cell-wall antigen from virulent brucellae has been recognised which inhibits the bactericidal action of bovine phagocytes as evidenced by promotion of increased survival and growth of an attenuated strain of brucellae in phagocytes pretreated with preparations containing the virulence antigen; this antigen is different from the one (see above) that interferes with the bactericidal action of bovine serum (Frost *et al.*, 1972). Although the nature of the aggressins is unknown it appears that virulent tubercle bacilli inhibit intracellular bactericidal mechanisms by preventing the phagocytic granules discharging into the phagosomes whereas leprosy bacilli allow the discharge but are unaffected by the liberated bactericidins (D'Arcy Hart *et al.*, 1972). The biochemical basis for these important recent observations should be elucidated before long.

2.5 Damaging the host

There are two methods whereby pathogenic bacteria can damage the host, directly by the production of poisons or toxins and indirectly by sensitising the host to bacterial products so that subsequent immunological reaction of the host results in tissue damage (so called immunopathology).

2.5.1 Bacterial toxins

This is the one area in microbial pathogenicity where biochemical studies in depth have been made both on the nature of the toxins and their mode of action at cellular and subcellular levels. Only the main outlines and a few high-lights can be covered here.

The toxic activities of bacteria can be divided into five categories:

1. Toxins responsible for non-infectious disease because they are produced outside the host. Bacteria can produce poisons in foodstuffs. The toxin of *Clostridium botulinum* and the enterotoxin of staphylococci are the main examples of these poisons. The disease that occurs on ingestion of the infected food material is a chemical poisoning comparable to that seen on eating a poisonous plant, fungus or shell fish. It is not an infectious process but clearly the microbial toxin is responsible for disease. The botulinum toxin has the general characteristics outlined in the next section; it is a protein neurotoxin acting on the autonomic system by interfering with acetylcholine synthesis or release. The staphylococcal enterotoxin is the one microbial toxin for which a full amino acid sequence is available.

2. Toxins of over-riding importance in infectious disease. Clostridium tetani and *Corynebacterium diphtheriae* produce *in vitro* powerful exotoxins which have been well characterised. These toxins are produced *in vivo* and are responsible for almost the whole disease syndrome. Immunisation with toxoid (formalin treated, detoxified toxin) protects against disease. Both are proteins with no toxic moieties or abnormal amino acids. Tetanus toxin is a neurotoxin acting on the central nervous system possibly by interfering with synaptic inhibitors. Diphtheria toxin interferes with protein synthesis and consists of two parts, one determining entry into a cell and the other inhibition of a transferase which halts protein synthesis (Uchida, Pappenheimer and Harper, 1972).

3. Toxins which are significant but not the only factors responsible for infectious disease. These toxins were originally recognised in cultures *in vitro* and then shown to be produced *in vivo*. They can be responsible for some pathological effects of infection. However, they are not the sole determinants of disease for often as much toxin is produced by avirulent as by virulent strains, sometimes injection of toxin does not produce all the pathological effects of disease and usually immunisation with toxoid does not confer solid protection against infection. Examples are the α-toxin of staphylococci which alters membrane permeability and the rash-forming toxin of streptococci. But the most important representatives of these toxins are the endotoxins, lipopolysaccharides intimately associated with the cell walls of many different Gram-negative bacteria. The basic structure consists of lipid moieties containing 2-keto-3 deoxy-octanoic acid and polysaccharide side chains attached to a heptose polymer. Their detailed chemistry is known more in relation to serological activity than toxicity. When extracted from cell walls by fairly drastic means (treatment with trichloracetic acid or warm aqueous phenol) and injected into animals they produce toxic manifestations – pyrexia, diarrhoea, prostration and death. In some infections, there is little doubt that endotoxins are liberated from the cell wall of the invading bacteria and are responsible for pathological effects, such as pyrexia, leucopenia, shock and death in typhoid fever or, pyrexia and shock in brucellosis of man and abortion in brucellosis of domestic animals. On the other hand, factors other than endotoxins must contribute to the pathology of many Gram-negative infections. First, endotoxins have the same biological properties no matter from which species they are obtained, yet pathological effects of Gram-negative infections vary enormously. Thus, in dysentery and cholera, bacteria remain in the gut and there is much diarrhoea, whereas in typhoid fever organisms invade the bloodstream and there is relatively little diarrhoea. Second, endotoxin is sometimes never released in significant amounts *in vivo*, since avirulent strains of Gram-negative species including the *E. coli* of normal gut contain much endotoxin yet when growing enterically they do not harm the host. We shall see later that in some diseases the pathological effects are due to an easily liberated toxin and not the cell-wall endotoxin which is never liberated from the cell wall in significant quantities.

4. Toxins produced in vitro but of unknown importance in disease. Many substances, producing toxic effects related or unrelated to disease syndromes, have been isolated from cultures. Some of these products may be laboratory artefacts having no relevance to disease *in vivo*. Even if formed *in vivo*, the question is whether they play significant roles in infection. Examples are some of the many enzymic and haemolytic products of staphylococci, strepto-cocci and other organisms.

5. Hitherto unknown toxins recognised by studying bacterial behaviour in vivo or in biological tests relevant to the disease. The reasons for this approach to problems of pathogenicity were given above and several important bacterial toxins have been recognised in this way. The first was the anthrax toxic complex now generally accepted as the cause of death in anthrax; it was found originally in the plasma of guinea pigs that had just died of anthrax. Later it was reproduced *in vitro*, purified and shown to consist of three synergistically acting components, two proteins and one a metal chelating agent containing protein, some carbohydrate and phosphorus. Its mode of action is unknown; the third component could not be replaced by another metal chelating agent such as versine. A second toxin which might be important in death from plague was recognised using *Pasteurella pestis* obtained from infected guinea pigs. Live virulent *P. pestis* killed both guinea pigs and mice but a former product, 'murine' toxin, obtained from cultures *in vitro*, killed only mice. However, an extract of *P. pestis* isolated directly from guinea pigs killed guinea pigs as well as mice. Later, guinea pig toxin was found in *P. pestis* grown *in vitro* and fractionated into two synergistically acting components, both proteins and unconnected with either the 'murine' toxin or endotoxin. Whether the guinea pig toxin or the 'murine' toxin or both are involved in death of man from plague is unknown.

In the past decade there have been spectacular advances in our understanding of the role of toxins in the acute diarrhoeal diseases of man and animals the so-called 'enterotoxic enteropathies'. This has been due largely to discarding mouse toxicity tests for other biological and animal tests in which organisms and their products were put into the gut lumen and their effects in this site observed. Investigations on cholera formed the template for those on other diseases.

An enterotoxin from *Vibrio cholerae*, responsible for the gross fatal fluid loss from the intestine which occurs in cholera, was recognised by using two tests. First a ligated segment of small intestine in a living rabbit would fill with fluid following intraluminal injection of *V. cholerae* and its products. Second *V. cholerae* and its products caused fluid accumulation and diarrhoea in suckling rabbits when introduced into the gut lumen by a gastric tube. Other animal tests followed. The extracellular enterotoxin has been purified and its mode of action studied. It is a heat labile, trypsin resistant, antigenic protein with a molecular weight of 90 000; it is different from the cell-wall endotoxin. It acts by increasing the normal secretion of the small intestine, possibly by activating adenyl cyclase present in the intestinal epithelial membrane, thereby raising intracellular cyclic-AMP levels which in turn would affect electrolyte transport. Simple replacement of fluid and electrolytes usually leads to rapid and complete recovery from the disease.

Using similar 'gut reaction' tests, enterotoxins now have been demonstrated for the following diarrhoea producing organisms: *E. coli* (scours in young pigs and calves and diarrhoea in babies; enterotoxin production is plasmid transmitted and the enterotoxin activates adenyl cyclase); *Vibrio parahaemolyticus* (a marine vibrio which in the summer causes the majority of the food poisoning cases in Japan); *Clostridium perfringens* (food poisoning in man) and *Shigella dysenteriae* (human dysentery). These enterotoxins have not been investigated as thoroughly as that from *V. cholerae* but all appear to be heat labile proteins.

These recent discoveries of relevant toxins in several important diseases, are a warning against attributing damage in other microbial diseases to causes other than direct toxicity until the possible production of toxins has been throughly investigated using realistic biological tests.

2.5.2 The role of immunopatholgy in bacterial disease

Anyone who has suffered hay-fever or asthma will know that the immunological reactions of the host although usually protective, can sometimes have unpleasant consequences. And classical work with *Mycobacterium tuberculosis* in guinea pigs has shown that hypersensitivity to bacterial products can be dangerous and even fatal for the host. Furthermore, skin tests indicate that hypersensitive states occur in many bacterial diseases such as tuberculosis, staphylococcal infections, streptococcal infections, pneumococcal infections, brucellosis, tularaemia, glanders, leprosy and salmonellosis. The hypersensitive reactions are usually of the delayed type indicating that cellular mechanisms are involved, but, antibody mediated, Arthus-type reactions can also occur, e.g. against bacterial polysaccharides. Thus, potentially in many diseases, non-toxic bacterial products could produce harm by evoking hypersensitivity reactions. But, just as production of a toxin *in vitro* does not necessarily mean that it is relevant *in vivo*, mere demonstration of a state of hypersensitivity by a skin test is no proof of the implication of hypersensitivity reactions in the main pathological effects of the disease. More extensive investigations are needed; the main systemic and local effects of the disease must be simulated by hypersensitivity reactions evoked in a sensitised host by products of the appropriate microbe. The prolonged work on tuberculosis and rheumatic fever emphasises the difficulties of obtaining precise biochemical knowledge in this field. There seems little doubt now that the pathology of tuberculosis is largely due to hypersensitivity to the products, particularly the waxes, of *M. tuberculosis*. Hypersensitivity also appears to play a role in the cardiac and kidney lesions following infection with streptococci, although the influence of direct toxicity of streptococcal products on the susceptible tissues is still advocated by some.

2.6 Tissue and host specificity in bacterial infections

Why, in man, do diphtheria bacilli, pneumococci and meningococci show predilections for the throat, lung and meninges respectively? Why is gonorrhoea confined to man and Johne's disease to cattle and related species? The two most likely explanations for differences in susceptibility to infection between different tissues or hosts, are differential distributions of bactericidal mechanisms and differential distributions of nutrients for which the metabolism of the parasite is especially adapted. Despite much effort attempts to lay the responsibility for specificities of single infections unequivocally on variation of defined bactericidal mechanisms have so far failed. There is some evidence, however, that kidney tissue is prone to a number of infections due to inhibition of complement and of phagocytosis by the high pH and salt concentrations respectively in this site.

More success has been achieved in investigations of the influence of nutrition. *Coryne-bacterium renale* and *Proteus mirabilis* persist in and cause severe damage to the kidney of cattle and man respectively. These localisations appear to be due to the possession of ureases which enable the bacteria to use urea for growth and for the production of ammonia which damages the tissue. Brucellosis in many animals (e.g. humans, rats, guinea pigs and rabbits) is a relatively mild and chronic disease; the causative organisms do not grow prolifically and have no marked affinity for particular tissues. However, in pregnant cows, sheep, goats and sows there is an enormous growth of brucellae in the placentae, the foetal fluids and the chorions, leading to the characteristic climax of the disease – abortion. Recent investigations have shown that the presence of erythritol, a growth stimulant for brucellae, in the susceptible tissues of susceptible species explains this tissue specificity in brucellosis. Thus, in cattle the foetal placentae, the chorion and foetal fluids contained more erythritol than the other foetal tissues and the maternal tissues contained no erythritol. Furthermore, the foetal placentae of cattle, goats, sheep and sows contained erythritol but not those of man, rats, guinea pigs and rabbits. Male genitalia of susceptible species contained erythritol thus correlating its presence with localisation of infection in the male. During growth *Brucella abortus* has a great affinity for erythritol; in a complex medium containing glucose at a concentration 1 000 times that of erythritol, it uses 1·5 times its own weight as a general energy source (the carbon from labelled erythritol was found in all cell constituents). Analogue studies showed the stimulation of growth by erythritol to be specific. Finally the S19 vaccine strain of *Br. abortus* which has been used safely without significant abortion in the field was inhibited by erythritol.

Studies on the same pattern as those on brucellosis have indicated that *Vibro fetus*, the cause of vibrionic abortion in domestic animals localises in the placental tissues for nutritional reasons but erythritol is not the growth stimulant. Also *Leptospira* species may localise in kidney tubules because of the high concentration there of glucose. A recent fascinating example of host resistance being determined by nutritional influences involves Brazilian strains of the plague bacillus *P. pestis* (Burrows and Gillett, 1971). These did not kill guinea pigs which are normally killed by *P. pestis* because, unlike normal strains, they required asparagine to grow well and guinea pig serum contains a powerful asparaginase. Clearly the biochemistry of tissue and host specificity is advancing.

3 Viruses

The mechanisms of virus pathogenicity are not clear. The first essential for studying the subject, quantitative comparison of the virulence of different strains, is inaccurate. Disease effects in animals (L.D.$_{50}$; lesion size) must be related to amounts of virus particles indicated by plaque formation or egg infection. These assays detect only a small proportion of the total virus particles and therefore may not measure the number of particles capable of multiplying in the experimental animals. Hence, only strains, for which the presently available tests have indicated the greatest possible difference in virulence, should be compared to recognise virulence markers and determinants.

Early virus attack of the respiratory tract has received some attention and the site of membrane lesions and the preferentially attacked cell types have been observed in Newcastle disease of chickens and influenza of experimental animals and man. Nevertheless, there appear to be few deeper investigations, for example, on the influence of oxygen tension, temperature, pH, commensals and mucus on virus survival on and replication in membranes; and also on the biochemical mechanisms of entry of various viruses into the particular epithelial cells they select for attack.

A virulent virus must be able to replicate in host cells and to spread from one to another; any change in these abilities will almost certainly result in changes in virulence. Investigations of the biochemistry of virus replication in animals are complicated by the obligate parasitism involved which not only entails complexity of the factors required for replication but increases the difficulty of distinguishing the influence of their absence from that of host factors (defence mechanisms) which actually destroy virus or inhibit replication. The ability of a virus to replicate in a particular cell depends on inherent features of that cell. These features can be involved in one or more stages of replication; attachment and penetration of virus, uncoating, provision of energy and precursors of low molecular weight, synthesis of viral nucleic acid and proteins, assembly and release. The characteristics of the host cell which determine these stages of replication might be called 'replication factors' and seem to be the counterparts in virology of the environmental factors such as low molecular weight nutrients, necessary for bacterial multiplication in host tissues or fluids. Tissue culture experiments show that 'replication factors' vary from cell type to cell type and are influenced by changes in environment of the cell. In animal infection, variation of availability of 'replication factors' in particular hosts or tissues and under different conditions will affect virus pathogenicity. Attenuated viruses may have a decreased capacity to use the factors. But few investigations of the influence of such factors comparable in depth to the experiments in tissue culture have been conducted either in animals or organ culture. Nevertheless, there are signs of this influence in the available studies. With regard to receptors, there was some parallel between the ability of homogenates of various primate tissues to bind polio-virus and their susceptibilities to virus replication and damage in infection and some attenuated strains adhered to susceptible nerve tissues less strongly than virulent strains. The effect of temperature on virus virulence probably reflects a temperature sensitivity of the enzymes used in virus synthesis rather than an influence on host defence mechanisms. And virus replication *in vivo* can be affected by low molecular weight materials; vaccinia virus infection of mice, like that in cell culture, was enhanced by injection of leucine.

As for bacterial infections, there are practically no detailed examinations of the early stages of viral infections. Host defence mechanisms appear to include inhibitors in serum, interferon (a virus inhibitor induced in host cells), antibody and macrophages. Polymorphonuclear leucocytes may be less important than in bacterial infections. Virulent viruses inhibit or do not induce host defence mechanisms. Virulent strains of influenza virus resist serum inhibitors more than avirulent strains and virulent strains of some but not all viruses induce less interferon than avirulent strains. Mouse macrophages are more readily infected by virulent than avirulent strains. But what virion constituents or virus induced products determine the superior resistance to host defence of virulent strains? Is some peculiarity of the envelope protein responsible for resistance to, or inhibition of, serum and phagocyte viricidins? Could host nucleases (which could destroy an invading virus) be inhibited by a particular folding of the viral nucleic acid or by an internal protein? What chemical difference determines induction of interferon by an avirulent strain in contrast to a virulent strain? Some viruses such as Newcastle Disease virus appear to produce interferon antagonists, and these may be the viral counterparts of the bacterial aggressins.

Viruses produce cytopathic effects in cells both in tissue culture and in disease. These effects could result from a passive role of the virus; cells might burst after acting merely as hosts for virus replication or die from the profound re-direction of metabolic processes required by such replication. Cell damage might also occur as a consequence of cessation of host cell macro-molecular synthesis, and even by direct 'toxic' action of viral products. Thus massive doses of some viruses such as influenza virus and pox viruses produce rapid toxic effects in animals and virion components such as the penton of adenovirus produce morphological damage in tissue culture. Furthermore in tissue cultures of polio-virus, influenza virus, mengo-virus, Newcastle Disease virus and vaccinia virus, replication of virus is not essential for cytopathic effects which require, however, virus-induced protein synthesis. Evidence is increasing that the virus-induced proteins which accompany these cytopathic effects are cytotoxins. Virus-free products from vaccinia virus-infected Hela cells have produced cytotoxic effects in fresh uninfected Hela cells in the presence of magnesium sulphate solution to increase membrane permeability thus allowing entry of the virus products (Woodward, Birkbeck and Stephen, 1972). Hypersensitivity also enters into the pathology of virus infections and auto-allergic (sensitisation of the host to host components) phenomena are more likely to occur in virus infections than in bacterial disease because many viruses incorporate host components into their structure. Deciding whether evocation of hypersensitivity reactions or direct toxicity is responsible for the main pathological effects of virus disease is not made easier by the present lack of knowledge of virus toxicity. Nevertheless, clinical observations and experimental studies in which pathological effects of virus diseases have been provoked or made worse by introducing into an infected host antibody or immune cells, indicate that hypersensitivity or auto-allergic reactions are involved in the pathology of some rashes, lymphocytic chorio-meningitis, dengue and viral encephalitides.

Tissue and host specificities occur in virus infections and are largely unexplained. Variation of host defence mechanisms and variation of 'replication factors' from tissue to tissue and host to host are the most likely reasons for the phenomena. Only in the classical work of Holland have we clear cut evidence; the susceptibility of appropriate (brain, anterior horn of spinal chord, intestine) primate tissues and the resistance of non-primate and some primate tissues to infection with polio-virus was correlated with the presence and absence of surface receptors. If the receptors were by-passed by using infectious polio-virus nucleic acid the virus grew in 'resistant' tissues.

4 Mycoplasmas

Mycoplasmas (micro-organisms without a rigid cell wall) such as *Mycoplasma pneumoniae* cause respiratory and urogenital tract infections in man and animals. The host-defence mechanisms which may operate against them on the surfaces are unknown. However, they may be adversely affected by products from other microbial commensals because of their extreme sensitivity to lysis by many factors. Estimations of relative virulence of mycoplasma strains are, like those of viruses, hampered by inaccuracy of viable counts. Nothing is known of factors determining multiplication *in vivo*. Materials lethal for mycoplasmas are found in tissue extracts and may include lysolecithin. Phagocytosis and intracellular killing of mycoplasmas by polymorphonuclear and mononuclear cells has been observed *in vitro* and *in vivo*. But whether mycoplasmas produce aggressins which interfere with these host defences is not clear.

Mycoplasmas may damage host tissues by similar mechanisms to those of bacteria and viruses, passively by excessive usage of nutrients such as arginine, or directly, by the production of hydrogen peroxide or neurotoxins such as those of *Mycoplasma gallisepticum* and *Mycoplasma neurolyticum*. Also close adherence of mycoplasmas to cells may change surface antigens leading to hypersensitivity or auto-immune effects. Nothing is known of the basis for tissue and host specificity of mycoplasmas.

5 Fungi

Many fungi produce animal diseases such as dermatitis, 'farmers lung', and mycotic abortion. Some have two morphological forms differing in virulence and they seem particularly appropriate for studies of mechanisms of virulence. Nevertheless we know very little about fungal pathogenicity.

Antifungal mechanisms operative on mucous and other body surfaces include inhibitory activity of bacterial commensals (since antibiotic treatment can result in fungal infection) and fungistatic materials in various secretions such as those of the conjunctiva and the fatty acids in teat secretions. Humoral antifungal factors including complement have been noted and killing by phagocytosis also appears important in some mycoses. The factors which determine the survival of fungi on mucous membranes are not known, nor those which determine replication in animal tissues. Many fungi are larger than phagocytes and mere size may prevent ingestion thus rapid growth alone can be an aggressive mechanism. A capsular polysaccharide prevents the phagocytic ingestion of virulent strains of *Cryptococcus neoformans* and like brucellae and tubercle bacilli, some fungi, e.g. *Histoplasma capsulatum, Coccidiodes immitis* and *Candida albicans* can survive and grow within phagocytes. Inhibition of intracellular killing may be due to fungal aggressins comparable to those produced by brucellae. Production of aggressins may be related to yeast forms; the non-invasive mycelial dermatophytes probably lack the powerful aggressins of the yeast-like fungi causing the deeper mycoses.

In some mycotic diseases, mechanical blockage by large mycelia could damage the host. Fungi produce in foodstuffs powerful toxins such as aflatoxin and sporidesmin but these have yet to be demonstrated in infection. Peptidases, collagenase and elastase appear to be involved in dermatophyte damage but hypersensitivity probably explains to a large degree the pathology of most fungal skin diseases.

Recently one example of tissue specificity in fungal infections, namely the growth of *Aspergillus fumigatus* in placental tissue which causes mycotic abortion in ewes and cattle may have been shown to have a nutritional basis although the fungal growth stimulant is yet unknown (White and Smith, 1973).

6 Protozoa

The microbial factors responsible for the pathological effects of protozoal infections have received little attention. Quantitative comparisons of virulence present no insuperable difficulties. The major difficulty in identifying virulence factors appears to lie in the versatility of protozoa. The different phases of their life cycles, their different morphological forms and their antigenic plasticity all have profound influences on pathogenicity. Keeping one strain in one form for repeated experiments on a particular problem in pathogenicity is a major operation.

The stimulating effect of host p-amino benzoic acid on malarial infections indicated that microbial nutrition influences protozoal pathogenicity. Host defence mechanisms against protozoal attack include humoral factors (antibody and complement) in some infections, e.g. trypanosomiasis, and phagocytosis in others, e.g. malaria and leishmaniasis. The mechanisms whereby protozoa avoid or inhibit host defence mechanisms are obscure. Trypanosomes appear to avoid lysis by antibody and complement and malarial parasites destruction by macrophages by changing their antigens *in vivo* as infection proceeds. The high mobility of some protozoa may prevent phagocytosis and entamoebae appear to produce factors which kill leucocytes. But the author is not aware of a protozoal product which prevents ingestion by phagocytes. *Toxoplasma gondii* produces a factor which promotes penetration of host cells and, as for bacterial aggressins, enhances its virulence for mice. Because *Leishmania* species survive and grow within phagocytes they may produce as yet unidentified aggressins similar to those of brucellae.

The overall toxic effects of protozoa have been investigated, especially those of *Plasmodium* species; for example, red cell destruction, anuria and circulatory failure. Clearly cell destruction by intracellular growth of protozoa could contribute to damage in such diseases as malaria and leishmania. A microbial toxin has not yet been recognised as being unequivocally responsible for the main pathological effects of any protozoal disease. In malaria increase of capillary permeability leading to shock and brain damage appear to be mediated by kallikrein, kinins and adenosine but a malarial toxin responsible for the release of such host products has not been demonstrated. The lytic effects of *Entamoeba* species are known but the mechanisms of cell destruction are not clear; lysosomal enzymes may be transferred from the entamoebae to the host cells by tubules formed between the two cells on contact. Toxic products of trypanosomes and toxoplasmas are known but their importance in disease is not clear. Hypersensitivity and auto-allergic phenomena occur in protozoal infections but their responsibility for important pathological effects is hard to judge. In malaria antibodies to host antigens changed by red blood cell parasitisation may, by opsonisation, promote phagocytosis and destruction of unparasitised red blood cells. Tissue and host specificity in protozoal infection is so far unexplained in biochemical terms.

In conclusion I hope this survey over the field of microbial pathogenicity will provide the student of biochemistry:

1. With the evidence from bacteriology that the phenomena which cause so much trouble to mankind can be explained in biochemical terms; and

2. With the feeling that much biochemical work is needed on the pathogenicity of microbes other than bacteria.

References

Further reading

AJL, S. J., KADIS, S. and MONTIE, T. C. (1970). *Microbial Toxins.* Academic Press, London and New York.

DUBOS, R. J. and HIRSCH, J. G. (1965). *Bacterial and Mycotic Infections of Man*, 4th ed. Lippincott, Philadelphia.

DUNLOP, R. H. and MOON, H. W. (1970). *Resistance to Infectious Disease.* Modern Press, Saskatoon.

HOWIE, J. W. and O'HEA, A. J. (1955). *Mechanisms of Microbial Pathogenicity.* Cambridge University Press, Cambridge.

MIMS, C. A. (1964). 'Aspects of the pathogenesis of virus diseases', *Bact. Rev.* **28**, 30-71.

SMITH, H. (1960). 'The biochemical response to bacterial injury', in *Biochemical Response to Injury.* Blackwell, Oxford and Edinburgh, p. 341.

SMITH, H. (1968). 'The biochemical challenge of microbial pathogenicity', *Bact. Rev.*, **32**, 164-84.

SMITH, H. (1972). 'Mechanisms of virus pathogenicity', *Bact. Rev.*, **36**, 291-310.

SMITH, H. and PEARCE, J. H. (1972). *Microbial Pathogenicity in Man and Animals.* Cambridge University Press, Cambridge.

SMITH, H. and TAYLOR, J. (1964). *Microbial Behaviour in vivo and in vitro.* Cambridge University Press, Cambridge.

SMITH, W. (1963). *Mechanisms of Virus Infection.* Academic Press, London and New York.

TAMM, I. and HORSFALL, F. L. (1965). *Viral and Rickettsial Infections of Man*, 4th ed. Lippincott, Philadelphia.

WILSON, G. S. and MILES, A. A. (1964). *Topley and Wilson's Principles of Bacteriology and Immunity*, 5th ed. Edward Arnold, London.

Recent references cited

D'ARCY HART, ARMSTRONG, J. A., BROWN, C. A. and DRAPER, P. (1972). 'Ultrastructural study of the behaviour of macrophages towards parasitic mycobacteria', *Inf. Imm.*, **5**, 803-7.

BURROWS, T. W. and GILLETT, W. A. (1971). 'Host specificity of Brazilian strains of *Pasteurella pestis*', *Nature, Lond.*, **229**, 51-52.

FROST, A. J. SMITH, H., WITT, K. and KEPPIE, J. (1973). 'The chemical basis of the virulence of *Brucella abortus.* X. A surface virulence factor which facilitates intracellular growth of *Brucella abortus* in bovine phagocytes', *Brit. J. Exp. Path.* In press.

UCHIDA, T., PAPPENHEIMER, A. M. Jr. and HARPER, A. V. (1972). 'Reconstitution of diphtheria toxin from two non-toxic cross-reacting mutant proteins', *Science*, **175**, 901-3.

WHITE, L. O. and SMITH, H. (1973). 'Placental localisation of *Aspergillus fumigatus* in bovine mycotic abortion', *J. Gen. Microbiol.* In press.

WOODWARD, C. G., BIRKBECK, T. H. and STEPHEN, J. (1972). 'Possible vaccinia virus cytotoxic factors', *J. Gen. Microbiol.*, **71**, V.

16
Protozoa as Tools for the Biochemist

B. A. Newton
Medical Research Council Biochemical Parasitology Unit, The Molteno Institute,
University of Cambridge

1 Introduction

Protozoa are, in the author's opinion, the most exciting of all microorganisms a biochemist could choose to study; they offer him an almost infinite variety of experimental material and pose challenging problems in every field of his science. The phylum includes most eukaryotic unicellular organisms which move; they range in size from a few microns to several centimetres and in complexity from relatively undifferentiated amoebae to highly specialised ciliates. The latter may contain, within the confines of a single cell, specialised feeding apparatus, light sensitive organelles, digestive and contractile vacuoles, both a micro and macro nucleus, locomotory structures, a rudimentary nervous system and defensive weapons. Some protozoa are ideally suited to microsurgical techniques, some undergo dramatic changes in structure and metabolism in the course of a complex life cycle in two hosts; such organisms provide ideal experimental systems for the investigation of factors initiating and controlling differentiation in single cells. Other protozoa (about 20 different species) are important pathogens of man and domestic animals, many cause diseases which have an almost worldwide distribution and, at any one time, probably infect a quarter of the human race: these organisms still pose many problems for the chemotherapist and immunologist.

In the last 20 years protozoa have attracted the attention of biochemists to an increasing extent and some have beome established as the 'organisms of choice' for certain types of investigation (e.g. the use of amoebae for the study of nuclear/cytoplasmic relations), but in general, knowledge of their biochemistry still lags far behind that of other microorganisms. The aim of this chapter is to discuss some of the problems of working with these organisms, to emphasise some of their virtues and outline some of the ways in which they are being employed to investigate basic biochemical problems.

2 Classification and general characteristics of protozoa

Protozoologists have had a particularly difficult task in devising a satisfactory scheme for classifying protozoa due to the diversity of the organisms which make up the phylum. A dozen major schemes have been proposed since 1900, all have numerous limitations and some taxonomic confusion still exists. However, a detailed discussion of these problems is not relevant to this chapter; it will be sufficient to describe some of the general characteristics of protozoa which help to distinguish them from other microorganisms. Readers interested in more detailed aspects of protozoal taxonomy are referred to Hall (1953) and Honigberg *et al.* (1964).

If we accept the views of the German zoologist, Haeckel (1866) — and it should be stressed that not all biologists do! — protozoa form a phylum in the kingdom Protista. This kingdom, according to Haeckel, is divided into two major sections on the basis of the presence or absence in cells of a discrete nucleus bounded by a nuclear membrane. All bacteria and blue green algae (i.e. prokaryotic cells which lack a discrete nucleus) form the lower protista and other algae,

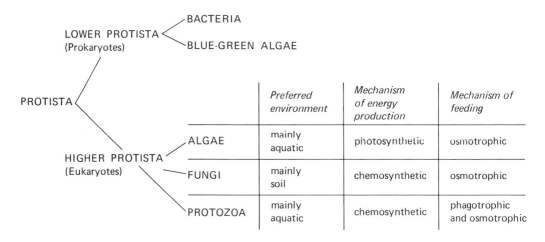

Fig. 16.1 Constitution of the kingdom Protista and some characteristics of higher protista.

protozoa and fungi (all eukaryotic cells with a well defined nucleus) form the higher protista. If we consider the most common habitats of higher protista, together with their mechanisms of feeding and of obtaining energy and whether or not they are motile (Fig. 16.1), it is possible to arrive at a description of protozoa which will distinguish them from other higher protista. They can be defined as motile, unicellular, chemosynthetic, phagotrophic, eukaryotes which are generally found in an aquatic environment.

It is of course impossible to formulate a definition of protozoa which will exclude all other microorganisms. Some photosynthetic organisms can also feed phagotrophically; some parasitic protozoa are osmotrophes; many protozoa form cysts under adverse conditions which, in contrast to the endospores of some bacilli, are bounded by a protective wall external to the cell's boundary membrane; however the myxobacteriales form external cystic membranes like protozoa. Many other examples which do not fit the above definition will be found in any textbook of protozoology. However, all protozoa are motile at some stage of their life cycle and movement may be amoeboid or by means of flagella or cilia. The type of motility shown by these organisms has formed a basis for the division of the phylum into four major groups: the sarcodina, the mastigophora, the sporozoa and the ciliophora. Before going on to discuss the biochemistry of protozoa it is essential that the reader should be familiar with a few of the general characteristics of these major groups of protozoa.

2.1 The Sarcodina

The group includes the simplest amoeboid protozoa which are of indefinite form and capable of extending pseudopodia in any direction. They feed by engulfing smaller microorganisms, particularly bacteria, which are digested within a food vacuole. There is no particular area in these cells which is organised for the purpose of feeding: ingestion of food can occur at any point on the cell surface and a food vacuole can form at any point in the cytoplasm. A conspicuous feature of these organisms is the contractile vacuole which is believed to control the water balance of the cell. This vacuole fills with fluid and discharges its contents into the environment, a new vacuole then forming elsewhere in the cytoplasm. Most of the Sarcodina are free living forms and are found in soil or water; some of these (*Hartmanella* and *Naegleria*) can be pathogenic to man and if they have the opportunity to invade the nasal mucosa and become established in the brain they produce acute amoebic meningoencephalitis. Other amoebae have beome obligate parasites of higher animals, an example being *Entamoeba histolytica* which causes amoebic dysentery in man, a disease which is widespread in tropical regions. In most simple amoebae reproduction is exclusively by binary fission but in some of the more complex marine forms such as the Foraminifera sexual reproduction occurs. The Foraminifera are able to secrete a simple protective shell around themselves while another marine organism, *Radiolaria* supports its cytoplasm on an elaborate siliceous endoskeleton.

2.2 The Mastigophora

All members of this group move by means of flagella; they range from forms with a single nucleus and flagellum, some of which are colourless counterparts of photosynthetic algae, to highly specialised multinucleate (polymastigotes) and multiflagellated (hypermastigotes) organisms. Some hypermastigotes lead a symbiotic existence, e.g. *Trichonympha* which occurs in the gut of termites where it provides a vital service to its host by converting cellulose into utilisable carbohydrate. Many flagellates are parasitic, some (monogenetic species) have one host in their life cycle, others (digenetic species) have two. Hosts of monogenetic flagellates are generally arthropods, usually insects and the parasites inhabit their gut. The second hosts of digenetic species are vertebrates of all classes in which the parasites inhabit the blood and other tissues. Digenetic species of *Trypanosoma* and *Leishmania* are of great medical and economic importance causing diseases such as sleeping sickness (African trypanosomiasis), Chagas disease (South American trypanosomiasis), oriental sore and visceral leishmaniasis. In all these flagellates sexual stages are unknown and multiplication is by binary fission.

Many schemes of classification split the mastigophora into plant-like flagellates which possess chlorophyll (*Phytomastigina*) and animal-like flagellates which may contain a glycogen-like rather than a starch-like polysaccharide as storage carbohydrate (*Zoomastigina*). The general definition of protozoa given in the previous section excludes photosynthetic flagellates from this discussion and allows the botanists to claim them as algae!

2.3 Sporozoa

All species of sporozoa are obligate parasites, they all have complex life cycles and they vary widely in morphology and size (*Plasmodium rouxi*, a parasite of birds is amongst the smallest with a diameter of 1–2 μm and *Porospora gigantea*, a parasite of crustacea and molluscs, may exceed 12 mm in diameter). Like the mastigophora mono- and digenetic species of sporozoa are known, also like the flagellates some sporozoa are of great medical and economic importance: malaria parasites (plasmodia), it is estimated, cause disease in one-eighth of the world's population at any one time; the coccidia are lethal to certain domestic animals, some cnidosporidia cause serious fish diseases and others are highly pathogenic to bees and silk-worms.

It is difficult to define the sporozoa in terms of morphology, they vary greatly and different stages in the life cycle of a single species may show little resemblance. One general characteristic of the subphylum is a lack of locomotory organelles in the adult stages; however, many species are capable of independent movement at certain periods of their developmental cycles. For example, microgametes are often flagellated, zygotes of some species may move even though they have no locomotory organelles, some stages in the life cycle of myxosporida show active amoeboid movement and gregarines have an ability to glide by a mechanism which is not fully understood. Life cycles of all sporozoa have both sexual and asexual phases of development.

2.4 Ciliophora

Free swimming ciliated protozoa move by a co-ordinated beating of large numbers of short cilia which generally cover their whole surface. They are further characterised by a unique type of nuclear organisation, each cell containing two types of nucleus which differ in size and function. The small 'micronucleus' is responsible for genetic continuity through the sexual phase of the life cycle. During this phase the larger 'macronucleus' breaks down and disappears, a new one subsequently being formed from the micronucleus. During asexual division both types of nuclei divide synchronously with cell division. The macronucleus is polyploid and may contain more than a hundred times the amount of DNA present in the micronucleus; it is genetically active in the sense that genes localised in it determine the phenotype of the ciliate but it divides amitotically. Amicronocleate strains of ciliates occur and in these the macronucleus ensures normal function of the cell with the exception that it loses its power to conjugate. Amacronucleate strains do not occur. Some genera of this subphylum include reproductively isolated groups or varieties made up of mating types (e.g. *Paramecium*).

In most ciliates the mouth or cytostome is a conspicuous feature. It may have associated with it highly developed undulating membranes and membranelles the movement of which guides food particles into the cytostome. Food enters the cytoplasm by becoming enclosed into a digestive vacuole which may be seen to follow a definite path through the cytoplasm, eventually reaching a pore or anal opening through which undigested material is excreted. Many ciliates contain specialised organelles such as the dischargeable trichocyst of *Paramecium* which may serve as a defence mechanism. Not all ciliates are free living throughout their life cycle. In some forms (suctorians) the mature stages are sessile and capture food by means of tentacles. The surface or cortical layers of ciliates (e.g. *Stentor coeruleus*) may contain longitudinal or spiral myonemes which are capable of rapid contraction. Sessile forms reproduce by budding off small ciliated cells which swim to a new site and anchor themselves.

As in the other three major groups of protozoa described there are many parasitic species of ciliates; *Balantidium coli* may produce a serious or fatal infection in man, there are important and deadly fish parasites amongst the genus *Ichthyophthirus*. Certain parasites of crustacea (e.g. Astomatida) are remarkable for the fact that some stages in their life cycles, such as division or encystment, are determined by physiological changes and sexual activity in their hosts. In all there are thought to be some 6000 species of ciliates.

3 The problems of obtaining a pure culture

Meaningful studies on the nutrition or metabolism of a microorganism only become possible when it is available in pure (axenic) culture. Obtaining such cultures is one of the major factors limiting the development of biochemical studies of protozoa. The nature of the problem varies from one protozoan species to another and is greatly influenced by the organisms natural habitat and feeding habits. Trypanosomes develop in the bloodstream of their mammalian hosts, generally in the absence of other microorganisms, and axenic cultures of these flagellates can often be established relatively easily by aseptic transfer of blood from an infected host to a suitable growth medium. A diphasic preparation composed of a blood nutrient–agar slope covered by a liquid layer composed of a suitably buffered salts solution containing glucose, is the classical medium for such isolations. Growth of trypanosomes in such a diphasic medium results in changes in their morphology and metabolism so that the organisms differ significantly from the 'bloodstream forms' from which they were derived. This and other aspects of differentiation in these flagellates will be discussed in detail in a later section. Some species of trypanosomes grow to such high densities in the blood stream of laboratory rodents that it is possible to obtain sufficient quantities for biochemical studies by simply separating them from blood constituents either by differential centrifugation or by passing infected blood through a suitable ion-exchange column (e.g. DEAE-cellulose) which adsorbs the more negatively charged components (Lanham 1968). Other parasitic protozoa, for example those found in the intestinal tract such as *Entamoeba histolytica*, and free living protozoa pose a much more difficult problem of isolation since they invariably have to be completely separated from other types of microorganisms which share their environment. Unfortunately techniques developed for the isolation of bacteria, such as the use of selective growth media or cloning on agar plates, are not generally applicable to protozoa but the following techniques have been used with varying degrees of success.

3.1 Isolation of single cells

Single cells of many protozoa have been isolated by micromanipulation and have been used to establish an axenic culture by transferring them to a suitable sterile growth medium. However, this technique requires sophisticated equipment and considerable technical skill, also the failure rate is high either because the single cells fail to divide in their new environment or because bacteria adhere to the surface of the protozoal cell.

3.2 Bulk separation of protozoa from other microorganisms

Differential centrifugation and filtration through columns of powdered quartz or, more recently, ion-exchange resins have been used to separate large numbers of protozoal cells from other microorganisms. Motile protozoa have been separated from non-motile contaminants by encouraging them (by application of a potential difference to a suspension so that all the protozoa move in the same direction) to move through a capillary tube bent into a series of 'S' shapes; the non-motile organisms settle under gravity and collect in the lower bends. However, all of these procedures may fail to remove adherent contaminants.

3.3 Differential killing

The technique now most widely used is based on selective killing of contaminating organisms. A study of the sensitivities of protozoa and other microorganisms to heat, desiccation, X-rays, heavy metals and other toxic chemicals has allowed selective treatments to be developed for particular organisms. The most successful procedure relies on the fact that protozoa in general are not sensitive to antibiotics which kill bacteria. Treatment of contaminated protozoal suspension with a 'cocktail' of penicillin, streptomycin and chloramphenicol removes most

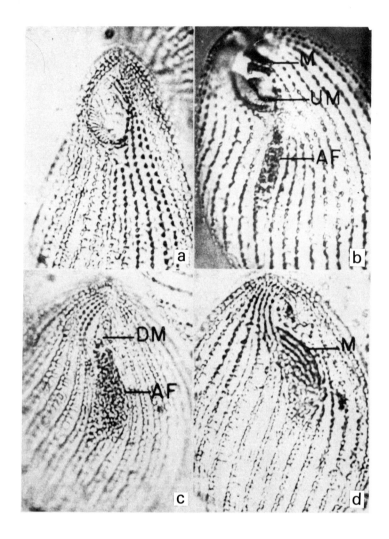

common bacterial contaminants. Unfortunately this approach is not satisfactory if yeast or moulds are present as some protozoa are sensitive to antifungal antibiotics. Under these circumstances a combination of the methods described has to be used.

4 Effects of *in vitro* cultivation on the morphology and morphogenesis of protozoa

Once an axenic culture of a protozoan has been established in a growth medium which will yield large numbers of organisms the way is clear for detailed biochemical and nutritional studies of what protozoologists generally refer to as 'culture forms'. Much basic information can be obtained from such studies but if we begin to enquire into the relationship of the forms growing in a laboratory culture to the forms found in their natural environment a whole range of new problems are uncovered. For example when some parasitic protozoa which have a

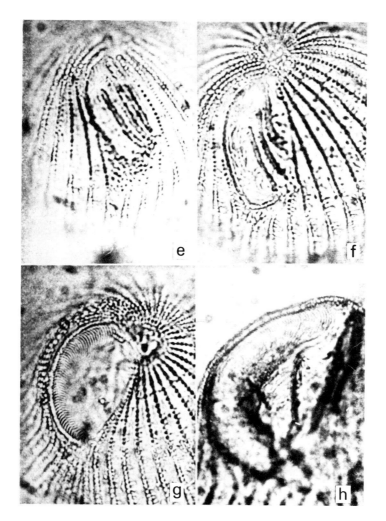

Plate 16.1 Successive stages in oral replacement in *Tetrahymena patula* (reproduced from Stone 1963). Cells have been stained by the Chatton–Lwoff silver technique. (*a*) Shows oral area of a microstome before transfer from 25° to 15°C. Note small size of the oral area; (*b*) and (*c*): kinetosomal field of forming mouth parts (AF) with reabsorption of old mouth parts (DM) including old membranelles (M) and undulating membrane (UM); (*d*): longitudinal organisation of membranelles (M); (*e*): organisation of undulating membrane now almost complete; (*f*) and (*g*) show a deepening of the oral cavity and elongation of the oral area; (*h*): fully formed microstome with large buccal cavity.

complex life cycle, perhaps involving more than one host, are cultured in the laboratory the normal life cycle is curtailed and only one particular developmental form grows. Comparison of such 'culture forms' with other developmental forms and investigation of factors influencing transformation of one form into another is now providing biochemists with model systems for the investigation of a number of morphogenetic problems. Two examples of such systems with which considerable progress has been made in recent years are provided by the free living ciliate *Tetrahymena* and the parasitic flagellate *Trypanosoma brucei*.

4.1 Microstome–macrostome transformation in ciliates

Ciliates of the genus *Tetrahymena* can occur in three distinct morphological forms. In pond water they are usually 200–250 μm in length and have a large oral opening. These forms are called macrostomes and they lead a carnivorous existence feeding on other ciliates. When macrostomes are adapted to grow in axenic culture in a liquid medium they develop as much smaller cells, about 30 μm long, with a small oral opening (Kidder *et al.* 1940); these are called microstomes. The third distinct developmental form of these ciliates is a reproductive cyst. Under certain conditions in culture microstomes can be induced to transform into macrostomes without passing through the reproductive cyst stage or undergoing cell division. This transformation is a function of the population growth cycle (Stone 1963); during the logarithmic phase of growth the ciliates are all in the microstome form but when division ceases (i.e. the stationary phase of the growth cycle) they become predominantly macrostomes. Examination of silver-stained preparations shows that this transformation involves a process of oral replacement in which the existing mouth parts and associated undulating membrane are reabsorbed and kinetosomes (basal granules of ciliates, capable of division and thought to contain DNA, which can give rise to cilia, fibrils, trichocysts and mouthparts) below the original opening become organised into longitudinal rows ultimately giving rise to new enlarged mouth parts and an undulating membrane. (Plate 16.1 (a)–(h)). These polymorphic properties would make *Tetrahymena* an attractive organism for biochemical and physiological studies of morphogenesis providing populations of one type of ciliate could be obtained and induced to transform under controlled conditions. Stone (1963) achieved this by controlling the temperature and pH of cultures; in stationary phase cultures grown at 25°C with pH maintained at a value of 7·5 only 30 per cent of the ciliates formed macrostomes; lowering the pH value to 6·5 or raising it to 8·5 resulted in 100 per cent transformation. Complete transformation of a population was also obtained by lowering the temperature of cultures to 20°C. Under either of these conditions microstomes transformed into macrostomes in the absence of cell division, the whole process taking about 3–6 h. When macrostomes from a stationary phase culture are transferred to fresh growth medium they divide giving two microstomes. This reverse transformation however has not been observed to occur in the absence of cell division. It has been suggested that in axenic culture transformation may be an overcrowding phenomenon or in the presence of suitable food organisms, a response of a predator to its prey, the latter producing a substance which triggers the transformation process. Separation of microstomes from food organisms by a semipermeable membrane stimulates transformation and a transforming principle (called stomatin) has been isolated from *Tetrahymena pyriformis* (the prey) which will induce transformation in *Tetrahymena vorax* (the predator) (Claff 1947; Buhse 1967; Buhse and Cameron 1968). The chemical nature of stomatin remains unknown; it is water soluble, has a low molecular weight and is not digested by a number of lytic enzymes. Incorporation experiments using radioactive precursors of nucleic acids and proteins have shown that both stomatin and temperature induced transformation are preceded by increased RNA and protein synthesis but there is no evidence that nuclear DNA synthesis occurs. Transformation is prevented by inhibitors of RNA and protein synthesis such as cyclohexamide, puromycin and actinomycin (Frankel 1970) and a study of the sequence of events suggests that there is a clear cut dissociation between RNA synthesis and the visible reorganisation of cell structures associated with the formation of a macrostome. It seems that assembly of microtubules plays an important role in transformation since colchicine, which binds specifically to microtubule protein, inhibits the process. Now that microstome–macrostome differentiation can be controlled in axenic culture it provides the biochemist with an ideal system for investigating the molecular basis of a complex cellular reorganisation process which proceeds in the absence of nuclear DNA synthesis and cell division.

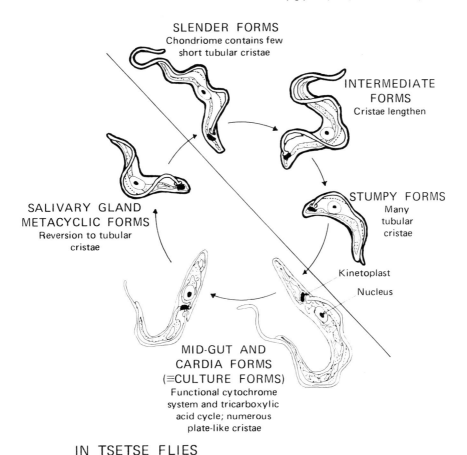

IN MAMMALS

BLOODSTREAM FORMS: Metabolise glucose mainly to pyruvate; terminal respiration by glycerophosphate oxidase system

SLENDER FORMS
Chondriome contains few short tubular cristae

INTERMEDIATE FORMS
Cristae lengthen

STUMPY FORMS
Many tubular cristae

SALIVARY GLAND METACYCLIC FORMS
Reversion to tubular cristae

Kinetoplast

Nucleus

MID-GUT AND CARDIA FORMS (≡CULTURE FORMS)
Functional cytochrome system and tricarboxylic acid cycle; numerous plate-like cristae

IN TSETSE FLIES

Fig. 16.2 Diagram of developmental stages of *Trypanosoma brucei* to show changes in the surface coat, state of development of the mitochondrion and position of the nucleus and kinetoplast (based on Vickerman 1971). The slender bloodstream form lacks a functional cytochrome system and tricarboxylic acid cycle (TCAC). Some TCAC enzymes have been detected in stumpy forms but cytochromes are still missing. Terminal respiration in bloodstream forms involves a glycerophosphate oxidase system. Forms which occur in the tsetse midgut and in culture have an active cytochrome system and TCAC; they also have a well-developed mitochondrial network containing many plate-like cristae. The forms which occur in the salivary gland have a less extensive mitochondrial structure. Forms which have a surface coat are indicated by a thick outline. The transformation from slender to stumpy forms occurs within each relapse population in pleomorphic strains of *T. brucei*. When trypanosomes enter the tsetse all variants lose their antigenic identity and revert to a basic antigenic type; they re-acquire a surface coat and become metacyclic forms in the salivary glands. (Reproduced from Newton 1972.)

4.2 Differentiation in *Trypanosoma brucei* ssp.

Trypanosoma brucei ssp. develop in the tsetse fly (*Glossina* sp.) and mammalian host (man, cattle or game animals in Africa) during its natural life cycle. The mammalian host becomes infected when the fly takes a blood meal; metacyclic forms of the flagellates pass from the salivary glands of the fly to blood and tissue fluids via the fly's proboscis. In the early stages of an infection the majority of trypanosomes found in the bloodstream are long slender forms (Fig. 16.2) with the kinetoplast close to the posterior end; later there is a preponderance of shorter stumpy forms. If such trypanosomes are removed from the bloodstream and incubated at 37°C in a diphasic medium of the type already described they do not continue to divide, but lowering the incubation temperature to 25°–28°C permits growth and the 'culture' forms which develop resemble not the bloodstream forms from which they were derived but the stage found in the midgut of the tsetse fly, with the kinetoplast adjacent to the nucleus. These forms are generally non-infective to mammals.

Comparison of 'bloodstream' and 'culture' forms of *T. brucei* has shown that they differ not only in morphology but also in intermediary metabolism chemical composition and ultra-structure. Long slender 'bloodstream' forms oxidise glucose very rapidly (as much as their own dry weight per hour) but incompletely, 80 per cent of the carbon being recoverable as pyruvate; this is due to a lack of pyruvate oxidase and a non-functional tricarboxylic acid cycle. Uptake of oxygen by these organisms is insensitive to cyanide and they do not contain detectable amounts of cytochromes. The reduced NAD produced during glycolysis is reoxidised by an

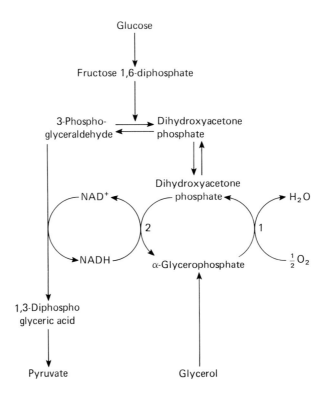

Fig. 16.3 Role of L-α-glycerophosphate oxidase (1) and dehydrogenase (2) in the metabolism of glucose by *Trypanosoma brucei* spp. (from Grant 1966).

unusual mechanism involving L-α-glycerophosphate oxidase (Fig. 16.3). In contrast the respiration of 'culture' forms is sensitive to cyanide; they contain cytochromes and a functional tricarboxylic acid cycle and so are able to oxidise glucose to carbon dioxide and water. Cytochemical studies and electron microscopy have shown that these differences in respiratory metabolism are associated with changes in the structure of the kinetoplast–chondriome complex of the flagellates. This complex is now generally regarded as a giant mitochondrion in which the mitochondrial DNA is localised in one small area. Electron microscopy shows that in slender bloodstream forms this is a simple tubular structure lacking normal mitochondrial cristae but in 'culture' forms it appears as a highly developed network containing many 'platelike' cristae (Fig. 16.2). These findings raise the question: do the changes in respiratory activity and mitochondrial structure occur in response to the change in environment encountered when bloodstream forms are transferred to a culture medium or, in the natural life cycle, to the tsetse fly gut in a blood meal? Evidence from studies of ultrastructure and respiratory activity of both long slender and short stumpy bloodstream forms indicate that this is not so (Vickerman 1965); at least partial activation of mitochondrial enzymes occurs while the flagellates are still in the bloodstream and short stumpy forms are able to metabolise α-ketoglutarate. What initiates these changes in the bloodstream remains unknown. It seems that acquisition of a functional tricarboxylic acid cycle and cytochrome system by stumpy forms preadapts them for life in the tsetse gut or in culture – slender forms will not grow in either situation. The presence of kinetoplast DNA is thought to be mandatory for the 'respiratory switch', organisms which have lost this DNA, either as a result of spontaneous mutation or treatment with mutagenic drugs such as acridines do not infect tsetse, cannot be cultured *in vitro* and have never been shown to contain a functional cytochrome system. However, these observations do not establish that kinetoplast DNA contains the genetic information required for synthesis of some or all of the oxidative enzymes present in culture forms but absent in bloodstream forms of *T. brucei*. Kinetoplast DNA has been isolated and purified; like mitochondrial DNA from other cell types it contains small circular molecules, but these are about one-tenth the size of other mitochondrial DNA's; they range from 0·3 to 0·7 μm in contour length in different species of trypanosome (Simpson 1972). Physical studies indicate that the circular molecules of a given species are very homogeneous in base composition, thus the genetic information encoded in a kinetoplast may only be sufficient to determine the amino-acid sequence of three or four small proteins. This would be a very minor contribution to the enzyme requirement for a functional tricarboxylic acid cycle and respiratory chain. However, it is premature to do more than speculate about this at present since the possibility that some kinetoplast DNA exists as long linear molecules cannot yet be ruled out (for further discussion see Newton, Cross and Baker 1973).

Clearly *T. brucei* provides a good illustration of the point made at the beginning of this section, namely that the form and physiology of a protozoan developing in a laboratory culture may differ strikingly from other forms which occur in the natural developmental cycle. These trypanosomes also provide the biochemist with a unique experimental system in which there is a cyclical development and regression of mitochondrial structures and enzymes: further study will certainly yield valuable information about initiation and control of mitochondriogenesis.

5 Interrelationships between DNA, RNA and histone synthesis in a ciliate macronucleus

Protozoa display a greater variety of nuclear organisation than any other cell type. In discussing the general characteristics of protozoa it was pointed out that a unique feature of ciliates is the presence of a macronucleus which plays an important role in the determination of phenotype and yet divides amitotically. The macronucleus of hypotrich ciliates has attracted particular attention because it undergoes unusual structural changes during the 'S phase' of the cell cycle. Gall (1959) and Prescott and Kimball (1961) made use of this feature to investigate whether the processes of RNA and DNA synthesis are mutually exclusive. They used *Euplotes eurystomus*, a ciliate commonly found in pond water. In common with other ciliates this organism has a small micronucleus and a large densely staining macronucleus, but the latter is

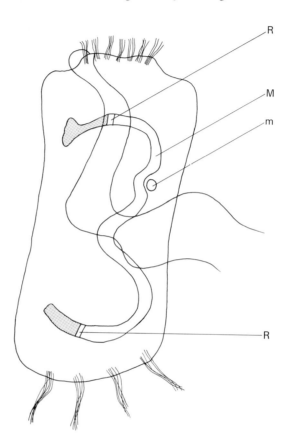

Fig. 16.4 Ventral or oral view of *Euplotes eurystomus* showing the small micronucleus (m) located in a notch in the W-shaped macronucleus. Near each end of the macronucleus (M) reorganisation or replication bands (R) are shown. The stippled regions distal to the bands have completed the duplication of DNA and histone.

unusual in being a long ribbon-like structure, about 140 x 7 μm, which frequently assumes a C- or W-shape in the cytoplasm (Fig. 16.4). The macronucleus contains about 200 times as much DNA as the micronucleus. Cell division of this ciliate involves the macronucleus pinching in half amitotically, the micronucleus dividing mitotically and finally transverse fission of the cell. Several hours prior to cell division the macronucleus undergoes a process which has been termed 'reorganisation'; the beginning of the process is indicated by the appearance, at each end of the nucleus, of a transverse band which stains very weakly with methyl green and other nuclear stains. Over a period of 8–12 h these two 'reorganisation or replication bands', as they

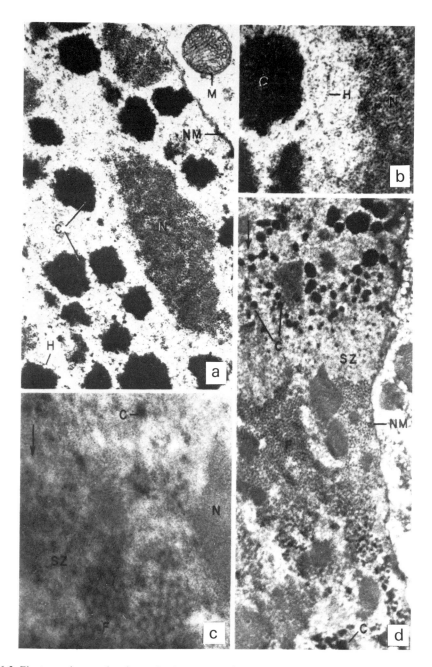

Plate 16.2 Electron micrographs of areas in the macronucleus of *Euplotes eurystomus*. (*a*): the small very dense particles are due to lanthanum acetate used in staining, it deposits selectively on the chromatin bodies (c) and not on the nuclear bodies of lower density (N). M, mitochondrion in cytoplasm; NM, nuclear membrane; H rows of particles thought to be sectioned helices; (*b*): Similar area to (*a*) but a higher magnification; (*c*): a preparation stained with ferric chloride showing 'replication or reorganisation band' SZ, i.e. the zone of DNA synthesis. The band is moving down the macronucleus in the direction of the arrow. The chromatin bodies (c) at the bottom of the figure have begun to break up and form a network of twisted microfibrils (F); (*d*): higher magnification micrograph of the zone of DNA synthesis (SZ). This zone is immediately followed by a region where minute microfibrils appear to be aggregating to form small chromatin bodies (C). (Reproduced from Kluss 1962.)

are called, move slowly towards one another and finally meet near the centre of the nucleus. Marked differences in the appearance of the chromatin on either side of the reorganisation band have been detected by electron microscopy (Plate 16.2); initially it is finely granular but in the areas through which the bands have passed the granules become much larger. Nuclei stained selectively for DNA (Feulgen) and histone (alkaline fast green) show that the concentrations of both these nuclear components is much increased in the regions of the nucleus distal to the reorganisation bands. This observation suggests that both DNA and histone synthesis commences at the tips of the nucleus and proceeds towards the middle. Photometric measurements and autoradiography of cells pulse-labelled with tritiated thymidine have shown that, in the case of DNA this hypothesis is correct; ^{3}H-thymidine is only incorporated in a small area immediately distal to the reorganisation bands and the average amount of Feulgen dye bound by the nucleus rises as the replication bands move towards the centre and is doubled by the time they meet. These results suggest that the replication band is an area in which there is a temporary reorganisation of chromatin accompanied by DNA synthesis. Electron microscopy has shown the presence of two regions in the replication band, one at the forward border where chromatin granules become broken down and the DNA organised into a regular array of 150 Å fibrils, and a second zone behind this area where the fibrils become more dispersed and so finely divided that they cannot be resolved as individual structures in the electron microscope. It is thought that the DNA in this region is in a very hydrated form. This change in molecular configuration may account for the low affinity of this area for stains such as Feulgen and methyl green.

Further experiments in which nuclei were stained for histone or labelled with ^{3}H-basic amino acids indicate that basic proteins undergo duplication in a similar manner to DNA and that the two processes are intimately associated. But whether histones are in fact synthesised at the site of DNA synthesis or are accumulated from the cytoplasm and subsequently bound to newly synthesised DNA cannot be decided from these experiments. There is now evidence from other cell systems that most, if not all, histone is of cytoplasmic origin and a histone messenger-RNA has been isolated. Experiments using ^{3}H cytidine have established that RNA is destroyed or eliminated from the macronucleus just ahead of the advancing reorganisation band. Neither RNA nor RNA synthesis can be detected in the zone of DNA synthesis but in the distal area, where reaggregation of chromatin has occurred, RNA synthesis is resumed. At the time these findings were made the direct involvement of DNA in RNA synthesis was still a matter for speculation and these experiments provided an elegant demonstration that nuclear RNA synthesis is 'switched off' during DNA synthesis. Many further questions about DNA synthesis in ciliate macronuclei remain to be answered, particularly those concerning the regulation of the amount of DNA synthesised in a nucleus which divides by an amitotic mechanism. It is well known that exactly equal distribution of macronuclear DNA between daughter cells is not essential to viability. During division of these nuclei some DNA forms a bridge between two daughter nuclei and may become detached into a separate vesicle and ultimately degraded in the cytoplasm. There is evidence that underendowed daughter cells can make up their macronuclear DNA to the correct level before the next cell division. Whether partial DNA synthesis is achieved by restricting replication to a limited number of the many genomes in the polyploid macronucleus is unknown. Clearly protozoa with macronuclei like *Euplotes* permit both cytologists and biochemists to carry out experiments of a type quite impossible with other cell systems and as we have seen the results can be of considerable significance to biology as a whole.

6 Synchronously dividing *Tetrahymena*: a model system for studying control of cell division

During exponential growth only a small percentage of cells in a culture are dividing at any one time and the majority are generally completely out of phase with respect to any particular activity which takes place during only a fraction of the generation time. Thus if we wish to study biochemical events related to cell division we must either devise techniques for the study of single cells or study populations in which growth and cell division are synchronised. If we begin with a single cell and observe its division and the division of its daughter cells we would find, initially, that the cell division of the offspring was reasonably well synchronised, but, as time proceeded, the synchrony would be lost because of random variations in the time taken for a cell to divide. The cause of these random variations is unknown; perhaps they are due to the non-equivalence of the products of cell division, for example, in ciliates, daughter cells do not have identical cytoplasmic structures, one of each pair will retain the original oral apparatus and the other will have a newly formed one. Randomisation of division in cultures derived from a single cell would occur long before the cell density reached a level which would provide adequate material for biochemical analysis.

Cell biologists have attempted to overcome this problem by searching for conditions which will bring all the cells of a culture into 'step'. A number of effective techniques have now been developed including temperature shocks, separation of cells at a particular stage of mitosis or cells of a certain size and the use of specific and reversible inhibitor of DNA synthesis. One of the first successful systems was heat-shock synchronisation of the ciliate *Tetrahymena*

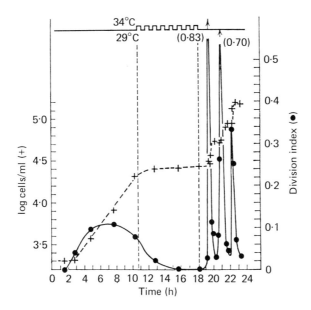

Fig. 16.5 Synchrony of *Tetrahymena pyriformis* induced by repetitive heat shocks. ●——● indicates division index and x – – – x cell number (from Zeuthen and Scherbaum 1954).

pyriformis (Scherbaum and Zeuthen 1954). Axenic cultures of an amicronucleate strain of *Tetrahymena* used by these workers have a generation time of 2·5 h when growing in a complex medium at 29°C. If the culture is given a series of heat shocks by raising the temperature to the just sub-lethal level of 34°C for 30 min, then returning it to 29°C for 30 min and if this procedure is repeated 8 or 10 times the cells stop dividing and the division index (i.e. the proportion of dividing cells) falls to zero. If such a culture is then maintained at 29°C a period of synchronous division commences, over 80 per cent of the population dividing after about 85 min, and a second synchronous division occurs 2 h later with about 70 per cent of the population dividing (Fig. 16.5). After a further 2 h or so the division index has fallen to 0·3 and in due course the culture reverts to normal exponential and asynchronous growth.

Although this system does not produce a perfectly synchronised population of cells it does permit the biochemist to study changes in cell composition and metabolism associated with a cell division cycle. Many detailed studies have been made since the system was developed and readers are referred to a comprehensive review by Zeuthen and Rasmussen (1972) and to a discussion by Mitchison (1971) for an assessement of the value of heat synchronised *Tetrahymena* as a model for studies on the control of cell division. The reason heat shocks produce synchrony in this ciliate is not completely understood but it is clear that the response of individual cells to elevated temperature varies during their division cycle. After a particular point in this cycle (termed the transition point and occurring about 25 min before division) cells are relatively insensitive to the raised temperature, also the division of young cells (i.e. those which have just divided) suffers little delay. However the delay induced by the higher temperature increases progressively from the time of division up to the transition point. Zeuthen and Williams (1969) have suggested a molecular model to explain these observations. They propose that cell division requires the assembly of some structure from a number of components at least two of which are assumed to be proteins (called division proteins). One of these proteins has to be synthesised continuously if assembly is to proceed. These protein components, it is suggested, are linked to form a strand which connects larger subunits to form a highly labile intermediate structure which will break down if the supply of its components is interrupted. When this intermediate structure is complete it develops, perhaps by some change in tertiary structure, into a stable final structure which plays some essential role in cell division. Temperature shocks, it is suggested, interrupt the assembly of division proteins either by accelerating their breakdown or slowing their assembly, and once this happens the whole intermediate structure breaks down and the cell has to start again from the beginning. A cell at the start of the division cycle will have only a small amount of this 'division protein' and thus will be delayed for only a short time by a heat shock but an older cell will have more of the intermediate structure assembled and so will suffer a greater delay from the same heat shock. The transition point is where the unstable intermediate structure is stabilised to give the final structure. Pulse labelling experiments with an amino-acid analogue, *p*-fluorophenylalanine and the use of cyclohexamide, a specific inhibitor of protein synthesis, have provided some supporting evidence for the existence of 'division proteins' in *Tetrahymena* and it is interesting that the oral apparatus of this ciliate, which develops in interphase as the precursor of the new mouth for one daughter cell, behaves in the same way as the postulated 'division protein' by showing regression after a heat shock and a transition point. At present it is not clear how far the concepts that have developed from studies of the *Tetrahymena* system can be applied to a wider range of cells. However there is evidence from the use of heat shocks and inhibitors of protein synthesis that a transition point exists in a wide range of cell types, from *Escherichia coli* to mammalian cells (Mitchison 1971).

7 The use of amoebae in studies of nuclear–cytoplasmic relationships

The unusual properties of the amoebal cell surface have allowed cell biologists to develop microsurgical techniques and to successfully transplant nuclei from one amoeba to another without affecting cell viability. The success of these methods opened up an entirely new experimental approach to problems of nuclear function. The earliest experiments (Comandon and de Fonbrune 1939) showed that co-ordinated movement generally ceases when the nucleus is removed from an amoeba but returns within minutes of introducing a nucleus from another individual even when the initial enucleation had been carried out one or two days previously. In the early 1950s Brachet and his collaborators made many comparative studies of nucleated and enucleated halves of amoeba and demonstrated that the respiration of each half was identical for up to 10 days after sectioning. In similar experiments the RNA content of enucleated halves gradually decreased whereas the amount of RNA in nucleated halves remained constant for as long as 12 days. This was the first clear cut demonstration that a nucleus exerts some control over cytoplasmic RNA and Brachet pointed out that these results were compatible with nuclear RNA being the precursor of cytoplasmic RNA.

Following the development of these techniques amoebae were used in a number of laboratories to investigate the effects of enucleation on various biosynthetic processes. Unfortunately results of many of these experiments are difficult to interpret because the amoebae were not grown in axenic culture; they were maintained on cultures of live bacteria and, although they were generally starved for a period before carrying out an experiment, the possibility that bacteria in the cytoplasm were contributing to the observed biosynthetic abilities of enucleated portions could not be eliminated. However, in 1955 Goldstein and Plaut described experiments which overcame this difficulty and provided evidence of the transfer of RNA from nucleus to cytoplasm. They found that the nucleus and cytoplasm of *Amoeba proteus* became radioactive if they were fed on the ciliate of *Tetrahymena* which had been grown in the presence of ^{32}P-labelled inorganic phosphate. The majority of the radioactivity could be removed by digestion of amoebae with ribonuclease. The next stage of the experiment was to remove a radioactive nucleus from a labelled amoeba by micromanipulation, wash it free of radioactive cytoplasm by passing it through the cytoplasm of one or more unlabelled amoebae and, finally, implant it into the cytoplasm of another unlabelled amoeba from which the nucleus had been removed. After 42 h it was found by autoradiography that much of the radioactivity had passed from the nucleus to the cytoplasm; this activity was again removed by digestion with ribonuclease. The experiment was then modified by leaving the unlabelled nucleus in the cell which received the radioactive nucleus. Under these circumstances labelled RNA was again found in the cytoplasm after two days but none could be detected in the initially unlabelled nucleus. These results imply that transfer of RNA from nucleus to cytoplasm can occur and it is a one-way process. The second experiment also indicated that the RNA in the transplanted labelled nucleus was probably not broken down and then resynthesised in the cytoplasm; if this occurred the unlabelled nucleus would have become labelled with ^{32}P.

The great potential of this experimental technique is obvious: it has been well exploited in the last 20 years and is still yielding valuable results. One of the most interesting findings concerns a group of proteins which appear to shuttle between nucleus and cytoplasm during the cell division cycle. In contrast to the experiments just described, when nuclei containing proteins labelled with radioactive amino acids are transferred to non-radioactive host cells which still contain their own unlabelled nucleus it is found that some radioactive protein (about 40 per cent of the total activity) appears in this nucleus and that there is little net transfer of labelled protein to the host cell cytoplasm. This redistribution occurs rapidly and is unaffected by excess unlabelled amino acids added to the incubation medium. The migratory protein is thought not to be a histone. Further studies have shown that transfer of certain proteins from nucleus to cytoplasm and back again to the nucleus is restricted to certain periods of the mitotic cycle. These experiments were carried out as follows: amoebae were labelled in their cytoplasm and nucleus by growth in the presence of ^{3}H-amino acids for a period of one generation time. After transfer to a non-radioactive medium about half of the cytoplasm of amoebae was removed with a glass microneedle every 36 h. These amputations prevented the cells from entering mitosis and between operations the cells regained their lost cytoplasm by growth in the unlabelled medium. After 15 operations 99·98 per cent replacement of the original radioactive cytoplasm had occurred in the absence of mitosis and autoradiography showed that the nucleus of these cells was still highly radioactive. If 30 operations were performed radioactivity was also lost from the nucleus, presumably due to a gradual transfer to the cytoplasm. If amputations were stopped after 15 operations and cells were allowed to enter mitosis autoradiography at various stages showed that suddenly, during prophase, the radioactive nuclear protein was lost into the cytoplasm. Little or no radioactivity could be detected in metaphase chromosomes or spindle, but in late telophase, the labelled proteins began to re-enter the nucleus, the transfer being completed in about half an hour. The function of these migrating proteins is not yet known. Goldstein has speculated that the shuttling of protein during interphase may be connected with the transfer of RNA to the cytoplasm. Another possibility is that these proteins, which are originally synthesised in the cytoplasm, may be involved in some sort of regulatory activity such as gene repression. It is interesting that no similar effect can be demonstrated in the macronucleus of ciliates, which it will be recalled, divide amitotically.

An obvious criticism of experiments involving such drastic procedures as nuclear transplantation and microsurgery is that the techniques might cause injuries to the cell which would lead to invalid conclusions being drawn about normal cellular processes. However, the possibility that the results described are due to artefacts has been carefully investigated (Legname and Goldstein 1972) and is considered small. The existence of a class of proteins that shuttles non-randomly between nucleus and cytoplasm in amoebae seems to be well established; whether these proteins provide a means of communication between these two parts of the cell remains to be established. A second question which must be asked is: how significant are these observations to other cell types? It has recently been demonstrated by the use of fluorescent antibodies against nuclear antigens that proteins from HeLa cell nuclei move into reactivated chick erythrocyte nuclei following fusion of the two cell types (Ege, Carlsson and Ringertz 1971). These results clearly demonstrate that proteins shuttle between nuclei against a concentration gradient in vertebrate cells which have not been subjected to micromanipulation.

Further reading

†CHEN, T. T. (1967–72). *Research in Protozoology*, vols. 1–4. Pergamon Press, Oxford.
†LWOFF, A. and HUTNER, S. H. (1951–64). *Biochemistry and Physiology of Protozoa*, vols. 1–3. Academic Press, New York.

References

†BRACHET, J. (1961). 'Nucleocytoplasmic interactions in unicellular organisms,' pp. 771–842. In *The Cell*, vol. II. Eds. J. Brachet and A. E. Mirsky. Academic Press, New York.

BUHSE, H. E. (1967). 'Microstome–macrostome transformation in *Tetrahymena vorax* strain vz type S induced by a transforming principle, stomatin.' *J. Protozool.*, **14**, 608–13.

BUHSE, H. E. and CAMERON, I. L. (1968). 'Temporal pattern of macromolecular events during the microstome–macrostome cell transformation of *Tetrahymena vorax* strain vz S[1].' *J. Exp. Zool.*, **169**, 229–36.

CLAFF, C. L. (1947). 'Induced morphological changes in *Tetrahymena vorax*.' *Biol. Bull.*, **93**, 216–7.

COMANDON, J. and FONBRUNE, P. de (1939). 'Ablation du noyau chez une amibe.' *C.r. Soc. Biol.*, Paris **130**, 740–44.

EGE, T., CARLSSON, S. A. and RINGERTZ, N. R. (1971). 'Immune microfluorimetric analysis of the distribution of species specific nuclear antigens in the HeLa-chick erythrocyte heterokaryons.' *Exp. Cell Res.*, **69**, 472–77.

FRANKEL, J. (1970). 'The synchronization of oral development without cell division in *Tetrahymena pyriformis*.' *J. Exp. Zool.*, **173**, 79–100.

GALL, J. G. (1959). 'Macronuclear duplication in the ciliated protozoan *Euplotes*.' *J. Biophys. Biochem. Cytol.*, **5**, 295–308.

GOLDSTEIN, L. and PLAUT, W. (1955). 'Direct evidence for nuclear synthesis of cytoplasmic ribose nucleic acid.' *Proc. Natn. Acad. Sci. U.S.A.*, **41**, 874–80.

GRANT, P. T. (1966). 'The selective inhibitors of energy-yielding reactions', pp. 281–93. In *Biochemical Studies of Antimicrobial drugs*. Eds: B. A. Newton and P. E. Reynolds. 16th Symp. Soc. Gen. Microbiol. University Press, Cambridge.

HALL, R. P. (1953). *Protozoology*. New York, Prentice-Hall.

HONIGBERG, B. M., BALAMUTH, W., BOVEE, E. C., CORLISS, J. O., GOJDICS, M., HALL, R. P., KUDO, R. R., LEVINE, N. D., LOEBLICH, A. R., WEISER, J. and WENDRICH, D. H. (1964). 'A revised classification of the phylum protozoa.' *J. Protozool.*, **11**, 7–20.

KIDDER, G. W., LILLY, D. M. and CLAFF, C. L. (1940). 'Growth studies on ciliates. The influence of food on structure and growth of *Glancoma vorax* sp. nov.' *Biol. Bull*, **78**, 9–23.

KLUSS, B. C. (1962). 'Electron microscopy of the macronucleus of *Euplotes eurystomus*,' *J. Cell Biol.* **13**, 462–65.

LANHAM, S. M. (1968). 'Separation of trypanosomes from the blood of infected rats and mice by ion-exchangers.' *Nature, Lond.*, **218**, 273–74.

LEGNAME, C. and GOLDSTEIN, L. (1972). 'Proteins in nucleocytoplasmic interactions.' *Exp. Cell Res.*, **75**, 111–21.

† MITCHISON, J. M. (1971). *The Biology of the Cell Cycle*. University Press, Cambridge.

† NEWTON, B. A. (1972). 'Protozoal pathogenicity,' pp. 269–301. In *Microbial Pathogenicity in Man and Animals*. Eds. H. Smith and J. H. Pearce. 22nd Symp. Soc. Gen. Microbiol. University Press, Cambridge.

† NEWTON, B. A., CROSS, G. A. M. and BAKER, J. R. (1973). 'Differentiation in trypanosomatidae.' pp. 339–73. In *Microbial Differentiation*, 23rd Symp. Soc. Gen. Microbiol. Eds. J. M. Ashworth and J. E. Smith, University Press, Cambridge.

PRESCOTT, D. M. and BENDER, M. A. (1963). 'Synthesis and behaviour of nuclear proteins during the cell life cycle.' *J. Cell. Comp. Physiol.*, **62**, 175 (Supplement).

PRESCOTT, D. M. and KIMBALL, R. F. (1961). 'Relation between RNA, DNA and protein synthesis in the replicating nucleus of *Euplotes*.' *Proc. Natn. Acad. Sci. U.S.A.*, **47**, 686–93.

SCHERBAUM, O. and ZEUTHEN, E. (1954). 'Induction of synchronous cell division in mass cultures of *Tetrahymena pyriformis*.' *Exp. Cell Res.*, **6**, 221–27.

† SIMPSON, L. (1972). 'The kinetoplast of Hemoflagellates.' *Int. Rev. Cytol.*, **32**, 139–207.

STONE, G. E. (1963). 'Polymorphic properties of *Tetrahymena patula* during growth in axenic media.' *J. Protozool.*, **10**, 74–84.

VICKERMAN, K. 'Polymorphism and mitochondrial activity in sleeping sickness trypanosomes.' *Nature, Lond.*, **208**, 762–66.

† VICKERMAN, K. (1971). 'Morphological and physiological considerations of extracellular blood protozoa,' pp. 58–91. In *Ecology and Physiology of Parasites – A Symposium.* Ed. A. M. Fallis, University Press, Toronto.

† ZEUTHEN, E. and RASMUSSEN, L. (1972). 'Synchronized cell division in protozoa,' vol. 4, pp. 11–145. In *Research in Protozoology.* Ed. T. T. Chen. Pergamon Press, Oxford.

ZEUTHEN, E. and SCHERBAUM, O. H. (1954). 'Synchronous division in mass cultures of the ciliate protozoan *Tetrahymena pyriformis*, as induced by temperature shock.' *Colston Pap.*, **7**, 141–55.

† ZEUTHEN, E. and WILLIAMS, N. E. (1969). 'Division-limiting morphogenetic processes in *Tetrahymena*,' pp. 203–16. In *Nucleic Acid Metabolism, Cell Differentiation and Cancer Growth.* Eds. E. V. Cowdry and S. Seno. Pergamon Press, Oxford.

---

† Major reviews or texts with comprehensive reference lists.

17
Lysosomes
(With a short note on Peroxisomes)

T. F. Slater
Department of Biochemistry, Brunel University

1 Introductory outline

1.1 Historical background

A lucid and entertaining account by de Duve (1969a) has described the sequence of events in the early 1950s leading up to the recognition of the lysosomes as a distinctive class of intracellular particle. De Duve and colleagues in the department of Physiological Chemistry at the University of Louvain had been actively interested in the metabolic behaviour of the substrate–specific enzyme glucose-6-phosphatase in rat liver, and in distinguishing such enzyme activity from the unspecific liver acid phosphatase acting, for example, on β-glycerophosphate or p-nitrophenylphosphate. To understand how such interests developed, such that the existence of a hitherto unrecognised class of cellular particles could be predicted with confidence on biochemical ground alone requires some preliminary remarks concerning the historical development of the technique of differential centrifuging, and the application of that technique to the study of the intracellular localisation of enzymes and coenzymes.

The successful separation of classes of intracellular particles from each other by differential centrifuging depended on the development of a number of ancillary techniques and procedures: (a) a satisfactory method of homogenising the starting material such that artefactual damage is minimal; (b) the development of a suspending medium in which to homogenise the tissue and which does not produce significant particle agglutination or enzyme inhibition; (c) and the availability of reliable high speed centrifuges. Until these were all available the biochemical study of the intracellular localisation of enzyme activities and of coenzyme content was slow to develop and to attract the attention of many biochemists.

A satisfactory technique for homogenising soft tissues such as liver had been described by Hagan in 1922 although the later description of a similar technique by Potter and Elvehjem (1936) is usually accepted as the breakthrough in this context. Commercially supplied 'high speed' centrifuges became widely available shortly post-war and although lacking the great sophistication of present generation machines they enabled the routine isolation of mitochondrial and microsomal fractions to be carried out. Early studies on intracellular localisation were performed with water or dilute salt solutions as the suspending media (see Claude 1969) and the distributions of enzymes during the various intracellular fractions were adversely affected as a consequence. The use of aqueous sucrose as a suspending medium was pioneered by Hogeboom, Schneider and Palade (1948) and 0·25 M sucrose has been widely used ever since. Further details concerning the isolation of cellular fractions by centrifuging can be obtained by reading Campbell (1966) and Hillman (1972).

512

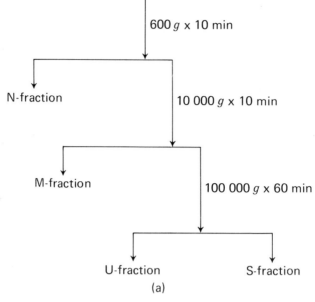

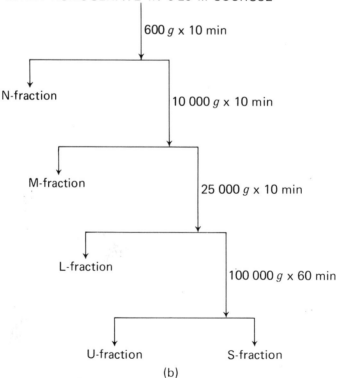

Fig. 17.1 (*a*) Simple centrifuging scheme for liver homogenate in 0·25 M sucrose to isolate a nuclear fraction (N), a mitochondrial fraction (M), a microsomal fraction (U) and cell sap (S). In each centrifuging stage the centrifugal acceleration in *g* and the time in minutes on the plateau at top speed are given. Each centrifuging stage results in a pellet shown to the left of the vertical lines and a supernatant suspension shown on the right in each case.

(*b*) A modified centrifuging scheme to allow for the separation of a light mitochondrial fraction (L).

1.2 Centrifuging analysis

A typical and simple scheme for separating intracellular fractions from a suspension of rat liver is 0·25 M sucrose as shown in Fig. 17.1(a). It was soon found that fraction N contained nuclei, unbroken cells, erythrocytes and membranous debris; it is a grossly impure fraction, but almost all of the DNA in the tissue sediments in it. Fraction M, the mitochondrial fraction, was found to contain the major part of the respiratory activity of the homogenate including such enzymes as succinate dehydrogenase and cytochrome oxidase; the organised cyclic sequence of enzymes constituting the tricarboxylic acid cycle also was located in this fraction. The smallest particulate fraction U was found to be rich in RNA and this fraction was given the name of 'microsomes' for the component particles (see Claude 1969 for review on the origin of the name 'microsomes' and early studies on their composition). The microsomes were later identified as the disrupted membranes of the endoplasmic reticulum by Porter (1953) and Palade and Porter (1954). The clear supernatant fraction S was rich in glycolytic activity and in various cofactors such as NAD^+. These and other results led most if not all investigators at that time to think of the intracellular distribution of enzymes in simple terms: a particular enzyme normally having a specific single location within the cell. This was the position in the early 1950s when de Duve and colleagues started their now classical studies on the intracellular localisation of glucose-6-phosphatase and of non-specific acid phosphatase.

The substrate-specific enzyme glucose-6-phosphatase was found to sediment wholly with the microsomal fraction U; indeed, it is now generally used as a microsomal enzyme marker in liver tissue. The substrate-unspecific enzyme acid phosphatase sedimented largely with fraction M. However, some unexpected peculiarities of the activity of acid phosphatase in sucrose suspensions were soon recognised. Firstly, the activity in 0·25 M sucrose homogenates was much less than previously observed in water homogenates but increased on storing at $+2°C$.

This phenomenon whereby the enzyme activity was much increased by prolonged storage, or by freezing and thawing the particle suspension, is known as latency; in the undamaged, fresh particle suspension the enzyme activities are said to be latent. Latency was explained by the suggestion that the acid hydrolases exist inside the (lysosomal) particle, separated by a membrane from the outside environment containing the substrate for the enzyme. The membrane was normally largely if not totally impermeable to substrate but the accessibility of substrate to enzyme was greatly increased by membrane damage. The second unexpected observation was that the acid phosphatase activity did not sediment with the normal mitochondrial fraction M when an apparently minor modification in centrifugal procedure was made (see de Duve 1969a) and this prompted the introduction of an additional centrifuging step between the mitochondrial (M) and the microsomal (U) sedimentations. The modified scheme (Fig. 17.1(b)) allowed for the separation of a light mitochondrial fraction L in which acid phosphatase activity was preferentially concentrated (see Appelmans et al. 1955). Thus acid phosphatase showed a distribution intermediate between the classical mitochondrial and microsomal fractions. Similar differences in the sedimentation behaviour of acid phosphatase, respiratory enzymes and glucose-6-phosphatase had been noted independently by Novikoff's group in 1954.

Before long other hydrolytic enzymes were found to exhibit centrifuging behaviour and latency similar to that found for acid phosphatase; for example, β-glucuronidase, cathepsin, acid ribonuclease and acid deoxyribonuclease. The enzymes having this distinctive centrifuging behaviour were all most active at acid pH and are classified as acid hydrolases. In 1955 de Duve and colleagues published their data to show that acid hydrolases sediment together in fraction L and also exhibited the phenomenon of latency. The different centrifuging behaviour and the property of latency led to the conclusion that the enzymes were located within a new type of cytoplasmic particle where enzyme-substrate interaction was modified by the integrity of an outer limiting membrane. This process, involving the separation of enzyme and substrate by a membrane, made sense in that the destructive acid hydrolases were kept compartmented away from the rest of the cell. In view of the potentially damaging lytic nature of the acid hydrolases de Duve and colleagues suggested the name lysosome for the new species of intracellular organelle in 1955. The concept of the lysosome was at that time both elegant and simple.

De Duve's work on liver lysosomes was soon extended to show that particles with similar properties occurred in tissues other than liver, for example in mammary gland (Greenbaum *et al.* 1960). It is now known that lysosomes have a very wide distribution in nature, being present in almost all types of mammalian cells for example. As the studies concerning the tissue distribution of lysosomes were being carried out other investigations were showing that numerous other acid hydrolases sedimented with the lysosomes in various tissues, including liver. More than forty lysosomal acid hydrolases have now been reported (see section 2.4).

Electron microscopy of the semi-purified fractions of liver lysosomes was carried out by Novikoff in 1956 and the lysosomes were identified with particles concentrated around the biliary canaliculi in rat liver; such particles had been previously seen by Rouiller in 1954 in electron micrographs and called peribiliary bodies. Following the identification of liver lysosomes with peribiliary bodies, it was natural to try to ease the problems of identifying the lysosomal particles in tissue sections by the use of histochemical procedures for acid hydrolases; the most frequently used assay being for acid phosphatase. With light microscopy the technique gives good results and in liver the stained particles show a clear-cut peribiliary localisation. However, when electron microscopy was applied to such stained sections it was found, somewhat paradoxically, that the increased resolution of the electron microscope blurred the simple image of the lysosome particle. For under the electron microscope and after appropriate staining procedures for acid hydrolases it was found that all sorts of bizarre and unexpectedly complex bodies showed acid phosphatase activity. As a result of such studies the concept of the lysosome as a single particle with no significant internal structure was changed and the lysosomes were recognised to include a family of particles of dissimilar shapes and structure but having in common the property of possessing an outer limiting membrane and a high acid hydrolase activity.

1.3 The concept of lysosomes as a family of particles

Lysosomal acid hydrolase proteins are synthesised by the rough endoplasmic reticulum and there is strong evidence that they are then packaged within small particles or vesicules through an interaction with the Golgi region (see Novikoff and Holtzmann 1970). The resultant collection of acid hydrolases compartmented away from the rest of the cytoplasm by a bounding membrane are called primary lysosomes.

Many types of cell (e.g. liver parenchymal cells, Kupffer cells, kidney tubular cells) can actively take in extraneous material either as solid particles or dissolved in solution. Other cell types have this property to a greater or lesser extent. This intake of material is called endocytosis; the reverse process in which material is pushed out from a cell is called exocytosis. It is a convenient and usual practice to consider endocytosis in two sub-divisions: when the material ingested is in solid form, the process is called phagocytosis; when the material is in solution, it is named pinocytosis. Material taken into cells in this way is present within the cytoplasm in a membrane-limited compartment as indicated in Fig. 17.2. Phagocytosed material is within a phagosome and pinocytosed material is within a pinocytotic vesicle. The bounding membrane of such particles arises from the invagination of the exterior cell membrane itself.

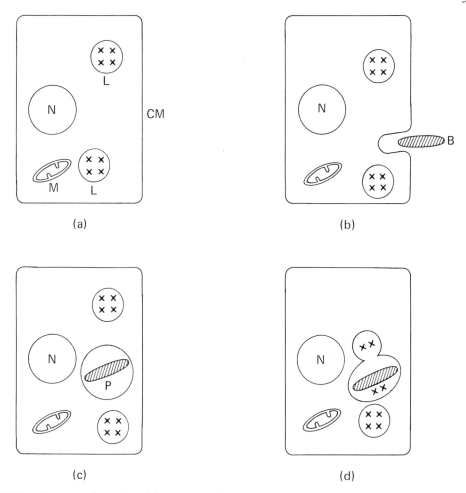

Fig. 17.2 A diagrammatic version of the events in phagocytosis.

In (a) a cell is shown with its outer membrane (CM), lysosomes (L), a mitochondrion (M) and a nucleus (N).

In (b) a bacterium (B) is being engulfed as a consequence of invagination of the cell membrane.

In (c) the bacterium is now totally enclosed by a portion of cell membrane and exists within a phagosome (P) in the cytoplasm.

In (d) the lysosomes are discharging their acid hydrolases (shown as x) into the phagosome and are thereby converting it to a secondary lysosome.

The intake of material by endocytosis is clearly distinguished from the intake of other materials (e.g. ions, nutrients) which are transported across the exterior cell membrane directly into the cytosol without passing through an intermediate stage of being compartmented within a membrane-limited particle.

The phagosome or pinocytotic vesicle formed by endocytosis are not lysosomal particles: they do not generally possess acid hydrolase activity to any significant extent although it is conceivable that a minute amount of enzyme activity present in extracellular fluid may be ingested with the other material. Within the cytosol, however, it is the general rule for primary lysosomes to undergo fusion with the pinocytotic vesicles or phagosomes and this process results in secondary lysosomes which now contain a rich variety of acid-hydrolases such that digestion of the ingested material may occur. The fusion of primary lysosomes with a membrane-limited vacuole containing particulate material is not confined to the situation resulting from phagocytosis. Under certain conditions a volume of the cytoplasm containing mitochondria, endoplasmic reticulum, etc., may become enclosed by an outer limiting membrane and this structure is called an autophagosome (i.e. a self-created phagosome) or, in more recently accepted nomenclature, a cytosegresome. This particle may then fuse with primary lysosomes to yield a secondary lysosomal structure. Another route leading to secondary lysosomes by a completely intracellular mechanism arises from the fusion of primary lysosomes with secretory granules which are normally associated with the exocytosis of protein material from the cell. Such a process is well marked in endocrine tissue where excess hormone in secretory granules undergoes intracellular digestion within a secondary lysosome; this process is given a special name, crinophagy.

The material within a secondary lysosome is attacked by the secondarily acquired acid hydrolases and is digested at a rate which obviously depends on many factors including the chemical nature of the ingested material itself. If this material is elementary carbon in the form of soot particles, then obviously no digestion can occur. Biological material however, is digested in general, although some types of lipid are broken down very slowly. As a consequence lipid material may gradually concentrate within a secondary lysosome as other substances are digested rapidly and this lipid may oxidise to give coloured products long-known as lipofuschin pigments. The particle containing such a lipid-rich inclusion undergoing very slow degradative change is called a residual body.

From the above comments and an inspection of Fig. 17.3 it becomes evident that there occur several forms of intracellular particle which contain acid hydrolase activity. This group of particles constitutes the family of lysosomal particles and it can be seen how diverse the morphological features may be within this group, ranging from the simple primary lysosome to the very complex structures derived from cytosegresomes (see for example, Helminen and Ericsson 1968; Kerr 1969). Having defined what we mean by the lysosomal family of particles we can now consider general properties of this group of intracellular organelles including the procedures used in separating them from other cellular components.

2 General properties

2.1 Separation procedures

Early studies on liver lysosomes revealed their fragility to strenuous homogenising techniques (e.g. the Waring blender) and to hypo-osmotic conditions. This potential fragility during *in vitro* preparation is characteristic of lysosomes from tissues generally and is not a special feature of liver lysosomes alone. As a consequence considerable care must be exercised in the choice of a suspending medium and in the homogenisation of the tissue under study if extensive damage to the lysosomes is to be avoided.

Aqueous sucrose is generally used as the suspending medium for preparing tissue dispersions; sometimes a buffer is also included to produce a desired initial pH value, and other additions such as metal chelating agents (e.g. EDTA) may also be included where it is considered necessary to protect sensitive enzymes against metal inactivation during the preparative procedures. The molarity of the sucrose (or other solute to be used) in the suspending medium should be that most suited to the osmotic characteristics of the lysosomal particles present in the tissue under investigation. Aqueous sucrose (0·25 M) is suitable for many tissues, however, including rat liver which will be specifically considered in the rest of this section.

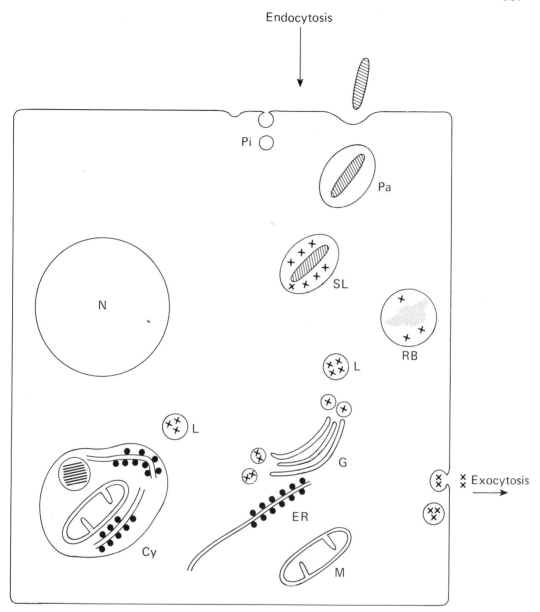

Fig. 17.3 A representation of the major stages of lysosomal formation and function within a cell.

Endocytosis can lead to pinocytotic vesicles (Pi) or a phagosome (Pa). A cytosegresome containing a mitochondrion (M), some endoplasmic reticulum and a peroxisome is shown as Cy. Endoplasmic reticulum (ER) and Golgi apparatus (G) are shown participating in the synthesis of acid hydrolases (x) and their packaging into primary lysosomes (L). Fusion of L with Pa gives a secondary lysosome (SL) which can lead to a residual body (RB) containing masses of autoxidised fat material. Some acid hydrolase activity is shown being discharged from the cell by exocytosis. The cell nucleus is shown at N.

If the tissue to be homogenised is relatively friable as is rat liver then satisfactory homogenates can be prepared using a standard Potter–Elvehjem technique (see Campbell 1966 for description). For comparative purposes from experiment to experiment the actual process of homogenising should be controlled as closely as possible in relation to the speed at which the pestle is revolved (generally about 1000 r.p.m.), the time of homogenising, and the clearance between the pestle and the barrel. If the clearance is too small (say 0·1 mm) then damage to mitochondria and lysosomes may occur; if the clearance is too large then a high proportion of the cells in the tissue sample may remain intact. In the author's experience a clearance of approx. 0·5 mm with three passes of the piston up and down the barrel is satisfactory for rat liver in 0·25 M sucrose. In addition, it is essential to maintain the temperature of the material as near to 0°C as possible during the preparative procedures to prevent enzyme denaturation.

Some tissues are much harder to homogenise by the Potter–Elvehjem technique than is rat liver and repeated passes of the piston up and down the barrel may be necessary to obtain satisfactory dispersion. With tissues rich in connective tissue (e.g. mammary gland) it may be necessary to stop frequently to remove and cut up the strands of connective tissue which wrap around the piston and may cause jamming. As a general rule the isolation of lysosomal fractions from such tough tissues by the Potter–Elvehjem technique is unsatisfactory as lysosomal damage occurs during the over-vigorous homogenising process required. In rat mammary gland, for example, the use of the Potter–Elvehjem procedure for 3 min resulted in 75 per cent of the total acid ribonuclease activity being in the free form (Greenbaum *et al.* 1960); when a more gentle procedure (the Chaikoff–Emanuel tissue press) was used this percentage was considerably reduced.

Other homogenising techniques more suited to the dispersion of tough tissues than the Potter–Elvehjem procedure that have been used in connection with lysosomal studies include (a) the Chaikoff–Emanuel press already mentioned, (b) the nitrogen-bomb and (c) enzymic tissue dispersion. These techniques will be outlined briefly below.

Emanuel and Chaikoff (1957) described a procedure suited for the dispersion of tough tissues such as thyroid gland. The tissue is initially chopped up finely with scissors in an appropriate cold medium and then placed inside a stainless steel press. By applying a considerable pressure through a hydraulic ram the tissue and suspending medium is forced through a narrow orifice across a very high pressure difference. This sudden drop in pressure results in tissue dispersion. By controlling the clearance between the central rod and the orifice selective damage can be produced to the cell membrane (i.e. dispersion of cell contents into suspension) with minimal damage to intracellular organelles.

A rather similar mechanism underlies the use of the nitrogen bomb. Here the tissue in a finely chopped state is exposed to an inert gas at very high pressure until equilibrium has been achieved. The external gas pressure is then rapidly dropped to normal and the sudden release of gas dissolved in the tissue causes cell breakage (see Contractor 1969).

Tissue dispersion by enzymic digestion has been described by Bullock *et al.* (1970) principally in connection with the isolation of morphologically well-preserved and biochemically active mitochondria from muscle; the technique involves exposing the finely minced tissue to proteolytic enzyme digestion for 30 min at 0°C followed by straining through muslin to remove strands of connective tissue and any remaining lumps of tissue.

Having obtained a satisfactory tissue dispersion the next stage in procedure is to separate lysosomal particles as cleanly as possible from other cellular organelles; centrifuging procedures have been used for this purpose in nearly all of the published work. In outlining general features of the centrifuging procedures the following discussion will be restricted to describing the isolation of lysosomes from rat liver homogenates.

If a relatively crude lysosomal preparation is required, and this may be quite adequate for many investigations, then simple differential centrifuging separations in 0·25 M sucrose as given in Figs. 17.1(a) and (b) may be used. Using the procedure of Fig. 17.1(b) the acid hydrolases are concentrated in fraction L about 10-fold over their concentration in the initial homogenate. Even so, fraction L is a highly impure fraction and the majority of the protein in it is due to contamination by mitochondria, microsomes and peroxisomes (for a review of lysosomal isolation procedures see Beaufay 1969). In fact it can be calculated with some fair degree of accuracy that although the proportion of total homogenate protein sedimenting in fraction L is about 8 per cent (de Duve et al. 1955) the contribution to this from the lysosomal particles is probably less than 0·7 per cent. This suggests that if purified lysosomal suspensions could be prepared they would possess a specific activity of acid hydrolases some 150-fold greater than in the initial homogenate. However, even quite complex centrifuging procedures (see Sawant et al. 1964) based on differential rates of sedimentation have not consistently resulted in a concentration of acid hydrolases of more than about 20-fold. A modification of Sawant's procedure involving a more complex fractionation scheme has been described by Ragab et al. (1967) but does not greatly increase the overall purity of the lysosomal preparations obtained.

Similar difficulties in obtaining relatively pure suspensions of lysosomes by isopycnic centrifuging have been found in numerous studies (see Beaufay 1969 for reference) and it must be concluded that the wide variability in lysosomal size in rat liver, and the overlapping of the density range with mitochondria, microsomes and peroxisomes rules out clean-cut separations based on centrifuging methods currently in use, including those employing the relatively recently introduced zonal rotors (Scheul et al. 1968; Brown 1968). A new approach to lysosome isolation based on a combination of electrophoretic and sedimentation characteristics has been outlined by Stahn, Maier and Hannig (1970). Only preliminary studies have been reported so far with this apparatus which at present is likely to find restricted usage due to its high cost.

In attempts to overcome the difficulties outlined above (e.g. overlapping size and density) that restrict the resolution obtainable by centrifuging much use has been made of the participation of lysosomes in fusing with endocytotic vesicles to form secondary lysosomes. Exogenous materials introduced into the parenteral circulation of an animal may become concentrated in the secondary lysosomes of, for example, the liver. One material that is of special interest in this respect is the non-ionic detergent Triton-WR-1339. When injected intraperitoneally into rats it accumulates progressively in secondary lysosomes in liver and the liver parenchymal cells can then be seen to contain large electron transparent vacuoles. These vacuoles contain high concentrations of the detergent which confer to them a much lower density than other intracellular organelles. As a consequence, isopycnic centrifuging of such liver suspension results in a clean-cut separation of such secondary lysosomes (or 'tritosomes') from other particles whose density remains unaffected by Triton administration. By this procedure it has been found possible to obtain very pure preparations of secondary lysosomes containing Triton-WR-1339 and this has helped in preparative procedures for studying lysosomal acid hydrolase activities in the virtual absence of other intracellular organelles. It must be emphasised however, that such 'tritosomes' are obviously artificial structures and that the high concentration of detergent in them may cause substantial changes in membrane properties in comparison with the membrane of primary lysosomes. 'Tritosomes' for example show a much less marked degree of latency than do normal lysosomes prepared from control rat livers when incubated at 37°C in 0·25 M sucrose.

2.2 Chemical composition

The composition of the lysosomal particles can be considered under the separate headings of (a) membrane composition and (b) matrix composition; the data obtained in these contexts should be interpreted bearing in mind the discussions in sections 1.3 and 2.1, where it was emphasised that the lysosomal family of particles is heterogeneous in character and that preparations of lysosomes may be grossly impure.

Membrane subfractions have been isolated after the rupture of parent lysosomal suspensions and the chemical composition examined (Tappel 1969a). Such analyses are complicated by the heterogeneity of the fractions studied which include not only contamination by mitochondria, microsomes and peroxisomes but also cytosegresomes containing such other organelles in various stages of digestion. In fact the phospholipid fraction extracted from liver lysosomes showed no qualitative differences compared to the values reported for mitochondrial or microsomal membrane phospholipid. The overall composition may indeed be greatly affected by the presence of even quite minor (on a protein basis) components as contaminants. For example, the Golgi membranes sediment with the microsomal fraction in standard centrifuging schemes, and it has been shown that ubiquinone previously thought to be associated with the membranes of the endoplasmic reticulum is in fact, present in the Golgi membranes (Nyquist *et al.* 1970). Ubiquinone is also found to a limited extent in the membranes of a tritosome preparation (Henning and Stoffel 1972). The latter finding has been interpreted as evidence for the participation of the Golgi apparatus in the formation of lysosomes.

The matrix components of lysosomal particles isolated from a number of different tissues (e.g. kidney, liver, brain) have been studied in respect of their anionic (negatively charged) and cationic (positively charged) species. Density gradient centrifuging enables two major fractions to be distinguished: (a) a lipoprotein fraction with density less than 1.35 g/ml and (b) a protein fraction containing most of the acid hydrolase activity with a density greater than 1.35 g/ml (Goldstone and Koenig 1970). The majority of the acid hydrolase enzyme activity is associated with neutral and cationic material that is relatively rich in bound carbohydrate; the isoelectric points of the components of the enzyme fraction were generally greater than 7.0. It is probable that many of the lysosomal acid hydrolases are glycoproteins that normally carry a positive charge. The lipoprotein fraction largely devoid of enzyme activity also contains some carbohydrate-bound material and is negatively charged at neutral pH (Goldstone *et al.* 1970). It has been suggested that the close association of the cationic and anionic proteins within the lysosome controls the enzyme activity of the numerous acid hydrolases through the interaction of the 'macro-anionic lipoprotein' with the protonated regions of the active sites on the hydrolases (Goldstone and Koenig 1970). More studies on this area are awaited with considerable interest since the basic questions (a) what limits the self-digestion of the lysosomal enzymes within a primary lysosome and (b) what limits their attack on the surrounding membranes in both primary and secondary lysosomes whilst permitting digestion of enclosed membranous structures such as endoplasmic reticulum in cytosegresomes remain unanswered at present.

The acidic anionic lipoprotein fraction described above has been studied in relation to its role in the uptake of positively charged material by the lysosomes. It has been known for some years that lysosomes both *in vivo* and *in vitro* will readily concentrate positively charged substances such as metal ions, dyes and drugs. In human clinical conditions involving disturbances of iron and copper metabolism large excesses of these metals accumulate in the body. In haemochromatosis (iron-overload of parenchymal tissues) and in Wilson's disease (associated with copper storage in the liver and brain) the excess metal ions have been shown to concentrate in liver lysosomes to a very marked extent. When lysosomal fractions are mixed *in vitro* with iron and copper salts a similar uptake of metal ions occurs. This concentrating property of the lysosomes for positively charged material is dependent on the lysosomal acidic lipoprotein. A particularly interesting example of such concentrating ability has been studied by Dingle and Barrett (1969a) and involves the uptake of fluorescent dyestuffs (e.g. acridine orange). This binding also can be readily demonstrated *in vivo* and *in vitro*. Since other cytoplasmic organelles do not bind the dye to any significant extent the uptake of fluorescent material can be used to monitor the purity of tissue fractions by simple fluorescence microscopic viewing after exposure to acridine orange (the procedure is fully described by Dingle and Barrett 1969b).

As noted already the studies involving the positively and negatively charged matrix components of the lysosomal particles have been carried out on a number of different tissues including liver and kidney. A more specialised non-enzymic component of the lysosome matrix has been reported for polymorphonuclear leucytes where Zeya and Spitznagel (1966) found evidence for the occurrence of a cationic component that possessed antimicrobial activity.

2.3 Stability

The reaction of lysosomal particles in suspension to changes in their local environment has been studied carefully on numerous occasions. Only a brief account of the general conclusions can be given here; a summary of procedures that damage the lysosomal membrane, leading to decreased latency of the enclosed acid hydrolases is given in Fig. 17.4.

Lysosomes respond to changes in the osmotic strength of the external medium in a manner similar to mitochondria: lysosomes swell and may rupture when placed in water; they are well preserved in 0·2–0·25 M sucrose.

Lysosomal suspensions are rather unstable at pH values outside the range 4–9. For example, incubation at pH 2·0 and 37°C even in the presence of an enzyme substrate for cathepsin leads

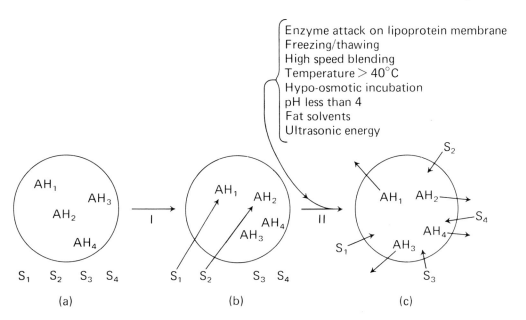

Fig. 17.4 In (*a*) an intact lysosome is shown with the bounding membrane preventing interaction between the internal acid hydrolases (AH$_i$) and the corresponding substrates S$_i$. Minor damage to the membrane as in Stage I leads to (*b*) an increased permeability of the membrane to some substrates but the high molecular weight enzymes are retained within the lysosome. In (*c*) the membrane has been severely damaged such that the enzymes can now leak out and substrates can freely penetrate the lysosome. Examples of procedures that can cause extensive damage to the lysosomal membrane are shown above Stage II.

522

to spontaneous membrane rupture. The numerous hydrolases within the lysosomal particles have optimal pH values around 5–6. This is more acid than the pH generally acknowledged for the cell sap or cytosol. Measurement with micro-electrodes or with pH-indicators on cytosol pH have given values of about 6·3–6·8; at 37°C the neutral point is pH 6·8 and the blood is

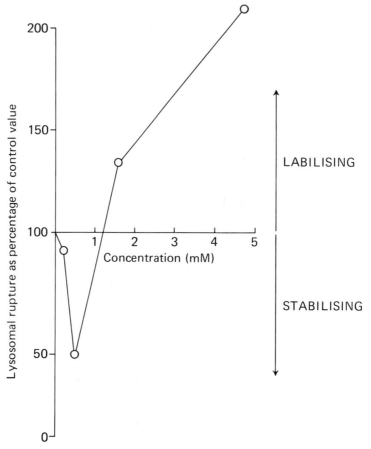

Fig. 17.5 The effects of the antihistamine drug promethazine on the release of β-glucuronidase from rat liver lysosomes incubated at 37°C for 30 min at pH 5·0. It can be seen that low concentrations of promethazine stabilise the lysosomes whereas high concentrations have a labilising action. Data of T. F. Slater and B. C. Sawyer.

approx. 7·4. In a recent study Reijngoud and Tager (1973) have suggested that the lysosomal membrane can maintain a pH gradient such that the lysosomal matrix is more acid than the cytosol. Using a derivation based on the Henderson–Hasselbalch equation they have obtained values for the intralysosomal pH of tritosomes. The pH (internal) was always more acid than pH (external) and with a pH (external) value of 7·5 they obtained a pH (internal) of 6·5. With a pH (external) of 6·3 their data suggests the pH (internal) would be approx. 5·8. One enigma that has surrounded the lysosome concept has been how the hydrolases which have acid pH optima can work efficiently inside the cell unless local and significant variations in pH occur within cells. The approach of Reijngoud and Tager (1973) may well have resolved this problem.

    Lysosomes are adversely affected by temperatures in excess of about 40°C which result in spontaneous damage to the membrane. Lysosomal suspensions may be stored for several hours at 0°C without significant damage to the membrane but freezing the suspension and subsequent thawing causes rupture of the particles. As shown by some early experiments of Gianetto and de Duve (1955) several successive cycles of freezing and thawing are necessary to liberate all of the enzyme initially restricted within a particle into a free or soluble form.

Since the membrane of the lysosome is lipoprotein in character any chemical procedure that causes dissociation or degradation of the lipoprotein structure will result in a decrease in enzyme latency, i.e. in increased interaction between enzymes and substrates. For example, enzyme attack by trypsin or lecithinase causes lysosomal membrane damage as does exposure to surface active materials such as detergents. The study of Wattiaux and de Duve (1956) with the non-ionic detergent Triton-X-100 (NOT the same as Triton-WR-1339 used for preparing 'tritosomes') showed clearly that when the concentration of the detergent reached a critical value of 0·08 per cent all of the acid hydrolases studied were released simultaneously in a free and soluble form. Many drugs also possess detergent-like properties when present in high concentration and this may result in biphasic effects on lysosomal stability. For example a drug may operate as a stabiliser of the lysosomal suspension when in low concentration due to its chemical properties as a metal binding agent or a free radical scavenger. At a higher concentration however its physical properties as a detergent may overwhelm the chemical protection and as a result lysosomal damage occurs. An illustration of this behaviour, which has been called 'the cross-over effect' (Slater 1969), is given in Fig. 17.5.

Since the lysosomal membrane contains unsaturated fatty acids, and the external medium contains dissolved oxygen, the possibility exists for membrane damage to occur through a peroxidative attack on the unsaturated membrane lipids. This process is called lipid peroxidation and involves the initial formation of a free radical derivative of the unsaturated fatty acid by interaction with a free radical (R$^\bullet$) present in the adjacent medium. Thus, lipid peroxidation requires free radicals (R$^\bullet$) for initiation of the process (see Fig. 17.6). Such a radical R$^\bullet$ can itself arise in a number of ways (a) by exposing the suspension to high energy

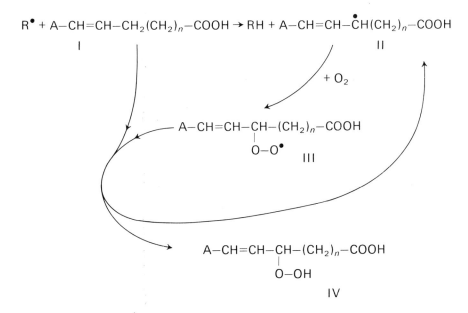

Fig. 17.6 Mechanism of attack by a free radical R$^\bullet$ on an unsaturated fatty acid (I) to give the unsaturated fatty acid radical (II), the peroxy-radical (III) and the hydroperoxide (IV) with re-cycling through (II).

radiation leading to the formation of hydroxyl radicals for example; (b) by a metal-catalysed reaction such as the Fenton reaction involving ferrous ions and hydrogen peroxide. The reaction of such an extraneous radical $R^{\bullet}$ with an unsaturated lipid in the presence of O_2 may result in a chain sequence which can cause extensive disruption of the organised structure of the lysosomal membrane. As a result enzyme latency is decreased. Tappel and colleagues have demonstrated this aspect of lysosomal damage by exposing lysosomal suspensions to γ-radiation, to ultra-violet radiation, and to a Fenton type reaction. All three types of free radical producing system resulted in release of acid hydrolase activity into a free or soluble form. A review of lipid peroxidation and its effects on biological membranes is given by Slater (1972).

Free radical attack on the lysosomal membrane can also occur following exposure to low energy radiation as with visible light provided that a photosensitiser is present. In such an arrangement the photosensitiser (e.g. a porphyrin) absorbs the visible light and is itself raised to an excited state analogous in chemical reactivity to a free radical; this may then attack the lysosomal membrane initiating the process of lipid peroxidation. This type of reaction is of primary importance in photosensitised injury to the skin and will be considered in more detail later.

2.4 Enzymic composition

The early studies of de Duve and colleagues on liver lysosomes concerned the acid hydrolases: acid phosphatase, β-glucuronidase, cathepsin, acid deoxyribonuclease and acid ribonuclease. This list of hydrolytic enzymes, having an acid pH optimum, and exhibiting latency was soon

Table 17.1   A representative list of acid hydrolases that occur in lysosomal particles.
(For additional enzymes and further details see Tappel, 1969a)

β-glucuronidase	β-galactosidase
α-glucosidase	β-N-acetyl-glucosaminidase
Cathepsins, A,B,C,D	α-mannosidase
Acid phosphatase	Phosphoprotein phosphatase
Aryl sulphatase	Acid deoxyribonuclease
Acid ribonuclease	Acid lipase
Phospholipase A	Sphingomyelinase
Glucocerebrosidase	Hyaluronidase
Lysozyme	

extended to include a number of glucosidases, galactosidases, aryl sulphatases and phosphoprotein phosphatases. By 1964 the number of acid hydrolases shown to be lysosomal in localisation had reached 12 (see de Duve 1963) and by 1969 it had reached 40 (see Tappel 1969a). In recent years the most important general finding in relation to the diversity of acid hydrolases has been the recognition of lipid degrading enzymes within the lysosome. It can be seen (Table 17.1) that the acid hydrolases known to be lysosomal in location are potentially capable of breaking down all of the major classes of biologically important material: proteins, fats and carbohydrates. A particularly wide variety of enzymes is present to degrade the complex lipids and mucosaccharides; a large number of serious inborn errors of metabolism are now recognised as having their origin in the absence of one or more enzymes of this type as will be mentioned in section 4.5.

Although a large number of acid hydrolases is now known to be associated with lysosomal particles this does not mean that all the known acid hydrolases necessarily occur within the lysosomes of every type of tissue. Some of the acid hydrolases are more or less restricted to particular types of cell; for example lysozyme is found particularly in the neutrophil leucocytes and hyaluronidase is found to a high concentration in the acrosomal cap of the spermatozoon and in kidney. With other acid hydrolases the specific activity may vary widely from tissue to tissue. In addition it is possible that even in a particular tissue, say liver, the acid hydrolases that exhibit latency may not all be contained together within a uniform set of particles. Data obtained by zonal centrifuging have suggested that some particles may be relatively enriched in some acid hydrolases whereas other particles from the same tissue may be relatively enriched in other acid hydrolases. In liver cathepsin C and acid phosphatase sedimented together whereas acid ribonuclease and cathepsin D were found in a different subfraction (Rahman *et al.* 1967).

Not all the acid hydrolase activity of any particular tissue is necessarily located in the lysosomal particle fraction. In rat liver, for instance, β-glucuronidase occurs to a considerable extent on the membranes of the endoplasmic reticulum as well as in the lysosomes and the activities in the two locations are controlled from separate genetic loci. Acid phosphatase activity occurs in the cytosol as well as in the lysosomal fraction and the two enzymes show different responses to a number of experimental procedures.

2.5 Enzyme analysis

Provided that reasonably large amounts of tissue material are available (e.g. about 10 mg or more of rat liver) then many acid hydrolases can be estimated simply and routinely by colorimetric or spectrophotometric methods. The principles underlying the assay of several acid hydrolases are outlined below.

1 β-glucuronidase

Phenolphthalein glucuronide $\xrightarrow{\text{hydrolysis at pH 5·0}}$ phenolphthalein which can be measured by adding alkali to form a red colour at the end of the incubation period.

2 Acid phosphatase

p-Nitrophenylphosphate $\xrightarrow{\text{hydrolysis}}$ p-nitrophenol which is yellow in alkali.

or

phenyl phosphate $\xrightarrow{\text{hydrolysis}}$ phenol which can be measured by the Folin reaction.

3 Acid ribonuclease

Highly polymerised RNA $\xrightarrow{\text{hydrolysis}}$ low molecular weight nucleotides which can be measured by absorption of light at 260 nm after precipitating the unhydrolysed RNA with acid.

The sensitivity of the estimations can be greatly increased by using fluorescent techniques. A number of procedures have been developed in which hydrolysis of the substrate is accompanied by liberation of a fluorescent material; most of these procedures are based on the use of derivatives of umbelliferone (Fig. 17.7). With these fluorescent procedures enzyme activity has been determined in samples containing small numbers of cells. The use of fluorescent substrates in enzyme estimation has been described by Leaback (1970).

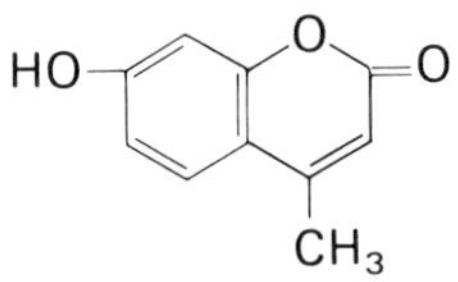

Fig. 17.7 The structure of methyl umbelliferone.

The histochemical study of acid hydrolase activity in tissue sections with either the light or electron microscope has mainly involved acid phosphatase. The most widely used procedure is that described by Gomori 1952. In this technique the phosphate liberated by the action of the phosphatase is precipitated as an insoluble lead salt. The lead salt is then converted to the insoluble and highly coloured lead sulphide which can be observed directly as a coloured deposit with light microscopy or, because of the absorbing effects of lead in an electron beam, can be seen as dense areas in electron micrographs. A short review of the histochemical staining reactions for lysosomal enzymes has been given by Beck and Lloyd (1969).

3 Physiological functions of the lysosomes

3.1 General comments

The lysosomes have been demonstrated to be involved in a wide variety of natural phenomena of fundamental biological importance. In general terms the activity of lysosomes in these diverse processes is dependent upon the digestive activity of the constituent acid hydrolases. Figure 17.8 outlines the types of biological behaviour in which an important role for lysosomes have either been firmly established or for which suggestive evidence exists. The discussion in the rest of this section will illustrate the processes listed in Fig. 17.8 in the order of sections A to E. Other procedures which result in accentuation of or interference with the processes shown in Fig. 17.8 and which reflect abnormal events will be discussed in section 4: pathological conditions involving lysosomal disturbances.

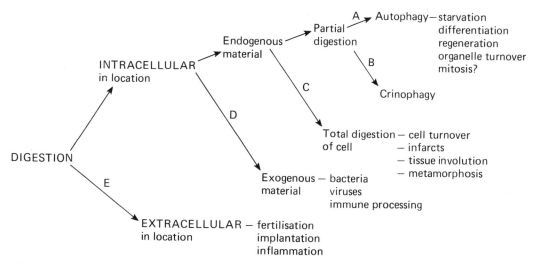

Fig. 17.8 The diagram illustrates the participation of lysosomes in various types of digestive phenomena. Examples of sub-divisions A to E are discussed in the text.

3.2 Autophagy

In this process (see Arstila and Trump 1968; Ericsson 1969) a partial digestion of intracellular material occurs within the confines of a secondary lysosome. In acute starvation, for example, it has been found that numerous cytosegresomes arise in rat liver and that digestion of enclosed material occurs within the subsequently formed secondary lysosome. The process which is considerably accelerated by starvation probably operates to provide a continuing supply of low molecular weight building blocks for the biosynthetic maintenance of key components.

A similar increase in autophagy occurs in the liver when maintained in an isolated perfused condition; Fig. 17.9 shows an electron micrograph of a section of isolated perfused rat liver with numerous lysosomal structures.

In the remodelling of tissues during differentiation, food stores within the cell may be broken down by acid hydrolases to yield components necessary for the sudden spurts that occur in biosynthesis. In fresh water sponges for instance hibernation occurs in the form of gemmules which consist of totipotent cells, the archaeocytes, surrounded by a protective outer casing. The archaeocytes contain masses of food reserves in inclusion bodies that are digested by lysosomal enzymes during histogenesis.

Lysosomes have been implicated in teratogenesis (see review by Lloyd and Beck 1969) by modifying the amount and nature of materials to which the embryo has access. As Lloyd and Beck (1969) write: 'congenital defects are a paediatric problem of growing importance in developed countries since, unlike infections and malnutrition, their incidence in the population (possibly about 2 per cent of all live births in Britain) and their importance as a cause of perinatal death have not decreased significantly in recent decades'. One important way in which lysosomes are involved in early embryonic development is at the stage of embryotrophic nutrition where the very early embryo relies on the absorption of nutrient materials from the endometrial tissue of the mother's uterus. This absorptive stage is dependent on an actively functioning lysosomal system in the embryonic surface layers. Interference with the activity of the lysosomes at this stage may plausibly be imagined to cause potentially serious disturbances in the patterned sequence of development. In this context it can be said that Trypan blue which

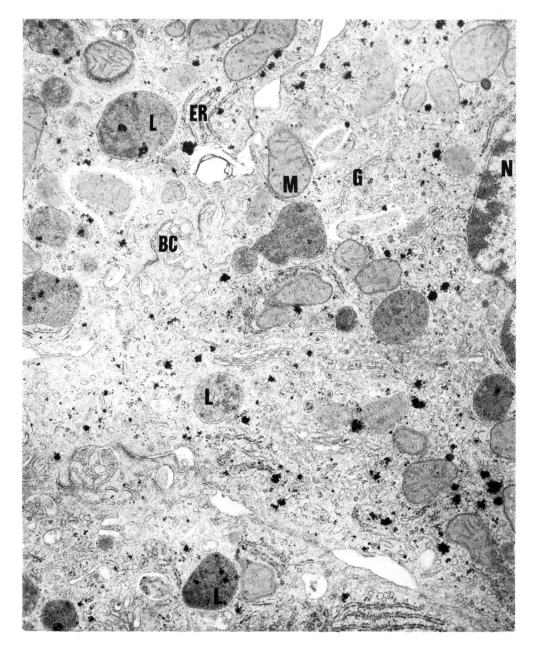

Fig. 17.9 Electron micrograph of a section of rat liver; the liver had been perfused in an isolated state for 3 h prior to fixing and embedding. The perfusion system and experimental details involved in the use of the isolated perfused rat liver technique are described by Bullock *et al.* (1973). The photograph shows a variety of lysosomal structures (L) which are much more frequent in perfused liver than in control liver samples taken from intact rats. Other structures indicated are mitochondria (M); endoplasmic reticulum (ER); Golgi zone (G); bile canaliculus (BC); and a portion of the nucleus (N). Unpublished data of Dr G. Bullock and the author.

is an active teratogen has teratogenic activity only during the stage of embryotrophic nutrition. Trypan blue has been found to decrease the activity of a number of lysosomal hydrolases when added *in vitro* and to inhibit pinocytosis in the yolk sac endothelium. It seems possible to conclude that some at least of the spontaneous malfunctions that are seen in neonates may have arisen from lysosomal disturbances at an early stage of embryogenesis perhaps due to the intake of medicines or dietary factors that contain as yet unrecognised lysosomal inhibitors.

During mitosis the lysosomal population undergoes an intracellular movement to a juxtanuclear orientation (see Kent *et al.* 1965). This and other evidence has led some workers to suggest that lysosomes may have an important functional role in mitosis perhaps in terms of dissolution of the nuclear membrane, or in destruction of the spindle. Evidence that changes in lysosomal membrane permeability occur during mitosis in Hela cells has been given by Maggi (1965) who briefly outlines other contributions to this interesting area of study.

3.3 Crinophagy

The importance of crinophagy in digesting excess hormone within the cell of origin has already been outlined and will not be discussed further here.

3.4 Total digestion

Under this heading will be considered some examples where whole cells undergo digestion by their component lysosomal enzymes.

Many types of cell within the adult human body undergo relatively rapid turnover; obviously to preserve the status quo in terms of tissue size then the number of new cells formed must average out with the number of cells dying-off. Liver cells for example have life times of several hundred days (say 300). Thus on average $\frac{1}{300}$ of the total liver cells turnover every day and in an adult human liver there are about 10^{11} parenchymal cells. It is easy to calculate that on average some 10^3–10^4 liver cells are dying every second and are being replaced by an equivalent number of young cells. The cells that die in such a manner are removed in a variety of ways. They may undergo shrinkage necrosis in which large volumes of the cell are digested within cytosegresome structures (see Kerr 1971): in such a case the acid hydrolases are still restricted within an intracellular vacuole. The cell may be digested by a process of liquifaction necrosis in which lysosomal hydrolases are released into the cell and autolysis takes place outside the confines of a lysosomal particle. The dying or dead cell or cell debris may be engulfed by another phagocytic cell, a macrophage, and then be digested within a secondary lysosome of the macrophage.

If the blood supply to a piece of tissue is cut off or becomes severely restricted by a blood clot then the tissue peripheral to the clot becomes ischaemic and anoxic. If the condition persists then the affected tissue which is deprived of nutrients and oxygen becomes necrotic. Such a mass of necrotic tissue arising from a circulatory block is called an infarct. The tissue in the infarct undergoes autolysis in the process of which lysosomal enzymes carry out massive tissue degradation. Considerable amounts of material from such necrotising tissue may find its way back into the circulation where it can result in low grade fever. The release of such cellular material into the systemic circulation due to such uncontrolled tissue breakdown can be used to monitor the progress of the tissue destruction as for example in the use of serum enzyme assays in myocardial infarction.

In certain physiological conditions the activity of an entire organ may change dramatically from a high to a low activity; such changes are under tight physiological control. An example is

530

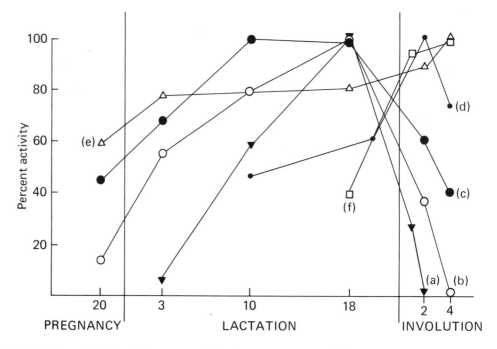

Fig. 17.10 Changes in RNA and in enzyme activities in rat mammary gland through pregnancy, lactation and early involution. Data are from Greenbaum, Slater and Wang (1965), Slater (1961), Greenbaum and Slater (1957), McLean (1958) and Greenbaum and Greenwood (1954). Enzymes: (*a*) glucose-6-phosphate dehydrogenase; (*b*) succinate oxidase; (*c*) RNA; (*d*) cathepsin; (*e*) acid ribonuclease; (*f*) β-glucuronidase.

the involution of the mammary gland when the suckling stimulus is removed. The gland rapidly changes from a highly active functional structure to a much smaller, functionally inactive tissue. In the rapidity of the changes at involution one sees the reverse to the processes occurring at parturition where metabolic activity rapidly builds up. The size and rapidity of such changes are illustrated in Fig. 17.10 for rat mammary gland. During early involution almost every component of a large number so far examined shows a precipitous decrease; acid hydrolase activity however increases during this period (Fig. 17.10). The increase is partly due to a concentration effect (due to decreases in other material) thus leaving the acid hydrolases present in larger proportion; it also results partially from the invasion of the involuting gland by leucocytes and macrophages which play a role in the removal of cell debris.

During the early stage of mammary involution there is an apparent decrease in the latency of acid hydrolases as measured by the bound/free ratio of lysosomal enzymes in mammary gland homogenates (Greenbaum *et al.* 1965). This suggested that involution involves an autolytic type of cell destruction similar to that observed in infarcts. However, unlike the situation with infarcts, involution is not accompanied by release of material indiscriminately into the circulation so that presumably the process is under some tight physiological control. The explanation for these apparently discrepant observations was provided by a morphological assessment of the involuting gland using the electron microscope. Ericsson found that increasingly large cytosegresomes appear in the cytoplasm of the mammary alveolar cells and that a process akin to shrinkage necrosis takes place within the confines of a secondary lysosome. These structures may be enormous in size and contain numerous mitochondria and other cytoplasmic structures (Fig. 17.11). In view of their size it is not difficult to imagine that

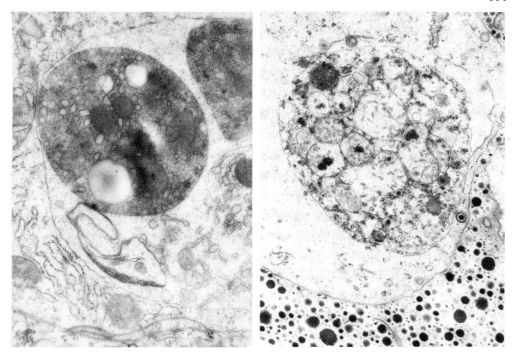

Fig. 17.11 Electron micrographs of sections of involuting rat mammary gland. In Fig. 11 (*a*) a complex cyto-segresome can be seen containing mitochondria, endoplasmic reticulum and fat droplets; the sample was taken from a rat on the second day of mammary involution. In Fig. 17.11 (*b*), an even more complex cytosegresome can be seen and the sample was taken on the third day of mammary involution. The figures were kindly supplied by Professor J. L. E. Ericsson and are reproduced by permission of Academic Press Inc. and Editors of *Journal of Ultrastructure Research* in which the illustrations were published (*J. Ultrastructure Research,* **25**, 214 (1968)).

they rupture more readily than the simpler primary lysosomes during homogenisation. The increased proportion of free enzyme found in mammary gland homogenates, even when prepared by the relatively gentle procedure of Emanuel and Chaikoff (1957) probably represents an artefact of tissue preparation rather than intracellular release under *in vivo* conditions.

Other examples of lysosomal involvement in tissue remodelling or involution include processes of metamorphosis (e.g. the regression of the tadpole tail), the involution of the Mullerian duct in the male chick embryo (see Scheib 1963), and the post-partum involution of the uterus (Wiessner 1969).

3.5 Intracellular digestion of exogenous material

This type of digestion has already been discussed generally in connection with the process of endocytosis. Some aspects of this process are of immediate and prime importance in relation to coping with infectious agents. For example, bacteria can be taken into cells by phagocytosis and many types of bacteria are readily digested by the acid hydrolases of the secondary lysosome; a discussion of the bactericidal properties of lysosomal components has been given by Allen (1969).

Some viruses also get taken into the lysosomal system via phagocytosis (see Dales 1969). In some cases the viral coat-protein is digested and the active nucleic acid is released and replicates in the cell. In other cases the virus particles within the phagosome appear to prevent fusion of the phagosome with a primary lysosome by an unknown mechanism.

Table 17.2   Compounds that have been reported to stabilise or labilise lysosomes under particular conditions

As pointed out in the text a particular compound may stabilise at one concentration or under one set of experimental conditions whereas under other conditions or at a different concentration it may labilize.

STABILISERS	LABILISERS
Cortisone and Hydrocortisone	Steroids such as progesterone, and etiocholanolone
Promethazine and other phenothiazine drugs at low concentration	Ultra-violet, X- and γ-radiation
Chloroquine	Visible light with a photosensitiser
Antioxidants such as vitamin E at low concentration	Triton WR-1339
	Detergents Silica Free radical reactions such as Iron catalysed lipid peroxidation

The lysosomal system has an important function in dealing with the numerous types of infectious microorganisms which enter our bodies. Interference in the proper functioning of the lysosomal system may thus result in increased susceptibility to infection and to decreased resistance to infection. For example the known increased susceptibility to infections of patients on long-term cortisone treatment may be a reflection of the stabilising action of cortisone on the lysosomal membrane (Table 17.2). In patients with the Chediak–Higashi syndrome an inborn error of the lysosomal system results in abnormally large and functionally inactive lysosomes in the leucocytes. Such patients are very susceptible to infectious disease.

Still within the context of the body's resistance to infectious agents, the lysosomes have been suggested to participate in the processing of antigenic material in the macrophages prior to stimulation of the lymphoid tissue. Agents that labilise the lysosomal system such as silica particles and vitamin A increase the immune responses through an adjuvant-like effect. A discussion of the participation of lysosomes in immune processes is given by Weissman and Dukor (1970).

3.6 Digestion by acid hydrolases in extracellular sites

The acrosomal cap of the spermatozoon can be considered as a single modified lysosome. Release of acid hydrolases during the early stage of sperm-egg contact is believed to facilitate penetration of the sperm through the outer wall of the egg. In a recent report Zaneveld *et al.* (1972) have studied the proteinase activity of human spermatozoa and have reported that it has an alkaline pH-optimum; this is in contrast to the acidic pH-optima of lysosomal proteinases generally.

In humans the fertilisation of the egg normally occurs in the Fallopian tube and rapid cell divisions occur during the next day or so to form a blastocyst; part of the outer layer of the blastocyst is specialised to form the trophoblast. The blastocyst attaches to the uterine wall through the trophoblast which rapidly multiplies and invades the uterine epithelium to intitiate implantation. It is believed that the invasion of one tissue by another as occurs in the trophoblast–uterus interaction involves the secretion of extracellular hydrolases that facilitate cell movements into a previously organised tissue structure.

Lysosomal enzymes have also been found to be released extracellularly in a number of other natural conditions. For example in inflammatory conditions resulting from diverse stimuli a number of lysosomal components have been detected in the extracellular fluid; these include various proteins that cause histamine release from mast cell granules and factors that are chemo-attractive for leucocytes. Another type of example of the extracellular release of acid hydrolases concerns the release of adrenal medullary hormones where the concomitant release of acid hydrolases during exocytosis is thought to reflect the activity of lysosomal granules during hormone release.

4 Lysosomes in pathology

4.1 General comments

Ever since the lysosomes were first recognised as a distinct class of intracellular organelle it was attractive, indeed natural to think that disturbances in normal lysosomal function might be associated with severe abnormalities (i.e. pathological conditions) of cellular behaviour. There is no doubt that the release of substantial quantities of destructive acid hydrolases into the cytoplasm would be accompanied by non-specific damage associated with the partial or complete hydrolysis of essential macromolecules. Direct evidence on this point has been obtained by exposing suspensions of cell organelles to lysosomal extracts. Tappel's group (1963) for example found that exposure of liver mitochondria to lysosomal acid ribonuclease resulted in the rapid loss of mitochondrial respiratory control. Further, human diploid cell chromosomes are grossly affected with many apparent chromosomal breaks when the lysosomes are damaged by a photosensitisation reaction (see section 4.7). The chromosomal damage was suggested to result from the release of acid deoxyribonuclease from the lysosomes with subsequent interaction with chromosomes in the nucleus; it was also found that purified acid deoxyribonuclease causes chromosomal breaks when incubated with isolated chromosomes. There is an interesting article by Allison (1969) on the possible close involvement of chromosomal damage by acid hydrolases in experimentally induced cancer, and of the role of lysosomes in carcinogenesis generally.

534

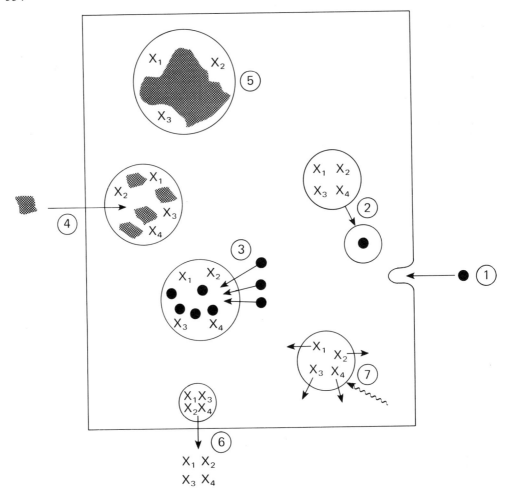

Fig. 17.12 Different mechanisms for causing lysosomal disturbances. (1) Disturbance of uptake;
(2) disturbance in fusion of phagosome and primary lysosome; (3) excessive accumulation of material
normally present in low concentration; (4) excessive accumulation of material not normally present;
(5) deficiency of an enzyme leading to build-up in a component normally metabolised; (6) disturbance
in exocytosis; (7) damage to lysosomal membrane by external agent such as radiation or chemical toxins.

It is clearly important therefore to consider how stable the lysosomal membrane is with
respect to any given treatment, either natural or unnatural in origin. Sensitivity of the
lysosomal fraction to any particular procedure can be considered as a potential initiator of
cellular injury. If the lysosomes are particularly sensitive to a given situation, such that they are
preferentially damaged in comparison to other organelles or enzyme systems, then we can say
that the disturbance to the lysosomal fraction is the primary locus of the cell injury in that
particular situation. The damage to the lysosomes may be sufficient to result in extensive
enzyme release with subsequent cell death; it is in relation to such situations that de Duve
coined the phrase 'suicide particle' in 1956. It was not intended to mean, however, that this was
a normal or frequent function of these organelles but simply that in some situations they may
be responsible for the death of the parent cell itself.

A wide variety of conditions is now known to affect lysosomal function and it seems
probable that the reaction of the lysosomes to any particular treatment varies widely with the

tissue of origin. This tissue variability in lysosomal response may indeed help to explain the known tissue selectivities in response to a number of injurious processes.

Disturbances to the normal functions of lysosomes may occur through interference with any one of a number of separate stages in lysosomal physiology. Fig. 17.12 illustrates a number of ways in which normal lysosomal function may be affected.

In mechanism (1) there is a disturbance in the normal process of endocytosis so that material normally taken in to the cell is retained in the extracellular compartments. In mechanism (2) although material is taken in normally there is a subsequent failure in the fusion of phagosomes with primary lysosomes so that the ingested material is not digested or, at best, digested very slowly within the lysosomes.

Mechanisms (3) and (4) concern disturbances in uptake of materials by the lysosomal system; in (3) there is an excessive accumulation of materials normally present in low concentrations, whereas in (4) there is an accumulation of material not normally present in the lysosomes. As a result of such excessive accumulations the normal processes of digestion may be overwhelmed by the presence of excess substrate or inhibited by the presence of abnormal and toxic materials. The digestion of materials within the lysosomes can also be affected as the result of the absence of a particular enzyme due to a genetic defect; this is covered in mechanism (5). It is evident that absence of a particular enzyme can result in the progressive accumulation of the material normally hydrolysed by the missing enzyme.

In contrast to the abnormalities of accumulation mentioned above there may also be abnormality in the process of extrusion or exocytosis (mechanism 6). In such a situation excessive amounts of acid hydrolases may be liberated outside the cell and cause extracellular disruption.

Another mechanism (7) resulting in lysosomal disturbance includes lysosomal membrane damage resulting from physical changes in the environment (e.g. anoxia, acidity, irradiation) or the direct action of toxic agents. In such cases the attack on the membrane may vary from barely detectable to gross disruption of the lysosomes with subsequent cell death.

Examples of each of the above seven types of lysosomal disturbance will now be given.

4.2 Disturbances of endocytosis

The details governing the biochemical basis of endocytosis are not yet worked out but some broad generalisations can be made:

(a) Materials are known that increase the endocytotic activity of a given type of cell (e.g. the effect of serum on endocytosis in Hela cells; Ahearn *et al.* 1966) as well as (b) decrease it; (c) the process of endocytosis is dependent upon the supply of metabolic energy currency in the form of ATP (or cyclic AMP); (d) the process of endocytosis most probably involves the active participation of microfilaments in the cell periphery. It is in relation to the last point that the material cytochalasin B, is believed to exert its inhibitory action on endocytosis. Cytochalasin B, is a complex natural product that has been found to produce a wide variety of unusual effects on cells including nuclear extrusion and the production of polynucleate varieties (see Carter 1967). It has also been found to inhibit the uptake of bacteria by polymorphonuclear leucocytes (Malawista *et al.* 1971); the effect reported by Malawista and colleagues is clear cut and whereas normal leucocytes will take up numerous bacteria into the lysosomal system the leucocytes exposed to cytochalasin B keep the bacteria outside the cell. In contrast to the inhibitory effects found with cytochalasin B, Najjar and Nishioka (1970) have described a peptide 'Tuftsin' formed from a γ-globulin which stimulates phagocytosis in neutrophil leucocytes; four patients studied who had recurring infections were found to have much reduced levels of 'Tuftsin'.

4.3 Disturbance in fusion

In some instances although exogenous material can gain access to a cell inside a phagosome the normal fusion of the phagosome with a primary lysosome is absent or greatly diminished. Again, not much is known of the detailed biochemistry governing such particle fusions but it is believed to be a process dependent upon ATP or cyclic AMP. Some viruses (reoviruses) can gain access to the cell by normal phagocytotic mechanisms but no fusion with primary lysosomes occurs. Do the viruses secrete some material that inhibits the fusion process by affecting the interactions of the membranes, or do they affect a possible chemotactic mechanism which bring lysosome and phagosome into close juxtaposition? Much interesting work is being done currently on this problem but so far the basic mechanisms have escaped detection.

Another interesting example where fusion apparently is inhibited has come from the work of D'Arcy Hart and colleagues (see Armstrong and D'Arcy Hart 1971). These workers have found that cultured macrophages take up *Mycobacterium tuberculosis* into phagosomes but fusion with lysosomes only appears to occur if the bacterium is in a non-viable state. This suggests that an active living mycobacterium can produce some component that is inhibitory to the lysosome-phagosome fusion process.

It is possible to envisage a similar type of cell dysfunction in which there is a failure of fusion between secretory granules and primary lysosomes as in crinophagy. In such a situation the particular hormone may accumulate to large excess and spill over into the systemic circulation to produce a clinically observable hormone imbalance.

4.4 Disturbance of uptake phenomena

4.4.1 Excessive accumulation of materials normally ingested in small amounts

Two examples will be considered under this heading: the excessive accumulations of copper in Wilson's disease and of iron in haemochromatosis.

Copper is a potent inhibitor of many important enzymes when tested under conditions *in vitro;* the oral ingestion of large amounts of copper salts leads to abdominal irritation, prostration and eventually death from circulatory failure. In hepatolenticular degeneration (also known as Wilson's disease) there is an inherited disturbance of copper metabolism and large amounts of this metal accumulate in the liver, brain and other tissues. The changes in the liver are associated with an increased susceptibility to cirrhosis of the liver where a proliferation of fibrous tissue gradually throttles the liver parenchymal cells into a state of failure; in addition there is a marked increase in the amount of lipid pigment (lipofuschin) in the liver cells. The morphological changes in the liver in Wilson's disease are accompanied by an accumulation of copper ions in the lysosome fraction. Bearing in mind the general toxic properties of copper ions on enzyme systems (including ATP-ases and other ATP dependent reactions) it is possible to see that the intralysosomal accumulation of copper may result in serious lysosomal dysfunction. The mechanism by which patients with Wilson's disease accumulate high concentrations of copper (they are in positive copper balance) is unknown but there seems little doubt that the accumulation within the lysosomes could be responsible for the increased lipid peroxidation that occurs and which results in the presence of much lipofuschin pigment in the livers of such patients.

It is known that iron salts administered by mouth are partially absorbed in the gut and are stored mainly in the reticulo-endothelial system; in liver, the iron is normally confined largely to the Kupffer cells. Under certain conditions, however, the iron can be taken up by the liver parenchymal cells and this condition in the human is associated with a tendency to liver fibrosis; it is called haemochromatosis. The aetiology of haemochromatosis has been the subject of a searching analysis by MacDonald (see MacDonald 1963; MacDonald and Pechet 1965) who was led to the conclusion that, on the one hand, there was an association between haemochromatosis and the daily intake of iron and, on the other hand, between haemo-chromatosis and chronic alcoholism. The analysis of a number of alcoholic beverages showed that some were very rich in iron content. Thus the intake of large volumes of alcohol *may* be associated with a large input of iron salts. MacDonald found, for example, that in the neighbourhood of Rennes (in Brittany), where it was known that a high incidence of haemochromatosis occurred, the local population consumed large volumes of wine and cider. The volume consumed averaged 5 litres per day per person; the drinks were rich in iron. As a result, large amounts of iron may accumulate in the parenchymal cells over a period of years as the high alcohol intake causes a slight disturbance of the parenchymal cell membrane enabling the iron to enter. The iron accumulates within the lysosomes of the parenchymal cells in the same way as mentioned for copper. Iron salts catalyse lipid peroxidation reactions so that it is no surprise to find that patients with haemochromatosis may have increased amounts of lipofuschin pigment in their livers. A further aspect of the excessive accumulation of iron and copper within the lysosomes is that these metal ions have a labilising action on the lysosomal membrane so that a certain amount of acid hydrolase release may occur chronically during the course of Wilson's disease or in haemochromatosis. Whether such lysosomal overload and increased fragility is responsible for the development of fibrosis is not known but is an interesting point for further study.

4.4.2 Excessive accumulation of materials not normally ingested.

Under this heading we can consider the examples of (a) silica, (b) asbestos, and (c) dyes.

Continued exposure to silica dust or asbestos particles in the atmosphere is known to be associated with increased susceptibility to the chronic lung conditions of silicosis and asbestosis. Both diseases involve a progressive fibrosis of the lungs which may greatly affect normal respiratory function; with asbestosis there is an additional change that may develop; a rapidly growing invasive tumour of the cells forming the outer (pleural) membrane covering the lung, a mesothelioma.

Experimental studies have shown that silica dust (size 2–5 μ) is readily taken in by phagocytosis into macrophages present in the lung. Later, the silica particles can be seen within secondary lysosomes which are less stable to environmental conditions than are secondary lysosomes in the absence of silica. The labilisation has been suggested to result from the numerous hydrogen bonds that silica can form within the secondary lysosome and which may result in distortion or disruption of the outer membrane. In fact the compound polyvinyl pyrrolidone which is a very good acceptor for hydrogen-bonding systems, and which also gets into the lysosome system when added to the external medium, protects the lysosomes against the damaging action of silica (see Allison *et al.* 1966; Nadler and Goldfischer 1970; an interesting brief account of the role of silica compounds in biological systems is given by Allison 1968).

Lung damage may be caused by the intake of asbestos fibres, and every attempt is now made to limit exposure to this particular material which has been found to be associated with fatal lung damage in man after variable time periods, often of many years. Exposure of animals to asbestos dust has been found to result in the uptake of the mineral into the lysosomal particles (Smith and Davis 1971). The mechanism by which asbestos can cause lysosomal disturbance, if indeed it does, has not yet been worked out. Although both silica and asbestos can be shown to accumulate within the secondary lysosome particles of lung macrophages it is not yet established whether lysosomal disturbance is the primary stimulus for the chronic development of lung dysfunction. The hypothesis that lysosomal uptake and lung damage are related is, however, an attractive one and shares some similarity with the disturbances mentioned in the previous section (Wilson's disease and haemochromatosis) where chronic lysosomal overload with metal ions is associated with chronic liver damage that progresses towards fibrosis. An interesting development in relation to asbestosis is the finding by Rajan *et al.* (1972) that when human pleura explants maintained in organ culture are exposed to the asbestos mineral crocidolite they show marked proliferation of mesothelial cells within a few days. This technique should prove valuable in the analysis of how asbestos fibres can cause such specific stimulation of the mesothelium. Although crocidolite appears to be the most active of the asbestos varieties in terms of producing mesotheliomas, other types of asbestos also possess toxic activity towards the lung. A report on a recent conference on asbestos toxicity can be found in *Nature, Lond.*, **240**, 256 (1972).

Finally in this section we can mention the uptake by lysosomes of dyestuffs. It has been mentioned already that lysosomes bind positively charged materials as a result of their content of negatively charged glycoproteins. Many dyes are positively charged at neutral pH and accumulate within the lysosomes when added to the external medium either *in vitro* or *in vivo*. For example, acridine orange binds tightly to lysosomes and can be used to monitor the increasing purity of fractions during a differential centrifuging separation since it shows a vivid orange fluorescence when irradiated at 400–450 nm. If acridine orange is added to the medium bathing fibroblasts in culture then within a short time the lysosomes fluoresce bright orange against a dark cytoplasmic background (see Allison and Young 1969). Moreover, when acridine orange is injected intraperitoneally into rats the dye can be localised in tissue lysosomes (e.g. in liver) some hours later (Beeken and Roessner 1972). Other dyes that have been used to demonstrate the lysosomal population in this way include neutral red and tetrazolium salts. The concentration of such materials within the lysosomes by ionic binding to intra-lysosomal anions has been reviewed by Dingle and Barrett (1969a).

4.5 Inborn errors of lysosomal enzyme activity

There are three main mechanisms resulting in lysosomal disturbance which may have a genetic character: (a) an inherited defect in the composition of the lysosomal membrane resulting in abnormal leakage of acid hydrolases into the surrounding cytoplasm; probably only relatively minor disturbances of this kind are compatible with cell viability; (b) an inherited disturbance in some extra-lysosomal metabolic route leading to the presence of abnormally high concentrations of a material that may concentrate within the lysosomes and cause enzyme disturbance; the example of Wilson's disease discussed in section 4.4.5 is relevant in this context; (c) an inherited disturbance in the structure of one of the lysosomal acid hydrolases such that material normally degraded within the lysosomes is not attacked and as a consequence accumulates. This last type of inborn lysosomal disturbance gives rise to the lysosomal storage diseases and there are now numerous examples known that fall under this heading. The remainder of the discussion in this section will concern such storage diseases.

The general features to be expected in such lysosomal storage diseases were outlined by Hers (1965) who has pioneered the development of this area of human biochemical genetics. Hers pointed out clearly that examination of tissue samples from a patient suspected of having a lysosomal storage disease should demonstrate the following characteristics: (a) the storage of the abnormal material is within a membrane-limited particle possessing at least some enzymically active acid hydrolases; (b) several tissues may be affected by the storage phenomena but often to a widely varying degree; (c) the storage product accumulates progressively; (d) the storage product concerns a particular class of substrate normally degraded by the missing enzyme or enzymes. Since some of the acid hydrolases have a broad substrate specificity the accumulated material may be rather heterogeneous in character; (e) the relationship between the storage disease and the clinical syndrome may be difficult to establish; (f) due to the relationship of the lysosomes with endocytosis generally, there is a theoretical possibility for enzyme replacement therapy where the missing enzyme can be supplied in an extra-cellular form and may get taken into the lysosomes by endocytosis.

A few examples of lysosomal storage diseases will now be discussed in order to illustrate the major features outlined above. Although the examples will concern the accumulation of macromolecules within the lysosomes this is not universally the case; small molecules may also accumulate under abnormal conditions. For example, Patrick and Lake (1968) have described the localisation of large crystals of cystine within particles that possess acid hydrolase activity; cystinosis is now classed as a lysosomal storage disease. A short account of inborn diseases involving lysosomes, which also provides numerous background references is given by Raivio and Seegmiller (1972); for a more detailed review, see Hers and Van Hoof (1969).

A number of diseases are associated with the storage of glycogen; in glycogen storage disease type II (also known as Pompe's disease) the glycogen accumulates within the lysosomes mainly in skeletal muscle, heart and liver. Patients with the type II disease have a normal phosphorylase enzyme sequence for degrading glycogen in the cytosol, they respond to epinephrine and glucogen normally and are not hypoglycaemic. They are deficient, however, in the enzyme α-glucosidase which is a lysosomal acid hydrolase and which normally degrades glycogen and oligosaccharides to monosaccharide. It is believed that glycogen particles are trapped normally within the lysosomes as a consequence of autophagy; under normal circumstances such glycogen is degraded but in patients with Pompe's disease it slowly accumulates. Eventually the lysosomes are so engorged with the glycogen that the membrane ruptures, liberating acid hydrolases into the cell and this may cause local cell death. A feature of Pompe's disease is the damage to muscle tissue by such lysosomal enzymes and which is progressively more severe in its consequences. Many of the affected children die in the first years of life but a less severe variant has also been found and is associated with much longer survival. Attempts have been made to treat the disturbance by the administration of an extract containing the missing acid hydrolase. For example, Hug and Schubert (1967) treated a 3-month old patient with an extract prepared from *Aspergillus niger* and possessing acid glucosidase activity. They observed a decrease in the intra-lysosomal glycogen in the patient's tissues following daily treatment with the extract over a prolonged period. Unfortunately, the patient died from an immunological complication. Other possible ways of introducing the deficient enzyme into patients with lysosomal storage diseases and which may be more satisfactory in relation to the immune response will be outlined later in this section.

The second example of storage disease to be discussed concerns the closely related syndromes known as the Hurler syndrome and the Hunter syndrome. In both conditions there is a disturbance in mucopolysaccharide metabolism and products such as dermatan and heparansulphate are stored. Hurler's syndrome, known originally as gargoylism, is associated with skeletal abnormalities, dwarfism, thickening of blood vessel walls, increased sizes of liver and spleen (hepatosplenomegaly), and mental retardation. Hunter's syndrome is clinically similar but is inherited in a recessive X-linked manner whereas the Hurler's disturbance shows an autosomal recessive nature. Despite this genetic difference in the mode of transmission the clinical and biochemical features of the two conditions are very similar. In the tissues of such patients the stored material is localised within lysosomal particles which in liver appear as transparent areas, whereas in nervous tissue there is a zebra-like banding due to increased lipid content. The accumulated material appears heterogeneous in nature and this may explain why variable data have been obtained in relation to the nature of the enzyme defect. It may be that the overall clinical syndrome can be produced by a number of different routes involving separate enzyme deficiencies and where the end result is the accumulation of mucopolysaccharide and glycolipid material. In some patients a marked deficiency in acid α-fucosidase was observed (Hers and Van Hoof 1969); in others, a decreased level of β-galactosidase (see Gerich 1969). An interesting experiment on the basic biological defect in Hurler's and Hunter's syndromes has been reported by Fratantoni et al. (1969). Fibroblasts cultured from the skin of affected individuals accumulate excessive quantities of mucopolysaccharides. When such fibroblasts from a Hurler patient are mixed with normal fibroblasts or with fibroblasts from a patient with Hunter's disease the mucopolysaccharide abnormality in both the affected samples is relieved. A heat-labile high molecular weight material was shown to be responsible for this effect. Presumably a protein enzyme that is deficient in Hurler's syndrome is present in normal cells and in cells from patients with Hunter's syndrome and is released into the culture medium and is subsequently taken up by the deficient cells. Similarly, with the case of cells taken from patients with Hunter's syndrome. In fact the mucopolysaccharide abnormality in, for example, the Hurler's condition can be normalised by the culture medium alone in which normal fibroblasts have been growing, by homogenates of normal cells, and by a factor extractable from normal human urine. However, as Ravivio and Seegmiller (1972) write: 'the specific nature of the corrective factors and their relationship to normal mucopolysaccharide degradation are unknown'.

In a related type of mucopolysaccharide storage, the San Filippo syndrome, there is an accumulation of heparan sulphate. As with the cases of Hurler's and Hunter's syndromes mentioned above the fibroblasts derived from patients with the San Filippo syndrome can be normalised in terms of their mucopolysaccharide storage by the addition in vitro of a protein corrective factor. The San Filippo syndrome has, in fact, been separated into two variants (A and B) as a result of such experiments. San Filippo variant A lacking A-factor and the variant B lacking B-factor; the two conditions showing cross-correction ability. With San Filippo variant B the corrective factor has now been shown to possess N-acetyl-a-D-glucosaminidase activity (Figura and Kresse 1972); a deficiency of this enzyme has been proposed to be the basic defect in the type B disease.

The possible treatment of such inborn lysosomal diseases has attracted increasing attention in terms of how best to attempt the restoration of normal enzyme activity within the lysosomal particles. Obviously, the enzyme administered should produce minimal immune response and this had led to studies aimed at isolating normal human acid hydrolases. Normal plasma contains some acid hydrolase activities and can be used as a replacement source in certain instances. In Fabry's disease, for example, a disturbance in glycosphingolipid metabolism leads to the accumulation of ceramide trihexoside. Patients with Fabry's disease, which is associated with renal malfunction, have been treated with plasma from cross-matched human donors and substantial decreases in the plasma concentration of ceramide trihexoside were observed in the patients (Mapes et al. 1970).

Another and most interesting development is to enclose the enzyme to be administered within a lipid-rich wrapping which will be taken up by tissues into the lysosomal system but which, it is hoped, will have a greatly decreased immunogenic activity when injected into the systemic circulation. Sessa and Weissmann (1970) for example have incorporated the acid hydrolase lysozyme within lipid-rich liposome particles.

4.6 Abnormalities in exocytosis

In the majority of normal situations the digestive action of the acid hydrolases is carried out within the cell whilst contained within the lysosomal system. In some normal situations however, as in fertilisation or in the process of embryonic implantation, an extracellular role for hydrolases is observed. In a number of pathological conditions we can also see acid hydrolases at work outside the cell and in such cases abnormal cell reaction may result at sites distant to the original locus of injury.

Serious damage to the liver, for example, as follows a high dose of CCl_4 which produces extensive liver necrosis, is accompanied by sharp rises in the levels of acid hydrolases in the serum (Slater et al. 1963) in company with the release of other enzymes from the cytosol and mitochondria (see Fig. 17.13). The acid hydrolases released into the circulation are obviously capable of causing widespread degradative changes to blood components and possibly the endothelial cells of the vessels but the extent and significance of such effects have not been quantitatively assessed in such types of tissue injury.

Another type of tissue damage which has been studied very extensively in relation to the release of acid hydrolases out of damaged cells is arthritis. It was suggested by Dingle that in such conditions the local release of lysosomal enzymes caused destruction of neighbouring articular cartilage. This idea was extended by Weissman who considered the role of lysosomes in autoimmune disturbances in general, and rheumatoid arthritis in particular. Extension of such ideas to measurements on the synovial membrane cells have required great sophistications in enzymic techniques and in quantitative cytochemistry (see Altman 1971). The latter

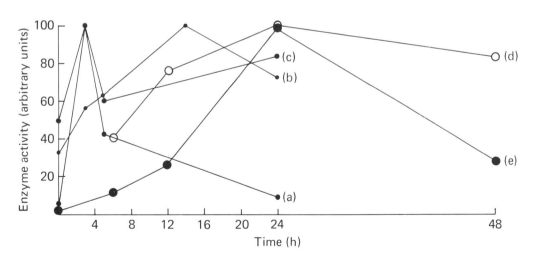

Fig. 17.13 Changes in the activities of enzymes in the serum after poisoning a rat with CCl_4 at time 0. Data are from Slater and Greenbaum (1965) and Rees and Sinha (1960).
Enzymes: (a) β-glucuronidase; (b) acid phosphatase; (c) acid ribonuclease; (d) malic dehydrogenase; (e) glutamate–oxaloacetate transaminase.

developments have been pioneered by the painstaking and elegant approaches of Chayan and Bitensky who have reviewed the role of lysosomal enzyme release in inflammation (Chayan and Bitensky 1971); another interesting account of lysosomal mechanisms in arthritis is that given by Weissmann (1972).

The effects of many anti-inflammatory drugs on lysosomal stability both *in vitro* and *in vivo* have been widely studied. Ignarro (1972) has found that a number of anti-inflammatory agents when administered *in vivo* stabilised the liver lysosome fraction in relation to enzyme release *in vitro* resulting from a number of experimental conditions. He concluded that the results were consistent with the hypothesis that anti-inflammatory drugs inhibit the extrusion of enzymes from lysosomes. An extension of this work (Ignarro *et al.* 1972) reported that the release of lysosomal enzymes is inhibited by cyclic-AMP or ATP which suggests that membrane stability is dependent upon these agents.

4.7 Physical damage to the lysosomal membrane

The effects of high temperature, low osmotic strength of the surrounding medium on lysosomal particle stability have already been mentioned and will not be discussed further.

If lysosomes are incubated at $37°C$ in a medium having a pH value less than 4 they progressively release acid hydrolases into an unsedimentable form (i.e. they lose their property of latency). This effect is believed to be at least partly responsible for the increased fragility of lysosomes in tissues that are rendered anoxic by, for example, a blood clot producing an infarct. Under such conditions the respiratory activity decreases rapidly in parallel with the degree of hypoxia, and lactic acid begins to accumulate; as a result the pH in severely hypoxic tissues begins to decrease. Different tissues may react at widely differing rates to such oxygen deficits. In liver De Duve and Beaufay (1959) found that significant release of acid hydrolases from the lysosomes into a non-sedimentable form did not occur until several hours after imposing total ischaemic anoxia on a specific lobe of rat liver; liver lysosomes are thus rather resistant to anoxic conditions despite very rapid changes in the respiratory activity and levels of components such as ATP which show large alterations within a few seconds of producing anoxia.

The damaging effects of lipid peroxidation on lysosomal membranes have already been mentioned. The production of a suitable type of free radical in the surrounding medium is required to initiate lipid peroxidation; one way of producing substantial quantities of free radicals within cells is to subject them to high energy irradiation such as γ-radiation, X-radiation or ultra-violet radiation. Exposure of lysosomal suspensions to such types of radiation has been shown clearly by Desai *et al.* (1964) to lead to lysosomal membrane damage with subsequent acid hydrolase release. One of the mechanisms by which high radiation can cause tissue injury can thus be seen to result from damaging the integrity of the lysosomal membranes; other mechanisms involve direct chemical modification of protein and nucleic acid structure.

Free radical production normally requires a high energy radiation input since the primary mechanism for free radical production involves the scission of a covalent bond; another possible route is by electron capture (Fig. 17.14). Under certain conditions, however, relatively low radiation energy (e.g. the radiation energy incoming as sunlight) may be sufficient to cause free radical type reactions to occur; such cases require the presence of a photosensitiser (P) such as a porphyrin. The photosensitiser absorbs at the appropriate wavelength of the sun's spectrum and is thereby raised to an excited state of relatively long half-life. In the excited state the molecule acts similarly to a free radical in being able to initiate processes such as lipid peroxidation. For example, the peroxidation of unsaturated fatty acids can be initiated by the addition of a dilute porphyrin solution in the presence of oxygen and exposing to light of wavelength about 410 nm. This mechanism of membrane damage initiated by the process of lipid peroxidation appears to be the cause of epidermal cell injury in photosensitive patients and in animals who are severely affected by sunlight. In 1965 Slater and Riley proposed that in cases of

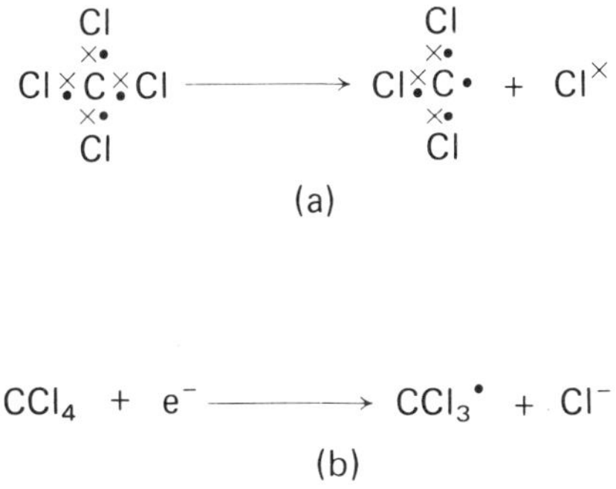

$$
\underset{\text{Cl}}{\overset{\text{Cl}}{\text{Cl} \overset{\times \bullet}{\underset{\times \bullet}{\text{C}}} \text{Cl}}} \longrightarrow \underset{\text{Cl}}{\overset{\text{Cl}}{\text{Cl} \overset{\times \bullet}{\underset{\times \bullet}{\text{C}}} \bullet}} + \text{Cl}^{\times}
$$

(a)

$$
\text{CCl}_4 + e^- \longrightarrow \text{CCl}_3{}^\bullet + \text{Cl}^-
$$

(b)

Fig. 17.14 Formation of free radicals. In (a) the carbon tetrachloride undergoes covalent bond scission at one carbon–chlorine bond to yield the trichloromethyl radical $\text{CCl}_3^{\bullet}$ and a chlorine atom. In (b) the carbon tetrachloride undergoes electron capture with subsequent bond scission to yield $\text{CCl}_3^{\bullet}$ and chloride ion.

photosensitisation due to the presence of excess porphyrin pigment in the skin and other tissues the primary attack of the excited photosensitiser was on the lysosomal membranes. In fact, very rapid changes in the lysosomes of rat epidermal cells could be demonstrated when low concentrations of porphyrins (10^{-6} M) were present and short exposure to light performed. In this type of tissue damage it is probable that disruption of the lysosomal membrane by a free radical motivated attack is a primary mechanism leading to subsequent cell injury. As already mentioned free radical reactions may be modified by the inclusion of scavengers in the system. This suggests one way of developing a rational therapy for photosensitive patients: find a non-toxic free radical scavenger which when taken by mouth concentrates in the skin. The natural pigment β-carotene possesses such properties and in a limited trial of patients with severe photosensitisation due to erythropoietic protoporphyria it was reported (Matthews-Roth et al. 1970) that administration of β-carotene produced substantial improvement in skin reaction to sunlight. Further details of the reactions involved in photosensitisation can be found in the general account of free radical mechanisms on tissue injury by Slater (1972).

Lysosomal membrane damage could also be produced by the direct or indirect action of toxic agents. For example, CCl_4 causes liver necrosis in many species but before it causes such serious liver cell damage it has first to undergo metabolism to a more toxic form. The most likely toxic product is a free radical, $\text{CCl}_4^{\bullet}$ (see Slater 1966, 1972 for reviews). It has been suggested by a number of investigators that such activated molecules could cause lysosomal membrane damage, possibly by lipid peroxidation, and that lysosomal enzyme release is an important feature of such injuries. However, in a number of types of liver injury that were studied (Slater 1969) lysosomal injury appeared as a late and secondary feature of the liver damage rather than as a primary or initiating locus. These data do not exclude the possibility that direct attack on the lysosomal membrane by toxic agents can occur, for each type of cellular injury has its own special features and peculiarities. Indeed, a recent report by Dolara et al. (1971) has shown that phalloidin, a toxic principle from Amanita phalloides, causes very early disturbance to the liver cell lysosomes.

5. Peroxisomes

5.1 General comments

Only a very brief account of the discovery, properties and possible biological functions of the peroxisomes will be given here since recent reviews adequately cover these particles (De Duve and Wattiaux 1966; De Duve 1969; Avers 1971).

In the early centrifuging studies of De Duve and colleagues with rat liver the enzymes urate oxidase and D-amino acid oxidase were found to sediment with the lysosomal acid hydrolases. However, the oxidases did not display latency and, later, when purified 'tritosomes' were separated after administering Triton-WR-1339 (see section 2.1) it was observed that the particles containing urate oxidase were now clearly distinguished from lysosomes. Further studies confirmed the existence of a new class of cell organelle in rat liver containing a number of oxidase enzymes producing hydrogen peroxide as an end product of reaction; this feature led to the suggestion of 'peroxisome' as the name for these organelles by De Duve in 1965.

In contrast to the widespread distribution of lysosomes in higher animal tissues, peroxisomes have only been demonstrated so far in a relatively small number of cell types. Peroxisome particles do occur, however, in a number of other species such as in protozoa and in some plant tissues. It has been reported that peroxisomes occur in the leaves of plants that perform photo-respiration but not in other leaves that do not carry out this function. This particular

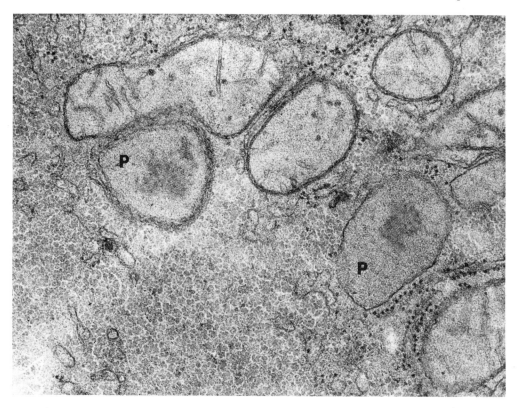

Fig. 17.15 Electron micrograph of section of rat liver; the rat had been injected 2 h previously with sodium salicylate (400 mg/kg body wt.) as described by Bullock *et al.* (1970). Two peroxisomes (P) can be seen with limiting membranes and central crystalline inclusions. Unpublished data of Dr G. B. Bullock and the author.

association of peroxisomes with photo-respiration has been suggested to be relevant to nitrate and nitrite reduction via nitrate – and nitrite – reductases (Lips and Avissar 1972) present in the same particles.

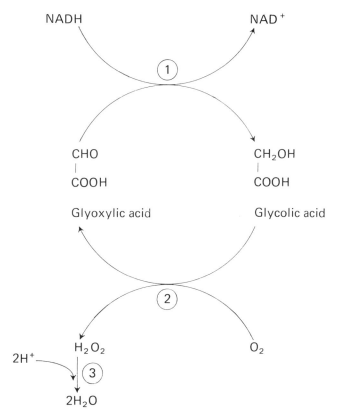

Fig. 17.16 Oxidation of NADH by trace amounts of glyoxylate in leaf peroxisomes using (1) glyoxyate reductase and (2) glycolate oxidase; the hydrogen peroxide produced can be metabolised by catalase (3).

In rat liver the peroxisome has a characteristic appearance under the electron microscope, having a bounding membrane and an inner crystalline core or inclusion body which has been shown to be crystalline uricase (see Fig. 17.15). In rat kidney the crystalline core is absent, as is the uricase enzyme. Although the number of peroxisomes in rat liver is generally small certain treatments cause considerable increases in the relative proportion of these particles. For example, chronic feeding with the hypo-cholesterolemic drug 'Atromid' results in increased numbers of peroxisomes as does a large dose of salicylate (Bullock *et al.* 1970).

5.2 Composition

The enzyme content of peroxisomes varies rather widely from tissue to tissue. In rat liver the major enzymes so far characterised are uricase, catalase, L-α-hydroxyacid oxidase, D-amino acid oxidase, glycolate oxidase and NADP-isocitrate dehydrogenase. In the 'glyoxysomes' of germinating castor-oil bean and other fatty seedlings which contain catalase the enzymes capable of carrying out the glyoxylic acid cycle are present: citrate synthase, aconitase,

546

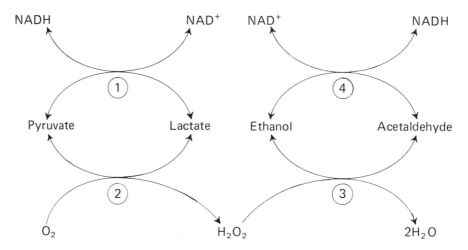

1. Lactate dehydrogenase
2. L-α-Hydroxyacid oxidase
3. Catalase
4. Alcohol dehydrogenase

Fig. 17.17 Possible inter-actions of cytosol enzymes with peroxisomal enzymes in the control of NAD$^+$/ NADH reactions. Cell sap enzymes (1) and (4) carry out either the oxidation of NADH or the reduction of NAD$^+$; enzymes (2) and (3) catalyse the transfer of hydrogen from lactate or ethanol to form water.

isocitrate lyase, malate synthase and malate dehydrogenase. In the course of each turn of this cycle, two molecules of acetyl coenzyme A are converted to succinate. In leaf peroxisomes two enzymes, glyoxylate reductase and glycolate oxidase are present which can catalyse the oxidation of NADH by oxygen using catalytic amounts of glyoxylate (Fig. 17.16). Although this system is apparently not present in rat liver it is possible to imagine similar oxidative schemes for NADH using dehydrogenases present in the cell sap (Fig. 17.17). In such an arrangement lactate and pyruvate would institute a cell sap–peroxisome shuttle. More direct evidence on the significance of peroxisomes in affecting or even controlling cytoplasmic NAD$^+$/NADH levels will be awaited with interest. Interesting speculations on the possible role of peroxisomes in cell respiration at different oxygen tensions are discussed by De Duve (1969).

References

AHEARN, M. J., HAMILTON, T. H. and BIESELE, J. J. (1966) 'Serum-induced formation of lysosomes in Hela cells: a process sensitive to actinomycin-D'. *Pro. Natn. Acad. Sci. U.S.A.,* **55,** 852-57.

ALLEN, J. M. (1969) 'Lysosomes in bacterial infection', vol. 2, chapter 3 of *Lysosomes in Biology and Pathology.* Eds.: J. T. Dingle and Honor B. Fell, North-Holland, Amsterdam.

ALLISON, A. C., HARRINGTON, J. S. and BIRKBECK, M. (1966) 'An examination of the cytotoxic effects of silica on macrophages', *J. Exp. Med.,* **124,** 141-53.

ALLISON, A. C. (1968) 'Silica compounds in biological systems'. *Proc. Roy. Soc. B.,* **171,** 19-30.

ALLISON, A. C. and YOUNG, M. R. (1969) 'Vital staining and fluorescence microscopy of lysosomes', vol. 2, chapter 22 of *Lysosomes in Biology and Pathology,* loc. cit.

ALLISON, A. C. (1969) 'Lysosomes and cancer', vol. 2, chapter 8 of *Lysosomes in Biology and Pathology,* loc. cit.

ALTMAN, F. P. (1971) 'The use of a recording microdensitometer for the quantitative measurement of enzyme activities inside tissue sections.' *Histochemie, 27,* 125-36.

APPELMANS, F., WATTIAUX, R. and DE DUVE, C. (1955) 'The association of acid phosphatase with a special class of cytoplasmic granules in rat liver.' *Biochem. J., 59,* 438-45.

ARMSTRONG, J. A. and D'ARCY-HART, P. (1971) 'Response of culture macrophages to *Mycobacterium tuberculosis,* with observations on fusion of lysosomes with phagosomes.' *J. Exp. Med., 134,* 713-40.

ARSTILA, A. U. and TRUMP B. F. (1968) 'Studies on cellular autophagocytosis.' *Amer. J. Pathol, 53,* 687-733.

BEAUFAY, H. (1969) 'Methods for the isolation of lysosomes,' vol. 2, chapter 18 of *Lysosomes in Biology and Pathology,* loc. cit.

BECK, F. and LLOYD, J. B. (1969) 'Histochemistry and electron microscopy of lysosomes,' vol. 2, chapter 21 of *Lysosomes in Biology and Medicine,* loc. cit.

BEEKEN, W. L. and ROESSNER, K. D. (1972) '*In vivo* labelling of hepatic lysosomes by intragastric administration of acridine orange.' *Lab. Invest., 26,* 173-77.

BROWN, D. H. (1968) 'Separation of mitochondria, peroxisomes and lysosomes by zonal centrifugation in a Ficoll gradient.' *Biochim. Biophys. Acta, 162,* 152-53.

BULLOCK, G., CARTER, E. E. and WHITE, A. M. (1970) 'The preparation of mitochondria from muscle without the use of a homogeniser.' *FEBS Letters, 8,* 109-11.

BULLOCK, G., DELANEY, V. B., SAWYER, B. C. and SLATER, T. F. (1970) 'Biochemical and structural changes in rat liver resulting from the parenteral administration of a large dose of sodium salicylate.' *Biochem. Pharmacol., 19,* 245-53.

CAMPBELL, P. N. (1966) *The Structure and Function of Animal Cell Components.* Pergamon Press, Oxford.

CARTER, S. B. (1967) 'Effects of cytochalasin on mammalian cells.' *Nature, Lond., 213,* 261-64.

CHAYEN, J. and BITENSKY, L. (1971) 'Lysosomal enzymes and inflammation.' *Ann. Rheum. Dis., 30,* 522-36.

CLAUDE, A. (1969) 'Microsomes, endoplasmic reticulum, and interactions of cytoplasmic membranes,' pp. 3-39 in *Microsomes and Drug Oxidations.* Eds.: J. R. Gillette, A. H. Conney, G. J. Cosmides, R. W. Estabrook, J. R. Fouts and G. J. Mannering. Academic Press, New York.

CONTRACTOR, S. (1969) 'Lysosomes in human placenta.' *Nature, Lond., 223.* 1275.

DALES, S. (1969) 'Role of lysosomes in cell-virus interactions,' vol. 2, chapter 4 of *Lysosomes in Biology and Pathology,* loc. cit.

DE DUVE, C., PRESSMAN, B. C., GIANETTO, R., WATTIAUX, R. and APELMANS, F. (1955) 'Intracellular distribution patterns of enzymes in rat liver tissue.' *Biochem. J., 60,* 604-17.

DE DUVE, C. and BEAUFAY, H. (1959) 'Influence of ischaemia on the state of some bound enzymes in rat liver.' *Biochem. J. 73,* 610-16.

DE DUVE, C. (1963) 'The lysosome concept,' pp. 1-31 in *Lysosomes* CIBA Foundation Symposium. Eds.: A.V.S. de Reuck and M.P. Cameron. J. & A. Churchill, London.

DE DUVE, C. and WATTIAUX, R. (1966) 'Peroxisomes,' *Physiol. Rev., 46,* 323-57.

DE DUVE, C. (1969a) 'The lysosome in retrospect,' vol. 1, chapter 1 of *Lysosomes in Biology and Pathology,* loc. cit.

DE DUVE, C. (1969b) 'The peroxisome: a new cytoplasmic organelle,' *Proc. Roy. Soc. B.,* **173,** 71-83.

DESAI, I. D., SAWANT, P. L. and TAPPEL, A. L. (1964) 'Peroxidative and radiation damage to isolated lysosomes.' *Biochim. Biophys. Acta,* **86,** 277–85.

DINGLE, J. T. and BARRETT, A. J. (1969a) 'Uptake of biologically active substances by lysosomes.' *Proc. Roy. Soc. B.,* **173,** 85-93.

DINGLE, J. T. and BARRETT, A. J. (1969b) 'Some special methods for the study of the lysosomal systems' in *Lysosomes in Biology and Pathology,* loc. cit.

DOLARA, P., BUIATTI, E. and GEDDES, M. (1971) 'Hydrocortisone protection of phalloidin-induced rat liver lysosome damage,' *Pharmacol. Res. Commun.,* **3,** 1-12.

ERICSSON, J. L. E. (1969) 'Mechanism of cellular autophagy,' vol. 2, chapter 12 of *Lysosomes in Biology and Pathology*, loc. cit.

FIGURA, K. von and KRESSE, H. (1972) 'The San Filippo B corrective factor: an *N*-acetyl-α-D-glucosaminidase'. *Biochem. Biophys. Res. Commun.,* **48,** 262-69

FRATANTONI, J. C., HALL, C. W. and NEUFELD, E. F. (1969) 'The defect in Hurler and Hunter syndromes.' *Proc. Natn. Acad. Sci. U.S.A.,* **64,** 360-66.

GERICH, J. E. (1969) 'Hunter's syndrome,' *New Engl. J. Med.,* **280,** 799-802.

GIANETTO, R. and de DUVE, C. (1955) 'Comparative study of the binding of acid phosphatase, β-glucuronidase and cathepsin by rat liver particles.' *Biochem. J.,* **59,** 433-38.

GOLDSTONE, A. and KOENIG, H. (1970) 'Lysosomal hydrolases as glycoproteins.' *Life Sciences,* **9,** part II, 1341-50.

GOLDSTONE, A., SZABO, E. and KOENIG, H. (1970) 'Isolation and characterisation of acidic lipoprotein in renal and hepatic lysosomes.' *Life Sciences,* **9,** part II, 607-16.

GREENBAUM, A. L. and GREENWOOD, F. C. (1954) 'Some enzymic changes in the mammary gland of rats during pregnancy, lactation and mammary involution.' *Biochem. J.,* **56,** 625–31.

GREENBAUM, A. L. and SLATER, T. F. (1957) 'The relationship between enzyme activity and particle counts in mammary gland suspensions.' *Biochem. J.,* **66,** 161-66.

GREENBAUM, A. L., SLATER, T. F. and WANG, D. Y. (1960) 'Lysosomal-like particles in rat mammary gland.' *Nature, Lond.,* **188,** 318-20.

GREENBAUM, A. L., SLATER, T. F. and WANG, D. Y. (1965) 'Lysosomal enzyme changes in enforced mammary-gland involution.' *Biochem. J.,* **97,** 518-22.

HAGAN, W. A. (1922) 'The value of heat-killed cultures for the prevention of the *Bacillus abortus* inoculation disease of guinea pigs.' *J. Exp. Med.,* **36,** 711-25.

HELMINEN, H. J. and ERICSSON, J. L. E. (1968) 'Studies on mammary-gland involution.' *J. Ultrastruct. Res.,* **25,** 214-27.

HENNING, R. and STOFFEL, W. (1972) 'Ubiquinone in the lysosomal membrane fraction of rat liver.' *Hoppe-Seyler's Z. Physiol. Chem.,* **353,** 75-8.

HERS, H. G. (1965) 'Inborn lysosomal diseases,' *Gastroenterology,* **48,** 625-33.

HERS, H. G. and VAN HOOF, F. (1969) 'Genetic abnormalities of lysosomes,' vol. 2, chapter 2 of *Lysosomes in Biology and Pathology,* loc. cit.

HILLMAN, H. H. (1972) *Certainty and Uncertainty in Biochemical Techniques.* Surrey University Press.

HOGEBOOM, G. H., SCHNEIDER, W. C. and PALADE, G. (1948) 'Cytochemical studies of mammalian tissues: isolation of intact mitochondria from rat liver; some biochemical properties of mitochondria and submicroscopic particulate material.' *J. Biol. Chem.,* **172,** 619-35.

HUG, G. and SCHUBERT, W. K. (1967) 'Lysosomes in Type II glycogenosis.' *J. Cell Biol.*, **35**, C1-C6.

IGNARRO, L. J., KRASSIKOFF, N. and SLYWKA, J. (1972) 'Inhibition of lysosomal enzyme release by adenosine triphosphate and dibutyryl-cyclic AMP.' *Life Sciences,* **11**, part II, 317-22.

IGNARRO, L. J. (1972) 'Lysosome membrane stabilisation *in vivo:* effects of steroidal and non-steroidal anti-inflammatory drugs on the integrity of rat liver lysosomes.' *J. Pharmacol. Exp. Therapeut.,* **182**, 179-88.

KENT, G., MINICK, O. T., ORFEI, E., VOLINI, F. I. and MADERA-ORSINI, F. (1965) 'The movement of iron-laden lysosomes in rat liver cells during mitosis.' *Amer. J. Pathol.,* **46**, 803-27.

KERR, J. F. R. (1969) 'An electron microscopic study of giant cytosegresomes in acute liver injury due to heliotrine.' *Pathology,* **1**, 83-94.

KERR, J. F. R. (1971) 'Shrinkage necrosis: a distinct mode of cellular death.' *J. Path.,* **105**, 13-20.

LEABACK, D. H. (1970) *An Introduction to the Fluorimetric Estimation of Enzyme Activities.* Koch-Light Laboratories Ltd.

LIPS, S. H. and AVISSAR, Y. (1972) 'Plant leaf microbodies as the intracellular site of nitrate reductase and nitrite reductase.' *Eur. J. Biochem.,* **29**, 20-24.

LLOYD, J. B. and BECK, F. (1969) 'Teratogenesis,' vol. 1, chapter 16 of *Lysosomes in Biology and Pathology,* loc. cit.

MACDONALD, R. A. (1963) 'Idiopathic hemochromatosis,' *Arch. Int. Med.,* **112**, 184-90.

MACDONALD, R. A. and PECHET, G. S. (1965) 'Experimental hemochromatosis in rats.' *Amer. J. Pathol.,* **46**, 85-109

MAGGI, V. (1965) 'A study of lysosomal acid phosphatase during mitosis in Hela cells.' *J. Roy. Microscop. Soc.,* **85**, 291-95.

MALAWISTA, S. E., GEE, B. L. and BENSCH, K. G. (1971) 'Cytochalasin B reversibly inhibits phagocytosis: function, metabolic and ultrastructural effects in human blood leukocytes and rabbit alveolar macrophages.' *Yale J. Biol. Med.,* **44**, 286-300.

MAPES, C. A., ANDERSON, R. L., SWEELEY, C. C., DESNICK, R. J. and KRIVIT, W. (1970) 'Enzyme replacement in Fabry's disease, an inborn error of metabolism.' *Science,* **169**, 987-89.

MATHEWS-ROTH, M. M., PATHAK, M. A. FITZPATRICK, T. B., HARBER, L. C. and KASS, E. H. (1970) 'β-carotene as a photoprotective agent in erythropoietic protoporphyria.' *New Engl. J. Med.,* **282**, 1231-34.

MCLEAN, P. (1958) 'Carbohydrate metabolism of mammary tissue.' *Biochim. Biophys. Acta,* **30**, 316-24.

NADLER, S. and GOLDFISCHER, S. (1970) 'The intracellular release of lysosomal contents in macrophages that have ingested silica.' *J. Histochem. Cytochem.,* **18**, 368-71.

NAJJAR, V. A. and NISHIOKA, K. (1970) 'Tuftsin: a natural phagocytosis stimulating peptide.' *Nature, Lond.,* **228**, 672-73.

NOVIKOFF, A. B. and HOLTZMAN, E. (1970) *Cells and Organelles,* Holt, Rinehart and Winston, New York.

NYQUIST, S. E., BARR, R. and MORRE, D. J. (1970) 'Ubiquinone from rat liver Golgi: apparatus fractions.' *Biochim. Biophys. Acta,* **208**, 532-34.

PALADE, G. E. and PORTER, K. R. (1954) 'Studies on endoplasmic reticulum: its identification in cells *in situ.*' *J. Exp. Med.,* **100**, 641-56.

PATRICK, A. D. and LAKE, B. D. (1968) 'Cystinosis: electron microscopic evidence of lysosomal storage of cystine in lymph node.' *J. Clin. Path.,* **21**, 571-75.

PORTER, K. R. (1953) 'Observations on a submicroscopic basophilic component of cytoplasm.' *J. Exp. Med.,* **97**, 727-49.

POTTER, VAN R. and ELVEHJEM, C. A. (1936) 'A modified method for the study of tissue oxidations.' *J. Biol. Chem.,* **114**, 495-504.

RAGAB, H., BECK, C., DILLARD, C. and TAPPEL, A. L. (1967) 'Preparation of rat liver lysosomes.' *Biochim. biophys. Acta,* **148**, 501-5.

RAJAN, K. T., WAGNER, J. C. and EVANS, P. H. (1972) 'The response of human pleura in organ culture to asbestos.' *Nature, Lond.,* **238**, 346-47.

RAHMAN, Y. E., HOWE, J. F., NANCE, S. L. and THOMSON, J. F. (1967) *Biochim. Biophys. Acta,* **146**, 484-92.

RAIVIO, K. O. and SEEGMILLER, J. E. (1972) 'Genetic diseases of metabolism,' p. 543 in *Annual Reviews of Biochemistry,* vol. 41, Annual Reviews, Palo Alto, California, U.S.A.

REES, K. R. and SINHA, K. P. (1960) 'Blood enzymes in liver injury,' *J. Path. Bact.,* **80**, 297-307.

REIJNGOUD, D. J. and TAGER, J. M. (1973) 'Measurement of intralysosomal pH,' *Biochim. Biophys. Acta,* **297**, 174-178.

SAWANT, P. L., SHIBKO, S., KUMTA, U.S. and TAPPEL, A. L. (1964) 'Isolation of rat liver lysosomes and their general properties.' *Biochim. biophys. Acta,* **85**, 82-92.

SCHEIB, D. (1963) 'Properties and role of acid hydrolases of the Mullerian ducts during sexual differentiation in the male chick embryo,' p. 264 in *Lysosomes* CIBA Foundation Symposium, loc. cit.

SCHUEL, H., SCHUEL, R. and UNAKAR, N. J. (1968) 'Separation of rat liver lysosomes and mitochondria in the A − XII zonal centrifuge.' *Analyt. Biochem.,* **25**, 146-63.

SESSA, G. and WEISSMANN, G. (1970) 'Incorporation of lysozyme into liposomes.' *J. Biol. Chem.,* **245**, 3296-301.

SLATER, T. F. (1961) 'Ribonuclease activity in the rat mammary gland during pregnancy, lactation and mammary involution.' *Biochem. J.,* **78**, 500-4.

SLATER, T. F., GREENBAUM, A. L. and WANG, D. Y. (1963) 'Lysosomal changes during liver injury and mammary involution,' p. 311 in *Lysosomes* CIBA Foundation Symposium, loc. cit.

SLATER, T. F. and GREENBAUM, A. L. (1965) 'Changes in lysosomal enzymes in acute experimental liver injury.' *Biochem. J.,* **96**, 484-91.

SLATER, T. F. and RILEY, P. A. (1965) 'Photosensitisation and lysosomal damage.' *Nature, Lond.,* **209**, 151-54.

SLATER, T. F. (1969) 'Aspects of cellular injury and recovery,' vol. 1, chapter 11 in *Biological Basis of Medicine.* Eds. E. E. Bittar and N. Bittar, Academic Press, New York.

SLATER, T. F. (1972) *Free Radical Mechanisms in Tissue Injury.* Pion Ltd., London.

SLATER, T. F. (1969) 'Lysosomes and experimentally induced tissue injury,' vol. 1, chapter 18 of *Lysosomes in Biology and Pathology,* loc. cit.

SLATER, T. F. (1966) 'The necrogenic action of CCl_4 on the rat; a speculative hypothesis based on activation.' *Nature, Lond.,* **209**, 36–40.

SMITH, B. A. and DAVIS, J. M. G. (1971) 'The association of phagocytosed asbestos dust with lysosome enzymes.' *J. Pathol.,* **105**, 153-57.

STAHN, R., MAIER, K-P., and HANNIG, K. (1970) 'A new method for the preparation of rat liver lysosomes.' *J. Cell Biol.,* **46**, 576-91.

TAPPEL, A. L., SAWANT, P. L. and SHIBKO, S. (1963) 'Lysosomes: distribution in animals, hydrolytic capacity and other properties,' p.78 in *Lysosomes* CIBA Foundation Symposium, loc. cit.

TAPPEL, A. L. (1969a) 'Lysosomal enzymes and other components,' vol. 2, chapter 9 of *Lysosomes in Biology and Pathology*, loc. cit.

TAPPEL, A. L. (1969b) 'Methods for the study of lysosomal function,' vol. 2, chapter 19 of *Lysosomes in Biology and Pathology*, loc. cit.

WATTIAUX, R. and DE DUVE, C. (1956) 'Release of bound hydrolases by means of Triton X-100.' *Biochem. J.*, **63**. 606-8.

WEISSMANN, G. and DUKOR, P. (1970) 'The role of lysosomes in immune responses.' *Adv. Immunol.*, **12**, 283-331.

WEISSMANN, G. (1972) 'Lysosomal mechanisms of tissue injury in arthritis.' *New Engl. J. Med.*, **286**, 141-47.

WOESSNER, J. F. (1969) 'The physiology of the uterus and mammary gland,' vol. 1, chapter 11 of *Lysosomes in Biology and Pathology*, loc. cit.

ZANEVELD, L. J. D., DRAGOJE, B. M. and SCHUMACHER, G. F. B. (1972) 'Acrosomal proteinase and proteinase inhibition of human spermatazoa.' *Science*, **177**, 702-4.

ZEYA, H. I. and SPITZNAGEL, J. K. (1966) 'Antimicrobial specificity of leukocyte lysosomal cationic proteins.' *Science*, **154**, 1049-51.

18
Mitochondrial Oxidative Phosphorylation

A. P. Dawson and M. J. Selwyn

School of Biological Sciences, University of East Anglia

1 Introduction

The mechanism of energy transduction by mitochondria remains one of the outstanding problems of modern biochemistry. The object of this review is to attempt to summarise the experimental evidence in favour of the various hypotheses of oxidative phosphorylation and to indicate how evidence for one hypothesis may be interpreted in terms of another. Since the relative merits of the chemical hypothesis, the chemiosmotic hypothesis and the conformational hypothesis have been argued at length for several years, it should not be expected that a definite decision is possible here. We have attempted to be as unbiased as possible, but allowance should perhaps be made for our personal preference for the chemiosmotic hypothesis.

There are certain observations with which all theories of oxidative phosphorylation must be consistent. To explain the stoichiometry of phosphorylation relative to oxygen uptake, there must be three sites along the electron transport chain at which energy can be conserved. The phenomenon of reversed electron transport shows that energy from one site can be used at another and hence must ultimately be convertible into a common form other than ATP,

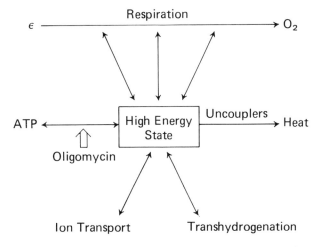

Fig. 18.1 The general relationship between energy-linked functions of mitochondria and electron transport.

whichever coupling site it arises from. The common intermediate can be utilised in several ways, e.g. ATP synthesis, ion transport, transhydrogenation as well as reversed electron transport. All these energy linked functions are inhibited by uncoupling agents, while only ATP synthesis is blocked by oligomycin, which also inhibits ion transport, transhydrogenation and reversed electron transport when these are ATP-supported. The high energy intermediate must under normal circumstances be relatively stable, since the phenomenon of respiratory control indicates that under conditions where use of the intermediate is minimised, its breakdown is rate limiting for electron transfer.

All theories must therefore conform to the idea of a high energy intermediate, which may be chemical or physical in nature. In addition they must provide, as well as a mechansim of synthesis of the intermediate, convincing mechanisms by which the energy of the intermediate state can be utilised for ATP synthesis, ion transport or electron transport (Fig. 18.1).

2 Hypotheses

2.1 The chemical hypothesis

The chemical theory as originally proposed (Slater 1953) was based on a mechanism of substrate level phosphorylation. The central feature of the theory is that there exist discrete chemical compounds between the respiratory carriers and hypothetical coupling factors. Using the notation of Chance and Williams (1956):

$$\left. \begin{array}{l} AH_2 + B + I \; \rightleftharpoons \; A \sim I + BH_2 \\ A \sim I + ADP + P_i \; \rightleftharpoons \; A + I + ATP \end{array} \right\} \; \textit{Scheme I}$$

AH_2 and B are two adjacent respiratory carriers. There is, however, no *a priori* reason for the formation of $A \sim I$ rather than $BH_2 \sim I$ and the scheme is flexible in this regard. When the requirement for a common factor between coupling sites became apparent, together with the necessity of explaining the absence of any effect of the redox state of the respiratory carriers on the uncoupler–stimulated ATPase and the ATP-P_i exchange reaction, the hypothesis was elaborated by the introduction of a further intermediate.

$$\left. \begin{array}{ll} AH_2 + B + I \; \rightleftharpoons \; A \sim I + BH_2 & (1) \\ A \sim I + X \; \rightleftharpoons \; A + X \sim I & (2) \\ X \sim I + P_i \; \rightleftharpoons \; X \sim P + I & (3) \\ X \sim P + ADP \; \rightleftharpoons \; X + ATP & (4) \end{array} \right\} \; \textit{Scheme II}$$

$X \sim I$ in this scheme can be regarded as the common energy transducing intermediate. This is considered as being the target of action of uncoupling agents, which, directly or indirectly, catalyse the reaction:

$$X \sim I \; \rightarrow \; X + I$$

Oligomycin is thought to block reaction (3) and transhydrogenation and ion transport are both driven by utilisation of $X \sim I$, produced either by respiration or by ATP.

The utilisation of $X \sim I$ for ion transport adds a considerable complication to the chemical theory. As will be seen (section 4), mitochondrial ion transport involves H^+ movement across the inner membrane, and it has been suggested (Chappell and Crofts 1965) that $X \sim I$ could be used to drive a proton translocating pump. If this were the case, it would provide an explanation for one of the biggest stumbling blocks of the chemical hypothesis – the failure over many years to find a suitable candidate for X ~ I. Fragmentation of the system would lead to short circuit of the pump and breakdown of X ~ I.

2.2 The chemiosmotic hypothesis

The chemiosmotic hypothesis, proposed by Mitchell in 1961, circumvented two big problems which the chemical hypothesis had been unable to explain convincingly up to that time. The first of these was the failure to produce any convincing candidates for the roles of X and I after many years intensive search. The second was the absence of any explanation for the necessity of an intact, closed, membrane system for oxidative phosphorylation.

Mitchell pointed out that, since the basis of oxidative phosphorylation was a dehydration reaction to form an acid anhydride link, a system which caused the removal of water from the active centre of an ATPase would favour the formation of ATP:

$$ADP + P_i \rightleftharpoons ATP + H_2O$$

To poise this equilibrium such that ADP = ATP at a P_i concentration of 10 mM requires the water concentration to be reduced from 55 M to 5 μM. The system Mitchell suggested for causing this dehydration consisted originally of two parts, an anisotropically organised respiratory chain consisting of alternating electron carriers and hydrogen carriers and an anisotropic membrane ATPase. The respiratory chain caused the transport of H^+ from one side

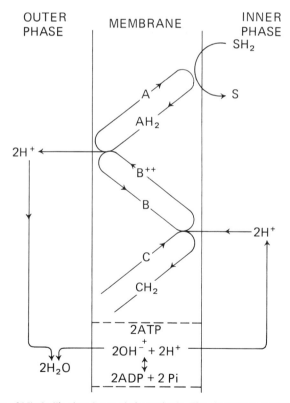

Fig. 18.2 Basic scheme of Mitchell's chemiosmotic hypothesis. The alternation between hydrogen carriers (A, C) and electron carriers (B) causes translocation of H^+ from the inner to the outer phase. The reentry of H^+ is coupled to the dehydration reaction at the active centre of the ATPase system.

of the membrane to the other, and the resulting pH gradient and membrane potential caused the removal of H^+ in one direction and OH^- in the opposite direction from the active centre of the ATPase (Fig. 18.2). AH_2 and CH_2 are hydrogen-carrying members of the respiratory chain and B is an electron carrier. For this system to work successfully the membrane itself has to be impermeable to H^+ since otherwise the pH gradient generated by respiration would be short circuited. Also the active centre of the ATPase has to be inaccessible to water and organised such that OH^- is removed in one direction and H^+ removed in the other.

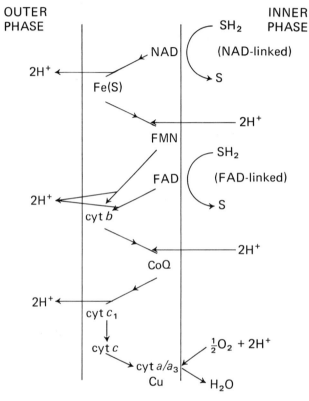

Fig. 18.3 Detailed arrangement of the respiratory carriers according to Mitchell. The scheme provides for the translocation of $6H^+/O$ for the oxidation NAD-linked substrates and $4H^+/O$ for succinate oxidation.

The detailed arrangement of respiratory carriers as proposed by Mitchell is shown in Fig. 18.3. There are three so called translocating loops from NAD-linked substrates and two from succinate, so that the predicted H^+/O ratios are 6 from NAD-linked substrates and 4 from succinate. The positioning of ubiquinone on the oxygen side of cytochrome b is acceptable on the basis of redox potential, but there is as yet no good evidence for placing it there rather than in the more conventional position on the substrate side of cytochrome b.

As originally postulated (Fig. 18.2) the chemiosmotic hypothesis required that the ATPase system straddled the mitochondrial inner membrane. However, it is now thought that the mitochondrial ATPase which is implicated in the phosphorylation process is in fact located in the stalked particles (Fernandez-Moran particles or inner membrane sub-units) projecting from the inner membrane into the matrix space (Racker 1967). A further consideration is the stoichiometry of the system. The scheme shown in Fig. 18.2 predicts that for each pair of H^+ ions pumped out by respiration, 2 ATP molecules would be synthesised. Since the arrangement of the respiratory chain shown in Fig. 18.3 predicts that $6H^+$ are pumped out per pair of electrons passing from NADH to oxygen, the expected P/O ratios for NAD-linked substrates would be 6, instead of the observed value of 3. To account for these discrepancies, Mitchell has produced a modified system (Fig. 18.4) in which the pH gradient is utilised to convert X^- and IO^- into $X \sim I$ in the membrane, $2H^+$ being translocated from outside to inside in the process. The intermediate is seen as existing in two forms; on the outside of the membrane X–I is in equilibrium with water, while on the inside of the membrane $X \sim I$ is involved in a reversible reaction with ADP, P_i and ATP. The energy difference is thought of as arising partly from the difference in H^+ activity on the two sides of the membrane and partly from the low concentration of X^- and IO^- on the inside of the membrane due to the negative electrical potential. This negative potential arises directly from the outward pumping of H^+ carried out by the respiratory chain.

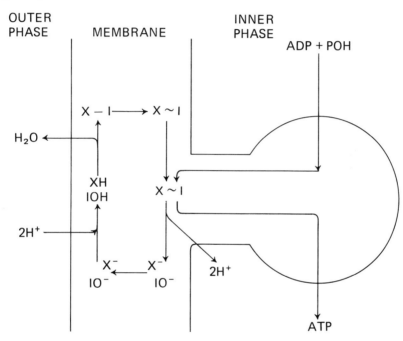

Fig. 18.4 Modified scheme for the anisotropic ATPase system, allowing for the translocation of $2H^+$/ATP synthesised.

Although an alternative explanation is possible to explain the stoichiometry, and this will be dealt with later, the requirement for X and I is at the moment absolute. Their chemical nature is however rather more flexible than is required on the chemical hypothesis. On the chemiosmotic hypothesis neither X nor I have to enter into combination with a redox carrier in the respiratory chain, and this is an important distinction between the two theories. Mitchell has suggested that X and I should be regarded as ionisable groups in the proton translocating part of the ATPase system.

In terms of the basic requirements of the coupling mechanism, the common high energy intermediate of the chemiosmotic hypothesis is best thought of as the combination of pH gradient and electrostatic membrane potential built up by the operation of the respiratory loops. This 'proton motive force' (the difference in electrochemical potential of protons across the membrane) is considered to be the driving force for ion transport. It was proposed that uncoupling agents increased the permeability of the mitochondrial membrane to H^+, thereby dissipating the pH gradient. Oligomycin could be thought of as being an inhibitor of the proton translocating ATPase system.

2.3 Conformational hypothesis

As distinct from the chemical hypothesis which proposes a discrete covalent compound as the high energy intermediate and the chemiosmotic hypothesis which suggests a pH gradient and membrane potential as being the primary energy conserving system, the conformational hypothesis introduces the idea of energy being stored in a conformational state of the macromolecules comprising the inner membrane. The original concept (Boyer 1965; King, Kuboyama and Takemori 1965) was of energy being conserved in a conformational state of a respiratory carrier

$$AH_2 + B \rightleftharpoons A^* + BH_2$$
$$A^* + ADP + P_i \rightleftharpoons A + ATP$$

Thus, like the chemical hypothesis, it directly predicts high energy states of respiratory carriers.

Various further elaborations of the hypothesis have appeared. Green's group (Green *et al.* 1968) observed a variety of conformational states of mitochondria using electron microscopy and suggested that these gross structural changes arose from changes in conformation of the sub-units making up the membrane. The changes were well correlated with the energy state of the mitochondria. A more recent model proposed by Green and Ji (1972) suggests that the conformational change is induced in the ATP synthesising system by an electric field developed by the respiratory carriers. This electric field can be considered as a dipole within the membrane, as distinct from the transmembrane proton gradient of the chemiosmotic hypothesis. It is however, quite difficult to distinguish from the latter, since in both cases the primary energy conserving event involves charge separation. The main difference between the two hypotheses lies in the proposed mechanism for the utilisation of the primary high energy state. The conformational hypothesis proposes that this is done via induced conformational changes while the chemiosmotic hypothesis suggests that it involves chemical intermediates produced by proton flow through the membrane.

Distinction between covalent compounds and conformation changes or charge separations is clear but there are many intermediate situations, for example co-ordination compounds or charge-transfer complexes, many of which are capable of storing energy.

In our view the problem is whether there is a direct connection between $X \sim I$ (or $\sim X$) and respiration, with ATP synthesis and H^+ pumping as alternative uses, or whether the connection is by the proton gradient.

3 High energy states and electron transfer

3.3 Spectral states of redox carriers

Both the chemical hypothesis and the chemiosmotic hypothesis make precise, though different, predictions about the properties of the respiratory chain carriers. In terms of the chemical hypothesis, three of the respiratory carriers, either in the reduced or oxidised state, should have the capacity to form covalent high energy intermediates with one or more coupling factors. The chemiosmotic hypothesis on the other hand, predicts that the respiratory chain is organised in a sided fashion, so that in three different regions protons are taken up on the inside of the membrane and expelled on the outside. The conformational theory is rather more adaptable in this regard, since on the original formulation (Boyer 1965) high energy states of cytochromes should exist, while later proposals (Green and Ji 1972) suggest that the respiratory chain should be able to carry out charge separation, inducing conformational changes in neighbouring proteins.

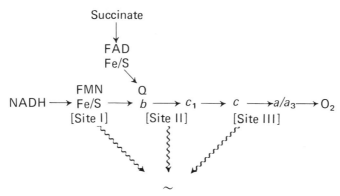

Fig. 18.5 Approximate positions of coupling sites on the respiratory chain. The position of site II between the cytochrome b region and cytochrome c_1 is quite well established. The precise points of coupling at sites I and III are rather more vague.

The three points along the respiratory chain at which coupling sites are located are approximately known, firstly from the crossover studies of Chance and Williams (1956), secondly from measurements of P/O or $P/2e$ ratios for particular regions of the chain and thirdly from redox potential considerations. The results of these studies are shown in Fig. 18.5. Any findings on the nature of the coupling mechanism have to be compatible with these results.

The formation of a high energy state of a respiratory carrier might be expected to alter its redox potential. The redox potential of

$$AH_2 + I \rightleftharpoons A \sim I + 2H\cdot$$

would be higher than the non-energised couple

$$AH_2 \rightleftharpoons A + 2H\cdot$$

In other words $A \sim I$ would be more readily reduced than A.

If, on the other hand, the reduced form of the cytochrome was the high energy form, the redox potential of the couple

$$B + I + 2H\cdot \rightleftharpoons BH_2 \sim I$$

would be lower than for the couple

$$B + 2H\cdot \rightleftharpoons BH_2$$

To be involved in a coupling mechanism, the reactions

$$BH_2 \sim I \rightarrow BH_2 + I$$
$$\text{or } A \sim I \rightarrow A + I$$

would have to have $\Delta G_0'$ of about -8 kcal/mole, which would be equivalent to an alteration of redox potential of 175 mV for a two-electron reaction and 350 mV for a one-electron reaction. Formally equivalent reactions can be written for the conformational hypothesis, involving A* (or possibly BH*$_2$).

Mitchell (1968), however, has pointed out that, in terms of the chemiosmotic hypothesis, if a membrane potential is developed, the activity of electrons on one side of the membrane is altered compared with the activity on the other side. Thus electron carriers which span the membrane will show effects of a membrane potential on their redox state. Hydrogen carriers which span the membrane will similarly show effects due to changes in hydrogen ion activity in their locality. On a thermodynamic basis, it can be seen that if electron transport is responsible for the development of a membrane potential and hydrogen ion gradient, these parameters, must, in return, affect the equilibria of the redox carriers.

Recently, Wilson and Dutton (1970a) have developed very refined methods for the estimation of redox potentials of cytochromes *in situ* in intact mitochondria, in the absence of respiratory inhibitors. The redox states of the cytochromes are measured spectrophotometrically under strictly anaerobic conditions when they are brought into equilibrium with a variety of redox couples such as ferri/ferrocyanide and the oxidised and reduced forms of tetramethyl-*p*-phenylene diamine. Using these techniques, Wilson and Dutton and their co-workers have found that the redox potentials of a cytochrome *b* component and cytochrome a_3 depend on the energy state of the mitochondria. For rat liver mitochondria, Wilson and Dutton found that cytochrome *a* had a midpoint potential of 190 mV, and this value was not changed by the addition of ATP. Cytochrome a_3, however, had a midpoint potential of 395 mV but on the addition of ATP this was reduced to 290–300 mV. On the basis of two electrons having to pass through cytochrome a_3 to yield the energy for the synthesis of one ATP molecule, a change in midpoint of 100 mV is still only about half what is required on thermodynamic grounds. However, in pigeon heart submitochondrial particles, Lindsay, Dutton and Wilson (1972) report that the midpoint potential of cytochrome a_3 shifts from 350 mV to 130 mV on the addition of ATP, which would be more consistent with the formation of cytochrome $a_3^{2+} \sim I$.

Erecinska *et al.* (1972) have reported that the addition of ATP to pigeon heart mitochondria causes a change in the spectrum of oxidised cytochrome oxidase. This change, which is blocked by uncouplers and oligomycin, indicates that there may be an energy dependent equilibrium between two forms of oxidised cytochrome oxidase, differing in the environment of the haem iron. The relationship between this phenomenon and the energy dependence of the midpoint potential is not clear, since the decrease in midpoint potential observed on energisation is more in accord with a high energy form of the reduced rather than the oxidised cytochrome. More complex schemes, involving high and low potential forms of both oxidised and reduced cytochrome, are required to account for these data.

In the case of cytochrome *b*, Wilson and Dutton (1970b) report two midpoint potentials in the absence of ATP, one at −55 mV and one at 35 mV. When ATP is added, the latter remains at 35 mV, while the former increases to 245 mV, a change of 300 mV. Cytochrome *b* can also be resolved into two components spectrophotometrically (e.g. Sato, Wilson and Chance 1971) the two being present in roughly equal quantities. If mitochondria are allowed to go anaerobic in the presence of succinate or glutamate, the cytochrome *b* of midpoint potential 35 mV (named cytochrome b_K by Chance) becomes reduced while the cytochrome *b* of variable midpoint potential (cytochrome b_T) remains oxidised. Addition of ATP raises the midpoint of cytochrome b_T and it becomes reduced. Cytochrome b_K has a symmetrical α band at 561 nm while cytochrome b_T has a double α band, with maxima at 558 and 566 nm. The spectrum of reduced cytochrome b_T is the same, whether it is in the high or low potential form.

The interaction of Antimycin A with the cytochromes b has added a further dimension of complication to the picture. Slater and his co-workers (see Slater 1971) found for example that the binding of Antimycin A to mitochondria, measured by the red shift which it induces in the spectrum of cytochrome b, is cooperative in uncoupled anaerobic mitochondria, but is non-cooperative in anaerobic mitochondria in the presence of ATP. Later investigations by Dutton *et al.* (1972) showed that Antimycin caused a red shift of about 1 nm in the α band of cytochrome b_K and also blocked the effect of ATP on the midpoint potential of cytochrome b_T. The Antimycin binding site thus appears to be very closely related to energy conservation at site II.

These pieces of evidence appear to support the chemical hypothesis fairly strongly. However, Caswell (1972) has criticised the techniques used for measuring redox potentials, since if there is a coupling site between the region of the respiratory chain at which the applied redox couple feeds in and the cytochrome being measured there will be artifactual effects of the ATP/ADP ratio on the redox states of the cytochromes. Wilson and co-workers have, however, used a large number of different artificial redox couples with identical results.

There are also the considerations of the effects of membrane potential and pH gradient predicted by the chemiosmotic hypothesis (Mitchell 1968). By the use of valinomycin, to make the mitochondrial membrane permeable to K^+, in the presence of a K^+ concentration gradient between the inside and outside of the mitochondria, it is possible to set up artificial membrane potentials across the inner membrane. Hinkle and Mitchell (1970) showed, using this technique, that the midpoint potential of cytochrome a depends on the applied membrane potential, which can be varied by altering the external K^+ concentration. A membrane potential of 100 mV (negative inside) caused the apparent midpoint potential of cytochrome a to go 50 mV more negative. The chemical hypothesis can however, defend itself against this experiment. The

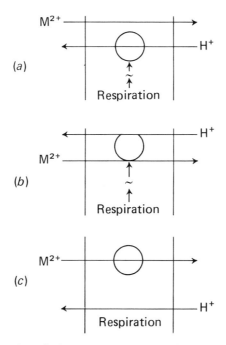

Fig. 18.6 Proposed mechanisms for a divalent cation accumulation by the chemical hypothesis (a and b) or the chemiosmotic hypothesis (c). In (a), the divalent cation movement is secondary to an $X \sim I$ driven H^+ pump. In (b), the H^+ movement is secondary and coupled to an $X \sim I$ driven cation translocator.

use of $X \sim I$ for ion pumping (Fig. 18.6) means that ion flux can generate $X \sim I$, i.e. the K^+ gradient can drive $X \sim I$ formation thereby altering the redox potential of the cytochromes.

The effect of pH on the midpoint potentials of cytochromes is at present rather confused. Azzi and Santato (1971) found that the effect of ATP in bringing about the reduction of cytochrome b_T after anaerobiosis could be mimicked by raising the pH from 7·5 to 8·7. This implies that either the midpoint of cytochrome b_T increases with increasing pH, or the midpoint potential of its electron donor decreases with increasing pH. In support of the latter idea, Urban and Klingenberg (1969) found that the midpoint potentials of both ubiquinone and cytochrome b (probably cytochrome b_K) decreased with increasing pH. Lee and Slater (1972) suggest, however, that there is a pH dependent accessibility barrier between cytochrome b_T and the substrate end of the respiratory chain.

The only clear case of pH affecting cytochrome redox potentials is therefore the one described by Urban and Klingenberg, where above pH 6·8, cytochrome b is thought to lose a proton on oxidation:

$$b^{2+}H^+ \rightleftharpoons b^{3+}H^+ \rightleftharpoons b^{3+} + H^+$$

This does indicate that oxidation and reduction of cytochromes need not necessarily be pure electron transfer and that some cytochromes may be, in effect, hydrogen carriers.

In summary it appears that at the moment, although high energy states of respiratory carriers can be most conveniently explained by the chemical hypothesis, the chemiosmotic hypothesis can provide explanations for most of the observed effects in terms of alterations in membrane potential and H^+ activity within the membrane. The effects of ATP on the spectrum of oxidised cytochrome oxidase and the complex effects of Antimycin on site II appear at the moment to favour the chemical hypothesis since they do not seem to be associated with any transmembrane electrical phenomena. Effects of membrane potential on ligand binding to the haem cannot, however, be ruled out.

3.2 H^+ pumping and electron transfer

The relationship of H^+ pumping across the inner mitochondrial membrane to electron transfer is the crucial difference between the chemical hypothesis and the chemiosmotic hypothesis. To account for ion transport phenomena, all theories have to include H^+ pumping, but several formulations are possible (Fig. 18.6). Under the chemical hypothesis, proton movements might be either primary (Scheme (a)) or secondary (Scheme (b)), in the latter case resulting from the movement of another ion such as Ca^{++}. For reasons to be discussed in section 4, the $X \sim I$ driven proton pump (a) is more satisfactory than (b) as an alternative to the chemiosmotic hypothesis.

When anaerobic mitochondria are given a small pulse of oxygen the external medium shows a temporary acidification. If the amount of H^+ expelled from the mitochondria is expressed in terms of the amount of oxygen delivered in the pulse, it is found that for succinate as substrate H^+/O is 4, while for β hydroxybutyrate it is 6. In accordance with the idea that uncouplers allow the passage of H^+ across the membrane, addition of the uncoupler FCCP after the oxygen pulse causes rapid re-equilibration of the external and internal pH. While these observations are entirely in accord with the chemiosmotic hypothesis, there is a problem connected with them. Calculation reveals that transfer of 1 nmol H^+/mg protein across the membrane should result in a membrane potential of 250 mV, which is in theory sufficient to drive ATP synthesis. However, in experiments such as these, the amount of H^+ expelled is proportional to the amount of oxygen added up to about 10 nmol H^+/mg protein. In the absence of any other ion movements to compensate for the charge carried by the H^+ ions, this would result in a membrane potential of 2·5 V, which is energetically impossible. The suggestion is that during the prolonged anaerobic period (about 30 min to avoid drift in the pH of the system) before the addition of oxygen, endogenous Ca^{++} which does in fact amount to about 10 nmol/mg protein, leaks out. On addition of oxygen, H^+ is pumped out and Ca^{++} re-enters the mitochondria, compensating for the negative potential inside and therefore allowing more H^+ to be expelled.

Although this very satisfactorily explains the observations, it opens the way to the chemical hypothesis to account for them also. If other ions move, this may be as a result of $X \sim I$ utilisation via a pump such as that described in Fig. 18.6. Thus, although the observations were predictable on the basis of the chemiosmotic hypothesis, they are also entirely explicable by the chemical hypothesis.

In an attempt to discriminate further between the two hypotheses, it is necessary to look at the other prediction made by Mitchell's scheme, that the respiratory carriers are organised in a sided fashion in the membrane. This is absolutely required by the chemiosmotic hypothesis, but is not at all necessary on the basis of the chemical coupling mechanism. As a basis for observations of this sort, it is found that submitochondrial particles prepared by sonicating whole mitochondria have the Fernandez-Moran (ATPase) particles on their outer surfaces, and are therefore inside out in comparison with whole mitochondria. Submitochondrial particles prepared by digitonin treatment, however, are orientated in the same way as whole mitochondria. Studies on ion transport using submitochondrial particles show that the sideness of the transporting systems follows the orientation of the membrane, e.g. sonicated particles tend to accumulate H^+ on initiation of respiration while digitonin particles behave like intact mitochondria, expelling H^+.

By the use of electron donors and acceptors which will not penetrate the membrane it has been possible to study the accessibility of their points of reaction with the respiratory chain when the membrane is orientated either normally or inside out. It is found, for example, that externally added reduced cytochrome c will sustain reversed electron transport in digitonin particles, but sonicated particles will use only endogenous cytochrome c. This strongly suggests that normally in whole mitochondria cytochrome c is in the outside of the membrane and is accessible there to its electron donor (cytochrome c_1) and electron acceptor (cytochrome oxidase).

There is some evidence also that cytochrome oxidase reacts with oxygen on the inside of the membrane. Azide, an inhibitor of cytochrome oxidase which can be accumulated by whole mitochondria, is a far more potent inhibitor under conditions where it accumulates than under those where there is no accumulation. However, to say that the inhibitor binding site and the site of reaction with oxygen are the same represents a very considerable assumption. Mitchell (1971) has shown that in mitochondria inhibited with Antimycin A the oxidation of cytochrome c by cytochrome oxidase (brought about by the addition of oxygen to anaerobic mitochondria) apparently results in the uptake of protons from the matrix space of the mitochondria and no change in the external pH. This is deduced from the observation that in the presence of FCCP to allow entry of hydrogen ions, about 0·41 ng ions H^+/mg protein disappear from the external medium compared with a cytochrome $c + c_1$ content of 0·38 nmol/mg. Similar deductions are made using ferrocyanide as an impermeant electron donor on the oxygen side of the Antimycin A block. Although this is what would be predicted on the basis of Mitchell's arrangement of the respiratory chain (Fig. 18.3), the chemical theory is also capable of explaining the H^+ movements associated with these results, using the $X \sim I$ driven proton pump (Fig. 18.7).

564

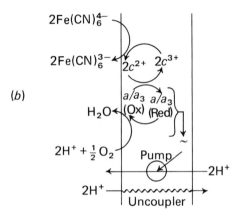

Fig. 18.7 Explanation of the proton movements associated with ferrocyanide oxidation by the terminal portion of the respiratory chain from the point of view of (*a*) the chemiosmotic hypothesis and (*b*) the chemical hypothesis.

The observation is that during ferrocyanide oxidation there is only a very slight pH change in the external medium over the first 30 s. If an uncoupler (e.g. FCCP or DNP) is present, there is an immediate alkaline drift in the external pH which corresponds to the uptake of $2H^+/O$.

Investigations of the orientation of the respiratory carriers in the membrane are still in rather a rudimentary state. Although the data on cytochrome c and the available data on cytochrome oxidase are in accord with Mitchell's predictions, the case for the chemiosmotic hypothesis rests rather heavily on the azide inhibition experiments. The chemical hypothesis interpretation of the hydrogen ion movements associated with electron transfer through site III involves cytochrome oxidase reacting with protons in the outer surface of the membrane. If it could be shown conclusively that oxygen reduction took place on the inner face of the membrane this possibility is made less likely.

Despite this, the way in which hydrogen ion translocation is associated with electron flow in a manner predicted by Mitchell's basic scheme is impressive. As well as the experiments outlined above, other investigations have used ferricyanide as an electron acceptor on the substrate side of the Antimycin A block, and the results in terms of $H^+/2e$ ratios are in accord with prediction. The inner mitochondrial membrane has been shown to be impermeable to ferro- and ferricyanide so that these compounds can be assumed to react in the outer surface of the membrane. However, the chemical hypothesis, by the invocation of the $X \sim I$ driven H^+ pump is able to explain all the observations in terms similar to those of Fig. 18.7. Greville (1969) has recently reviewed all these findings in detail.

3.3 Conformational changes

Hackenbrock (1966) and Green *et al.* (1968) using electron microscopy showed that large structural differences exist between mitochondria in different energy states. These differences take the form of alterations in the morphology of the cristae and in the size of the matrix space. The changes seem to correspond to the low amplitude swelling and shrinking phenomena observed by Packer (1963) using light scattering techniques. The original suggestion by Green was that the large changes in the configuration of the cristae reflected conformational changes in the protein sub-units making up the inner membrane. Using light scattering measurements it was found that the configurational transitions were rather slow. For instance the change from the energised to the energised-twisted state on addition of phosphate to beef heart mito-chondria showed a half life of about 7 s. The possibility exists however, that the secondary changes in the configuration of the cristae might be slower than any primary conformational changes in membrane proteins.

A different interpretation of the morphological changes which occur on changing the mito-chondrial energy state is that they are a result of pH and osmotic effects due to ion transport. Evidence for this has been presented by Hunter *et al.* (1969) and Stoner and Sirak (1969). Since all the changes described by Green *et al.* involve the presence of permeable ionic species in the experimental medium (e.g. substrate anions, phosphate, ADP and ATP) it is almost certain that movements of these species take place on energisation or de-energisation. The resultant osmotic and electrical effects could very plausibly account for the observed changes.

The other other line of evidence for conformational changes in mitchochondria comes from the use of fluorescent probes. The most commonly used probe has been 1 anilino naphthalene 8-sulphonic acid (ANS). The fluorescence of this compound depends on the hydrophobicity of its environment. An increase in hydrophobicity results in a higher quantum yield and a shift in the emission maximum to shorter wavelengths. The addition of ANS to sub-mitochondrial particles prepared by sonication results in enhanced fluorescence as it binds to or dissolves in the membranes. Energisation of the particles, either by addition of ATP or oxidisable substrate, causes an increase in fluorescence which can be reversed by adding uncoupling agents. Brocklehurst *et al.* (1970) found that this enhancement of fluorescence can be accounted for by two effects. Firstly, there is an increase in binding of ANS to the particles. Since ANS is negatively charged the membrane potential resulting from H^+ pumping would cause the accumulation of ANS inside sonicated particles, so that this effect does not require the invocation of any conformation changes. However, the second contributory factor to the increase in fluorescence is apparently an increase in quantum yield. This can be deduced from experiments in which the fluorescence resulting from the addition of ANS to energised or de-energised particles is extrapolated to infinite particle concentration. At infinite particle con-centration, all the ANS should be bound at all times, so that any fluorescence change must be due to a change in quantum yield. The observed increase in quantum yield on energisation presumably reflects an increase in the hydrophobicity of the environment of the ANS, an indication of a conformation change. Unfortunately for the conformational hypothesis, Brocklehurst *et al.* (1970) and Freedman *et al.* (1971) have found that the induction of the fluorescence increase is apparently too slow to be a primary event in the coupling process, being very much slower than changes in the redox state of the respiratory chain components under similar conditions. The quantum yield increase on the addition of NADH to sonicated particles, for example, shows a half-life of about 3 s, and the decrease in quantum yield on addition of uncoupler shows a similar time response. It is possible, therefore, that the changes measured by ANS fluorescence may be similar to the structural changes observed in the electron microscope.

At the present time there is no convincing evidence for the existence of conformation changes which are sufficiently rapid to be the primary energy conserving event. It must be remembered however, that high energy conformation states of respiratory carriers would probably appear to the impartial observer as indistinguishable from carrier $\sim$ I, so that to that extent the conformation hypothesis is not distinguishable from the chemical hypothesis.

4. Ion and substrate transport

4.1 Transport processes

During preparation mitochondria retain ions such as K^+, Mg^{2+} and Ca^{2+} as well as small amounts of substrates. Two types of retention can be distinguished: firstly that in which the ion is retained by virtue of the impermeability of the membrane or by a Donnan effect as a counter ion to an impermeable ion of opposite charge; secondly, retention of a permeable ion by the continual utilisation of energy provided by respiration. Under appropriate conditions the latter process can be extended so that the mitochondria accumulate ions in an energy dependent fashion.

Substances can move across membranes in several ways and their effects on, and responses to, pH, electrostatic potential and other ion concentration differences across the membrane, depends very much on their mode of transport. It is very necessary to be absolutely clear about the type of transport involved (both in this section and in the section on ionophoretic uncoupling, 5.3) and to prevent confusion we give below an outline of the classification of transport processes which we use (Mitchell 1967).

1 Uniport. The substance crosses the membrane either alone or via a carrier (which for continued transport must be able to return empty) but in either case there is no *net* transport of any other substances. When the substance is an ion there is a flow of charge, hence the terms electro-active, electrogenic and electrophoretic uniport.

2 Antiport. Otherwise known as exchange diffusion. The passage of a substance in one direction across the membrane is compulsorily linked to the passage of another substance in the reverse direction. Depending on the charges on the two compounds this process may be electro-active or electro-neutral. The compulsory nature of the exchange is important because a non-compulsory exchange is equivalent to, and behaves as, two uniport systems.

3 Symport. In this case two substances cross the membrane in the same direction. As with antiport the process may be electrically neutral or not and again it is preferable to restrict the term to the condition where the linkage is compulsory.

Features of transport processes which require emphasis are:

a In aqueous media proton symport is equivalent to hydroxide antiport, and vice versa, and measurements of net transfer cannot distinguish between the alternatives.

b In coupled transport where one of the ions is a proton or hydroxyl ion the distribution of the other substance will depend on the pH differential across the membrane.

c If significant ion movement is to occur, an electro-active transport must be balanced either by counter flow of the same charge or parallel flow of the opposite charge. This condition for electro-neutrality can cause two uniport processes to appear to be antiport or symport. The distinction between the types of linkage (molecular and electrical) is important because under some conditions their behaviour is very different.

d The work involved in a transport process depends on the chemical potentials of all the substances which are compulsorily linked in the transport (and only those substances) on both sides of the membrane and, if the process involves a net transfer of charge, on the electrostatic potential difference across the membrane. Secondary active transport of a substance (movement of that *one* substance against its electrochemical gradient) can occur as a result of the condition for electro-neutrality, by symport or by antiport. Sets of symport or antiport systems can be linked by common substrates.

4.2 Divalent cation uptake

Although it is clear that mitochondria can retain and under some conditions accumulate K^+ and Mg^{2+} ions against a concentration gradient by an energy linked process, the mechanism is not well defined and the nature of the carriers or even their existence is not known. At present the effect is probably best left as an empirical observation and the same may be said of the Na^+/H^+ antiport across the inner mitochondrial membrane.

The uptake of Ca^{2+}, Mn^{2+} and Sr^{2+} has proved much more interesting and much of the work on these ions has been reviewed by Lehninger, Carafoli and Rossi (1967). Early reports on Ca^{2+} ions were that they acted as uncouplers, stimulated respiration and could be accumulated by mitochondria. Subsequent work confirmed these observations: although under some conditions Ca^{2+} can cause uncoupling by damage to the mitochondrial structure, under more controlled conditions what appears to be uncoupling is better thought of as energy-linked uptake of the Ca^{2+} ions. The amount of Ca^{2+} accumulated bears a stoichiometric relationship to the 'X $\sim$ I' generated by respiration, the ratio Ca^{2+} to 'X $\sim$ I' approaching 2:1. Addition of Ca^{2+} ions to respiring mitochondria results in a short burst of increased respiration and limited uptake of Ca^{2+} except in the presence of ions such as phosphate or acetate (which enter the matrix by hydroxide antiport or proton symport) in which case increased respiration is maintained until virtually all the Ca^{2+} has been accumulated. During Ca^{2+} accumulation protons appear in the medium. The ratio of H^+ ejected to Ca^{2+} taken up varies with the conditions but in many circumstances is about 1:1. Uptake of Ca^{2+} is prevented by uncouplers such as DNP but not by oligomycin. Ca^{2+} uptake can also be driven by hydrolysis of ATP. This again is abolished by DNP and also by oligomycin.

Similar uptake has been observed with Sr^{2+} and Mn^{2+} ions and there may be a rather feeble uptake of Ba^{2+} ions. This specificity strongly suggests the presence of a carrier for these ions in the mitochondrial membrane and this is confirmed by the finding that certain trivalent rare earth ions, e.g. Pr^{3+} (Mela 1969) are very potent inhibitors of Ca^{2+} transport and allow titration of the carrier sites.

Two main schools of thought developed. Firstly, there was a proposal that Ca^{2+} was exchanged for protons on a carrier driven by X $\sim$ I. There are variations of this proposition but a common prediction is a stoichiometric relationship between Ca^{2+} entry and H^+ ejection. The other main hypothesis was based on Mitchell's chemiosmotic hypothesis but is equally applicable to the chemical hypothesis with an outwards proton pump driven by X $\sim$ I. In either case the inside of the mitochondrion becomes electrically negative and alkaline. If Ca^{2+} can enter by electro-active uniport then it will do so in response to the electrical potential difference across the membrane. Collapse of this component of the proton motive force will allow development of an increased pH differential and this has been experimentally observed.

Two further findings support the view that the proton pump and Ca^{2+} carrier are separate. Firstly, the ratio of Ca^{2+}: H^+ is variable. With a separate pump and carrier this is permissible since the ratio depends on the initial and final states of the mitochondria. Secondly, it has been possible to demonstrate passive entry of Ca^{2+} by electro-active uniport. This entry is inhibited by Pr^{3+} ions but does not require the presence of an uncoupler as would be predicted by the $X \sim I$ driven Ca^{2+} pump (Selwyn, Dawson and Dunnett 1970).

It should be appreciated that on this basis neither Ca^{2+} nor phosphate is actively transported since both ions enter down electrochemical gradients. Entry of the ions discharges the gradients and hence energy is required to maintain the gradients and produce accumulation of significant quantities of the ions.

4.3 Redox substrate and phosphate uptake

Using the techniques of osmotic swelling of mitochondria in isotonic ammonium salts and reduction of intra-mitochondrial pyridine nucleotides, Chappell has been able to show the presence of carriers for several respiratory substrates and for phosphate. The properties of, and evidence for, these carriers has been admirably reviewed by Chappell (1968). Phosphate enters either as the acid, H_3PO_4 or by an equivalent antiport process such as exchange of HPO_4^{2-} for two hydroxyl ions. Malate, and succinate, also cross the membrane rapidly but only in the presence of catalytic traces of phosphate. The very similar substrates fumarate and oxalo-acetate do not enter which suggests the presence of a highly specific carrier, a view which is supported by the inhibitory action of butylmalonate. Oxoglutarate, citrate, isocitrate and cis-aconitate are also transported across the membrane but only in the presence of both phosphate and malate. Chappell postulated a series of exchange diffusion carriers in which phosphate has a prime position, malate and succinate a secondary position exchanging for phosphate, and the other substrates exchanging for malate. This interpretation has been confirmed by more direct methods for measuring substrate entry. The properties of these carriers are summarised and their mode of operation shown in Fig. 18.8.

According to this mechanism phosphate and the respiratory substrates should be accumulated in response to a pH gradient (alkaline inside) and, since the exchanges are electro-neutral, should not be affected by the electrical potential difference across the membrane. Harris and Berent (1970) have measured the distribution of these substrates and also acetate and pyruvate and, because the ratio of internal to external concentrations when reduced to the $1/n$th root (where n is the charge on the substrate anion) is constant, have proposed that the anions are distributed in a Donnan type of equilibrium. However the response to the pH gradient depends on the number of hydroxyl ions for which the anion is exchanged and this gives exactly the same numerical relationship for the distribution ratios as the Donnan equilibrium. Reagents such as nigericin or trialkyltins which collapse the pH gradient lead to depletion of intra-mitochondrial substrate whereas reagents which render the inside more negative, such as valinomycin in a medium of low K^+ concentration, do not lead to loss of substrate. On the present evidence, accumulation of phosphate and respiratory substrates in an electro-neutral fashion in response to a pH gradient is a satisfactory hypothesis.

4.4 Adenine nucleotide translocase

The initial evidence for this carrier, which catalyses the exchange of ATP for ADP across the inner mitochondrial membrane, came from work on the inhibitor atractyloside. In early investigations this appeared to be an inhibitor of the energy conversion process similar to

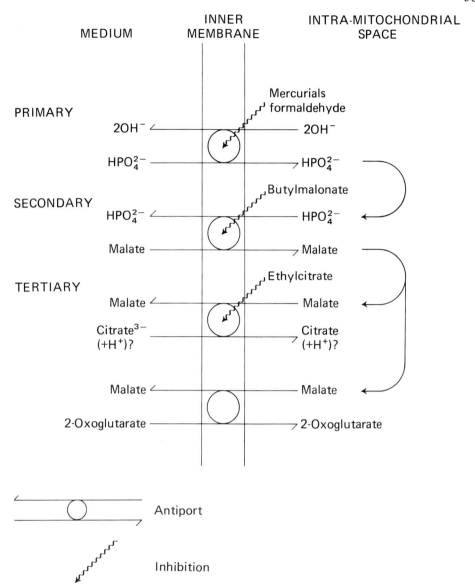

Fig. 18.8 Antiporters for phosphate and substrate anions.

    Note how malate entry is catalysed by phosphate which can re-enter by the phosphate carrier. Citrate or 2-oxoglutarate exchange for malate and thus require the presence of catalytic concentrations of both malate and phosphate.

    Succinate appears to enter on the same carrier as malate and the malate-citrate antiporter is also effective with iso-citrate and *cis*-aconitate.

    Note that in effect all these ions exchange for hydroxyl ions.

oligomycin. On further investigation it was found that atractyloside inhibited ADP-stimulated uncoupling by arsenate but not the uncoupling by arsenate in the absence of ADP (both are equally inhibited by oligomycin). Also in contrast to oligomycin, atractyloside failed to inhibit phosphorylation of endogenous ADP or hydrolysis of endogenous ATP. These observations suggested that atractyloside inhibited the passage of ADP and/or ATP across the mitochondrial membrane. In a very thorough investigation Klingenberg (1970) and his colleagues demonstrated that while *net* entry of all adenine nucleotides was slow, exchange of ATP for ADP across the membrane was very rapid and that this antiport was inhibited by atractyloside.

ADP and ATP are not electrically equivalent at physiological pH but the exchange could be, if, for example, the carrier carried $ATPH^{3-}$ and ADP^{3-}. Klingenberg has estimated the net charge and proton transfer under passive conditions and finds that the exchange is approximately $ATPH_{0.5}^{3.5-}$ for ADP^{3-}. The exchange is therefore affected by both pH and electrostatic potential differences across the membrane, a negative potential inside favouring ADP entry and relative alkalinity inside favouring retention of ATP. The experimental circumstances suggest that 3·5 may be an underestimate of the magnitude of the charge on the ATP and on a molecular basis the transfer of a fractional charge seems improbable. It is probable that the exchange is really ATP^{4-} for ADP^{3-} in which case it will respond only to the electrostatic potential difference across the membrane.

4.5 Ion movements during phosphorylation

The phosphorylation of ADP is often written as a straightforward dehydration reaction but under physiological conditions the reaction involves the uptake of around 0·8 protons per ATP formed and is therefore better written as in equation (a). The alternative shown in (b) should not

(a) $ADP^{3-} + HPO_4^{2-} + H^+ \leftrightarrow ATP^{4-} + H_2O$

(b) $ADP^{3-} + HPO_4^{2-} \leftrightarrow ATP^{4-} + OH^-$

be forgotten since the reaction may actually proceed by the elimination of an hydroxyl ion. Mitochondrial oxidative phosphorylation can be measured by the pH increase in the external medium. It is clear from the site of the coupling factor ATPase (F_1) and from the effects of inhibitors of the phosphate and adenine nucleotide transporters that phosphorylation of ADP takes place on the inner side of the coupling membrane. In view of the small volume and limited buffering capacity of the intra-mitochondrial compartment there must be some means whereby the base production from the internal phosphorylation is compensated, for otherwise this space would become very alkaline indeed.

One way to overcome this problem would be to have the membrane permeable to protons (or hydroxyl ions) but it has been found experimentally that the membrane has a very low permeability to protons. (See sections 5.2 and 5.3). Since the transporters for phosphate, ADP and ATP involve movement of charge and hydroxyl ions across the membrane and, under more or less steady state conditions, ADP and phosphate disappear from the medium and ATP appears there in equivalent amounts, it is necessary to consider the carriers as well as the phosphorylation. This is shown in Fig. 18.9. Figure 18.9(*a*) shows that an exchange of ADP^{3-} for $ATPH^{3-}$ would together with the phosphate hydroxyl ion antiport lead to a balance of both charge and hydroxyl ions inside the mitochondrion. However the experimental evidence is against such an electro-neutral adenine nucleotide antiport (discussed in the previous section). Moreover such an exchange would have the unfortunate effect that conditions which favour phosphate entry, i.e. a pH gradient alkaline on the inner side, would favour retention of ATP and expulsion of ADP. The situation with the electro-active exchange is shown in Fig. 4.2(*b*). In this case it will be seen that the system is not compensated and requires compensation by an outwardly directed proton pump. Since the evidence favours this mode of adenine nucleotide antiport this seems to provide a fundamental necessity for some form of proton pump regardless of other aspects of the process. An outwardly directed proton pump would make the inside alkaline and produce an electrostatic potential difference, negative inside, across the coupling membrane. Thus phosphate accumulation would be favoured by the pH gradient while entry of ADP and exit of ATP would be promoted by the electrostatic potential difference.

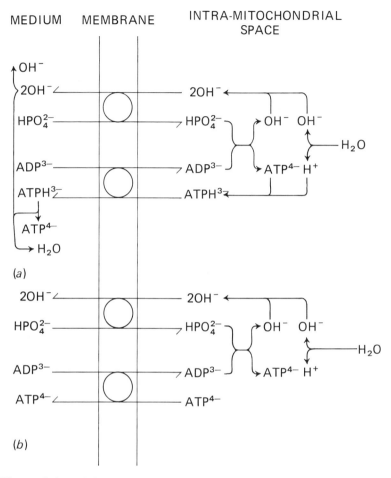

MEDIUM   MEMBRANE   INTRA-MITOCHONDRIAL SPACE

Fig. 18.9 Acid-base and charge balance across the mitochondrial membrane during phosphorylation of ADP. (a) Assuming that the adenine nucleotide exchange is ADP^{3-} for $ATPH^{3-}$. Note the balance of both acid-base and charge. (b) Assuming that the adenine nucleotide exchange is ADP^{3-} for ATP^{4-}. Note the imbalance; one H^+ remaining inside for each ATP formed.

5 Uncoupling

5.1 Uncoupling and inhibition of energy conversion

It is necessary to be precise in the names used for substances affecting oxidative phosphorylation. Respiratory chain inhibitors by blocking electron transport also inhibit phosphorylation and, in tightly coupled mitochondria, inhibitors of the phosphorylation process also inhibit electron flow. An uncoupler of oxidative phosphorylation produces a decrease (inhibition) of phosphorylation but as it releases respiratory control it often *stimulates* respiration. Uncouplers are sometimes referred to as inhibitors of oxidative phosphorylation but this term can lead to confusion and should be avoided. Those who like analogies with familiar objects may like to compare respiratory chain inhibitors, uncouplers and phosphorylation inhibitors with the throttle, clutch and brakes of an automobile.

A distinction which is not usually made clear is that between catalytic or kinetic uncoupling, where neither the uncoupler nor anything other than respiratory substrate is consumed, and thermodynamic uncoupling where work is done on the uncoupler, or more often on an ion in the medium, but this work is done at very low efficiency. Again the analogy with an automobile may be helpful; efficient, energy conserving operation being represented by the car going uphill slowly, kinetic uncoupling by the clutch being depressed or by ice on the road, and thermodynamic uncoupling by the car travelling fast along a level road. It will be appreciated that in thermodynamic uncoupling the mechanism may be working faultlessly on a molecular basis even though no energy is conserved.

In a system as complex as the mitochondrial coupling mechanism it is not surprising that compounds acting at different sites can produce similar effects. Extreme examples are atractyloside and certain organo-mercurials which appear to act as inhibitors of the coupling system but which actually inhibit the adenine nucleotide and phosphate transporters. Trialkyltin compounds appear to act at a very similar site to oligomycin but aurovertin, which has a similar effect to oligomycin on phosphorylation, is known to act at a different site since it inhibits the solubilised ATPase whereas oligomycin does not (Roberton *et al.* 1967). Several types of uncouplers are known which produce uncoupling in very different ways.

5.2 Classical uncouplers

1 Phosphorylation-site by-pass uncouplers. These are compounds, such as menadione, which can react with electron carriers before and after a region (site 1 for menadione) where electron transport is coupled to phosphorylation. Their mode of uncoupling is self-evident but they have not as yet yielded any very valuable information on the mechanism of phosphorylation.

2 Arsenate. Arsenate is a potent uncoupler of glyceraldehyde-3-phosphate dehydrogenase and when it was found to uncouple mitochondrial oxidative phosphorylation it appeared by analogy to provide good evidence for the $X \sim P$ intermediate. It was postulated that arsenate acted by forming an unstable $X \sim As$ intermediate since the pyroarsenates and acyl-arsenates are much less stable than the corresponding phosphate compounds. However the uncoupling effect of arsenate is markedly stimulated by addition of ADP. Since ADP and ATP cross the membrane on a strictly coupled exchange and only slowly by non-exchange uniport, it is difficult to eliminate the possibility that even in the absence of added ADP the endogenous ADP is arsenylated and that the unstable arsenate compound is ADP-As. This of course does not preclude $X \sim As$ and $X \sim P$ as intermediates and quantitative estimates of the K_i for arsenate and the inhibition of its uncoupling activity by phosphate suggest differences in the mechanism of action in the presence and absence of ADP. Even so the interpretation of these effects as evidence for $X \sim As$ is not conclusive.

3 2,4-Dinitrophenol (DNP). Localisation of the site of action of 2,4-dinitrophenol stems from the work of Lardy *et al.* (1958) who used it in conjunction with the inhibitor oligomycin. In the presence of dinitrophenol oligomycin has little effect on respiration but it does inhibit the dinitrophenol-stimulated ATPase. It is thus possible to write the sequence:

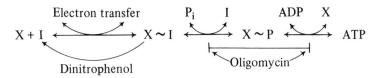

Estabrook (1961) extended this work by including arsenate, and concluded that arsenate acted after the site of action of DNP and at a similar site to oligomycin. This type of work, together with the lack of requirement for ADP or phosphate (Borst and Slater 1961) supports the hypothesis that DNP acts by catalysing the breakdown of X ~ I.

The nature of this catalytic breakdown of X ~ I remained vague on the chemical hypothesis but it received encouraging support from the finding that a preparation of a soluble ATPase from mitochondria was stimulated by DNP (Lardy and Wellman 1953; see also section 6.2).

By the time Mitchell proposed his chemiosmotic hypothesis a great variety of compounds had been found to act in a similar fashion to DNP. Mitchell rationalised their structures as weak acids in which the negative charge of the anionic form would be delocalised. This property allows the compound, both in acid and anionic states, to be soluble and mobile in the lipid phase of the membrane. Movement of these forms of the compound across the membrane

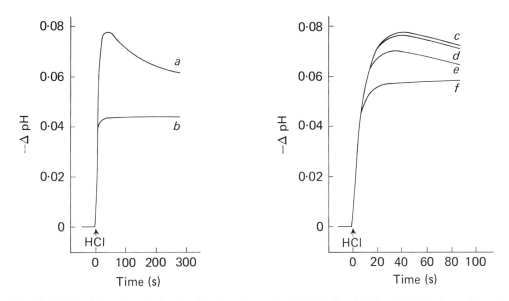

Fig. 18.10 Pulsed titration of mitochondria. Experiment by P. Mitchell and J. Moyle (1967). Recordings of the time course of pH after adding 2 μl of air-free 50 mM HCl in 100 mM KCl to anaerobic rat liver mitochondria in 150 mM KCl in the absence of added buffer but in the region pH 7·0–7·1.

(a) No further addition. Note the overshoot – initially the buffering capacity is that of any buffering by the medium, the outer membrane, the inter-membrane space and the outside of the inner membrane. The intra-mitochondrial space is titrated slowly as the acid has to cross the inner membrane.

(b) Plus 0·1% Triton X-100. The total buffering capacity is titrated rapidly.

(c) As (a) but note the different time scale for this and the following curves.

(d) Plus 10 μg valinomycin/g protein.

(e) Plus 50 μM DNP.

(f) Plus 10 μg valinomycin/g protein and 50 μM DNP.

Note that DNP alone (e) allows rapid entry of some acid but, since DNP effects H$^+$ uniport, this is limited by charge imbalance. In the presence of valinomycin as well as DNP (f) charge balance is permitted by the exit of K$^+$ and complete titration of the contents is rapid.

catalyses proton uniport. The effect of DNP on the coupling factor ATPase could be regarded as being a misleading and unfortunate coincidence, since DNP is known to have effects on several other enzymes.

Investigation, for example by pulsed titrations, Fig. 18.10, has shown that the mitochondrial membrane has a low permeability to protons. This type of experiment (Mitchell and Moyle 1967) and more direct measurements on model membranes (Liberman and Topaly 1968 and Hopfer *et al.* 1970) have shown that uncouplers such as DNP do catalyse proton uniport across lipid membranes. Several studies have been made correlating their activity as proton conductors in artificial membranes with their activity as uncouplers and usually, but not always, good correlation has been found. In view of the simple composition of the model membranes and obvious differences from the mitochondrial membrane system perfect correlation is not to be expected. Since, for other reasons, a proton pump driven by $X \sim I$ is included in the chemical hypothesis (and also the conformational hypothesis) proton uniport should be sufficient by itself to confer uncoupling activity on a compound. Thus these uncouplers would act in the way originally proposed in the chemical hypothesis but it is now possible to be more precise about the way in which breakdown of $X \sim I$ is produced.

5.3 Ionophoretic and permeant ion uncouplers

From the previous section it will be realised that the favoured mechanism for uncouplers of the dinitrophenol type is that they act as proton uniporters. It is not necessary for compensation of proton translocation to be made by actual proton uniport and the process can be dissected into two parts, neutralisation of charge and pH equilibration. Information about these processes has come from a study of ionophoretic antibiotics (Lardy, Johnson and McMurray 1958; Pressman *et al.* 1967; Henderson *et al.* 1969; see also the review by Chance and Montal 1971), anion antiport by organometals (Selwyn *et al.* 1970) and the use of permeant ions and weak electrolytes (Skulachev *et al.* 1969). Skulachev *et al.* have divided these types of uncouplers into four classes; weak acids, weak bases, cations and anions. This is useful in so far as it goes but it does not cover the situation completely. For example, acetic acid is a weak acid but is not normally an uncoupler; the permeable cations only uncouple under certain circumstances, that is when there is an appropriate counter ion. In the experiments which Skulachev used to illustrate his classification he used a medium which contained phosphate. Figure 18.11 shows that the permeable cation dimethyldibenzylammonium (DMDBA) produces little uncoupling until phosphate is added. When the experiment is performed in this way phosphate, rather than DMDBA, appears to be the uncoupler. The situation is clarified if we remember that the uncoupling can be split into charge and pH neutralisation. It is also necessary to consider the differences between kinetic and thermodynamic uncoupling, the uncoupling shown in Fig. 18.11 being of the latter sort. The continued accumulation of electrolytes from the medium is indicated by the light scattering trace. Applying these ideas leads to the more complete classification of ionophoretic uncouplers given in Table 18.1.

It should be noted that thermodynamic uncoupling, where continued movement of electrolytes in one direction across the coupling membrane takes place, is sensitive to the polarity of the proton pump. Thus, uncoupling of the type II(c) (Table 18.1) is very effective with sub-mitochondrial particles or chloroplasts where the proton pump is directed inwards and permeant anions can move inwards from the medium to compensate for the positive charge. It is ineffective with whole mitochondria owing to the unavailability of a continuing source of permeant anions within the mitochondrion.

The multiplicity of ways of producing uncoupling by electrolyte movement, together with

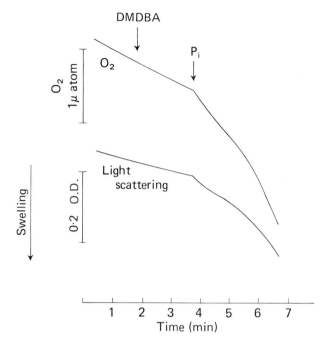

Fig. 18.11 Uncoupling by DMDBA (dimethyldibenzylammonium).

The figure shows the simultaneous recording of oxygen uptake and light scattering changes (a decrease in scattering indicating swelling of the mitochondria) of mitochondria suspended in 100 mM KCl, with 10 mM K-HEPES buffer, pH 7·5, at 30°C with succinate as substrate. The arrows show the addition of 50 μmoles DMDBA and 50 μmoles of inorganic phosphate (P_i). Note that there is no release of respiratory control until phosphate is added. The rapid swelling indicates accumulation of ions from the medium.

the need for both charge and pH neutralisation as their common feature and the polarity requirements of thermodynamic (class II) uncouplers leave little doubt that it is ion transport resulting in the net movement of a proton with its charge which is the condition for uncoupling. Arguments that uncoupling is caused by side effects of these compounds are very tenuous since side effects have not been demonstrated in every case. Furthermore, in cases where an appropriate second component is required, uncoupling based on the other effects does not explain the requirement for the second component (e.g. SCN^- as opposed to Cl^- with tri-alkyltins). Similarly, arguments that these compounds are activating natural pumps or carriers are unnecessarily contrived since the compounds have been shown to act on artificial membranes which do not have such pumps or carriers.

6 Enzymological approaches

6.1 Exchange reactions

Ultimately a description of oxidative phosphorylation in biochemical terms will entail knowledge of the enzymes which catalyse the process. Two approaches to this problem have been made, firstly the extraction of such enzymes, which is dealt with in section 6.2, and secondly the study of partial reactions of oxidative phosphorylation of which the exchange reactions have proved most valuable.

Table 18.1 Types of Uncoupling by Ionophores and Permeant Electrolytes

Type		Description	Example	Mechanism
I		*Kinetic*		
	(a)	Weak acid and anion uniport	2,4-Dinitrophenol FCCP, etc.	DNPOH → DNPOH; H^+ ↑; DNPO⁻ ← DNPO⁻ ; ↳H^+
	(b)	Anion-hydroxide exchange plus anion uniport	Trialkyltins with SCN⁻ (Stockdale *et al.* 1970)	OH^- ← (TAT) ← OH^-; SCN^- → SCN^-
	(c)	Weak base and cation uniport	Tributylamine (Skulachev *et al.* 1969) NH_3/NH_4^+ + Gramicidin (Chappell and Crofts 1965)	BH^+ —(GRAM)→ BH^+; H^+ ↑; B ← B; ↳H^+
	(d)	Cation-proton exchange plus cation uniport	K^+ + Nigericin + Valinomycin (Montal *et al.* 1970)	H^+ → (NIG) → H^+; K^+ → (VAL) → K^+
II		*Thermodynamic*		
	(a)	Cation and weak acid uniport	K^+ + Valinomycin/ acetate or dimethyl dibenzylammonium/ acetate (mitochondria) (Skulachev *et al.* 1969)	K^+ —(VAL)→ K^+; HAc → HAc ⟨Ac⁻ / H^+⟩
	(b)	Cation uniport and anion-hydroxide exchange	K^+ + Valinomycin, or dimethyldibenzylam- monium with phos- phate or chloride + trialkyltin (mitochondria)	$DMBA^+$ → $DMBA^+$; Cl^- → Cl^-; (TAT); OH^- ← OH^-
	(c)	Anion and weak base uniport	NO_3^- or tetraphenylboron plus NH_3 (sub-mitochondrial par- ticles) (Skulachev *et al.* 1969; Montal *et al.* 1970)	NO_3^- → NO_3^-; NH_3 → NH_3 ; ↳H^+ → NH_4^+
	(d)	Anion uniport and exchange cation-proton	NO_3^- with K^+ + Nigericin (sub- mitochondrial particles) (Montal *et al.* 1970)	NO_3^- → NO_3^-; K^+ → K^+; (NIG); H^+ ← H^+

Exchange reactions using isotopically labelled compounds have proved valuable in the investigation of enzyme reaction mechanisms not only by virtue of the indication of mechanism given by the presence or absence of a particular exchange but also because the exchange reaction is a more simple system than the overall reaction. However, it has become clear that effects such as compulsory order of addition of substrates or substrate-induced conformation changes of enzymes may lead to confusing observations and the lack of an exchange reaction does not conclusively eliminate a covalent enzyme-substrate intermediate. Unfortunately the presence of an exchange reaction does not conclusively prove the existence of a covalent intermediate. Nevertheless, taken with other methods the exchange reactions can be very valuable. This topic has been excellently reviewed by Boyer (1967) and references to original papers and fuller discussion should be sought in his review.

The situation in mitochondria is further complicated by the presence of many enzymes besides those of oxidative phosphorylation which utilise adenine nucleotides and/or phosphate as substrates. The sensitivity of an exchange reaction to dinitrophenol and oligomycin is often used as a diagnostic test for its relevance to oxidative phosphorylation. The early observation on exchange reactions gave very promising results and four were considered to be part of the coupling process.

a P_i^{32}-ATP exchange

$$ATP + P_i^{32} = ADP\text{-}P_i^{32} + P_i$$

Involves the last two reactions of Scheme II given in section 2.1.

b C^{14}-ADP-ATP exchange

Involves only the last reaction of the scheme.

c H_2O-PO_4^{18}

P_i labelled with O^{18} loses its label to water but this process is greatly accelerated by mitochondria during oxidative phosphorylation or in the presence of ATP. This would appear to be a reversal of reaction 3 of Scheme II, ATP or respiration being necessary to allow formation of the $X \sim I$ necessary to form $X \sim P$ from P_i.

d H_2O^{18}-ATP

Examination of the previous reaction in the opposite direction, i.e. incorporation of O^{18} from water into P_i led to the discovery of this fourth exchange reaction, incorporation of O^{18} from water into ATP. While this is qualitatively perfectly in accord with the scheme, it was found that the exchange was faster than the H_2O^{18}-P_i exchange. To fit with the scheme either a second site of entry of O^{18} or some compartmentation of phosphate, nucleotides or water has to be proposed.

Later work has on balance decreased the value of the exchange reactions as evidence for the existence of $X \sim P$. Thus, in sub-mitochondrial particles (from which endogenous adenine nucleotides can be effectively removed, unlike intact mitochondria) ADP is required for the O^{18}-P_i-H_2O exchange. In mitochondria ADP can be removed by an ATP-generating system and then both the H_2O^{18}-ATP and the P_i^{32}-ATP exchanges are inhibited. The involvement of ADP weakens the evidence for the existence of $X \sim P$ as an intermediate since it suggests either that ADP must be bound to the enzyme before $X \sim P$ is formed (cf p. 572) or that a concerted reaction takes place and hence that ATP is the first and only high energy phosphate compound. An intermediate with a pentacovalent terminal phosphorus atom could account for the observed exchange reactions:

One feature which has emerged from a study of these exchange reactions is the lack of a very marked inhibition of the exchange by oxidation or reduction of the electron carries. This is further support for the separation of phosphorylation from the redox carriers, and argues against the schemes which involve this, many of which are based on model reactions in which high energy phosphate bonds are formed by oxidation of quinones or quinol phosphates.

6.2 Protein factors

Two approaches to the isolation of proteins from the energy conserving systems have been used. The first is the extraction of enzymes catalysing partial reactions of the process, such as the exchange reactions, mitochondrial ATPase and the electron transporting proteins. The second is the purification of soluble factors, the coupling factors, which restore or enhance phosphorylation or other energy linked functions in sub-mitochondrial particles.

It is worth noting that even though the electron carriers catalyse readily detectable reactions and often have well characterised spectra, their purification has had a lengthy and tortuous history. In some cases really satisfactory preparations are still not available. It is therefore hardly surprising that purification of the ill-defined energy-conserving enzymes has proved even more difficult. Coupling factors are known in great profusion but in many cases with only poor characterisation. The number of different factors is of some interest since on a theoretical basis the chemical or conformational hypotheses, involving site specific coupling factors, require more than does the chemiosmotic hypothesis which has only one coupling system. However, it is worth noting that if there are three distinct coupling factors, there are seven possible qualitatively different combinations of them. If allowance is made for quantitative differences, different sources and modification of activity during extraction and purification, a system comprising three or four coupling factors could account for the large number of different ones reported. This problem has been succinctly reviewed by Beechey (1970). The properties of coupling factors have been reviewed extensively by Lardy and Ferguson (1969) and by Racker (1965, 1970), and Van Dam and Meyer (1971) have published a brief review of recent developments. Full details should be sought in these reviews.

A protein catalysing the ADP–ATP exchange was purified by Lehninger (1960). The activity was not sensitive to DNP but became so when the soluble enzyme was recombined with digitonin sub-mitochondrial particles. Under some conditions this preparation could produce enhancement of phosphorylation. Another partially purified protein, called the M factor, enhanced the DNP sensitivity of the recombined exchange enzyme under some conditions, but has no other known activity. Neither of these factors appears to have revealed much information and the ADP–ATP exchange enzyme may not be part of the coupling system (Lardy and Ferguson 1969).

Linnane and Titchener (1960) partially purified a factor, obtained by sonicating mito-chondria, which improved phosphorylation in sub-mitochondrial particles. This factor was studied in D. E. Green's laboratory and it appeared to have been separated into factors specific for the three coupling sites. This latter work has not been substantiated and the relationship of the original factor to other coupling factors is not clear.

One of the earliest reported extractions of a protein of the energy conserving system was the DNP stimulated ATPase extracted from acetone-dried rat liver mitochondria by Lardy and Wellman (1953). Purification of a similar enzyme from beef heart mitochondria (Selwyn, 1967) has shown that this enzyme is fundamentally the same as the coupling factor ATPase reported by Penefsky et al. (1960). In this paper, the coupling factor activity was shown to reside in a Mg^{2+}-dependent, DNP stimulated ATPase which was insensitive to oligomycin. This ATPase, called the F_1 coupling factor, is the most satisfactory of all coupling factors. It has been extracted and purified by a variety of methods in many laboratories, and although there are variations in the properties of the different preparations, there is a fundamental similarity which leaves no doubt that there is a protein in mitochondria which, under appropriate conditions, has potent ATPase and coupling factor activity. In some preparations, such as the F_1-X of Vallejos et al. (1968), the factor $A-D$ complex of Fisher et al. (1971) and the oligomycin-sensitive ATPase of Kagawa and Racker (1966) and Tzagaloff et al. (1968), the ATPase is complexed with some other factor. MacLennan and Tzagaloff (1968) were able to separate and purify a soluble oligomycin sensitivity conferring protein (OSCP) of molecular weight 18 000. Bulos and Racker (1968) have isolated a similar protein which they call F_c. Racker had earlier prepared a factor designated F_0 which conferred oligomycin sensitivity on the F_1 ATPase but this was found to contain respiratory enzymes. F_0 was resolved to yield CF_0, which is a particulate preparation of hydrophobic proteins that inhibits the ATPase activity of F_1. Addition of phospholipids to the CF_0-F_1 complex restores ATPase activity and this activity is sensitive to oligomycin. The OSCP or F_c appears to be a well established entity which in appropriate test conditions also has coupling factor activity. Racker and his collaborators (Racker 1965, 1970) proceeded with further fractionation of sub-mitochondrial particles by a variety of techniques such as sonication, exposure to high pH and treatment with urea, trypsin or phospholipase and in this way obtained evidence for the involvement of other factors in the restoration of phosphorylation or other energy linked functions to these particles. These have been given numbers, F_2, F_3, F_4, F_5 and F_6, and are characterised by their method of preparation, the type of particles to which they restore activity, the type of activity they restore and the need for other factors. F_4 is now thought to be a mixture of F_2, F_3 and F_5 and the status of the remaining factors as individual proteins is not satisfactorily established. Lam et al. (1967) prepared a factor B which was originally thought to correspond to F_3 but is now thought to be more akin to F_2. This group have also reported a factor C and a factor D (Fisher et al. 1971), the latter being extracted as an appendage to factor A in the $A-D$ complex. This complex has been renamed the ATP synthetase complex since it catalyses ATP-P_i and ADP-ATP exchange reactions which are inhibited by uncouplers, oligomycin and antiserum to F_1. The $A-D$ complex preparation greatly enhances the ATP-P_i exchange activity of 'urea' sub-mitochondrial particles and its specific activity in this enhancement is 600 nmol/mg/min. Without the particles its activity is much lower and rather variable, being in the range 2-30 nmol/mg/min. The possibility that the activity is due to contamination by membrane fragments cannot be eliminated and the inhibition by uncouplers may well be a result of stimulation of ATPase activity.

Kagawa and Racker (1971) have reported the reconstruction, from partially purified coupling factors, phospholipids and a preparation of hydrophobic proteins from mitochondria, of a system which catalyses the ATP-P_i exchange and exhibits a fluorescent probe response on addition of ATP. With the incorporation of cytochrome c and cytochrome oxidase, oxidative phosphorylation was also observed (Racker and Kandrach 1971). Without belittling this achievement it must be admitted that it has more value as a promising technique than in providing information about the mechanism. However, the finding that the reconstituted system is vesicular is in line with the predictions of the chemiosmotic hypothesis.

One final report of the use of purified preparations is that by Yong and King (1972) in which a complex, based on detergent micelles, between purified cytochrome c and cytochrome oxidase shows phosphorylation, respiratory control, inhibition by oligomycin and uncoupling by FCCP. The absence of any requirement for coupling factors or added phospholipids may well be a significant breakthrough but the alternative possibility, that cytochrome oxidase even in highly purified form is contaminated with coupling factors, seems at least as likely.

Further work on these reconstructed systems and investigation of the need for vesicular structures for coupled phosphorylation is obviously a valuable but difficult approach to the problem. An outstanding need is for the characterisation of coupling factors in terms of catalysis of a specific reaction and in other identifiable properties.

7 Conclusion

The value of an hypothesis lies in its ability to explain existing results and suggest new and profitable lines of enquiry. In historical terms, the chemical hypothesis has served these functions admirably, as has, more recently, the chemiosmotic hypothesis. The conformational hypothesis has, perhaps, not been as useful as the other two, since it appears to be a special case of the chemical hypothesis.

The chemiosmotic hypothesis has led directly to the discovery of the ion pumping properties of mitochondria and chloroplasts. It has suggested:

a The impermeability of the membrane to H^+.
b The movements of H^+ associated with respiration, ATP hydrolysis and ion transport.
c The effects of classical uncouplers on the proton conductance of the membrane.

The chemical hypothesis can explain these findings only in terms of a secondary hypothesis – the $X \sim I$ driven H^+ pump. On the other hand, the original simplicity of the chemiosmotic hypothesis has been lost by the need to introduce some form of $X \sim I$. The exact nature of $X \sim I$ in terms of the chemiosmotic hypothesis has much greater flexibility than on the chemical hypothesis, where it represents a discrete covalent compound. It could, for instance, represent the ionisation state of amino acid residues in a coupling protein. However, the necessity for its existence means that formally (Fig. 18.12) the two hypotheses are so similar that there is probably no immediately practicable method of distinguishing between them. The chemical hypothesis predicts, in contrast to the chemiosmotic hypothesis, the possibility of reconstructing an ATP synthesising system without an H^+ pump, but there are as yet no signs of this being achieved. Perhaps the brightest hope for the future lies in the further study of the organisation of the respiratory chain to see whether it continues to fulfil the predictions of Mitchell's hypothesis.

In section 4 it was pointed out that the likelihood is that phosphorylation itself leads to electrical and pH imbalance across the membrane due to the operation of the ADP–ATP exchanger and the P_i–OH^- exchanger. For phosphorylation to continue, this means that about $1H^+$/ATP synthesised must be pumped outwards in compensation. In Fig. 18.4 the suggestion was that $2H^+$/ATP synthesised are translocated through the ATPase system. On the basis of a H^+/O ratio of 6 for NAD-linked substrates and 4 for succinate, this would leave the ion exchanger systems uncompensated. Another suggestion from Mitchell is that shown in Fig. 18.13 in which only one proton is translocated through the ATPase, the other proton compensating for the ion transporter systems.

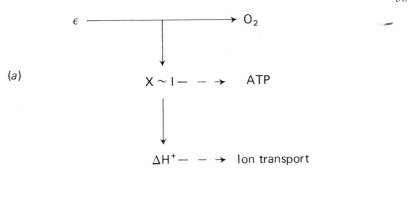

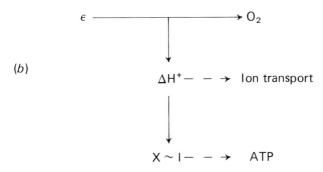

Fig. 18.12 The relationship between electron transfer, proton translocation and the high energy chemical state (a) according to the chemical hypothesis and (b) according to the chemiosmotic hypothesis.

These findings are also explicable in terms of the chemical hypothesis. One possibility is that the $X \sim I$ driven proton pump operates in parallel with ATP synthesis. Alternatively the use of $X \sim I$ for ATP synthesis might cause the expulsion of H^+ from the mitochondria.

A major criticism which has been levelled at the chemiosmotic hypothesis (Slater 1971) is made on energetic grounds. The proton-motive force (the combination of pH gradient and membrane potential) required to poise the ATP/ADP ratio at unity is of the order of 210 mV, and values of at least 230 mV have been demonstrated. However, respiration is capable of driving ATP synthesis when the ATP/ADP ratio is very high, and the suggestion is that the demonstrated proton motive force is inadequate under these conditions. It should be pointed out that this also creates a problem for the chemical hypothesis since if the proton motive force is produced by an H^+ pump driven by $X \sim I$, the proton motive force reflects the energy available in $X \sim I$. It seems likely that the value for the free energy of hydrolysis of ATP (9·4 kcal/mol) assumed by Slater is too large. The presence of Mg^{2+} reduces the value of $\Delta G_0'$ to about 6·8 kcal/mol and with this reduction the proton motive force is sufficient to drive the synthesis of ATP against the adverse phosphorylation potential. A word of caution may also be sounded about the direct application of values and equations of equilibrium thermodynamics to a steady state situation.

The mechanism, in enzymological terms, of oxidative phosphorylation will probably have to wait upon detailed knowledge of the nature of the high energy entity, since this is, in a sense, a substrate for the ATP synthesising enzymes. Apart from the F_1 ATPase nothing is known of the chemical activities of the coupling factors and examination of these is unlikely to produce

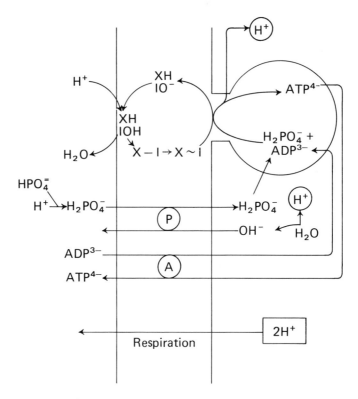

Fig. 18.13 Proton movements during phosphorylation according to the chemiosmotic hypothesis. For every ATP synthesised, approximately 1 H$^+$ is consumed. In addition, respiration ejects 2H$^+$ from the inside of the mitochondria. Of these, one is thought of as re-entering through the ATP synthesising system (upper part of diagram) and one by a combination of the phosphate transporter (P) and the adenine nucleotide carrier (A).

the answer, since usually enzymes have been found for substrates and not vice versa. From time to time reports of X ~ P, detected by incorporation of radioactive phosphate, appear and the latest is that by Cross *et al.* (1970). Lack of further reports about it suggest that like earlier findings it may not have lived up to its initial promise.

On balance the chemiosmotic hypothesis appears to accommodate the experimental observations with fewer secondary assumptions than the chemical hypothesis or the conformational variant of the chemical hypothesis. Unfortunately neither hypothesis provides any basis for predicting the chemical nature of X ~ I which both require.

It is also worth pointing out that the chemical and chemiosmotic hypotheses are not as yet mutually exclusive. Schemes in which electron transport leads to proton translocation and X ~ I production in parallel are by no means impossible. However it would be very difficult in practical terms to distinguish such schemes from the alternatives shown in Fig. 18.12.

Finally, with regard to the failure to find X ~ I or X ~ P it may be worth remembering the story by G. K. Chesterton about the crime committed by a man dressed as the postman — unnoticed because he was always there.

8 Bibliography

As has, we hope, become clear in the above chapter the investigation of mitochondrial oxidative phosphorylation is a rapidly advancing field and the theories of the mechanism of this process are often the subject of considerable contention. Because of this we have felt it necessary to include original references to most of the points discussed so that the reader may assess the information for himself. However this bibliography also contains a number of recommended reviews which are suitable for more general reading and these are marked with an asterisk.

AZZI, A. and SANTATO, M. (1971). 'ATP and pH induced spectral changes of cytochrome b in rat liver mitochondria', *Biochem. Biophys. Res. Commun.*, **45**, 945-54.

* BEECHEY, R. B. (1970). 'Mitochondrial coupling factors'. *Biochem. J., 116,* 6P-8P.

BORST, P. and SLATER, E. C. (1961) 'The site of action of 2,4-dinitrophenol on oxidative phosphorylation, *Biochim. Biophys. Acta,* **48**, 362-79.

BOYER, P. D. (1965). 'Carboxyl activation as a possible common reaction in substrate-level and oxidative phosphorylation and in muscle contraction'. In: *Oxidases and Related Redox Systems,* Vol. 2, pp. 994-1008. Eds. T. E. King, H. S. Mason and M. Morrison. New York, John Wiley.

BOYER, P. D. (1967), '^{18}O and related exchanges, in enzymic formation and utilization of nucleoside triphosphates', *Current Topics in Bioenergetics,* **2**, 99-149.

BROCKLEHURST, J. R., FREEDMAN, R. B., HANCOCK, D. J. and RADDA, G. K. (1970). 'Membrane studies with polarity-dependent and excimer-forming fluorescent probes', *Biochem. J.,* **116**, 721-31.

BULOS, B. and RACKER, E. (1968). 'Further resolution of the rutamycin-sensitive adenosine triphosphatase', *J. Biol. Chem.,* **243**, 3891-900.

CASWELL, A. H. (1971). 'The estimation of redox potentials of cytochromes in mitochondria', *Arch. Biochem. Biophys.,* **144**, 445-47.

* CHANCE, B. and WILLIAMS, G. R. (1956). 'The respiratory chain and oxidative phosphorylation', *Adv. Enzymol.,* **17**, 65-134.

* CHANCE, B. and MONTAL, M. (1971). 'Ion-translocation in energy-conserving membrane systems', *Current Topics in Membranes and Transport,* **2**, 99-156.

CHAPPELL, J. B. and CROFTS, A. R. (1965). 'Gramicidin and ion transport in isolated liver mitochondria', *Biochem. J.,* **95**, 393-402.

* CHAPPELL, J. B. (1968). 'Systems used for the transport of substrates into mitochondria', *Br. Med. Bull.,* **24**, 150-57.

CROSS, R. L., CROSS, B. A. and WANG, J. H. (1970). 'Detection of a phosphorylated intermediate in mitochondrial oxidative phosphorylation', *Biochem. Biophys. Res. Commun.,* **40**, 1155-61.

DUTTON, P. L., ERECINSKA, M., SATO, N., MUKAI, Y., PRING, M. and WILSON, D. F. (1972). 'Reactions of *b* cytochromes with ATP and Antimycin A in pigeon heart mitochondria', *Biochim. Biophys. Acta,* **267**, 15-24.

ERECINSKA, M., WILSON, D. F., SATO, N. and NICHOLLS, P. (1972). 'The energy dependence of the chemical properties of cytochrome *c* oxidase', *Arch. Biochem. Biophys.,* **151**, 188-93.

ESTABROOK, R. W. (1961). 'Effect of oligomycin on the arsenate and 2,4-dinitrophenol (DNP) stimulation of mitochondrial oxidations', *Biochem. Biophys. Res. Commun.,* **4**, 89-91.

FISHER, R. J., CHEN, J. C., SANI, B. P., KAPLAY, S. S. and SANADI, D. R. (1971). 'A soluble mitochondrial ATP synthetase complex catalyzing ATP-phosphate and ATP-ADP exchange', *Proc. Natn. Acad. Sci. U.S.A.*, **68**, 2181-84.

FREEDMAN, R. B., HANCOCK, D. J. and RADDA, G. K. (1971). 'The Design of Fluorescent Probes'. In: *Probes of Structure and Function of Macromolecules and Membranes*, Vol. 1 pp. 325-37. Eds. Chance, B., Lee, C-P. and Blasie, J. K. London and New York, Academic Press.

GREEN, D. E., ASAI, J., HARRIS, R. A. and PENNISTON, J. T. (1968). 'Conformational basis of energy transformations in membrane systems III. Configurational changes in the mitochondrial inner membrane induced by changes in functional states', *Arch. Biochem. Biophys.*, **125**, 684-705.

GREEN, D. E. and JI, S. (1972). 'Electromechanochemical model of mitochondrial structure and function', *Proc. Natn. Acad. Sci. U.S.A.*, **69**, 726-29.

* GREVILLE, G. D. (1969). 'A scrutiny of Mitchell's chemiosmotic hypothesis of respiratory and photosynthetic phosphorylation', *Current Topics in Bioenergetics*, **3**, 1-78.

HACKENBROCK, C. R. (1966). 'Ultrastructural bases for metabolically linked mechanical activity in mitochondria. I. Reversible ultrastructural changes, with change in metabolic steady state in isolated liver mitochondria', *J. Cell Biol.*, **30**, 269-97.

HARRIS, E. J. and BERENT, C. (1970). 'The applicability of the Donnan relation to the distribution of certain anions between mitochondria and medium', *FEBS Lett.*, **10**, 6-12.

HENDERSON, P. J. F., McGIVAN, J. D. and CHAPPELL, J. B. (1969). 'The action of certain antibiotics on mitochondrial, erythrocyte and artifical phospholipid membranes. The role of induced proton permeability', *Biochem. J.*, **111**, 521-35.

HINKLE, P. and MITCHELL, P. (1970). 'Effect of membrane potential on equilibrium poise between cytochrome *a* and cytochrome *c* in rat liver mitochondria', *Bioenergetics*, **1**, 45-60.

HOPFER, U., LEHNINGER, A. L. and LENNARZ, W. J. (1970). 'The effect of the polar moiety of lipids on bilayer conductance induced by uncouplers of oxidative phosphorylation', *J. Membrane Biol.*, **3**, 142-55.

HUNTER, G. R., KAMISHIMA, Y. and BRIERLEY, G. P. (1969). 'Ion transport by heart mitochondria XV. Morphological changes associated with the penetration of solutes into isolated heart mitochondria', *Biochim. Biophys. Acta*, **180**, 81-97.

KAGAWA, Y. and RACKER, E. (1966). 'Reconstruction of the oligomycin-sensitive andenosine triphosphatase', *J. Biol. Chem.*, **241**, 2467-474.

KAGAWA, Y. and RACKER, E. (1971). 'Reconstitution of vesicles catalyzing $^{32}P_i$-adenosine triphosphate exchange', *J. Biol. Chem.*, **246**, 5477-487.

KING, T. E., KUBOYAMA, M. and TAKEMORI, S. (1965). 'On cardiac cytochrome oxidase: a cytochrome *c*–cytochrome oxidase complex'. In: *Oxidases and Related Redox Systems*, Vol. 2 pp. 707-36. Eds. T. E. King, H. S. Mason and M. Morrison. New York. John Wiley.

* KLINGENBERG, M. (1970). 'Metabolite transport in mitochondria: an example for intracellular membrane function', *Essays in Biochemistry*, **6**, 119-59.

LAM, K. W., WARSHAW, J. B. and SANADI, D. R. (1967). 'Purification and properties of a second energy-transfer factor', *Arch. Biochem. Biophys.*, **119**, 477-84.

* LARDY, H. A. and FERGUSON, S. M. (1969). 'Oxidative phosphorylation in mitochondria', *A. Rev. Biochem.*, **38**, 991-1034.

LARDY, H. A., JOHNSON, D. and McMURRAY, W. C. (1958). 'Antibiotics as tools for metaboli studies. I. A survey of toxic antibiotics in respiratory, phosphorylative and glycolytic systems', *Archs. Biochem. Biophys.*, **78**, 587-97.

LARDY, H. A. and WELLMAN, H. (1953). 'The catalytic effect of 2,4-dinitrophenol on adenosinetriphosphate hydrolysis by cell particles and soluble enzymes', *J. Biol. Chem.*, **201**, 357-70.

LEE, I. Y. and SLATER, E. C. (1972). 'Effect of pH on cytochromes *b* in ATP-Mg submito-chondrial particles', *Biochim. Biophys. Acta,* **256**, 587-93.

LEHNINGER, A. L. (1960). 'Components of the energy-coupling mechanism and mitochondrial structure', In: *Biological Structure and Function,* Vol. II, pp. 31-51. Eds. T. W. Goodwin and O. Lindberg. London and New York, Academic Press.

* LEHNINGER, A. L., CARAFOLI, E. and ROSSI, C. S. (1967). 'Energy-linked ion movements in mitochondrial systems', *Adv. Enzymol.,* **29**, 259-320.

LIBERMAN, E. A. and TOPALY, V. P. (1968). 'Selective transport of ions through bimolecular phospholipid membranes', *Biochim. Biophys. Acta,* **163**, 125-36.

LINDSAY, J. C., DUTTON, P. L. and WILSON, D. F. (1972). 'Energy-dependent effects on the oxidation-reduction midpoint potentials of the *b* and *c* cytochromes in phosphorylating submitochondrial particles from pigeon heart', *Biochemistry,* **11**, 1937-43.

LINNANE, A. W. and TITCHENER, E. B. (1960). 'A factor for coupled oxidation in the electron transport particle', *Biochim. Biophys. Acta,* **39**, 469-78.

MacLENNAN, D. H. and TZAGALOFF, A. (1968). 'Purification and characterisation of the oligomycin sensitivity conferring protein', *Biochemistry,* **7**, 1603-10.

MELA, L. (1969). 'Inhibition and activation of calcium transport in mitochondria. Effect of lanthanides and local anaesthetic drugs', *Biochemistry,* **8**, 2481-86.

MITCHELL, P. (1961). 'Coupling of phosphorylation to electron and hydrogen transfer by a chemiosmotic type of mechanism', *Nature, Lond.,* **191**, 144-8.

MITCHELL, P. (1967). 'Translocations through natural membranes', *Adv. Enzymol.,* **29**, 33-87.

*MITCHELL, P. (1968). *Chemiosmotic Coupling and Energy Transduction,* Bodmin. Glynn Research Ltd.

* MITCHELL, P. (1971). 'Structure and function of the respiratory chain of mitochondria and bacteria'. In: *Energy Transduction in Respiration and Photosynthesis,* pp. 123-52. Eds. E. Quagliariello, S. Papa and C. S. Rossi. Adriatica Editrice, Bari.

MITCHELL, P. and MOYLE, J. (1967). 'Proton-transport phosphorylation: some experimental tests'. In: *Biochemistry of Mitochondria,* pp. 53-74. Eds. E. C. Slater, Z. Kaniuga and L. Wojtczak. London and New York, Academic Press. Warszawa, PWN–Polish Scientific Publishers.

MONTAL, M., CHANCE, B. and LEE, C-P. (1970). 'Ion transport and energy conservation in submitochondrial particles', *J. Membrane Biol.,* **2**, 201-34.

PACKER, L. (1963). 'Size and shape transformations correlated with oxidative phosphory-lation in mitochondria. I. Swelling-shrinkage mechanisms in intact mitochondria', *J. Cell. Biol.,* **18**, 487-501.

PENEFSKY, H. S., PULLMAN, M. E., DATTA, A. and RACKER, E. (1960). 'Participation of a soluble adenosine triphosphatase in oxidative phosphorylation', *J. Biol. Chem.,* **235**, 3330-36.

PRESSMAN, B. C., HARRIS, E. J., JAGGER, W. S. and JOHNSON, J. H. (1967). 'Antibiotic-mediated transport of alkali ions across lipid barriers', *Proc. Natn. Acad. Sci. U.S.A.,* **58**, 1949-56.

RACKER, E. (1965). *Mechanisms in Bioenergetics,* pp. 159-77. New York, Academic Press.

RACKER, E. (1967). 'Resolution and reconstruction of the inner mitochondrial membrane', *Fed. Proc.,* **26**, 1335-40.

RACKER, E. (1970). 'The two faces of the inner mitochondrial membrane', *Essays in Biochem.* **6**, 1-22.

RACKER, E. and KANDRACH, A. (1971). 'Reconstitution of the third site of oxidative phosphorylation', *J. Biol. Chem.,* **246**, 7069-71.

ROBERTON, A. M., BEECHEY, R. B., HOLLOWAY, C. T. and KNIGHT, I. G. (1967). 'The effect of aurovertin on a soluble mitochondrial adenosine triphosphatase', *Biochem. J.,* **104,** 54C–55C.

SATO, N., WILSON, D. F. and CHANCE, B. (1971). 'The spectral properties of the *b* cytochromes in intact mitochondria', *Biochim. Biophys. Acta,* **253,** 88–97.

SELWYN, M. J. (1967). 'Preparation and general properties of a soluble adenosine triphosphatase from mitochondria', *Biochem. J.,* **105,** 279–88.

SELWYN, M. J., DAWSON, A. P. and DUNNETT, S. J. (1970). 'Calcium transport in mitochondria'. *FEBS Lett.,* **10,** 1–5.

SELWYN, M. J., DAWSON, A. P., STOCKDALE, M. and GAINS, N. (1970). 'Chloride-hydroxide exchange across mitochondrial, erythrocyte and artificial lipid membranes mediated by trialkyl- and triphenyltin compounds', *Eur. J. Biochem.,* **14,** 120–26.

SKULACHEV, V. P., JASAITIS, A. A., NAVICKAITE, V. V., YAGUZHINSKY, L. S., LIBERMAN, E. A., TOPALI, V. P. and ZOFINA, L. M. (1969). 'Five types of uncouplers for oxidative phosphorylation'. In: *Mitochondria Structure and Function.* Eds. L. Ernster and Z. Drahota. *FEBS Symposium,* Vol. 17, pp. 275–84. London and New York, Academic Press.

SLATER, E. C. (1953). 'Mechanism of phosphorylation in the respiratory chain', *Nature, Lond.,* **172,** 975–82.

* SLATER, E. C. (1971). 'The coupling between energy-yielding and energy-utilizing reactions in mitochondria', *Q. Rev. Biophys.,* **4,** 35–71.

STOCKDALE, M., DAWSON, A. P. and SELWYN, M. J. (1970). 'Effects of trialkyltin and triphenyltin compounds on mitochondrial respiration', *Eur. J. Biochem.,* **15,** 342–51.

STONER, C. D. and SIRAK, H. D. (1969). 'Passive induction of the "energized-twisted" conformational state in bovine heart mitochondria', *Biochem. Biophys. Res. Commun.,* **35,** 59–66.

TZAGALOFF, A., BYINGTON, K. H. and MacLENNAN, D. H. (1968). 'The isolation and characterisation of an oligomycin-sensitive adenosine triphosphatase from bovine heart mitochondria', *J. Biol. Chem.,* **243,** 2405–12.

URBAN, P. F. and KLINGENBERG, M. (1969). 'On the redox potentials of ubiquinone and cytochrome *b* in the respiratory chain', *Eur. J. Biochem.,* **9,** 519–25.

VALLEJOS, R. H., VAN DEN BERGH, S. G. and SLATER, E. C. (1968). 'On coupling factors of oxidative phosphorylation', *Biochim. Biophys. Acta,* **153,** 509–20.

* VAN DAM, K. and MEYER, A. J. (1971). 'Oxidation and energy conservation by mitochondria', *A. Rev. Biochem.,* **40,** 115–60.

WILSON, D. F. and DUTTON, P. L. (1970a). 'The oxidation-reduction potentials of cytochromes *a* and a_3 in intact rat liver mitochondria', *Archs Biochem. Biophys.,* **136,** 583–85.

WILSON, D. F. and DUTTON, P. L. (1970b). 'Energy-dependent changes in the oxidation-reduction potential of cytochrome *b*', *Biochem. Biophys. Res. Commun.,* **39,** 59–64.

YONG, F. C. and KING, T. E. (1972). 'Respiratory control and oxidative phosphorylation of the cytochrome *c*–cytochrome oxidase complex', *Biochem. Biophys. Res. Commun.,* **47,** 380–86.

19
The Hormonal Control of Metabolism

M. C. Perry

Department of Biochemistry, Chelsea College, University of London

1 Introduction

In the face of various challenges from the environment, the organism maintains homeostasis in part through the actions of its hormones, which co-ordinate the distribution of substrates between, and the metabolism of substrates by, its tissues. Many tissues are now known to be at least partly endocrine in function. These include the pancreas, the anterior and posterior pituitary, the adrenal cortex and medulla, the thyroid and parathyroids, the ovary, testis and placenta, the kidney, stomach and small intestine. This wide range of endocrine organs and wider range of hormones they produce make a specific treatment of each impracticable within the context of this chapter. The approach adopted will be to consider the control of metabolism by, and the mechanism of action of, one selected hormone (insulin) as an example of hormone action in general. Other hormones will be introduced where they interrelate with insulin in the control of metabolism, or where they add to the general concepts of the hormonal control of metabolism.

The hormonal control of metabolism encompasses both rapid changes that occur following activation or inhibition of pre-existing enzymes or transport processes by hormones, and slower changes that occur in enzyme levels within the cell. It is difficult to classify individual hormones under either of these categories since one hormone may elicit both types of response within a sensitive tissue. Since the mechanism of regulation of gene expression by hormones is little understood, this chapter will concentrate on the rapid metabolic effects and will introduce long-term changes in enzyme levels only where they augment the rapid changes.

During recent years, the emphasis of research has moved from the determination of the effects of hormones on cell metabolism to investigations of the mechanisms by which hormones act. This chapter will attempt to reflect this change in emphasis and will concentrate on the mechanism of hormone action.

2 The general scheme of insulin action

Insulin is synthesised and stored within the beta cells of the islets of Langerhans in the pancreas. In response to a given stimulus, usually a rise in blood glucose concentration, the hormone is released into the extracellular space, ultimately entering the blood stream. At its target tissues, insulin combines with a specific receptor site, probably protein in nature, situated in the cell membrane. As a consequence of insulin binding to its receptor site the message carried by the hormone is translated into an intracellular message that can be 'read' by the enzyme complement of the cell. In response to this second message certain aspects of the metabolism of the target cell are changed, and the cell is then said to respond to the hormone. The hormone is finally inactivated. There is, in addition, a feedback control system linking the target tissues to the endocrine organ, usually by means of some product of hormone action on the target tissue. In the case of insulin changes in blood glucose concentration, produced by the action of the hormone on muscle or adipose tissue, control the rate of secretion of insulin and hence control the concentration of insulin in the blood perfusing the target tissues.

3 Insulin synthesis and storage

The insulin molecule consists of two polypeptide chains, the A chain and the B chain containing respectively 21 and 30 amino-acid residues. The chains are linked by two disulphide bonds with an additional disulphide bond in the A chain. The tertiary structure of crystalline zinc–insulin has been determined by X-ray crystallography (Blundel et al. 1971).

The question of whether insulin is synthesised as two separate chains that are subsequently linked by disulphide bond formation or as a single-chain precursor that undergoes maturation into the active hormone was resolved in 1967 by the discovery of proinsulin (Steiner et al. 1969). Reading from the N-terminal residue, this single chain precursor of insulin consists of the B-chain of insulin, a connecting peptide (the C-peptide) of 30–33 amino-acid residues length, and the A-chain of insulin. Proinsulin is synthesised on the ribosomes of the beta cell, and assumes the correct tertiary configuration for disulphide bond formation to occur. The subsequent maturation of the protein occurs by limited proteolysis which removes the C-peptide leaving the active insulin molecule. The enzyme(s) involved in this conversion are unknown but probably have trypsin- and carboxypeptidase-like activities. Trypsin can be used in vitro to convert proinsulin to a biologically active molecule, desalanino-insulin, which is insulin minus the C-terminal alanine of the B-chain. Proinsulin itself has little or no intrinsic biological activity.

The proteolytic conversion of proinsulin to insulin occurs either during the transfer of proinsulin from the ribosomes to the Golgi apparatus or within the Golgi apparatus during the formation of the insulin storage granules, and the hormone is stored in the beta cell in its active form. Analysis of storage granules isolated from rat islets shows they contain insulin and C-peptide in equimolar amounts, and very small amounts of proinsulin. The hormone is present as the zinc–insulin complex and each storage granule is surrounded by a membrane.

4 Insulin secretion

In response to an appropriate stimulus, such as a rise in the glucose concentration of the blood perfusing the islets of Langerhans, insulin is released from the beta cells.

4.1 Mechanism of secretion

Electron microscopy of the pancreas shows that each beta cell contains many membrane-bound insulin storage granules scattered throughout the cytoplasm. Secretion of insulin is accompanied by degranulation of the beta cell. Severe stimuli to secretion such as the injection of anti-insulin serum result in a marked degranulation of the beta cells. It seems probable that insulin secretion does not occur by dissolution of the insulin granule into the cytoplasm followed by diffusion of the hormone through the cell membrane, but by emiocytosis (Lacy and Howell 1970). In this process, when the beta cell is stimulated to secrete, the granules migrate to the cell periphery and there fusion occurs between the granule membrane and the beta cell membrane. The granule contents are liberated into the extracellular space, where they dissolve and diffuse into the blood stream. The fate of the granule membrane is not clear. The emiocytotic process has been convincingly demonstrated in studies by freeze-etching of rat islets stimulated to secrete by glucose (Orci *et al.* 1973). Both insulin and the C-peptide are secreted, but under normal conditions only very small amounts of proinsulin are released, although the precursor may be secreted in increasing amounts following prolonged stimulation of the pancreas *in vitro*.

It is not clear how the movement of secretory granules through the cytoplasm of the beta cell is controlled. A system of microtubules has been identified in electron micrographs of rat pancreatic beta cells (Lacy *et al.* 1968). Colchicine, a plant alkaloid that at low concentrations disrupts microtubules in a variety of tissues, suppressed the stimulation of insulin secretion by glucose, suggesting that the microtubular system may be involved in granule translocation within the beta cell.

4.2 Control of secretion

The control of insulin levels in the blood and hence of the amount of insulin presented to its target tissues is effected by control of the rate of secretion alone. There is no evidence for a control mechanism exerted through the rate of inactivation of the hormone. The feedback control of plasma insulin levels mediated by the glucose concentration of the blood is also exerted at the level of insulin secretion. The control of insulin secretion and its role in the hormonal regulation of metabolism has been discussed by Hales (1967), and by Mayhew, Wright and Ashmore (1969).

The major physiological stimulus to insulin secretion *in vivo* is the concentration of glucose in the blood perfusing the pancreas. Glucose stimulates both insulin synthesis and secretion by the islets, but it stimulates both processes independently. (Insulin secretion from islets incubated *in vitro* in the presence of puromycin to suppress insulin synthesis is stimulated by increasing the glucose concentration of the incubation medium.) Glucose initially stimulates the release of preformed, stored insulin and it is only after prolonged stimulation that appreciable amounts of newly synthesised insulin are released.

The beta cells of the pancreatic islets possess a mechanism by which the change in blood glucose concentration is translated into a change in the rate of insulin secretion. Below a threshold glucose concentration of approximately 40 mg% negligible insulin secretion occurs. Above this threshold concentration there is a linear relationship between the glucose concentration and the rate of insulin secretion until a plateau is reached when the glucose concentration attains 250–500 mg%. Increasing the glucose concentration above this level fails to elicit a further increase in insulin secretion. Thus the beta cell is capable of responding to changes in the physiological range of blood glucose concentrations with changes in the rate of insulin secretion. This relationship forms the basis of the feedback control of insulin secretion by blood glucose levels, a fall in plasma glucose concentration as a result of the insulin stimulation of glucose uptake and metabolism by adipose tissue and muscle decreasing the rate of insulin secretion.

4.3 Mechanism of glucose stimulation of insulin secretion

Although the pancreas responds rapidly and sensitively to changes in the blood glucose level, there is compelling evidence to suggest that glucose itself is not the primary stimulus to insulin secretion, and that it has to be metabolised first. Membrane transport is not a rate-limiting step for glucose metabolism in the beta cells. Inhibitors of glucose metabolism such as manno-heptulose (which inhibits glucose phosphorylation by hexokinase) and 2-deoxy-D-glucose (which following phosphorylation to 2-deoxy-D-glucose-6-phosphate is not metabolised further, and competitively inhibits glucose-6-phosphate metabolism) inhibit the stimulation of insulin secretion by glucose *in vitro*.

The stimulation of insulin secretion is not specific to glucose, and in the perfused rat pancreas mannose and fructose, which are metabolised by this preparation, stimulated insulin secretion but xylose and arabinose, which are not metabolised, did not. The normal pathways for glucose metabolism (glycolysis, the pentose phosphate pathway, the citric acid cycle and the enzymes of glycogen synthesis and degradation) are present in the beta cell, but there is little evidence directly implicating one or more of these pathways in the control of insulin secretion. The nature of the metabolite or metabolic state that acts as the direct stimulus to insulin secretion is still unknown.

Monosaccharides are not the only stimuli to insulin release. A wide variety of other compounds have been found to be effective. Many of these observations have been made on *in vitro* systems, and, in the absence of corroborative *in vivo* evidence, are of questionable physiological significance; but are of value for the light they shed on possible mechanisms for the control of insulin secretion.

Many amino acids have been reported to stimulate insulin secretion *in vivo* and *in vitro*. An oversensitivity of the beta cells to stimulation by leucine seems to be the cause of leucine-sensitive hypoglycaemia, a condition in which the intake of leucine in the diet results in a severe hypoglycaemia as a result of excessive insulin secretion.

Certain drugs are of clinical significance in the control of insulin secretion in disease states. Sulphonylureas, such as tolbutamide and chlorpropamide, promote insulin secretion *in vivo* and *in vitro*, and are used to treat certain mild forms of diabetes in which the pancreas still maintains its capacity to synthesise and secrete insulin. Diazoxide, which inhibits insulin secretion, is used in conditions in which there is excessive insulin release such as islet cell carcinoma and leucine-sensitive hypoglycaemia.

In addition to these stimuli, insulin secretion is modified by other hormones. The control of the secretion of one hormone by another is an important point of interrelation between hormones in the control of metabolism (Hales 1967). Effective hormonal stimuli to insulin secretion include ACTH, secretin, pancreozymin and glucagon. Glucagon, synthesised by the alpha cells of the islets of Langerhans, is secreted in response to a fall in plasma glucose concentration and is a potent stimulus to insulin secretion provided glucose is present. In the absence of glucose, glucagon is ineffective. Inhibitors of insulin secretion are less common, the most potent and important amongst the hormones being adrenaline. In liver and adipose tissue both adrenaline and glucagon act similarly through the adenylate cyclase/adenosine $3':5'$-cyclic monophosphate (cyclic AMP) system (see section 5.2) to promote glycogenolysis and lipolysis. In the beta cell their actions differ, glucagon increasing and adrenaline decreasing insulin secretion in response to glucose. An explanation for this apparent inconsistency was provided by Turtle and Kipnis (1967), who measured cyclic AMP levels in islets. When glucagon was present cyclic AMP levels rose, but when adrenaline was present there was no change or even a fall in the levels of cyclic AMP. Theophylline and caffeine, which also increase cyclic AMP levels in the islets, stimulate insulin secretion in response to glucose. It is probable that the increased cyclic AMP levels found in response to these stimuli promote glucose metabolism within the islets and so lead to enhanced insulin secretion. The possibility that insulin may regulate its own secretion by decreasing cyclic AMP levels in the beta cell merits consideration.

In common with secretion from many cells that contain secretory products in the form of granules, insulin secretion is calcium-dependent (Matthews 1970). Ca^{2+} itself is not stimulatory, but its presence is necessary for other stimuli to be effective. Insulin secretion by pancreas preparations stimulated *in vitro* by glucose is inhibited if Ca^{2+} ions are not present in the incubation medium. The role of Ca^{2+} in this system has many features in common with its role in the control of muscle contraction. In skeletal muscle, depolarisation of the muscle cell membrane results in the release of Ca^{2+} from the sarcoplasmic reticulum into the sarcoplasm where it triggers muscle contraction by relieving the inhibitory action of the troponin complex on myosin ATPase activity and actin-myosin interaction (see Chapter 21). Similarly the depolarisation of the beta cell membrane from its resting potential of about -20 mV induced by raising the extracellular concentration of K^+ ions enhances insulin secretion in a Ca^{2+}-dependent manner. Although the evidence for the involvement of Ca^+ in the emiocytotic process is strong, the site of its action has not been defined. Since insulin secretion in response to other stimuli can also be inhibited by the omission of Ca^{2+} from the incubation medium, it is probable that Ca^{2+} is involved in the later stages of emiocytosis, possibly in the control of the movement of the secretory granules to the cell periphery, or in the fusion of the granule and cell membranes at the moment of insulin release.

4.4 Summary

An attempt to summarise some of the findings relating to the control of insulin secretion by glucose is shown in Fig. 19.1. Such a scheme is tentative and will doubtless require modification in the light of future observations.

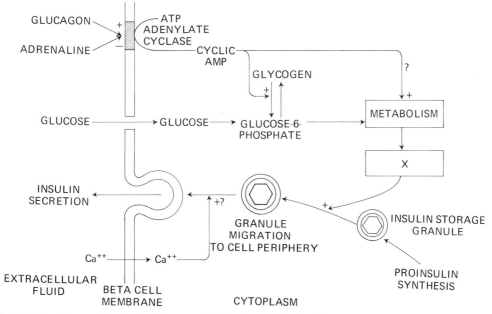

Fig. 19.1 Tentative mechanism for the control of insulin secretion by glucose.

In this scheme glucose enters the beta cell freely and is metabolised to an unknown metabolite or metabolic state (X) that constitutes a direct stimulus to insulin secretion. When the rate of glucose metabolism is increased, either by an increased supply of glucose to the beta cell or by increased cyclic AMP levels within the beta cell consequent upon stimulation of adenylate cyclase by glucagon, an increased rate of production of X results in an increased rate of insulin secretion. The preformed insulin storage granules migrate to the cell periphery and fuse with the cell membrane to liberate their contents. Ca^{2+} is essential for the later stages of emiocytosis although its site of action is not known.

5 Hormone action

In discussing hormone action, an *effect* of a hormone is defined as an observed consequence of the administration of that hormone to an animal *in vivo* or to a tissue *in vitro*. The *mechanism of action* of a hormone is a precise description of the molecular mechanisms in the chain of events initiated by the hormone and leading ultimately to the observed effects of the hormone on cell metabolism. The *function* of a hormone is a general conclusion as to the role of the hormone in the body. Being a subjective analysis on the part of the observer, a function can only be evaluated in the light of present knowledge.

For a hormone to modify cell metabolism requires that the cell:

1 is able to recognise the hormone as a molecular species and in some way to estimate the concentration of the hormone reaching the cell so as to produce a graded, rather than an all-or-none, response;
2 contains the appropriate means for transmitting the message carried by the hormone, in the same or a different language, to the intracellular enzyme systems that respond to the hormone.

The first criterion requires the presence of hormone-specific receptor sites at which discrimination between the different hormones occurs. The second concerns the concept of second messengers.

5.1 Receptors

The receptor concept was introduced into pharmacology to explain the specific effects of drugs and hormones. It envisages the presence in the target tissue of receptor sites that specifically bind, and hence identify, a hormone. They may also effectively concentrate the hormone within its target tissue. The combination of the hormone with its receptor triggers the chain of events that leads ultimately to changes in cell metabolism. The corollary to this concept is that there should exist within target tissues a definable macromolecular species with the properties expected of the receptor, namely (1) the ability to bind the hormone specifically, and (2) the ability to transmit the hormonal message into the cell. For the second of these functions, the receptor in all probability needs to exist in the correct spatial and functional relationships with the responsive metabolic systems of the cell. Any attempt at purification using classical biochemical techniques immediately disrupts these relationships. Thus potential receptor sites can only be identified at the present time by their specific hormone-binding ability. This consideration, together with the extreme scarcity of receptor sites within a tissue, has hindered progress in the identification and isolation of receptors, but important progress has been made in the isolation and characterisation of the probable insulin receptor from liver and fat cell membranes (Cuatrecasas 1972a).

5.1.1 The insulin receptor

The belief that insulin exerts its primary action, that is the interaction with its receptor, at the cell membrane of the target cell has developed from many lines of evidence (Levine 1966). Perhaps the most striking single piece of evidence is the recent demonstration that insulin covalently attached to large Sepharose beads through the alpha-amino group of the N-terminal phenylalanine of the B-chain or through the epsilon-amino group of lysine B29 was active in stimulating glucose uptake and suppressing ACTH-stimulated lipolysis in isolated fat cells (Cuatrecasas 1969). The Sepharose beads (60–300 μ in diameter) were larger than the isolated fat cells (50–100 μ) and could not penetrate the cells. These experiments indicated that insulin, although physically restricted to the extracellular space, was still capable of exerting its full hormonal activity. To do so the hormone had to be interacting with a cell membrane component that had at least part of its structure superficially located on the outside of the cell membrane.

Specific binding of insulin to isolated fat cells and to membranes prepared from isolated fat cells was identified and characterised using lightly iodinated ^{125}I-insulin (Cuatrecasas 1971a). ^{125}I-iodine can be introduced into the insulin molecule at one or more of the four tyrosines present, but ^{125}I-insulin containing less than one atom of ^{125}I-iodine per insulin was used since the introduction of more leads to progressive loss of biological activity.

^{125}-I-insulin specifically bound to cell membranes was displaced by native insulin, but not by oxidised or reduced insulin, glucagon, ACTH, vasopressin, oxytocin, prolactin or growth hormone (Cuatrecasas 1971a). Proinsulin binds specifically but with a 20-fold lower affinity than insulin. Somatomedin, a low molecular weight peptide that mediates the action of growth hormone on cartilage, is the only other hormone-related compound so far reported to compete with insulin for binding to the insulin receptor (Hintz *et al.* 1972). This observation is of particular interest since somatomedin possesses many insulin-like actions. ^{125}I-insulin binding was saturable with respect to the ^{125}I-insulin concentration, a maximum of 11 000 insulin molecules binding per fat cell. A previous estimate based on a study of the concentration dependence of the biological actions of insulin was that the antilipolytic action of insulin required the binding of between 50 and 1000 insulin molecules per cell, and the stimulation of glucose transport 100–3000 insulin molecules per cell (Crofford, Minemura and Kono 1970). The association of insulin (I) with its receptor (R) is a simple bimolecular reaction:

$$I + R \rightleftharpoons IR$$

with a dissociation constant of 5×10^{-11} M, which correlated well with the concentration of insulin ($6\cdot1 \times 10^{-11}$ M) giving half-maximal stimulation of glucose oxidation by isolated fat cells (Cuatrecasas 1971a). This correlation was important since it indicated that the ^{125}I-insulin was binding to the true insulin receptor in the cell membrane.

The binding of insulin to isolated fat cell or liver membranes did not result in a chemical change in the insulin molecule suggesting that binding of the hormone and its inactivation are two independent processes. The inactivation of insulin may be the function of a cytoplasmic protease identified in liver and muscle, which hydrolyses insulin with a K_m for the hormone of $0\cdot1$ μM and shows specificity for insulin over proinsulin (Burghen, Kitabchi and Brush 1972). Insulin-receptor binding did not require heavy metal ions and was non-covalent, thus invalidating earlier hypotheses that insulin might bind to its receptor by a disulphide-sulphydryl interchange reaction. Specific insulin binding was present only in the cell membrane and no binding was detectable in the nuclear, mitochondrial or cytoplasmic fractions. Treatment of liver and isolated fat cell membranes with Triton X-100, a non-ionic detergent, abolished specific ^{125}I-insulin binding, and all the insulin-binding activity was solubilised (Cuatrecasas 1972a). The insulin-binding characteristics of the solubilised protein were very similar to those of the protein *in situ* in the fat cell membrane. The binding protein was found to be highly asymmetrical with a molecular weight of about 300 000 daltons (Cuatrecasas 1972b). Its insulin-binding capacity was destroyed by trypsin but not by phospholipase C or neuraminidase, indicating that phospholipids and sialic acid were not involved in insulin binding. The receptor has been purified 250 000-fold from a rat liver homogenate by extraction followed by affinity chromatography on insulin–agarose columns (Cuatrecasas 1972c).

Digestion of isolated fat cell membranes with phospholipase C, which in other membranes hydrolyses up to 80 per cent of the phospholipid, caused a 3–6-fold increase in their capacity to bind ^{125}I-insulin (Cuatrecasas 1971b). The binding properties of the receptors exposed by phospholipase treatment were similar to those of the receptors normally exposed in the membrane. The role of this second, major, population of insulin receptors present in sites inaccessible to the extracellular environment is unclear.

5.1.2 Other polypeptide hormone receptors

Evidence has been presented for the specific binding of ^{125}I-glucagon to rat liver membranes (Rodbell *et al.* 1971) and of ^{125}I-ACTH (adrenocorticotrophic hormone) to fragmented adrenal cortex membranes (Lefkowitz *et al.* 1970). Although both glucagon and ACTH activate adenylate cyclase in rat liver, the independence of the initial binding of each hormone to its receptor was shown by the failure of ACTH to displace ^{125}I-glucagon from liver membranes.

5.1.3 Steroid hormone receptors

While it is probable that the receptors for most if not all the polypeptide and catecholamine hormones are located in the cell membrane, steroid hormones combine with specific steroid-binding proteins found in the cytoplasm. These proteins are both hormone- and tissue-specific and have been identified in the target tissues for several steroid hormones. Combination of the steroid with the specific cytoplasmic receptor protein precedes transfer of the steroid hormone–receptor complex into the nucleus where binding to the chromosomes occurs in such a way as to promote the expression of specific genes within the cell.

5.1.4 Summary

A protein that specifically binds insulin has been isolated from liver and fat cell membranes. This property and its binding characteristics suggest that it is in all probability the biologically significant insulin receptor. Evidence for the specific binding of other hormones to receptor sites in the cell has been obtained. The mere existence of such an insulin-binding protein does not however prove that it is the insulin receptor. Conclusive proof requires the demonstration that combination of the hormone with the binding protein is an essential step in the mechanism of insulin action.

5.2 The second messenger hypothesis

In no case has it been possible to demonstrate effects of physiological concentrations of a hormone on the activity of a purified enzyme *in vitro*, even though the hormone is known to control the activity of the enzyme *in vivo*. Although the polypeptide and catecholamine hormones probably do not penetrate the cell membrane of their target cells to influence cell metabolism directly, the information carried by the hormone penetrates, usually after translation into a form that can be 'read' by the enzymic machinery of the cell. This translation of information from one form, the hormone or first messenger, into another, the second messenger inside the cell, forms the basis of a general theory of polypeptide hormone action proposed by Sutherland and his co-workers (reviewed in Robinson, Butcher and Sutherland 1971), as a result of their investigations of the role of cyclic AMP in the action of the glycogenolytic hormones.

The only second messenger unambiguously identified so far is cyclic AMP. Many enzyme systems in the cell respond to this cyclic nucleotide where they are insensitive to the direct application of the hormone itself to the enzyme *in vitro*. In response to changes in the intracellular concentration of cyclic AMP the activity of these enzymes changes in ways characteristic of the administration of the hormone itself to the intact tissue.

Since the original discovery of cyclic AMP in its role as the second messenger for adrenaline and glucagon in the control of glycogen metabolism, much evidence has accumulated that cyclic AMP is involved in the actions of a wide variety of hormones. A comprehensive list of all hormones believed to act via changes in the levels of cyclic AMP within their target cells can be found in the review by Robinson, Butcher and Sutherland (1971). It has also been shown that insulin lowers cyclic AMP levels in liver and adipose tissue.

If cyclic AMP is to be implicated as the second messenger for a particular hormone, four experimental criteria have been proposed:

a. The hormone should increase cyclic AMP levels in its target tissues and not in unresponsive tissues, with the increase in cyclic AMP levels preceding or at least paralleling in time the earliest observed metabolic response. A response occurring before any change in cyclic AMP levels clearly cannot be mediated by the nucleotide. The change in cyclic AMP levels should be proportional to the concentration of the hormone over the range of hormone concentrations producing a metabolic response.

b. The hormone should stimulate adenylate cyclase activity in homogenates or broken cell preparations.

c. The action of the hormone should be mimicked by inhibitors of cyclic $3'5'$-nucleotide phosphodiesterase, the enzyme that inactivates cyclic AMP by hydrolysis to $5'$-AMP. The methylxanthines (e.g. caffeine and theophylline) are inhibitors of the phosphodiesterase, raising cyclic AMP levels by altering the balance between the basal unstimulated activity of adenylate cyclase and the destruction of cyclic AMP by the phosphodiesterase. In the presence of a methylxanthine the rise in cyclic AMP levels in response to hormonal stimulation of adenylate cyclase is accentuated. The methylxanthine should therefore potentiate the effects of sub-maximal concentrations of the hormone.

d. Theoretically, addition of cyclic AMP to the tissue *in vitro* should mimic the effects of the hormone, but practical difficulties limit the value of this approach. Cyclic AMP is inactive in many cases, either because it cannot penetrate the cell membrane or because once inside the cell it is rapidly inactivated by the phosphodiesterase. Acylated derivatives of cyclic AMP, in particular $N^6,2'$-O-dibutyryl cyclic AMP (dibutyryl cyclic AMP) have been synthesised and shown to mimic the actions of many hormones with cyclic AMP as their second messenger. The reason for the activity of the dibutyryl derivative where cyclic AMP itself is inactive is unclear, but may relate to resistance to inactivation by phosphodiesterase or to a greater membrane permeability on the part of the acylated nucleotide.

5.2.1 Adenylate cyclase

Adenylate cyclase, the enzyme catalysing the formation of cyclic AMP from ATP, has been localised to the plasma membrane in liver, adipose tissue, avian erythrocytes and other tissues. The mechanism of the enzyme reaction is uncertain. Mg^{2+} is an obligatory cofactor, and in some cases Ca^{2+} is inhibitory, although in the adrenal cortex Ca^{2+} is essential for ACTH stimulation of adenylate cyclase activity. Fluoride also stimulates the enzyme in homogenates but not in intact tissues.

Early attempts to prepare a solubilised adenylate cyclase were accompanied by loss of hormonal sensitivity, but recently preparations have been described in which a partially solubilised adenylate cyclase has retained a degree of hormone responsiveness. Disruption of an ACTH-responsive adrenal tumour in a French pressure cell in the presence of phosphatidyl-ethanolamine yielded a soluble fraction with adenylate cyclase activity that was sensitive to ACTH (Lefkowitz et al. 1970). This fraction also bound ^{125}I-ACTH specifically.

Adenylate cyclase is distributed asymmetrically in the cell membrane (Øye and Sutherland 1966). When ATP was added to the exterior of intact turkey erythrocytes, it was hydrolysed to ADP by an ATPase in the cell membrane, but conversion to cyclic AMP did not occur until the cells were haemolysed. Treatment of the intact erythrocytes with proteolytic enzymes reduced the activity of the ATPase without affecting adenylate cyclase activity. When haemolysed cells were treated with trypsin both the ATPase and adenylate cyclase were inactivated. These results indicated that adenylate cyclase occupies a position in the membrane facing the interior of the cell with no functionally important part of the enzyme accessible to proteolytic attack from the outside. This contrasts with the situation of the hormone receptors which are orientated towards the outside of the membrane. The asymmetric distribution of these two entities in the membrane provides a sequential coupling system for transmitting the hormonal message across the cell membrane from the free hormone in the extracellular fluid to the second messenger in the cytoplasm. The mechanism of the coupling between the hormone receptors and adenylate cyclase is unknown, although there is evidence that phospholipids and a high degree of membrane integrity are necessary for efficient information transfer to occur. Digestion of rat liver membranes with phospholipase A reduced the binding of ^{125}I-glucagon and inhibited the stimulation of adenylate cyclase activity by glucagon without affecting stimulation by fluoride, showing that the phospholipolytic digestion had not inactivated adenylated cyclase (Pohl et al. 1971). Both glucagon binding and adenylate cyclase activation were in part restored by exposing the treated membranes to aqueous dispersions of phosphatidylserine, phosphatidyl-choline and phosphatidylethanolamine. The noradrenaline-responsiveness of cat heart adenylate cyclase solubilised with a nonionic detergent [Lubrol PX] was restored by the addition of phosphatidylinositol, but not by phosphatidylserine or phosphatidylethanolamine (Levey 1971a), whereas addition of phosphatidylserine restored sensitivity to glucagon and histamine (Levey 1971b).

5.2.2 The cyclic AMP receptor and protein kinase activity

The mechanism by which cyclic AMP itself modifies cell metabolism has remained obscure until recently when the basis of a possible general mechanism for cyclic AMP action has been elucidated. Two cyclic-AMP dependent activities have been identified in and isolated from several tissues:

a   Equilibrium dialysis studies indicated the presence of a protein that bound cyclic AMP specifically in homogenates of adrenal cortex (Garren et al. 1971). This protein was found mainly in the cytoplasm and endoplasmic reticulum. It has been purified and binds cyclic AMP noncovalently with a dissociation constant of 3×10^{-8} M. This protein has strong claims to being the cyclic AMP receptor.

b   Cyclic AMP has been shown to stimulate the activity of protein kinases in several tissues. These kinases catalyse the ATP-dependent phosphorylation of a wide variety of proteins *in vivo* and *in vitro* including histone, casein, protamine, phosphorylase 'b' kinase and glycogen synthetase. The cyclic-AMP dependent phosphorylase 'b' kinase kinase from rabbit muscle also phosphorylated, and simultaneously activated, partially purified hormone-sensitive lipase from rat epididymal adipose tissue *in vitro* (Huttunen, Steinberg and Mayer 1970).

The cyclic-AMP binding and cyclic-AMP dependent protein kinase activities from adrenal cortex migrated together as a single band during polyacrylamide gel electrophoresis at several gel concentrations in the absence of cyclic AMP (Gill and Garren 1971). When cyclic AMP was present the two proteins ran independently and the resulting protein kinase freed from the binding protein was fully active and not stimulated by cyclic AMP. The kinase did not bind cyclic AMP. Separation of the two activities from rabbit skeletal muscle has been accomplished using affinity chromotography on casein-Sepharose in the presence of cyclic AMP (Reimann *et al.* 1971). When the protein kinase and binding protein were recombined in the absence of cyclic AMP, kinase activity was depressed and could once again be stimulated by cyclic AMP. Comparison of the molecular weights of the cyclic-AMP binding protein (92 000) and the protein kinase (60 500) with that of the complex (144 000) as determined by polyacrylamide gel electrophoresis suggested the binding occurred with a 1:1 stoicheiometry (Gill and Garren 1971). Similar observations have been made with cyclic-AMP stimulated protein kinases from beef heart, rabbit reticulocytes and rat liver. It appears from these observations that in the absence of cyclic AMP, the cyclic-AMP binding protein (B) binds to and inhibits the activity of the protein kinase (K). Elevation of the intracellular levels of cyclic AMP following hormonal stimulation of adenylate cyclase leads to binding of cyclic AMP to its receptor protein followed by dissociation of the binding protein–kinase complex. The kinase in the free state phosphorylates protein substrates:

$$\text{cAMP} + \text{B–K} \rightleftharpoons \text{cAMP–B} + \text{K}$$

          (inactive   (active
          kinase)   kinase)

Thus cyclic AMP relieves the protein kinase from the action of its inhibitor, the cyclic-AMP binding protein. A fall in the levels of cyclic AMP displaces the equilibrium of the above dissociation in favour of the inactive complex, with suppression of protein kinase activity.

Stimulation of protein kinase activity may be the means by which cyclic AMP controls many intracellular enzyme activities, but the nucleotide has other actions which, as yet, are not easy to reconcile with the involvement of a protein kinase. Foremost amongst these are the changes in levels of certain enzymes such as tyrosine transaminase and phosphoenolpyruvate carboxykinase in liver in response to glucagon or dibutyryl cyclic-AMP administration. It is possible that stimulation of a protein kinase in the nucleus is involved in the control of gene expression since Langan (1969) has observed phosphorylation of rat liver histone *in vivo* in response to dibutyryl cyclic AMP injected intraperitoneally. Similarly cyclic AMP is the second messenger for the control of water permeability in the mammalian kidney by vasopressin. In this case a cyclic-AMP stimulated protein kinase, which phosphorylates kidney cell membrane proteins, and a protein phosphatase, which catalyses dephosphorylation of the membrane proteins, have been identified (Dousa, Sands and Hechter 1972). Clearly these are areas of great interest in terms of the possible involvement of protein kinases in all cyclic-AMP mediated control systems, but such a universal role for protein kinases seems unlikely since, for example, the cyclic-AMP receptor protein involved in the regulation of gene expression in *Escherichia coli* seems not to be related to a protein kinase (Pastan *et al.* 1971).

The cyclic-AMP stimulated protein kinases may be closely integrated with particular enzyme systems or they may be non-specific in that one protein kinase may phosphorylate and control the activity of more than one enzyme system within the cell. The lack of substrate specificity *in vitro* might suggest the latter, but this takes no account of possible binding of the protein kinases to specific enzyme systems within the cell.

5.2.3 Cyclic 3'5'-nucleotide phosphodiesterase

For effective control of enzyme activities by cyclic AMP there needs to be in addition to a means of producing cyclic AMP in response to a hormone a method for inactivation of the nucleotide to maintain low intracellular levels in the absence of hormonal stimulation. This inactivation is brought about by hydrolysis to 5'-AMP catalysed by cyclic 3'5'-nucleotide phosphodiesterase. There are two phosphodiesterase activities in rat adipose tissue with K_m values for cyclic AMP of 0·88 μM and 41 μM respectively (Loten and Sneyd 1970). Insulin increased the V_{max} of the low K_m enzyme and decreased the K_m of the high K_m enzyme, raising the possibility that insulin may lower intracellular cyclic-AMP levels by activation of phosphodiesterase.

5.2.4 Summary

The status of cyclic AMP as the second messenger for many hormones is well established. The possible existence of other second messengers has been suggested, e.g. Ca^{2+} (Rasmussen 1970), but not proved. The intracellular concentration of cyclic AMP reflects a balance between production by adenylate cyclase and inactivation by cyclic 3'5'-nucleotide phosphodiesterase. An increase in adenylate cyclase activity following hormone administration distorts this balance in favour of an increased concentration of cyclic AMP. The cyclic nucleotide binds to a specific receptor protein which is also an inhibitor of a protein kinase. Its binding promotes dissociation of the receptor protein–protein kinase complex with activation of the kinase, which catalyses the phosphorylation of certain key enzymes and possibly other physiologically important proteins within the cell, thus controlling their activity.

6 Effects of insulin on metabolism

6.1 Carbohydrate metabolism

The one unequivocal action of insulin is the ability of the hormone to increase the facilitated transport of glucose across the plasma membranes of muscle and fat cells. There is strong evidence that insulin acts at the membrane transport step rather than at the subsequent phosphorylation step (Randle and Morgan 1962; Park, Crofford and Kono 1968). A comparison between the distribution of glucose and of sorbitol in rat hearts perfused with media containing glucose and sorbitol showed the spaces occupied by each compound in the tissue to be the same in the absence of insulin (Morgan *et al.* 1961). Sorbitol, being impermeant, was restricted to the extracellular space, so the sorbitol space, calculated as

$$\frac{\text{tissue sorbitol content}}{\text{medium sorbitol concentration}}$$

was a measure of the extracellular volume. Since the glucose space was the same as the sorbitol space, it was likely that free glucose was also restricted to the extracellular volume in the absence of insulin. In the presence of insulin, the glucose space was greater than the sorbitol space, which was interpreted to mean that free glucose was present within the cells. This increase in the glucose space was most marked in hearts from alloxan diabetic rats which have a depressed rate of glucose phosphorylation. Insulin had increased glucose transport until the rate of entry of glucose into the cells exceeded the rate of its phosphorylation and free glucose had accumulated in the cells. This technique and the use of transported but non-metabolised analogues of glucose, such as 3-*O*-methylglucose, have shown that insulin increases glucose transport in adipose tissue and muscle, but not in liver and the brain.

Although the increased flux of glucose into a tissue in response to insulin leads to a general increase in glucose metabolism, insulin preferentially diverts an increased proportion of the glucose into the pathways for glycogen and triglyceride synthesis. Glycogen synthetase (UDP-glucose glycogen transglucosylase) exists in two forms, the 'D' form dependent on glucose-6-phosphate for activity, and the 'I' form independent of the presence of glucose-6-phosphate. The interconversion of the two forms is brought about by the synthetase 'I' kinase, which phosphorylates and inactivates the 'I' form (converting it to the 'D' form), and synthetase 'D' phosphatase, which catalyses the dephosphorylation and activation of the 'D' form (converting it to the 'I' form). Insulin administration to muscle and liver leads to an increase in the amount of the synthetase in the more active 'I' form. Glycogen phosphorylase also exists in two interconvertible forms. The glycogenolytic hormones, adrenaline and glucagon, stimulate the conversion of the less active 'b' form into the more active 'a' form. This effect is mediated by cyclic AMP which activates phosphorylase 'b' kinase kinase. The cascade effect initiated by this activation leads to the sequential activation of phosphorylase 'b' kinase and phosphorylase resulting in an increased rate of glycogenolysis. Adrenaline and glucagon also stimulate the conversion of the synthetase 'I' form into the less active 'D' form. The identification of glycogen synthetase 'I' kinase with phosphorylase 'b' kinase kinase has provided a simple rationale for the parallel activation of glycogenolysis and inhibition of glycogen synthesis observed following administration of adrenaline or glucagon. A rise in tissue levels of cyclic AMP activates phosphorylase 'b' kinase kinase leading to stimulation of the synthetase 'I' to 'D' conversion and of the phosphorylase 'b' to 'a' conversion. This avoids a futile cycling mechanism between glycogen and glucose-1-phosphate and provides for either net synthesis or net glycogenolysis to occur.

In view of the role of cyclic AMP in the control of glycogen metabolism by the glycogenolytic hormones, the possibility that the insulin antagonism of the action of these hormones is mediated by a decrease in the intracellular levels of cyclic AMP within a stimulated tissue merits serious consideration. However, this interpretation seems to be an oversimplification, for, in rat diaphragm incubated *in vitro*, insulin had no effect on the levels of cyclic AMP in the presence or absence of adrenaline, but increased the amount of glycogen synthetase in the 'I' form without at the same time having a significant effect on glycogen phosphorylase (Craig, Rall and Larner 1969). This suggests that the insulin effect on glycogen synthetase activity in muscle at least does not depend on changes in cyclic AMP concentrations. Whether the same is true for liver is less certain since in this tissue insulin decreases the adrenaline-stimulated cyclic AMP levels.

Gluconeogenesis is stimulated by the catecholamines, glucagon and the adrenal corticosteroids. It is increased in diabetes and during starvation, two conditions associated with an increased level of free fatty acids in the blood. In diabetes it appears that the major sources of substrate for gluconeogenesis are amino acids released from the peripheral tissues by proteolysis. Glycerol, released from adipose tissue during lipolysis, and lactate are relatively minor sources. *In vivo* the stimulation of gluconeogenesis by glucagon and adrenaline may be in part mediated by an increased supply of gluconeogenic substrates to the liver, but a more direct effect of the hormones is argued for by the observation that in the perfused rat liver they have been shown to have a stimulatory effect on the conversion of these substrates to glucose in the absence of any change in substrate concentration (Exton *et al.* 1970). This direct effect of glucagon was rapid in onset and was mediated by cyclic AMP, since this nucleotide, when added to the medium perfusing the liver, produced changes in the rate of gluconeogenesis and in the levels of the gluconeogenic intermediates in the liver cells that closely paralleled those brought about by the administration of glucagon or adrenaline. Crossover point analysis of the levels of intermediates in the perfused liver before and after hormonal stimulation has identified a control site for these hormones between pyruvate and phosphoenolpyruvate in the gluconeogenic pathway, suggesting activation of either pyruvate carboxylase or phosphoenolpyruvate carboxykinase. In agreement with this analysis glucagon stimulated gluconeogenesis from lactate and alanine, substrates that are converted to glucose via pyruvate, but not from glycerol or fructose which enter gluconeogenesis at stages beyond phosphoenolpyruvate.

Since glucagon stimulates lipolysis in the liver both *in vivo* and *in vitro*, its gluconeogenic potential may be augmented by an increased supply of endogenous free fatty acids. Similarly the increased levels of circulating free fatty acids in diabetes may stimulate gluconeogenesis, since any condition leading to increased oxidation of free fatty acids by the liver will promote gluconeogenesis. Oleate increased the synthesis of glucose from alanine, lactate and pyruvate when added to the medium perfusing rat liver *in vitro* (Williamson, Browning and Scholz 1969). In these experiments, crossover point analysis identified a major control site at pyruvate carboxylase, which when stimulated generated increased levels of oxaloacetate and malate. The stimulation of pyruvate carboxylase activity may be caused by the increased levels of acetyl CoA, the allosteric activator of the enzyme, resulting from increased fatty acid oxidation. Competition for pyruvate by pyruvate dehydrogenase (leading into the citric acid cycle) is depressed under conditions of increased fatty acid oxidation because pyruvate dehydrogenase is inhibited by the raised acetyl CoA/CoASH and NADH/NAD$^+$ ratios that result, and because under these conditions the enzyme exists mainly in the phosphorylated inactive form (Wieland, Patzelt and Löffler 1972; see also section 6.2). The result is that pyruvate is channelled through pyruvate carboxylase into gluconeogenesis rather than into the citric acid cycle. The equilibrium of the malate dehydrogenase reaction, particularly when the NADH/NAD$^+$ ratio in the mitochondrion is high, greatly favours malate formation from oxaloacetate. Malate transport from the mitochondrion into the cytoplasm provides a supply of NADH by acting as substrate for cytoplasmic malate dehydrogenase. (A supply of NADH is necessary for reversal of the glyceraldehyde-3-phosphate dehydrogenase step.) Moreover, the generation of oxaloacetate in the cytoplasm by the oxidation of malate originating from the mitochondrion provides a source of substrate for gluconeogenesis. It has not proved possible to identify precisely the control site(s) stimulated by cyclic AMP in gluconeogenesis since the nucleotide is without effect on the activities of the relevant enzymes *in vitro*. The possibility also exists that cyclic AMP may stimulate the transfer of gluconeogenic substrate across the mitochondrial membrane.

These aspects of the integration of carbohydrate and fat metabolism in diabetes and starvation are discussed in more detail by Greville and Tubbs (1968), and by Gumaa, McLean and Greenbaum (1971). Further consequences of increased fatty acid oxidation by the liver during diabetes are described in section 6.2.1.

The rapid control of gluconeogenesis by hormones is augmented in diabetes and prolonged starvation by changes in the levels of certain of the glycolytic and gluconeogenic enzymes. Various authors have reported that the levels of pyruvate carboxylase, phosphoenolpyruvate carboxykinase, fructose-1:6-diphosphatase and glucose-6-phosphatase increase in liver under gluconeogenic conditions (Scrutton and Utter 1968). The reversal of the gluconeogenic state of diabetes by insulin has been reported to be associated with a fall in the levels of these enzymes and an increase in the levels of glucokinase, phosphofructokinase and pyruvate kinase, enzymes that catalyse irreversible steps in glycolysis. A futile cycle between glucose and glucose-6-phosphate is avoided by this reciprocal relationship between glucokinase and glucose-6-phosphatase, and possibly also by compartmentation of glucose-6-phosphate within the liver cell.

6.2 Fat metabolism

When incubated with adipose tissue *in vitro*, insulin preferentially stimulates the conversion of glucose into fatty acids, doubling the rate of metabolism of glucose to pyruvate, yet at the same time stimulating the conversion of pyruvate to fatty acids five-fold. The hormone stimulates the conversion of glucose to fatty acids at a point between pyruvate and the fatty acids. A possible explanation for this lies in the observation that insulin activated pyruvate dehydrogenase in rat adipose tissue (Coore *et al*. 1971), the activation being sufficient to account for the observed degree of stimulation of fatty acid synthesis by insulin. The activation of pyruvate dehydrogenase increases the supply of acetyl CoA within the mitochondrion and hence (through the activity of citrate synthase in the mitochondrion and ATP-citrate lyase in the cytoplasm) in the cytoplasm. The activity of ATP-citrate lyase is depressed in diabetes and increased following insulin administration to a diabetic animal.

Pyruvate dehydrogenase exists in two interconvertible forms, an inactive phosphorylated form and an active dephosphorylated form. The phosphorylation/dephosphorylation cycle catalysed by pyruvate dehydrogenase kinase and pyruvate dehydrogenase phosphate phosphatase respectively is similar to those by which the activity of glycogen synthetase and phosphorylase are regulated, with the important difference that it appears to be insensitive to cyclic AMP (Coore *et al*. 1971). Insulin administration to intact adipose tissue, but not to isolated mitochondria, led to enhanced levels of the active dephosphorylated form by a mechanism as yet unknown. Insulin could act either by inhibition of pyruvate dehydrogenase kinase or by activation of pyruvate dehydrogenase phosphate phosphatase. Pyruvate dehydrogenase kinase was inhibited (and pyruvate dehydrogenase activated) by ADP and pyruvate (Martin *et al*. 1972). Pyruvate dehydrogenase phosphate phosphatase was activated by low concentrations of calcium (Denton, Randle and Martin 1972), raising the intriguing possibility that the action of insulin on this enzyme may be related to changes in intracellular calcium levels.

In addition to promoting fatty acid synthesis in adipose tissue by activation of pyruvate dehydrogenase, insulin by its enhancement of glucose uptake and metabolism in this tissue provides a supply of glycerol-1-phosphate for esterification of the newly synthesised fatty acyl CoA.

On feeding a high-carbohydrate, low-fat diet to starved rats there was a co-ordinated increase in the levels of several of the enzymes of fatty acid synthesis in rat liver (Gibson *et al*. 1972). The levels of glucose-6-phosphate dehydrogenase, malic enzyme, ATP-citrate lyase and, to a lesser extent, acetyl CoA carboxylase and the fatty acid synthetase itself, increased on refeeding the high-carbohydrate diet. These changes were paralleled by an increase in the concentration of insulin in the blood but it was not possible from these experiments to decide whether insulin was directly implicated in the increase in activity of these enzymes.

In addition to its effects on triglyceride synthesis, insulin decreases the stimulation of lipolysis in adipose tissue by a variety of hormones including adrenaline, ACTH and glucagon. This is accompanied by a decrease in hormone-stimulated cyclic AMP levels in the tissue, an indication that this fall may mediate the antilipolytic effect of insulin. This decrease may be produced by inhibition of adenylate cyclase and/or by activation of phosphodiesterase. Jungas

(1966) has reported that adenylate cyclase activity in homogenates of rat epididymal adipose tissue was reduced by exposure of the tissue to insulin before homogenisation, and a recent report (Loten and Sneyd 1970) suggested activation of adipose tissue cyclic $3'5'$-nucleotide phosphodiesterase by insulin.

6.2.1 Ketone body formation

Ketosis, an increase in the concentrations of the three ketone bodies (beta-hydroxybutyrate, acetoacetate and acetone) in blood, is common in severe untreated diabetes. In mild conditions of starvation and diabetes, ketone bodies are produced by the liver and can be used as metabolic substrates by other tissues. The heart, for instance, metabolises ketone bodies in preference to glucose, and during prolonged fasting ketone bodies become a major metabolic substrate for the brain (Krebs 1966). In acute diabetes a severe ketosis can occur, which, if untreated, can lead to acidosis, diabetic coma and ultimately death. In diabetes increased rates of gluconeogenesis and ketogenesis occur together and for much the same metabolic reasons. The raised plasma free fatty acid levels in diabetes result in an increased uptake and metabolism of free fatty acids by the liver. This in turn leads to elevated levels of fatty acyl CoA which inhibit acetyl CoA carboxylase and strongly suppress lipid synthesis. The depression of citric acid cycle activity, probably by the inhibition of citrate synthase by elevated mitochondrial ATP levels and by the inhibition of isocitrate dehydrogenase and possibly the other dehydrogenases of the cycle by elevated NADH levels, decreases the rate of oxidation of acetyl CoA by the cycle. The loss in ATP yield due to depressed cycle activity is compensated for by the increased oxidation of fatty acids as far as acetyl CoA. The resultant overproduction of acetyl CoA could also deplete the cell of coenzyme A, thus inhibiting fatty acid oxidation, were it not for the presence of the ketogenic pathway which provides for disposal of the acetyl units and regeneration of coenzyme A. In mild conditions of starvation and diabetes ketone body production by the liver rises, but is balanced by metabolism in muscle. In severe diabetes, this balance is disturbed and excessive ketone body production leads to ketosis.

6.3 Protein metabolism

6.3.1 Amino acid transport

Insulin stimulates the transport of most amino acids into muscle (Scharff and Wool 1967), by a mechanism that is independent of any action of the hormone on protein synthesis since it occurred in the presence of puromycin. Stimulation of amino acid transport also occurred in the absence of glucose, indicating independence of the effects of insulin on these two membrane transport processes.

6.3.2 RNA synthesis

Although insulin stimulated the incorporation of radioactive uridine into RNA in muscle, this effect was not essential for its stimulation of amino acid incorporation into protein, since the stimulation of amino acid incorporation occurred in the presence of actinomycin D, an inhibitor of DNA transcription.

6.3.3 Amino acid incorporation into protein

Insulin has been shown to stimulate the incorporation of radioactive amino acids into protein in several tissues. The subcellular location of this effect has proved difficult to determine but evidence has been presented that diabetes is accompanied by a defect in ribosomal function

(Wool *et al.* 1968). Ribosomes from the skeletal muscle of diabetic animals catalysed the incorporation of amino acids less effectively than ribosomes from normal animals. Treatment of diabetic animals with insulin immediately prior to isolation of the ribosomes restored their synthetic capacity to normal (Wool and Cavicchi 1967). This restoration of activity was rapid in onset, an increase in efficiency of ribosomal function being apparent within 5 min of the administration of the hormone *in vivo*. Insulin was without effect when added to the ribosomes *in vitro*, confirming the suggestion that the original message carried by insulin has to be translated into a form that can be 'read' by the ribosomes before it is effective. Treatment of normal animals with insulin failed to enhance ribosomal function. In the normal state the circulating levels of insulin are sufficient to maintain the protein synthetic machinery in a fully active state. The role of insulin in this respect would be one of maintenance rather than as a mediator of rapid responses in protein synthesis.

Insulin also has a general antiproteolytic action. Untreated diabetics often show a negative nitrogen balance in which protein catabolism exceeds protein anabolism, and insulin administration can reverse this condition.

7 Summary and the pleiotypic response

The main conclusion to be drawn is that almost all the effects of insulin lie in the direction of anabolism. The hormone promotes a wide range of anabolic processes, including the synthesis of glycogen, fat and protein, while at the same time decreasing the corresponding catabolic processes, glycogenolysis, lipolysis and proteolysis. Insulin may be regarded as the main anabolic hormone of the body, providing the general anabolic background against which the catabolic hormones, e.g. adrenaline and glucagon, can act. A constant concentration of insulin in the blood perfusing the target tissues maintains these in a general anabolic state that can be modified by catabolic hormones mediating the minute-to-minute needs of the animal. In the absence of a biologically active insulin in the blood, the anabolic background is lost and a general catabolic state ensues. This is diabetes mellitus, in severe untreated forms of which the body may literally waste away.

It is most improbable that insulin directly affects each metabolic process known to respond to insulin administration. More probable is the suggestion that insulin has a single primary action, which can be equated with the combination of the hormone with its receptor in the cell membrane. This interaction in ways as yet undefined translates the message carried by the hormone into a form that can be 'understood' by the enzyme complement of the cell. In response to this message specific changes occur in cell metabolism resulting in an overall change characteristic of the action of insulin on the particular tissue concerned. The hormone switches on a co-ordinated series of reactions within the cell leading finally to the observed effects of insulin on cell metabolism.

A striking analogy has been drawn between, on the one hand, the control by serum or insulin of growth and metabolism in untransformed mouse fibroblasts in tissue culture, and on the other, stringent control in bacteria (Hershko *et al.* 1971). In stringent control, the absence of an essential amino acid initiates a series of apparently unrelated biochemical processes that include inhibition of protein and RNA synthesis, and polysome formation, inhibition of the uptake of glucose and nucleic acid precursors and stimulation of proteolysis. Mouse fibroblasts require serum or insulin to grow, and when these growth promoting factors are withdrawn a negative pleiotypic response ensues, characterised by changes similar to those that occur in stringent control. (A pleiotypic response is a series of superficially unrelated metabolic reactions that respond in a co-ordinated fashion to an environmental change.) Conversely when growth is stimulated by serum or insulin, a positive pleiotypic response is initiated resulting in changes in the opposite direction. Thus glucose and nucleic acid precursor uptake is stimulated, protein and RNA synthesis and polysome formation are enhanced and proteolysis is inhibited. The authors further suggest, by analogy with the possible mediation of the stringent response in bacteria by guanosine tetraphosphate, that the resting, unstimulated cell membrane produces a pleiotypic mediator leading to the expression of the negative pleiotypic response. Insulin is viewed as inhibiting the formation of the pleiotypic mediator, thus allowing expression of the anabolic reactions of a positive pleiotypic response.

Although this hypothesis was based on experiments with fibroblasts it is readily extended to other tissues, and it suggests that instead of looking for a second messenger that increases in concentration following insulin administration, a search for a messenger that decreases in concentration in response to insulin might be rewarding. Cyclic AMP is an obvious candidate, but as has been discussed previously, not all the effects of insulin can be correlated with changes in intracellular cyclic AMP levels.

8 References

Key references and reviews

CUATRECASAS, P. (1969). 'Interaction of insulin with the cell membrane: the primary action of insulin.' *Proc. Natn. Acad. Sci. U.S.A.*, **63**, 450-57. Demonstration that the primary action of insulin is exerted at the cell membrane which led to the characterisation and isolation of the probable insulin receptor described in section 5.1.

GARREN, L. D., GILL, G. N., MASUI, H. and WALTON, G. M. (1971). 'On the mechanism of action of ACTH.' *Recent Prog. Horm. Res.*, **27**, 433-74. Discussion of the role of protein kinases in the mechanism of action of cyclic AMP (section 5.2.2).

GREVILLE, G. D. and TUBBS, P. K. (1968). 'The catabolism of long chain fatty acids in mammalian tissues.' *Essays Biochem.*, **4**, 155-212.

GUMAA, K. A., MCLEAN, P. and GREENBAUM, A. L. (1971). 'Compartmentation in relation to metabolic control in liver.' *Essays Biochem.*, **7**, 39-86.

These last two references discuss the integration of fat and carbohydrate metabolism in diabetes and starvation (sections 6.1 and 6.2).

HALES, C. N. (1967). 'Some actions of hormones in the regulation of glucose metabolism.' *Essays Biochem.*, **3**, 74-104. A review of metabolic control generally and the regulation of carbohydrate metabolism by insulin (section 6.1).

HERSHKO, A., MAMONT, P., SHIELDS, R. and TOMKINS, G. M. (1971). 'Pleiotypic response.' *Nature, New Biol.*, **232**, 206-11. Development of a theory to correlate the many effects of insulin on cell growth and metabolism (section 7).

LEVINE, R. (1966). 'The action of insulin at the cell membrane.' *Am. J. Med.*, **40**, 691-94. A brief review of the earlier evidence that the primary action of insulin is exerted at the cell membrane (section 5).

MAYHEW, D. A., WRIGHT, P. H. and ASHMORE, J. (1969). 'Regulation of insulin secretion.' *Pharmac. Rev.*, **21**, 183-212. Review of the various factors that control insulin secretion from the beta cell (section 4).

ROBISON, G. A., BUTCHER, R. W. and SUTHERLAND, E. W. (1971). *Cyclic AMP*. Academic Press, London. A comprehensive survey of the discovery, properties and role of cyclic AMP as a second messenger (section 5.2).

STEINER, D. F., CLARK, J. L., NOLAN, C., RUBENSTEIN, A. H., MARGOLIASH, E., ATEN, B. and OYER, P. E. (1969). 'Proinsulin and the biosynthesis of insulin.' *Recent Prog. Horm. Res.,* **25**, 207-72. Description of the discovery and function of proinsulin in the synthesis of insulin (section 3).

WOOL, I. G., STIREWALT, W. S., KURIHARA, K., LOW, R. B., BAILEY, P. and OYER, D. (1968). 'Mode of action of insulin in the regulation of protein biosynthesis in muscle.' *Recent Prog. Horm. Res.,* **24**, 139-208. Review of the control of protein synthesis by insulin (section 6.3).

Specific references

BLUNDELL, T. L., DODSON, G. G., DODSON, E., HODGKIN, D. C. and VIJAYAN, M. (1971). 'X-ray analysis and the structure of insulin.' *Recent Prog. Horm. Res.,* **27**, 1-34.

BURGHEN, G. A., KITABCHI, A. E. and BRUSH, J. S. (1972). 'Characterisation of a rat liver protease with specificity for insulin.' *Endocrinology,* **91**, 633-42.

COORE, H. G., DENTON, R. M., MARTIN, B. R. and RANDLE, P. J. (1971). 'Regulation of adipose tissue pyruvate dehydrogenase by insulin and other hormones. *Biochem. J.,* **125**, 115-27.

CRAIG, J. W., RALL, T. W. and LARNER, J. (1969). 'The influence of insulin and epinephrine on adenosine-3′,5′-phosphate and glycogen transferase in muscle.' *Biochim. Biophys. Acta,* **177**, 213-19.

CROFFORD, O. B., MINEMURA, T. and KONO, T. (1970). 'Insulin–receptor interaction in isolated fat cells.' *Adv. Enzyme Reguln.,* **8**, 219-38.

CUATRECASAS, P. (1971a). 'Insulin–receptor interactions in adipose tissue cells: direct measurement and properties.' *Proc. Natn. Acad. Sci. U.S.A.,* **68**, 1264-68.

CUATRECASAS, P. (1971b). 'Unmasking of insulin receptors in fat cells and fat cell membranes. Perturbation of membrane lipids.' *J. Biol. Chem,* **246**, 6532-42.

CUATRECASAS, P. (1972a). 'Isolation of the insulin receptor of liver and fat cell membranes.' *Proc. Natn. Acad. Sci. U.S.A.,* **69**, 318-22.

CUATRECASAS, P. (1972b). 'Properties of the insulin receptor isolated from liver and fat cell membranes.' *J. Biol. Chem.,* **247**, 1980-91.

CUATRECASAS, P. (1972c). 'Affinity chromatography and purification of the insulin receptor of liver cell membranes.' *Proc. Natn. Acad. Sci. U.S.A.,* **69**, 1277-81.

DENTON, R. M., RANDLE, P. J. and MARTIN, B. R. (1972). 'Stimulation by calcium ions of pyruvate dehydrogenase phosphate phosphatase.' *Biochem. J.,* **128**, 161-63.

DOUSA, T. P., SANDS, H. and HECHTER, O. (1972). 'Cyclic AMP-dependent reversible phosphorylation of renal medullary plasma membrane protein.' *Endocrinology,* **91**, 757-63.

EXTON, J. H., MALLETTE, L. E., JEFFERSON, L. S., WONG, E. H. A., FRIEDMANN, N., MILLER, T. B. and PARK, C. R. (1970). 'The hormonal control of hepatic gluconeogenesis.' *Recent Prog. Horm. Res.,* **26**, 411-57.

GIBSON, D. M., LYONS, R. T., SCOTT, D. F. and MUTO, Y. (1972). 'Synthesis and degradation of the lipogenic enzymes of rat liver.' *Adv. Enzyme Reguln.,* **10**, 187-204.

GILL, G. N. and GARREN, L. D. (1971). 'Role of the receptor in the mechanism of action of adenosine 3′5′-cyclic monophosphate.' *Proc. Natn. Acad. Sci. U.S.A.,* **68**, 786-90.

HINTZ, R. L., CLEMMONS, D. R., UNDERWOOD, L. E. and VAN WYK, J. J. (1972). 'Competitive binding of somatomedin to the insulin receptors of adipocytes, chondrocytes and liver membranes.' *Proc. Natn. Acad. Sci. U.S.A.,* **69**, 2351-53.

HUTTUNEN, J. K., STEINBERG, D. and MAYER, S. E. (1970). 'Protein kinase activation and phosphorylation of a purified hormone-sensitive lipase.' *Biochem. Biophys. Res. Commun.*, **41**, 1350–52.

JUNGAS, R. L. (1966). 'Role of cyclic 3'5'-AMP in the response of adipose tissue to insulin.' *Proc. Natn. Acad. Sci. U.S.A.*, **56**, 757–63.

KREBS, H. A. (1966). 'The regulation of the release of ketone bodies by the liver.' *Adv. Enzyme Reguln.*, **4**, 339–53.

LACY, P. E. and HOWELL, S. L. (1970). 'The mechanism of emiocytotic insulin release.' In: *The Structure and Metabolism of the Pancreatic Islets*, pp. 171–78. Eds. S. Falkmer, B. Hellman and I.-B. Täljedal. Pergamon Press, Oxford.

LACY, P. E., HOWELL, S. L., YOUNG, D. A. and FINK, C. J. (1968). 'New hypothesis of insulin secretion.' *Nature, Lond.*, **219**, 1177–79.

LANGAN, T. A. (1969). 'Action of adenosine 3'5'-monophosphate-dependent histone kinase *in vivo*.' *J. Biol. Chem.*, **244**, 5763–65.

LEFKOWITZ, R. J., ROTH, J., PRICER, W. and PASTAN, I. (1970). 'ACTH receptors in the adrenal: specific binding of ACTH-^{125}I and its relation to adenyl cyclase.' *Proc. Natn. Acad. Sci. U.S.A.*, **65**, 745–52.

LEVEY, G. S. (1971a). 'Restoration of norepinephrine responsiveness of solubilised myocardial adenylate cyclase by phosphatidylinositol.' *J. Biol. Chem.*, **246**, 7405–07.

LEVEY, G. S. (1971b). 'Restoration of glucagon responsiveness of solubilised myocardial adenyl cyclase by phosphatidylserine.' *Biochem. Biophys. Res. Commun.*, **43**, 108–13.

LOTEN, E. G. and SNEYD, J. G. T. (1970). 'An effect of insulin on adipose tissue adenosine 3'5'-cyclic monophosphate phosphodiesterase.' *Biochem. J.*, **120**, 187–93.

MARTIN, B. R., DENTON, R. M., PARK, H. T. and RANDLE, P. J. (1972). 'Mechanisms regulating adipose tissue pyruvate dehydrogenase.' *Biochem. J.*, **129**, 763–73.

MATTHEWS, E. K. (1970). 'Calcium and hormone release.' In: *Calcium and Cellular Function*, pp. 163–82. Ed. A. W. Cuthbert. Macmillan, London.

MORGAN, H. E., HENDERSON, M. J., REGEN, D. M. and PARK, C. R. (1961). 'Regulation of glucose uptake in muscle. I. The effects of insulin and anoxia on glucose transport and phosphorylation in the isolated perfused heart of normal rats.' *J. Biol. Chem.*, **236**, 253–61.

ORCI, L., AMHERDT, M., MALAISSE-LAGAE, F., ROUILLER, C. and RENOLD, A. E. (1973). 'Insulin release by emiocytosis: demonstration by freeze-etching technique.' *Science*, **179**, 82–83.

ØYE, I. and SUTHERLAND, E. W. (1966). 'The effect of epinephrine and other agents on adenyl cyclase in the cell membrane of avian erythrocytes.' *Biochim. Biophys. Acta*, **127**, 347–54.

PARK, C. R., CROFFORD, O. B. and KONO, T. (1968). 'Mediated (nonactive) transport of glucose in mammalian cells and its regulation.' *J. Gen. Physiol.*, **52**, 296s–313s.

PASTAN, I., PERLMAN, R. L., EMMER, M., VARMUS, H. E., DE CROMBRUGGHE, B., CHEN, B. P. and PARKS, J. (1971). 'Regulation of gene expression in *Escherichia coli* by cyclic AMP.' *Recent. Prog. Horm. Res.*, **27**, 421–30.

POHL, S. L., KRANS, H. M., KOZYREFF, V., BIRNBAUMER, L. and RODBELL, M. (1971). 'The glucagon-sensitive adenyl cyclase system in plasma membranes of rat liver. VI. Evidence for a role of membrane lipids.' *J. Biol. Chem.*, **246**, 4447–54.

RANDLE, P. J. and MORGAN, H. E. (1962). 'Regulation of glucose uptake by muscle.' *Vitams. Horm.*, **20**, 199–249.

RASMUSSEN, H. (1970). 'Cell communication, calcium ion and cyclic adenosine monophosphate.' *Science*, **170**, 404–12.

REIMANN, E. M., BROSTROM, C. O., CORBIN, J. D., KING, C. A. and KREBS, E. G. (1971). 'Separation of regulatory and catalytic subunits of the cyclic 3'5'-adenosine monophosphate-dependent protein kinase(s) of rabbit skeletal muscle.' *Biochem. Biophys. Res. Commun.,* **42**, 187-94.

RODBELL, M., KRANS, H. M. J., POHL, S. L. and BIRNBAUMER, L. (1971). 'Glucagon-sensitive adenyl cyclase system in plasma membranes of rat liver. III. Binding of glucagon: method of assay and specificity.' *J. Biol. Chem.,* **246**, 1861-71.

SCHARFF, R. and WOOL, I. G. (1967). 'Accumulation of amino acids in muscle of perfused rat heart. Effect of insulin in the presence of puromycin.' *Biochem. J.,* **97**, 272-76.

SCRUTTON, M. C. and UTTER, M. F. (1968). 'The regulation of glycolysis and gluconeo-genesis in animal tissues.' *A. Rev. Biochem.,* **37**, 249-302.

TURTLE, J. R. and KIPNIS, D. M. (1967). 'An adrenergic receptor mechanism for the control of cyclic 3'5'-adenosine monophosphate synthesis in tissues.' *Biochem. Biophys. Res. Commun.,* **28**, 797-802.

WIELAND, O. H., PATZELT, C. and LOFFLER, G. (1972). 'Active and inactive forms of pyruvate dehydrogenase in rat liver.' *Eur. J. Biochem.,* **26**, 426-33.

WILLIAMSON, J. R., BROWNING, E. T. and SCHOLZ, R. (1969). 'Control mechanisms of gluconeogenesis and ketogenesis. I. Effects of oleate on gluconeogenesis in perfused rat liver.' *J. Biol. Chem.,* **244**, 4607-16.

WOOL, I. G. and CAVICCHI, P. (1967). 'Protein synthesis by skeletal muscle ribosomes. Effect of diabetes and insulin.' *Biochemistry,* **6**, 1231-42.

20
The Structure and Genetics of Immunoglobulins

A. J. Munro

Department of Pathology, University of Cambridge

Vertebrates possess a mechanism which protects them against the effects of invading organisms. This defence mechanism is based in part on the formation of protein molecules known as antibodies, as well as of specialised cells. Both molecules and cells can react with foreign substances or antigens. This short chapter cannot deal in detail with the whole subject of immunobiology but will describe the structure of antibody molecules and the consequent genetic problems.

1 The structure of immunoglobulins

If an animal is injected with antigen, after a few days protein molecules which will specifically bind to the antigen are found in the serum. The protein molecules are immunoglobulins known as antibodies and all have the same basic structure (Fig. 20.1). Determination of this structure has been advanced by the study of multiple myeloma tumours in which the cells responsible for making antibodies have become malignant. These malignant cells continue to secrete antibody, and the serum of animals carrying such tumours therefore contains high concentrations of a unique homogeneous immunoglobulin known as a myeloma protein. Investigation of these myeloma proteins and of naturally occurring antibodies (heterogeneous) shows that the unit from which the immunoglobulins are built is a protein containing two pairs of polypeptide chains. There are two identical light chains with a molecular weight of 25 000 each and two identical heavy chains each with a molecular weight of 50 000. Each light chain is usually attached to a heavy chain by a disulphide bridge which varies in location; the heavy chains too are joined by disulphide bridges. The amino-terminal regions of the light and of the heavy chains contain variable sequences of amino acids; this is shown in Fig. 20.1.

Cleavage of the molecule under limiting conditions with the proteolytic enzymes papain or pepsin divides it into antigen-binding (Fab) or crystallisable (Fc) fragments (Fig. 20.1). The region which is subject to attack by papain and pepsin is flexible and acts as a hinge. The antibody-combining site is in the Fab region which contains both light and heavy chains. The structure shown in Fig. 20.1 has therefore two identical combining sites for antigen.

1.1 Classes of human immunoglobulins

There are five classes of human immunoglobulin molecules (Table 20.1) as defined by the heavy chain amino acid sequence from about the amino acid in position 110 onwards to the

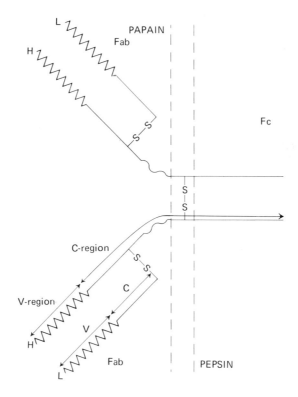

Fig. 20.1 Diagrammatic representation of the basic structure of immunoglobulins. It contains two light chains (L) and two heavy chains (H). The amino-terminal regions of each chain contain variable sequences (V) of amino acids. See text for further details.

carboxy-terminal end. Two types of light chain, known as kappa and lambda, can be associated with each class of immunoglobulin molecule. Comparison of sequences of light chains of the same type, or heavy chains of the same class, shows in each case that approximately the first 110 amino acids from the amino-terminal end of the polypeptide chains are variable. The amino acid sequence of the rest of the chain is relatively constant. These parts of the chains are designated the V- and C-regions respectively.

Thus, the diversity of antibody structure arises from not only the different combinations of C-regions which are possible, but also from the permutations of sequences in the V-region of each chain. So far no two human myeloma proteins have been found with the same amino acid sequence in the V-region.

1.2 Molecular structure

1.2.1 Combining sites

The combining site for antigen comprises the V-regions of both the light and heavy chains. The wide variation in V-region sequences is potentially able to provide for an immense diversity of structure of combining sites. This allows for a great number of highly specific antibody–antigen reactions; for example, the antibodies formed in response to polio virus will not react with influenza virus. The nature of the binding of antigen to the antigen binding site on an antibody is in every way analogous to the initial binding of substrate to an enzyme.

1.2.2 Polymers

Not all the immunoglobulins are made up of only one of the units shown in Fig. 20.1. IgM antibody molecules contain five such units (Fig. 20.2(b)); consequently the molecule has a molecular weight of approximately 900 000 and has ten combining sites, each of which is again

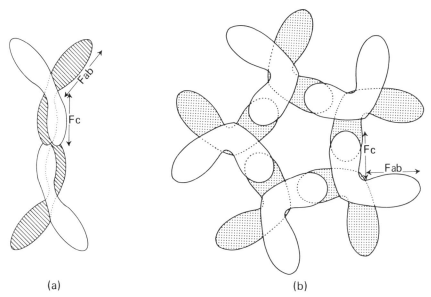

Fig. 20.2 Diagrammatic representation of IgA (a) and IgM (b) molecules. The units are identical but in each case the hatched units have been rotated through 180° with respect to the unhatched unit lying on top.

identical. With IgA the degree of polymerisation is variable. The usual aggregate is a dimer although higher aggregates are also found in serum. All the combining sites on each molecule are similarly identical (Fig. 20.2(a)).

1.2.3 J Chain

An additional polypeptide chain called the J chain of molecular weight approximately 25 000 is found in both IgM and in the polymeric forms of IgA. The initial suggestion that the J chain was necessary for polymerisation is now disputed although it may stabilise the polymeric forms *in vivo*. It is interesting that in the cells secreting IgM and polymeric forms of IgA the bulk of

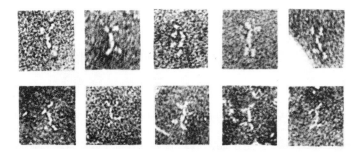

Fig. 20.3 Electron micrographs of single IgA molecules. (From Munn *et al.* (1971). *Nature, Lond.*, **231**, 527–29.)

the intracellular material is in the unpolymerised form. The same cells synthesise the J chain and polymerisation occurs shortly before secretion.

1.2.4 Three-dimensional structure

It is possible to see the different immunoglobulin molecules by electron microscopy (Figs 20.3, 20.4). This and other studies show that the molecule is a flexible structure with freedom of movement around the region of attachment of the Fab part to the rest of the heavy chain (see

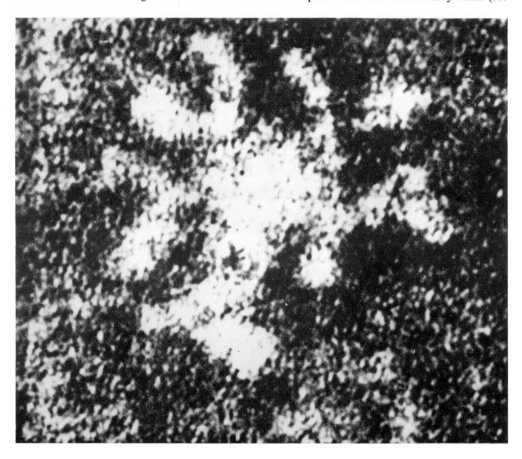

Fig. 20.4 Electron micrograph of a single mouse IgM molecule. (From Parkhouse *et al.* (1970). *Immunology*, **18**, 575–84.)

Fig. 20.1). X-ray crystallographic studies have not yet produced the structure of a complete immunoglobulin molecule with sufficient resolution to show individual amino acids but the results obtained at low resolution show a molecule with an open configuration (see Fig. 20.5).

1.3 Significance of polyvalency

The fact that immunoglobulin molecules are at least divalent and often have a higher valency is very important. First, some structures, such as viruses or bacterial surface membranes, have

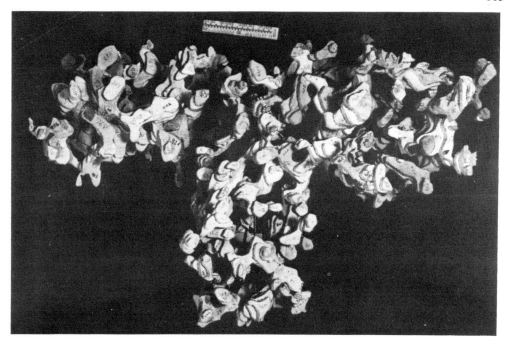

Fig. 20.5 A balsa wood model of an immunoglobulin molecule constructed from the electron density maps obtained by X-ray crystallography. The vertical region corresponds to Fc and the two arms to the Fab portion of the molecule. (From Sorna *et al.* (1971). *J. Biol. Chem.*, **246**, 3753.)

repeating antigenic determinants. In such cases, the immunoglobulin molecule will be able to react with the antigen at two or more sites. This will greatly stabilise the interaction between antigen and antibody molecules; it has been calculated that for IgG molecules the use of two combining sites rather than one increases the binding equilibrium constant (affinity constant) by about three orders of magnitude. For the IgM molecule this effect would be much greater. The second consequence of the polyvalency of antibody molecules is that they can agglutinate antigen molecules into a three-dimensional lattice (Fig. 20.6). If the reactants are in the right proportions this causes precipitation; this phenomenon is employed in many immunological techniques. If either antigen or antibody is present in excess, the antibody–antigen complexes formed are too small to precipitate from solution (Fig. 20.6) but nevertheless can initiate reactions such as complement fixation.

1.4 Effects of constant region heterogeneity

The various amino acid sequences in the constant region of the heavy chains that differentiate the classes of antibody molecule give each class or subclass different biological properties (Table 20.1). For example the constant region sequence of IgE immunoglobulin molecules is such that these molecules will bind preferentially to mast cells and basophils in humans. On the other hand the IgA molecule will combine with an additional polypeptide chain known as the secretory piece, enabling the IgA molecules to be secreted into the saliva and other mucous secretions in the gut and respiratory tract. Unlike the J chain the secretory piece is not produced by the IgA-synthesising cells.

614

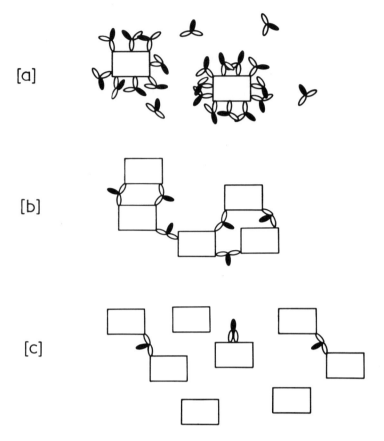

Fig. 20.6 Diagrammatic representation of antibody reacting with antigen. (a) Excess antibody; (b) equivalence; (c) excess antigen.

1.5 Problems for geneticists

The precise antigen specificity of antibody molecules implies the existence of a large number of different combining sites in the pool of immunoglobulin molecules. The exact number of such sites is not known. As mentioned earlier, immunoglobulin molecules may have many different permutations of amino acid sequence in the V-region and thus have the potential for great specificity of reaction. Despite the wide variety possible, the pair of combining sites in each individual immunoglobulin molecule is nevertheless found to be identical. This poses considerable problems for molecular geneticists. Some answers have already been found; for example the number and distribution of the genes responsible for the constant region parts of the different polypeptide chains has been determined. However, there is still considerable controversy over the number of genes required to code for all the V-region sequences that are needed to explain the high degree of antigen specificity seen in immunoglobulin molecules. The immunoglobulins also display two further genetic features which seem to be unique to these molecules. First, a contradiction to the dogma 'one gene one polypeptide chain' (since there must be separate genes for the variable and constant regions); second, an example of the use of one of two almost identical chromosome regions in a diploid cell (i.e. allelic exclusion; see section 1.8).

Table 20.1 Some properties of human immunoglobulins

Class	IgM	IgG				IgA	IgD	IgE
Sub-class		IgG_1	IgG_2	IgG_3	IgG_4			
Serum level (mg/ml)	1·2	9·0	1·8	0·8	0·4	2	0·03	0·00004
Half-life in serum (days)	10			23		6	2·8	2·3
Molecular weight approx.	900 000			150 000		$(150\,000)^n$	178 000	187 000
Heavy chains	μ	γ_1	γ_2	γ_3	γ_4	α	δ	ϵ
Number of combining sites for antigen	10			2		$2n$	2	2
Placental transfer	−	+	+	+	+	−	−	−
Reaginic antibody binds to mast cells, etc.	−	−	−	−	−	−	−	+
Complement fixation classical	+	+	±	+	−	−	−	−
Complement Alternative pathway	+	+	+	+	+	+	−	+
Found in saliva and other mucous secretion	−	−	−	−	−	+	−	−

1.6 Pools of genes

Much information on the genetic requirements for antibody production has been obtained from comparison of the amino acid sequence of light chains of the same type, or of heavy chains. This shows that there are sufficient constant amino acids in the variable region to define the type of variable region without reference to the constant part of the molecule. In other words there is a pool of sequences for the variable regions of kappa light chains which are always associated with kappa constant regions and never with lambda light chains or with heavy chains. The same is true for lambda and for heavy chains. This implies that there are three separate pools of genes, one coding for kappa, the second for lambda and the third for heavy chain variable sequences. Further variable sequences can be allocated to 'subgroups': the similarity between sequences of one subgroup is greater than that between sequences in different subgroups. For example the sequence in a particular subgroup may be longer than the sequences in another subgroup; the sequences in human kappa subgroup I (κI) comprise 107 amino acids while the sequences in subgroup III (κIII) comprise 108 amino acids. Further each subgroup contains many subgroup-specific residues, the first three amino acids of human κI are usually Asp-Ile-Gln-; κII, Asp-Ile-Val-; and κIII, Glu-Ile-Val-. The simplest explanation for this is that there is at least one gene for each subgroup; it is generally accepted that the *minimum* number of genes for the V-region of each chain is the number of well-defined subgroups. If one examines the sera of many different normal individuals one finds light chains from each subgroup in each individual. This means that the subgroups do not represent allelic forms of each other; that is, they are not different forms of the same stretch of DNA carried on maternal or paternal chromosomes. Instead, each haploid set of chromosomes contains at least one gene representing each of the subgroups found in human light chains. In summary, for humans the minimum number of genes for the V-regions of each type of chain is 4 for kappa chains, 4 for lambda chains and 3 for heavy chains. In other species the same argument gives different minimal numbers of genes, and this will be discussed later.

Comparison of the amino acid sequence of the constant region of kappa light chains shows that there is a single amino acid substitution, leucine to valine, in position 191. The genetic segregation of this substitution (which can be followed by a serological reaction) shows that unlike the kappa V-regions, the two forms of kappa C-regions are true alleles of each other. There is no evidence of more than one kappa C-region gene in the haploid set of chromosomes. This leads to the paradoxical situation where the information for a single polypeptide chain is stored in the DNA not as one continuous sequence but rather as two sequences, one for the V-region and one for the C-region. For example, it is argued that for human kappa chains there are at least four genes for the V-region and only one gene for the C-region. Since DNA is a linear array of nucleotides with information stored in a linear structure it is impossible for all four V-genes to be adjacent to a single C-gene. Theoretically it is possible that the bringing together of the genetic information held in the two separate pieces of DNA could have occurred at any time during the process of transcription and translation of the DNA into the final single polypeptide chain. It now seems clear that this does not occur during protein synthesis since there is, both for light and for heavy chains, only one growing point, and there is probably in each case a single molecular messenger RNA species, one for light chains and one for heavy chains. The least implausible alternative is that rearrangement of the DNA occurs before transcription, in such a way that the sequence corresponding to the variable region of the polypeptide chain is brought into juxtaposition with the region that codes for the constant part.

Comparison of the C-region of lambda chains in humans also reveals certain amino acid substitutions but in this case the substitutions are found to occur in all individuals. In other words they represent recent duplications of the lambda gene. In humans there seem to be at least three genes, each coding for slightly different forms of the lambda C-region. On the other hand all the different lambda subgroups of the V-regions are found with all the three different C-regions. These three lambda C-region genes represent such a recent gene duplication that very few variants have accumulated in the duplicated genes. In the case of the heavy chain C-region genes one can also see evidence for relatively recent gene duplications, because the homology

Table 20.2 The minimum number of genes for human immunoglobulins. The V-genes are found expressed in association with the C-genes in each row. It is predicted that for each chain the set of V-genes will be closely linked to the corresponding C-genes, and that each represents a different chromosome.

		V-genes				C-genes										
Light chains	κ	$\overline{\kappa I_a}$	$\overline{\kappa I_b}$	$\overline{\kappa II}$	$\overline{\kappa III}$	$\overline{\kappa}$										
	λ	$\overline{\lambda I}$	$\overline{\lambda II}$	$\overline{\lambda III}$	$\overline{\lambda IV}$	$\overline{\lambda lys}$	$\overline{\lambda arg}$	-----								
Heavy chains		$\overline{\overline{HI}}$	$\overline{\overline{HII}}$	$\overline{\overline{HIII}}$		$\overline{\gamma 4}$	$\overline{\gamma 2}$	$\overline{\gamma 3}$	$\overline{\gamma 1}$	$\overline{\alpha 1}$	$\overline{\alpha 2}$	$\overline{\mu 2}$	$\overline{\mu 1}$	$\overline{\delta}$	$\overline{\epsilon}$	

between the different subclasses of gamma chains is greater within a species than the homology between the gamma chains of two different species. As in lambda chains, the V-regions for heavy chains are found with all the different classes and subclasses of heavy chain C-regions (Table 20.2).

1.7 Genetic markers

In rabbits there is fortunately a genetic marker which correlates with amino acid changes in the heavy chain V-region. This marker behaves like a true Mendelian gene and occurs in three apparently allelic forms known as a1, a2, a3. As would be expected, this genetic marker occurs in all the different classes of immunoglobulin molecules because the same V-regions are employed in each of the different classes. There is also a genetic marker in the heavy chain C-region of rabbit immunoglobulin molecules. Using the genetic markers in the V- and C-regions one can study the linkage of the genes for these structures. Recombinants have been found between these two genes which give a recombination frequency of approximately 0·5 per cent. This suggests that the V-region genes are close to, but not adjacent to, the C-region genes that they serve. Using these markers it is also possible to see whether the V-region from one chromosome can interact with the C-region from the other chromosome in diploid cells. In fact such interactions are rare; the majority of the heavy chains formed have the same arrangement of genetic markers as those found in the chromosomes inherited from each parent. This means that the mechanism which brings together the information for the V-region and the C-region usually operates for the parts held on the same chromosome. It has been suggested that the V-region genes for the two types of light chain are also closely linked to the C-region genes for these chains.

Genetic markers have also been detected in the C-regions of the two types of light chain in rabbits. It can be demonstrated that the C-regions for light chains of each type and for the heavy chains probably occur on three separate chromosomes. For each of the three types of polypeptide chain – kappa, lambda and heavy chains – found in immunoglobulin molecules, there are three separate sets of genes. These comprise a pool of V-region genes of known minimum but unknown maximum size, and one or more C-region genes. In each case all the V-regions in each set of genes can become associated with all the C-region genes in each set (Table 20.2).

1.8 Allelic exclusion

A problem posed by the structure shown in Fig. 20.1 is that there must be some mechanism to ensure that light chains and heavy chains of only one particular V-region are used to assemble the structure in order that the combining sites are identical. Since there are many possible V-regions for each chain it would be highly likely that the V-region sequence that could be employed in the two different chromosomes would not be identical. Consequently if each chromosome were active two chains with different V-region sequences would be produced. This would lead to a molecule having two different combining sites. This difficulty is solved by a process that allows only one of the two chromosomes to be active at any one time and which is known as allelic exclusion. It is apparently restricted to immunoglobulin genes apart from the complete inactivation of the second X chromosome seen in female cells. In antibody-forming cells there is no evidence for allelic inactivation of any other autosomal genes. The exact mechanism for this very precise inactivation of a particular gene has yet to be discovered.

2 The origin of antibody diversity

2.1 Introduction

For many years the nature of the action of antigen in eliciting immune response has been much discussed. The first of two extreme views proposes that the antibody which reacts with the antigen does not exist until the antigen appears in the system. This is known as the instructive model and implies that the nature of the antigen helps to generate the correct antibody structure. The second proposes that the antibody structure exists before the antigen appears in the system, and that the role of antigen is to select the structure with which it can react. This is known as a selective theory and has been most fully developed by Burnet into a clonal selection theory. According to this, cells pre-exist which carry recognition elements for antigen, and antigen selects these cells for proliferation, and secretion of antibody. The combining site for antigen on the antigen recognition element, and on the final antibody produced would be identical.

It is difficult to imagine how instructive models work in terms of molecular biology and it is generally accepted that a clonal selection model is the correct description. So far all the necessary elements for such a model have been found; for example, lymphocytes from foetuses carry immunoglobulin determinants on their surfaces.

Since it seems that both the light and heavy chains are constructed in the same way as any other polypeptide chain it is generally accepted that the cell producing a particular immunoglobulin possesses one sequence of nucleotides in the DNA which corresponds to the amino acid sequence of the light chain, and another sequence for the heavy chain. The diversity of antibody specificity calls for a large number of different polypeptide chains in order that the correct combining sites may be constructed. This presupposes the existence of a considerable number of genes to produce a wide range of antibody structures. The origin of this necessarily large pool of genes is still a controversial subject. It can be discussed under three headings: first, the number of genes necessary; second, the role of somatic mutation in the origin of diversity; and third, the need (if any) for additional mutational processes.

2.2 The number of V-region genes

There is obviously some correspondence between the number of antibody structures and the number of V-region genes. However, we do not know how many different antibody structures any one animal can produce. Further we do not know how efficient is the mechanism of pairing different light and heavy chains in giving a diversity of different structures. For example, theoretically, if there were P light chains and Q heavy chains then $P \times Q$ different antigen-combining sites could be constructed from these chains if all possible combinations gave useful antibody molecules.

Estimates of the number of necessary antibody structures vary from 10^4 to more than 10^{12}. Consequently numerological arguments from this starting point are not particularly useful. All processes for generating diversity require genes to be transmitted in the germ line. In a pure germ line model, it is argued that the genes for the production of all the antibody structures that an entire animal can employ, are carried in every cell of that animal. In other words all the genes for antibodies are passed on in the germ line. The difficulties inherent in the germ line model are twofold. First, the evolution of the V-region genes is hard to explain in terms of a simple evolutionary tree. For example, all the kappa light chains of rabbit have valine at position 11 while leucine is the only amino acid in human κ chains independent of subgroup. Consequently it is necessary to postulate a recent massive gene duplication to produce the set of V-region genes for rabbits from one V-gene in an ancestral animal which had valine in this position. This type of problem is also seen in the V-genes of heavy chains in many species including man, mice, rabbits and guinea pigs, cats and dogs where one finds other species-specific amino acids. The situation is very similar to the evolution of the C-region genes where one finds species-specific amino acids in the subclasses of particular classes of immunoglobulin. It is hard to imagine a rapid gene expansion process giving a set of identical genes followed by a rapid mutational process generating the necessary diversity in the short time available for these many species. It would be possible to imagine that such an event might occur occasionally but that it should happen in all species would seem to be particularly unlikely.

The second argument against the germ line model is based on the apparent failure to get recombinants between the V-genes as seen in rabbits. As mentioned earlier, rabbits contain an allelic marker in the V-regions of the heavy chains, and this marker behaves in a strictly Mendelian fashion. If a set of 100–1000 genes are necessary for the heavy chain, and they all contain this allelic marker which has arisen by a process of massive gene duplication, then it is difficult to see why a recombinant has not occurred between this large set of identical genes. This is particularly striking when one considers that in other systems where large numbers of genes are known to occur, for example with the genes for ribosomal RNA, there is a high frequency of mismatched recombination between these genes, leading to deletion and consequent failure to survive.

2.3 The role of somatic mutation

Somatic mutation models present an attractive alternative to a germ line model. In this case a small number of genes would be inherited in the germ line and the generation of diversity would rely on the accumulation of useful somatic mutations.

Mutations can occur with every cell division. The rate of this somatic mutation has not been clearly measured in eukaryotic cells. However various estimates give a value between 10^{-6} to 10^{-5} mutations per gene per cell division. Taking a value of 10^{-6} this would mean that if 10^7 cells were to divide then there would be ten cells carrying a mutation in a particular gene. Normally these mutants are not seen, because they have no selective advantage and are masked by the large numbers of non-mutated forms of that particular gene product. With immuno-globulins this is not the case, since an antigen selects positively for cells with which it can specifically react. This means that if a mutant occurs which will react with a particular antigen more avidly, then there is more chance of that cell and its progeny being stimulated to produce antibodies during the course of a response than cells which bear immunoglobulin receptors that react less strongly with the antigen. It is known that during the course of a response the affinity of binding of the antibody to the antigen increases; this may well be due to antigen-selecting mutants which can react with the antigen with greater affinity. It is hard to see how antigen can fail to select somatic mutants and thus increase the range of antibody diversity.

2.4 The possibility of special somatic mutation mechanisms

The drawback of relying too heavily on somatic mutation as a means of generating diversity is the problem of explaining how an immune system is set up during embryonic development. For this reason several different mechanisms have been suggested as a possible means of increasing the diversity of V-genes during development. Some of these mechanisms rely on selection against unmutated forms of the germ line V-genes, while others propose specific processes for increasing mutation in the V-genes. These mechanisms have not been shown to be essential. One example suggests that there are hypermutatable regions in the germ line V-genes which correspond to the hypervariable regions seen in the V-gene structures. It should be stressed that this may not be an essential feature; the appearance of the hypervariable regions in the V-gene structures may only represent antigen selection, since it is thought that the amino acids in this part of the chain are those most intimately involved in the construction of the antigen combining site.

2.5 Relative importance of the factors involved

The solution to the question of the origin of antibody diversity is likely to differ between different species. The three variables that one can play with are the nature of the V-gene, the number of the V-genes, and the time allowed before a competent immune system has to arise. The germ line V-genes may be selected to encourage a hypermutatable region or to code for antibodies which react with common pathogens of the species. This latter type of germ line gene would have the dual advantage of giving early protection to the newborn and would also be suitable for early antigen stimulation and proliferation. This would lead to an increasing number of mutant cells in the population through the natural process of somatic mutation. It is clear from the amino acid sequence data that the number of V-genes found in mice is considerably larger than that in man. This may reflect the short time during ontogeny between the initial appearance of lymphocytes and the time that the creature is born as an immunologically competent though not fully responsive animal.

In conclusion, I feel that at present somatic mutation by itself acting on an evolutionarily selected, small number of germ line V-genes is adequate to account for antibody diversity, and that the number of these V-genes and their nature will vary from species to species.

3 Additional Comments

In this chapter we have considered only the structure and synthesis of antibody molecules and the genetic problems related to these phenomena. We have not at all discussed the triggering mechanisms involved in inducing the cells to divide when an antigen appears. It is now clear that in many cases this process requires co-operation between three cell types: lymphocytes derived from thymus (T cells), lymphocytes which are not derived from the thymus (B cells) and macrophages. Co-operation between these involves soluble factors which are at present being studied intensively at a biochemical level. After the correct presentation of antigen to the cell surface, further signals must then be transmitted from the cell membrane to the nucleus of the cell so that proliferation and differentiation may occur. The biochemical nature of these signals has not yet been elucidated. In addition, the consequences of antibody–antigen reactions in the whole animal are profound, ranging from the activation of complement to the release of chemical mediators as in acute anaphylaxis. Some of these phenomena are understood at the biochemical level.

I thank Mrs H. Bateman for help in the preparation of this chapter.

4. Bibliography

General immunology

ROITT, I. M. (1971). *Essential Immunology.* Blackwell, Oxford.

DAVID, B. D., DULBECCO, R., EISEN, H. N. GINSBERG, H. S. and WOOD, B. W. (1968). *Principles of Microbiology and Immunology,* Harper, New York and London.

PORTER, R. R. (Ed.) (1974). *Defence and Recognition.* MTP Series. Butterworth, Oxford.

BURNET, M. (1969). *Cellular Immunology.* Cambridge Univ. Press.

Structure

METZGER, H. (1970). 'Structure and function of γM macroglobulins.' *Adv. Immun.* **12**, 57–116.

GREEN, N. M. (1969). 'Electron microscopy of the immunoglobulins.' *Adv. Immun.* **11**, 1–30.

PARKHOUSE, R. M. E., ASKONAS, B. A. and DOURMASHKIN, R. R. (1970). 'Electron microscopic studies of mouse immunoglobulin M; structure and reconstitution following reduction.' *Immunology.* **18**, 575–84.

21
The Molecular Basis of
Muscular Contraction

Gerald Offer

Department of Biophysics, King's College, London

1 Introduction

Whether it is for the movement of a limb, the snapping of a jaw, the flapping of a wing, the beating of a heart or the peristalsis of the gut, muscles are needed by animals to produce force and movement. We, as biochemists, should find muscles intriguing. They convert chemical energy directly into mechanical work and there are few examples of such machines in the non-biological world. The production of force and movement is caused at the molecular level by a conformational change in one of the protein components of the contractile machinery — a conformational change literally being put to work. Muscles also provide us with excellent examples of biological control mechanisms, for the contractile machinery must be capable of being rapidly switched on and off. Without this a Chopin prelude could not be played nor a dragonfly take flight. The allosteric interactions between the protein components which achieve this control are subtle and interesting.

Apart from their interest as machines, muscles provide us with splendid examples of structures intermediate in size between the particles traditionally studied by the biochemist (with molecular weights, say, of less than 10^6) and the very much larger structures (nuclei, mitochondria, etc.) traditionally studied by the cell biologist. The study of how protein molecules assemble into the filaments that comprise the contractile machinery should give us an important insight into how the more complicated organelles in the cell may be built up.

The principles by which muscles contract are potentially of wider significance, since protein components similar to those in muscle occur in other cells which show contractile activity, e.g. in the platelets which are responsible for the retraction of blood clots.

In this chapter we shall discuss the molecular make-up of the contractile machinery of muscle, and see how the protein components interact with one another to convert the free energy of hydrolysis of ATP into mechanical work. But before we can do this we need to be familiar with the elementary physiology and structure of muscle.

2 The physiological properties of muscle

2.1 Classification of muscle

It is convenient to classify the muscles of vertebrate animals according to the degree of structural order they possess.

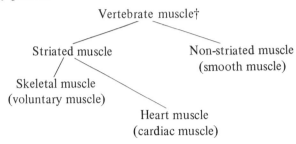

The striated muscles have a high degree of structural regularity, so much so that the fibres, of which the muscle is composed, display a regular transverse striation (Plate 21.1). Skeletal muscle is so named because it works on the skeleton and therefore is largely responsible for determining the external shape of a vertebrate. Typically about 40 per cent or more of the mass of an animal such as man is skeletal muscle. The alternative name, voluntary muscle, indicates that the contraction is under voluntary control. Heart muscle is very similar in structure to skeletal muscle and much of what we say about skeletal muscle is directly applicable to it. In contrast vertebrate non-striated muscle (e.g. muscles of the walls of the gut and of blood vessels) has a much less regular structure and although it is now clear that it operates by a mechanism similar to that of skeletal and heart muscle, the structural details of the contractile elements are not as well understood. We shall therefore concentrate on vertebrate skeletal muscle on which most work has been done. The reader should, however, be aware that many clues to the mechanism of muscular contraction can be picked up by studying other types of muscles: some of these show fascinating specialisations, e.g. insect flight muscle which can undergo very rapid oscillatory contractions (up to 10^3 s^{-1}), and certain muscles in molluscs like the oyster which can sustain tension for days with little expenditure of energy.

2.2 Muscle and movement

Skeletal muscles are usually attached to the bones of the skeleton via inelastic tendons. These serve not only to concentrate the pull of the muscle on a small area, but also to keep the bulk of the muscle away from the joint operated on. The attachment is usually quite close to the joint (Fig. 21.1); this means that a small shortening of the muscle produces a large movement at the end of the bone; conversely, the muscle needs to develop a greater tension than the applied load.

Muscles can only pull and not push. That is, they can shorten actively, developing a large tension but they cannot extend actively. It is true that some excised muscles will return to their original length after a contraction, but only a very weak force is produced. Hence for every movement of the skeleton, at least two opposing muscles are involved (Fig. 21.1). When movement in one direction is required, one muscle shortens and the other (the antagonist) pays

---

† The muscles of invertebrates are similarly classified into striated and non-striated muscles but vertebrate and invertebrate non-striated muscles are structurally quite distinct.

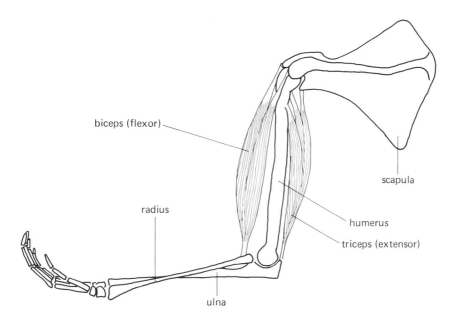

biceps (flexor)

scapula

radius

humerus

triceps (extensor)

ulna

Fig. 21.1 Action of antagonistic muscles at the elbow joint. The triceps acts to extend the arm, the biceps muscle to flex it. (From *Introduction to Biology* by D. G. Mackean (John Murray.)

out (the tension in the latter being anything from zero to nearly the same as the shortening muscle, depending on the movement). This is rather an oversimplified view, since more than two muscles are often involved.

Examination of a muscle shows that it is composed of numerous fibres (Plate 21.1 and Fig. 21.2). These lie parallel to the long axis in some muscles but do not necessarily extend from one end of the muscle to the other. In other muscles the fibres are arranged at a considerable angle to the tendon. This allows the muscle to exert larger tensions than a parallel-fibred muscle of similar size but the possible extent of shortening is correspondingly decreased. Parallel-fibred muscles, such as the frog sartorius, are obviously more suitable for physiological studies of tension.

2.3 The nerve impulse and contraction

When the central nervous system decrees that a muscle contracts, an action potential passes along the motor nerve and, because the latter is branched, simultaneously reaches several muscle fibres within the muscle (the motor unit). The junction between nerve and muscle fibre (the motor end plate) is usually situated about half-way along the muscle fibre. The arrival of an action potential at this junction causes the release of acetylcholine which depolarises the membrane of the muscle fibre so causing a new action potential to be propagated along this membrane. Experimentally a muscle can be made to contract either by electrical stimulation of the motor nerve or by direct electrical stimulation of the muscle itself.

A muscle fibre responds to a single action potential after a short delay (about 1–3 ms) by contracting and then quickly relaxing. This response, called a twitch, takes 20–200 ms at 37°C depending on the muscle but it may be slowed down by being made to contract at a lower temperature. The muscle membrane is capable of conducting a further action potential a short interval (∼3 ms) after the first, but because the mechanical response is much slower, the muscle responds to a series of closely spaced action potentials by a sustained contraction, the tetanus. In physiological experiments the muscle may variously be allowed to shorten at constant load (isotonic contraction), be allowed to shorten at constant velocity (isovelocity contraction), be prevented from shortening so that the tension rises (isometric contraction)† or actually be pulled out by a greater force (stretch experiment). All these have counterparts *in vivo*. An excised muscle can usually shorten to about 60 per cent of its resting length on stimulation, but *in vivo*, the amount of shortening permitted by the skeleton is usually less than this.

The velocity with which a skeletal muscle fibre can shorten depends on the load which opposes the motion. With no load it shortens at maximum velocity (about $2-20$ lengths s^{-1} at $37°$C) depending on the muscle. As the load increases, the velocity of shortening decreases until a load is reached at which shortening does not occur (isometric tension). If the load is higher still, the muscle fibre will be pulled out.

2.4 Relaxed, contracting and rigor muscle

A muscle fibre can exist in three states, two of them natural, the third not. In the relaxed state, the fibre is pliable and can be stretched with little resistance. In the contracting state, force is obviously required to pull the muscle out, but this can be done without damage and indeed this happens every time we walk downstairs.

The rigor state is not physiological, but it has been extremely useful in understanding the mechanism of contraction. It is produced when the contractile machinery is deprived of ATP, as in rigor mortis. Rigor muscle is very resistant to being stretched. If sufficient force is applied to it to cause lengthening, irreparable damage is done. The rigor state can be reversed by restoring the supply of ATP to the muscle; it then contracts or relaxes depending on the conditions. We shall see the cause of the inextensibility of rigor muscle later.

3 The coarse structure of the contractile apparatus and how it shortens

3.1 The skeletal muscle cell and its contents

A skeletal muscle may be composed of as many as a million fibres; these fibres are immense multinucleate cells. Their diameter, typically between 20 and 100 μm, is larger than that of most cells and their length may be millimetres or as much as half a metre. The function of muscle fibres is to contract and their contents reflect that specialisation. They contain a very large number of long, parallel, approximately cylindrical contractile elements, the myofibrils, which run from one end of the muscle fibre to the other (Plates 21.1, 21.2; Fig. 21.2). The myofibrils are about $1-2$ μm in diameter, so a typical fibre would contain a thousand or more myofibrils.

Typically then, about two-thirds of the dry weight of a muscle fibre is accounted for by the myofibrils. These are bathed in the soluble fraction of the cell (the sarcoplasm) which contains the glycolytic enzymes and glycogen particles. It is important to know the ionic composition of the sarcoplasm so that the appropriate medium can be designed for *in vitro* experiments. As with the interior of most cells the major cation is K$^+$ ($\sim$120 mM) but there is in addition a

---

† The term contraction is used to mean mechanical activity of the muscle and does not necessarily imply shortening. Note also that the volume of the muscle remains almost exactly constant; any shortening or lengthening of the muscle is compensated by a change in cross-sectional area.

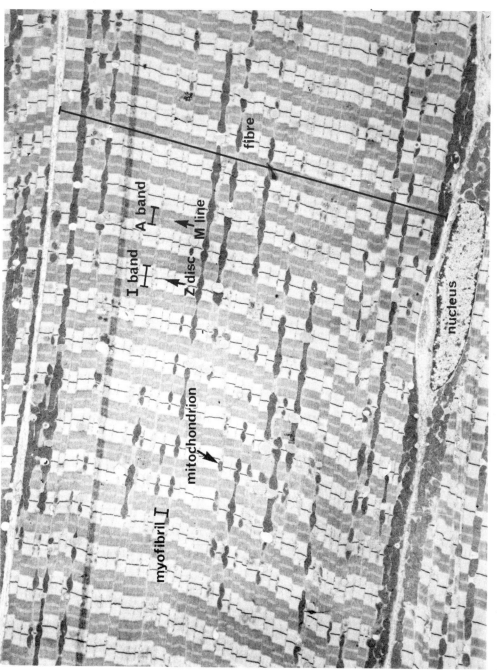

Plate 21.1 Low power electron micrograph (longitudinal section) of rabbit masseter muscle showing parts of three adjacent fibres. The cross striation of the fibres is very apparent. The magnification is comparable to that given by the light microscope but much more detail is visible. Note the large number of myofibrils in the fibre and the narrow spaces between them. This is a fast-twitch red muscle and there are many mitochondria in these spaces. An A and I band, a Z disc and M line are marked. (Because the muscle has been stretched the H zones are nearly as wide as the A bands.) (Courtesy of S. Page.)

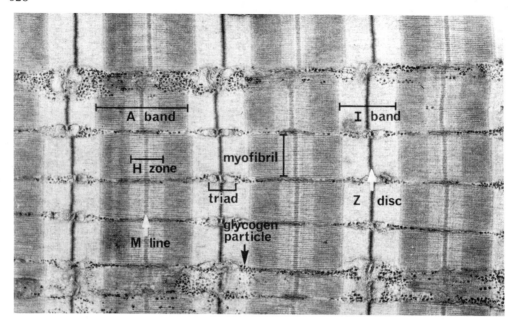

Plate 21.2 Medium power electron micrograph (longitudinal section) of frog sartorius muscle showing several myofibrils. The filamentous structure of the myofibrils may just be discerned. An A and I band, and H zone, a Z disc and M line are marked. Note the fine stripes in both the A and I bands. In the spaces between the myofibrils there are numerous glycogen particles and the membranes of the sarcoplasmic reticulum and T-tubules. A triad is marked. (Courtesy of R. Craig.)

substantial amount of Mg^{2+} (~10 mM). The anions (apart from protein) are predominantly organic phosphates, namely creatine phosphate (~24 mM) and ATP (~4 mM). Smaller amounts of Cl^-, HCO_3^- and inorganic phosphate are present. We could therefore roughly describe the ionic medium inside the muscle fibre as the potassium and magnesium salts of ATP and creatine phosphate.

We can now consider the other contents of the muscle cell. Around each myofibril there is a network of membrane-bounded channels, the sarcoplasmic reticulum (Plate 21.5) whose function we shall see later. The nuclei are squeezed to the edge of the cell and take up only a very small fraction of its volume. Ribosomes are present only in small numbers since adult muscle cells are not very active in protein synthesis. Mitochondria are present in numbers which vary between fibres.

Fibres which have many mitochondria will contain much cytochrome and often much myoglobin. Such fibres are therefore red in colour, in contrast to white fibres which have few mitochondria and little myoglobin. (The function of myoglobin appears to be to aid transfer of O_2 from the cell surface to the mitochondria.) Thus red fibres can oxidise muscle glycogen or fat to CO_2 but white fibres have a predominantly glycolytic activity and break glycogen down to lactic acid. Three different types of muscle fibre may actually be distinguished in mammals by their colour and their mechanical behaviour. Most muscles contain all three types but frequently one type predominates. Fast-twitch white fibres contract and relax quickly but cannot maintain tension for long. They are used for short bursts of activity (e.g. in the muscles of the back used for jumping). Fast-twitch red fibres also contract and relax quickly but they are not so easily fatigued. They are used for repetitive contractions (e.g. in the jaw muscles of a rabbit). Thirdly, the slow-twitch intermediate fibres contract and relax slowly but can maintain tension for long periods.

(They are called intermediate because they are less strongly red in colour than the fast red fibres.) They seem to be used for sustained contractions (e.g. for posture).

3.2 Simplified systems for studying the contractile apparatus

The presence of the cell membrane makes it experimentally difficult to alter the ionic composition of the sarcoplasm. Consequently experiments on intact muscle fibres do not straightforwardly provide information on what substances are necessary for contraction and relaxation. Three different preparations, which are still capable of contraction, have been used to get over the permeability problem.

Firstly, the cell membrane may be physically removed under oil by peeling it away from the fibre, rather like a stocking from a leg. Nothing else is altered in these skinned fibres.

A second type of preparation involves damaging the structure of all the membranes by osmotic shock in a medium of low ionic strength. Conventionally this is done by soaking a small bundle of fibres in 50 per cent aqueous glycerol. It may then conveniently be stored in the deep freeze. As a result of this treatment soluble proteins and metabolites diffuse out. The preparation, called glycerinated muscle by its originator A. Szent-Györgyi, is in the rigor state (because of the absence of ATP). It is no longer excitable by electrical stimulation, but if ATP is added under the appropriate ionic conditions it will contract even if it has been stored for many months. Since the tension it can produce is similar to that of intact muscle, the contractile apparatus itself is clearly undamaged. The breakdown of the membranes is sufficiently complete that not only can the effect of ions be studied but large protein molecules can diffuse into the contractile apparatus. This property has enabled antibody techniques to be used to locate muscle proteins in the myofibril.

Although the membrane has been removed from these preparations, a diffusion problem remains. For example, if ATP is used up by the contractile apparatus, the concentration of ATP in the centre of the fibre will be far lower than on the outside. Hence glycerinated muscles, while useful for studying the conditions required for development of tension, are not really suitable for measurements of ATP consumption.

This problem can be surmounted by using a third type of preparation. Myofibrils may be isolated from a homogenate of muscle by differential centrifugation (Plate 21.3). Because myofibrils constitute so great a proportion of the fibre, there is little difficulty in obtaining a moderately pure preparation. The main problem is to prevent them from shortening during isolation but there are ways of doing this. Naturally myofibrils are broken during the homogenisation and are far shorter than they are *in vivo*. Isolated myofibrils are useful, not only for studying the ionic conditions required for relaxation and contraction, but also for measuring ATPase activities. As we shall see in a moment, observation of their contraction in the light microscope also provides crucial information on how they contract.

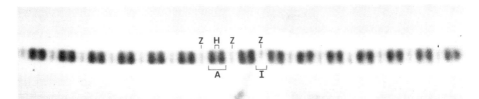

Plate 21.3 An isolated myofibril in the light (phase contrast) microscope. An A and I band, an H zone and several Z discs are marked. (Courtesy of R. Craig.)

3.3 The striation of the myofibril

We may now enquire into the structure of the myofibril. In the light (phase contrast) microscope, isolated myofibrils show transverse bands, alternately dense and less dense (Plate 21.3). It is because the myofibrils have this extremely regular structure and are arranged in register across the width of the fibre that the fibre appears cross-striated (Plates 21.1, 21.2; Fig. 21.2). The dense bands, which contain the higher concentration of protein, are also strongly birefringent and hence are called the A bands (A for anisotropic). The less dense bands are only

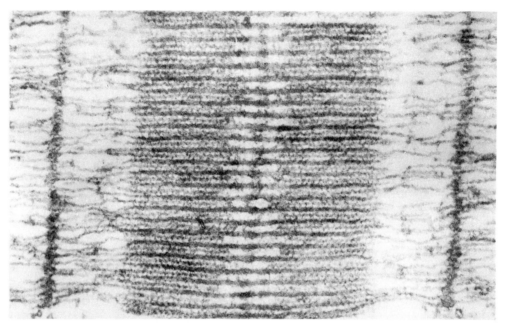

Plate 21.4 High power electron micrograph (thin longitudinal section) of rabbit psoas muscle showing the interdigitation of thick and thin filaments and the cross-bridges connecting them. To reveal the cross-bridges in this way requires very thin sections (~ 20 nm). (Examination of an edge of the myofibril shows that the cross-bridges belong to the thick filaments rather than to the thin filaments.) (Courtesy of H. E. Huxley and The Rockefeller University Press.)

weakly birefringent and are termed the I (for isotropic) bands. Each I band is bisected by a structure called the Z disc. The part of the myofibril between two successive Z discs is called a sarcomere; at rest length this is 2.4 μm or so long depending on the muscle. The sarcomere can be regarded as the mechanical unit of the myofibril since the length changes occurring when a muscle fibre shortens or is stretched are the sum of the length changes in the sarcomeres in series along the length of the fibre. All the sarcomeres in a muscle fibre change in length together; the regularity of sarcomere length is so well maintained that a muscle fibre (and indeed a whole muscle) behaves like a diffraction grating. Thus if a narrow beam of monochromatic light (e.g. from a laser) is shone on the muscle, the light is diffracted so that strong first, second and third order diffraction beams are produced. These change in position as the muscle is stretched or allowed to shorten, thus allowing us to measure accurately the sarcomere length in a muscle without destroying it.

3.4 The filament lattice of the myofibril

The origin of the striation of the myofibrils may be seen in electron micrographs (Plates 21.2 and 21.4). The myofibrils are composed of two interdigitating sets of filaments, the thick filaments about 15 nm in diameter and 1·6 μm in length and the thin filaments about 8 nm in diameter and 1·0 μm in length (Fig. 21.2). The A band consists of thick filaments, together with those parts of the thin filaments which overlap them. The I band contains those parts of the thin filaments not overlapping the thick ones. (A structure composed of interdigitating filaments was in fact suggested by X-ray diffraction before it was confirmed by electron microscopy.)

The thick filaments in an A band are packed side by side in a hexagonal lattice (Fig. 21.2). While the regularity of this lattice is not perfect, it is sufficently good for the X-ray diffraction pattern of living muscle to show well developed equatorial reflections. In transverse sections through the H zone (the region of the A band containing no overlapping thin filaments) only the thick filaments are present. At the centre of the A band the thick filaments are connected by a structure called the M line. In the overlap zone of the A band both thick and thin filaments are present with one thin filament placed midway between three thick ones. The geometry is such that there are four times as many thin filaments as thick filaments in a sarcomere. In the I band, the thin filaments are not so regularly arranged. This is because the thin filament lattice changes from hexagonal (in the overlap zone) to a square arrangement where the filaments are attached to the Z disc.

3.5 The sliding filament hypothesis

Undoubtedly the most important advance in our understanding of muscle contraction came in 1954 with two complementary studies of how the band pattern of the sarcomere changes with its length. A. F. Huxley and Niedergerke showed that when an intact fibre was stretched or stimulated to contract, the I bands changed in length while the A bands remained constant. Using myofibrils isolated from glycerinated muscle, Hanson and H. E. Huxley gained the advantage that the band pattern was clearer and in particular that H zones could be resolved (Plate 21.3). In rigor myofibrils, produced from muscle at varying degrees of stretch, the A bands were constant in length and changes in the length of the H zone equalled those of the I band. If low concentrations of ATP were added to the myofibrils they contracted slowly and the closing up of the H zone as the I band decreased in length was clearly visible.

On the basis of these experiments, the two groups of workers independently proposed the sliding filament hypothesis: contraction or extension of muscle took place by the sliding of thick filaments past thin filaments with no length changes in the filaments themselves (Fig. 21.3).

The hypothesis has been tested very thoroughly. Strong evidence supporting the hypothesis is the constancy of length of the thick and thin filaments in muscles fixed for electron microscopy at a variety of sarcomere lengths. Moreover the X-ray diffraction patterns of resting and contracting muscle indicate that the periodicities with which the components of the two types of filaments are packed remain essentially unchanged either during stretch or during contraction.

When the hypothesis was first proposed it surprised those who had been expecting large conformational changes in the protein components of muscle. As we shall see in section 10, constancy of length does not preclude conformational changes occurring in the filaments.

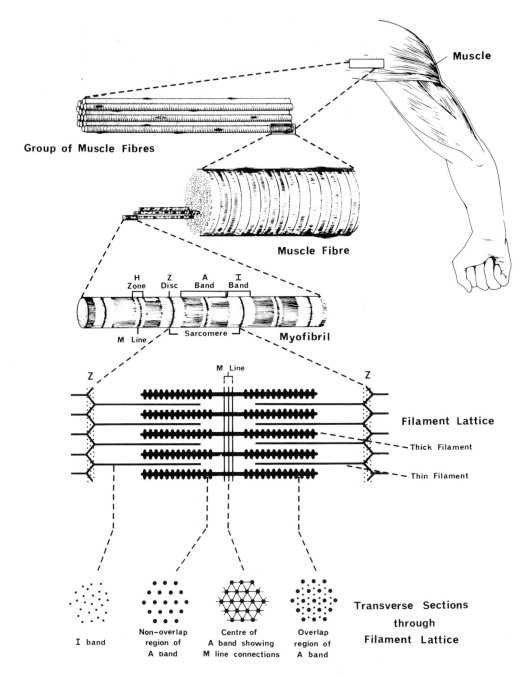

Muscle

Group of Muscle Fibres

Muscle Fibre

H
Zone

Z
Disc

A
Band

I
Band

M Line

Sarcomere

Myofibril

M Line

Z

Z

Filament Lattice

— Thick Filament

— Thin Filament

I band

Non-overlap
region of
A band

Centre of
A band showing
M line connections

Overlap
region of
A band

Transverse Sections
through
Filament Lattice

Fig. 21.2 Schematic diagram of the organisation of skeletal muscle from the gross to the filament level. The diagrams at the bottom represent transverse sections of the filament lattice at the levels indicated. (Courtesy of D. W. Fawcett and W. B. Saunders Co.)

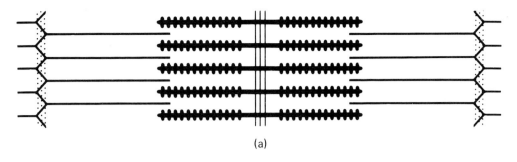

(a)

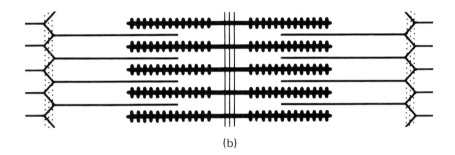

(b)

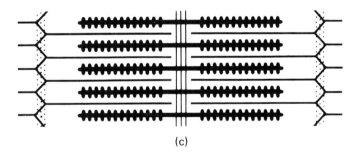

(c)

Fig. 21.3 Schematic representation of the sliding filament hypothesis. The degree of interdigitation of the filaments alters as the muscles are stretched or contract. (a) A highly stretched sarcomere (this situation probably does not occur *in vivo* but muscles stretched to this extent are not damaged and can still contract). (b) A moderately stretched sarcomere (this situation represents roughly the most stretched case occurring *in vivo*). (c) A shortened sarcomere. In this case all the cross-bridges are overlapped by thin filaments. Further shortening (which is still physiological) is possible by thin filaments from one half overlapping with thin filaments from the other.

3.6 The sites of tension production

If a myofibril made of discontinuous filaments is to produce tension, a longitudinal force must be developed between the thick and the thin filaments. *A priori*, the force acting between the filaments could be a long-range force (electrostatic or van der Waals) but there is now strong evidence that structural links called cross-bridges are formed between the filaments. These cross-bridges, although poorly preserved by electron microscope preparative methods, may be discerned in longitudinal sections at high magnification (Plate 21.4). The thick filaments have many projections along their length which make contact with the thin filament in the overlap zone. (None is however present in a region near the centre of the thick filament.) Note that the projections are called cross-bridges whether or not they are attached to the thin filament. Because of the constancy of filament length during contraction, it was natural to assume that the cross-bridges were like oars in a boat which would execute cyclic motions and 'row' the filaments past one another. A. F. Huxley was responsible for showing that a cross-bridge mechanism of this type could account for many of the mechanical and thermal properties of muscle. There is now abundant evidence that the cross-bridges are not artifacts and that the cross-bridges can indeed make contact with the thin filaments.

The following evidence implicates the cross-bridges as sites of tension production. The isometric tension exerted by a muscle fibre varies with the sarcomere length. If the muscle is stretched so that there is no overlap of filaments, no tension is produced. At shorter sarcomere lengths the tension developed is proportional to the amount of overlap between the filaments up to the point where the thin filaments have overlapped all the cross-bridges (compare Fig. 21.3). A further small shortening causes no increase in tension.

This and other evidence suggests that the cross-bridges are the sites of tension production but it should be appreciated that no-one has yet shown directly that *movements* of attached cross-bridges generate tension. Some workers, while not denying the existence of cross-bridges, are pursuing the possibility that the forces are long range. In this chapter we shall conform to the general view that the force is generated by cross-bridge movements and we shall return to the problem in section 10.

4 Energetics of contraction

4.1 The fuel used for contraction

When a muscle contracts work is done and heat given out. What fuel provides the energy for this? What we regard as the fuel will depend on the system we consider. From the standpoint of the whole animal, the fuel is of course the food consumed. From the standpoint of the whole muscle, the fuel is the glucose (or fatty acids) which are taken up from the blood. But we are here concerned with the fuel used *directly* by the contractile apparatus. It has been known for a long time that muscles, in which both glycolysis and respiration have been inhibited, can produce a large number of apparently normal contractions on stimulation. So neither of these metabolic processes is directly linked to contraction.

Under suitable ionic conditions, both glycerinated muscle and isolated myofibrils contract when ATP is added, hydrolysing it to ADP and P_i. This suggests rather strongly that ATP is the immediate fuel for contraction but it is desirable to see if this is also the case in living muscle. The difficulty here is the occurrence of recovery processes which cause the resynthesis of ATP from ADP and P_i. These must be prevented if we are to observe the primary events. A good way of suddenly stopping metabolic processes is to plunge the muscle into a bath of a fluorohydrocarbon refrigerant cooled to the temperature of liquid nitrogen. Liquid nitrogen itself is unsatisfactory because the thin film of gas formed between the muscle and liquid acts as an insulating layer. To ensure very rapid freezing, the muscle should be thin and it is desirable to have it already near $0°C$. An even faster method of freezing is to squeeze the muscle rapidly between two metal hammers cooled to the temperature of liquid nitrogen. It is thereby possible to cool the muscle to below $-10°C$ within $0·1$ s. Having frozen the muscle in this way, it may be fragmented into small pieces. These are then allowed to thaw into ice-cold perchloric acid so that the enzymes are denatured as thawing occurs. After removing insoluble proteins, the solution is neutralised and the extracted low molecular weight components (such as creatine phosphate, creatine, ATP, ADP, AMP, P_i), can be estimated, often by enzymic assays.

Now this procedure is satisfactory for measuring the level of these components in any particular muscle at some instant of time, but we want to measure the *change* in the levels consequent on contraction. Unfortunately there can be no 'before' and 'after' experiments on the same muscle for obvious reasons. The only course is to compare muscles that are as similar as possible (e.g. muscles from the left and right sides of the same animal). One muscle of the pair can be allowed to rest and the other made to contract. Both are then frozen. There are of course natural variations even in such paired muscles and it may be necessary to examine a large number of pairs to obtain statistically useful results. It is particularly difficult to measure changes in creatine phosphate levels because the starting level is so high. It is also obviously easier to measure the larger changes arising from a series of twitches or a long tetanus than the small change arising from a single twitch or part of a twitch.

If a muscle is stimulated for a long time, the conversion of muscle glycogen to lactic acid can be demonstrated (as shown in the classical experiments of Fletcher and Hopkins and of Meyerhof). We are here concerned, however, with the changes occurring at short time intervals. For about the first 15 s (at $0°C$) from the start of a contraction, no changes occur in the level of glycolytic metabolites. Hence changes in ATP and creatine phosphate can be followed during this period without having to take these slower recovery processes into account. If a longer contraction or series of contractions is used, glycolysis may be inhibited by treating the muscles beforehand with iodoacetate, and oxidative processes may be prevented by working in nitrogen.

All experiments of this nature show that no change in the ATP level occurs as a result of contraction. Instead, the level of creatine phosphate falls and there is a corresponding rise in creatine and phosphate. It is now realised that this is because the sarcoplasm contains large amounts of the enzyme creatine kinase which causes the rephosphorylation of ADP to ATP as soon as it is formed:

$$ATP \rightarrow ADP + P_i$$

$$ADP + CP \rightleftharpoons ATP + C$$

(Creatine phosphate and creatine kinase are of course often used in the laboratory when we want to maintain a constant level of ATP in a system in which ATP is consumed; this is presumably their function *in vivo*.)

To demonstrate ATP hydrolysis *in vivo*, it is necessary to inhibit creatine kinase. Fluorodinitrobenzene (the Sanger reagent) has been shown to inhibit purified creatine kinase and fortunately does not upset the ability of muscles to be stimulated and to contract. In fluorodinitrobenzene-poisoned muscles a drop in ATP and concomitant rises in ADP, AMP and IMP levels can be demonstrated after a contraction with no change in creatine phosphate. AMP and IMP are formed because adenylate kinase and adenylate deaminase remain active:

$$2ADP \rightleftharpoons AMP + ATP$$

$$AMP \rightarrow IMP + NH_3$$

The poisoned muscle is unfortunately therefore not a simple system but the results do suggest that ATP is the fuel used by the poisoned muscles, and creatine phosphate that used by the unpoisoned muscle whose creatine kinase enzyme is still active. There are, however, more critical standards to be satisfied. The first law of thermodynamics demands that the heat plus work produced by the muscle should equal the total change in enthalpy of all the chemical reactions which have occurred in the muscle:

$$\text{Heat} + \text{work} = -\sum n_i \Delta H_i \tag{1}$$

where n_i is the number of moles reacted and ΔH_i is the molar enthalpy change of the ith reaction.

If a substance is to be regarded as the fuel for contraction, the amount of the substance used should *quantitatively* account for the heat and work produced, under circumstances where there is no overall change in the system under study, other than the consumption of fuel and the production of heat and work. This means that we should examine the changes after a complete contraction–relaxation sequence.

The systems we can study are either unpoisoned muscles or fluorodinitrobenzene-poisoned muscle. Both systems include the activating mechanism as well as the contractile apparatus proper.

As a result of the pioneer work of A. V. Hill, the heat output of a muscle can be accurately measured with a sensitive thermopile, and the measurement of external work (load x distance lifted) is straightforward. However the values used for the molar enthalpy of hydrolysis of ATP or of creatine phosphate are not simply obtained. We have to remember that H^+ ions are liberated when ATP is hydrolysed, and are taken up when creatine phosphate is hydrolysed. We must therefore include a term due to the enthalpy of neutralisation of these H^+ ions by the cell buffers (assumed to be mainly histidine residues). We also have to take into account the enthalpies of binding of Mg^{2+} to creatine phosphate, ATP and ADP. The best current estimates of the molar enthalpy change when creatine phosphate and ATP are hydrolysed under conditions similar to those in the muscle cell are −33 and −42 kJ/mol respectively.

If unpoisoned muscles are frozen after a tetanus or series of twitches, the value of the (heat plus work) term is slightly greater than the calculated enthalpy change for creatine phosphate hydrolysis. It is possible that this small discrepancy arises because creatine phosphate in the muscle cell is partially bound to protein and yet another term (the enthalpy of binding of creatine phosphate) has to be included in the expression for the total enthalpy. Although it is thus clear that creatine phosphate is the major fuel in the unpoisoned muscles, we lack proof that it is the only one.

The definitive experiments on fluorodinitrobenzene-poisoned muscles have yet to be done. These are more difficult because, especially in long tetani, the ATP hydrolysed forms a variety of products (ADP, AMP, IMP). There are however indications that the amount of ATP hydrolysed approximately accounts for the energy liberated at the end of a tetanus. This is consistent with the view that the fuel used by the contractile apparatus (and activating mechanism) is indeed ATP.

Of course equation (1) must apply at *all* stages of a contraction but in general the $\Sigma n_i \Delta H_i$ term will include *changes in the system under study* as well as those due to consumption of fuel. Thus if unpoisoned muscles are frozen in the very early stages of an isometric tetanus, the heat output greatly exceeds the calculated enthalpy change for creatine phosphate hydrolysis. This arises because the rate of creatine phosphate hydrolysis is constant but there is an early burst of heat production before the tension has built up. Evidently some exothermic reaction is occurring during the early stages of contraction which is very likely reversed during or immediately after relaxation. This exothermic reaction probably involves some part of the activating mechanism but it is not yet clear which.

A further effect is observed in isotonic contractions. Heat is given out during the shortening which cannot be accounted for by the amount of free ATP (or creatine phosphate) hydrolysed. Again this exothermic reaction must be reversed at the end of a complete contraction–relaxation sequence (possibly driven by ATP hydrolysis). This exothermic reaction presumably involves a change in the state of the contractile apparatus, perhaps of the population of the cross-bridges in each step of their cycle. Heat measurements are clearly useful for revealing the existence of reactions which would otherwise go undetected.

To summarise, taken together, the study of intact muscles and the study of the simplified muscle systems shows that ATP is the fuel used directly by the contractile apparatus. *In vivo* the creatine kinase enzyme maintains the ATP level. The breakdown of muscle glycogen by glycolysis and respiration, which restore the level of creatine phosphate (via ATP), are very much slower recovery processes.

4.2 The consumption of ATP in contracting muscle

An important finding is that the amount of ATP hydrolysed in a contraction is not constant, but depends on the duration of the contraction and on how much work is done. So muscle is not like a stretched rubber band which releases a fixed amount of energy on contraction. On the contrary, muscle, like a motor car, uses up more fuel when the demands made on it are greater (the Fenn effect). However it should be noted that ATP is hydrolysed to an appreciable extent even when the muscle does no external work (i.e. when it is not loaded or when it is not allowed to shorten). This explains why we feel tired when we hold a heavy suitcase for a long time. Evidently there is no tight coupling between the hydrolysis of ATP and the performance of work.

The consumption of large amounts of ATP (up to 3 μmol ATP/g wet weight muscle) can be demonstrated when a frog sartorius muscle is made to perform a large amount of work, e.g. if the muscle is stretched and then allowed to contract slowly over a large distance with a high load. We shall see later that the sites of ATP hydrolysis are on the cross-bridges and that there are two ATPase sites per cross-bridge. The concentration of cross-bridges is about 0·16 μmol/g wet weight muscle. Thus during a contraction a cross-bridge may hydrolyse many ATP molecules and the only reasonable explanation is that the cross-bridges act many times; that is they act *cyclically*.

It is important to know the maximum rate at which muscle can hydrolyse ATP. This occurs during an isovelocity contraction at moderate speed. The value obtained for frog sartorius muscle is 1·4 μmol ATP/g wet weight muscle/s at 0°C. The rate at 20°C would be about four times higher. Neglecting the small amount of ATP consumed by the activating mechanism, we can calculate that at 20°C the ATP consumed by the cross-bridges is about 35 mol ATP/s/mol cross-bridges.

4.3 The efficiency of muscle contraction

We obviously want to know how efficient muscles are at converting the free energy of hydrolysis of ATP into work. The thermodynamic efficiency is given by work/$(-n\Delta F)$ where n is the number of moles of ATP hydrolysed and ΔF is the molar free energy of hydrolysis of ATP in the muscle cell. Unfortunately we do not know the precise concentrations of ATP, ADP and P_i in the neighbourhood of the contractile apparatus, so although the *standard* molar free energy of hydrolysis of ATP is known, the molar free energy change in this neighbourhood can only be estimated. Using the most reasonable figure (-50 kJ/mol) for the latter, the highest thermodynamic efficiency (which occurs at moderate velocities of shortening) has been

638

estimated to be about 60 per cent. If we make allowances for the ATP consumed by the activating mechanism, the estimated efficiency is higher still. The efficiency in whole muscle *in vivo* is much lower. This has been estimated to be about 20 per cent based on the extra O_2 breathed during bicycling exercise. The low net efficiency is presumably therefore mainly due to the inefficiency of glycolysis rather than to the contractile apparatus itself.

5 The control of muscular contraction

5.1 Excitation–contraction coupling

A muscle which contracted all the time would be as useless as one which never contracted. What causes the passage of an action potential along the membrane of a muscle fibre to elicit a contractile response, so causing something like a 2000-fold increase in the rate of consumption of ATP? And why does the fibre stop contracting and relax again?

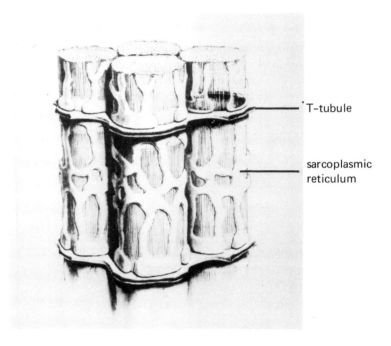

Plate 21.5 Diagram of the arrangement of the sarcoplasmic reticulum and T-tubules in the spaces between the myofibrils in frog muscle. In this muscle T-tubules occur at the level of every Z-disc and make contact with the sarcoplasmic reticulum at structures called triads. (Courtesy of K. R. Porter and Lea & Febiger.)

It was early noted that a muscle fibre contracts almost simultaneously across its width. The outside does not contract first, followed much later by the interior. If a substance were released just inside the cell membrane in response to the action potential, it would take at least 0·5 s for diffusion to cause the concentration of the centre of a 100 μm fibre to reach half the peripheral concentration. It was thus argued that the inward spread of excitation must be electrical. This was substantiated by the discovery that the cell membrane was not simply a cylindrical sleeve but formed numerous tubular pockets at regular intervals. These pockets form a branched network in the spaces between myofibrils and thus penetrate to the centre of the fibre. In the frog, these transverse (T) tubules, as they are called, occur at the level of every Z disc (Plate 21.5). Since they are continuous with the cell membrane and have similar electrical properties, it seems reasonable to assume that the action potential is transmitted along the tubules to the interior of the fibre. (It is not yet quite certain that the T-tubules transmit an action potential rather than merely a passive inward spread of current.)

But the stimulus felt by the contractile apparatus cannot be electrical because glycerinated muscle (in which no action potentials can be propagated) can still contract perfectly normally. This suggests that the final stimulus is a chemical one.

5.2 The chemical agent responsible for triggering contraction

When ATP is added to glycerinated muscle or to myofibrils (in say 0.1 M KCl, 5 mM $MgCl_2$, pH 7.0), and no special precautions are taken to purify the reagents, contraction with concomitant splitting of ATP occurs. However, if ATP is added to a homogenate of fresh muscle in a similar medium, the myofibrils present in the homogenate relax. Evidently there is some factor in the homogenate which induces this relaxation. The factor responsible is present in the low speed supernatant after centrifuging the homogenate to pellet the myofibrils, mitochondria and nuclei. If ATP is added to myofibrils in the presence of this supernatant, relaxation again occurs. The supernatant was therefore at one time believed to contain a relaxing factor which played a *direct* role in relaxation.

However, it was noted that quite small amounts of Ca^{2+} ions could reverse the effect of the relaxing factor preparation. Moreover, chelating agents which could bind Ca^{2+} ions tightly would replace the natural relaxing factor and cause relaxation. For example, a chelating agent called ethylene glycol bis-(β amino-ethylether) NN'-tetraacetate (abbreviated EGTA) is very effective at causing relaxation and this binds Ca^{2+} tightly but Mg^{2+} very weakly. This suggested rather strongly that the contractile apparatus was sensitive to traces of Ca^{2+} ion (10^{-5} M). Proof that Ca^{2+} ions are *required* for contraction came with the discovery that if scrupulous care was taken to remove traces of Ca^{2+} impurities from the reagents, the addition of ATP to glycerinated muscle or to myofibrils caused relaxation and not contraction! The conditions required for relaxation and a low rate of ATP splitting are simply the presence of ATP and Mg^{2+} ions and the absence of Ca^{2+} ions; neither the chelating agents nor the relaxing factor are necessary provided Ca^{2+} ions are absent.

The levels of Ca^{2+} required for contraction are best studied using a metal buffer system such as a mixture of CaEGTA and free EGTA. This stabilises the Ca^{2+} level in much the same way as a mixture of acetic acid and acetate ions stabilises the H^+ concentration. Using this system we can show that the development of full tension in glycerinated muscle (or a high ATPase rate in myofibrils) requires about 10^{-5} M Ca^{2+}, while relaxation (and a low ATPase rate in myofibrils) is produced at a Ca^{2+} level of less than about 10^{-7} M Ca^{2+} (Fig. 21.4).

We can now discuss the source of the Ca^{2+} *in vivo* and its fate during relaxation. Local application of a solution containing Ca^{2+} to skinned fibres gives a localised contraction followed rapidly by a relaxation. This suggests that the fibre contains some mechanism for removing Ca^{2+} ions. It is obviously important to examine the nature of the relaxing factor preparation mentioned earlier to see if this could be part of the mechanism. The active principle in the preparation turned out to be particulate and could be spun down at high centrifugal speeds. The particles contain an ion pump which can concentrate Ca^{2+} to an astonishing extent, so that in the presence of Mg^{2+} and ATP the external concentration of Ca^{2+} is reduced to 10^{-9} M. This explains their ability to cause relaxation. The energy required for this active transport comes from the hydrolysis of ATP; two Ca^{2+} ions can be pumped per ATP molecule hydrolysed. The number of the particles present in a muscle homogenate is quite large and examination by electron microscopy shows that they are vesicles bounded by membrane material. The only reasonable source is the sarcoplasmic reticulum. This was verified by a histochemical demonstration that the intact sarcoplasmic reticulum could indeed concentrate Ca^{2+}. Thus the sarcoplasmic reticulum, which surrounds every myofibril, is a structure designed to take up Ca^{2+} from the enclosed myofibril.

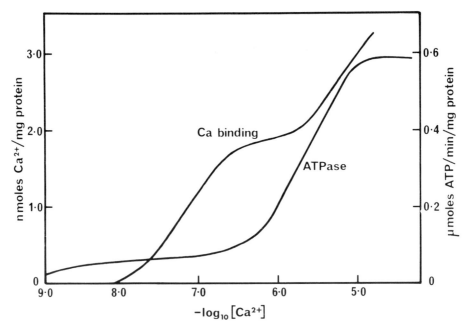

Fig. 21.4 The effect of Ca^{2+} concentration on (a) the ATPase of myofibrils (b) the amount of Ca^{2+} bound. Note the steep dependence of the ATPase on the Ca^{2+} concentration. (Courtesy of A. Weber and Prentice-Hall.)

We can see therefore why muscle can relax, but we have not explained how the action potential can cause the release of Ca^{2+}. The transverse tubules which conduct the action potential inwards from the cell surface form junctions with the sarcoplasmic reticulum at regular intervals. In the frog the tubules form a sandwich (called a triad) between two sacs of the sarcoplasmic reticulum at the level of the Z disc (Plate 21.5). We can reasonably infer that the arrival of an action potential at this junction causes the depolarisation of the membrane of the sarcoplasmic reticulum and the release of Ca^{2+}. This Ca^{2+} is bound by the myofibrils causing them to contract. If no further action potential arrives at the triad, the sarcoplasmic reticulum abstracts Ca^{2+} from the sarcoplasm causing the myofibrils to release their bound Ca^{2+} and relax.

A direct demonstration of the release of Ca^{2+} resulting from the passage of an action potential has been obtained in certain large invertebrate striated fibres, injected with aequorin. Aequorin is a protein from a jellyfish which forms a transitory luminescent complex with Ca^{2+}. When the injected fibres are stimulated they emit light! The time course of the light production shows that the Ca^{2+} is released prior to tension production. So far technical difficulties have prevented a similar demonstration in vertebrate muscle.

An interesting question is what triggers the greatly increased rate of glycolysis which occurs a few seconds after the start of a contraction. The obvious candidate is the Ca^{2+} ion itself and indeed it now appears that phosphorylase *b* kinase (the enzyme that catalyses the phosphorylation of phosphorylase *b* by ATP) is situated on the surface of the sarcoplasmic reticulum and is activated by low concentrations of Ca^{2+} ion. This would account for the increase of phosphorylase *a* (the active form of phosphorylase) observed during contraction. Thus Ca^{2+} seems to be the common trigger for activating both the contractile apparatus itself and the glycolytic pathway which generates the ATP required for contraction.

We may now summarise the general features of muscular action. Ca^{2+} acts as the trigger for contraction, the energy for contraction comes from the hydrolysis of ATP and contraction occurs by the sliding of filaments past each other. We are now in a position to discuss the molecular structure of the contractile machinery and the mechanism of contraction that drives the filaments.

6 How the molecular structure of the myofibril is investigated

6.1 The components of the myofibril

It has already been mentioned that the contractile organelles of the muscle fibre, the myofibrils, can be isolated. A myofibril is made up of protein (concentration about 180 mg/ml). Traces of RNA and lipid are present but they very likely arise from contamination by

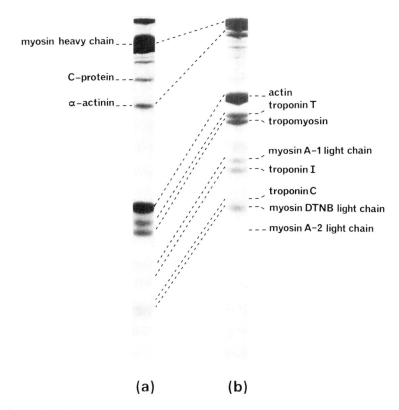

(a)   (b)

Plate 21.6 Separation of the components of the myofibril by SDS polyacrylamide gel electrophoresis, (a) on 5 per cent (b) on 8 per cent polyacrylamide gels. The separation of the high molecular weight components is best achieved on gels of lower concentration, those of low molecular weight on gels of higher concentration.

Table 21.1 The components of the myofibril

Protein	Location	Approx. % of total myofibrillar protein	MW	Number of polypeptide chains and chain wt.
Myosin	Thick filaments	55	460 000	2 × 190 000 (heavy chain) 1.2 × 21 000 (A-1 light chain)‡ 2 × 18 000 (DTNB light chain) 0.8 × 17 000 (A-2 light chain)‡
C-protein	Thick filaments (located at nine sites spaced about 43 nm apart in each half of the A band)	2	140 000	1 × 140 000
M-line protein	M line†	?	88 000	2 × 44 000
Actin	Thin filaments	23	41 700	1 × 41 700
Tropomyosin	Thin filaments	6	70 000	2 × 35 000
Troponin	Thin filaments (located at sites spaced 38·5 nm apart throughout the thin filament)	6	80 000	1 × 37 000 (TN-T) 1 × 23 000 (TN-I) 1 × 18 000 (TN-C)
α-Actinin	Z disc†	1	180 000	2 × 90 000

† Probably other proteins are also present in the M line and Z disc.
‡ See text for explanation of non-integral quantities of A-1 and A-2 light chains.

other organelles. Obviously we want to know what protein components the contractile machinery contains. One of the most powerful analytical techniques the muscle biochemist now has at his disposal is the technique of gel electrophoresis in sodium docecyl sulphate (SDS). Muscle proteins tend to associate very strongly with themselves and with each other and the resulting complexes are usually much too large to run on gels. But myofibrils dissolve almost completely in SDS and the protein components dissociate into the individual polypeptide chains. When the mixture is subjected to electrophoresis, the polypeptide chains migrate according to their molecular weights. Patterns obtained at two gel concentrations are shown in Plate 21.6. They show that the number of polypeptide species present in myofibrils in appreciable amounts is at least 13 (and is probably a little greater because two different polypeptide species may overlap). It is encouraging that all but two of the bands (and they are fairly minor) can be attributed to well-characterised proteins. A list of these is given in Table 21.1. At the moment the quantities of the various components present in the myofibril are known only approximately, but better methods of estimating the protein present in electrophoretic bands should improve this situation.

6.2 Location of the components

In some cases, information about the location of a protein component may be obtained by selectively extracting it from isolated myofibrils and noting the accompanying change in the band pattern in the light microscope. For example, myosin can be extracted from myofibrils with high ionic strength solvents and the A band is then seen to disappear.

Alternatively, one can elicit antibodies to a purified myofibrillar protein, couple them to a fluorescent dye and label the myofibrils with the fluorescent antibody. The location of the component in the labelled myofibrils can then be observed by fluorescence microscopy. In some favourable cases the coupling to a fluorescent dye is not necessary and the extra electron density produced by attaching the unmodified antibody to the myofibril is detectable in the electron microscope.

A third powerful method is to synthesise artificial filaments from the purified protein components and to compare them with the natural filaments. We shall now discuss how the latter may be prepared.

6.3 Fragmentation of the myofibril into segments and filaments

Determination of the structure of the myofibril is facilitated by fragmentation into its components. This can be at two levels. Homogenisation of relaxed fibrils causes the two types of filament arrays to separate from one another, so producing a mixture of assemblies of thin filaments attached to either side of a Z disc (I segments), and assemblies of thick filaments held together by the M line (A segments). It is not yet possible to arrest the dissection at this stage and a large number of individual filaments are also found. Attempts to separate the thin filaments from thick filaments have been only partly successful; were this to be achieved, the location of the myofibrillar components would be unequivocal. Nevertheless these homogenates have proved to be extremely useful in electron microscope studies of the structure of the filaments.

6.4 Techniques for examining the molecular structure of filaments

The molecular structure of the filaments has been investigated both by X-ray diffraction and by electron microscopy. Neither can be singled out as the better technique. Each has distinct advantages over the other and the two are best used to complement one another. On the one hand electron microscopy does give an image of the object. We have seen that sections of muscle have given useful information about the arrangement of the filaments and filament length, but finer details of structure are best examined by the very simple technique of negative staining, in which the outline of the shape of the object is delineated by an electron dense stain such as uranyl acetate. A great advantage of electron microscopy is that individual objects (e.g. a single filament or molecule) may be picked out. However the object examined in the electron microscope is a very dead object inevitably distorted by the treatment it has received. For example, no-one has yet found a method of preserving the details of the cross-bridges of the thick filaments. On the other hand, X-ray diffraction allows us to study living muscle with little damage; the muscle is still able to contract at the end of the X-ray exposure. It is even possible to study the structure of muscle in the contracting state by stimulating the muscle at intervals, each time opening the shutter over the X-ray source. In this way the X-ray exposure may be built up in short increments over a period of many hours. The spacings recorded may then be expected accurately to reflect the true spacings in the living muscle. A disadvantage of X-ray diffraction is that the pattern obtained is not an image of the object. Phase information in the diffraction pattern is lost, and it is often difficult to interpret the pattern and therefore to determine the structure of the object giving rise to the pattern. Increasing use is being made of optical diffraction patterns of the corresponding electron micrographs to assist the interpretation of the X-ray patterns. Another disadvantage of the X-ray method is that a very large number of molecules is required to give a diffraction pattern and unless these are well aligned, the information obtained is very limited. X-ray diffraction has been used mainly on whole muscle but it is possible to obtain highly ordered gels of some of the components of myofibrils and the diffraction of these should give even greater structural detail.

We are now in a position to discuss the individual proteins of the myofibril and the details of molecular structure of the filaments. The proteins are normally prepared from rabbit muscle and the properties we shall mention refer to the rabbit proteins.

7 Myosin

7.1 The main characteristics of myosin

Myosin comprises about half the protein of the myofibril and is the major protein of the thick filaments. It can be extracted from a muscle mince by solutions of high ionic strength. In an impure state it was first isolated by Kühne in 1859 but the first detailed study was by Edsall in 1930. It was not until the early 1940s that Straub and Szent-Györgyi noted that preparations of 'myosin' extracted for different times behaved differently on addition of ATP (see below). They concluded that these preparations were contaminated by greater or smaller amounts of a second protein which they termed actin. Actin could be obtained from an acetone-dried powder made from the residue after extracting myosin. Nowadays myosin can be obtained in a highly purified state and the old nomenclature (myosin A for moderately pure myosin obtained from a short extraction, and myosin B for very impure myosin obtained from a long extraction) is no longer used.

The discovery in 1939 by Engelhardt and Lyubimova that myosin had ATPase activity was a crucial event. Myosin had been recognised as a structural protein for many years; this discovery provided a link between structural studies and energetics.

The second important characteristic of myosin is that it binds strongly to F-actin. (F-actin is the polymerised fibrous form of actin, the main protein of the thin filament.) If a moderately viscous solution of myosin is added to one of F-actin at high ionic strength (to bring the myosin in solution), the viscosity of the mixture becomes very high indicating the formation of a complex (sometimes called actomyosin). But if ATP is now added, the viscosity of the mixture falls to the value expected for the two non-interacting components, i.e. the complex has dissociated. When the ATP is all hydrolysed the viscosity rises again. Here in this simple experiment we have clear evidence that links could be formed between the major proteins of the thick and the thin filaments, and that these links are sensitive to ATP.

The third main characteristic of myosin is its ability to associate with itself at salt concentrations similar to those in the cell. In high salt concentrations (e.g. 0·5 M KCl, pH 7·0) and low protein concentrations, myosin exists as molecules of molecular weight 460 000. At higher protein concentrations reversible dimerisation occurs. If the ionic strength of myosin is lowered to between 0·1 and 0·3, the solution becomes noticeably turbid and the analytical ultracentrifuge shows the presence of a rapidly sedimenting species. Clearly large structures have been formed. If the ionic strength is lowered still further (to 0·04) visible precipitation occurs; this property is made use of in the purification of myosin.

7.2 Shape of the myosin molecule and the location of its several functions

The shape of the myosin molecule can be visualised in the electron microscope when contrasted by rotary shadowing (which shows up more detail than unidirectional shadowing). The molecule is very long (160 nm) and thin, with two globular heads attached to one end of a long tail (Fig. 21.5). It is not known how flexibly the heads are joined to the tail and whether the two heads can make contact. The molecule thus combines features of a fibrous protein with

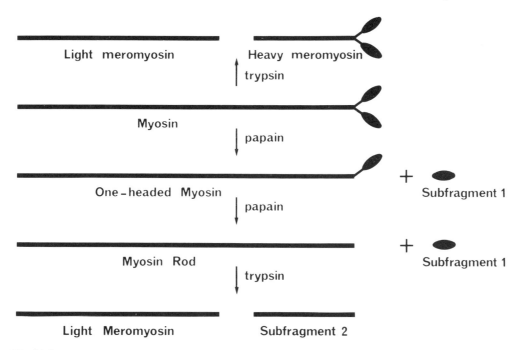

Fig. 21.5 The myosin molecule and the products of cleavage by proteolytic enzymes. (After S. Lowey.)

features of a globular protein. The asymmetry of the molecule explains why solutions of myosin are rather viscous; this is noticeable even at concentrations of 5 mg/ml.

It is possible to chop the myosin molecule up into fairly well-defined fragments by the use of proteolytic enzymes, and hence to find where the important properties of the molecule are located. The isolated fragments may be visualised in the electron microscope and they may also be tested for ATPase and actin binding activity. For example, a low concentration of trypsin attacks a large number of peptide bonds but the susceptible bonds lie within a region of only 20 nm or so, about half way along the molecule (Fig. 21.5). The ATPase activity is not decreased during this digestion. The truncated tail (molecular weight 120 000) is called light meromyosin (Greek: *meros* = part of). Light meromyosin retains the ability to associate and it forms large ordered aggregates at low ionic strength. It can neither hydrolyse ATP nor bind to actin. It has an astonishingly high α-helical content (>90 per cent) The other fragment, heavy meromyosin (MW 350 000), which contains the two globular heads attached to the other part of the tail, retains both the ATPase activity and the ability to bind actin. However it has no ability to associate and remains in solution at low ionic strength. It should be emphasised that heavy and light meromyosin are *not* subunits of myosin; they are proteolytic fragments. They must not be confused with the heavy and light polypeptide chains referred to later.

The myosin molecule can be broken in another way. Papain preferentially attacks near the globular heads (Fig. 21.5). The products are thus the full-length tail (called the myosin rod, MW 200 000) and the separated heads (each called subfragment 1, MW 120 000). Shorter digestion yields a myosin molecule with only one head chopped off. The ATPase activity and actin binding capacity of myosin are fully retained in subfragment 1 and are absent in myosin rod, so we know that these activities lie exclusively in the globular heads of the myosin molecule. The myosin rod, like light meromyosin, has a high α-helical content and associates at low ionic strength.

A combination of treatments with trypsin and papain allows us to isolate the short piece of tail (known as subfragment 2) found in heavy meromyosin (Fig. 21.5). Although it has a high α-helical content, it is important to note that subfragment 2 remains freely soluble at low ionic strength and neither associates with itself nor binds to light meromyosin. It is also possible to produce a one-headed heavy meromyosin fragment by shorter digestion with papain.

Dissection of the myosin molecule in this way allows us to build up a general idea about the packing of myosin molecules in the thick filament. The part of the tail corresponding to light meromyosin is believed to associate to build up the backbone of the filament leaving the remainder of the molecule (equivalent to heavy meromyosin) free from the backbone to function as a cross-bridge. The globular heads (containing the ATPase and actin binding sites) are thus able to make contact with actin in the thin filament.

7.3 Subunit structure of myosin

The subunit structure of myosin, once the subject of much controversy, may now be readily established by SDS gel electrophoresis. It appears that rabbit skeletal myosin has two heavy polypeptide chains (with a MW of about 190 000 they are the largest polypeptide chains known) and four light polypeptide chains (with MW ranging from 17 000 to 21 000) (see Table 21.1). Of the four light chains, two are termed DTNB light chains (MW 18 000), and the other two are made up from A-1 light chains (MW 21 000) and A-2 light chains (MW 17 000). This nomenclature needs some explanation. If the sulphydryl groups of myosin are allowed to react with Ellman's reagent, 5,5'-dithiobis-(2-nitrobenzoate) (abbreviated DTNB), two of the light chains are dissociated from the parent molecule and hence are known as DTNB light chains.

Somewhat surprisingly the removal of these light chains has no discernible effect on the properties of myosin. If the denuded myosin is now exposed to alkali (pH 11), the two remaining light chains are dissociated from the parent. This is accompanied by a loss in the ability of myosin to bind actin and to hydrolyse ATP.

The A-1 light chain is longer than the A-2 light chain but much of their amino-acid sequence is common to both. There has been frequent speculation that myosin might exist as isoenzymes. For example it is not yet clear whether some myosin molecules have one A-1 light chain and one A-2 light chain or whether most myosin molecules have two A-1 light chains and the remainder two A-2 light chains. As for the two heavy chains it has been established that two slightly different amino-acid sequences are present, but again it is not known whether both are present in the same myosin molecule.

We must consider how the polypeptide chains of myosin are arranged within the molecule. Light meromyosin and myosin rod have very high α-helical contents. Now in the α-helical

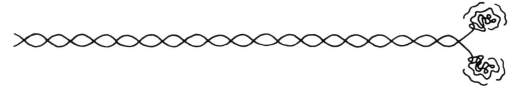

Fig. 21.6 Diagram of the arrangement of polypeptide chains in the myosin molecule. The two heavy chains stretch from one end of the molecule to the other. The four light chains are present only in the heads.

structure, the amino-acid residues follow one another at intervals of 0·15 nm. Taking the average molecular weight of a residue as 110, we can calculate that if molecules of light meromyosin (MW 120 000) and myosin rod (MW 200 000) contained a single α-helix, their lengths would be about 160 nm and 270 nm respectively. These are close to twice the observed lengths and it is therefore deduced that light meromyosin and myosin rod contain *two* α-helices running in parallel. X-ray diffraction of fibres of light meromyosin suggests that the α-helices are would round each other in a coiled-coil conformation.

SDS gel electrophoresis shows that the light chains are present in subfragment 1, so the two polypeptide chains involved in the tail of the myosin molecule must be the heavy chains. The heavy chains have too high a molecular weight to be present only in the tail and so must make up a substantial part of the heads as well. The arrangement of the polypeptide chains in the myosin molecule is shown in Fig. 21.6. The two acetylated N-terminal ends of the heavy chains are retained in heavy meromyosin so we know these two chains run parallel to one another, the direction of the chains being from head to tail as we pass from the N-terminal to the C-terminal end.

Since it is possible that in the same myosin molecule one head might contain an A-1 light chain and the other head an A-2 light chain, we do not know whether or not the two heads of a myosin molecule are structurally and functionally identical.

7.4 Assembly of myosin into thick filaments

I have already mentioned that if a solution of myosin is dialysed against a medium with a composition resembling that of the muscle cell ($I = 0·15$, pH 7·0), the solution becomes turbid, due to the presence of aggregates. In the electron microscope these aggregates are seen to be filaments that resemble the natural thick filaments released by blending myofibrils (Plate 21.7). In particular they have a similar diameter, their ends taper and the surface is not smooth but is covered with numerous projections except at the centre. These projections we have seen are believed to act as the cross-bridges. Thus myosin molecules have the innate ability to assemble into filaments without the direction of any template. However it should be noted that the length of the synthetic filaments varies widely unlike the natural filaments and it is not known how the extremely constant length of the latter is determined. Unfortunately the filaments are not well enough preserved in the electron microscope for us to say whether the packing of the molecules in the natural and synthetic forms are precisely the same. To explain the observed

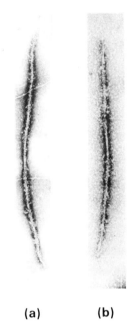

(a)   (b)

Plate 21.7 Comparison of electron micrographs of (a) a natural thick filament obtainedy by blending muscle in a relaxing medium, (b) a synthetic thick filament obtained by polymerising myosin. Note the similarity of the two structures, especially the presence of a bare central shaft. (Courtesy of H. E. Huxley and Academic Press.)

structures, H. E. Huxley suggested that the myosin molecules were packed in opposite directions in each half of the thick filament. This would account for the zone in the middle of the thick filament which does not have projections (the bare zone) and the tapering of the ends (Fig. 21.7). The reversal in polarity at the centre of the filament would help to explain how thin filaments overlapping the two ends of a thick filament are propelled in opposite directions, towards the centre of the thick filament (Fig. 21.3).

The precise way in which the myosin molecules are packed is not known, but useful information on the arrangement of the cross-bridges has been obtained from X-ray diffraction of living resting muscle (Plate 21.8). There is a strong meridional reflection at 14·3 nm and there are layer lines at 42·9 nm and its orders. The interpretation is that the cross-bridges in resting muscle are arranged on n helical strands of pitch 42·9 n nm and axial repeat 14·3 nm. The number (n) of helical strands is however still controversial and could be two, three or four, that is there could be two, three or even four cross-bridges every 14·3 nm; (models with two and three cross-bridges every 14·3 nm are shown in Fig. 21.8). Assuming that one myosin molecule gives rise to one cross-bridge, the figure of three would fit best with current estimates of the amount of myosin in the myofibril and corresponds to about 300 myosin molecules in the thick filament. It should not be thought that the cross-bridges are arranged with very precise helical order (as for example are the subunits in tobacco mosaic virus). The part of the X-ray pattern due to myosin fades out at high angles indicating that the cross-bridges are not precisely fixed and may move around a bit.

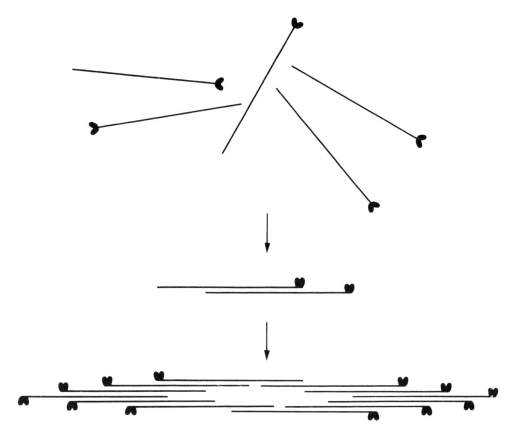

Fig. 21.7 Schematic diagram to show how myosin molecules are arranged with opposite polarity in each half of a thick filament. A dimer is shown as a possible intermediate. (After H. E. Huxley.)

How do the tails of the myosin molecules pack? The simplest model of the thick filament would require the tails to be arranged with the same helical symmetry as the heads so that all the myosin molecules had an equivalent environment. Since the length of light meromyosin is much greater than 42·9 nm, the periodicity along the filament, each myosin molecule must be tilted slightly with respect to the filament axis so that the end of the molecule farthest from the heads lies near to the axis of the filament, while the junction between the light meromyosin and subfragment 2 moieties lies on the surface. Any such arrangement would leave a hollow core down the axis of the filament. We do not know whether such a hollow core exists or whether it is filled by another protein; the amount of such a core protein would be small.

In a number of invertebrate muscles the thick filaments are wider than those of vertebrate skeletal muscle. In certain molluscan muscles, for example, the diameter of the thick filaments may reach 150 nm. In these invertebrate muscles the myosin molecules are packed on the outside and the centre of the filament is taken up by a core protein called paramyosin. The ratio of paramyosin to myosin may vary from about 1:10 to about 10:1 depending on the

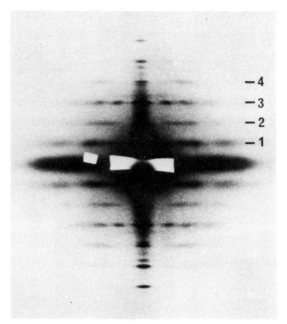

Plate 21.8 Low angle X-ray diffraction pattern of resting muscle. The set of layer lines at 42·9 nm and its orders (arising from the arrangement of projections about the thick filament) are marked. Note the strong third order meridional reflection at 14·3 nm. (Courtesy of H. E. Huxley and Academic Press.)

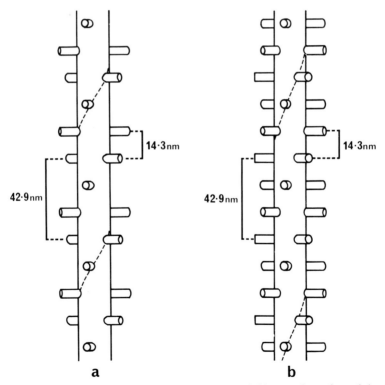

Fig. 21.8 Simplified models of two possible arrangements of cross-bridges on the surface of the thick filament. (a) Two-strand helix with pitch 2 × 42·9 nm and two cross-bridges emerging every 14·3 nm. (b) Three-strand helix with pitch 3 × 42·9 nm and three cross-bridges emerging every 14·3 nm. The dotted lines trace the path of one of the helical strands. (Courtesy of H. E. Huxley and of J. Squire.)

muscle. Paramyosin is another two chain coiled-coil α-helical protein with a molecular weight of about 200 000 and a molecular length of about 140 nm. Like light meromyosin, it associates to form ordered aggregates at low ionic strength.

It has been noted in section 7.1 that the myosin molecule is in reversible equilibrium with a dimer. The intrinsic viscosity of the dimer is 311 ml/g, which is only a little greater than that of the monomer (245 ml/g). This suggests that in the dimer the two myosin molecules are arranged neither end to end nor with total side to side overlap. Detailed hydrodynamic measurements favour a model in which the two myosin molecules point the same way and are arranged side to side with a stagger of about 43 nm (Fig. 21.7). This dimer may well turn out to be the building block of the filament.

Myosin is not the only protein of the thick filaments. In each half of an A-segment there is a set of 11 transverse stripes at about 43 nm intervals. A recently described protein called C-protein has been shown by antibody methods to be responsible for most of the stripes. Its function is unknown.

7.5 The ATPase activity of myosin

I have already mentioned that myosin and two of its fragments, heavy meromyosin and subfragment 1, have ATPase activity. Many of the earlier studies of this activity were made in all kinds of unphysiological media and under these circumstances myosin by itself may be shown to have quite a high ATPase activity. For example, in a medium devoid of all traces of Mg^{2+} ions and in the presence of 0·6 M K^+ ions, the ATPase activity of myosin can reach 30 s^{-1} at 25°C. However if we are concerned with myosin as part of the contractile apparatus we should examine its activity in a medium which resembles the interior of the muscle cell (e.g. 0·15 M K^+ ions, 5 mM Mg^{2+} ions, 5 mM ATP, pH 7·0) It is then found that the ATPase activity of myosin, while detectable, is very low (about 0·02 s^{-1} per head).† Although initially surprising, this is exactly what we should expect if the hydrolysis of ATP is coupled to the performance of mechanical work. The system with myosin alone should correspond to the resting state of muscle in which it would be wasteful to have appreciable hydrolysis of ATP. The resting ATPase of muscle is not accurately known but heat measurements suggest that the energy turnover in contracting muscle is about 2000 times that in resting muscle. The ATPase in resting muscle is thus roughly comparable with that of myosin.

We should now consider the transient kinetics of hydrolysis of ATP by myosin. ATPase measurements are often performed by mixing myosin or heavy meromyosin with ATP in a suitable medium, stopping the reaction at various times with trichloroacetic acid, and then measuring the inorganic phosphate liberated. Since the enzyme is denatured by the acid, the phosphate measured will include not only the free inorganic phosphate but any inorganic phosphate bound non-covalently to the enzyme. Using apparatus which will rapidly mix myosin or heavy meromyosin with ATP, one can demonstrate the rapid release of about 1-2 mol P_i/mol enzyme at short time intervals (the 'early burst') (Fig. 21.9). This is followed by the slow constant release of phosphate corresponding to the steady state rate (0·02 s^{-1}). The rate of the early burst of phosphate depends on the concentration of ATP, being about 2×10^6 [ATP] s^{-1} at low ATP concentrations but reaching a plateau of approximately 100-150 s^{-1} per site at saturating ATP concentrations.

On the other hand, if only the *free* inorganic phosphate liberated is measured (for example by using a coupled enzyme assay with glyceraldehyde 3-phosphate dehydrogenase and

---

† Since the molecular weight of myosin and its fragments is known, it is useful to state activities in units of s^{-1} rather than the more traditional μmole ATP/min/mg enzyme. The K_m for ATP is very low for all myosin systems and the velocities given are V_{max} values obtained at saturating ATP concentrations.

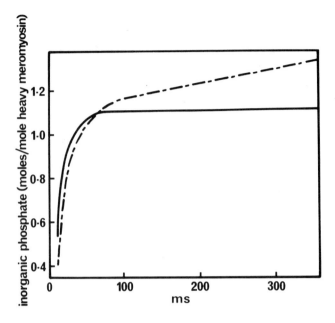

Fig. 21.9 The early burst of inorganic phosphate produced by stopping the reaction with acid, ——— with heavy meromyosin **alone** – – – – with heavy meromyosin plus actin (molar ratio 1:4·8). (Courtesy of E. W. Taylor and The American Chemical Society.)

phosphoglycerate kinase) then the time course is linear and shows no early burst. The early burst, then, is due to the rapid formation of *enzyme-bound* phosphate, not of *free* phosphate. Although a covalent phosphorylated enzyme intermediate has been sought, it has not been found. Thus the enzyme-bound phosphate is apparently bound by secondary forces only. To explain these findings Taylor proposed the following scheme:

$$\text{M} + \text{ATP} \underset{}{\overset{\text{fast}}{\rightleftharpoons}} \text{M . ATP} \underset{}{\overset{\text{fast}}{\rightleftharpoons}} \text{M . ADP . P}_i \underset{}{\overset{\text{slow}}{\rightleftharpoons}} \text{M} + \text{ADP} + \text{P}_i \qquad (2)$$

On this scheme myosin (or heavy meromyosin) combines rapidly with ATP and rapidly hydrolyses it to ADP and P_i. The ADP and P_i however dissociate only slowly from this complex and, until they have done so, no more ATP can bind at the active site. Thus the low steady-state ATPase activity is not due to the fact that myosin is a poor catalyst of the hydrolytic step. Rather it is due to the slow rate at which the products leave the active site. Thus in the steady state myosin is predominantly in the M . ADP . P_i state.

To measure the rate of decay of the myosin ADP . P_i complex directly, myosin was mixed with stoichiometric amounts of radioactive ATP (in which the adenine ring was labelled with ^{3}H and the terminal phosphate with ^{32}P); myosin. ^{3}HADP . ^{32}P$_i$ was thus formed very quickly. The mixture was then subjected to rapid gel filtration. Protein is eluted ahead of low molecular weight components. If none of the myosin ^{3}HADP . ^{3}P$_i$ complex were broken down, all the ADP and P_i would be eluted with the protein peak. If some of the ADP or P_i were dissociated in the time required to run the column, some ADP or P_i would be eluted after the protein peak. From the distribution of radioactivity in the eluted fractions the rate of dissociation of products can be calculated. The rate of dissociation of ADP and P_i were indeed found to be similar to the steady state rate of ATP hydrolysis.

Further details of the breakdown of the M . ADP . P_i complex are now available. The rate of dissociation of ADP at 20°C from an M . ADP complex (formed by mixing M and ADP) is several times the steady state ATPase rate. The dissociation of ADP is not therefore the rate-limiting step at this temperature and the slow decomposition of the M . ADP . P_i complex has to be explained by another cause. It is not yet clear whether the slow decomposition involves the ordered slow release of P_i before the dissociation of ADP, or whether an additional M . ADP . P_i intermediate is involved (see Taylor 1973).

Attempts are being made to see whether any of the kinetic steps are associated with conformational changes in the myosin molecule. The addition of ATP or ADP to myosin causes no detectable change in secondary structure as shown by circular dichroism and absorption studies in the far ultra-violet. However, there is now evidence both from ultra-violet spectroscopy and fluorescence that the tertiary structures of M, M . ADP (formed by mixing M and ADP) and M . ADP . P_i (formed by mixing M and ATP) are different. Unfortunately it is not known whether in these intermediates there is any change in the position of the heads with respect to each other and to the tail.

Myosin has two globular heads and each has an ATPase site. We know this because myosin binds 2 mol ADP/mol, and because subfragment 1 (or single headed heavy meromyosin) has about half the ATPase activity of myosin and heavy meromyosin on a molar basis. It is not yet clear whether the two heads function independently. As we shall see this is of considerable importance in deciding how the cross-bridges operate.

Plate 21.9 Electron micrograph of F-actin filaments revealed by negative contrast. Note the periodic change in the apparent width of the filaments (compare Fig. 22.10). This is best seen by viewing each filament obliquely along its length. In some areas the subunits can be **discerned. (Courtesy of J. Hanson.)**

8 Actin

8.1 Structure of G-actin and F-actin

Actin is normally extracted from the acetone-dried powder of muscle at very low salt concentrations. Under these conditions it exists in the form of globular molecules (hence called G-actin) of MW 41 700. The complete amino-acid sequence of the single polypeptide chain of G-actin is now known. The molecule of G-actin contains one bound Ca^{2+} ion and one non-covalently bound ATP molecule, but their function is obscure.

Owing to the globular shape of the molecule, solutions of G-actin are only slightly more viscous than water. However, a very dramatic effect occurs when salt (e.g. 0·1 M KCl) is added

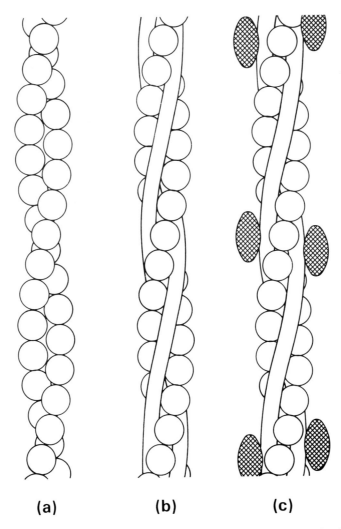

(a)　　　**(b)**　　　**(c)**

Fig. 21.10 Diagram of the structure of the components of the thin filament. (a) F-actin, showing the double helical arrangement of subunits (represented as spheres). (b) F-actin plus tropomyosin. The tropomyosin strands are drawn lying in the long pitched grooves of the actin helix. For clarity their diameter has been exaggerated. (c) F-actin plus tropomyosin plus troponin. The troponin molecules are represented as the hatched ellipsoids.

to such solutions. The viscosity rises very greatly and indeed at only moderate protein concentrations (e.g. ~5 mg/ml) the solution may set as a gel so that the vessel containing it may be inverted without loss of the contents. The only reasonable explanation for this rise in viscosity is that the globular molecules have polymerised to give a fibrous structure; this product is therefore called F-actin. During polymerisation the bound ATP molecules are converted to ADP. The polymerisation can be reversed by dialysis against low ionic strength solutions containing ATP.

The structure of F-actin has been investigated both by electron microscopy and by X-ray diffraction. F-actin consists of long filaments of undefined length made up of globular subunits in a helical arrangement (Plate 21.9 and Fig. 21.10). Although we can regard the structure as consisting of two 'strands' of subunits wound helically around one another (with a pitch of about 74 nm), there is no reason to think the 'strands' can exist independently of each other. Each subunit of the structure is presumably joined by non-covalent bonds to its two neighbours in the same 'strand' and to its two neighbours in the adjacent 'strand'. The size of each subunit suggests that it comes from a single G-actin molecule.

The importance of these filaments made from purified actin is that in the electron microscope they look very similar indeed to natural thin filaments. Moreover part of the X-ray diffraction pattern of whole muscle can be attributed to the F-actin structure. F-actin is thus thought to be the basis of the thin filament structure (Fig. 21.10). So here we have an excellent example of self-assembly: the subunits are able to assemble independently of any template to give a polymer similar to the natural one. But like synthetic myosin filaments, filaments of F-actin are of indeterminate length and we do not know what determines the precisely defined lengths of natural filaments.

One point worth noting about the steric arrangement of the subunits in F-actin is that the number per turn of each strand (about $13\frac{1}{2}$) is not an integral multiple of 3. Thus although each thin filament is surrounded by three thick filaments, the arrangement of the subunits does not take account of this. We shall consider the direction of the two strands later (section 8.2)

8.2 Interaction of F-actin with myosin

We have already noted in sections 7.1 and 7.2 that myosin, heavy meromyosin and subfragment 1 all bind strongly to F-actin. In each case the complex is dissociated by ATP. It is obviously desirable to know more about the stoichiometry and structure of these complexes in order to understand how the cross-bridges function.

The binding of subfragment 1 (or heavy meromyosin) to F-actin can be measured in the ultracentrifuge. It is generally agreed that when subfragment 1 is in excess, one subfragment 1 molecule binds per actin subunit and that the binding at each site is independent of that at the other sites. Thus in this simple system there is no interaction between neighbouring actin subunits. In the electron microscope the structure of the complex may be seen (Plate 21.10). The details of the structure are difficult to make out, partly because we are seeing a two-dimensional projection of the original three-dimensional object. The two-dimensional distribution of density in the electron micrographs may however be used to compute the three-dimensional structure if it is assumed that the structure has helical symmetry. This is shown in Plate 21.11. The underlying F-actin structure can be discerned. The elongated subfragment 1 molecules are seen to be attached to the actin subunits at an angle of about 50° to the filament axis. Viewed along the filament axis they do not emerge radially but are slewed at an angle of about 60° to a radius (see also Fig. 21.12). Incidently this reconstruction tells us a lot about the shape of the subfragment 1 molecule which was not previously known; the shape

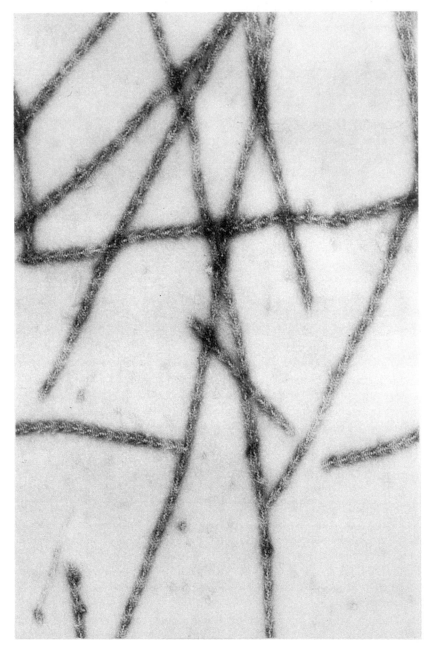

Plate 21.10 Electron micrograph of complexes of thin filaments and subfragment 1. Note the appearance of arrowheads indicating the polarity of the structure. Compare with Plate 21.11. (Courtesy of H. E. Huxley and Academic Press.)

is not unlike a bent finger. The length is about 15 nm. This picture gives us a first exciting glimpse of what cross-bridges might look like attached to the thin filament at some point in their cycle of operation. A further important feature to note is that the structure of the complex has polarity and this must mean that the underlying F-actin is also polarised. The actin subunits in Fig. 21.10 are represented as spheres only because we do not yet know their shape, but if we were to represent them realistically, the strands would be seen to have a direction. The polarity of the structure means that the direction of the two strands is the same. When labelled with subfragment 1, thin filaments emerging from either side of a Z disc show opposite polarities; the attachment point of each subfragment 1 molecule lies farther from the Z disc than the rest of the molecule. This reversal of polarity is to be expected from the necessity of drawing thick filaments in from opposite directions on either side of the Z disc (Fig. 21.3).

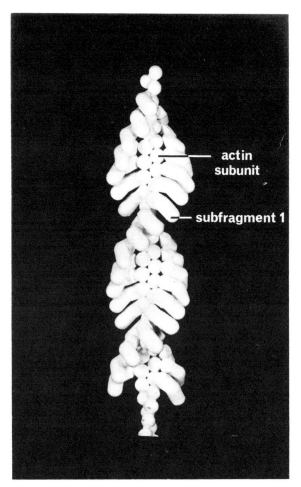

Plate 21.11 A model of the complex of F-actin and subfragment 1 based on information present in micrographs such as Plate 21.10 and assuming that the structure has helical symmetry. Note that the subfragment 1 molecules follow the helical symmetry of the actin subunits to which they are attached. Compare with the transverse section of Fig. 21.12. (Courtesy of H. E. Huxley and Academic Press.)

The structure of the F-actin/subfragment 1 complex is very characteristic and recently it has been possible to identify the F-actin filaments which occur at low concentrations in some non-muscular cells by allowing subfragment 1 to diffuse into the cell. This is one of the important indications that the contractility displayed by some cells may be related to muscle contraction.

So far we have discussed the binding of subfragment 1 to F-actin. Since heavy meromyosin has two heads, we might have expected that one heavy meromyosin molecule could bind simultaneously to two neighbouring actin subunits. Unfortunately the stoichiometry of binding of heavy meromyosin to F-actin is a matter of controversy at present and it has not been agreed whether or not the two heads bind simultaneously. Since the cross-bridge is thought to consist of a heavy meromyosin moiety this is a large gap in our knowledge.

8.3 Effect of F-actin on myosin ATPase

We have already noted that the activity of myosin ATPase is low. If the cross-bridge cycle is associated with the hydrolysis of ATP, then the presence of actin should stimulate myosin ATPase. This is indeed found. F-actin stimulates the ATPase of myosin by a factor of about 20 at low ionic strength. When ATP is added to a slightly turbid solution of F-actin and myosin at

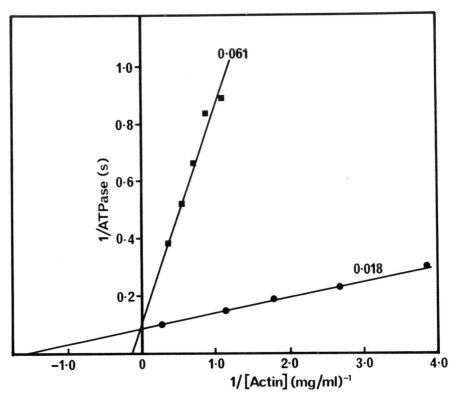

Fig. 21.11 Double reciprocal plots to show the activation of heavy meromyosin ATPase by actin at two ionic strengths (0·018 and 0·061). Both plots intersect the 1/ATPase axis at 0·09 s indicating that the maximum ATPase of heavy meromyosin at saturating F-actin concentrations is 11 s⁻¹ (per site). (Courtesy of C. Moos and the American Chemical Society.)

low ionic strength, the high ATPase activity is associated with the formation of numerous clusters of filaments which cause the turbidity to increase. These clusters grow and tend to settle to the bottom of the vessel over a period of a minute or so. This phenomenon is known as superprecipitation. While often studied as a model of muscular contraction, the movement of filaments giving rise to the clusters has not yet been explained.

Although actin produces a large activation, the ATPase rate (of the order of 3 s⁻¹) is still much smaller than the maximum rate of splitting of ATP in intact muscle. This might be expected since when we mix myosin and actin in the filamentous forms, we are not duplicating the organised array of the myofibril. Thus many of the globular heads of the myosin molecules may be sterically prevented from reaching the actin filament. Moos therefore suggested that a more suitable system for *in vitro* studies was F-actin plus heavy meromyosin (or subfragment 1) in which there is no such steric restriction.

Under most experimental conditions, the presence of ATP would cause the F-actin-heavy meromyosin complex to be largely dissociated. Maximum activation is therefore not achieved with stoichiometric quantities of F-actin; high concentrations of F-actin are needed. It is convenient to treat F-actin as a modifier of the enzyme and plot the reciprocal of the increase in ATPase against the reciprocal of the F-actin concentration. The resulting plot is a straight line (Fig. 21.11). The intercept on the $1/(v\text{-}v')$ axis is the value of $1/(v\text{-}v')$ at saturating F-actin concentrations, the intercept on the $1/[\text{F-actin}]$ axis gives the concentration of F-actin required for half maximum activation. It is thereby found that the maximum rate of ATP splitting for rabbit heavy meromyosin is between 10 and 20 s^{-1} per site at 25°C and is independent of ionic strength. This is comparable to the rates of ATP splitting in intact frog muscle (about 17 s^{-1} per site) (section 4.2). So we feel justified in thinking that in contracting muscle the high ATPase is explicable in terms of the actin-activated myosin ATPase.

The rate-limiting step in the hydrolysis of ATP by heavy meromyosin is the release of products ADP and P$_i$ from the enzyme (section 7.5). Since F-actin accelerates the ATPase, this step must be bypassed. The possible steps for the F-actin activated enzyme can be written out by adding to equation (2) a further set of steps with the F-actin–myosin complex replacing myosin:

$$
\begin{array}{ccccccc}
\text{M + ATP} & \underset{}{\overset{1}{\rightleftharpoons}} & \text{M . ATP} & \underset{}{\overset{2}{\rightleftharpoons}} & \text{M . ADP} \cdot \text{P}_i & \underset{}{\overset{3}{\rightleftharpoons}} & \text{M + ADP + P}_i \\
\Big\updownarrow{}^{8} & & \Big\updownarrow{}^{5} & & \Big\updownarrow{}^{6} & & \Big\updownarrow{}^{8} \\
\text{AM + ATP} & \underset{}{\overset{4}{\rightleftharpoons}} & \text{AM . ATP} & \underset{}{\overset{9}{\rightleftharpoons}} & \text{AM . ADP} \cdot \text{P}_i & \underset{}{\overset{7}{\rightleftharpoons}} & \text{AM + ADP + P}_i
\end{array}
\tag{3}
$$

Not all the steps will occur at appreciable rates. The following analysis by Taylor allows us to eliminate several of the steps as being too slow. The rate of formation of the early burst phosphate is slightly slower in the presence of actin (Fig. 21.9). This means that step 9 can be no faster than step 2. As we have mentioned before, the addition of ATP to a complex of actin and myosin causes dissociation of the complex. The rate of dissociation (measured by following the decrease in turbidity) is very fast indeed (>1000 s^{-1}). This means that the AM . ATP intermediate must break down via step 5 rather than via step 9. So the actual hydrolysis of ATP *even in the presence of F-actin* is believed to occur on myosin alone. Step 3 is however very slow, so the release of products is supposed to occur after recombination of myosin with actin (step 6). The cause of the activation by actin is thus the acceleration of the release of products, not the acceleration of the hydrolytic step. Since the rate of dissociation of AM to A + M is very slow we are left with the scheme:

$$
\begin{array}{ccc}
\text{M . ATP} & \overset{2}{\rightleftharpoons} & \text{M . ADP . P}_i \\
\Big\updownarrow{}^{5} & & {}_{+A}\Big\updownarrow{}^{6} \\
\text{AM + ATP} \underset{}{\overset{4}{\rightleftharpoons}} \text{AM . ATP} & & \text{AM . ADP . P}_i \overset{7}{\rightleftharpoons} \text{AM + ADP + P}_i
\end{array}
\tag{4}
$$

(which is rearranged in Fig. 21.15 with different numbering to emphasise the cyclic nature of the scheme).

We shall see in section 10 that it is of considerable importance to know the rate of step 7 but so far no direct estimates are available for this, nor indeed is it certain that the rate constant for step 7 is less than that of step 2 (~100 s^{-1}). This is mainly because the value for the turnover rate of the cycle is controversial. Hence we do not know what is the rate-determining step in the presence of excess actin.

It will be noted that for each ATP molecule hydrolysed, a myosin head associates once with actin and the complex breaks down once. In this simple scheme no mention has been made of the possibility that the two sites in myosin might interact.

The beauty of this scheme is that it integrates many of the known facts about the interaction of actin, myosin and ATP. We shall return to it in later sections.

9 The molecular basis for regulation

9.1 The regulatory proteins

We have previously seen that Ca^{2+} ions switch the contractile machinery on. We should now discuss how this is achieved. The actin plus myosin system we have just mentioned is quite unaffected by the presence or absence of 10^{-5} M Ca^{2+} ions. Something clearly is missing from the system. Before actin preparations were as fully characterised as they are now, it was noted that a system made up of impure F-actin and myosin *did* show Ca^{2+} sensitivity. A long analysis by Ebashi in the early 1960s of the missing factor present in the crude actin, showed it was made up of two proteins, tropomyosin (which had already been discovered in 1947 by Bailey, but for which there was no known function) and a new protein called troponin. As we shall see these two proteins together function as a regulatory system.

9.2 Tropomyosin

Tropomyosin is a long thin rod-like molecule about 40 nm long and 2 nm wide with a molecular weight of 70 000. Like light meromyosin, it contains two α-helical polypeptide chains wound in a coiled-coil. A substantial part of the amino-acid sequence is now known, and it is found that hydrophobic residues occur alternately at intervals of three and four residues in agreement with Crick's predictions for a coiled-coil.

Tropomyosin is the only fibrous protein which has been crystallised. The crystal is made up of a network of strands formed by end to end polymerisation of the tropomyosin molecules. The viscosity of tropomyosin solutions greatly increases at very low ionic strength also suggesting end to end polymerisation. We shall see that this property is used in the construction of the thin filament.

Antibody to tropomyosin labels the region of the myofibril occupied by the thin filaments. Moreover tropomyosin binds strongly to F-actin and the structure of the complex resembles the natural thin filament. Therefore tropomyosin is present in the thin filament. Unfortunately no accurate figures are available for the stoichiometry of binding of F-actin and tropomyosin. However, electron microscopy and X-ray diffraction of complexes of F-actin and tropomyosin suggest a model in which molecules of tropomyosin lie end to end in the two long-pitched helical grooves of the F-actin structure (Fig. 21.10(b)). The tropomyosin helix obviously has the same pitch as that of the F-actin helix. The axial repeat of the tropomyosin molecules is 38·5 nm which within experimental error is seven times the axial repeat (5·46 nm) of the subunits in one strand of the actin structure. If there is an exact integral relation this would imply that there are specific bonds between the two proteins.

One interesting point is that F-actin complexed with tropomyosin actually activates the ATPase of subfragment 1 more than F-actin alone, but the plots of $1/v$ against $1/[\text{F-actin}]$ are now curved suggesting that in the presence of tropomyosin the actin subunits no longer function independently of each other.

9.3 Troponin

As isolated, troponin has a molecular weight of 80 000 and is thought to be a slightly elongated molecule. Although it binds to tropomyosin it does not bind to actin. So tropomyosin provides the means of attachment of troponin to the thin filament. The location of troponin in the thin filament has been determined by antibody staining. Antibody to troponin is not bound continuously along the thin filament but at discrete sites spaced at intervals of 38·5 nm. The stripes seen in the I band of sectioned muscle (Plate 21.2) are therefore attributed to troponin. By densitometry of SDS gels of whole myofibrils, the molar ratios of actin, troponin and tropomyosin are found to be about 7:1:1. Thus in the thin filament each troponin molecule is thought to be bound to one tropomyosin molecule in the structure shown in Fig. 21.10(c). It is not yet clear how the long axis of the troponin molecule is disposed with respect to the axis of the thin filament, nor how close the troponin molecules are to the ends of the tropomyosin molecules, nor how close are neighbouring troponin molecules in the two long pitched grooves.

Of all the myofibrillar proteins, troponin alone binds Ca^{2+} ions strongly but reversibly in the presence of Mg^{2+} ions. Present indications are that the troponin molecule has two Ca^{2+} binding sites of high affinity (association constant of about 2×10^6 M^{-1}) and two with low affinity (about 4×10^4 M^{-1}). When troponin is bound to tropomyosin and actin these constants are but little changed. From its spatial distribution the content of troponin in the myofibril can be deduced to be 0·8 μmol/g protein. Since the myofibril binds about 3 μmol Ca^{2+}/g protein at 10^{-5} M Ca^{2+} when maximum tension is reached (Fig. 21.4), a large part (if not all) of the binding can be attributed to troponin. Hence the only myofibrillar component which picks up Ca^{2+} released by the sarcoplasmic reticulum appears to be troponin, and all four sites on troponin must be filled for maximum tension to be achieved. It will be of considerable interest to determine the rate constants for association and dissociation of Ca^{2+} from troponin to see whether the rates of development and decay of contractile activity are due primarily to the properties of troponin or whether they are mainly determined by the sarcoplasmic reticulum.

Certain invertebrate muscles do not have troponin. Such muscles have instead a special kind of myosin containing a light chain which binds Ca^{2+} reversibly in the presence of Mg^{2+}. Interaction between such myosin and actin requires binding of Ca^{2+} to this light chain. Thus in these muscles the control lies on the thick filaments rather than on the thin filaments.

The subunit composition of troponin is revealed by SDS electrophoresis. Three types of polypeptide chain are present in an equimolar ratio. TN-T (MW 37 000) is so-called because it can bind to tropomyosin; TN-I (MW 23 000) is so-called because in high concentrations it inhibits actomyosin ATPase; TN-C (MW 18 000) is so-called because it can bind Ca^{2+}. The components of troponin are exceptionally stable and they may be separated from each other by chromatography in 8 M urea without impairment of activity. The role of the subunits in the troponin molecule is still controversial (see Weber and Murray 1973).

9.4 Regulation by tropomyosin and troponin

An equimolar ratio of troponin (or a mixture of its three components) and tropomyosin confers on the actomyosin system the ability to be regulated by Ca^{2+}; in the absence of Ca^{2+} the ATPase activity is low and the actin filaments are dissociated from the myosin filaments, while in the presence of Ca^{2+} the ATPase activity is high and the system superprecipitates. Both proteins are necessary for this regulation. The greatest regulation is achieved when the molar ratio of tropomyosin, troponin and actin is about 1:1:7. However, even in this case, in the absence of Ca^{2+} the ATPase activity does not usually decrease below about a tenth of that in the presence of (say) $10^{-5} M$ Ca^{2+}, whereas in intact muscle the difference in ATPase activity between the resting and contracting muscle is thought to be about 2000. The cause of this discrepancy is not yet clear.

The cause of the inhibition by the tropomyosin–troponin system in the absence of Ca^{2+} has been studied kinetically. When these two proteins are added in a constant ratio to F-actin they do not affect the *maximum rate* at which subfragment 1 can hydrolyse ATP in the presence of saturating actin; what they alter is the affinity of actin for subfragment 1 so that a higher concentration of actin is required to achieve the same activation. It has recently been shown that these two regulatory proteins act on step 6 of equation (4), slowing the rate at which the (myosin ADP . P_i) complex binds to actin.

To summarise, in the absence of Ca^{2+} ions, tropomyosin and troponin reduce the ability of actin to interact with myosin, thereby decreasing the rate of hydrolysis of ATP and causing relaxation in intact muscle.

9.5 Structural explanation of the regulation

It seems at first difficult to understand how Ca^{2+} ions binding at troponin sites spaced 38·5 nm apart could affect the reactivity of *all* the actin subunits in the thin filament. Some recent X-ray diffraction results on whole muscle have helped to explain this. In resting muscle the second layer line of the 37 nm period of the actin–tropomyosin structure is weak, whereas in contracting muscle (even in the case of muscle stretched out so that there is very little overlap between the thick and thin filaments) it is relatively strong. This difference can be explained if in the contracting muscle the strings of tropomyosin molecules lie closer to the centre of the grooves of the actin double helix, whereas in resting muscle they lie much closer to one strand of actin subunits than the other (Fig. 21.12).

Electron micrographs of actin–tropomyosin complexes labelled with subfragment 1 suggest how this large-scale movement of tropomyosin molecules is able to switch the activity of the thin filament on and off. The position of tropomyosin in the relaxed state appears to overlap the binding site of subfragment 1 (Fig. 21.12). (We say 'appears' because the position of subfragment 1 is taken from electron micrographs, that of tropomyosin in the resting position calculated from X-ray patterns.) It therefore seems very plausible that tropomyosin and subfragment 1 compete with one another for sites on the F-actin subunits. If tropomyosin is constrained by troponin to occupy this site in the absence of Ca^{2+} ions, then the globular heads of myosin cannot bind and the muscle relaxes. Conversely in the rigor state, where the heads bind very strongly to the actin because of the absence of ATP, the tropomyosin molecule is pushed over in the 'contracting' position.

No evidence is available to explain how Ca^{2+} binding at troponin molecules is able to cause conformational changes in this protein and hence in tropomyosin.

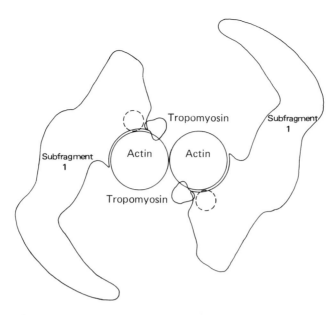

Fig. 21.12 Diagram of a transverse section of a thin filament–subfragment 1 complex in the contracting state showing the positions occupied by subfragment 1 and tropomyosin. The position that tropomyosin is thought to occupy in the *relaxed* state is shown by dotted lines. (Courtesy of H. E. Huxley and The Cold Spring Harbor Laboratory.)

9.6 Heterotropic interactions in muscle

We should now realise that in muscle we have a set of interacting proteins. We can consider subfragment 1–actin–tropomyosin–troponin as an allosteric system with ATP as a ligand for subfragment 1, and Ca^{2+} as a ligand for troponin. We have seen that the binding of Ca^{2+} to troponin affects the binding of subfragment 1 to actin. Conversely ATP (which affects the binding of subfragment 1 to actin) alters the affinity of troponin for Ca^{2+}. The reader is referred to the review by Weber and Murray (1973) for a detailed account of these heterotropic interactions.

10 The cross-bridge cycle and the generation of force

10.1 The simplest cross-bridge cycle

It should now be obvious to the reader that the central problem in muscular contraction is how the globular heads of the myosin molecule interact with sites on actin to produce force. The simplest version of a cross-bridge cycle (ignoring for the moment details of structure) would involve four steps

    a. The attachment of the cross-bridge to actin.
    b. The movement of the attached cross-bridge producing tension (the working stroke).
    c. The detachment of the cross-bridge.
    d. A recovery stroke.

We can now consider the evidence to support this scheme.

10.2 Evidence about the movements of cross-bridges

It is obviously difficult to follow the structural changes in the cross-bridge which might occur in contracting muscle because it is not a static system. One way round the problem is to regard the resting and rigor positions of the cross-bridge as arrested states of the cross-bridge cycle. Elucidation of the structure of these two states is a much simpler problem.

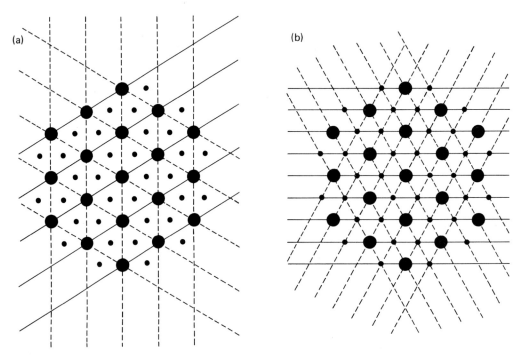

Fig. 21.13 End-on view of the hexagonal lattice of thick and thin filaments. The three sets of lattice planes in the (10) crystallographic directions are shown in (a), those in the (11) crystallographic directions in (b). (Courtesy of H. E. Huxley and The Royal Society.)

The inextensibility of rigor muscle has long been attributed to the attachment of cross-bridges to the thin filaments. The association of actin and myosin is strong and in the absence of ATP, the cross-bridges would presumably be unable to execute cyclic motions. X-ray diffraction of rigor muscle has confirmed that many cross-bridges are indeed attached and has given an indication of the kind of movements the cross-bridges might make. This can be simply explained. The side-to-side spacing of the filaments in the hexagonal lattice gives rise to equatorial reflections. The two strongest reflections arise from the crystallographic (10) and (11) planes of this lattice (Fig. 21.13) and are hence called the (10) and (11) reflections. Both the thick and the thin filaments lie on the (11) planes and therefore contribute in phase to the (11) reflection. While the thick filaments also lie on the (10) planes, the thin filaments lie between the (10) planes and therefore reduce the amplitude of the density variation between these planes. In rigor muscle the ratio of the intensity of the (11) reflection to that of the (10) reflection is much greater than in resting muscle. This suggests that in rigor there has been a very substantial transfer of mass away from the thick filament and towards the thin filament. This is, of course, consistent with the cross-bridges moving out from positions close to the backbone and approaching the thin filament. Further information can be obtained from changes in the remainder of the X-ray pattern. In rigor, additional intensity appears on the layer lines associated with the actin helix. The most obvious change is the very great intensification of the first layer line at 37 nm but in addition the maxima of the intensity on the 5·1 nm and 5·9 nm layer lines move nearer the meridian. These changes indicate that mass has attached itself to the periphery of the actin helix and by implication that cross-bridges have actually become *attached* to actin. These effects on the X-ray pattern are enhanced when subfragment 1 is used to label the vacant sites in the I bands of glycerinated muscle. This lends further support to the above conclusions.

What has happened to the structure of the thick filament to allow the myosin heads to be arranged in conformity with the actin helix (compare Plate 21.11)? In the resting state, the spatial arrangement of the myosin heads is quite different from that of the actin subunits. The pitches of the helices are unrelated (42·9 n and 74 nm) and so are the axial periodicities of the myosin heads and actin subunits (14·3 and 2·73 nm respectively) (Figs. 21.8 and 21.10). If the same symmetry were preserved in rigor, very few heads would be disposed at the precise position required for attachment. If a large number of cross-bridges are to attach in rigor, the arrangement present in resting muscle must be disturbed; that this occurs is shown by the total disappearance in the rigor X-ray pattern of the layer lines at 42·9 nm and its orders. Moreover the meridional reflection due to the axial repeat of the cross-bridges is much weaker in the rigor pattern. Thus the helical arrangement of cross-bridges present in the resting state is destroyed in rigor. The extent of the longitudinal and azimuthal movements involved, whether or not the new arrangement of cross-bridges about the thick filament axis is regular, what fraction of the cross-bridges are attached, and whether or not the movements of the cross-bridges are accompanied by a conformational change in the backbone of the thick filament are all unsolved questions.

Although the cross-bridges of vertebrate skeletal muscle are poorly preserved in electron micrographs, this is not the case for oscillatory insect flight muscle. In the resting state of this muscle, the myosin heads are clearly seen projecting out at right angles from the thick filament backbone. In the rigor muscle the myosin heads are attached to the thin filaments but are tilted at an angle of about 45° to the thin filament axis with the points of attachment of the heads lying farthest from the Z disc. Thus there is good reason for thinking that in the rigor state the myosin heads label the thin filament in the same manner as in the subfragment 1-F-actin complexes (section 8.2). Thus we have clear evidence from rigor muscle not only that myosin heads can move outward and attach to the thin filament but can change their angle of tilt with respect to the backbone.

But rigor is an unnatural state; do we have any evidence that cross-bridges attach *during contraction*? Fortunately X-ray diffraction patterns may be obtained from isometrically contracting muscle. The main change observed in contracting muscle is a very considerable decrease in intensity of the myosin layer lines. Unlike the case in rigor they do not disappear entirely, but this is probably because it is difficult to activate all the fibres in a muscle simultaneously. The intensity of the 14·3 nm meridional reflection is also weakened indicating a decrease in axial order of the cross-bridges. Thus in contracting muscle there is clear evidence that the cross-bridges have moved from their resting position. They have not all moved in the same direction by the same amount for then the helical order would be preserved. The change is consistent with continuous cycling motion of the cross-bridges (but equally consistent with a static structure in which all the cross-bridges have moved randomly from their resting positions). Further information is obtained from the equatorial reflections. The ratio of intensities of the (11) and (10) reflections is greater than in resting muscle. The magnitude of the change indicates that there has been an outward movement of the cross-bridges of roughly half the extent of the rigor case. (It should be emphasised that this piece of evidence in itself does not tell us whether *attachment* of cross-bridges to the thin filament has occurred. The change could be explained by about half of the cross-bridges moving out and attaching, but it could equally well be explained by all of the cross-bridges moving half-way!) The crucial evidence required to show that cross-bridges *attach* in contracting muscle is a change in the actin-like layer lines which indicated that mass had attached to the periphery of the actin helix. It is technically difficult to detect the presence of a small number of attached cross-bridges because the expected intensity increase in the layer lines concerned depends on the *square* of the fraction of sites occupied. At present the changes observed are too small to be conclusive but they do put an upper limit of about one fifth on the fraction of cross-bridges attached at any instant. This is, of course, perfectly reasonable if the cross-bridges are undergoing cycles of attachment and detachment, and spend the greater part of their time detached. Considerable effort will no doubt be expended in attempting to prove that cross-bridges do attach in contracting muscle and in measuring the fraction attached under various physiological conditions.

To summarise, the cross-bridges must be rather flexibly attached to the backbone of the thick filament. In rigor they move out and become attached to the thin filament, the helical arrangement about the backbone being disrupted. In contracting muscle there is also good evidence that the cross-bridges have moved from the resting position but there is as yet no conclusive evidence for attachment. Nor do we know how far a myosin head is able to move from its resting position in order to attach to an actin subunit at the appropriate angle. Most seriously, we lack *direct* evidence that the cross-bridges execute cyclic motions and that when attached to the thin filaments a mechanical movement produces tension.

10.3 The mechanics of the working stroke

In the original conception of the cross-bridge it was likened to a rigid oar. But now we know a lot more about the structure of the cross-bridge. We have seen it consists of two globular heads (subfragment 1) joined to a tail (subfragment 2). How can such a structure interact with the thin filament in such a way as to produce a sliding force?

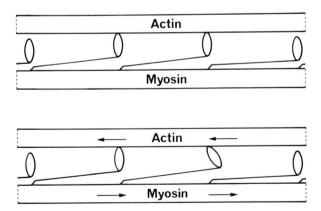

Fig. 21.14 Schematic diagram to show how a change in the angle of attachment of a myosin head to a thin filament could produce tension in the link between it and the backbone. In the upper diagram two cross-bridges ae shown attached and two detached. In the lower diagram one of the attached cross-bridges has tilted. For simplicity only one head is shown for each myosin molecule. (After H. E. Huxley.)

H. E. Huxley pointed out that subfragment 2 was unlikely to be rigid enough to act like an oar. It would not sustain the necessary couple; try pushing an object along a table with one end of a hair (which after all contains a high percentage of α-helices) while holding the far end! The other important point is that ATP is hydrolysed at sites in the heads, not in the tails; in other words energy for contraction is generated near the point of attachment of the cross-bridge to the thin filament rather than in the subfragment 2 tail.

The oar analogy is thus unsatisfactory. To replace it, H. E. Huxley suggested the model shown in Fig. 21.14. The main feature of his model is that the working stroke involves a change in angle between the actin filament and the attached myosin heads. This movement causes tension to be exerted on the subfragment 2 tail (to which it is flexibly hinged) and thence on the backbone of the thick filament. A nice feature is that the subfragment 2 is not required to be very rigid but is instead under tension; a hair will support a surprisingly large weight! We do not know whether the change in tilt of the heads is caused by a change in conformation of the actin subunits, or of the heads, or of both, and also what provides the driving force for the change. This is a major area of ignorance.

The simplest way to regard the change in tilt of the myosin heads is as the transition between two conformational states with no intermediates. (A good analogy is the haemoglobin molecule which passes from one quaternary state to another on oxygenation.) Thus in the above model, the heads are supposed to rotate through a constant angle† (Fig. 21.14). If this is so, the work done by the cross-bridge (and hence the efficiency) will not be fixed but will depend on the tension which opposes motion. If there is no tension opposing motion, no work is done and the free energy of hydrolysis of ATP is wasted. This is presumably what happens in free solution when actin and heavy meromyosin hydrolyse ATP.

An important question is whether attached cross-bridges rotate together. It should be obvious from Fig. 21.14 that if the subfragment 2 tails were inextensible this would have to occur. Rotation of the heads would *directly* cause the filaments to slide. If however the tails could be stretched, the cross-bridges could move independently.

It is not yet clear whether the tails are sufficiently elastic for this to occur. We have seen that the tail contains a high proportion of coiled-coil α-helical structure, and a stretch of the required amount (about 8 nm, see section 10.4) would presumably disrupt the structure. However an element with the required elasticity may be present in the link joining the tail to the backbone or the tail to the head. It is even possible that the myosin heads themselves are not rigid. If rotation of the heads causes the tails of the cross-bridges to be stretched, then the energy produced by the rotation is stored as potential energy in the stretched tails. Rotation of the heads would not therefore be directly linked to sliding of the filaments.

A further proposal of the H. E. Huxley model is that the subfragment 2 tails are flexibly hinged to the backbone. This allows the myosin heads to operate over the same angle irrespective of the separation of the thick and thin filaments. Since the myofibril acts as a constant volume system, the thick and thin filaments get farther apart as the muscle shortens.

10.4 The size of the working stroke

One important fact that we want to know is the size of the working stroke, i.e. the axial distance the ends of the myosin heads move when they rotate. This quantity is not known with certainty but some clues about it can be obtained from several approaches:

---

† It may be necessary to add the proviso that if the tension in the subfragment 2 tail opposing the tilt of the head is excessive, the link between actin and myosin may be mechanically broken before the head has completed its transition.

1. We have seen that in insect flight muscle the myosin heads can adopt two angular positions (90° and 45°). The simplest hypothesis is that there are only two stable configurations of the heads. If that is so, the myosin heads would initially attach at right angles to the thin filaments and in the working stroke would rotate by 45° to the rigor position. If this is also true of vertebrate skeletal muscle, reference to Fig. 21.14 shows that the axial distance moved by the end of the myosin head in the working stroke would be about 15 sin 45° = 11 nm, if the length of the myosin head, like subfragment 1, is 15 nm.†

2. Information about the size of the working stroke can be obtained from the physiological quick release experiments of A. F. Huxley. In an isometrically contracting muscle some of the attached cross-bridges will be in the position of attachment and others in the fully rotated position (Fig. 21.14). In the latter case the tails will be stretched. What happens if such a muscle is suddenly allowed to shorten by a small distance? If we look at short enough times (~1 ms), there should not be time for cross-bridges to attach or detach, nor even for the attached cross-bridges to rotate. So by recording the instantaneous fall in tension we will be following the elasticity of the tails with the heads frozen in position. Experimentally there are difficulties in completely resolving the instantaneous fall in tension from the subsequent partial recovery of tension due to cross-bridge rotation, but the amount of shortening required to reduce the instantaneous tension to zero is about 8 nm per half-sarcomere. Provided our assumption that the cross-bridge can exist in only two configurations is correct, this distance corresponds to the size of the working stroke. (If the cross-bridge can adopt intermediate positions this figure will be an underestimate.)

3. An approximate idea of the size of the working stroke can be obtained from energetic considerations. A mammalian muscle at 37°C in which all the fibres are contracting can exert an isometric tension of about 40 N/cm^2 cross-sectional area. This corresponds to a tension of about 8×10^{-10} N per thick filament. The number of cross-bridges in each half of the thick filament is about 100 (if there are two cross-bridges every 14·3 nm of length) (see section 7.4). If a fraction *n* of these are attached at any instant, the average longitudinal tension in the tails of the cross-bridges is $(8 \times 10^{-10}p)/100n$ N when the muscle is exerting a fraction *p* of the isometric tension. (It is interesting to compare this tension with the force required to break a hydrogen bond. This is about 3×10^{-10} N. Thus provided *n* is not too small, a single hydrogen bond should be able to sustain the tension.) Suppose on average a cross-bridge remains attached while the filaments slide *x* nm. Then the work done by the cross-bridge is $(8 \times 10^{-10} px)/100 n \times 10^{-9}$ J. Now the free energy of hydrolysis of ATP in the muscle cell is about 50 kJ/mol. If the thermodynamic efficiency is *E* and *z* molecules of ATP are split per cross-bridge cycle

$$\frac{8 \times 10^{-10}px \times 10^{-9}}{100\,n} = \frac{50\,Ez \times 10^3}{N} \quad \text{(where } N \text{ is Avogadro's number)}$$

Hence

$$\frac{xp}{nzE} = 10 \text{ (or 15 if there are } \textit{three} \text{ cross-bridges every 14·3 nm)}$$

We have as yet insufficient knowledge to exploit this equation fully. Unfortunately we have information on *n* only under isometric conditions from X-ray measurements and we cannot assume that the fraction of cross-bridges attached is proportional to the load. Suppose we assume that *n* lies in the range 0·05–0·2 when the tension is one-third of the isometric tension;

---

† This argument is rather over-simplified. The direction at which the heads are attached has to be specified by an azimuthal angle as well as by the angle between the head and the long axis of the thin filament. A full treatment would therefore require the change in *both* angles to be known.

under these conditions E is high ($0.5 < E < 1$). Since z could plausibly be 1 or 2, reasonable values for the size of the stroke would lie between about 1 and 20 nm. Although this range is wide it is important to note that it does not conflict with the other estimates.

In summary, there is considerable uncertainty about the axial distance moved by the ends of the myosin heads in the working stroke but the most reasonable estimate is about 8 nm.

10.5 The rate of the steps in the cross-bridge cycle

A complete description of the cross-bridge cycle would define the rates of each of the steps (both in the forward and back directions) and how these are influenced by factors such as the load or the rate of sliding of the filaments past each other. There are obviously several problems which need answering. Firstly the velocity with which the heads rotate should be as

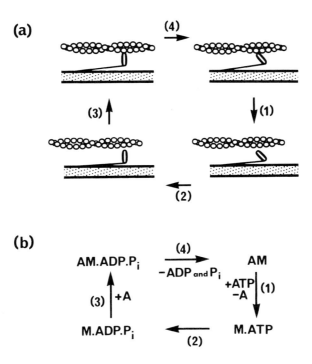

Fig. 21.15 Comparison of the mechanical steps of the cross-bridge cycle with the steps of the actomyosin ATPase cycle. Note that the numbering of the steps is different from that used earlier in equation (4). (After E. W. Taylor.)

fast or faster than the velocity of sliding of filaments past each other (about 10 μm s^{-1} in an unloaded contraction). Thus the time required for the ends of the heads to move 8 nm should be 0.8 ms or less in an unloaded contraction (corresponding to a rate constant of ~1000 s^{-1}). Secondly the heads should rotate soon after attachment for if they do not the tails will be slack when they do rotate due to the movement of the filaments, and the energy will be wasted. Finally after they have rotated they should not remain attached for too long or they will ultimately interfere mechanically with further movement. Nor should they detach too quickly if the tails can be stretched or the potential energy stored in the stretched tails will not be converted into movement of the backbone.

But is it not yet possible to measure the rates of the mechanical steps. An alternative approach used by Taylor is to correlate the mechanical steps of the cross-bridge cycle with the kinetic steps of F-actin activated heavy meromyosin ATPase which *are* experimentally measurable. It is assumed that heavy meromyosin and F-actin in free solution mimic exactly the conformational changes which occur *in vivo*. The only difference, of course, is that the working stroke in muscle may be opposed by tension, which is absent in free solution. In Fig. 21.15 the working cycle is compared with the cycle of ATP hydrolysis. As we have said before, each cycle of ATP hydrolysis involves the attachment of heavy meromyosin to actin and its later detachment.

Starting with the cross-bridge at the end of its working stroke, the detachment of the cross-bridge (step 1) can be correlated with the binding of ATP and the subsequent very fast dissociation of the actin–heavy meromyosin complex. The recovery stroke can be correlated with the hydrolysis of ATP by heavy meromyosin alone. The attachment of the cross-bridge to the thin filament corresponds to the combination of actin with the heavy meromyosin–ADP . P_i complex. Finally the working stroke corresponds to the dissociation of the products from the actin–heavy meromyosin–ADP . P_i complex.

If this interpretation is correct, the rate of turnover of cross-bridges in unloaded contractions should be similar to the ATPase rate of heavy meromyosin at high concentrations of F-actin. Unfortunately we can only measure the cross-bridge turnover by measuring ATP hydrolysis. We have already seen in section 8.3 that the maximum rate of ATP splitting by muscle is comparable to the maximum ATPase of F-actin activated heavy meromyosin. A second prediction of the scheme is that the rate of the working stroke in lightly loaded contractions ($\sim$1000 s^{-1}) should agree with the rate of dissociation of products from the actin–heavy meromyosin–ADP . P_i complex. We have noted in section 8.3 that we do not yet know how fast this step is and it will be important to measure this.

The model predicts the state of the cross-bridge in resting, contracting and rigor muscle. In the resting state, where actin is prevented from reacting, the cross-bridge will spend most of its time in the M . ADP . P_i state. This has been verified by examining the amount of ^{14}C ADP bound to protein when glycerinated muscle is incubated in a relaxing solution containing ^{14}C ATP.

In the rigor state, due to the absence of ATP, the cross-bridges will be arrested as the AM intermediate.

In the contracting state, none of the steps is prevented and the population of cross-bridges in the various states will depend on the rate constants for the forward and back reactions. It is important to note that the rate constants for the tension-producing step will depend on the work produced and hence on the load applied to the muscle (see Taylor 1973).

The scheme shown in Fig. 21.15 is obviously only a start and many details need to be filled in. Each of the steps in the cycle very likely includes several more fundamental reactions, the nature and velocity of which are as yet unknown. The scheme shown does not explain in what way actin competes with ATP (or products) for sites on the myosin heads nor has the possible interaction between the two heads of myosin been included. It does however provide a very useful framework on which further experimentation can be based.

10.6 Conclusion

The grosser features of muscular contraction are now well-established and an elegant model is available to explain the generation of tension at the molecular level in mechanical terms. But we are still in a state of very considerable ignorance about the details (both structural and kinetic) of the cross-bridge cycle and further advances will require the combined approaches of structural studies, physiological studies, physical chemistry of muscle proteins and enzyme kinetics.

References

AIDLEY, D. J. (1971). *The Physiology of Excitable Cells.* Cambridge University Press. (An excellent complement to this article covering the physiology of muscle and nerve.)

BENDALL, J. R. (1969). *Muscles, Molecules and Movement.* Heinemann Educational Books, London. (The only inexpensive and reasonably up-to-date student textbook on the subject. A useful source of references to key papers.)

EBASHI, S. and ENDO, M. (1968). 'Calcium ion and muscular contraction.' *Prog. Biophys.,* **18**, 125.

HANSON, J. (1968) 'Recent X-ray diffraction studies of muscle.' *Quart. Rev. Biophys.* **1**, 177.

HOLMES, K. C. and BLOW, D. M. (1965). *The Use of X-ray Diffraction in the Study of Protein and Nucleic Acid Structure.* Interscience, New York and London. (For background reading.)

HUXLEY, H. E. (1960). 'Muscle cells.' In *The Cell,* Vol. IV, p. 365 eds. J. Brachet and A. D. Mirsky. Academic Press, New York and London. (Particularly useful for the description of the hierarchies of structure and the sliding filament hypothesis.)

HUXLEY, H. E. (1963). 'Electron microscope studies of natural and synthetic protein filaments from striated muscle.' *J. Mol. Biol.,* **7**, 281.

HUXLEY, H. E. and BROWN, W. (1967). 'X-ray diffraction studies of muscle.' *J. Mol. Biol.,* **30**, 383.

HUXLEY, H. E. (1969). 'The mechanism of muscular contraction.' *Science,* **164**, 1356. (Very clear and stimulating.)

LOWEY, S. (1971). 'Myosin: molecule and filament.' In *Subunits in Biological Systems,* Part A, eds. S. N. Timasheff and G. D. Fasman, Marcel Dekker, New York.

McNEIL ALEXANDER, R. (1968). *Animal Mechanics.* Sidgwick & Jackson, London. (For background reading. Includes a fascinating account of how muscles act on joints in living animals.)

NEEDHAM, D. M. (1971). *Machina Carnis.* Cambridge University Press. (A very detailed work on the biochemistry of muscular contraction from a historical viewpoint. Includes a very comprehensive list of references.)

TAYLOR, E. W. (1972). 'Chemistry of muscle contraction.' *Ann. Rev. Biochem.,* **41**, 577. (Reviews the recent literature. Particularly useful for the sections on kinetics and the sarcoplasmic reticulum.)

TAYLOR, E. W. (1973). 'Mechanism of actomyosin ATPase and the problem of muscular contraction.' In *Current Topics in Bioenergetics* Vol. 5, ed. D. R. Sanadi, Academic Press, New York & London. (A penetrating analysis of the cross-bridge cycle.)

WEBER, A. and MURRAY, J. (1973). 'Molecular Control Mechanisms in Muscle Contraction.' *Physiological Reviews.*

WILKIE, D. R. (1960). 'Thermodynamics and the interpretation of biological heat measurements.' *Prog. Biophys.,* **10**, 259.

WOLEDGE, R. C. (1971). 'Heat production and chemical change in muscle.' *Prog. Biophys.,* **22**, 37.

Also see the *Cold Spring Harbor Symposium for Quantitative Biology,* Vol. 37 on the Mechanism of Muscle Contraction for very recent work.

Index

673

electrophoresis: diagonal separations of modified peptides by, 102–3, 104, 105; of proteins, 126

elongation factor (bacterial) EF-G, 61; amount of, in bacterial cell; 63; GTPase activity of, 62, 67, 69; involved in translocation of peptidyl-tRNA on ribosome, 4, 66–7; mutants of, 68; translocation product complex of 50S ribosome unit, GDP, and, 67–8

elongation factor (bacterial) EF-T, 61; amount of, in bacterial cell, 63; in elongation process, 64, 66

elongation factor (bacterial) EF-Ts, subunit of EF-T, 61; identical with subunit IV of RNA replicase of Qβ phage, 70; present in subunit ψ of bacterial RNA polymerase, 70; releases GDP from EF-Tu complex, 64

elongation factor (bacterial) EF-Tu, subunit of EF-T, 61; complex of aminoacyl-tRNA, GTP, and, binds to ribosomal A site, 4, 31, 50, 64, 70; complex of GDP and, 64; identical with subunit III of RNA replicase of Qβ phage, 70; methionine-tRNA$_m$ transferred by, not -tRNAf, 60; present in subunit ψ of bacterial RNA polymerase, 70

elongation factors of *Bacillus stearothermophilus*, 65

elongation factors, eukaryotic, EF-1 and EF-2; form complexes with GTP, 69

embryo: lysosomes in absorption of nutrients from endometrium by, 527, 529; problem of development of immune system in, 620

emiocytosis (fusion of storage granule with cell membrane, and liberation of granule contents outside cell), 589; calcium in, 591

emission spectrum (wavelength distribution of fluorescence), 177

encephalomyocarditis virus (single-stranded RNA), and ascites tumour cell extracts, 43, 59

endocrine tissue, intracellular digestion of excess hormone in (crinophagy), 516, 529, 536

endocytosis (phagocytosis + pinocytosis), 514, 517; disturbances of, 535–6; intake of infectious organisms by, 292, 531–2

endonucleases: recognising and cleaving foreign DNA, 255; specific for regenerating linear from circular DNA of lambda phage, 254; tend to degrade mRNA not on ribosomes, 448

endopeptidase of *E. coli*, in mucopeptide synthesis, 351

endoplasmic reticulum: in cell cycle, 409–10; in cell division, 374, 381; in cytokinesis, 382; fragments of, sediment as microsomes, 513; lysosomal enzymes synthesised on, 514, 517; micro-tubules and, 385; in plant cells, 369, 371, 386, 393; poliomyelitis virus and, 302; ribosomes in (eukaryotes), 34; *see also* sarcoplasmic reticulum

Entamoebae, leucocyte-killing factors produced by, 488

Enterobacteriaceae: identification of, by lipopolysaccharides, 352

enterotoxins, 483

envelopes of bacteria, 343; *see also* cell walls

envelopes of viruses, derived from host, 285–6, 302, 306; in attachment of virus to cell, 288

enzymes: absorption spectra of, after binding of substrate or inhibitor to, 170, 210–12; active sites in, 109–17, 211; allosteric, 249; chromogenic substrates for, in isolation of mutants, 450; cleavage of polypeptides by, at specific sites, 94, 97–101; commercial production of, 472; complex formation between ligands and, 212–13, 223, 224; control of synthesis of, in micro-organisms, 437–9; inducibility of bacterial, confined to one phase of cell cycle, 407; latency of, in lysosomes, 513, 523; of lysosomes, 524–6, (inborn errors in) 538–41; of necrosing tissue, released into serum, 529; new, in virus-infected cells, 299–300, 306; in oxidative phosphorylation, 575, 577–8; purified, insulin and, 594, 595; rapid reaction techniques applied to, 197, 224; relaxation spectra of reactions of, 222–4; synthesis of, in cell cycle, 406–7; transient kinetics of reactions of, 205–13; in virions, 286, 306

epimerases, in change from primary to secondary cell-wall formation in plants, 388

equatorial reflections, in X-ray diffraction analysis, 311, 312

erythritol of bovine placenta, etc., as growth-stimulant for *Brucella abortus*, 484

Escherichia coli: accumulation of guanosine phosphates in, on deprivation of an essential amino acid, 69–70; active transport into, 468; cAMP receptor protein in regulation of gene expression in, 597; antiphagocytic activity of, 480; arabinose system in, 458–9; conjugation of, 451–2; defective cell division in mutants of, 469; DNA of, 255, 257, (repair of) 464–5; endonuclease of some strains of, cleaves DNA of lambda phage, 255; enterotoxin of, 483; growth of, with restricted oxygen, 433–4, and with suboptimal carbon dioxide, 432; growth patterns of, 431; lactose system in, 256, 456–8; lipopolysaccharide synthesis in, 354; mapping of genes of, 453; mucopeptide synthesis in, 346; rate of protein synthesis in, 433–4; rod and round forms of, 363; strain of, with 25% of protein as β-glucosidase, 472; tryptophan system in, 460–1

ethanol, in *in-vitro* system for simulating peptide-bond formation, 65–6

ethidium, blocks nuclear division but allows cytokinesis, 381

ethylmethanesulphonate, mutagen, 449

Euplotes eurystomus (ciliate): interrelations of DNA, RNA, and histone syntheses in macronucleus of, 501–4

evolution: chemostasis in approach to processes of, 441; of immunoglobulins, 619

excimer, exciplex, in fluorescence, 176

excitation energy of molecules, 176–7

excitation spectrum, 177

exciton systems, 173–5, 176; optical activity of disymmetric, 186–7; of peptide-bond absorption band, 189

respiration: and muscle contraction, 634, 635, 637; not affected in amoeba by removal of nucleus, 507

restriction enzyme (endonuclease), cleaving foreign DNA, 255

reticulocytes: polyribosomes in, 36; mRNA from, 43; synthesis of mutated haemoglobin by, when supplied with modified aminoacyl-tRNA, 20

rhabdoviruses (RNA), 280

L-rhamnose, in lipopolysaccharide of *Salmonella*, 353, 356, 357

rhamnose units, at kinks in pectin chains, 334

ribitol phosphate, in teichoic acids, 358, 359, 360

ribonuclease, 36; A and S forms of, 154-5; acid, in lysosomes, 513, 518, 524, 530, 533, 541; adsorbed on ribosomes, 37; assay of, 526; H form of, in tumour viruses, 286; histidine at phosphate-binding site in, 115; lysine in, 114, 133; NMR spectra of, 153-8; pancreatic form of, specific for two bases, 258; mRNA degraded by, 42, unless protected by ribosomes, 42, 44, 58; synthetic, effects of modifications on, 127-8; T form of, specific for one base, 258

ribosomal ambiguity mutations, 471

ribosomes, bacterial: binding sites for, (on mRNA) 44, (on phage DNA) 257, (on phage RNA) 268, 270-1, 274; binding sites on (A and P), (for aminoacyl-tRNAs) 2-3, 10, 34, 50, (for elongation factors) 68-9, (for initiator tRNA) 49, 52 (puromycin at P site displaces polypeptidyl-tRNA from A site) 49-51; molecular weight of, 2; number of, per cell, 3; protect mRNA from ribonucleases, 42, 44, 58; proteins of, 3, 34, 37, 39-40, 69; rate of protein synthesis proportional to number of, 435; relative movement of mRNA and, 5, 44, 62; streptomycin binds to, 471; subunits of, (50S and 30S) 5, 34, 35, 36, 44, 55, 67, (23S and 5S) 37, 38, 39; unit of (70S), 3, 34, 35, 36; in virus-infected cells, 264, 272, 300

ribosomes, eukaryotic of cytoplasm: defective in diabetes, 602-3; few in muscle cells, 268; subunits of (60S and 40S), 34; unit of (80S), 34

ribosomes, eukaryotic of mitochondria and chloroplasts, 34

L-ribulokinase, in *E. coli*, 458

L-ribulose-5-phosphate epimerase, in *E. coli*, 458

rifampicin, rifamycin, inhibit DNA-dependent but not RNA-dependent RNA polymerase, 267, 296, 305

rigor in muscle (in absence of ATP), 626, 629; cross-bridges in, 662, 664-5, 670

ristocetin, antibiotic: inhibits synthesis of bacterial mucopeptide, 349, 352

RNA (ribonucleic acid): cell content of, in cell cycle, 404; constant in nucleated half of amoeba, decreases in enucleated half, 507; 'extra', in slowly growing bacteria, 436; in mammary gland, 530; methods for determining nucleotide sequences of, 262; non-coding sequences in, 253, 274; passes from nucleus to cytoplasm, 507; stable, of egg cell, 410; synthesis of, *see* RNA synthesis

RNA, of bacteriophages, 262-3; control of transcription and translation of, 271-2; genes in, 268; host specificity resides in, 288; replication of, 263-5, 297-8; ribosome-binding sites on, 268, 270-1, 274; specificity of replicases for, 265-6; structure of, 272-4, (secondary) 262, 269, 271; translation of, 266-7

HnRNA (heterogeneous nuclear), 261

mRNAs (messenger): bound to 30S subunit of ribosome, 44, 53, 56, 59-60; degradation of, 42, 448; discovery of, 40-1; of eukaryotes, are monocistronic, 43; length of life of, 41; molecular weight of, 4; number of molecules of, per bacterial cell, 41; nucleotide sequence of, for coat protein of MS2 phage, 13, 42; as percentage of total RNA, 19, 36, 41; polycistronic, 41, 256, 446; post-transcriptional processing of, 261; precursor of, 45; rapid turnover of, 405; relative movement of ribosomes and, 5, 44, 62; ribosomes strung along, 36; secondary and tertiary structure of, 44-5; spacer sequences separate coding sequences of, 253; synthesised after phage infection, 34; synthesised and translated from 5′ end to 3′ end, 5, 6, 42; synthetic, in solution of genetic code, 9, 11, 41; transcription of DNA into, 5, 6, 70, 256, 445; transcription of viral DNA into, 296, 299; translation of, into protein, 5, 6, 35, 41, 266-7, (in vitro) 43; transport of, from nucleus to cytoplasm, 45, 261; viral RNA acting as, 262, 299

rRNA (ribosomal), 36-9; bulk of cellular RNA in, 405; DNA cistrons for, 259-60, 619; non-translated, 253; as percentage of total RNA, 19; secondary structure of, 38; smaller ratio of, to protein, in eukaryotes than in bacteria, 428

tRNAs (transfer), 18-21; anticodons in, 14, 15, 23, 25, 28-9, 30; attachment of amino acids to, 15-18; cytokinins bound to?, 375; as determining part of aminoacyl-tRNA molecule, 14, 20; fractionation of, 11, 20-1; inactivation of, 31; minor nucleotides in, 21-3; molecular weight of, 2; new species of, in phage-infected cells, 300; non-translated, 253; percentage of DNA coding for, 19, 33; as percentage of total RNA and of cell dry weight, 19; precursors of, 33; recognition of aminoacyl-tRNAs by, 29-30; repair enzyme for, 31-2; required for synthesis of glycine-peptide bridge in mucopeptide, 350; sequence determination in, 21; sequence the same at 3′ end of all, 2, 25, 31; structure of, 23, 24, 25-8; suppressor (decoding chain-terminating codons), 12, 29; two forms of, for methionine, 47-8

RNA bacteriophages, 262-3; absorb only to pili of male bacteria, 288, 290, 292

RNA polymerase (DNA-dependent): antibiotics inhibiting, 267; binding of, to promoter sites on DNA, 256, 446, 457; of *E. coli*, altered on infection by T₄ phage, 299; inhibited by MS1 (guanosine tetraphosphate), 70; subunit ψ of, includes elongation factors, 70; syntheses by, 11, 12, 33, 42, 265; in transcription of viral DNA, 286, 296

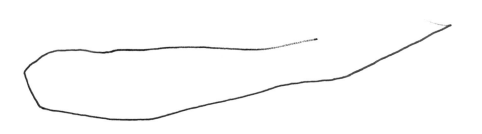